Advances in Intelligent Systems and Computing

Volume 622

Series editor

Janusz Kacprzyk, Polish Academy of Sciences, Warsaw, Poland
e-mail: kacprzyk@ibspan.waw.pl

About this Series

The series "Advances in Intelligent Systems and Computing" contains publications on theory, applications, and design methods of Intelligent Systems and Intelligent Computing. Virtually all disciplines such as engineering, natural sciences, computer and information science, ICT, economics, business, e-commerce, environment, healthcare, life science are covered. The list of topics spans all the areas of modern intelligent systems and computing.

The publications within "Advances in Intelligent Systems and Computing" are primarily textbooks and proceedings of important conferences, symposia and congresses. They cover significant recent developments in the field, both of a foundational and applicable character. An important characteristic feature of the series is the short publication time and world-wide distribution. This permits a rapid and broad dissemination of research results.

Advisory Board

More information about this series at http://www.springer.com/series/11156

Elena G. Popkova
Editor

The Impact of Information on Modern Humans

 Springer

Editor
Elena G. Popkova
Volgograd State Technical University
Volgograd
Russia

ISSN 2194-5357   ISSN 2194-5365   (electronic)
Advances in Intelligent Systems and Computing
ISBN 978-3-319-75382-9   ISBN 978-3-319-75383-6   (eBook)
https://doi.org/10.1007/978-3-319-75383-6

Library of Congress Control Number: 2018934335

Printed on acid-free paper

This Springer imprint is published by the registered company Springer International Publishing AG
part of Springer Nature
The registered company address is: Gewerbestrasse 11, 6330 Cham, Switzerland

Contents

Legal Foundations of Human Society

Philosophy of Modern Humans

The Role of Project Management of the Innovative Activities of Large Industrial Structures

Innara R. Lyapina[1]([✉]) [ID], Tatyana N. Ivashchenko[2],
Olga O. Komarevtseva[2], Oksana V. Leonova[2],
and Alexander V. Shchegolev[2]

[1] Orel State University, Orel, Russia
innara_lapina@mail.ru
[2] Russian Presidential Academy of National Economy and Public
Administration, Moscow, Russia
itn-57@mail.ru, komare-91@mail.ru,
leonova-o-v@yandex.ru, alexander_shchegolev@bk.ru

Abstract. Changes in economic conditions of development of economic subjects require application of new tools that improve functioning of entrepreneurial structures in the system of adaptation to global processes and challenges. These tools are especially important for implementation of innovative activities in large industrial structures that form the main part of the country's GDP. One of such tools is project management, which is a targeted process of development and implementation of decisions aimed at successful execution of works within innovative activity. According to this, the purpose of the research is consideration of the models of project management of large industrial structures' innovative activities that are used in the Russian and foreign practice. Within the research, it is necessary to solve the following tasks: consider the theoretical foundations of the essence of project management of innovative activities in large industrial structures; determine the peculiarities of the model of project management of innovative activities in large industrial structures of the Russian Federation; generalize foreign experience of project management of innovative activities of large industrial companies; form the proprietary model of project management of innovative activities of large industrial structures. The scientific novelty of the research consists in generalization of peculiarities of project management of innovative activities in the Russian and foreign realia of development with possible creation of the proprietary model of project management. The methodology of the research includes the following methods: theoretical analysis, used for generalization of existing theoretical and methodological studies in the sphere of project management of large industrial structures; generalization, used for determining peculiarities within the models of project management of innovative activities; comparison, used for comparing facts for the purpose of determining the common and different features within the Russian and foreign models of project management of innovative activities of large industrial structures; modeling, used for creating the proprietary model of project management of innovative activities within large industrial structures. The main conclusions and results could be used for studies in the sphere of project management of large industrial structures in the conditions of digital economy.

© Springer International Publishing AG, part of Springer Nature 2018
E. G. Popkova (Ed.): HOSMC 2017, AISC 622, pp. 3–15, 2018.
https://doi.org/10.1007/978-3-319-75383-6_1

Keywords: Innovations · Project management · Industrial structures
Modeling · Innovative activities · Intellectual property · Technological cities
Technological parks · Industrial centers · Industrial node · Agglomeration

1   Introduction

The key to effective development of a company in the conditions of domination of the technological paradigm is organization of innovative activities, which is oriented at development of unique innovations and their implementation at the competitive market of technologies. Organization of innovative activities is based not on random and non-systemic processes of production of innovations but is a structured model within the planned framework of methodological rules, related to implementation of ideas, knowledge, and project offers. Very often this model of innovative activities includes the stage-by-stage steps, based on development of final innovational product. According to this, innovative activities are organized in the aspect of project management.

Project management is one of the important tools in the modern system of market economy. This circumstance is related to application of project management to solving complex and specific tasks. The main attributes of project management are as follows:

- creation of an innovational product within limited time and resources;
- dynamic development of the R&D system as the institute of project management;
- focus at achievement of the set goal and final result;
- specification of project management events that stimulates formation of the final result on the basis of application of more accessible methods;
- using the interdisciplinary approach to management of innovative activities.

Project management of innovative activities is related to organization of works on creation of innovational goods and technologies and intellectual products over a limited time. As a tool, project management is topical for large entrepreneurial structures that conduct innovative activities on the basis of partnership agreements. On the whole, actuality and urgency of this topic are obvious. Firstly, modern realia of development of economy require application of new tools and methods in management of innovative activities. Secondly, innovative activities of large entrepreneurial structures may lead to the quickest technological development of the whole economy of the Russian Federation, which is related to large scale and capital of the projects that are implemented within these economic activities.

2   Theoretical Foundations of Studying the Essence of Project Management of Innovative Activities in Large Entrepreneurial Structures

The theoretical foundations of project management are viewed in scientific works of the Russian and foreign scholars. It should be noted that the essence of project management consists in the systematized offer of the complex of measures with technological, scientific, financial, and commercial character, mutually connected to the main criteria (terms, means, and resources) and performed until receipt of the final result. The essence of managing the projects is studies in the works by Bryde (1995), Hutchins (2001), Kaab (2016), Lyapina et al. (2017), Mirzoeva (2015), Yaluner and Chernysheva (2016), etc. These authors have formed certain theses that open the essence of project management in the production and economic activities.

Project management of innovative activities is reflected in scientific articles of Cooke-Davies and Arzymanow (2003), Komarevtseva (2017), Rose (2002), Smyth (2014), and Danchenko (2011). Having generalized the studies of the above authors, it is possible to distinguish the peculiar features of project management within innovative activities:

1. Targeted orientation of innovative activities that consists in creation of the planned innovational product.
2. Formed budget of implementation of innovational projects for provision of business planning within innovative activities.
3. Limited terms of implementation of innovative activities for formation of attractiveness of the developed innovational product among the investors.
4. Novelty of the company that develops innovational product within the implementation of joint innovative activities of the companies.

According to limited resources within project management, its main elements are selected subjects of innovative activities. In this research, the subjects are large industrial structures. Description of development of large entrepreneurial structures is generalized in the works by Yee-Pagulayan (2003), Clark and Colling (2005), Simonova et al. (2017), Arslanov (2015), Kalabina (2017), Lukmanova and Khurshudyan (2017), Pavlovskaya (2016). The large entrepreneurial structures, according to the authors, include the following:

1. Technopolis and technological parks created on the basis of R&D organizations of the industrial type.
2. Industrial centers – large production structures separated from technological interconnections of the main company, located in the Oblast centers of the subjects of the RF.
3. Industrial node – a group of large companies located at the adjacent territories.
4. Industrial agglomeration – territorial entity with high concentration of large productions of various economic spheres and high population density.

Despite the significance of the study of large entrepreneurial structures, most of the works are devoted to implementation of project management at a specific industrial company. We think that only total development of these subjects will allow forming a certain level of competitiveness of entrepreneurial structures and realizing the technological initiative within the Russian Federation. This thesis is supported by the scholars involved with development of the models of project management of innovative activities of industrial structures in the RF and abroad.

Among the studies devoted to project management of innovative activities of large entrepreneurial structures, it is possible to distinguish the scientific articles of

Anderson and Merna (2003), Lechler and Dvir (2010), Anastasova et al. (2016), Furta et al. (2016), and Khovaev and Kozhevnikova (2016).

Based on the theoretical evaluation of the essence of the research topic, it is necessary to form the models of project management used in foreign and Russian industrial structures. This circumstance is related to the fact that development of the proprietary model requires generalization of the material that opens the peculiarities of project management in the conditions of increase of significance of innovational & technological development of business structures.

3   The Model of Project Management of Innovative Activities in Large Entrepreneurial Structures of the RF

Project management of innovative activities in large entrepreneurial structures of the RF is used for solving complex tasks in the sphere of innovational developments. This model is a closed system with directions of input and output (Fig. 1).

The model of project management of innovative activities of large industrial structures in the RF is based on five stages.

Stage 1. Goal of innovative activities. Determining the goal allows substantiating the directions of interaction within innovative activities. The main goals of project management are as follows:

- solving the tasks in the sphere of current functioning of large entrepreneurial structures;
- solving the tasks in the sphere of long-term functioning of large entrepreneurial structures;
- domination of national subjects of business over the foreign ones.

According to the above goals, the subjects of large entrepreneurial structures determine the category within which they will conduct innovative activities.

Stage 2. Categories of innovative activities. The mono-category of innovative activities supposes execution of scientific and technological and R&D works within one direction, which stimulates solving the tasks in the sphere of current functioning of the specific subject of a large industrial structure. In the aspect of this category, execution of multiple actions on creation of technologies is coordinated, which allows improving the production process of the given industrial structure. The multi-category of innovative activities is related to implementation of the technological process within joint activities of large entrepreneurial structures aimed at solving the tasks in the sphere of long-term development of these structures' subjects.

Applying this category of innovative activities, large industrial structures try to receive the intellectual product that could be further used for development of their own production and for selling in the market of intellectual resources. The mega-category of innovative activities is aimed at domination of the national large entrepreneurial structures over foreign industrial corporations and conglomerates. Using this category in the Russian realia is minimum. Firstly, the mega-category character supposes creation of innovational business within the industrial structures. Secondly, production of innovational is of the constant character.

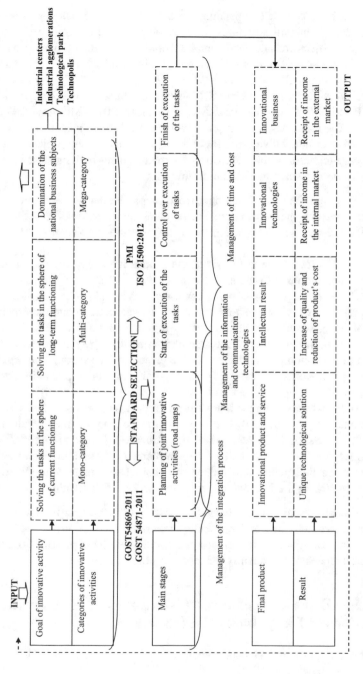

Fig. 1. The model of project management of innovative activities of large industrial structures in the RF.

It should be noted that depending on selection of the category of innovative activities, large industrial structures have to observe the national and international standards of project management. Implementation of project management in the structure of mono-category or multi-category requires adaptation of innovative activities for the standards GOST 54869-2011 and GOST 54871-2011. The main standards within the mega-category are Project Management Institute (CAPM, PfMP, PgMP, PMI-SP, PMI-RMP, PMI-PBA, PMI-AC), ISO 21500:2012 (Moscow 2017).

Stage 3. The main stages of project management. Project management is based on implementation of innovative activities in the form of investment project. According to this, this stage includes peculiar measures for management in the sphere of investment planning. These measures include the following:

1. Planning of joint innovative activities (road maps). Recently, development of a road map has become the main document of planning of innovative activities, which includes the system of measures and stage-by-stage execution of algorithmic actions.
2. Start of execution of the tasks within the innovative activities. Implementation of this measure supposes execution of works within the specific innovational project that includes management of resources for achieving planned indicators and coordination of activities between all members of large entrepreneurial structures.
3. Control over execution of tasks within the innovative activities. This measure allows conducting analysis of the state that is realized by the project according to the indicators announced at the start of the project and correcting and regulating the directions of works in view of the changes that take place in the external environment.
4. Finish of the works within the innovative activities. This measure leads to formation of the final product (stage 4) and determination of the received result from innovative activities of large entrepreneurial structures (stage 5). According to this model, large industrial structures try to implement the unique technological solution, increase the quality and reduce the cost of the manufactured product, and receive income in the internal and external market of innovations.

Project management is implemented somewhat differently in foreign countries, in which innovative activities of large entrepreneurial structures are related to detailed analysis of external (competitive) environment and selection of priorities of activities in view of preferences of the country's economic development.

4   Foreign Experience of Project Management of Innovative Activities of Large Industrial Structures

The foreign model of project management of innovative activities of large industrial structures is presented in Fig. 2.

According to this figure, the model of project management of innovative activities is based on seven stages. Each of the stages is formed within roadmapping – the effective method of planning, management, and forecasting of innovative activities, which are based on road maps.

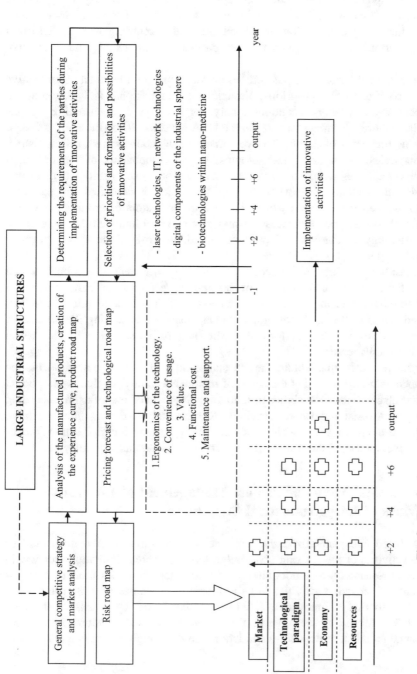

Fig. 2. The foreign model of project management of innovative activities of large industrial structures

The initial stage of innovative activities of large entrepreneurial structures is formation of the general strategy of business subjects on the basis of competitive advantages and market analysis. This analysis is based on evaluation performed at the moment of compilation of the product strategy and creation of the experience curve. The experience curve will allow determining how the scale of issue of innovational products influences the formation of market demand. According to this, industrial structures determine the requirements of the parties for conduct of joint innovative activities.

The priorities are formed and possibilities are determined for conduct of innovative activities according to the top-priority directions of innovational and technological development of the country's economy. Analyzing the stage of determining the priorities of large entrepreneurial structures in the USA for 2007–2015 (Economics 2017), the following directions of economic development of the country were distinguished: laser technologies, IT, network technologies; digital components of the industrial sphere; biotechnologies within nano-medicine. Based on these priorities and formed technological innovations, the cost forecast of the project is compiled and the technological road map is developed which includes the indicators of ergonomics of the developed technologies; conveniences of using technological innovations; value of the created technology or innovational product; functional cost of the issued product; service and support.

At the last but one stage of project management, the risk road map is built, which determines the positions of the change of the market, technological paradigm, economy, and resources over the period of existence of the created innovational product. The formed stages of the model of project management allow starting the process of innovative activities, which is performed on the basis of the standards of the Project Management Professional.

It should be noted that unlike the Russian model of project management, the foreign model is more oriented at the needs of national economy. At that, an important advantage is elaboration of the choice of innovational products' manufacture according to the mega-categorial innovative activities. Russian industrial structures are more oriented at the internal market, which does not allow forming the global positions of economic development of the state in the sphere of innovational technologies.

5 The Authors' Model of Project Management of Innovative Activities of Large Industrial Structures

Having analyzed and generalized the main peculiarities of project management of innovative activities of the Russian and foreign large entrepreneurial structures within the performed research, the authors attempted to develop their own model of project management of innovative activities in large industrial structures (Fig. 3, Table 1).

According to Fig. 3, innovative activities are conducted by means of the total system of R&D, formed by the subjects of large entrepreneurial structures in the form of total capital (conduct of financing) and labor resources (highly-qualified personnel).

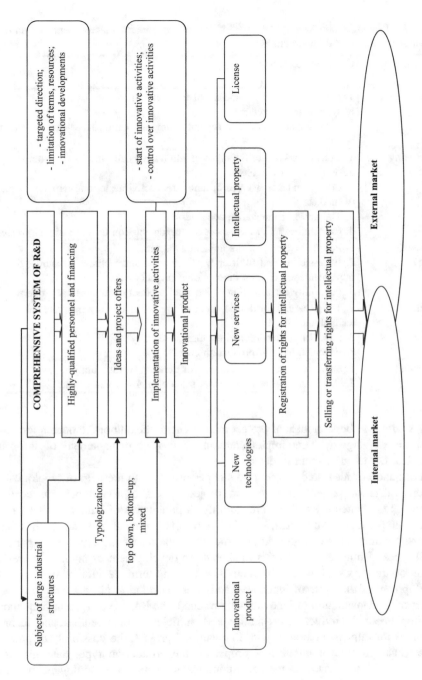

Fig. 3. The authors' model of project management of innovative activities of large industrial structures

Table 1. Typologization within the authors' model of project management of innovative activities of large industrial structures

Type of criterion	Group indicators
Generalized type	1. Compatibility of parameters of innovative activities of large entrepreneurial structures 2. Indicator of risk of each separate company within large entrepreneurial structures
Marketing type	1. Correspondence of the types of innovative activities to short-term development of markets 2. Correspondence of types of innovative activities to long-term development of markets 3. The cycle of innovative activities 4. Cost indicators in the aspect of implementation of innovational projects
Type of R&D	1. Patent frequency of the implemented product 2. Accessibility and sufficiency of scientific and technical resources for implementation of joint innovational project 3. Possibility for patenting of the final product of intellectual property
Production type	1. The necessity for development of the corresponding innovational technologies 2. Level of production and ecological security 3. Expected level of production costs
Financial type	1. Sufficiency of financial assets for creation of total budget 2. Distribution of revenues from sales of innovational product

As with the above models of project management, the authors' model is formed on the basis of stage-by-stage implementation of project management of innovative activities. Let us view them in detail.

Formation of ideas and project offers is performed with the help of typologization. At this stage of project management, project offers of five types are evaluated (Table 1): the generalized type (compatibility of parameters of innovative activities of large entrepreneurial structures, calculation of the indicator risk for each separate company within large entrepreneurial structures); the marketing type (correspondence of all types of innovative activities to short-term development of markets, correspondence of the types of innovative activities to long-term development of markets, development of the cycle of innovative activities, calculation of cost indicators in the aspect of implementation of innovational projects); the R&D type (patent frequency of the implemented product, accessibility and sufficiency of scientific and technical resources for implementation of joint innovational project, the possibility of patenting of the final product of intellectual property); the production type (determining the necessity for development of the corresponding innovational technologies, evaluating the level of production and ecological security, forecasting the expected level of production costs); the financial type (sufficiency of financial assets for creation of the total budget, distribution of the revenues from future sales of the innovational product). Based on evaluation of group indicators of the project within innovative activities, the approaches to project management are formed. Three main approaches are viewed within this model: top-down, bottom-up, and mixed. The centralized approach to project management top-down is oriented at formation of domestic technologies that stimulate creation of innovational products and services. The decentralized approach bottom-up includes orientation at using foreign technologies that stimulate emergence of intellectual property similar to foreign samples. The mixed approach allows correcting the approaches top-down and bottom-up depending on emerging tasks within innovative activities.

Based on the selected approaches, implementation of innovative activities is begun, similar to the algorithm of the Russian model of project management of innovative activities of large industrial structures. The final stage of implementation of innovative activities is development of innovational and/or licensed product, new technologies and services, and intellectual property. In the Russian and foreign models of project management of innovative activities, this stage is the final one. We think that it is important to set further movement of the developed product, for which the right of intellectual property is to be formed, sale is to be performed, and/or transition of rights for intellectual property in the internal and/or external markets of intellectual resources is to be conducted.

6 Conclusions

According to the determined theses, the authors have studied the Russian and the foreign models of project management of innovative activities in large entrepreneurial structures. Thus, the Russian model of project management is largely aimed at implementation of the tasks of domestic economic activities of industrial structures. On the contrary, the foreign model of project management of innovative activities is connected to the technological development of the country's economy. These conclusions allowed offering the proprietary model of project management of innovative activities of large industrial structures in which the mechanism of the measures aimed at implementation of the final innovational product is determined. The presented authors' model allows determining the level of priority of the projects implemented within innovative activities; forming the most probable results of innovative activities; thirdly, modeling the optimal structure of innovational works in the aspect of selected approaches.

References

Bryde, D.J.: Establishing a project organization and a project-management process for telecommunications project management. Int. J. Proj. Manage. **13**(1), 25–31 (1995)

Hutchins, G.: Strategic planning for project management using a project management maturity model. Proj. Manage. J. **32**(4), 61 (2001)

Kaab, M.: Construction project management and risk management. Int. Sci. Rev. **8**(18), 20–26 (2016)

Lyapina, I., Stroeva, O., Vlasova, M., Konobeeva, O., Konobeeva, E.: Approaches to organization of project management in Russia. In: Popkova, E.G., et al. (eds.) Integration and Clustering for Sustainable Economic Growth (Ser. Contributions to Economics), pp. 91–99. Springer (2017)

Mirzoeva, S.M.: Peculiarities of definitions "management of projects" and "project management". Econ. Entrepreneurship 10-1(63-1), 1118–1120 (2015)

Yaluner, E.V., Chernyshev, E.A.: Project management as a creative element of managing the projects and systems of management. Perspect. Sci. 12(87), 88–91 (2016)

Cooke-Davies, T.J., Arzymanow, A.: The maturity of project management in different industries – an investigation into variations between project management models. Int. J. Proj. Manage. 21(6), 471–478 (2003)

Komarevtseva, O.O.: The construction of the "Edgeworth box" to determine the effectiveness of changes in the municipality. Sci. Soc. 2(1), 36–46 (2017)

Rose, K.: Modern project management: successfully integrating project management knowledge areas and processes. Proj. Manage. J. 33(1), 60 (2002)

Smyth, H.: Relationship Management and the Management of Projects, 290 p (2014)

Danchenko, E.B.: Strategic management of business through the prism of managing innovational projects and the program. East Eur. J. Lead. Technol. 1(6(49)), 31–33 (2011)

Yee-Pagulayan, P.: Communicating project management: the integrated vocabulary of project management and systems engineering. Softw. Qual. Prof. 5(4), 44 (2003)

Clark, I., Colling, T.: The management of human resources in project management-LED organizations. Pers. Rev. 34(2), 178–191 (2005)

Simonova, E.V., Lyapina, I.R., Kovanova, E.S., Eldyaeva, N.A., Sibirskaya, E.V.: Characteristics of interaction between small innovational and large business for the purpose of increase of their competitiveness. In: Popkova, E.G. (ed.) Russia and the European Union, Development and Perspectives, pp. 407–413. Springer (2017)

Arslanov, S.D.: Regarding the issue of modernization of the knowledge-intensive sector of the modern entrepreneurship. Int. J. Appl. Fundam. Res. 10-1, 123–128 (2015)

Kalabina, E.G.: New industrialization, technological changes, and the labor sphere of industrial companies. Bull. OMSK Univ. Ser. Econ. 1(57), 72–81 (2017)

Lukmanova, I.G., Khurshudyan, G.V.: The strategy of development of large production structures in the crisis conditions. Econ. Entrepreneurship 3-2(80-2), 813–815 (2017)

Pavlovskaya, I.G.: The methodological approaches to formation of the organizational and economic mechanism of innovational development of large entrepreneurial structures. Drucker Bull. 2(2(10)), 249–255 (2016)

Anderson, D.K., Merna, T.: Project management strategy-project management represented as a process based set of management domains and the consequences for project management strategy. Int. J. Proj. Manage. 21(6), 387–393 (2003)

Lechler, T.G., Dvir, D.: An alternative taxonomy of project management structures: linking project management structures and project success. IEEE Trans. Eng. Manage. 57(2), 198–210 (2010)

Anastasova, A.S., Nikushina, A.N., Pavlova, A.S., Sarafanov, A.D.: Management of projects: peculiarities of project management in Russia. Theory Pract. Modern Sci. 11(17), 29–31 (2016)

Furta, S.D., Solomatina, T.B., Popova, M.M.: Similarities between risk management and management of project's stakeholder. In: Initiatives of the 21st Century, no. 3-4, pp. 13–19 (2016)

Khovaev, S.Y., Kozhevnikov, A.D.: From management of projects to the project mechanism of implementation of the corporate strategy: development of the conceptual model. Econ. Entrepreneurship **6**(71), 429–438 (2016)

Moscow branch of Project Management Institute (2017). https://pmi.ru

Economics & Statistics Administration United States Department of Commerce (2017). http://www.esa.doc.gov

Increase of Effectiveness of Public Welfare Through Improvement of the System of Investment Management

Alla L. Lazarenko[✉], Svetlana A. Orlova, Irina A. Rykova,
Irina M. Golaydo, and Elena E. Uvarova

Orel State University of Economics and Trade, Orel, Russia
orel-osu@mail.ru, super-ya-57@mail.ru, osuet@mail.ru,
my-orel-57@mail.ru, 1278orel@mail.ru

Abstract. Changeability of economic processes and formation of the modern reality of socio-economic development requires the search for new tools that increase the level of public welfare. The process of consumption and production provides less influence on the improvement of economic and social welfare of population and state. This process is being replaced by investment management. According to this, the purpose of this article is to form the main approaches to increase of effectiveness of public welfare through improvement of the system of investment management. According to this, it is necessary to solve the following tasks: study the theoretical basis of scientific ideas in the sphere of increase of effectiveness of public welfare by means of investment management; determine certain macro-economic approaches to increase of public welfare trough investment management; characterize the public-private and project approach to investment management; form the authors' idea of improvement of the system of investment management for possible increase of public welfare effectiveness. The research tools include the method of theoretical analysis, deduction, comparative evaluation, specification and generalization, and object tracking. The novelty of the research consists in the attempt to substantiate the interconnection between increase of effectiveness of public welfare and improvement of the system of investment management. This research will be of interest to a rather wide circle: scientists dealing with the issues of increasing the level of public welfare, researchers studying the issues of modernization and transformation of the system of investment management, and practical specialists conducting strategic research in the sphere of improvement of socio-economic development of territorial subjects.

Keywords: Investments · Management · Individual welfare · Public welfare
Effectiveness · Efficiency · Project approach · Turbulence
Public private approach

1 Introduction

Over the course of humanity's development, individuals and society tried to improve their welfare. The tools for increase of public welfare included taxes, factors of economic growth, accumulation of savings, growth of wages, and increase of labor

© Springer International Publishing AG, part of Springer Nature 2018
E. G. Popkova (Ed.): HOSMC 2017, AISC 622, pp. 16–29, 2018.
https://doi.org/10.1007/978-3-319-75383-6_2

efficiency. These tools allowed creating the "illusory" idea of increase of public welfare effectiveness by means of increase of individual indicators of welfare. However, the system of investment management as a tool of increase of public welfare effectiveness has not been studied sufficiently. Investments – as capitalizes and non-capitalized money assets – form the welfare of an individual consumer and create public goods. This thesis could be viewed on the basis of three targeted aspects of investment management. Firstly, effective investment management at the company's level maximizes private profit of investors and owners of business. Increase of profit is the indicator of welfare of an economic subject. The tax and insurance interest of obtained profit forms public welfare on the basis of state's taking fixed liabilities.

Secondly, investment management supposes re-investing of profit for further increase of internal norm of profitability within the investment projects. Investment business projects are a complex of interconnected measures, aimed at the achievement of a specific goal by means of formed investment resources (including human resources – individual subjects of welfare). Realization of investment projects leads to increase of material welfare of these subjects (from the received investment revenues) and non-material welfare of the public sector (from the effect of feedback within the investment project).

Thirdly, investment management of the public sector stimulates the process of reproduction of the main funds that ensure sectorial increase of public welfare. This process is primarily caused by sectorial priority within the state investment programs. As of now, these sectors include agriculture, processing industry, chemical production, machine building and transport complex, communications and telecommunications, and production and distribution of electric energy. These spheres are provided with investments on the basis of implementation of investment and innovational projects in the sphere of production technologization. Thus, growth of internal indicators of development of top-priority sector of economy stimulates the increase of welfare of the population which directly participates in the process of implementation of investments in these directions.

According to this, the role of investment management within the increase of public welfare effectiveness is rather large. Topicality and significance of this are confirmed by the fact that the search for new directions of the country's economic growth requires application and approbation of various tools in the issues related to increase of the level of gross domestic product. Public welfare is an important indicator that reflects these directions. At that, increase of effectiveness of public welfare stimulates the positive structural shift within the economic growth, while reduction of welfare leads to the opposite process. The substantiated position of the authors within this topic of the research set the goal of forming the main approaches to increase of effectiveness of public welfare through improvement of the system of investment management. According to this, it is necessary to solve the following tasks:

– consider the theoretical basis of scientific ideas in the sphere of increase of public welfare effectiveness by means of investment management;
– determine certain macro-economic approaches to increase of effectiveness of public welfare through investment management;

- characterize the public-private and project approach to investment management;
- form the proprietary idea of improvement the system of investment management for possible increase of public welfare effectiveness.

The main tools of this research include the method of theoretical analysis, deduction, comparative evaluation, specification and generalization, and object tracking.

2   The Theoretical Basis of Scientific Ideas in the Sphere of Increase of Public Welfare Effectiveness by Means of Investment Management

The preconditions of the theoretical basis of increase of effectiveness of welfare of society and individual were studied in the 18th century. This period started the macro-economic studies of the influence of various factors on the level of population's welfare. The founders of the theories and concepts of welfare are such scholars as L. Abalkin, T. Veblen, J. Galbraith, H. Gossen, J. Keynes, N. Nakamura, D. North, W. Eucken, V. Pareto, A. Pigou, D. Ricardo, W. Rostow, A. Rubinstein, P. Samuelson, A. Smith, A. Sen, J. Heckman, J. Hicks, and L. Errhard. Macro-economic ideas of these authors are studies in the works of Blaug (1991; 2002). Based on the theoretical evaluation of these articles, the main ideas of the fundamentalist scholars in the sphere of priority of individual welfare over public welfare and vice versa were formed (Table 1).

Table 1. Ideas of selection of priorities of individual welfare over public welfare

Stages	Priority of public welfare	Priority of individual welfare	Average priority of individual and public welfare
Idea	Public welfare is viewed as a totality of individual benefits and aggregation of the total level of public welfare	Individual welfare is viewed as the volume of consumption of various benefits (material and non-material) by individual, which forms a competitive market of demand and offer	At various stages, priorities shift from individual welfare to public welfare and vice versa
1750–1950	A. Smith D. Ricardo A Pigou J. Keynes	H. Gossen	V. Pareto
1951–1980	W. Eucken L. Erhard	T. Veblen J. Galbraith	J. Hicks P. Samuelson
1980–2000	L. Abalkin	W. Rostow	N. Nakamura
2000–2016	A. Rubinstein	A. Sen	D. North J. Heckman

In the modern conditions, these concepts allowed forming the groups of scholars that study the issue of increase of public welfare effectiveness through:

1. Effective consumption - Bateman and Thorp (2007), Deli and Varma (2002), Lens (2002), Kushnareva (2008), Titova (1999). The higher the effectiveness of consumption, the higher the growth of the rates of public welfare.
2. Effective production - Irani (2010), Rutkauskas and Žilinskij (2010), Xiaomeng (2016). Only production can increase the effectiveness of public welfare. Firstly, on the basis of total employment; secondly, with the help of increase of wages; thirdly, through implementation of state programs for formation and restoration of production funds.
3. Effective distribution - Flint-Ashery (2015), Mackenzie (2000), Kolmakov (2013), Shkirenko (2014). A manufactured product must be effectively distributed among the consumers. Effective distribution stimulates production of a new commodity, which leads to constant employment of the population.
4. Formation of the progressive scale of taxation - Besley and Coate (2003), Dzhetpisova (2011), Chigvintseva (2011). Public welfare can be "fed" only by the tax resources which are used for restoration of ineffective expenditures of the state.
5. Improvement of the system of investment management - Blekesaune (2007), Komarevtseva (2017), Lyapina et al. (2017), Yeo and Qiu (2003), Litvinenko (2008), Romanova (2011), Khachaturyan (2010). This concept is based on studying the process of investment management, namely evaluation of the algorithm of investments' movement in various sectors of economy. It should be noted that this concept includes the elements of consumption, production, and distribution. For a more detailed study if this issue, let us determine certain macro-economic approaches to increase of effectiveness of public welfare through investment management.

3 Certain Macro-economic Approaches to Increase of Effectiveness of Public Welfare Through Investment Management

Within the concept of improving the system of management for increase of public welfare effectiveness, there are four main approaches which are presented in Fig. 1.

1. The market approach, based on interaction between manufacturers and market capitalists in the process of formation of individual and public welfare. An important resource of this approach is investment capital, which is implemented in the competitive market by means of the production process, not demand and offer. The main subject that receives welfare is the capitalist individual. The effectiveness of public welfare could be raised only when the capitalist individual receives the necessary level of revenues from the investment capital.
2. The equilibrium approach reflects social inequality of society in the aspect of welfare. It is possible to note that the direction of this approach is related to the scientific ideas of K. Marx. Increase of effectiveness of public welfare can be achieved in the conditions of equal possibilities and revenues.

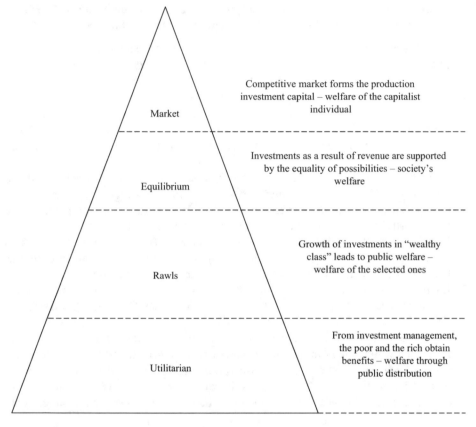

Fig. 1. Certain macro-economic approaches to increase of public welfare through investment management

The main subject – recipient of welfare – is society. The provided equilibrium approach does not take into account the factors related to reduction of motivation in struggle for resources and their effective use. We think that this approach can be implemented only if the representatives of this direction make an assumption related to existence of a certain error in social inequality.

3. The Rawls approach divides the whole society into two unequal classes: rich and poor. Growth of investments in the "rich class" will lead to public welfare. The existing approach is based on priority of the subject – recipient of welfare – which is the "rich class", or "selected class". This approach is a pure product of the capitalist society in which only the rich people can possess and use the limited resources. At that, an important condition is maximization of the received revenues of the "poor class". It is possible to increase effectiveness of public welfare on the basis of growth investments of the former and maximization of the revenues of the latter.

4. The utilitarian approach is a continuation of the previous approach – at that, it is viewed from the position of usefulness. The poor and the rich receive various levels

of usefulness from investment management. According to this, society has to have distribution of revenues, the subject of which is the state. This distribution will lead to growth of labor efficiency and will artificially increase effectiveness of society's welfare.

The above approaches open the theoretical significance of the macro-economic concepts in the sphere of increase of public welfare effectiveness. However, with formation of a new economic environment and change of public formations over a long period of time, different approaches to studying the issue of increase of public welfare effectiveness through investment management appeared. These approaches were built on the country's peculiarities of managing certain types of investments. Therefore, we deem it necessary to characterize the public-private and project approaches to investment management.

4 The Public-Private Approach to Redistribution of Investments Within Top-Priority Projects

The public-private approach (Fig. 2) of redistribution of investments within top-priority projects is based on studying the Russian market of direct and alternative investments, as well as analysis of the implemented investment programs in 2014–2016. The main drawbacks of the implemented approach are reflected in the following theses:

- investment market of public-private partnership is non-systemic;
- interaction between the state and investors is ineffective;
- the process of investment management within this approach is ineffective.

Interaction between the state and private investors within the federal targeted investment programs and projects is formed between certain subjects of management (Fig. 3).

Despite the fact that small structures of business are presented in the system of investment management, they do not participate within the interaction between state structures. Of course, limitation of the subjects of small business does not allow for direct interaction with the federal level in large-scale projects (the main share belongs to large business structures – 44.80% and corporations – 26.40%). At that, the lack of interaction between the municipal level and subjects of small business within investment management is not clear. Thus, the aspect of inefficiency of interaction within the public-private approach to redistribution of investments is obvious.

Non-systemic character and lack of coordination of the public-private approach is one of the critical drawbacks. According to Fig. 2, the goals of the management subjects within this approach are different (the state – additional investments, investors – additional revenues, population – infrastructural and labor effect). Despite the sustainability and regularity of these goals, redistribution of investments within top-priority projects should be based on at least one common goal for all territorial subjects. The process of investment management in the public-private approach is

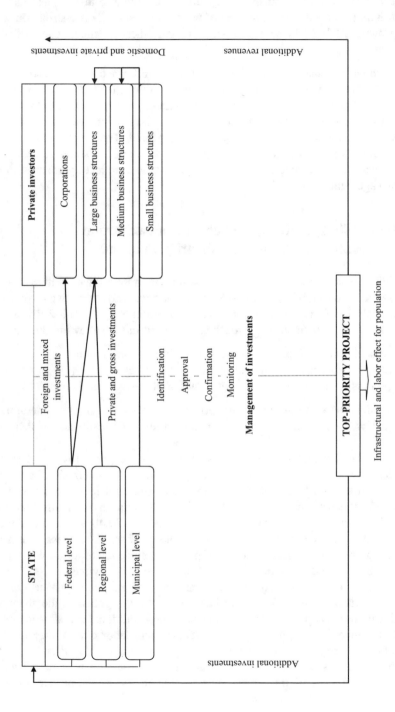

Fig. 2. The public-private approach to redistribution of investments within top-priority projects

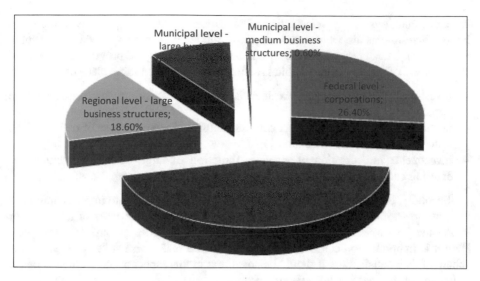

Fig. 3. Interaction between the state and investors (per cent ration) (Overviews 2017)

implemented on the basis of the five-stage algorithm of formation of the investment project (a set of steps (stages) within the investment project could be larger). At that, investment management should be conducted at all stages of interaction between the state and private investors. In the Russian realia, this process starts only after identification, approval, confirmation, and monitoring of the investment interaction. Besides, it should be noted that in the process of interaction between the state and private investors, the priority is given to private (44.2%), foreign (36.5%), gross (12.2%), and mixed (8.4%) investments (Overviews 2017). We think that limitation of choice also influences the investment management within increase of public welfare effectiveness.

5 The Project Approach to Investment Management (The West European Approach)

The project approach to investment management, which is often called in foreign literature the West European approach, is somewhat different (Fig. 4). According to this approach, investment management is conducted through interaction between three main subjects of the project environment: population, state, and economic & entrepreneurial sector. It should be noted that the economic & entrepreneurial sector is business structures that function in the internal and external sectorial environment of the state on the basis of the general principles of interaction. These structures include representatives of transnational corporations and subjects of small business. This peculiarity is distinguished by the authors in order to show the difference of the project approach from the public-private approach. The presented subjects have a range of criterial limitations for implementation of investment projects together with the state. Firstly, investments are viewed as a project of the state. According to this, investment management takes place within the state investment projects and programs. Increase of effectiveness of public welfare within investment management takes place by means of:

- investments for solving complex financial tasks (for specific financial issues of the population);
- investments for absorption of economic structures (for the purpose of revival of the state-controlled sector of economy);
- investments into creation of alliances (creation of large business structures for entering external markets).

Secondly, increase of effectiveness of public welfare is related to implementation of the state investment strategy. For this, the state does not set the tasks of attraction or taking the investment resources from the economic activities of business structures. Here it is important to determine the state needs and make wise state decisions in the sphere of their satisfaction. It should be noted that in this aspect public welfare is raised by means of the measures of effective economy.

The subjects presented in Fig. 4 interact in the process of managing various types of investments. This division is caused by the final goals of the management subjects:

- population (low-yielding and low-risk operations) – are used for preserving the current level of welfare;
- state (zero-yielding investments and disinvestments) – are used for receiving the economic and social effect, as well as supporting own functioning on the basis of taking the state capital from the investment turnover;
- economic & entrepreneurial sector (high-, medium-yielding investments, high-, medium-risk investments, reinvestments) – are used for receipt of additional and excess profit within investment capital, as well as the repeated process of investing the received profit.

On the whole, it is possible to note that this approach to investment management is more rational than the public-private approach. At that, the West European approach is based on the postulates of the capitalist system, developed investment market with the elements of state regulation and participation of the population in the investments management. In the Russian Federation, the practice of application of the project approach to investment management requires adaptation to economic conditions of the country's development. We think that investment management should form in the system that has clear interconnections between population and business, and the state is just an intermediary between these subjects. For that, let us form the proprietary approach to investment management for increase of public welfare effectiveness.

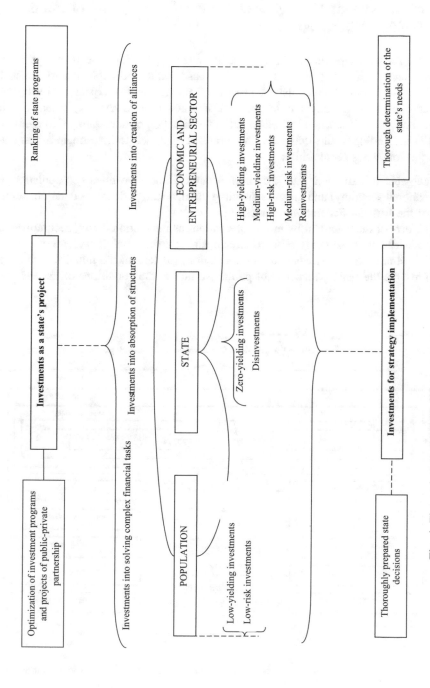

Fig. 4. The project approach to investment management (the West-European approach)

6 The Proprietary Idea of Improving the System of Investment Management for Possible Increase of Public Welfare Effectiveness

The proprietary approach to investment management for increase of public welfare effectiveness, which is presented in Fig. 5, is based on the principle of turbulence. The principle of turbulence includes the process of management of objects and subjects of specific phenomena that are in constant interaction with the implemented initial functions and parameters of development in the situation of uncertainty (Komarevtseva 2017). According to this definition, the turbulent approach to investment management has the following peculiarities:

– subjects are divided into the ones accumulating investments (population and national economy) and the ones directing the investment flows (Federation, subjects of the Russian Federation, municipal entities);
– objects of management by means of continuous interaction direct, accumulate, and redistribute investments between all subjects that form public welfare;
– investments that participate in the interaction, implement the initial functions, and preserve the main parameters of public welfare in the conditions of uncertainty.

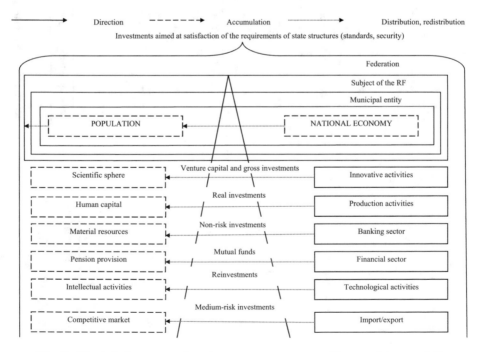

Fig. 5. The turbulent approach to investment management (the proprietary model)

Within the turbulent approach, the process of increase of public welfare effectiveness is implemented on the basis of interaction between the national economy and the population. The state is only an intermediary in this interaction, performing the functions of the initial control, monitoring, and preferential help. The management objects, presented in the turbulent approach, are the spheres in which direction, accumulation, and redistribution of investment resources takes place.

The investment resources are completely different forms of investments than the ones presented in Figs. 2 and 4:

- venture capital and gross investments – free investment resources, aimed at formation, creation, and purchase of new innovational means that allow reinvesting the capital;
- real investments – investment resources used in the process of investing into the object of intellectual property, material and non-material assets;
- low-risk investments – used for preservation of the current level of the society's welfare on the basis of short-term fixed investments;
- mutual funds – allow forming long-term accumulations on the basis of shared investments;
- reinvesting is the repeated process of investing the received profit for the purpose of maximization of the received initial income;
- medium-risk investments – investment resources that form the profitable part of the margin, corrected for the average value of risk within the investment market.

Management of presented resources allows increasing the effectiveness of welfare of certain sectors of national economy and directions of professional activities of the population. Thus, for example, venture capital and gross investments, which are used in the innovational sphere, allow increasing the effectiveness of public welfare in the scientific sphere; real investments from the production sector increase welfare within the development of human capital; low-risk investments preserve the formed financial capital of the population, etc.

On the whole, the turbulent approach distinguishes the most important investment tools that are to be used during improvement of the management system and ensures the rational interaction between all subjects that participate in the increase of effectiveness of public welfare.

7 Conclusions

The performed research, aimed at consideration of the possibility of increase of public welfare effectiveness through improvement of the system of investment management allowed forming a range of important conclusions. Increase of effectiveness of public welfare is related to the search for new sources of economic growth. The viewed process over the recent decades allowed forming certain factors that influence the effectiveness of public welfare. Theoretical analysis of the scientific works of foreign and Russian scholars allowed determining that these factors include the process of effective consumption, production, and distribution. Certain scholars viewed the interaction between the increase of public welfare effectiveness depending on implementation of the progressive scale of taxation. However, limitation of resources of economic development took the issue of increase of public welfare effectiveness into the sphere of improvement of the investment management. At that, the process of investment management is based on certain approaches. The authors determined the theoretical macro-economic approaches to investment management (market, equilibrium, Rawls, and utilitarian) and practical territorial approaches (the public-private and the project approaches). The main drawbacks of these approaches are lack of adaptation to the Russian realia of increase of public welfare effectiveness, non-systemic character of consideration of the investment market, inefficiency of interaction between subjects in the management system, and ineffectiveness of the process of investment management. According to these drawbacks, the proprietary approach to increase of effectiveness of public welfare through improving the system of investment management was formed. This approach is based on the principle of turbulence, which allowed dividing the subjects of increase of public welfare effectiveness into the ones cumulating investments and the ones directing investment flows; grouping the management objects on the basis of continuous interaction of the processes of direction, accumulation, and redistribution of investments; implementing the initial functions and preserving the main parameters of public welfare in the conditions of uncertainty. We think that this approach distinguishes the most important investment tools that are to be used during improvement of the management system and ensures the most rational interaction between all subjects that participate in the increase of public welfare effectiveness.

References

Blaug, M.: Economic Theories, True or False? Essays in the History and Methodology of Economics. Manchester School of Economic and Social Studies, vol. 59, pp. 91–92 (1991)

Blaug, M.: The classical theory of economic growth. Economica 69(274), 340 (2002)

Bateman, H., Thorp, S.: Decentralized investment management: an analysis of profit pension funds. J. Pension Econ. Fin. 6(1), 44 (2007)

Deli, D.N., Varma, R.: Contracting in the investment management industry: evidence from mutual funds. J. Fin. Econ. 63(1), 79–98 (2002)

Lens, V.: Public voices and public policy: changing the societal discourse on 'Welfare'. J. Sociol. Soc. Welfare 29(1), 137–154 (2002)

Kushnareva, O.: Regulating public welfare: a regional level. In: Problems of the Theory and Practice of Management, vol. 2, pp. 29–35 (2008)

Titova, N.E.: Individual and public welfare in West economic concepts. Econ. Sci. 1, 46–56 (1999)

Irani, Z.: Investment evaluation within project management: an information systems perspective. J. Oper. Res. Soc. 61(6), 917–928 (2010)

Rutkauskas, A.V., Žilinskij, G.: Financial leverage usage for active management of the investment portfolio. Bus. Theory Pract. 11(3), 194–203 (2010)

Xiaomeng, Zh.: Institutional transition from welfare enterprise to social enterprise: the localization of legislation and in Chinese context. J. Inst. Stud. 8(1), 121–131 (2016)

Flint-Ashery, S.: Public welfare or sectarianism: a new challenge for planning. Plann. Theory Pract. 16(3), 299–318 (2015)

Mackenzie, R.: Protecting transnational investment in intellectual property rights: legal issues and risk management strategies. Eur. Bus. Law Rev. **11**(2), 105–111 (2000)

Kolmakov, A.N.: The theoretical aspects of development of the system of public welfare. Bull. Tambov Univ. Ser. Humanit. Sci. **8**(124), 27–32 (2013)

Shkirenko, G.A.: Individual and public welfare – modern tendencies. Bull. Voronezh State Univ. Ser. Econ. Manage. **1**, 102–107 (2014)

Besley, T., Coate, S.: On the public choice critique of welfare economics. Pub. Choice **114**(3–4), 253–273 (2003)

Dzhetpisova, A.B.: The conceptual foundations of investment management. Bull. Astrakhan State Tech. Univ. Ser. Econ. **2**, 25–34 (2011)

Chigvintseva, E.S.: Public welfare in the context of a new quality of socio-economic relations. Terra Economicus **9**(3–2), 23–26 (2011)

Blekesaune, M.: Economic conditions and public attitudes to welfare policies. Eur. Sociol. Rev. **23**(3), 393 (2007)

Komarevtseva, O.O.: Smart city technologies: new barriers to investment or a method for solving the economic problems of municipalities? R-Econ. **3**(1), 32–39 (2017)

Lyapina, I., Stroeva, O., Vlasova, M., Petrukhina, E.: Strategic approach to investment activities management in russian industry [Электронный ресурс]. In: Popkova, E.G. (ed.) Russia and the European Union, Development and Perspectives (Ser. Contributions to Economics), pp. 333–339. Springer (2017)

Yeo, K.T., Qiu, F.: The value of management flexibility a real option approach to investment evaluation. Int. J. Proj. Manage. **21**(4), 243–250 (2003)

Litvinenko, V.A.: Directions and methods of intensification of the state investment policy in Russia's regions. Audit Fin. Anal. **6**, 314–324 (2008)

Romanova, L.A.: The issues of the investment process management. Sci. Anal. J. "Obozrevatel" – "Observer" **11**(262), 92–100 (2011)

Khachaturyan, N.M.: The potential of integration in managing the strategic investment development. Bull. V.N. Tatishchev Volga Univ. **20**, 37–43 (2010)

Overviews of the Russian market of alternative investments for 2013–2016. The official web-site of the managing investment company "RWM Capital" (2017). http://mfd.ru/commentfiles/2009/130813110056.pdf

Effective Import Substitution in the Agro-industrial Complex: Competition or Monopoly?

Aleksei V. Bogoviz[1]([⊠]) [iD], Svetlana V. Lobova[2] [iD],
and Yury A. Bugai[3] [iD]

[1] Federal State Budgetary Scientific Institution "Federal Research Center
of Agrarian Economy and Social Development of Rural Areas—All Russian
Research Institute of Agricultural Economics", Moscow, Russia
aleksei.bogoviz@gmail.com
[2] Altai State University, Barnaul, Russia
barnaulhome@mail.ru
[3] Altai State Agricultural University, Barnaul, Russia
yrbugai@mail.ru

Abstract. The purpose of this article is to determine the preferential market structure in the agro-industrial complex for provision of higher effectiveness of its import substitution and to develop practical recommendations for formation and support for optimal structure of this market. The methodology is based on the method of comparative analysis, which is used for comparing competition and monopoly as the variants of the market structure of the Russian agro-industrial complex. The authors use the method of analysis of statistical information for studying the state of the Russian agro-industrial complex. The authors have developed the method of evaluating the effectiveness of import substitution, which is also used in this research. The article studies the key factors that restrain the development of the process of import substitution of the Russian agro-industrial complex in 2016 and proves that oligopoly provides a very high quality of products with acceptable prices, as the market is presented by large companies that are easy to control, including the price control. In addition to this, they possess large resources, which, with the right management, allows achieving their high innovative activity and sustainability. Due to this, the market structure of oligopoly is the most effective as to import substitutions in the agro-industrial complex. As a result, the authors recommend application of flexible and preferential tools for stimulating the development of the process of import substitution in the agro-industrial complex, which should be based on the agro-industrial special economic areas.

Keywords: Effective import substitution · Agro-industrial complex
Competition · Monopoly · Special economic areas · Modern Russia

1 Introduction

At present, the model of economic development of modern Russia is reconsidered. On the path of creation of a theoretically perfect model of the post-industrial economy, the Russian economic system faced a range of problems that predetermined its

© Springer International Publishing AG, part of Springer Nature 2018
E. G. Popkova (Ed.): HOSMC 2017, AISC 622, pp. 30–36, 2018.
https://doi.org/10.1007/978-3-319-75383-6_3

inaccessibility in practice. The main such problem is violation of the international division of labor in the conditions of disintegration of economic systems under the influence of crisis phenomena of various nature.

Non-optimality of the conditions – primarily, in the external environment – for development of the post-industrial economy, which made the Russian economic system unstable and vulnerable against the influence of the smallest changes in the geo-political situation, is a substantial basis for change of the model of development. Mixed (diversified) economy was selected for such model, based on industry, with highly-developed service sphere which forms a half of gross domestic product.

In the process of post-industrialization of the Russian economy, the tertiary sector of economy has been developing, while the real sector was characterized by the deficit of investments and low innovational activity, which led to critical reduction of its competitiveness and high dependence of Russia on the import of industrial products.

If the dependence on import of most industrial products does not contradict the new model of Russia's economic development, import of agro-industrial products is unacceptable due to its strategic significance for provision of the national food security. The most important tool of overcoming the dependence on import of agro-industrial products, which is an essential aspect for creation of a new model of development of the Russian economy, is import substitution.

This actualizes the problem of search for the means of achieving high effectiveness of import substitution in agro-industrial complex (AIC). An important role here belongs to determining the optimal market structrure. The purpose of the article is to determine the preferential market structrure in AIC for provision of high effectiveness of its import substitution and to develop practical recommendations for formation and support of the optimal structure of this market.

2 Materials and Method

Methodology of this research is based on the method of comparative analysis, with the help of which competition and monopoly are compares, as variants of the market structure of the Russian agro-industrial complex. The authors also use the method of analysis of statistical information for studying the state of the Russian agro-industrial complex. The basic indicators and their values for the recent years are given in Table 1.

We developed the proprietary method of evaluating the effectiveness of import substitution, which is also used in this research. It supposes the usage of the following formula:

$$Eis = (QP/PP) * Cinn * Csus \tag{1}$$

where Eis   – indicator of effectiveness of import substitution in the market;
QP   – average market quality of products;
PP   – average market level of prices for products;
Cinn   – coefficient of innovativeness of enterprises in the market;
Csus   – coefficient of sustainability of enterprises against crises.

Table 1. Dynamics of indicators of development of the Russian agro-industrial complex in 2011–2016

Indicators	Values of indicators for the years					
	2011	2012	2013	2014	2015	2016
The volume of supplied goods of own production, RUB billion	3,262	3,602	4,001	4,272	4,840	5,861
Volume of import, RUB billion	2,184	2,550	2,442	2,598	2,394	1,590
Total volume of consumption, RUB billion	5,446	6,152	6,443	6,870	7,234	7,451
Share of domestic production, %	60	59	62	62	67	79
Level of using production capacities, %	66	70	69	66	67	74
Share of innovations-active enterprises, %	9.5	9.6	9.3	9	10.3	10.2
Number of enterprises	52,266	51,464	50,848	49,985	49,992	51,387
Concentration of production, %	15.1	16.3	14.6	21.2	22.6	18

Source: compiled by the authors on the basis of: (Federal State Statistics Service 2016), (Voronin 2017).

As is seen from Formula (1), the basis of measuring the effectiveness of import substitution is the ratio of products' quality in the market to the average level of prices. The indicators QP and PP are measures in points. The indicator QP can take the values from 1 (minimum value) to 10 (maximum value), and the indicator PP – from 10 (maximum price) to 1 (minimum price).

This ratio is multiplied by two coefficients. The coefficients Cinn and Csus are measures in points (in tenths of 1) and take values from 0.1 (minimum innovativeness/ sustainability) to 2.0 (maximum innovativeness/sustainability). Therefore, the indicator Eis can take values from 0.001 to 10. The higher its value, the higher the effectiveness of import substitution in the studied market. The developed method may be applied to any market, but in the context of this article it is applied for determining the effectiveness of various market structures in the market of AIC.

3   Discussion

The issues of effectiveness of import substitution in the AIC are studied in the works by (Bogoviz and Mezhov 2015), (Popkova et al. 2016), (Sadovnikova et al. 2013), (Popova et al. 2015), (Dudukalov et al. 2016), (Bogoviz et al. 2017), (Sandu et al. 2017), and (Przhedetskaya and Akopova 2015). However, despite the high level of elaboration of these issues, such aspects and the optimal sectorial structure, which stimulates maximization of effectiveness of import substitution in AIC, are not sufficiently studied in the existing publications and require deeper consideration.

4 Results

The results of analysis of data from Table 1 showed that in 2016 the share of the Russian products in the structure of offer in the AIC market constituted 79%, growing by 31.6% as compared to 2011 (in 2011 it constituted 60%). The tendency for strengthening the positions of the Russian enterprises in the AIC market is accompanied by the growth of the level of using the production capacities, which constituted 74% in 2016, growing by 12.1% as compared to 2011 (66%).

In the quantitative expression, growth of the number of enterprises of AIC was negative in the recent years. In 2016, their number constituted 51,387, reducing by 2.7% as compared to 2011 (52, 266). This is accompanied by growth of concentration of products in the AIC market – i.e., its monopolization. The level of concentration of 2016 constituted 18%, growing by 19.2%, as compared to 2011 (15.1%).

That is, at present, the structure of the Russian market of AIC can be characterized as monopolistic competition. The results of the performed comparative analysis of various market structures from the point of their effectiveness in the sphere of AIC are given in Table 2.

Table 2. Results of comparative analysis of various market structures from the point of view of effectiveness of import substitution in AIC

Criteria of comparison	Oligopoly	Monopolistic competition	Perfect competition
QP	8	6	5
PP	4	2	1
Cinn	1.0	1.5	0.3
Csus	2.0	0.2	0.1
Eis	4.0	0.9	0.15

Source: compiled by the authors.

As is seen from Table 2, we viewed three market structures: existing monopolistic competition in the Russian market of AIC, oligopoly as a manifestation of monopoly, the extreme level of which cannot be achieved in this market, and perfect competition. The existing structure of the AIC market – monopolistic competition – showed average effectiveness as to import substitution in this market.

Due to the complete action of the mechanism of competition and the enterprises' possession the minimum set of resources – as with the perfect competition – the monopolistic competition ensures the acceptable quality of products with low prices. However, due to the fact that market players represent a lot of small and medium companies, the accessible set of resources is not minimal but very small, so their innovational activity and sustainability is rather small. At this market structure, the indicator of effectiveness of import substitution in the AIC took the value of 0.9 points.

Despite the minimum level of prices, the market structure of competition leads to low quality of products, and the companies are characterized by low innovational capabilities and low sustainability die to minimum set of resources. Due to that, the market structure of competition leads to low effectiveness of import substitution in the AIC – 0.15 points.

Oligopoly provides a very high quality of products with acceptable prices, as the market is represented by large companies, which are easy to manage, including the price control. In addition to this, they possess large resources, which, with right management, allows achieving their high innovative activity and sustainability. Due to this, the market structure of oligopoly became most effective as to import substitution in AIC.

That's why for the purpose of achieving high effectiveness of import substitution in AIC we recommend to stimulate the realization of the started tendency of monopolization of this market, up to establishment of oligopoly. At that, it is necessary to form and support favorable conditions for getting maximum advantages from this market structure. In order to determine such conditions, let us use Fig. 1.

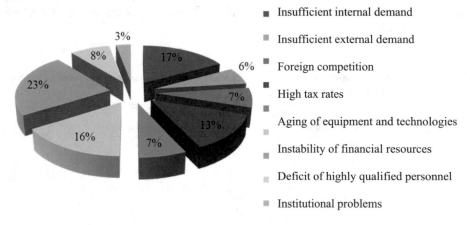

Fig. 1. The system of factors that restrain the development of the process of import substitution of AIC in Russian in 2016. Source: compiled by the authors on the basis of: (Federal State Statistics Service 2016).

As is seen from Fig. 1, the key factors that restrained the development of the process of import substitution of AIC in Russia in 2016 were high tax rates, instability of economic situation, inaccessibility of financial assets, etc. We recommend creation of agro-industrial special economic areas as a measure of influencing these factors in the interests of their transformation into the sources of growth and development of the Russian entrepreneurship in the AIC.

The special economic areas have tax preferences, financial assets are more accessible, and the economic situation is characterized by higher stability. At present, there are no special economic areas for enterprises of the AIC in Russia, which, in our opinion, hinders their development. This market is of high priority for the Russian economy, so the companies of the AIC have to receive access to the preferences provided within the special economic areas.

As compared to the current Russian policy for supporting the AIC companies, which supposes high expenses of the federal and territorial budgets for subsidizing, special economic areas are based on private investments. Moreover, unifying the companies of the AIC into special economic areas, the government will receive wider possibilities for their regulation, as, according to the Russian law, such possibilities do now allow setting strict requirements to the quality of products, innovative activity, and prices, which differs special economic areas.

5 Conclusions

It should be emphasized that AIC is not just a market, but has a strategic role for the Russian economy – especially in the conditions of unfavorable geo-political situation. This is a basis for application of flexible and preferential tools for stimulating the development of the process of import substitution into AIC, which should be based on agro-industrial economic areas.

The performed analysis showed that monopoly which is expressed in the form of oligopoly is the most preferable market structure in the AIC. The perfect competition, which is traditionally viewed as the most perspective market structure, showed the lowest effectiveness as to import substitution in the AIC. The perspectives of further scientific research in this sphere are seen in the context of specifying the tools of managing the process of restructuring the AIC market in favor of establishment of oligopoly.

References

Bogoviz, A., Mezhov, S.: Models and tools for research of innovation processes. Modern Appl. Sci. **9**(3), 159–172 (2015)

Dudukalov, E.V., Rodinorova, N.D., Sivakova, Y.E., et al.: Global innovational networks: sense and role on development of global economy. Contemp. Econ. **10**(4), 299–310 (2016)

Popkova, E.G., Shakhovskaya, L.S., Abramov, S.A., et al.: Ecological clusters as a tool of improving the environmental safety in developing countries. Environ. Dev. Sustain. **18**(4), 1049–1057 (2016)

Popova, L., Popova, S.A., Dugina, T.A., Korobeynikov, D.A., Korobeynikova, O.M.: Cluster policy in agrarian sphere in implementation of concept of economic growth. Eur. Res. Stud. J. **18**(Special Issue), 27–36 (2015)

Przhedetskaya, N., Akopova, E.: Institutional designing of continuous education in Russia under the conditions of neo-economy and globalization. Reg. Sectoral Econ. Stud. **15**(2), 115–122 (2015)

Sadovnikova, N., Parygin, D., Gnedkova, E., Kravets, A., Kizim, A., Ukustov, S.: Scenario forecasting of sustainable urban development based on cognitive model. In: Proceedings of the IADIS International Conference ICT, Society and Human Beings 2013, Proceedings of the IADIS International Conference e-Commerce 2013, pp. 115–119 (2013)

Bogoviz, A.V., Ragulina, Y.V., Shkodinsky, S.V., Babeshin, M.A.: Factors of provision of food security. Econ. Russ. Agric. **2**(1), 2–8 (2017)

Voronin, B.A.: Russian AIC – from import of agricultural products to export-oriented development. Fields Russ. **4**(148), 5–12 (2017)

Federal State Statistics Service. Industrial production in Russia. 2016: statistical collection. Moscow: Federal State Statistics Service (2016)

Sandu, I.S., Bogoviz, A.V., Ryzhenkova, N.E., Ragulina, Y.V.: Formation of the innovational infrastructure in the agrarian sector. AIC Econ. Manage. **1**(1), 35–41 (2017)

Import Substitution in the Agro-Industrial Complex in the Interests of Provision of Food Security: Option or Necessity?

Aleksei V. Bogoviz[1](✉) ⓘ, Yury A. Bugai[2] ⓘ,
and Vladimir S. Osipov[3,4,5] ⓘ

[1] Federal State Budgetary Scientific Institution "Federal Research Center
of Agrarian Economy and Social Development of Rural Areas—All Russian
Research Institute of Agricultural Economics", Moscow, Russia
aleksei.bogoviz@gmail.com
[2] Altai State Agricultural University, Barnaul, Russia
yrbugai@mail.ru
[3] Lomonosov Moscow State University, Moscow, Russia
vs.ossipov@gmail.com
[4] Russian State Agrarian University - Moscow Timiryazev Agricultural
Academy, Moscow, Russia
[5] All-Russian Research Institute of Potato Farming by A.G. Lorh,
Moscow, Russia

Abstract. The purpose of the research is to determine the value and specifics of
the need for provision of national food security in modern Russia, to evaluate
the level of necessity for using import substitution in the agro-industrial complex for solving this problem, and to search for the alternative means that allows
developing the agro-industrial complex without the critical load on the federal
budget. The authors offer the methodological approach to determining the level
of national food security and determine the expedience of application of import
substitution in the modern Russia's agro-industrial complex. It is proved that
import substitution in the agro-industrial complex in the interests of provision of
food security in modern Russia is a voluntary choice of the Russian government,
not the objective necessity. This choice contradicts the existing situation in the
agro-industrial complex and, instead of stimulating the development of the
Russian entrepreneurship in the agro-industrial complex, it leads to increase of
its dependence on the state financial support and the atrophy of the capability for
independent development. Clustering of agro-industrial complex is much more
effective for provision of the national food security in modern Russia, as an
alternative to import substitution. For its practical application, the authors offer a
cluster model of development of the Russian agro-industrial complex for the
purpose of provision of food security.

Keywords: Import substitution · Clustering
Companies of agro-industrial complex · National food security
Russia

© Springer International Publishing AG, part of Springer Nature 2018
E. G. Popkova (Ed.): HOSMC 2017, AISC 622, pp. 37–43, 2018.
https://doi.org/10.1007/978-3-319-75383-6_4

1 Introduction

In the crisis conditions, when dependence on import of strategic goods is unacceptable, the problem of the national economic security grows. An important role in solving it belongs to provision of the national food security, by eliminating the dependence on import of products of the agro-industrial complex. The main method of protecting the national interests in the food sphere is import substitution in the agro-industrial complex.

According to the existing scientific ideas and the international experience in provision of food security, the Russian government announced the necessity for import substitution in the agro-industrial complex and started to implement measures for stimulating this process. Due to specifics of the Russian economic system, related to domination of financial barriers on the path of development of entrepreneurship in the agro-industrial complex, these measures are brought down to provision of subsidies, which supposes large load on the federal budget.

Due to the long crisis, the budget resources are limited, and realization of measures for stimulating import substitution in the agro-industrial complex makes the Russian government reduce other expenditures – in particular, for socially important programs. In this article, the authors offer a hypothesis that these measures are not necessary, and provision of national food security of Russia does not require emphasis on import substitution and large expenditures of the federal budget.

The purpose of the research is to determine the value and specifics of the needs for provision of the national food security in modern Russia, to evaluate the level of necessity for using import substitution in the agro-industrial complex for solving this problem, and to search for an alternative means that allows developing the agro-industrial complex without the critical load on the federal budget.

2 Materials and Method

From the methodological point of view, this study offers the following approach to determining the level of the national food security:

$$\text{NFS}_t = (\text{Vaic}_t/\text{Caic}_t) * [(\text{Vaic}_t/\text{Caic}_t)/(\text{Vaic}_{t-5}/\text{Caic}_{t-5})] \tag{1}$$

where
NFS   – indicator of the level of national food security;
Vaic   – volume of national production of products of the agro-industrial complex
Caic   – volume of consumption of the product of the agro-industrial complex;
T   – time period (calendar year)

As is seen from Formula (1), the offered methodological approach is based on determining the current share of domestic production in the structure of consumption of the products of the agro-industrial complex and finding its products with its growth for the last five years. Both the static situation in the agro-industrial complex and the dynamics of its change are taken into account. The indicators Vaic and Caic are measured in the monetary items, and the indicator NFS – in the points, the tenths of 1. For the purpose of treatment of its values, we developed a special scale (Table 1).

Table 1. The scale for treatment of the value of the indicator of the level of national food security

Intervals of values of the indicator NFS and their economic sense			
$0.60 \geq$ NFS	$0.60 <$ NFS ≤ 0.75	$0.75 <$ NFS < 0.90	NFS ≥ 0.90
National food security under threat	Low level of national food security	Acceptable level of national food security	High level of national food security

Источник: составлено автором.

As is seen from Table 1, we distinguished four intervals of values of the indicator NFS. If its value is below or equals 0.60, this shows that national food security is under a threat. Import substitution is critically necessary in this case. If the value of this indicator is in the interval 0.60–0.75, it reflects the low level of national food security. The need for import substitution is rather strong.

If the value of the indicator NFS is in the interval 0.75–0.90, it shows the acceptable level of the national food security. The need for import substitution is preserved in this case, but it is at a low level. If this indicator takes the value that equals or exceeds 0.90, the level of national food security is high. Import substitution is not necessary.

3 Discussion

The theoretical and methodological issues of provision of national food security are studied in the works of such authors as (Bogoviz and Mezhov 2015), (Popkova et al. 2016), (Sadovnikova et al. 2013), and (Popova et al. 2015). The necessity for import substitution in the agro-industrial complex for provision of food security of modern Russia has been emphasized in the works of such authors as (Bogoviz et al. 2017), (Sandu et al. 2017), (Przhedetskaya and Akopova 2015), and (Gulyayeva et al. 2016). The theoretical problems of import substitution are viewed in the works (Osipov 2016), (Kosov et al. 2016), and (Gnezdova et al. 2016).

The performed overview of the modern scientific literature on the studied problem showed insufficient elaboration of the alternative tools of import substitution for provision of national food security and determination of the conditions at which import substitution or alternative tools are expedient to use.

4   Results

According to the official statistical information of the Federal State Statistics Service of the RF, the volume of consumption of the products of the agro-industrial complex in Russia in 2016 constituted RUB 7,451 billion, and the volume of the Russian production of the products of the agro-industrial complex constituted RUB 5,861 billion. In 2012, the values of these indicators constituted RUB 6,152 billion and RUB 3,602, respectively (Federal State Statistics Service 2016). Bases on these data, let us perform evaluation of national food security in 2016 with the help of the developed methodological approach:

$$\text{NFS}_{\text{Rus}(2016)} = (5,861/7,451) * ((5,861/7,451)/(3,602/6,152))$$
$$= 0.79 * (0.79/0.59) = 0.79 * 1.34 = 1.06.$$

The received value of the indicator NFS in 201 exceeded 0.9, which showed high level of national food security and the lack of necessity for import substitution. Thus, the share of national production of the agro-industrial complex products constitutes 79% - i.e., the share of import is minimal. At that, there has been growth of the share of the domestic production of the agro-industrial complex by 34% over the recent five years, which shows that import of the agro-industrial complex products is not a problem in modern Russia.

In this case, in the interests of provision of national food security, we recommend to leave the policy of import substitution, which supposes the active role of the state, related to supporting the domestic the agro-industrial complex manufacturers, and to realize the policy of clustering, which gives the state the secondary role of a passive observer and referee which creates and supports favorable conditions for independent development of domestic entrepreneurship in the agro-industrial complex.

That is, from the marketing point of view, it is expedient to implement the transition from the struggle for leadership, which is characterized by high resource intensity, to the strategy of keeping the leadership positions that have already been conquered by the Russian companies in the agro-industrial complex. This could be done with the help of clustering that allows strengthening the market positions of the Russian enterprises in the agro-industrial complex and providing them with the possibilities of further growth and development.

The policy of clustering supposes stimulation of "healthy" competition in the market, and thus it differs from the policy of import substitution, oriented at fighting competition. At that, during clustering state regulation of the agro-industrial complex gives way to market self-regulation.

For activation of cluster processes in the Russian agro-industrial complex, we recommend the state to establish formation of clusters at the federal level as a top-priority direction of development of the agro-industrial complex and support realization of cluster initiatives in the agro-industrial complex with legal and consultation support. After the formation of clusters in the agro-industrial complex, the state should use the following tools for their regulation:

- anti-monopoly policy: supporting highly-competitive environment that stimulates unification of the agro-industrial complex companies into clusters;
- tax policy: provision of tax preferences to members of clusters in the agro-industrial complex – e.g., subsidies for added value tax. Within the cluster, such measure will be effective, as the companies will be integrated horizontally and vertically in the cluster, and the liabilities for payment of this tax will not be shifted to other participants of the added value chain;
- credit & investment policy: providing cluster members in the agro-industrial complex with the access to subsidized credit resources and provision of tax preferences to external (as to the cluster) investors into development of cluster companies.

The offered cluster model of development of the Russian agro-industrial complex in the interests of provision of food security is shown in Fig. 1.

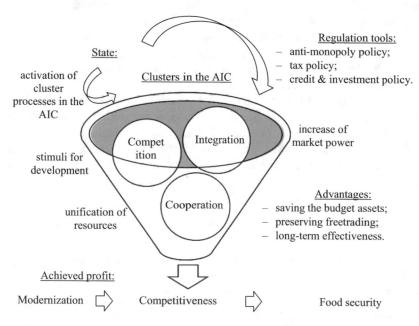

Fig. 1. Cluster model of development of the Russian the agro-industrial complex in the interests of provision of food security Source: compiled by the authors.

As is seen from Fig. 1, three forces are at work at the agricultural company: competition, which creates stimuli for their development, integration, which ensures the increase of their market power, and cooperation, which allows unifying their resources and realizing the potential for development. As a result, the modernization of the agricultural companies, their high competitiveness, and provision of the Russian food security are achieved.

An advantage of cluster policy, as compared to the policy of import substitution, is saving of budget assets, which is achieved by shifting the load for development of the national entrepreneurship in the agricultural sphere from the state to private business. Another advantage of clustering is preservation of freetrading, as clustering is a part of its concept, and import substitution is related to the opposite policy of protectionism. This allows Russia to correspond to the requirements of integration associations – such as the WTO.

Another advantage of clustering of the agro-industrial complex is the long-term effectiveness. If the subsidies, provided to the agricultural companies within the policy of import substitution are of short-term, cluster processes allow supporting high competitiveness of the agricultural companies in the long-term.

5   Conclusions

Thus, it was proved that import substitution in the agro-industrial complex in the interests of provision of food security in modern Russia is a voluntary choice of the Russian government, not the objective necessity. This choice is not substantiated and contradicts the existing situation in the agricultural complex; instead of stimulating the development of the Russian entrepreneurship in the agricultural complex, it leads to increase of its dependence on the state financial support and the atrophy of capability for independent development.

Clustering in the agro-industrial complex is much more effective as to provision of national food security in modern Russia – instead of the import substitution. During further scientific research, it is recommended to focus on the internal processes that take place in the agro-industrial cluster, as in this article attention is paid to the processes related to activation and regulation of clustering in the agro-industrial complex.

References

Bogoviz, A., Mezhov, S.: Models and tools for research of innovation processes. Modern Appl. Sci. **9**(3), 159–172 (2015)

Popkova, E.G., Shakhovskaya, L.S., Abramov, S.A., et al.: Ecological clusters as a tool of improving the environmental safety in developing countries. Environ. Dev. Sustain. **18**(4), 1049–1057 (2016)

Popova, L., Popova, S.A., Dugina, T.A., Korobeynikov, D.A., Korobeynikova, O.M.: Cluster policy in agrarian sphere in implementation of concept of economic growth. Eur. Res. Stud. J. **18**(Special Issue), 27–36 (2015)

Przhedetskaya, N., Akopova, E.: Institutional designing of continuous education in Russia under the conditions of neo-economy and globalization. Reg. Sectoral Econ. Stud. **15**(2), 115–122 (2015)

Sadovnikova, N., Parygin, D., Gnedkova, E., Kravets, A., Kizim, A., Ukustov, S.: Scenario forecasting of sustainable urban development based on cognitive model. In: Proceedings of the IADIS International Conference ICT, Society and Human Beings 2013, Proceedings of the IADIS International Conference e-Commerce 2013, pp. 115–119 (2013)

Bogoviz, A.V., Ragulina, Y.V., Shkodinsky, S.V., Babeshin, M.A.: Factors of provision of food security. Econ. Russ. Agric. **2**(1), 2–8 (2017)

Federal State Statistics Service. Industrial production in Russia. 2016: statistical collection. Federal State Statistics Service, Moscow (2016)

Sandu, I.S., Bogoviz, A.V., Ryzhenkova, N.E., Regulina, Y.V.: Formation of the innovational infrastructure in the agrarian sector. AIC Econ. Manage. **1**(1), 35–41 (2017)

Osipov, V.S., Skryl, T.V.: The strategic directions of the modern Russian economic development. Int. Bus. Manage. **10**(6), 710–717 (2016)

Kosov, M.E., Akhmadeev, R.G., Osipov, V.S., et al.: Socio-economic planning of the economy. Indian J. Sci. Technol. **9**(36), 102008 (2016)

Gulyayeva, T.I., Kuznetsova, T.M., Gnezdova, J.V., Veselovsky, M.Y., Avarskii, N.D.: Investing in innovation projects in Russia's agrifood complex. J. Internet Bank. Commer. **21**(6), 1–13 (2016)

Gnezdova, J.V., Kugelev, I.M., Romanova, I.N., Romanova, J.A.: Conceptual model of the territorial manufacturing cooperative system use in Russia. J. Internet Bank. Commer. **21**(Spec.issue 4) (2016)

Model of Innovational Development of Modern Russian Industry

Arutyun A. Khachaturyan[1](✉), Karine S. Khachaturyan[2],
and Arsen S. Abdulkadyrov[2]

[1] Institute of Market Issues of the Russian Academy of Sciences,
Moscow, Russia
karutyun@yandex.ru
[2] Military University of the Ministry of Defense of the Russian Federation,
Moscow, Russia
ars.rggu@mail.ru

Abstract. The purpose of the article is to develop the effective model of innovational development of the modern Russian industry. A proprietary model was developed for measuring the effectiveness of the model of industry's innovational development. With its help, the authors perform evaluation of the existing Russian model and determine the key problems of industry's innovational development in modern Russia. The results of the performed research showed that the current model of industry's innovational development in modern Russia, which is realized within modernization of the national economic system, is characterized by low effectiveness, and a new model should be developed. The offered new model of innovational development of modern Russian industry's innovational development allows restoring the logic of distribution of roles between participants of this process and ensuring its high effectiveness. The developed model is based on the action of market mechanism.

Keywords: Innovational development · Modern Russia · Modernization
Effectiveness · Competitiveness

1 Introduction

Economy of modern Russia is industry-oriented, or, as it is called, industrial. The attempts of following the modern model of production specialization according to the normative principle, that is on the basis of the set priorities of economy's development, made the Russian economy more susceptible to influence of crisis phenomena in the global economy and reduction of its global competitiveness.

This explains unreadiness and incapability of private business to implement the government plans for the country's post-industrialization. In order to restore the balance in the modern economic system, there's a need to return to the classic model of production specialization according to the positive principle, i.e., on the basis of possession of relative advantages in certain spheres of national economy.

Excess of natural resources provided Russia with relative advantages in the sphere of mineral industry, and existing production capacities and highly-qualified engineering personnel – in the sphere of processing industry. However, in the epoch of formation of knowledge economy in the global economic system, relative advantages are not sufficient for achieving high indicators of competitiveness – innovative activity also plays an important role.

This explains high topicality of studying the issues related to provision of innovational development of modern Russia's industry. Our hypothesis is that the current model of innovational development of modern Russia's industry, which is implemented within modernization of the national economic system, is characterized by low effectiveness, and it is required to build a new model – which is a purpose of this work.

2 Materials and Method

For measuring the effectiveness of the modern of industry's innovational development in this research, the following proprietary formula was developed:

$$Eidi = DAr(eff)/DAr(exp), \qquad (1)$$

где Eidi – indicator of effectiveness of the modern industry's innovational development;

DAr(eff) – direct average of growth rate of various indicators of efficiency of the model of industry's innovational development;

DAr(exp) – direct average of growth rate of various indicators of expenditures for provision of industry's innovational development.

If the estimate value of the indicator DAr(eff) or DAr(exp) is negative, it is equaled to zero. The value of the indicator Eidi is below or equals 1, which shows low effectiveness. The larger the indicator Eidi, the higher the effectiveness.

The indicators of efficiency of the model of innovational development of modern Russia's industry (type "R") are the number of organizations involved in research and development, number of personnel involved in scientific research and development, developed leading production technologies, implemented leading production technologies, and the share of organizations that conduct research and development.

The indicators of expenditures for provision of innovational development of modern Russia's industry (type "E") are financing of science from the assets of the federal budget, internal expenditures of organizations for R&D, and expenditures of organizations of the entrepreneurial sector for innovatons. Dynamics of values of these indicators for 2000–2016 according to the Federal State Statistics Service is shown in Table 1.

Table 1. Dynamics of indicators of innovational development of modern Russia's industry

Type	Indicator	2000	2016	Increase rate	Growth rate
P	Number of organizations that perform scientific R&D	4,099	3,604	−495	−0.12
R	Number of personnel involved in scientific R&D, thousand	887.7	732.3	−155.4	−0.18
R	Developed leading production technologies	1,138	1,398	260	0.23
R	Implemented leading production technologies	191,650	218,018	26,368	0.14
R	Share of organizations that perform R&D, %	11.1	10.9	−0.2	−0.02
E	Financing of science from the assets of the federal budget, RUB million	17,396.4	437,273.3	419,876.9	24.14
E	Internal expenditures for R&D, RUB million	76,697.1	847,527	770,829.9	10.05
e	Expenditures for innovations of the entrepreneurial sector's organizations, RUB million	14,326.2	145,836.9	131,510.7	9.18

Source: compiled by the author according to the materials: Federal State Statistics Service (2016).

3 Discussion

The sense and logic of the process of the industry's innovational development are studied in the works by (Khan 2017; Fujii and Managi 2016; Kinahan 2016; Anokhina et al. 2016; Bogoviz and Mezhov 2015; Duman and Kureková 2016). The peculiarities of innovational development of modern Russia's industry are studied in the works by (Przhedetskaya and Panasenkova 2014; Shakirtkhanov 2017; Trindade et al. 2016; Veselovsky et al. 2017; Korobkin, et al. 2015; Malyshkov and Ragulina 2014; Popkova 2017; Przhedetskaya and Akopova 2015).

4 Results

Based on the data of Table 1, let us perform calculation of effectiveness of the model of innovational development of modern Russia's industry according to the developed formula:

$$\text{Eidi} = [(-0.12 - 0.18 + 0.23 + 0.14 - 0.02)/5]/[(24.14 + 10.05 + 9.18)/3] = 0.01/14.46$$
$$= 0.0007(\rightarrow 0).$$

The received value of the indicator Eir (0.0007) strives to zero, which shows low effectiveness of the model of industry's innovational development, realized in modern Russia. In view of the fact that the indicators of expenditures are given in factual prices, it is expedient to take into account the inflation level, which constituted 200% in 2000–2016.

In this case, the growth rate of financing of science from the assets of the federal budget constituted 3.86, internal expenditures of organizations for R&D – −10.22, and expenditures for innovations of organizations of the entrepreneurial sector – −11.09. Let us perform calculation of effectiveness of the model of innovational development of modern Russia's industry in view of specified data for the expenditures:

$$\text{Eidi} = [(-0.12 - 0.18 + 0.23 + 0.14 - 0.02)/5]/[(3.86 - 10.22 - 11.09)/3] = 0.01.$$

Even with correction of expenditures in view of inflation, effectiveness of the applied model of industry's innovational development in modern Russia is low (0.01). Without mentioning the expenditures, it is possible to see that efficiency of application of this model is low. Thus, in 2016 – as compared to 2000 – the number of organizations that conduct scientific research and development reduced by 0.12%, the number of personnel involved in scientific research and development – by 0.18%, and share of organizations that conduct research and development – by 0.02%.

The number of developed leading production technologies grew by 23%, and the number of implemented leading production technologies – by 0.14%. The performed analysis allowed determining the following key problems of the industry's innovational development in modern Russia:

- deficit of leading production technologies, which increase dependence of Russian industrial enterprises on import of these technologies;
- industrial enterprises' founding on state-owned R&D organizations in the context of conduct of scientific R&D;
- low innovational activity of industrial enterprises, which is expressed in implementation of innovations being a rare thing for them.

The initiative in the sphere of innovational development of modern Russia's industry within the existing model comes from the state, which is the main reason for its low effectiveness. State financing of scientific R&D and stimulation of their implementation through the regulation measures should be replaced by the following market measures:

- stimulating development of competition in the sphere of R&D for institutes and industrial enterprises;
- reducing market barriers for R&D centers by simplification of the procedures of their registration and accountability, reduction of tax load, etc.;
- strengthening the cooperation between R&D institutes and industrial enterprises (e.g., through creation of industrial clusters);
- increase of requirements to innovations by stimulating the creation of completely new innovations.

The offered new model of innovational development of modern Russia's industry is shown in Fig. 1.

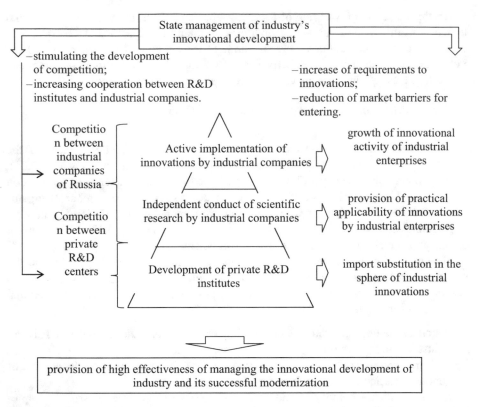

Fig 1. New model of innovational development of modern Russia's industry. Source: compiled by the author.

As is seen from Fig. 1, in the new offered model the state does not directly implement the strategy of industry's innovational development but only manages this process by creating conditions for private enterprises' implementing this strategy. As a result, growth of innovational activity of industrial enterprises, provision of practical applicability of innovations, and import substitution in the sphere of industrial innovations are achieved. This leads to provision of high effectiveness of managing the innovational development of industry and its successful modernization.

5   Conclusions

Thus, the offered hypothesis is proved – the existing model of managing the innovational development of industry in modern Russia has low effectiveness. The main reason for this is incorrect distribution of the roles of main subjects in this model. Instead of management, the state conducts scientific research and development, while private entrepreneurship does not see profits from innovational activity and does not show it.

The offered new model of innovational development of modern Russia's industry allows restoring the logic of distribution of roles between the participants of this process and ensuring its high effectiveness. Success of the developed model is based on the action of the market mechanism. However, as it is limited in modern Russia, the most important task on the path of realization of this model is achievement of high effectiveness of the anti-monopoly policy.

It should be noted that innovational development of industry is a necessary condition for provision of its global competitiveness in the long-term. The longer it takes to understand ineffectiveness of the existing model, the more will grow the underrun of the Russian industrial enterprises from foreign rivals and the more complex it will be to develop the action of the market mechanism.

In order to refuse from the protectionist measures, it is recommended to start the process of application of the new model of industry's innovational development in modern Russia. Though, it should be acknowledged that the change of the model is risky, which predetermines the necessity for controlling it. Determination of perspective directions of risk management of the process of transition to a new model of industry's innovational development in modern Russia is a perspective direction for further scientific research.

References

Anokhina, M., Zinchuk, G.M., Petrovskaya, S.A.: The development of the concept of economic growth of the agro-industrial complex. J. Internet Bank. Commer. 21(3), 224 (2016)

Bogoviz, A., Mezhov, S.: Models and tools for research of innovation processes. Mod. Appl. Sci. 9(3), 159–172 (2015)

Duman, A., Kurekovà, L.: The role of state in development of socio-economic models in Hungary and Slovakia: the case of industrial policy. Changing models of capitalism in Europe and the U.S., 99–120

Fujii, H., Managi, S.: Economic development and multiple air pollutant emissions from the industrial sector. Environ. Sci. Pollut. Res. 23(3), 2802–2812 (2016)

Khan, B.Z.: Human capital, knowledge and economic development: evidence from the British Industrial Revolution, 1750–1930. Cliometrica, 1–29 (2017)

Kinahan, K.L.: Design-based economic development: understanding the role of cultural institutions and collections of industrial and product design. Econ. Dev. Q. 30(4), 329–341 (2016)

Korobkin, D., Fomenkov, S., Kravets, A., Kolesnikov, S., Dykov, M.: Three-steps methodology for patents prior-art retrieval and structured physical knowledge extracting. Communications in Computer and Information Science 535, 124–136 (2015)

Malyshkov, V.I., Ragulina, Y.V.: The entrepreneurial climate in Russia: the present and the future. Life Sci. J. 11(6), 118–121 (2014)

Popkova, E.G.: Guest editorial. Int. J. Educ. Manage. 31(1), 2 (2017)

Przhedetskaya, N., Akopova, E.: Institutional designing of continuous education in Russia under the conditions of neo-economy and globalization. Reg. Sectoral Econ. Stud. 15(2), 115–122 (2015)

Przhedetskaya, N.V., Panasenkova, T.V.: Business education: concept and evolution of development in knowledge economy. World Appl. Sci. J. 2(1) (2014)

Shakirtkhanov, B.R.: Innovative development of industrial enterprises of Kazakhstan in the conditions of economic growth downturn. Int. J. Econ. Res. **14**(7), 121–133 (2017)

Trindade, J.R., Cooney, P., de Oliveira, W.P.: Industrial trajectory and economic development: dilemma of the re-primarization of the Brazilian economy. Rev. Rad. Polit. Econ. **48**(2), 269–286 (2016)

Veselovsky, M.Y., Khoroshavina, N.S., Bank, O.A., Suglobov, A.E., Khmelev, S.A.: Characteristics of the innovation development of Russia's industrial enterprises under conditions of economic sanctions. J. Appl. Econ. Sci. **12**(2), 321–331 (2017)

Federal State Statistics Service: Russia in numbers. 2016: statistical collection. Federal State Statistics Service, Moscow (2016)

Possibilities and Threats of Starting the Mechanism of Import Substitution in the AIC in the Context of Provision of Food Security

Aleksei V. Bogoviz[1]([⊠]) [iD], Svetlana V. Lobova[2] [iD],
and Yury A. Bugai[3] [iD]

[1] Federal State Budgetary Scientific Institution "Federal Research Center
of Agrarian Economy and Social Development of Rural Areas—All Russian
Research Institute of Agricultural Economics", Moscow, Russia
aleksei.bogoviz@gmail.com
[2] Altai State University, Barnaul, Russia
barnaulhome@mail.ru
[3] Altai State Agricultural University, Barnaul, Russia
yrbugai@mail.ru

Abstract. The purpose of the article is to study the contradiction of the mechanism of import substitution in the context of provision of food security and to develop recommendations for overcoming it in modern Russia. The authors use the method of SWOT analysis, which allows evaluating advantages and drawbacks of the mechanism of import substitution, as well as opportunities and threats, in the context of provision of national food security. The authors use the method of correlation and regression analysis to evaluate the results of application of import substitution in the modern Russia's AIC from the point of view of provision of national food security. The key conclusion of this research consists in the fact that the mechanism of import substitution provides wide possibilities for supporting domestic production of the AIC products and brings threats related to limitation of market self-management, which requires flexible application of this mechanism. In modern Russia, the threats of import substitution have not yet emerged, but it is not very effective either. Low innovational activity in the long-term may become a reason for reduction of competitiveness of domestic companies of the AIC. Thus, import will be a serious threat for the Russia's AIC. In order to prevent this scenario, we recommend using innovational policy with the policy of import substitution in the AIC. This will allow preserving and multiplying the effect that has been achieved due to successful policy of import substitution in Russia.

Keywords: Mechanism of import substitution · Agro-industrial complex
Food security

1 Introduction

Application of the mechanism of import substitution is a mandatory measure for provision of national food security. At that, confidence in high effectiveness of this mechanism is so strong that it is not put in doubt. Over the recent years, large

© Springer International Publishing AG, part of Springer Nature 2018
E. G. Popkova (Ed.): HOSMC 2017, AISC 622, pp. 51–57, 2018.
https://doi.org/10.1007/978-3-319-75383-6_6

experience of implementation of the mechanism of import substitution in the AIC was accumulated – it showed that it allows achieving food security not in all cases.

Thus, for example, the strategy of import substitution, together with development of export, is successfully implemented in the countries of East Asia, while the attempts of development of the AIC only on the basis of import substitution are not successful in the countries of Latin America. As any measures of protection of domestic production, import substitution belongs to the policy of protectionism, which takes the economic systems into the past, hindering them obtaining advantages from globalization and participating in the international division of labor.

At the same time, the problem of provision of national food security is very topical in many modern countries, including Russia. The age of stability and economic integration was replaced by disintegration processes that divide the participants of the global economic system and make them develop independent production of the AIC products, regardless of possession of absolute or relative competitive advantages.

The authors offer the hypothesis that the mechanism of import substitution provides wide possibilities for supporting domestic production of the AIC products and brings threats related to limitation of market self-management, which requires flexible application of this mechanism. The purpose of the article is to study contradictions of the mechanism of import substitution in the context of provision of food security and to develop recommendations for overcoming it in modern Russia.

2   Materials and Method

For verification of the offered hypothesis, the authors use the method of SWOT analysis, which allows assessing strengths and weaknesses of the mechanism of import substitution, as well as provided possibilities and threats in the context of provision of national food security. For evaluation of the results of application of the mechanism of import substitution in the modern Russia's AIC from the point of view of provision of national food security, we use the method of regression and correlation analysis. The initial statistical information is given in Table 1.

Table 1. Dynamics of indicators for evaluating the results of application of the mechanism of import substitution in the modern Russia's AIC in 2011–2016

Indicators	2011	2012	2013	2014	2015	2016
Share of domestic production of meat (except poultry), %	42	46	49	57	66	76
Share of domestic production of poultry, %	81	87	87	88	91	96
Share of innovations-active companies, %	9.5	9.6	9.3	9	10.3	10.2
Share of private investments	99.1	98.6	99.2	99.3	99.7	99.7
Share of subsidies to companies, RUB billion	52.8	53.5	54.9	56.1	57.2	58.8

Source: compiled by the authors on the basis of: (Federal State Statistics Service, 2017), (Ministry of Agriculture of the RF 2017).

For determination of consequences of implementing the policy of import substitution in the Russia's AIC, it is necessary to determine connection between various indicators of activities of the AIC companies (y_{1-6}) and the volume of subsidies to companies of the AIC (x). The large this connection, the more favorable influence of the policy of import substitution on development of the AIC.

3 Discussion

The principle of work of the mechanism of import substitution in the AIC in the context of provision of food security is studied and described in the works (Bogoviz and Mezhov 2015; Popkova et al. 2016; Sadovnikova et al. 2013; Popova et al. 2015; Bogoviz et al. 2017; Sandu et al. 2017; Przhedetskaya and Akopova 2015). This allows determining the high level of elaboration of the problem of this research. However, the focus on the possibilities of application of this mechanism without consideration of potential threats is a gap in the system of modern scientific knowledge, which is to be filled by this article.

4 Results

Based on studying the experience of the countries of the world in application of the mechanism of import substitution in the AIC and own logical comparisons of causal connections of development of this process, the authors conducted SWOT analysis, the results of which are given in Table 2.

Table 2. Results of SWOT analysis of application of the mechanism of import substitution in the AIC in the context of provision of national food security

(S) Strengths of the mechanism of import substitution:	(W) Weaknesses of the mechanism of import substitution:
- quick effect, which allows developing domestic production of deficit products of the AIC, which has been previously imported	- limitation of market self-management; - necessity for large state investments
(O) Opportunities of application of the mechanism of import substitution for provision of national food security:	(T) Threats of application of the mechanism of import substitution for provision of national food security:
- reduction of the share of import of the AIC products; - quick modernization of the AIC; - development of entrepreneurship in the segments of the AIC in which it is required for provision of food security	- deficit of the federal budget, which does not allow for long realization of the policy of import substitution; - reduction of the volume of private investments into development of the AIC; - reduction of innovational activity of the AIC companies

Source: compiled by the authors.

As is seen from Table 1, the key advantage of the mechanism of import substitution is quick effect, which allows developing domestic manufacture of deficit products of the AIC that was previously imported. Drawbacks of the mechanism of import substitution include limitation of market self-management and the need for large state investments.

Possibilities of application of the mechanism of import substitution for provision of national food security are related to reduction of the share of import of the AIC products, quick modernization of the AIC, and development of entrepreneurship in the segments of the AIC in which it is necessary for provision of food security.

Threats of application of the mechanism of import substitution for provision of national food security include deficit of the federal budget, which does not allow for long implementation of the policy of import substitution, reduction of the volume of private investments into development of the AIC, and reduction of innovational activity of the AIC companies.

The results of the regression and correlation analysis of the influence of the policy of import substitution on the Russia's AIC are given in Table 1.

As is seen from Table 3, in modern Russia the threats of import substitution have not yet been realized. However, it is not yet effective either. Thus, subsidizing has led to large growth of production of meat and poultry (+5.64% and +2.05%, accordingly, with increase of the volume of subsidies by RUB 1 billion) with acceptable level of correlation (77% and 88%, accordingly).

Table 3. Results of analysis of the influence of the policy of import substitution on the Russia's AIC

Indicators of AIC	Indicators of correlation/regression	
	Change with increase of the volume of subsidies by RUB 1 million	Strength of connection
Share of domestic production of meat (except for poultry), %	+5.64%	77%
Share of domestic production of poultry, %	+2.05%	88%
Share of innovations-active companies, %	+0.12%	31%
Share of private investments	+0.15%	73%

Source: compiled by the authors.

At the same time, the level of domestic production in these segments of the AIC constitutes 76% and 96%, accordingly. This shows the lack of necessity for fighting the import. In the segment of oil and wheat flour, Russia's provision with own production reaches 99%. That is, in the strategically important segments of the AIC, import is not a problem.

At that, due to absence of necessary geographical conditions in other segments of the AIC, especially production of exotic products, subsidizing will not allow reaching a significant effect, so it is not effective at all. The volume of private investments into

development of the AIC companies is not influenced by the policy of import substitution. Despite the acceptable level of correlation (73%), growth of the volume of
private investments constitutes 0.15%.

In view of the fact that in 2016 their share in the structure of investments constitutes
99.7, it is impossible to expect their large growth in percentage terms. This shows
absence of private business's habit for state support. At that, a negative aspect of the
policy of import substitution in the Russia's AIC is low innovational activity of
entrepreneurship.

The share of innovations-active companies constituted 10.2% in 2016. It does not
correlate with the state-provided subsidies and does not change with time. Such innovational activity in the long-term may become a reason for reduction of competitiveness
of the domestic companies of the AIC. Then, import will be a serious threat for Russia.

For prevention of this scenario, we recommend supplementing the policy of import
substitution in the AIC by innovational policy. This allows preventing the reverse cycle
and return to dependence on import of the AIC products, the existing demand for which
will not be satisfied by domestic companies. Innovations will allow preserving and
multiplying the effect achieved due to successful policy of import substitution in
Russia. The logic and the process of overcoming the contradiction of the mechanism of
import substitution in the context of provision of food security in modern Russia is
shown in Fig. 1.

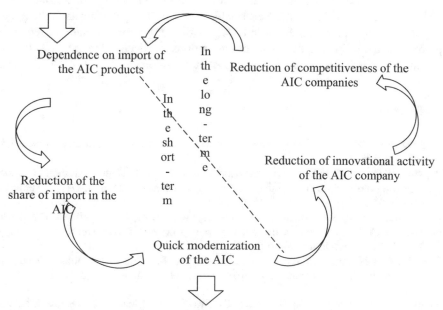

Fig. 1. Overcoming the contradiction of the mechanism of import substitution in the context of
provision of food security in modern Russia Source: compiled by the authors.

Development of the innovational policy in AIC is not a part of this work, being an independent study. It should be noted that the tools of subsidizing should not be used for stimulating the growth of innovational activity of the AIC companies, as well as similar financial tools for stimulating innovational activity of the AIC companies – e.g., tax subsidies.

Instead, we recommend concentrating on non-financial tools of stimulation of innovational activity of the AIC companies, aimed at stimulation of their own innovational initiatives and creation of common favorable conditions for that. It is recommended to strengthen the institutional system and develop close cooperation between state and business for exchange of information, which allows correcting the policy in the sphere of innovations management in the Russia's AIC.

5 Conclusions

Thus, contradiction of the mechanism of import substitution in the AIC in the context of provision of food security is manifested in the fact that application of this mechanism in the short term allows overcoming dependence on import and developing domestic manufacture of deficit products of the AIC. In case of long application of the policy of import substitution, domestic companies get used to state-provided subsidies, and lack of foreign competition leads to elimination of natural market stimuli for manifestation of innovational and investment activity.

In modern Russia, the policy of import substitution has been used for several years and has not yet led to serious negative consequences, ensuring profits related to provision of national food security. Nevertheless, for preventing the threats of the policy of import substitution, which are manifested in the long-term, the authors recommend performing a transition to the policy of innovations management in the AIC, which is a logical continuation of the policy of import substitution.

References

Bogoviz, A., Mezhov, S.: Models and tools for research of innovation processes. Modern Appl. Sci. **9**(3), 159–172 (2015)

Popkova, E.G., Shakhovskaya, L.S., Abramov, S.A., et al.: Ecological clusters as a tool of improving the environmental safety in developing countries. Environ. Dev. Sustain. **18**(4), 1049–1057 (2016)

Popova, L., Popova, S.A., Dugina, T.A., Korobeynikov, D.A., Korobeynikova, O.M.: Cluster Policy in Agrarian Sphere in Implementation of Concept of Economic Growth. Eur. Res. Stud. J. **18**, Special Issue, 27–36 (2015)

Przhedetskaya, N., Akopova, E.: Institutional designing of continuous education in Russia under the conditions of neo-economy and globalization. Reg. Sectoral Econ. Stud. **15**(2), 115–122 (2015)

Sadovnikova, N., Parygin, D., Gnedkova, E., Kravets, A., Kizim, A., Ukustov, S.: Scenario forecasting of sustainable urban development based on cognitive model. In: Proceedings of the IADIS International Conference ICT, Society and Human Beings 2013, Proceedings of the IADIS International Conference e-Commerce 2013, pp. 115–119 (2013)

Bogoviz, A.V., Ragulina, Y.V., Shkodinsky, S.V., Babeshin, M.A.: Factors of provision of food security. Agric. Econ. Russia **2**(1), 2–8 (2017)

Ministry of Agriculture of the RF: Regarding distribution of subsidies in the agro-industrial complex in 2017 (2017). http://government.ru/docs/26279/. Accessed 11 July 2017

Sandu, I.S., Bogoviz, A.V., Ryzhenkova, N.E., Ragulina, Y.V.: Formation of innovational infrastructure in the agrarian sector. AIC: Econ. Manage. **1**(1), 35–41 (2017)

Selection of the Key Spheres of Modern Russia's Industry: Prioritizing

Aleksei V. Bogoviz[1](✉) ⓘ, Vladimir S. Osipov[2,3] ⓘ,
and Elena I. Semenova[1] ⓘ

[1] Federal State Budgetary Scientific Institution "Federal Research Center
of Agrarian Economy and Social Development of Rural Areas—All Russian
Research Institute of Agricultural Economics", Moscow, Russia
aleksei.bogoviz@gmail.com
[2] Lomonosov Moscow State University, Moscow, Russia
vs.ossipov@gmail.com
[3] Russian State Agrarian University - Moscow Timiryazev Agricultural
Academy, Moscow, Russia

Abstract. The purpose of the article is to determine the key spheres of modern
Russia's industry in the interests of stimulating its strategic development in the
long-term. During determining the key spheres of Russia's industry, the authors
are guided by the Concept of long-term socio-economic development of the
Russian Federation until 2020, established by the Decree of the Government of
the RF dated November 17, 2008, No. 1662-r. The information and analytical
basis of the research includes the materials of the ranking for the values of the
global competitiveness index in 2016 according to the World Economic Forum
and the ranking for the knowledge economy index according to the World Bank.
In order to study the current state of Russia's industry and to determine the
priorities of its development, the authors use the method of structural and
functional analysis. In the course of the research, the authors come to the
conclusion that state's non-interference into the real sector of the Russian
economy is a reason for its development in strategically unprofitable direction; it
is necessary to set priorities of development of modern Russia's industry. The
authors substantiate the necessity for and present the logic of restructuring of
modern Russia's industry.

Keywords: Industry · Modern Russia · Priorities of development
National competitiveness · Sustainability

1 Introduction

The real (industrial) sector is a basis of provision of sustainability of economic systems'
development. For optimization of entrepreneurship (relocating industrial production
into countries with cheaper and more accessible resources and lower ecological standards) and post-industrialization, most developed and developing countries have been
preferring the development of the service sphere of economy.

© Springer International Publishing AG, part of Springer Nature 2018
E. G. Popkova (Ed.): HOSMC 2017, AISC 622, pp. 58–64, 2018.
https://doi.org/10.1007/978-3-319-75383-6_7

The long global economic recession, which was a result of this tendency, became an argument in favor of refusal from full-scale post-industrialization and transition to formation of knowledge economy or innovations-oriented economy, in which the preference is given not to the sector of national economy but to innovations-active companies in any sectors and high-tech spheres of industrial sector of economy.

Due to this, the interest to industry as to a means of overcoming the recession's consequences and prevention of new crises, as well as potential vector of growth of economy, grew. In modern Russia, industrial specialization still survives. However, the basis of economy and export orientation at the mineral extraction spheres are the reasons for unprofitable position of the Russian entrepreneurship at the beginning of the added value chain of industrial goods and for instability caused by high volatility of the global prices for raw materials.

Our hypothesis is that state's non-interference with the real sector of the Russian economy is a reason for its development in strategically unprofitable direction; it is necessary to put priorities of development of modern Russia's industry. The purpose of the article is to determine the key spheres of modern Russia's industry in the interest of stimulating its strategic development in the long-term.

2 Materials and Method

While determining the key spheres of Russia's industry in this research, we use the Concept of long-term socio-economic development of the RF until 2020, established by the Decree of the Government of the Russian Federation dated November 17, 2008, No. 1662-r (Government of the Russian Federation 2008).

The authors use the materials of the World Economic Forum and the World Bank for 2016 and the official statistical information of the Federal State Statistics Service for 2016, which is shown graphically in Figs. 1 and 2.

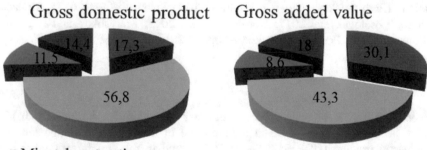

Fig. 1. Sectorial structure of internal industrial production of Russia in 2016. Source: compiled by the authors based on materials: (Federal State Statistics Service, 2016).

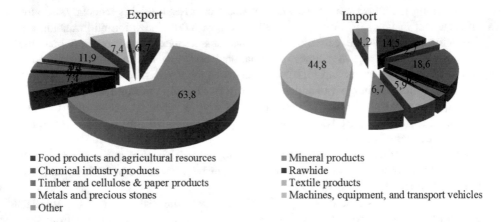

Fig. 2. Sectorial structure of export and import of Russia in 2016. Source: compiled by the authors based on: (Federal State Statistics Service, 2016).

For studying the current state of Russia's industry and determining the priorities of its development, this article uses the method of structural and functional analysis.

3   Discussion

The role and value of the real sector in development of modern economic systems are reflected in the works (Khan 2017; Fujii and Managi 2016; Kinahan 2016; Anokhi-naetal 2016; Bogoviz and Mezhov 2015; Duman and Kureková 2016; Osipov and Skryl 2016). The tendencies of development of modern Russia's industry are shown in publications (Przhedetskaya and Panasenkova 2014; Shakirtkhanov 2017; Trindade et al. 2016; Veselovskyetal 2017; Korobkin et al., 2015; Malyshkov and Ragulina 2014; Popkova 2017; Przhedetskaya and Akopova 2015; Kosov et al. 2016).

4   Results

As is seen from Fig. 1, the main sphere of industry in modern Russia (as of 2016) is processing production. Their share in the structure of the industrial components of gross domestic product (GDP) constitutes 56.8%, and in the structure of industrial component of gross added value – 43.3%. Minerals extraction is also an important sphere. Its share in the structure of the industrial component of gross domestic product (GDP) constitutes 17.3%, and in the structure of industrial component of gross added value – 30.1%.

The spheres of production and distribution of electric energy, natural gas, and water form 11.5% of the industrial components of gross domestic product (GDP) and 8.6% of the industrial component of gross added value. The construction sphere provides 14.4% of the industrial component of gross domestic product (GDP) and 18% of the industrial component of gross added value. The total share of industry in Russia's GDP constitutes 44.5%, and in the structure of gross added value – 32.6%.

In the structure of Russia's export in 2016, mineral products account for 63.8%. They are followed by metals and precious stones (11.9%), machinery, equipment, and transport vehicles (7.4%). The main imported products are machinery, equipment, and transport vehicles (44.8%), products of chemical industry (18.6%), food products, and agricultural resources (14.5%).

In the Concept of long-term socio-economic development of the RF until 2020, the strategic priorities are as follows: provision of national economic security, stability and sustainability of development of economic system, creation of innovations-oriented economy (knowledge economy), and provision of global competitiveness. In the ranking of global competitiveness index in 2016, Russia received 4.5 points and 43rd position (World Economic Forum 2017), and in the ranking of knowledge economy index – 5.78 points and 55th position (World Bank 2017).

Comparing the current situation in Russia's industry and strategic landmarks of economy's development, we determines a lot of mismatches. Firstly, high dependence on import of strategic goods (machinery, equipment, and transport means) and food products contradicts the idea of provision of national economic security. Secondly, export of mineral products contradicts the idea of stability and sustainability of development of economy, causing its dependence on the global prices for these products.

Thirdly, the fact that high-tech spheres are not separated in the structure of industry shows their insignificant role in development of the Russian real sector. This is confirmed by low value of knowledge economy index. Fourthly, domination of export into the CIS countries – on the partnership, not competitive, basis – reflects low competitiveness of the Russian industrial products, which is confirmed by low value of the global competitiveness index.

According to the Concept of socio-economic development of Russia, we determined the following key spheres of industry:

- export-oriented spheres of processing industry: their development is aimed at provision of stability of development of Russia's economic system and strengthening of its positions in the global arena as a supplier of finished products, not just mineral resources;
- high-tech spheres of industry: their development is orientes at stimulating creation of innovations-oriented economy (knowledge economy) in Russia;
- spheres of heavy machinery and food industry: their development is aimed at stimulating provision of national economic security of Russia through import substitution.

At the same time, it should be emphasized that while preferring the above top-priority spheres, one must remember the current main sphere of industry – processing industry, which provides a large share of GDP, is a basis of export and the key source of tax revenues into budgets of various levels.

That's why these priorities should be viewed as additions to the main sphere of Russia's industry, not as its replacement. In the long-term, when these top-priority spheres oust the processing industry from the leading position in the structure of gross national product, gross added value, and export, it will be expedient to take it from the list of top-priority spheres – still, until that time such measure has a risk of emergence of the national economic crisis. The offered logic of prioritization of the sphere of industry in modern Russia is shown in Fig. 3.

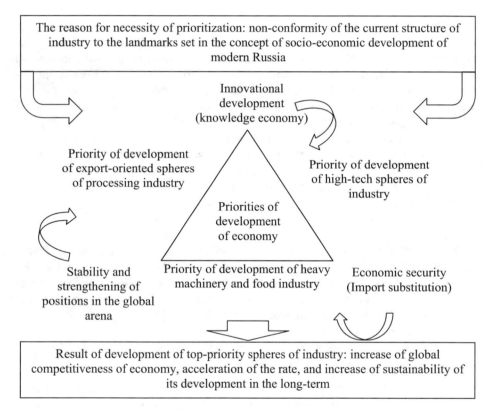

Fig. 3. Logic of prioritization of the spheres of industry in modern Russia. Source: compiled by the authors.

As is seen from Fig. 3, the reason for the necessity of prioritization consists in non-conformity of the current structure of industry to the landmarks, set in the concept of socio-economic development of modern Russia. As a result, development of top-priority spheres of industry achieves the increase of global competitiveness of economy, acceleration of rate, and increase of sustainability of its development in the long-term.

5   Conclusions

Thus, the results of the performed research confirm the offered hypothesis and provide the proofs of the necessity for restructuring of modern Russia's industry. The offered and substantiated priorities should be established in the national strategy of socio-economic development. The term of implementation of this strategy should be set for 2030-2035, as in the opposite case its implementation will be related to high level of risk and low probability of success.

It should be noted that selection of new key spheres of modern Russia's industry is necessary not only for achievement of internal goals of economic development but also for stimulating its establishment in the status of a leading economic state in the global level, as well as strengthening of international economic ties and relations. The offered logic of prioritization of the spheres of industry in modern Russia corresponds to the national program of modernization of economy, which is realized in modern Russia, which creates a basis for successful development of the distinguished top-priority spheres of industry.

However, prioritization is the first of the many steps on the way to successful restructuring of modern Russia's industry. An important step is also determination of effective tools of development of the above spheres of industry, as without the measures of state regulation these spheres will develop very slowly. Development of the system of regulation measures is a perspective direction for further scientific research in this sphere.

References

Anokhina, M., Zinchuk, G.M., Petrovskaya, S.A.: The development of the concept of economic growth of the agro-industrial complex. J. Internet Bank. Commerce **21**(3), 224 (2016)

Bogoviz, A., Mezhov, S.: Models and tools for research of innovation processes. Mod. Appl. Sci. **9**(3), 159–172 (2015)

Duman, A., Kurekovà, L.: The role of state in development of socio-economic models in Hungary and Slovakia: the case of industrial policy. Changing Models of Capitalism in Europe and the U.S.,pp. 99–120 (2016)

Fujii, H., Managi, S.: Economic development and multiple air pollutant emissions from the industrial sector. Environ. Sci. Pollut. Res. **23**(3), 2802–2812 (2016)

Khan, B.Z.: Human capital, knowledge and economic development: evidence from the British Industrial Revolution, 1750–1930. Cliometrica, pp. 1–29 (2017)

Kinahan, K.L.: Design-based economic development: understanding the role of cultural institutions and collections of industrial and product design. Econ. Develop. Q. **30**(4), 329–341 (2016)

Korobkin, D., Fomenkov, S., Kravets, A., Kolesnikov, S., Dykov, M.: Three-steps methodology for patents prior-art retrieval and structured physical knowledge extracting. Commun. Comput. Inf. Sci. **535**, 124–136 (2015)

Kosov, M.E., Akhmadeev, R.G., Osipov, V.S., et al.: Socio-economic planning of the economy. Indian J. Sci. Technol. **9**(36), 1–11 (2016)

Malyshkov, V.I., Ragulina, Y.V.: The entrepreneurial climate in Russia: the present and the future. Life Sci. J. **11**(6), 118–121 (2014)

Osipov, V.S., Skryl, T.V.: The strategic directions of the modern Russian economic development. Int. Bus. Manage. **10**(6), 710–717 (2016)

Popkova, E.G.: Guest editorial. Int. J. Educ. Manage. **31**(1), 2 (2017)

Przhedetskaya, N., Akopova, E.: Institutional designing of continuous education in Russia under the conditions of neo-economy and globalization. Reg. Sectoral Econ. Stud. **15**(2), 115–122 (2015)

Przhedetskaya, N.V., Panasenkova, T.V.: Business education: concept and evolution of development in knowledge economy. World Appl. Sci. J. **2**(1) (2014)

Shakirtkhanov, B.R.: Innovative development of industrial enterprises of Kazakhstan in the conditions of economic growth downturn. Int. J. Econ. Res. **14**(7), 121–133 (2017)

Trindade, J.R., Cooney, P., de Oliveira, W.P.: Industrial Trajectory and Economic Development: Dilemma of the Re-primarization of the Brazilian Economy. Rev. Radic. Political Econ. **48**(2), 269–286 (2016)

Veselovsky, M.Y., Khoroshavina, N.S., Bank, O.A., Suglobov, A.E., Khmelev, S.A.: Characteristics of the innovation development of Russia's industrial enterprises under conditions of economic sanctions. J. Appl. Econ. Sci. **12**(2), 321–331 (2017)

World Bank. Knowledge economy index (2017). http://web.worldbank.org/archive/website01030/WEB/IMAGES/KAM_V4.PDF. Accessed 25 Jul 2017

World Economic Forum. The Global Competitiveness Report 2016–2017 (2017). http://reports.weforum.org/global-competitiveness-index/. Accessed 25 Jul 2017

Government of the Russian Federation. The Concept of long-term socio-economic development of the Russian Federation until 2020, established by the Decree of the Government of the Russian Federation dated November 17, 2008, No. 1662-r (2008). http://www.consultant.ru/document/cons_doc_LAW_82134/28c7f9e359e8af09d7244d8033c66928fa27e527/. Accessed 25 Jul 2017

State Audit as a Mandatory Condition of Budget Policy Effectiveness

Aleksei V. Bogoviz[1](✉) ⓘ, Aleksei I. Bolonin[2],
and Svetlana V. Lobova[3] ⓘ

[1] Federal State Budgetary Scientific Institution "Federal Research Center
of Agrarian Economy and Social Development of Rural Areas—All Russian
Research Institute of Agricultural Economics", Moscow, Russia
aleksei.bogoviz@gmail.com
[2] Moscow State Institute of International Relations, Moscow, Russia
danrotten@yandex.ru
[3] Altai State University, Barnaul, Russia
barnaulhome@mail.ru

Abstract. The purpose of the work is to substantiate the necessity for applying
state audit for achieving high effectiveness of budge policy of economic systems
by the example of modern Russia. The authors develop a methodology of
evaluation of budget policy effectiveness and use it by the example of modern
Russia. The authors show that budget policy of modern Russia is not effective,
and systemic state audit is required for its increase. The authors determine the
main reasons for low effectiveness of the budget policy in modern Russia and
view the process of increase of budget policy effectiveness with the help of state
audit. As a result of research it is concluded that low effectiveness of budget
policy in modern Russia is caused by lack of control and independence of public
authorities' bodies that conduct the development and implementation of the
budget policy. Solving this problem requires introduction of state audit of the
budget policy. It should be noted that state audit leads to the interest of the public
authorities that conduct its development and implementation of the budget policy
in achieving its high effectiveness. That's why state audit is an inseparable
condition of effectiveness of the budget policy.

Keywords: State audit · Budget policy · Effectiveness · State management
Federal budget · Modern Russia

1 Introduction

Budget policy is a basis of implementation all other directions of state policy, as it
forms their resource provision. The effectiveness of budget policy determines the
economic systems' capability for achieving strategic goals of their development with
preservation of sustainability. This predetermines high topicality of scientific study of
the foundations of development and implementation of the budget policy from the
point of view of its effectiveness.

© Springer International Publishing AG, part of Springer Nature 2018
E. G. Popkova (Ed.): HOSMC 2017, AISC 622, pp. 65–70, 2018.
https://doi.org/10.1007/978-3-319-75383-6_8

At present, there is high interest to budget policy, as the global financial crisis that started in 2008 showed that ineffective budget policy may lead to serious negative consequences, related to multiple limitations on the economic system and become a reason for its default.

During the period of stability, the consequences of ineffective budget policy are manifested not so clearly, though in the long-term it increases the risk component of state management of economy. Despite the general acknowledgement of the necessity for provision of high effectiveness of budget policy of the state, the methodological foundations of its precise measuring are not strong enough, which is a reason for arguments among the experts.

The study is based on the hypothesis that budget policy of modern Russia is characterized by low effectiveness, and its increase requires systemic state audit. The purpose of the work is to substantiate the necessity for application of state audit for achieving high effectiveness of budget policy of economic systems by the example of modern Russia.

2 Materials and Method

We think that the most important indicator of effectiveness of budget policy is balance of the country's federal budget.

At the same time, the budget balance reflects mainly rationality of spending the budget assets, so it is necessary to take into account its results during assessment of effectiveness of budget policy. One of the most important results is social well-being in the country – reflected by index of happiness. In the conditions of creation of knowledge economy, another result is achievement of innovational development of economy, which is reflected by index of innovativeness of economy.

In the conditions of global competition, the result of budget policy is also competitiveness of economy, which is reflected by the corresponding index. At that, all these results are equal. We developed the methodology for evaluation of effectiveness of budget policy. Within this methodology, depending on the combination of balance of the federal budget and the results of implementation of budget policy, the indicator of its effectiveness (Pbe) is assigned the value from 1 point (minimum effectiveness) to 20 points (maximum effectiveness) in whole numbers according to the developed matrix (Table 1).

As is seen from Table 2, we distinguished five intervals of balance of the federal budget: zero balance, slight surplus of the budget (up to 1.5% of GDP), large surplus of the budget (more than 1.5% of GDP), and slight deficit of the budget (up to 1.5% of GDP) and large deficit of the budget (more than 1.5% of GDP).

We also distinguished four variants of combining the indicators of efficiency of budget policy: high values for all indicators (VVV), high values for two indicators, but low value for one of the indicators (VVN), high value for one indicator and low values for two other indicators (VNN), and low values for all indicators (NNN). High-quality assessment of the value of indicators of budget policy efficiency is conducted with the expert method depending on the national peculiarities of the economic system.

Table 1. Dynamics of indicators of implementation of budget policy in the Russian Federation in 2011-2017 on the basis of the federal budget

Indicators	2011	2012	2013	2014	2015	2016	2017 forecast
Revenues of the federal budget, RUB billion	11,366.0	12,853.7	13,019.9	14,496.8	15,082.4	16,271.8	17,088.6
Revenues of the federal budget, % of GDP	20.3	20.7	19.7	18.5	19.5	19.6	19.0
Expenditures of the federal budget, RUB billion	10,935.2	12,890.6	13,342.9	14,830.6	15,513.1	16,854.9	17,536.2
Expenditures of the federal budget, % of GDP	19.5	20.7	20.2	19.0	20.0	20.1	20.2
Balance of the budget, RUB billion	430.8	-36.9	-323.0	-333.8	-430.7	-583.1	-447.6
Balance of the budget, % of GDP	0.8	0.0	-0.5	-0.5	-0.5	-0.5	-1.2

Source: compiled by the authors on the basis of: Ministry of Finance of the Russian Federation (2017).

Table 2. The matrix of values, assigned to the indicator of effectiveness of budget policy

Balance of the federal budget (C), % of GDP	Values of index of happiness, index of innovativeness of economy, and index of competitiveness of economy			
	VVV	VVN	VNN	NNN
C = 0	18	13	9	5
0 < C ≤ 1.5	19	15	10	4
C > 1.5	20	14	8	3
-1.5 ≤ C < 0	17	12	7	2
C < -1.5	16	11	6	1

Source: compiled by the authors.

3   Discussion

Theoretical and methodological and applied issues of measuring and provision of effectiveness of budget policy of economic systems, including practical examples from the experience of modern countries of the world, are reflected in the materials of the research by Popkova et al. (2017); Bykanova et al. (2017); Przhedetskaya and Akopova

(2016); Sadovnikova (2013); Busemeyer and Garritzmann (2017); Osipov et al. (2017). However, not enough attention is paid to state audit of budget policy, which requires further research.

4 Results

Dynamics of balance of the federal budget of the RF in 2011–2017 is shown in Fig. 1.

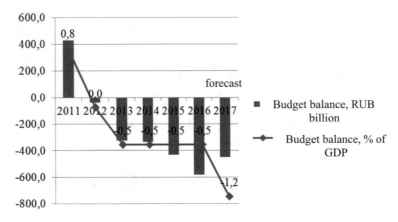

Fig. 1. Dynamics of balance of the federal budget of the RF in 2011–2017. Source: compiled by the authors

As is seen from Fig. 1, 2011 marked the surplus of the federal budget of the Russian Federation that equaled RUB 430.8 billion (0.8% of GDP), despite the post-crisis period. It was followed by deficit of the federal budget, which by 2017 reached RUB 447.6 billion (1.2% of GDP). In 2017, Russia received 5.963 points out of 10 (49th place in the ranking of the countries of the world) according to the index of population's happiness; 38.8 points out of 100 (45th place) according to the index of economy's innovativeness; 4.5 out of 10 (43rd place) according to the index of economy's global competitiveness.

For Russia, which strived to become a leading developed country of the world, such values of the indicators of efficiency of budget policy are rather low. Together with insignificant surplus of the federal budget, the indicator of effectiveness of budget policy is assigned with 2 points according to the developed estimate methodology. We determined the following main reasons of such low value of the indicator effectiveness of budget policy in modern Russia:

– strong influence of shadow economy on the process of implementation of budget policy, which leads to reduction of revenues into the federal budget (reduction of the revenue part) and inappropriate spending of its assets (exceeding the expenditure part);
– lack of responsibility, accountability, and mechanisms of regulation of the process of development and implementation of budget policy.

Both determined problems of achievement of high effectiveness of budget policy in modern Russia could be solved fully or partially with the help of state audit. In this research, we view state audit in a wide sense, seeing it not only as control over implementation of the budget by comparing revenues and expenditures but also as analysis of effectiveness of collecting the revenues into the budget and conducting the expenditures of the state budget.

State audit allows turning the developers of the budget policy from the last instance, which possesses unlimited state power, into controlled bodies of public authorities, which are responsible before the corresponding body (state auditor) and society. The sense of the process of increase of effectiveness of budget policy with the help of state audit is shown in Fig. 2.

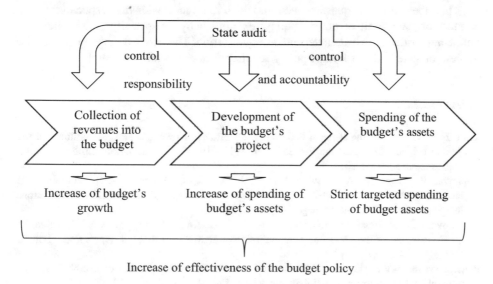

Fig. 2. The process of increase of effectiveness of budget policy with the help of state audit
Source: compiled by the authors.

As is seen from Fig. 2, state audit should be conducted over the whole budget process. At the stage of collection of revenues into the budget and spending of the budget assets, state audit allows reducing or overcoming shadow economy, ensuring the increase of revenues and strictly appropriate spending of budget assets. At the stage of development of the budget's project, state audit allows increasing rationality of spending of budget assets. In totality, this leads to increase of budget policy effectiveness.

State audit is a mandatory condition of effectiveness of budget policy, as it allows forming and supporting the stimulus for provision of its effectiveness. Lack of state audit leads to lack of control over developers of the budget policy, which contradicts the idea of high effectiveness of their work. That's why introduction of state audit of the budget policy is a necessary measure for providing its high effectiveness.

5 Conclusion

As a result of the research, it is possible to conclude that effectiveness of budget policy in modern Russia is low. This is caused primarily by lack of control and independence of public authorities' bodies, which conduct development and realization of the budget policy. In order to solve this problem, this work offers the state audit of budget policy.

At that, it should be noted that without state audit there's no interest from public authorities, which conduct development and realization of budget policy, to achievement of its high effectiveness. That's why state audit is a mandatory condition of budget policy effectiveness. The key condition of success of implementation of state audit in the interests of increase of budget policy effectiveness is observation of all principles of audit.

Thus there arises a question who should be assigned with the responsibility for conduct of state audit of budget policy: existing body or specially created body of public authorities or private independent organization. The detailed development of the process of state audit of budget policy requires further scientific research.

References

Ministry of Finance of the Russian Federation. Dynamics of revenues and expenditures of the federal budget of the Russian Federation in 2011–2017 (2017). http://info.minfin.ru/fbrash.php. Accessed 26 Jul 2017

Popkova, E., Bogoviz, A., Litvinova, T., Alieva, N., Gorbacheva, A.: Methodological recommendations for improvement of statistical accounting and assessment of innovations in agriculture. AIC: Econ. Manage. 7(1), 42–50 (2017)

Bykanova, O.A., Akhmadeev, R.G., Kosov, M.E., Ponkratov, V.V., Osipov, V.S., Ragulina, Y. V.: Assessment of the economic potential of sovereign wealth funds. J. Appl. Econ. Sci. 12 (1/47), 70–84 (2017)

Przhedetskaya, N.V., Akopova, E.S.: Imperative of state in the process of establishment of innovational economy in the globalizing world. Eur. Res. Stud. 9(2), 79–85 (2016)

Sadovnikova, N., Parygin, D., Gnedkova, E., Kravets, A., Kizim, A., Ukustov, S.: Scenario forecasting of sustainable urban development based on cognitive model. In: Proceedings of the IADIS International Conference ICT, Society and Human Beings 2013, Proceedings of the IADIS International Conference e-Commerce 2013, pp. 115–119 (2013)

Busemeyer, M.R., Garritzmann, J.L.: Public opinion on policy and budgetary trade-offs in European welfare states: evidence from a new comparative survey. J. Eur. Public Policy. 24 (6), 871–889 (2017)

Osipov, V.S., Bykanova, O.A., Akhmadeev, R.G., Bogoviz, A.V., Smirnov, V.M.: External debt burden and its impact on the countries' budgetary policy. J. Appl. Econ. Sci. 12(2), 342–355 (2017)

The Mechanism of Activation of the Process of Import Substitution in the Agro-Industrial Complex for Provision of Food Security

Aleksei V. Bogoviz[1]([✉]) [iD], Aydar M. Tufetulov[2],
and Denis A. Chepik[3] [iD]

[1] Federal State Budgetary Scientific Institution "Federal Research Center
of Agrarian Economy and Social Development of Rural Areas—All Russian
Research Institute of Agricultural Economics", Moscow, Russia
aleksei.bogoviz@gmail.com

[2] Institute of Management, Economics and Finance, Kazan (Volga) Federal
University, Kazan, Russia
ajdar-t@yandex.ru

[3] Russian Research Institute of Agricultural Economics, Moscow, Russia
denis_chepik@mail.ru

Abstract. The purpose of the work is to develop the practice-oriented mechanism of activation of the process of import substitution in the agro-industrial complex (AIC) for the purpose of provision of food security in modern Russia. The offered hypothesis is verified with the help of the regression and correlation analysis. In the course of the research, the authors prove that subsidizing, which is the only measure, used for stimulating import substitution in modern Russia, is not effective and leads to additional load on the deficit federal budget. In order to reduce the entering barriers and increase competition in the sphere of AIC in modern Russia, the authors offer to use the mechanism of public-private partnership. This will allow attracting flexible and adaptable small and medium business into this sphere due to provision of state production capacities, as well as private investments by means of state guarantees. The authors describe the functions of the state and private business, as well as logic of activation of the process of import substitution of the AIC for provision of food security by application of the mechanism of public-private partnership.

Keywords: Import substitution · AIC · National food security
Public-private partnership · Modern Russia

1 Introduction

Agro-industrial specialization is not a peculiar feature of the modern Russian economy – both due to the influence of geographical factors and to striving of the national economic system to quick post-industrialization, which is conducted not by means of development of the service sphere but by means of increase of its share in the structure of economy with reduction of business activity in other spheres.

© Springer International Publishing AG, part of Springer Nature 2018
E. G. Popkova (Ed.): HOSMC 2017, AISC 622, pp. 71–76, 2018.
https://doi.org/10.1007/978-3-319-75383-6_9

In view of high demand for agro-industrial products, caused by growth of the number of population and striving for increase of the living standards, this leads to the problem of provision of food security. This problem was traditional solved with the help of import. However, in recent years, this method of solving it became less accessible due to violation of international economic connections of Russia in the conditions of the financial crisis, which was replaced by the sanctions.

Change of suppliers of imported agro-industrial products did not lead to complete solution of the problem of provision of food security in modern Russia, as in the conditions of instability of the geopolitical situation there is a risk of violation of shipment of agro-industrial products. Therefore, the most effective means of provision of food security of modern Russia's economy is import substitution in the agro-industrial complex.

The course at import substitution in the AIC was officially accounces in the Strategy of socio-economic development of the agro-industrial complex of the RF until 2020, but over the five years of its implementation there were no substantial results, and the Russia's economy's dependence on import of agro-industrial products is preserved.

In this research, we offer the hypothesis that the reason for unsuccessfulness of implementation of the course for import substitution in the AIC in modern Russia consists in absence of practical measures of its implementation. The goal of this work is to develop the practice-oriented mechanism of activation of the process of import substitution in the AIC for provision of food security in modern Russia.

2 Materials and Method

The measures applied for implementation of the Strategy of socio-economic development of the agro-industrial complex of the RF until 2020 (Russian Academy of Agricultural Sciences 2017) are described in the program of state support for agro-industrial complex in the RF (Kuban Agricultural Information and Consultation Center 2017).

This program of state support for agro-industrial complex in the RF supposes exclusive subsidizing of enterprises of the AIC. Statistical information on the course of implementation of this strategy is given in the materials of the Ministry of Agriculture of the RF on distribution of subsidies in the agro-industrial complex in 2017 (Ministry of Agriculture of the RF 2017).

The offered hypothesis is verified by the authors with the help of the method of regression and correlation analysis. It is true if the coefficient of correlation of the number and turnover of enterprises of the AIC and the volume of their subsidizing by the state is below 90% - otherwise, the hypothesis will be overturned.

3 Discussion

The problem of provision of national food security is has been studied in multiple works of modern authors, which include Bogoviz and Mezhov (2015); Popkova et al. (2016); Sadovnikova et al. (2013); Popova et al. (2015); Bogoviz et al. (2017); Sandu et al. (2017); Przhedetskaya and Akopova (2015); Dudukalov et al. (2016).

The performed content analysis of scientific literature on the topic of the research showed that despite the statement on the necessity for applying import substitution for provision of national food security, practical recommendations and mechanisms of activation of this process are not offered, which is a basis for further scientific research in this direction.

4 Results

The results of regression and correlation analysis showed that increase of the volume of state subsidizing of enterprises of the AIC by RUB 1 billion leads to increase of their turnover by RUB 2.3 billion (correlation coefficient constitutes 81.4%), and their number decreases by 2 companies (correlation coefficient constitutes 0.1%).

This shows that the applied measures for implementation of the Strategy of socio-economic development of the agro-industrial complex of the RF influence the process of development of import substitution in an insignificant way and even slow it down. It should be noted that the connection between the volume of state subsidies and turnover of enterprises could be predetermined not so much by their real interdependence as by the influence of inflation.

Thus, subsidizing is characterized by low effectiveness as to stimulation of import substitution in modern Russia and leads only to additional load on the deficit federal budget. It is risky to rely on the market mechanism that works successfully in other spheres of economy, as the spheres of AIC is of top-priority from the point of view of provision of national food security, and high dependence on the geographical factor (which influence in unfavorable in Russia) due to close connection to agriculture reduces investment attractiveness of this sphere.

The companies of the AIC are represented mostly by large business, the important role in the work of which belongs to fixed assets and production technologies, which determines high entering barriers for new players. Current companies of the AIC have stationary organizational structure, which does not allow them to react to market signals and leads to their low competitiveness.

In order to reduce the entering barriers and increase competition in the sphere of AIC in modern Russia, the authors offer to use the mechanism of public-private partnership. This will allow attracting flexible and adaptable small and medium business on the basis of provision of state production capacities and attracting private investments by means of state guarantees.

The state's functions within partnership in the sphere of AIC should be brought down to the following:

– provision of production capacities: state should be the initiator of partnership with private business, providing premises, equipment, and technologies. These could be production capacities which the state already owns or the ones specifically purchased for this goal;

- establishment of national priorities of the AIC for import substitution: projects of public-private partnership should be created in the segments of the AIC in which entrepreneurship is least developed, but there's a large demand for its products, i.e., there's deficit;
- provision of guarantees for investors: in the conditions of deficit of the federal budget, state investments might be not enough for formation of necessary production capacities, and there would be a need for private investments, attracted by means of state guarantees of their return.

The functions of private business within partnership in the AIC should be the following:

- flexible and adaptable management of projects of public-private partnership in the AIC: effective management within the partnership with the state is provided by small and medium companies;
- placement of investments into the projects of public-private partnership in the AIC: the main investment load will be placed on private business;
- marketing in the AIC: determining perspective directions of development of the AIC and new tendencies of demand.

The offered mechanism of activation of the process of import substitution in the AIC for provision of food security is shown in Table 1.

Table 1. Dynamics of indicators of development of Russia's AIC in 2011-2016

Indicators	Values of indicators for the years					
	2011	2012	2013	2014	2015	2016
Turnover of enterprises, RUB billion	3,986.2	4,058.7	4,138.5	4,216.7	4,675.6	5,526.3
Number of enterprises, thousand	51.2	51.5	48.5	47.6	50.9	51.4
Volume of subsidies to enterprises, RUB billion	52.8	53.5	54.9	56.1	57.2	58.8

Source: compiled by the authors on the basis of: (Federal State Statistics Service, 2017), (Ministry of Agriculture of the RF 2017).

As is seen from Fig. 1, implementation of the mechanism of public-private partnership leads to growth of competition of enterprises in the AIC, as market barriers for players are reduced. This stimulates increase of flexibility and innovational activity of all enterprises in the AIC, as it makes them care for their competitiveness. This activate the process of import substitution in the AIC, as this leads to equal filling of its segments and manufacture of competitive products that satisfy the demand of domestic consumers. As a result, the problem of provision of national food security is achieved.

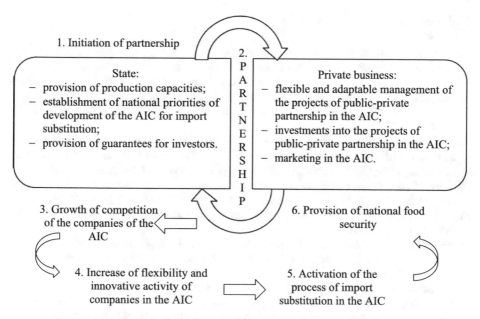

Fig. 1. Mechanism of public-private partnership for activation of the process of import substitution in the AIC for provision of food security. Source: compiled by the authors.

5 Conclusion

It should be concluded that only competitive environment can activate the process of import substitution in the AIC, as its absence became the reason for high dependence on import of products of the AIC in modern Russia. As it is difficult to achieve high level of competition in the AIC with private business only – due to high barriers for entering – it is necessary for the state to be involved.

Public-private partnership is an optimal variant of development of business activity in the AIC, as it allows reducing barriers for new players while preserving their marketing activity – which cannot be ensured by the measures for subsidizing the innovational activity of current enterprises in the Russia's AIC.

The results of the performed research are limited by underdevelopment of the mechanism of public-private partnership in domestic economic practice, which complicated the application of the offered authors' recommendations. That's why a perspective direction of further research should be development of the theoretical and methodological platform of public-private partnership and practical aspects of its implementation in modern Russia.

References

Dudukalov, E.V., Rodinorova, N.D., Sivakova, Y.E., et al.: Global innovational networks: sense and role on development of global economy. Contemp. Econ. **10**(4), 299–310 (2016)

Kuban Agricultural Information and Consultation Center. State support for agro-industrial complex in the RF (2017). http://www.kaicc.ru/gosudarstvennaja-podderzhka-apk/subsidirovanie-i-finansirovanie/gosudarstvennaja-podderzhka-agroprom. Accessed 11 Jul 2017

Russian Academy of Agricultural Sciences. Strategy of socio-economic development of the agro-industrial complex of the RF until 2020 (2017). http://www.vniiesh.ru/documents/document_9509_Стратегия%20АПК%202020.pdf. Accessed 11 Jul 2017

Ministry of Agriculture of the RF. Regarding distribution of subsidies in the agro-industrial complex in 2017 (2017). http://government.ru/docs/26279/. Accessed 11 Jul 2017

Bogoviz, A., Mezhov, S.: Models and tools for research of innovation processes. Mod. Appl. Sci. **9**(3), 159–172 (2015)

Popkova, E.G., Shakhovskaya, L.S., Abramov, S.A., et al.: Ecological clusters as a tool of improving the environmental safety in developing countries. Environ. Develop. Sustain. **18**(4), 1049–1057 (2016)

Sadovnikova, N., Parygin, D., Gnedkova, E., Kravets, A., Kizim, A., Ukustov, S.: Scenario forecasting of sustainable urban development based on cognitive model. In: Proceedings of the IADIS International Conference ICT, Society and Human Beings 2013, Proceedings of the IADIS International Conference e-Commerce 2013, pp. 115–119 (2013)

Popova, L., Popova, S.A., Dugina, T.A., Korobeynikov, D.A., Korobeynikova, O.M.: Cluster policy in agrarian sphere in implementation of concept of economic growth. Eur. Res. Stud. J. **18**(Special Issue), 27–36 (2015)

Bogoviz, A.V., Ragulina, Y.V., Shkodinsky, S.V., Babeshin, M.A.: Factors of provision of food security. Agric. Econ. Russ. **2**(1), 2–8 (2017)

Sandu, I.S., Bogoviz, A.V., Ryzhenkova, N.E., Ragulina, Y.V.: Formation of innovational infrastructure in the agrarian sector. AIC: Econ. Manage. **1**(1), 35–41 (2017)

Przhedetskaya, N., Akopova, E.: Institutional designing of continuous education in Russia under the conditions of neo-economy and globalization. Reg. Sectoral Econ. Stud. **15**(2), 115–122 (2015)

The Modern Methodology of Managing the Process of Import Substitution in the Agro-Industrial Complex for Provision of Food Security

Mikhail A. Babeshin(✉), Andrey S. Karpov, and Karina V. Karpova

Military University of the Ministry of Defence of the RF, Moscow, Russia
babeshin-78@mail.ru, karpof_a@mail.ru,
karpo-karina@yandex.ru

Abstract. The purpose of the article is to develop the modern methodology of managing the process of import-substitution in the agro-industrial complex for provision of Russia's food security. For assessing the effectiveness of the methodology of managing the process of import substitution for provision of food security, applied in modern Russia, the authors use the proprietary formula, which allows assessing the effectiveness of the methodology of managing the process of import substitution in the agro-industrial complex for provision of food security in statics. In order to build a full picture of its effectiveness, the authors determine the effectiveness of this methodology in dynamics with the help of regression analysis. As a result of the research, the authors prove the thesis that the methodology of managing the process of import substitution in the agro-industrial complex for provision of food security, which is used in modern Russia, is not effective. For solving the problem of provision of national food security, Russia requires a new modern managerial methodology.

Keywords: State management · Import substitution · Food security
Agro-industrial complex

1 Introduction

The problem of provision of national food is topical for a lot of countries, which implement the course of internal and foreign production specialization, which differs from the agro-industrial complex. This is true for developed and quickly developing countries, which want to conform to the global tendencies and conduct industrialization and then post-industrialization of economy, which is accompanied by slowdown of the rate of development of the agro-industrial complex and increase of their dependence on import of foods products.

In the period of global recession, the problem of provision of national food security in these countries grew due to failures in the system of international economic relations. This became a push for starting the processes of import substitution in the agro-industrial complex for provision of food security. In modern Russia, the methodology of managing the process of import substitution in the agro-industrial

© Springer International Publishing AG, part of Springer Nature 2018
E. G. Popkova (Ed.): HOSMC 2017, AISC 622, pp. 77–82, 2018.
https://doi.org/10.1007/978-3-319-75383-6_10

complex for provision of food security is of the integrated structure - that is it supposes usage of one managerial tool – financial support with small variations.

Our hypothesis within this article is that the methodology of managing the process of import substitution in the agro-industrial sphere for provision of food security in modern Russia is not effective. In order to solve the problem of provision of national food security of Russia, we need a new modern managerial methodology. In this article, we seek the goal of verification of this hypothesis and development of the modern methodology of managing the process of import substitution in the agro-industrial complex for provision of food security in Russia.

2 Materials and Method

For assessment of effectiveness of the methodology of managing the process of import substitution in the agro-industrial complex for provision of food security, applied in modern Russia, we offer to use the following proprietary formula:

$$Emu = \Delta Infs/\Delta VSaic \tag{1}$$

where Emu – effectiveness of methodology of managing the process of import substitution in the agro-industrial complex for provision of food security, applied in modern Russia;

$\Delta Infs$ – annual growth of the value of index of national food security of Russia;

$\Delta VSaic$ – annual growth of the volume of subsidies to the companies of the agro-industrial complex of Russia.

The value of Emu should be treated with a traditional method of treatment of indicators of effectiveness. If this indicator exceeds 1, effectiveness is positive, which shows the expedience of application of this methodology. If the value Emu is equal or below 1, effectiveness is negative, which shows inexpedience of application of this methodology.

The offered formula allows assessing the effectiveness of methodology of managing the process of import substitution in the agro-industrial complex for provision of food security, which is applied in modern Russia, in statics. In order to compile a fuller picture of its effectiveness, the authors determine effectiveness of this methodology in dynamics with the help of regression analysis.

This analysis is used for determining the presence and character (positive or negative) and strength (strong or weak) of the connection between the volume of subsidies to companies of the agro-industrial complex (VSaic) and the value of the index of national food security (Infs) and the place in the ranking of countries as to the index of national food security (Infs). The basic values of these indicators are shown in Table 1.

Table 1. Dynamics of values of the index of national food security and the volume of subsidies to the companies in Russia in 2012–2016

Indicators	Symbol	Values of indicators for the years				
		2012	2013	2014	2015	2016
Value of the index of national food security, points	Infs	68.9	60.9	62.7	63.8	63.8
Place in the ranking of countries as to the value of the index of national food security	Infs	29	40	40	43	43
Volume of subsidies to companies of the agro-industrial complex, RUB billion	VSaic	53.5	54.9	56.1	57.2	58.8

Source: compiled by the authors on the basis of: (The Economist Intelligence Unit 2017), (Ministry of Agriculture of the RF 2017).

3 Discussion

The issues of the methodological character, related to the issue of managing the process of import substitution in the agro-industrial complex for provision of national food security, are studied in the works (Bogoviz and Mezhov 2015), (Popkova et al. 2016), (Sadovnikova et al. 2013), (Popova et al. 2015), (Bogoviz et al. 2017), (Sandu et al. 2017), and (Przhedetskaya and Akopova 2015).

4 Results

In the course of application of the developed formula for assessing the effectiveness of methodology of managing the process of import substitution in the agro-industrial complex for provision of food security, applied in modern Russia, we obtained the following results (Table 2).

Table 2. Results of evaluation of effectiveness of the applied methodology

Indicators	Estimate values of indicators for the years			
	2013	2014	2015	2016
Infs	0.88	1.03	1.02	1.00
VSaic	1.03	1.02	1.02	1.03
Emu	0.86	1.01	1.00	0.97

Source: compiled by the authors

As is seen from Fig. 1, effectiveness of methodology of managing the process of import substitution in the agro-industrial complex for provision of food security, applied in modern Russia, was negative in 2013 (0.86), 2015 (1.00), and 2016 (0.97), and in 2014 it barely exceeded the threshold of positive value (1.01). This shows the inexpedience of application of this methodology. The results of regression analysis are given in Table 3.

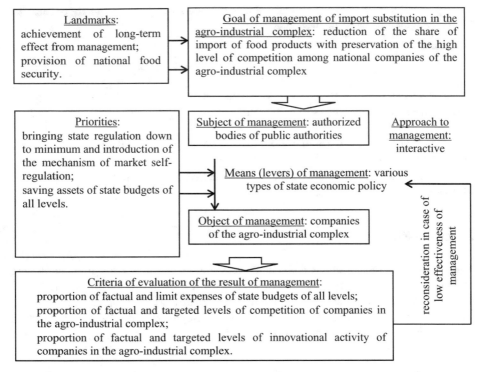

Fig. 1. Modern methodology of managing the process of import substitution in the agro-industrial complex for provision of food security в Russia. Source: compiled by the authors.

Table 3. Results of regression analysis

Characteristics of the model	Model 1: Infs = f (VSaic)	Model 2: Infs = f (VSaic)
Estimate coefficient	−0.58	2.39
Constant	96.87	−95.37
Standard error of deviation	0.76	0.87
Coefficient of determination	0.16	0.71
Coefficient of correlation	16%	71%
Fisher F criterion	0.57	7.52
Number of levels of freedom	3	3
F table	5.54	5.54

Source: compiled by the authors.

As is seen from Table 3, increase of the volume of subsidies to companies of the agro-industrial complex (VSaic) by RUB 1 billion leads to reduction of the value of the index of national food security (Infs) by 0.58 points and increase of the Russia's place in the ranking of countries as to the index of national food security (Infs) by 2.39 points up. At that, the connection between indicators of the first model is weak (correlation coefficient is 16%), and the model is statistically insignificant, as the estimate value of F-criterion is below the table value.

The second model is statistically significant, but the connection between indicators is rather large to state the possibility of explaining the change of one indicator with another (correlation coefficient equals 71%). The results of regression analysis coincide with the results of statistical analysis of effectiveness and confirm inexpedience of application of the methodology of managing the process of import substitution in the agro-industrial complex for provision of food security, applied in modern Russia, due to its negative effectiveness.

The better alternative is the modern methodology, developed specifically for Russia. Contrary to the applied methodology, it is polycomponent, i.e., the means (levers) of management are various types of state economic policy, including invest-ment, innovational, monetary, anti-monopoly, etc.

The choice of policy and the character of its application (activating or restraining) depends on the current situation in the agro-industrial complex, which supposes interactive approach to management. The offered methodology of managing the pro-cess of import substitution in the agro-industrial complex for provision of food security в Russia is shown in Fig. 1.

As is seen from Fig. 1, the landmarks of managing the process of import substi-tution in the agro-industrial complex within the modern methodology are achievement of long-term effect from management and provision of national food security. According to these landmarks, the goal of managing import substitution in the agro-industrial complex consists in reduction of the share of import of food products with preservation of a high level of competition among the national companies the agro-industrial complex.

The priorities of management are bringing state regulation down to minimum and realization of the mechanism of market self-regulation, as well as saving the assets of state budgets of all levels. A criterion of assessing the result of management is pro-portion of factual and limit expenditures of state budgets of all levels, proportion of the factual and targeted levels of competition of companies in the agro-industrial complex, and proportion of the factual and targeted levels of innovational activities of companies in the agro-industrial complex.

In case of low efficiency of management, the means (levers) of management are reconsidered. Due to high flexibility, the developed modern methodology is effective at different phases of the economic cycle. In addition to this, the developed methodology allows controlling the process of import substitution in the agro-industrial complex, which allows for self-improvement of management.

5 Conclusions

Thus, one of the most important causes of aggravation of the problem of provision of national food security in Russia has been application of ineffective mono-component methodology of managing the process of import substitution in the agro-industrial complex.

The modern poly-component and interactive methodology of managing this process is characterized by higher effectiveness and thus ensures full or partial solution to the problem of national food security in Russia. The developed methodology determined the general direction of optimization of the process of managing import substitution in the agro-industrial complex. Determination of specific methods and their application in various situations is a perspective direction of development of the performed scientific research.

References

Bogoviz, A., Mezhov, S.: Models and tools for research of innovation processes. Mod. Appl. Sci. **9**(3), 159–172 (2015)

Popkova, E.G., Shakhovskaya, L.S., Abramov, S.A., et al.: Ecological clusters as a tool of improving the environmental safety in developing countries. Environ. Dev. Sustain. **18**(4), 1049–1057 (2016)

Popova, L., Popova, S.A., Dugina, T.A., Korobeynikov, D.A., Korobeynikova, O.M.: Cluster policy in agrarian sphere in implementation of concept of economic growth. Eur. Res. Stud. J. **18**(Special Issue), 27–36 (2015)

Przhedetskaya, N., Akopova, E.: Institutional designing of continuous education in Russia under the conditions of neo-economy and globalization. Reg. Sectoral Econ. Stud. **15**(2), 115–122 (2015)

Sadovnikova, N., Parygin, D., Gnedkova, E., Kravets, A., Kizim, A., Ukustov, S.: Scenario forecasting of sustainable urban development based on cognitive model. In: Proceedings of the IADIS International Conference ICT, Society and Human Beings 2013, Proceedings of the IADIS International Conference e-Commerce 2013, pp. 115–119 (2013)

The Economist Intelligence Unit: The Global Food Security Index 2016 (2017). http://foodsecurityindex.eiu.com/. Accessed 29 July 2017

Bogoviz, A.V., Ragulina, Y.V., Shkodinsky, S.V., Babeshin, M.A.: Factors of provision of food security. Agric. Econ. Russ. **2**(1), 2–8 (2017)

Ministry of agriculture of the Russian Federation: Regarding the 2017 distribution of subsidies in the agro-industrial complex (2017). http://government.ru/docs/26279/. Accessed 11 July 2017

Sandu, I.S., Bogoviz, A.V., Ryzhenkova, N.E., Ragulina, Y.V.: Formation of the innovational infrastructure in the agrarian sector. AIC Econ. Manage. **1**(1), 35–41 (2017)

The Problem of Provision of Food Security Through Management of the AIC: Transnationalization Vs Import Substitution

Aleksei V. Bogoviz(✉) , Elena I. Semenova,
and Ivan S. Sandu

Federal State Budgetary Scientific Institution "Federal Research Center
of Agrarian Economy and Social Development of Rural Areas—All Russian
Research Institute of Agricultural Economics", Moscow, Russia
aleksei.bogoviz@gmail.com

Abstract. The purpose of the article is to substantiate the necessity and to determine the perspectives of modern Russia's transition to solving the problem of provision of food security through management of the agro-industrial complex (AIC) with the help of transnationalization instead of the traditional practice of import substitution. The indicator of potential of transnationalization of entrepreneurship in the AIC in this research is index of economic globalization according to the KOF. To evaluate the level of implementation of potential of transnationalization of entrepreneurship in the AIC, the authors the methods of regression and correlation analysis. The authors study the connection between the volume of export of products of companies of the AIC, share of export in the structure of sales of the companies of the AIC, and index of economic globalization. In the course of the research, the authors prove that import substitution is not the only method of solving the problem of provision of food security through management of the AIC. Modern Russia possesses the potential and the possibility for application of such tool of the AIC management as transnationalization, which allows achieving tactical and strategic results, which ensure elimination of the problem of provision of food security.

Keywords: Food security · Management of agro-industrial complex
Transnationalization · Import substitution

1 Introduction

Agro-industrial complex is the central element of the system of provision of national food security. Due to topicality of the problem, provision of food security in the AIC is often viewed by economic systems only in connection to the national food security – not as an independent sphere of economy. In this case, management of the AIC is brought down to stimulating development of the process of import substitution.

© Springer International Publishing AG, part of Springer Nature 2018
E. G. Popkova (Ed.): HOSMC 2017, AISC 622, pp. 83–88, 2018.
https://doi.org/10.1007/978-3-319-75383-6_11

Such economic practice is characterized by various drawbacks. Firstly, the main load within import substitution is laid on the state, which leads to high expenditures of budgets of various levels of the state system and dependence of the AIC business on state subsidies.

Secondly, import substitution, as any regulation tool, is not universal and possesses a limited specter of effective usage. That's why in certain situations its application is not effective, and can even show negative effectiveness; instead of strengthening the positions of national entrepreneurship, it leads to their weakening.

Thirdly, import substitution presupposes that positions of domestic entrepreneurship are weak, and it should take a defensive position as to foreign rivals. At that, the possibility of transition to attack is not viewed, which might be a reason for lost profit from unrealized potential in the sphere of development of entrepreneurship.

The working hypothesis of this research consists in the fact that import substitution is not the only method of solving the problem of provision of food security through management of the AIC. Modern Russia possesses the potential and possibility for application of such tool of management of the AIC as transnationalization, which allows achieving tactical and strategic results that ensure elimination of the problem of provision of food security as such.

Our purpose within this article is to substantiate the necessity for and to determine perspectives of transition of modern Russia to solving the problem of provision of food security through management of the AIC with the help of transnationalization instead of the traditional practice of import substitution.

2   Materials and Method

The indicator of potential of entrepreneurship's transnationalization в the AIC in this research in index of economic globalization according to the KOF. For evaluation of the level of implementation of the potential of entrepreneurship's transnationalization in the AIC, the authors use the methods of regression and correlation analysis. The authors study the connection between the volume of export of products of the companies of the AIC (y_1), share of export in the structure of sales of the AIC companies (y_2), and the index of economic globalization (x).

The logic of application of these methods to verification of the working hypothesis of the research is that if connection between the above indicators is strong (determination coefficient - $R^2 \geq 90\%$) and positive (estimate coefficient in the model of paired linear regression b with "+" sign), i.e., the potential of entrepreneurship' transnationalization in the AIC is fully implemented.

Accordingly, low value of the correlation coefficient and negative sign of the estimate coefficient b show incomplete realization of this potential. The values of potential of entrepreneurship's transnationalization in the Russia's AIC in 2011–2016 are shown in Table 1.

Table 1. Values of indicators of the potential of entrepreneurship's transnationalization in the Russia's AIC in 2011–2016

Indicators	Value of indicators for the years					
	2011	2012	2013	2014	2015	2016
Volume of supplied goods of own production, RUB billion	3,262	3,602	4,001	4,272	4,840	5,861
Volume of export of products, RUB billion	528	798	1,008	978	1,140	972
Share of export in the structure of production, %	16.19	22.15	25.19	22.89	23.55	16.58
Index of economic globalization, points from 1 to 100	61.67	52.79	50.13	51.96	53.74	52.06

Source: compiled by the authors on the basis of: (Federal State Statistics Service 2016), (Voronin 2017), (KOF 2017).

3   Discussion

The fundamental and applied issues of setting and solving the problem of provision of food security through management of the AIC are reflected in the studies of such scholars and experts as (Bogoviz and Mezhov 2015), (Popkova et al. 2016), (Sadovnikova et al. 2013), (Popova et al. 2015), (Bogoviz et al. 2017), (Sandu et al. 2017), and (Przhedetskaya and Akopova 2015).

At that, despite the high level of elaboration of this scientific problem, it has been studied only from one side – from the point of view of import substitution, while other methods of its solution were omitted by modern authors –which is a significant gap in the system of scientific knowledge.

4   Results

The data of Table 1 show that as of now (2017) the potential of entrepreneurship's transnationalization in the Russia's AIC is rather large, as the value of the index of economic globalization in 2016 constituted 52.06 points out of 100. This is also shown by large progress of domestic economic system in integration into the global economy, as well as existence of perspectives for development of this process. The results of regression and correlation analysis are given in Table 2.

Table 2. Results of regression and correlation analysis

Indicators of regression/correlation	Estimate values of indicators for the models	
	y_1	y_2
R^2	65%	43%
b	−0.70	−0.61

Source: compiled by the authors.

As is seen from Table 2, values of determination coefficients are low for both received models of paired linear regression (65% and 43%). Moreover, the estimate coefficient b is negative in both models. This shows that increase of the potential of entrepreneurship's transnationalization in the Russia's AIC (growth of the index of economic globalization) by 1 point leads to reduction of the level of its realization. Thus, the volume of export of products of companies of the AIC reduces by $ 70 million, and the share of export in the structure of sales of the AIC companies – by 0.61%.

This proves the offered hypothesis on perspectives of entrepreneurship's transnationalization in the Russia's AIC. The following arguments should be considered as substantiation of necessity for realization of these perspectives and refusal from the traditional practice of import substitution in the AIC for solving the problem of provision of national food security:

– deficit of the federal budget predetermines the impossibility of implementing the measures of import substitution in full;
– volume (absolute indicator) and share (relative indicator) of import in the structure of the Russia's AIC are too low for viewing it as a threat to the Russian entrepreneurship;
– foreign companies are a source of competition, the level of which reduced in the Russia's AIC, which might lead to elimination of natural market stimuli for its development. That's why foreign competition plays an important role in development of the Russia's AIC.

The most perspective directions of transnationalization of the Russian entrepreneurship in the AIC for solving the problem of provision of food security are the following: expansion of export of the AIC products; Russian AIC companies' integration with their foreign rivals and joint entering the world markets (this direction could be realized through formation of transnational cluster structures); transnationalization of entrepreneurship and placing the structural branches of Russian companies of the AIC in other countries for optimization of business processes. In order to achieve high effectiveness of management of the AIC with the help of transnationalization of entrepreneurship of modern Russia, we offer the following practical recommendations:

– provision of tax and customs preferences for domestic transnational companies of the AIC: this is necessary for provision of stimuli for transnationalization and for leveling the conditions in which Russian and foreign transnational companies with similar privileges in their countries are;
– active cooperation in international economic relations of Russian companies: a lot of decision on business cooperation can and should be taken and conducted at the level of states within partnership agreements.

The mechanism of solving the problem of provision of food security with the help of transnationalization of Russian business in the AIC is shown in Fig. 1.

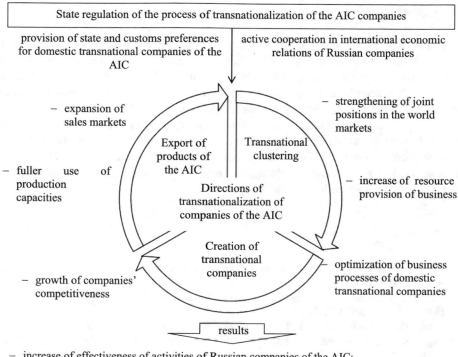

Fig. 1. Mechanism of solving the problem of provision of food security with the help of transnationalization of Russian entrepreneurship in the AIC. Source: compiled by the authors.

As is seen from Fig. 1, transnationalization of Russian entrepreneurship in the AIC provides a lot of advantages. This includes increase of effectiveness of the Russian companies of the AIC by means of optimization of business processes and simplification of the task of fronting of world markets, expansion of sales markets for Russian companies of the AIC, which allows using their production capacities to a full extent and the Russia's AIC transition from the strategy of survival to the strategy of attack and active development, which ensures more intense growth of business activity.

As a result, the problem of provision of national food security is solved and fully eliminated, as domestic entrepreneurship receives a strong and long-term impulse for development, which ensures its global competitiveness.

5 Conclusion

It is possible to conclude that economic globalization and integration brings new threats, violating the usual order of functioning of economic systems, and opens new possibilities and perspectives for their development. The Russia's AIC does not require import substitution – it is capable to go to a new level of the globally-oriented development: transnationalization of entrepreneurship.

It supposes domestic AIC companies' entering the global markets and their transformation from the ones vulnerable to the influence of globalization, which require state support, into highly effective and competitive players in the global arena, capable of independent functioning and bringing profit for the state. It allows solving the problem of provision of food security.

References

Bogoviz, A., Mezhov, S.: Models and tools for research of innovation processes. Mod. Appl. Sci. 9(3), 159–172 (2015)

KOF: Index of economic globalization: Russian Federation (2017). http://globalization.kof.ethz. ch/query/. Accessed 19 July 2017

Popkova, E.G., Shakhovskaya, L.S., Abramov, S.A., et al.: Ecological clusters as a tool of improving the environmental safety in developing countries. Environ. Dev. Sustain. 18(4), 1049–1057 (2016)

Popova, L., Popova, S.A., Dugina, T.A., Korobeynikov, D.A., Korobeynikova, O.M.: Cluster policy in agrarian sphere in implementation of concept of economic growth. Eur. Res. Stud. J. 18(Special Issue), 27–36 (2015)

Przhedetskaya, N., Akopova, E.: Institutional designing of continuous education in Russia under the conditions of neo-economy and globalization. Reg. Sectoral Econ. Stud. 15(2), 115–122 (2015)

Sadovnikova, N., Parygin, D., Gnedkova, E., Kravets, A., Kizim, A., Ukustov, S.: Scenario forecasting of sustainable urban development based on cognitive model. In: Proceedings of the IADIS International Conference ICT, Society and Human Beings 2013, Proceedings of the IADIS International Conference e-Commerce 2013, pp. 115–119 (2013)

Bogoviz, A.V., Ragulina, Y.V., Shkodinsky, S.V., Babeshin, M.A.: Factors of provision of food security. Agric. Econ. Russ. 2(1), 2–8 (2017)

Voronin, B.A.: Russian AIC – from import of agricultural products to export-oriented development. Fields Russ. 4(148), 5–12 (2017)

Federal State Statistics Service: Industrial production in Russia. Statistical collection. Federal State Statistics Service, Moscow (2016)

Sandu, I.S., Bogoviz, A.V., Ryzhenkova, N.E., Ragulina, Y.V.: Formation of innovational infrastructure in the agrarian sector. AIC Econ. Manage. 1(1), 35–41 (2017)

Elements of the Foresight Technology in Design Project-Oriented Training of Prospective Engineers

Raisa Petruneva$^{(\boxtimes)}$, Valentina Vasilyeva, and Olga Toporkova

Volgograd State Technical University, Volgograd, Russian Federation
raisa@vstu.ru, vasilyevavd2016@yandex.ru,
toporkova.vstu@gmail.com

Abstract. The article is devoted to one of the most important problems regarding the results of technical innovations implementation in people's life and to training future specialists to uncover those consequences during the preparation stage of technical projects. The objective of this paper is to analyze the problems of design project-oriented training of engineering students in Russia and find ways for its improvement. Methodology: Universal scientific research methods were used such as the methods of idealization and formalization. The experimental research methods of observation, modeling, abstraction, comparison, the analysis of students' educational activity results were also employed. Results: 1. The experience of some well-known organizations that use foresight technologies for predicting the consequences of innovative engineering and technical decisions implementation is analyzed. 2. The methodology of students training in forecasting skills making use of foresight technologies is developed. 3. The educational and expert procedure content is described. The results obtained are novel as for the first time foresight technologies are applied to the process of training of engineers. Previously the foresight technologies were used for elaborating large social and economic projects directed on the large territories development. The findings obtained comply with well-known technologies with similar focus, e.g. CDIO. This research contributes to the theory and methodology of professional education of engineers as it develops the understanding of their training content. Recommendations: The findings obtained can be useful for specialists of training and methodological associations, technical universities curricular designers and educators, specialists of various design organizations.

Keywords: Foresight technology · Project-oriented training · Engineers
Foresight · Project-based learning

JEL-code: A220

1 Introduction

Project-based learning is a core, systemically important component of educational process at a higher technical school because it integrates all the knowledge in humanities and sciences as well as social, mathematical, technical and other subjects

© Springer International Publishing AG, part of Springer Nature 2018
E. G. Popkova (Ed.): HOSMC 2017, AISC 622, pp. 89–96, 2018.
https://doi.org/10.1007/978-3-319-75383-6_12

learnt before. It is also the backbone and a guideline of all the future professional activities of an engineer because it defines his position as a designer and producer of new real products and systems (from the emergence of an idea to finished goods).

The importance of project-based learning in undergraduate education of engineers and other specialists is underlined by many Russian and foreign scientists (Lehmann et al. 2008), (Von Grabe et al. 2010), (Zamyatina et al. 2013) and (Shekar 2014).

The objective of this paper is to analyze the problems of design project-oriented training of engineering students in Russia and find ways for its improvement.

2 Background and Methodology

The following methods were employed: universal scientific methods, including methods of idealization and formalization. The experimental research methods of observation, modeling, abstraction, comparison, the analysis of students' educational activity results were also used.

Currently in Russia students are exposed to specific elements of project activities mainly through the study of unrelated subjects by solving physical, chemical, technical and other problems. Even an undergraduate degree paper cannot replace a full-fledged multidisciplinary project since the paper contains only some particular elements of a graduation paper: review of literature on the problem set by the teacher, choosing the way of solving the task and elaborating on its solution (doing some calculations, carrying out experiments). As a result, Bachelor of "Methods and technologies", being a graduate of the first tier of higher engineering education, cannot fully master the procedures and an entire array of engineering methods; consequently, it does not make it possible to qualify him as an engineer in the common sense of the word; it rather makes regard him as a vocational school graduate (Ministry of Labor and Social Protection of the Russian Federation 2013).

It should be noted that the problems of the quality of engineering education are not specifically Russian. Thus, in accordance with the study conducted by Australian researchers Nair et al. (2009).

Research on student-learning outcomes indicates that university graduates do not possess important skills required by employers, such as communication, decision-making, problem-solving, leadership, emotional intelligence, social ethics skills as well as the ability to work with people of different backgrounds.

At the 2014 Indianapolis Third International Forum for Engineering Education the problem of the dissatisfaction of employers with the quality of training of technical university graduates was also discussed: university graduates are not ready to work in modern environment and do not possess the required skills and knowledge (Ivanov and Zijatdinova 2014). With this, it was noted that employers demand very much of young specialists: manufacturing companies need a graduate possessing the skills of a project manager, a specialist in a specific field, a researcher, a designer and a talented manager with communication and leadership skills. Those findings comply with the earlier research conducted by Hesketh (2000), who has found out that:

Employers no longer seek graduates with the 'hard' technical or vocational skills required for the job... the 'softer' or interpersonal skills are the new vogue.

Manufacturing companies need a full-fledged specialist with all the necessary competences, including design skills, who is capable of getting down to work at once and making money for the company.

The Association for Engineering Education of Russia (AEER) held an international conference "Management of multidisciplinary projects in engineering education: planning and implementation" in the cities of Lisbon and Porto in Portugal in May 2014. The conference was aimed at the analysis of the best results in the organization and management of multidisciplinary projects which are undoubtedly implemented in Russian universities too.

Such matters as the methodology of planning and implementation of multidisciplinary projects, basic tendencies in and approaches to multidisciplinary aspects of engineering education as well as successful practices in the implementation of multidisciplinary projects, including the ones with the participation of undergraduates, were discussed (AEER 2015).

3 Discussion and Results

3.1 The CDIO Initiative and Instructional Engineering Design

The involvement of students in project activities requires new approaches to the project training of Bachelors of "Methods and technologies". The concept of project training, developed within the CDIO (Conceive, Design, Implement, Operate) Initiative international project, can be used as a guideline. Project competences, which are to be studied by undergraduates within educational programs in methods and technology, were included in the requirements of CDIO Syllabus (Crawley et al. 2007). The CDIO Standards, which have been developed, specify requirements to the Syllabus of Bachelors of "Methods and Technology" in the field of design education and stipulate the following:

- to introduce into an undergraduate curriculum an engineering orientation course which sets the basics to form relevant professional, personal and interpersonal competences of prospective bachelors;
- to introduce in the curriculum two or more projects to master practical design skills (one project is for a basic level, another is for an advanced level);
- to form working environment for practical engineering practice in order to create products and systems and to improve professional and social skills of students;
- to provide for an integrated education based on the use of active and effective practically targeted methods (CDIO Standards 2.0.).

A full-fledged instructional engineering design requires a revision of the entire philosophy of instructional engineering and the introduction of modern innovative approaches to the content of educational projects; the content shall be consistent both with the current technical practice and the requirements of international standards.

First of all, it is necessary to reconsider the content of the major subject which prepares an undergraduate for practical design work and, to a large extent, helps integrate the knowledge of various subjects into the entire "body" of the educational project. This subject, which gives an understanding of the principles of coordination and integration of various fields of engineering knowledge, has a multidisciplinary character and has the respective name – 'Fundamentals of Design …'.

The major element of educational engineering design process is the stage of the analysis and choice of a new technical decision which will be implemented in the project later. It is at this stage that a designer student can reveal all his professional knowledge and public stance.

The essential element of this stage should be the analysis and anticipatory estimates of mid-term and long-term risks and hazards for man and mankind. These risks and hazards arise due to the implementation of a complete life cycle of a technical product starting from its design and production and ending with its operation and disposal.

Any of the above stages can involve some hazards which can be critical, though they can be delayed in time and space. With this, these consequences can definitely be generated by technology, on the one hand, and caused by social and humane factors, on the other hand. This is due to the fact that in the modern global world a merge of technical, social and biological systems is observed, and the scope of a technical project has achieved royal dimensions never seen before. Nowadays mankind is in a fundamentally new social and cultural situation which can threaten mankind a real global ecological catastrophe should these consequences be neglected. Thus, a major task of a designer is to identify such consequences to society in the technological, social and humanitarian fields after the technical novelty has been implemented.

3.2 The Foresight Technology in Design Project-Oriented Training of Engineering Students

This forecast is impossible without using multidiscipline knowledge. The famous Russian engineer and philosopher P. Engelmeyer believed that an engineer had to rise above his creation and analyze related knowledge to understand the essence of the invention. Due to the tunnel vision of engineering or any other professional education a specialist cannot give an unbiased look at the results of his work and assess them objectively. In this situation a possible way out might be to bring in specialists (experts) from different (non-engineering) fields to engineering when analyzing and choosing the design concept (Engelmeyer 1898).

The idea of foresight might be very fruitful in this context; foresight is essentially a set of technologies which use expert reviews to determine some possible options in the future (Limonova 2015). The foresight technology implies the involvement of many experts from different walks of life who are related to this or that degree to the area of a particular project.

The idea of foresight is based on the assumption that the coming of the "desired" variant of the future heavily depends on today's activities. Hence, the choice of decision options is related to the choice of technologies which make it possible to minimize possible negative consequences of the project and foresee the most unexpected ways of the development of events as well as possible pitfalls. The work of experts is aimed not only at identifying possible options but at choosing the most advantageous ones in compliance with certain acceptability criteria.

In practice, work on a multidiscipline project can be organized in the form of students' expert board in the following way. A student designer receives a design engineering assignment and generates input data to choose the way to fulfill the project task.

The designer uses their knowledge of social sciences, humanities, natural and technical sciences as well as personal experience to formulate a hypothesis of possible man-induced, social and humanitarian consequences of the implementation of the proposed technical solution. On its basis the designer student makes up a list of alternatives to be studied. With this aim he or she forms, upon a project advisor's review, a group of experts. Each expert is assigned the task to scrutinize one of the alternatives and provide a scientifically grounded conclusion. The group of experts is formed of peer designer students. Each of them can act for their peers as an expert in a particular field. A number of experts can be invited to consider all possible options, scenarios of the development of events as well as to obtain a complete picture (Petruneva and Vasilyeva 2010; Petruneva et al. 2016).

The pattern of a future technical solution develops from the information which the experts share with the designer. Both traditional and comparatively novel expert methods are used during the discussion. With this, discussion methods are being constantly improved, procedures and practices are being perfected; generally, all this increases the validity of the scenario of the development of a man-induced situation.

The main aim of inviting experts is to use their knowledge in a specific scientific and applied field in order to resolve the engineering task assigned. An expert cannot rely only on common sense; they should make their judgment on the basis of scientifically grounded facts, opinions of respected authorities on the matter, results of additional investigations, including social studies, etc.

With this, experts should give answers to the following questions set by the chief performer of the project.

The proposed project is discussed by a board of experts. During the discussion various decision-taking methods can be used, including the well-known brain storming, scenario-building technique, expert panels, the Delphi method (when questionnaires are answered in two rounds by experts) and other modern technologies such as road maps, relevance trees, SWOT analysis, mutual influence analysis and others. Some of these methods require mathematical tools for data processing.

As a result, the board decides what physical, social and humanitarian consequences of various magnitude might follow the implementation of the proposed technical solution, and either recommends the design for the subsequent engineering implementation or suggests a further study of the problem with an appropriate technical solution in mind.

The necessity to assess prospective mid-term scenarios of the development of man-induced events is a necessity which has already been recognized by specialists. In this context foresight technologies are a fairly reliable and promising tool which is already being used for long-term forecasts of social and economic development of countries and regions, and an enormous number of experts, up to dozens of thousands, may be involved in the process (Stanovlenie Forsajta 2017).

In the countries, which are the most advanced ones in this aspect, there function and operate state institutions and organizations which carry out a comprehensive expert evaluation of technical objects which includes social and humanitarian aspects. In the USA, for example, Office of Technology Assessment was set up in 1972 which task is to provide senators and congressmen with objective information in this field.

At the same time Technology Assessment Board (TAB) was set up in the Congress with the main task to develop at the earliest stage guidelines for possible positive or negative consequences of technical projects as well as to provide the Congress with information required for making decisions (Stepin et al. 1999).

In German Bundestag in 1986 a similar board was set up (Enquete-Komission "Technikfolgenabschtzung"), which was later used as a basis for the Technology Assessment Bureau; the Bureau incorporates a group of multidiscipline scientists representing natural, social and technical sciences there. The initiatives of the German engineers' union which adopted the guidelines "Technology assessment: concepts and grounds" in 1991 are of particular interest to us. According to the guidelines, technical activities always necessarily assess technology, and not everything that is technically feasible must necessarily be created. Thus, technology should meet a number of requirements – not only technical expedience, functionality and cost effectiveness, but it should also improve the living standards, safety and health of people, the quality of natural and social environment and so on. Technology assessment means a systematic and orderly activity which assesses direct and indirect technological, ecological, humanitarian, social and other consequences of this technology and its possible alternatives; with this aim it works out pragmatic and creative possibilities to make justified decisions (Ibid.).

Thus, the assessment of technology and, respectively, technological processes, has become an indispensable part of engineering. Sometimes this assessment is also called social and humanitarian (social and economic, social and ecological, etc.) expert evaluation of technical projects. It goes without saying that this work is a multidiscipline task and requires training experts with encyclopedic knowledge both in technical and scientific fields, on the one hand, and social and humanitarian competences, on the other hand. Such responsibility means that it is necessary to build up the self-consciousness of engineers in terms of understanding the necessity of social, ecological etc. assessment of technology and technological processes.

4 Conclusions

In our opinion, good opportunities in engineering and design activities are provided by the implementation of the CDIO (Conceive, Design, Implement, Operate) ideas in technical universities. Those ideas form a basis for the implementation of new approaches in teaching and learning activities to designing technical objects, including the ones which require expert opinion technologies. Moreover, the CDIO ideas induce undergraduates to implement their first projects.

Thus, modeling a comprehensive multidiscipline expert opinion of new engineering and design solutions during education as well as incorporating such modeling in the syllabus of undergraduates seems to be a promising approach.

References

AEER: Mezhdunarodnaja konferencija «Upravlenie mezhdisciplinarnymi proektami v inzhen-ernom obrazovanii: planirovanie i vypolnenie». In: International Conference "Management of Multidisciplinary Projects in Engineering Education: Planning and Implementation" (2014). http://aeer.ru/ru/c_liss_2014-result.htm. Accessed 6 July 2017

CDIO Standards 2.0.: Chalmers University of Technology, Gothenburg. http://www.cdio.org/implementing-cdio/standards/12-cdio-standards. Accessed 6 July 2017

Crawley, E.F., Malmqvist, J., Östlund, S., Brodeur, D.R.: Rethinking Engineering Education: The CDIO Approach. Springer, New York (2007)

Engelmeyer, P.K.: Tehnicheskij itog XIX veka (The Technical Result of the 19th century). Tipografija K.A. Kaznacheeva, Moscow (1898)

Hesketh, A.J.: Recruiting an Elite? employers' perceptions of graduate education and training. J. Educ. Work 13(3), 245–271 (2000)

Ivanov, V.G., Zijatdinova, J.N.: Mezhdunarodnyj forum Amerikanskogo obshhestva po inzhenernomu obrazovaniju. In: International Forum of American Society for Engineering Education, Vysshee obrazovanie v Rossii (Higher Education in Russia), No. 8–9, pp. 65–75 (2014)

Lehmann, M., Christensen, P., Du, X., Thrane, M.: Problem-oriented and project-based learning (POPBL) as an innovative learning strategy for sustainable development in engineering education. Eur. J. Eng. Educ. 33(3), 283–295 (2008)

Limonova, M.: Chto takoe Forsajt? (What is the Foresight?), The Expert Club of Industry and Energy, Moscow (2015). http://www.expertclub.ru/sections/foresight/programm/0. Accessed 6 July 2017

Nair, C.S., Patil, A., Mertova, P.: Re-engineering graduate skills - a case study. Eur. J. Eng. Educ. 34(2), 131–139 (2009)

Petruneva, R.M., Vasilyeva V.D.: Jekspertiza inzhenerno-proektirovochnyh reshenij kak sovremennaja uchebnaja tehnologija. In: Examination of Engineering-Designing Decisions as Modern Educational Technology, Vysshee obrazovanie v Rossii (Higher Education in Russia), No. 8–9, pp. 122–128 (2010)

Petruneva, R.M, Vasilyeva, V.D., Toporkova, O.V.: «Sociokul'turnaja sreda universiteta i tradicii vospitanija budushhih inzhenerov» Socio-cultural Environment of the University, Traditions of Educational Work, and Extra-Curricular Activities of Future Engineers, Vysshee obrazovanie v Rossii (Higher Education in Russia), No. 7, pp. 135–141 (2016)

Ministry of labour and Social Protection of the Russian Federation: Prikaz Ministerstva truda I social'noj zashhity RF ot 12 aprelja 2013 g. № 148n «Ob utverzhdenii urovnej kvalifikacii v celjah razrabotki proektov professional'nyh standartov» (The Order of the Ministry of labour and Social Protection of the Russian Federation from Ministry of Labour and Social Protection of the RF, Moscow (2013). http://www.rosmintrud.ru/docs/mintrud/orders/48. Accessed 6 July 2017

Shekar, A.: Project-based learning in engineering design education: sharing best practices. In: ASEE Annual Conference and Exposition, Conference Proceeding, pp. 1–18 (2014)

"Stanovlenie Forsajta. Pervyeprecedenty Forsajta". (The Foresight Establishment. The First Cases of Foresight). http://foresight.sfu-kras.ru/node/9. Accessed 6 July 2017

Stepin, V.S., Gorohov, V.G., Rozov, M.A.: Filosofija nauki i tehniki (The Philosophy of Science and Technology). Izd-vo Gardariki, Moscow (1999)

Von Grabe, J., Dietsch, P., Winter, S.: Interdisciplinary design projects in the education of civil engineers. In: 11th World Conference on Timber Engineering 2010, WCTE, pp. 2635–2644 (2010)

Zamyatina, O.M., Mozgaleva, P.I., Solovjev, M.A., Bokov, L.À., Pozdeeva, À.F.: Realization of project-based learning approach in engineering education. World Appl. Sci. J. **27**(13A), 433–438 (2013)

Philosophical View on Human Existence in the World of Technic and Information

Anna Guryanova[1]([⊠]) [iD], Elmira Khafiyatullina[2] [iD],
Andrew Kolibanov[2] [iD], Alexander Makhovikov[1] [iD],
and Vyacheslav Frolov[1] [iD]

[1] Samara State University of Economics, Samara, Russia
annaguryanov@yandex.ru, shentala_sseu@inbox.ru,
frolov5070@yandex.ru
[2] Samara State Technical University, Samara, Russia
dek.fispos2009@yandex.ru, kort70@mail.ru

Abstract. The aim of the article is to study an impact of technic and information on human existence. In this context the concept of technic and a number of relative terms are explained. Main philosophical conceptions of technic describing its place and role in the social life are considered. The impact of information technic on the process of modern social development is analyzed. The social problems of information society are discussed from the philosophical point of view. This means special attention to the great changes brought by information and technical progress into the social life. Systems of culture, communication and media are seriously transformed. This makes problems of human existence in the world of technic and information really actual.

Keywords: Information · Technic · Human being · Philosophy of technic
Anthropology of technic · Information society · Mass culture · Mass media
Electronic communication

JEL Classification Codes: Z 10 · Z 13 · Z 19

1 Introduction

Human being is one of the main philosophical problems. Philosophy considers him as a complex system consisting of various universal characteristics. In general philosophy concerns everything that can be associated with the human being. For example objects of technics and information technology which are the result of human creative activity. If the human being thinks and produces them, he needs to understand these phenomena sooner or later - because the process of creation changes not only the world around him but also himself. That's why philosophy is closely connected with human activities in technics and information technology. It makes human being looking for an answer to the most difficult questions of his modern existence. These are the questions of reasonability and legality of his transformation activity.

The present stage of civilization development causes a great sense of alarm because of intensive growth of information and technical innovations. It threatens not only the

© Springer International Publishing AG, part of Springer Nature 2018
E. G. Popkova (Ed.): HOSMC 2017, AISC 622, pp. 97–104, 2018.
https://doi.org/10.1007/978-3-319-75383-6_13

human life but existence of humanity as a whole. All this makes actual philosophical analysis of human's being in the world of technic and information. We are sure in future this theme will be even more topical than in our days.

2 Materials and Methods

In the process of research we used philosophical, logical, dialectical and historical methods of analysis.

2.1 Philosophical Method of Analysis

As we have seen earlier modern philosophy pays special attention to the problem of human existence in the world of technic and information. Method of philosophical analysis gave us an opportunity to study this problem purposefully and completely (Guryanova et al. 2017). We searched and interpreted the rich philosophical material on this topic. These are conceptions from the field of existentialist philosophy and relative spheres of philosophical knowledge. They both concern nature and essence of the human being, technic and information.

2.2 Logical Methods of Analysis

The findings of the article were made with a help of the logical methods of cognitive activity. We can note among them methods of analysis and synthesis, method of deduction and method of comparison. In the process of consideration we also used logical methods of systematization and classification.

2.3 Dialectical Method of Analysis

We used dialectical method to consider an object we are interested (human existence in the world of technic and information) from different sides. Dialectical method gave us a chance to show the real impact of science and technology on modern life, to find out its positive and negative aspects. Today we can postulate that the very existence of humanity will depend finally on its decisions about advances in science and information technology.

2.4 Historical Method of Analysis

As a starting point for the analysis we accept a historical character of science and technology impact on the society. Science and technology are always determined by contemporary culture. So their influence on civilization must be considered in historical perspective, in the context of specific historical epoch. It's also important to understand in whose hands the technic is, who uses it and for what purposes.

3   Results

3.1   Modern Philosophy of Technic: Main Problems and Concepts

It's well known that information progress of humanity and the later appearance of information society became possible only at a high level of scientific and technological development. Technic accompanied humanity since ancient times. But civilization didn't immediately become technical and especially information. Intensive growth of technology took place in the XX century. Since that time technic has become an object of philosophical analysis.

3.1.1   Philosophy and Anthropology of Technic

The first scientists who began to discuss the problems of interaction between technology and society were the German philosopher E. Kapp and the Russian engineer P. Engelmeier. Technic and its importance for the future of humanity has become an object of systematic analysis in their works. Thus philosophy of technic was formed. Modern philosophy of technic explores phenomenon of technic, its place and role in the process of social development. It also pays special attention to the impact of technic on human existence.

In this context in the first half of the 20th century a new field of philosophical knowledge was formed. It was called "anthropology of technic". Anthropology of technic considers technical environment as a way of human existence. It explores technic as a necessary attribute of human life. But it often analyzes technic from the biological point of view. The sources of technical creativity are revealed exclusively in the biological activity of the human being. In other words technic is considered as a realization of some qualities and abilities typical for the world of nature. The human being compensates his biological failure with technic.

3.1.2   Concepts of "Technic" and "Technosphere" as the Units of Philosophical Analysis

Basic concept of the modern philosophy of technic is certainly "technic" itself. Today technic has become a force dominating the human being. Under "technic" we usually mean the following:

1. a set of technical devices, artefacts (from the simplest tools to complex technical systems);
2. a set of various types of technical activities for creation of these devices (from scientific and technological researches to their production and exploitation);
3. a set of technical knowledges (from specialized technical prescriptions to theoretical scientific researchers conducting in the field of technology) (Stepin et al. 1996).

As we can see the sphere of technic includes not only the using of scientific and technical knowledges but also their production. That's why modern technic is closely connected with development of science. It is also included into an independent field of life activity which is called "technosphere". Technosphere is a historically determined system of relations between humanity and technology, human being and nature, between different humans. It is based on a technical understanding of the world. Technosphere is consciously formed, maintained and perfected by humanity.

3.1.3   Concept of "Noosphere" as a Unit of Philosophical Analysis

In addition to technosphere philosophers also talk about "noosphere". The founders of noosphere conception V. Vernadsky and P. Teilhard de Chardin believed that human intellect is turning into a planetary geological force. It helps to regulate both natural and social reality and to create the more perfect forms of human existence. Noosphere is a result of systematic, consciously regulated transformation of the biosphere into a new qualitative state.

According to Vernadsky noosphere is a harmonic connection of nature and society, a triumph of intellect and humanism. Noosphere unites together science, social development and state policy for the benefit of the human being. It means a new world without weapons, wars and environmental problems. It is a wonderful dream, an important goal facing people of the good will. Noosphere is a great mission of science and an aim of humanity armed with the science.

It's clear that noosphere is directly connected with technosphere. But will the modern humanity move in direction of noosphere's construction in understanding of Vernadsky? Or technosphere and noosphere as the products of human activity will destroy humanity in future? It's significant that the greater part of modern philosophical conceptions interpret the project of noosphere as a utopia. They consider technic as a threat to human existence. The most illustrative in this context is philosophy of existentialism.

3.2   The Impact of Technic on Human Existence in Philosophy of Existentialism

Philosophers-existentialists look at the human future in tragic light. The reason is the achievements of scientific and technical progress. Existentialists interpret development of science and technology as a reason of public and personal standardization. Technic limits the human freedom, transforms the human nature and turns the human being into a soulless machine. This leads to the loss of human spirituality, morality and culture.

3.2.1   N. Berdyaev About Social Consequences of Technical Development

The problems of technical development and its social consequences are analyzed seriously by Russian philosopher-existentialist N. Berdyaev. He characterized technic as a turning point of human destiny. On the one hand technic liberates the human's spirit, but on the other - it conquers not only the world of nature but the human being himself.

In his article "The human being and the machine" Berdyaev wrote about the crisis of human being and humanity. This crisis is caused by the rapid development of technic. Philosopher considered technic as a factor determining human life and activity (Berdyaev 1989). But whether the humans are able to limit the power of technic? Berdyaev didn't give a final answer to this question.

3.2.2   M. Heidegger About the Present State of Technical Development

German philosopher M. Heidegger analyzed the nature of technic from existentialist point of view. He was sure that understanding of technic needs an appeal to the human being, creation of "human dimension" of technical progress. We mustn't only use technic as a tool, we must govern it as an instrument. An essence of technic is the way the human being discovers possibilities contained in nature.

In general Heidegger was dissatisfied with the present state of science and technic development. He thought that when the human being creates technic he doesn't pay the necessary attention to its nature. Philosopher wanted to see another form of technic existence. He hoped a new spiritual atmosphere will be formed around it. According to Heidegger technic must look like an art. In art the human being uses natural materials in such a way that an essence of art is fully determined by human nature.

3.2.3   K. Jaspers About Nature and Essence of Modern Technic

One of the leaders of modern existentialist philosophy K. Jaspers believed that technic has a dual character. It distances human being from nature, but at the same time it causes a new unity with it. Technic creates the beauty of technical products. it expands much the real vision of the world. But technic also has its limits. The human being must be afraid of technic because he can get lost in it and forget about himself. Technic is only an instrument of domination over the lifeless organic forces and people who sometimes look at it with horror.

As technic itself doesn't set goals it is on the other side of good and evil or precedes them. It can serve either good or evil for the people. It is neutral itself and opposes one and another. That's why technic must be directed (Jaspers 2014). The main sense of technic according to Jaspers is to transform the human being. This task is especially important in modern conditions associated with rapid technical progress and impressive growth of information technology.

3.3   The Impact of Information Technic on Social Development

Information technic has a great influence on modern humanity. In the 20th century it began to develop intensively. Today it impacts both human and social life. Ideas about the nature of technic, its place and role in social development are used by philosophers for understanding the new patterns and real processes of "information society" (Bell 1973). Technic causes many social changes there. It's clear that modern technical progress is impossible without radical changes of social life. This makes us looking for connections between technic and the new social processes.

3.3.1   Mass Culture in the Information Society

Modern society creates and consumes actively the mass culture. This process is closely related with development of information technic and technology such as TV and Internet.

Mass culture of the modern information society has a dual character. On the one hand it has a strong positive impact on the human beings. It enters into their everyday life practically everywhere - from villages to megacities. Its main consequences are democratic and accessible education, universal literacy, mass editions of newspapers and magazines, cheap color reproductions of paintings and high-quality recordings of musical compositions. This of course can be considered as positive results of mass culture's influence on information society.

But it's also necessary to mention the negative consequences of information technic development. Mass culture offers its users certain patterns and standards of behavior. The result is unification of mass consciousness in information society. People consume similar information products which have a global nature. They accept an active propaganda of the lifestyle typical for industrial civilization. This causes the loss of human individuality and national identity. The set of moral and ethical principles changes much. Degradation of language also takes place.

3.3.2   System of Communication in the Information Society

Appearance of computer networks is the most significant phenomenon of information society. It causes radical changes in the system of communication.

Information society is actively involved into a process of electronic communication. Real people there are replaced by "social agents", culture in its original forms - by social technologies, human interactions - by psychological manipulations. Certainly the main role in this process belongs to the Internet. It expands to the limit the practice of non-institutional relations. Thus electronic communication becomes practically free from the control of social structures.

The result is a number of psychological (and even psychic) problems connected with communication activity. Traditional types of communication based on a real contact between people disappear. Distant communication which is much more independent from the sphere of human emotions and feelings develops intensively. If the human being communicates mainly within information and network space he risks isolating himself from the society. This leads to disorder and deformation of his emotional sphere, to desocialization and alienation from social life (Guryanova 2015).

3.3.3   Mass Media in the Information Society

Modern media aren't already limited by capacities of common journalism, radio and cinematography. It's a wide system of interconnections between modern television and the Internet. Information transmitted by the satellites gets the planetary audience as an object of obsessive manipulation. In this case information technic is used to transmit all kinds of information including spam, gossips, intimate details of human private lives and conflict situations. The whole world is turning into a "global village" where nothing can be hidden from the neighbors (McLuhan 2003).

In the modern world the Internet plays a role of universal instrument and existence environment. It functions practically without and outside the society (if we understand the last one in its traditional sense - as a set of social structures and institutes). Society as a system and global structure of normative order doesn't perform its functions in the space of the Internet. Electronic communication gives the human being a full freedom of self-identification (virtual status and virtual name, virtual habits, psyche and body, even virtual advantages and disadvantages). At the same time communication through the Internet alienates and loses his real body, natural status, etc.

4 Discussions

Information technic plays a special role in the life of modern society. It changes not only the world around us (including nature and social environment) but the inner world of the human being. His mentality and traditional way of life are seriously transformed. The same can be said about the human society. Information here becomes a main criterion of human differentiation. This changes the very structure of social relations.

A quantitative growth of technical capacity takes place in the information society. When it reaches its critical point it gets a new quality. And technic moves into a new phase of its development. But this phase mustn't be dangerous to civilization. In these new conditions everyone must remember that it's necessary to be a human first of all, to honor morality, to follow the real human values. In this case technic won't be terrible for humanity. Because technic is only an instrument. It depends fully on the will of its owners and managers which are always the human beings.

5 Conclusions

An importance of technic and information technology in the modern world makes necessary their philosophical analysis. It concerns an impact of information technic on society and human being. A special role in this process belongs to philosophy and anthropology of technic. These disciplines describe in detail specifics of human existence in the world of technic and information.

The study of social problems of the information society finds out the following circumstance: information and technical achievements can't be considered from the positive point of view only. They have also a number of negative characters that can potentially destroy humanity in future. Both positive and negative consequences of human information development manifest themselves in modern systems of communication, mass media, mass culture.

Philosophical approaches to information technic show its existential dimension: modern technic is always a mirror image of the human being. That's why it's very contradictory (like a human being himself). But we mustn't be afraid of technic, give it a chance to destroy ourselves. We are sure there is the only one key to solving the problem of human existence in the world of technic and information. It is looking for the harmony between technic and the human being.

References

Berdyaev, N.A.: The human being and the machine. Questions Philos. **2**, 147–162 (1989)

Bell, D.: The Coming of Post-Industrial Society: A Venture in Social Forecasting. Basic Books, New York (1973). 507 p

Guryanova, A.: Phenomenon of homo informaticus: new dimension of human's being. In: Elyakov, A. (ed.) Problems of Human Existence in Information Society, pp. 203–220. Samara State University of Economics, Samara (2015)

Guryanova, A., Guryanov, N., Frolov, V., Tokmakov, M., Belozerova, O.: Main categories of economics as an object of philosophical analysis. In: Popkova, E. (ed.) Russia and the European Union: Development and Perspectives, pp. 221–228. Springer (2017)

Jaspers, K.: The Origin and Goal of History, Kindle Edition. Routledge Revivals (2014). 316 p

McLuhan, M.: Understanding Media. The Extensions of Man. Gingko Press (2003). 616 p

Stepin, V.S., Gorokhov, V.G., Rozov, M.A.: Philosophy of Science and Technic Gardariki, Moscow (1996). 399 p

Pedagogics as a Means of Knowledge Translation in Human Society

Implementation of the Information and Communication Technologies into Activities of a Pedagogue

Olga V. Dybina(✉)

Tolyatti State University, Tolyatti, Russia
dybinaov@yandex.ru

Abstract. This article is devoted to using the information and communication technologies in education and their implementation into activities of pedagogues of a pre-school educational organization. Topicality of the selected issue is related to the current process of informatization of the Russian education. A sphere of scientific sectors is outlined within which the informatization of education and organization of the process of pedagogues' acquiring the information and communication technologies are studied. Theoretical studies are viewed which shows that insufficient level of innovation in pre-school education is caused by the low level of informatization in a pre-school organization. It is shown that information society sets new requirements to educational systems – they have to become effective and innovational, which will allow each pedagogue to realize their potential. The indicators and results of pedagogues' acquiring the information and communication technologies are presented. The main attention is paid to the methodological provision of the process of pedagogues' acquiring the information and communication technologies and conditions of its implementation. The events held with pedagogues in "School of the information and communication technologies" are very interesting. The material has large practical value and allows using the offered program for pre-school educational organizations' pedagogues' acquiring the information and communication technologies. These issues are of a many-sided character.

Keywords: Informatization of education
Information and communication technologies
Pre-school educational organization · Pedagogues · Methodological provision
Socio-psychological · Functional conditions

1 Introduction

Modern development of the information and communication technologies leads to development of economy and large transformation of society, including informatization of the educational sphere [5]. New possibilities for implementing new methodologies and pedagogical developments into the pedagogical practice provide the pedagogues with scientific progress and informatization of pre-school education. The developments could be aimed at implementation of innovational ideas of instructional educational process, which is a necessary condition of development of pre-school education. That's why

© Springer International Publishing AG, part of Springer Nature 2018
E. G. Popkova (Ed.): HOSMC 2017, AISC 622, pp. 107–113, 2018.
https://doi.org/10.1007/978-3-319-75383-6_14

organization of work for creation of the information and educational space of a pre-school organization is an important factor of realization of the federal state educational standard of pre-school education.

The federal state educational standard of pre-school education supposes the modern pedagogue's possessing the skill to use a computer and modern multimedia equipment, as well as to create his own educational resources and use apply them in his pedagogical activities.

I.V. Robert distinguishes the main directions of implementing the means of new information and communication technologies into education. She states that the current acceleration of scientific and technical progress, which takes place against the background of implementation of automatized systems, microprocessors, program management, robots, and processing centers into production, leads to the necessity to teach the growing generation which would be able to participate in the new stage of the modern society's development, related to informatization. I.V. Robert speaks of uniqueness of the possibilities of new information and communication technologies [5, p. 30]. The information and communication technologies stimulate development of child's personality and preparation for living in the information society. Informatization of society makes us use information technologies, acquiring the child to the information and communication technologies, performing the social order of the society. All levels of the educational process take place with more intense and results [7].

An important advantage is the possibility to use the methods of the information and communication technologies as automatization of the processes of control and correction of the activities' results and computer pedagogical testing and psychological diagnostics. They allow for automatization of processing of the results of experiment (laboratory or demonstration). Application of the means of the information and communication technologies for organizing intellectual leisure and developing games is very interesting as well. It is impossible to overestimate the advantages of the methods of the information and communication technologies during studying.

A modern pedagogue must possess knowledge on the information and communication technologies and be a professional in using them in the educational process.

The order of the Ministry of Healthcare and Social Development of the RF dated August 26, 2010, No. 761n established the "Qualification guide for offices of managers, specialists, and public officers", which provides the requirements for pedagogues for using the information and communication technologies.

According to the normative documents, a pedagogue of pre-school educational organization should develop and use in his professional activities the information and communication technologies and be a guide into the world of new technologies for the child within the pre-school education.

Unfortunately, as the practice shows, not all pedagogues possess the information and communication technologies. There are no special programs and methodological guides that would help pedagogues in pre-school establishments to effectively use the information and communication technologies (hereinafter – ICT) in their work. This determines the topicality of the research at the socio-pedagogical level.

At the scientific and theoretical level in the process of studying the theoretical foundations of pedagogues' acquiring the ICT, the following provisions of L.S. Vygotsky on the necessity for integration of technical tools with psychological tools were determined, which provides children with wide possibilities of development in all spheres of studying. Theoretical analysis of studies [1, 4, 6, 7, 9] shows that using the ICT means in the modern pre-school education is one of the most important factors of increase of the educational process effectiveness.

Together with elaboration of the problem at the scientific and theoretical level, analysis of scientific studies and pedagogical practice shows insufficient theoretical and methodological elaboration of the foundations of applying the ICT in pre-school education, pedagogues' readiness for acquiring the ICT, and lack of methodological provision of this process.

The purpose of the article is to study the level of pedagogues' acquiring the information and communication technologies and to develop the methodological provision of this process that stimulates the effective professional activities of pedagogues in the sphere of pre-school education.

2 Description

Special attention is paid to organization of experimental work, its course, and the received results. For evaluating the level of pedagogues' acquiring the pre-school organization of the ICT, the following indicators were set:

1. Capability to perform information search. The skill to search for information is a basis for acquiring knowledge for a pedagogue. Information sources include professionals, documents, and communication means. At present, this takes place through the communication means – mass media, mobile devices, computer networks (blogs, social networks, etc.). This article emphasizes on the search for information in the Internet, which is an open educational space.
2. Capability to work with finished program and methodological complexes. The program and methodological complex is a complex of program and methodological means for supporting the process of teaching a certain course or its topic [8]. The program and methodological complexes allow pedagogues to plan their work in view of the federal state educational standard of pre-school education, take into account individual peculiarities of children, select games and exercises according to the educational goal, etc.
3. Capability to enter remote educational activities. Remote educational activities are becoming more popular in pre-school educational organizations (consultations, virtual exhibitions, etc.).

The research was performed on the basis of kindergartens of the autonomous non-profit organization "Planet of childhood 'Lada'" in Tolyatti; more than 150 pedagogues were surveyed. The following questions were asked: Do you use the search systems during solving the pedagogical tasks? Which ones? Do you use the ICT for communication with children and their parents (legal representatives)? Do you plan online activity? Do you use text editor in your work? What information can you offer for posting on the kindergarten's web-site?

Analysis of questionnaires showed that 67% of the respondents use the Yandex search engine, while the rest use Google. The survey showed lack of capability to use the information search, for 17% of the pedagogues are dissatisfied with the search results, and 33% cannot find the required information in the search systems (total 50%). Besides, only 10% use special symbols and thematic catalogues. 90% of the surveyed pedagogues do not use these resources. Pedagogues of the kindergarten face difficulties during search for information.

Analysis of questionnaires showed that only 50% of the respondents use the ICT for evaluation and monitoring, and the rest do not use such possibility and perform all routine work manually. However, 90% of the pedagogues have some skills with text editor and could have used it more often in their work. Only 10% of the respondents use for organization of educational activities the software means for presentations.

The results of the survey show that 90% of the respondents are ready to share their pedagogical experience with the help of an open educational resource – the kindergarten's web-site.

Most of pedagogues (67%) wish to conduct remote educational activities, but a lot of them lack technical knowledge on the method of information search and work with ready program and methodological complexes and require additional work for enlightenment on the possibilities of the ICT.

During analysis of digital catalogs, the catalogs for pedagogues attestation, catalogs of methodological materials, photo albums, catalog of videos, and archive of materials of the pre-school educational organization's web-site were found. With the methodological material for pedagogical activities with the use of the ICT there are no technical means for implementation of the ICT.

We supposed that the process of pedagogues' acquiring the information and communication technologies is possible if:

– the essence of the process of pedagogues' acquiring the information and communication technologies in the system of interconnected pedagogical notions in view of the work specifics is opened;
– the methodological provision of the process of pedagogues' acquiring the information and communication technologies is developed and implemented;
– the experimental substantiation of the criterial tools necessary for objective evaluation of the level of pedagogues' acquiring the information and communication technologies is given.

Based on analysis of ascertaining experiment, for pedagogues' acquiring the information and communication technologies during organization of open educational space, the following tasks are set:

1. Stimulating development of the pedagogues' skills to search for information with the help of search engines.
2. Developing pedagogues' skill to work with ready program and methodological complexes.
3. Preparing to conducting remote educational activities.

The Internet Day was organized in kindergartens, which helped to attract the pedagogues' attention to using the virtual space, its possibilities and perspectives. The exhibition stands with the history of the Internet, short information on its work and possibilities of use are very interesting.

A lot of attention was paid to the work of creative groups, while pedagogues got acquainted with the Internet, diversity of search systems, special symbols for optimal search of information, and acquired the basic skills of work with thematic catalogs.

The "School of the information and communication technologies" was organized, with regular "consultation lessons", the main topics of which are the following:

– "Learning text editor".
– "Compiling presentation".
– "Selecting web browser".
– "Optimal usage of search engines".
– "You've got mail! Acquainting with e-mail".
– "Kindergarten's web-site – my ticket to the open educational space".
– "Compiling diagnostics with the help of computer".
– "Social networks".
– "What is WEB 2.0?".
– "Program and methodological complexes", etc.

In the course of work of the "School of the information and communication technologies" the participants performed the tasks related to application of a new program in the educational activities. For example, after acquisition of text editor, the pedagogies had to compile a consultation for parents and prepare the planning of the educational process in the digital form.

Special attention was paid to the methodological provision aimed at implementation of the developed program. The program for pedagogues' acquiring the ICT supposes usage of the basic level of approbation and creation of the proprietary educational resources. The Program includes the following blocks:

1. "Implementing the ICT as a condition of realization of the federal state educational standard".
2. "Practical use of the PhotodexProShowProducer software for creation of professional presentations.
3. "Creating interactive publications at the Calameo service".
4. "Creation a personal web-site at the ucoz.ru web service".
5. "Creation a personal web-site at the wix.com web service".
6. "Studying MicrosoftPowerPoint – creating multimedia presentations".
7. "**Audacity** – sound editor with a wide set of professional capabilities".
8. "Vocalremover – online service for processing audio files".
9. "Nero – multifunctional multimedia package for working with CD and DVD, sound and video".
10. "PinnacleStudio – software for professional work with video".

The methodological provision was conducted with the following conditions: socio-psychological (solving the problems of formation of a pre-school organization's group, developing employees' activity, developing each member's personality, and coordinating various relations); functional (provision of activities of a pedagogical group, scientific work, increase of effectiveness, high quality of children training, and additional training of personnel) [2, p. 158].

3 Results

As a result of experimental study, large changes in the positions of pedagogues of pre-school educational organization as to using the information and communication technologies in their activities happens. They showed not only knowledge of software and specifics of technical work but also the skill to work with main applied software and multimedia programs, and to use the Internet. Pedagogues of a pre-school educational organization began to use the means of the ICT during planning of pedagogical activities, evaluation and creation of reports for children training, and learned new programs.

4 Conclusions

Implementation of the contents of the program blocks ensures the following: improvement of the process of acquiring the information and communication technologies by pedagogues; creation of integrated information environment: methodological guidebooks and recommendations for acquiring the information and communication technologies for pedagogues.

Using the information and communication technologies in a pedagogue's work allows for enrichment and renovation of the educational process in a pre-school educational organization and for increase of its effectiveness. The performed research has not analyzed all aspects of the studied issue due to its multi-aspect character.

References

1. Gorvits, Y.M., Chaynova, A.A., Poddyakov, N.N.: New information technologies in pre-school education, 328 p. Linka Press, Moscow (1998)
2. Dybina, O.V.: Individual managerial concept of manager of a pre-school educational organization. Vector Sci. Tolyatti State Univ. **2-1**(32-1), 154–160 (2015)
3. Vinnitsky, Y.A.: Informatizaton of education: problems and perspectives [E-source]: deputy director for educational work, School No. 169. https://edugalaxy.intel.ru/index.php?act=attach&type=blogentry&id=3409
4. Komarova, T.S., Komarova, I.I., Tulikov, A.V., et al.: Information and communication technologies in pre-school education, 31 p. Mozaika Sintez, Moscow (2011)
5. Robert, I.V.: Methodology of education informatization [E-source]: director of the Institute of Education Informatization of the RAE. http://ito.su/40/plenum/Robert.html?PHPSESSID=pts18etqpmcevg1cva1erI05I3
6. Morozov, K.A.: The information and communication technologies and their application in pedagogical activities. In: Personality, Family, and Society: Issues of Pedagogics and Psychology: Collection of Articles Based on Materials of the 30th International Scientific and Practical Conference. SibAK, Novosibirsk (2013)

7. Revnivtseva, R.M.: The information and communication technologies in a pre-school educational establishment. In: Pedagogics: Traditiona and Innovations: Materials of the 2nd International Scientific Confernece, Chelyabinsk, October 2012, pp. 67–69. Two Young Communists, Chelyabinsk (2012)
8. Robert, I.V.: Theory and methodology of education informatization (psychological & pedagogical and technological aspects)/IEI of RAE, 2nd edn., 274 p. (2007)
9. Shmakova, A.P., Khramova, L.V.: Informatization of pre-school education: advantages and problems. Siberian Sci. Bull. 1(7), 93–95 (2012)

Perspectives of Development of the Educational Services Market in Regions of Russia in the Conditions of the Knowledge Economy Formation

Irina V. Baranova[⊠]

Mikhailovka Vocational Training College Named After V.V. Arnautov,
Mikhaylovka, Russia
baranova_irina_v@mail.ru

Abstract. The purpose of the article is to study the perspectives of development of the educational services market in regions of Russia in the conditions of the knowledge economy formation. The methodological basis of the research consists of the method of structural analysis, which is used for studying the Russia's position in the ranking of the countries on development of knowledge economy, and the method of horizontal analysis, which is used for studying the dynamics of formation and development of knowledge economy in modern Russia. The information and analytical basis of the article includes the materials of the Federal State Statistics Service of the RF and the index of knowledge economy according to the World Bank. The authors also study the process of formation of knowledge economy in Russia and determine that it has been continuing for the last 25 years and will last for 8 more years in case of preservation of the current tendency. The state's efforts on management of this process are peculiar for low effectiveness. This is due to insufficient consideration of regional peculiarities and needs during development of the educational services markets in Russia's regions. A serious problem of knowledge economy development in Russia is weak connection between development of regional markets of educational services and the national course at formation of knowledge economy. For solving these problems, the authors offer practical recommendations and develop a perspective model of development of the educational services market in Russia's regions in the conditions of the knowledge economy formation.

Keywords: Educational service market · Regions of Russia
Knowledge economy

1 Introduction

Economic system's setting on the path of development of economy is a serious challenge for all other spheres of national economy, as the entrepreneurial structures that work in them have to change their business processes in the direction of increase of knowledge intensity, shift of emphasis of human resources, as compared to other types of resources, and orientation at innovational development. The educational services

E. G. Popkova (Ed.): HOSMC 2017, AISC 622, pp. 114–120, 2018.
https://doi.org/10.1007/978-3-319-75383-6_15

market is subject to most serious transformations, as it bears the main load for development of knowledge economy.

In the conditions of the knowledge economy formation, educational services market is not just an environment for preparation of necessary human resources but a source of knowledge – the newest technologies of production, organization, and management, innovational products (goods and services), etc. In the countries with large territory and high level of differentiation in the level of socio-economic development of economic systems, which are parts of the national economy, the conditions, intensity, and directions of development of regional educational services markets differ, and the level of diversity could be rather high.

On the one hand, this hinders their unification and formation of a non-structural national educational services market, which is easily predictable and manageable, and which structural elements are in close interconnection. On the other hand, development of educational services markets according to the peculiarities of the regions of their location allows accelerating the process of knowledge economy formation due to creation of knowledge and human resources that are necessary for each specific region, this leveling disproportions in development of the national economic system.

That's why regional models of development of educational services markets should take into account the peculiarities of regional economic systems. This explains the topicality of research of the regional aspect of development of the educational services market in the context of the knowledge economy development. The authors focus on the modern Russian economy and seek the goal of studying the perspectives of development of the educational services market in regions of Russia in the conditions of the knowledge economy formation.

2 Materials and Method

Peculiarities of functioning and development of regional economies in the national economic systems are viewed in the studies (Fleischmann et al. 2017; Otoiu et al. 2017; Ge and Zhao 2017; Guliak 2017; Anukoonwattaka 2016). The conceptual and applied issues of development of regional educational services markets of the countries of the world are reflected in the publications (Popkova et al. 2016; Ragulina et al. 2015; Bogoviz et al. 2017; Bogdanova et al. 2016; Popova et al. 2016). The essence of the process of the knowledge economy formation is studied in the works (Fathollahi et al. 2017; Amavilah et al. 2017; Kuleshov et al. 2017).

The performed content analysis of the scientific works on the set problem determined insufficient scientific elaboration of the influence of the processes, related to creation of knowledge economy, on development of regional educational services markets, which leads to the necessity for further research of this issue.

The methodological basis of the research consists of the methods of structural analysis, which is used by the authors to study the Russia's position in the ranking of the countries on creation of knowledge economy, and the method of horizontal analysis, which is used for studying the dynamics of formation and development of knowledge economy in modern Russia.

The information and analytical basis of the research includes the materials of the Federal State Statistics Service and the index of knowledge economy according to the World Bank, which is the basis for annual rankings of 140 countries of the world.

3 Results

The value of the knowledge economy index in Russia constituted 5.78 in 2016. It is ranked 55th, near Ukraine, Belarus, and Qatar. The key reason for such low position of Russia in the ranking of the countries is ineffective regional educational services market, the result of functioning of which is low innovative activity and low level of development of technologies (World Bank 2016).

The innovational system, which consists of companies, R&D centers, universities, consultation agencies, and other organizations, has not been formed in Russia yet. There's no connection between the sphere of science and education and the production sector of economy. As a result, the created innovations do not reach targeted consumers and remain at the level of dissertation studies.

Another important reason of insufficiently high level of development of science and innovations is the ineffective system of patenting the right for scientific inventions and innovational technologies. Complexity of the procedure of patenting of new knowledge is a serious barrier for the Russian scholars. The system of commercialization of innovations is not developed sufficiently.

Another reason is the lack of financial resources due to absence of assets with the state and due to low investment attractiveness of innovational projects for private investors. Scholars cannot perform research with their private assets, as the research requires a lot of money.

Thus, the studies are not performed at all, or are performed to a limited extent – which does not allow obtaining the desired results and compiling the results. Due to these reasons, the process of the knowledge economy formation in Russia is not yet completed and continues until now. The structure of this process is reflected in Fig. 1.

As is seen from Fig. 1, the process of the knowledge economy formation in Russia took place in three stages. The beginning of the first stage was the Russia's transition to the market path of development in 1991. After this, in 1992–2002, the share of expenditures for innovations in the structure of GDP was growing from 0.03% to 0.60%. The growth was very quick and constituted more than 250% per year. In this period, the volume of innovational goods, works, and services increased, constituting RUB 430 billion, and the share of the companies that implement innovations reached 11%.

During the second stage (2002–2013), the share of expenditures for innovations in the structure of GDP continued to grow, but slower – 13% per year, from 0.61% in 2000 to 0.83% in 2010. The volume of innovational goods, works, and services grew, consisting RUB 230–700 billion. The share of the companies that implement innovations grew insubstantially, constituting 11%–12%.

At the third stage, which began in 2014 and lasted until 2016, the shareo of expenditures for innovations in the structure of GDP grew very quickly – the growth

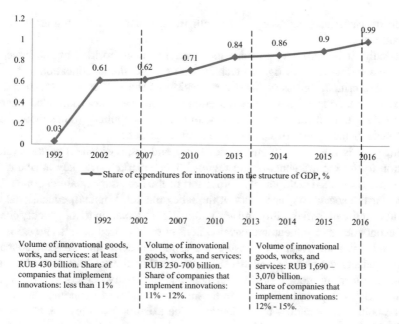

Fig. 1. The process of the knowledge economy formation in Russia. Source: compiled by the author with the use of the materials (Rosstat 2016).

constituted around 30% per year – from 0.84% in 2011 to 0.99% in 2014. The volume of innovational goods, works, and services grew more, constituting RUB 1,690–3,070. The share of the companies that implement innovations also grew, constituting 12%–15%.

At present, it is possible to suppose that the next stage of this process will begin in 2017 and will continue until full formation of the knowledge economy in Russia. In view of the modern state policy's orientation at formation of innovational economy and announcement of the strategy of innovational development of Russia until 2020, it is possible to suppose that this stage covers the period from 2017 to 2025.

At this stage, it is necessary to expect the increase of the share of expenditures for innovations in the structure of GDP up to 3%. The volume of innovational goods, works, and services might increase to RUB 5,000 billion. The share of the companies that implement innovations might constitute 35–40%. This will take Russia to the level of developed countries and will allow forming knowledge economy.

The most serious problems of the regional educational services market in Russia are low quality of education and weak connection between the educational market and the labor market. In order to created conditions for successful knowledge economy formation in modern Russia, it is necessary to solve these problems, which supposes increase of the quality of education and establishment of strong connection between the labor market and the educational services market. For this, we offer the following recommendations.

Firstly, it is necessary to raise the level of education. In addition to increase of the number of students, it is necessary to focus on improvement of quality and topicality of development of education and increase of qualification, in order to solve the problem of the lack of useful and necessary skills with the graduates of the higher educational establishments.

Secondly, it is necessary to raise the diversity of authorities of the educational establishments and the prestige of technical and professional education. At present, Russia has a critical necessity for diversification of the system of education for the purpose of reduction of the excessive emphasis on theoretical knowledge and for supplementing the educational programs with a range of applied studies that reflect the factual state of the sphere of national economy at present.

Thirdly, it is important to ensure flexibility and responsiveness of the system of education to the requirements of the regional labor market. The key attribute of the educational system's development is provision of the necessary qualification and competence for the specialists, necessary at the labor market. For that, educational establishments require more flexibility and adaptability, for adaptation to the quickly changing global economic environment and development of the required talents and skills.

Fourthly, it is important to ensure correspondence of the Russian educational establishments to the global standards. Higher educational establishments of the global standard and leading models of R&D centers and university centers are necessary for ensuring the development of talents in top-priority sectors of economy. Such models could be built from scratch – by modernizing the existing institutes due to partnership with the world-class universities.

Fifthly, it is necessary to create centers of leading experience in the sphere of scientific research. For this, it is necessary to invest large resources into educational and R&D centers. This will ensure conduct of original and potentially innovational research and creation of the society of scholars, technologists, and specialists in various spheres of national economy.

Sixthly, it is necessary to implement the information and communication technologies into the sphere of education. At present, it is clear that there will be a transition to the next generation of solutions in the sphere of the information and communication technologies in education. Recently, such technologies as platforms of online education, tools for joint virtual work, Internet resources, digital access to libraries, video courses, and creation of virtual training programs have been approbated in the Russian educational establishments with different levels of effectiveness.

Seventhly, it is necessary to stimulate implementation of the information and communication technologies as the main – not additional – means of teaching. The information and communication technologies in education will probably become the most powerful tool for raising the quality of education. The people who grew in the age of digital technologies – so called "digital natives" – perceive the surrounding environment differently, as compared to the past generations. That's why there's a necessity for new approaches to education for successful teaching of new generation.

Based on the above recommendations, we developed a perspective model of development of the educational services market in regions of Russia in the conditions of the knowledge economy formation, which is shown in Fig. 2.

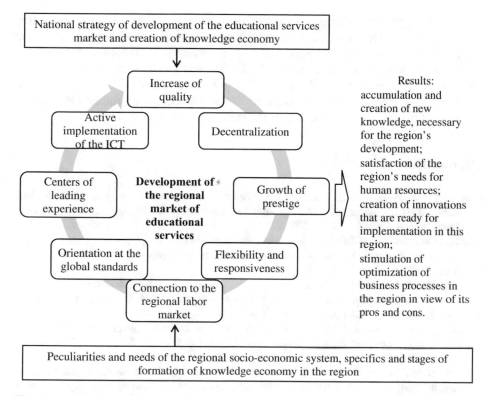

Fig. 2. A perspective model of development of the educational services market in regions of Russia in the conditions of the knowledge economy formation. Source: compiled by the authors.

As is seen from Fig. 2, according to the offered model, the development of the educational services market in a region takes place in connection to the national strategy of development of the educational services market and creation of knowledge economy and in view of the peculiarities and needs of the regional socio-economic system, as well as specifics and stage of the knowledge economy formation in a region. This allows achieving such results as accumulation and creation of new knowledge, necessary for development of the region, satisfaction of the region's needs for human resources, creation of innovations that are ready for implementation in this region, and stimulation of optimization of business processes in the region in view of its pros and cons.

4   Conclusions

Thus, modern Russia is in the process of the knowledge economy formation, which has been continuing for the last 25 years; with preservation of the current tendency it will last for at least 8 more years. The state's efforts on management of this process are peculiar for low effectiveness. According to the analysis performed in the research, the probable reason for this is insufficient consideration of regional peculiarities and needs during development of educational services markets in regions of Russia.

A serious problem of creation of knowledge economy in Russia is weak connection between development of regional educational services markets and the national course at formation of knowledge economy. The authors' recommendations and the offered perspective model of development of the educational services market in regions of Russia in the conditions of the knowledge economy formation allow solving the determined problems, increasing the effectiveness of management, and accelerating the process of the knowledge economy formation in modern Russia.

References

Amavilah, V., Asongu, S.A., Andrés, A.R.: Effects of globalization on peace and stability: implications for governance and the knowledge economy of African countries. Technol. Forecast. Soc. Chang. **122**, 91–103 (2017)

Anukoonwattaka, W.: Global value chains and competitiveness of the integrated regions: exchange rate issues. In: ASEAN Economic Community: A Model for Asia-Wide Regional Integration? pp. 127–151 (2016)

Bogdanova, S.V., Kozel, I.V., Ermolina, L.V., Litvinova, T.N.: Management of small innovational enterprise under the conditions of global competition: possibilities and threats. Eur. Res. Stud. J. **19**(2 Special Issue), 268–275 (2016)

Bogoviz, A.V., Ragulina, Y.V., Kutukova, E.S.: Ways to improve the economic efficiency of investment policy and their economic justification. Int. J. Appl. Bus. Econ. Res. **15**(11), 275–285 (2017)

Fathollahi, M.F., Elahi, N., Najafi, S.M.S.: Appropriate theoretical framework for understanding and analyzing economic issues in knowledge-based economy. J. Knowl. Econ. **8**(3), 957–976 (2017)

Fleischmann, K., Daniel, R., Welters, R.: Developing a regional economy through creative industries: innovation capacity in a regional Australian city. Creative Ind. J. **1**(2), 1–20 (2017)

Ge, Y., Zhao, X.: Regional economy and development: a viewpoint and application of spatial statistics. Spat. Stat. **2**(1), 126–129 (2017)

Guliak, R.: New resonance approach to competitiveness interventions in lagging regions: the case of Ukraine before the armed conflict. Rev. Econ. Perspect. **17**(1), 25–56 (2017)

Kuleshov, V.V., Untura, G.A., Markova, V.D.: Towards a knowledge economy: the role of innovative projects in the reindustrialization of Novosibirsk oblast. Reg. Res. Russ. **7**(3), 215–224 (2017)

Otoiu, A., Bere, R., Silvestru, C.: An assessment of the first round impact of innovation industries on Europe's regional economies. Amfiteatru Econ. **19**(44), 289–301 (2017)

Popova, L.V., Popkova, E.G., Dubova, Y.I., Natsubidze, A.S., Litvinova, T.N.: Financial mechanisms of nanotechnology development in developing countries. J. Appl. Econ. Sci. **11**(4), 584–590 (2016)

Ragulina, Y.V., Stroiteleva, E.V., Miller, A.I.: Modeling of integration processes in the business structures. Mod. Appl. Sci. **9**(3), 145–158 (2015)

World Bank: Building Knowledge Economies: Assessment Methodology. World Bank, Washington, DC (2016)

Popkova, E.G., Chechina, O.S., Abramov, S.A.: Problem of the human capital quality reducing in conditions of educational unification. Mediterr. J. Soc. Sci. **6**(3), 95–100 (2016)

Rosstat: Russian in Numbers: Short Statistical Collection. Federal State Statistics Service, Moscow (2016)

World Trends Impacting the Change of the Higher Education System in Russia in the Conditions of Global Crisis Management on the Basis of Entrepreneurship and Innovations

Svetlana M. Yakovleva[✉], Maria O. Suraeva,
and Alexander P. Zhabin

Samara State University of Economics, Samara, Russia
syakovleva80@mail.ru, marusyasuraeva@mail.ru,
zhabin@sseu.ru

Abstract. The current system of higher education in the Russian Federation is subject to changes at different levels: global, regional and local, especially in the context of Global Crisis Management. The goal of this research work is to identify and systematize world trends that influence the change of the system of higher education in the context of Global Crisis Management, including using the principles of entrepreneurship. At the global level, the system of higher education of the Russian Federation is affected by trends in political, economic, social, cultural and information technology. As early as in the twentieth century, it was possible to distinguish the influence of each of the trends on the education system and to characterize the impact of each of them. In the modern world, the processes of globalization and internationalization, which affect each of the macro-trends, come first. Accordingly, macro-trends are considered in the analysis of their impact on the education system. In the research, within the framework of the methodology for highlighting the main trends, the Russian education system is considered element-wise, which includes higher education institutions, lecturers, students, IT environment of the university, and their characteristics. Among the main results, trends shaping education of the present and near future are revealed, which change the system of higher education in the Russian Federation in the context of Global Crisis Management: internationalization of the educational environment, including programs and conditions for their implementation; simultaneous existence of two realities: on-line and off-line; Project education; Gaming. World trends shaping education are a necessity that should be taken into account when forming the development pathway of the higher education system in the Russian Federation, especially in the context of Global Crisis Management. Adherence to trends shaping education helps to form a stable education system, flexibly reacting to global and local changes. When developing programs for strategic and innovative development, universities need to draw on trends, which should be the criteria for selecting each management decision in a modern university. A special place is taken by the formation of the entrepreneurial university or transformation of the classical Russian university into it, and development characteristics and directions determine and will determine world trends.

© Springer International Publishing AG, part of Springer Nature 2018
E. G. Popkova (Ed.): HOSMC 2017, AISC 622, pp. 121–128, 2018.
https://doi.org/10.1007/978-3-319-75383-6_16

Keywords: Trends · Higher education · Crisis management
Entrepreneurial university · Innovation

JEL Classification: D8 · I23 · Q01

1   Introduction

There are various approaches to both the education system in the Russian Federation and the search for challenges facing the Russian education in the context of Global Crisis Management. In particular, Schedrovitsky (2016) distinguishes the following challenges: the disappearance of approximately 30% of traditional occupations, changes in infrastructure, changes in technology in the educational process. Challenges direct the further changes in education, form trends shaping education at the global and local levels.

The concept of the entrepreneurial university is more relevant than ever both for the Russian reality and the current development stage of higher education. Thus, Konstantinov and Filonovich (2007) define the entrepreneurial university as an institution that systematically makes efforts in three directions: generation of new knowledge, teaching and implementation of knowledge into practice, transformation of the internal environment and modification of relations with the external environment. The concept of the entrepreneurial university is disclosed in detail in the scientific works of foreign scientists, such as: Dabić and Švarc (2012). They explain entrepreneurial activities of universities and provide a reference to the position of entrepreneurial universities (in Croatia). The authors conclude that all universities are inevitably forced to change. Etzkowitz (2004) describes the evolution of the entrepreneurial university as well as Wong et al. (2007). They talk about "Entrepreneurial University" Model to Support Knowledge-Based Economic Development. Guerrero and Urbano (2012) are considering the development of the entrepreneurial university.

2   Methodology

To identify the main trends, it is necessary to consider the education system element-by-element. The authors distinguish the following elements of the education system in Russia:

- Higher education institutions,
- Lecturers,
- Students,
- IT-environment of the university.

Higher education institutions are, on the one hand, the most resilient elements of the system, and on the other, undergo a powerful impact from other elements. In particular, the approach to the activity of the university on the part of public authorities is changing. Instead of the city-forming institutions or institutions serving key enterprises in the region from the point of view of their supply of specialists and R&D

(historical aspect), modern universities are included in the competitive struggle both among themselves (participation in projects 5–100 and supporting universities) and internationally. Accordingly, the stability of universities is relative.

Lecturers of universities have also ceased to be a stable element of universities. If during the Soviet Union the stability of teaching activity was determined by strict compliance with necessary requirements, at present even the compliance with all teaching requirements and writing of articles does not guarantee the preservation of the lecturer's workplace. A colleague applying for this position may have higher-ranking articles, participation in international conferences, experience as a visiting lecturer, etc. Thus, the compliance with formal teaching requirements does not guarantee job stability, but rather creates the necessary minimum, which must be met if the lecturer plans to continue working.

The trends of non-departments universities, implemented in some universities, exacerbate lecturers' stability, the most successful lecturers will be invited to the programs, and the success criteria can vary from academic to bright personal traits. Speaking about competition as the main threat to lecturers' sustainability, it is necessary to understand competition in a broader format: starting from local within the university, continuing with competition in the inter-university space of the Russian Federation (invited lecturers, part-time work) and ending with global competition in the world educational space, for example, the replacement of full-time disciplines by on-line courses.

To increase the stability level of such an element of the education system as a lecturer, it is necessary, on the one hand, to create conditions for its "anchoring" within the main university, and on the other, to provide opportunities for the development. It is in this way, by constantly increasing the value and cost of the lecturer, the trend of maintaining lecturers' stability within the educational system can form.

Modern students as an element of the education system have a certain set of qualities and characteristics that distinguish them from the previous generation of students and from the lecturers with whom they interact. Modern students belong to the Generation Z, the birth years of which are the end of the 1990s - the beginning of the 2000s. Generation Z differs from the previous two generations: both the Generation Y, born in the early 1980s, and the Generation X, born in the 1960s and 70s, have more differences than similarities to the Generation Z. The distinctive feature of the Generation Z is the adequate evaluation of its generation parameters, on the one hand, and unwillingness to change their behavioral stereotypes, on the other. Thus, our focus groups with students of economic specialties of the 1st–2th year of SSEU (Samara State University of Economics) (3 focus groups were held, 15–18 students took part in each), highlight the main features of their generation in terms of attitude towards the learning process: high level of mobility, simplification of information, IT - as part of life, learning should be interesting, otherwise the material is not remembered. The analysis of the main features of the generation shows that in order to preserve students as an element of the education system, rather than as a list of higher education institutions, it is necessary to build new vectors of educational opportunities for the Generation Z, built on the use of modern trends.

The IT environment is taken out into a separate element, because the modern IT environment of the university with the development pathway cannot and should not be limited only to the formation of the university's website and access to lecture material in on-line mode. The implication of the Learning Management System, the use of MOOC (Massive Open Online Courses) of both Russian and foreign platforms, should become the basis for the development of the on-line environment of the university not as an alternative reality, but as a supplement to the off-line environment, which should develop in interaction and complementarity, create a unique information field within the university, which at the same time effectively interacts with information fields of other educational spaces: textbooks, libraries, on-line Incubators and projects.

3 Results

Analyzing the basic elements of the education system in Russia, the authors come to the conclusion about the main trends that will change the pathway of the educational space of the Russian Federation in the near future in the context of Global Crisis Management, if it seeks to the further development in accordance with global trends shaping education, or is aimed at their advance.

Among the main trends shaping education of the present and near future, changing the system of higher education in the Russian Federation can be identified:

- Internationalization of the educational environment, including programs and conditions for their implementation;
- The simultaneous existence of two realities: on-line and off-line, which is currently expressed in implementation of MOOC, LMS and access to on-line libraries of other HEIs, and as a trend will move from the format of information accessibility to the format of detailed selection of information that cuts off not the right material on the basis of specifying analytical queries;
- Project education - learning not in the classical format of the theoretical and applied interaction, but through the achievement of certain goals in the learning process through the use of a set of specific resources;
- Gaming as a trend has been formed and is still being formed on the basis of the request of the Generation Z - learning should be interesting. Interesting learning involves emotional involvement, shifting the emphasis from the useful in the future to the fascinating one in the present. Accordingly, Gaming, the educational game form becomes a modern trend shaping education.

Modern trends have an impact on each other in the further development of the educational environment and, perhaps, after a while the International Gaming, implemented in the project format, will become a new trend.

Each of the trends undoubtedly builds the future development pathway of higher education in the Russian Federation, while some pathways are already a reality for education systems of other countries.

If you look at the overview of trends shaping education made by the OECD (Trends Shaping Education 2016), the first trends also indicate globalization, which now covers all areas of activity, including education. In the context of education, the trend of globalization raises a large number of topical issues, affecting both international and national education systems:

- Do higher education institutions need to study the trends of the global labor market and prepare students for work in international and multinational companies?
- Should students who are leaving to study abroad feel responsible to their country and come back to work again? - What is the role of the OECD in reducing the brain drain?
- Internationalization of higher education has created many opportunities for students and higher education institutions - should all programs become averaged standard or should certain differences be taken into account when compiling them? (OECD, Trends Shaping Education 2016).

These and other questions asked when trying to determine trends in education make it possible to form strategic changes in both national and global education systems.

Internationalization of higher education as a form of manifestation of the trend of "globalization" has a separate effect on the formation of education pathways (Altbach and Knight 2007). According to Fiona Hunter, the current interpretation of internationalization is not the same as academic mobility. From the point of view of modern understanding, the main part of internationalization of higher education is the curriculum content development and search for talents, both employees and students.

Fiona Hunter gives the following definition of internationalization, relying on the development of recent trends: "Internationalization is the way to enrich the quality of education and research and the level of service to society" (de Wit and Hunter, 2016).

The majority of Russian universities, striving to implement a bundle of trends - globalization/internationalization are aimed at the development of incoming academic mobility. This trend was reflected in the rating of Russian universities (Methodology for calculating indicators to monitor the effectiveness of higher education institutions from April 3, 2014 N AK 39/05vn.) However, the quantitative analysis of academic mobility parameters no longer corresponds to new trends. Universities should reconsider their attitude towards internationalization on the basis of the updated definition given by Hunter (2016). Creation of conditions for the development of talents and internationalization of the substantive part of educational processes will contribute to the realization of the trend of globalization, which will lead to the further development of the education system.

Globalization is closely related to information technology, especially in the context of the transition from mobile to content ones.

Enriching the content of education implies the need for access to better educational resources, which are not always available in local universities. Modern information technologies give accelerated access to educational information from the best universities in the world.

Modern trends show that the availability of access to information does not contribute to the development of the education system, but is a certain basis for its stability.

The development of the education system is possible on the basis of combining two trends shaping education: active use of information technologies in education and their active implementation in project-oriented learning.

Project-oriented learning as a trend involves not only the practical implementation of knowledge gained, but rather the placement of knowledge in a certain framework: temporary, resource, targeted for its maximum effective use.

The trend of project-oriented learning shows that knowledge can become obsolete, can change, resources may be required to obtain it or to gain this knowledge. Accordingly, the trend of project-oriented learning allows us not only use it in a practical way, but also to give certain estimates for the effective use of knowledge.

The influence of these trends leads to the need and urgency to form and transform Russian universities in the entrepreneurial format through development programs. Naturally, the existing system of higher education has a number of limitations in the rapid implementation. Most of universities that remain active players on the market are budgetary and autonomous education institutions, whose activities are strictly regulated by the founder. One of the main problems while implementing the concept of "entrepreneurial university" are limited financial resources of the Russian university. In addressing this problem, Russia has established a program to increase the competitiveness of universities among the world's leading research and education centers "5top100", which really contributes to the establishment of the entrepreneurial university on the basis of support and financing of innovative activities in all areas of the university. One of the bright and effective projects while implementing the concept of the entrepreneurial university on the basis of support for the "5top100" project is St. Petersburg National Research University of Information Technologies, Mechanics and Optics.

Universities are of particular interest as sources of knowledge, transferring it to the business environment. Universities often act as intermediaries for the transfer of knowledge from researchers to industry.

The concept of the open innovation ecosystem is determined by a significant number of activities classified as open innovation initiatives. Interactions between actors and the implementation of policies suggest that goals are a distinctive feature of the ecosystem.

The role of universities as a source of open innovation for partner firms is becoming urgent, as technology is becoming a key factor for the firm to gain competitiveness in the world arena. This approach differs from the traditional approach that dominated universities and industry in the past.

As a consequence of three elements, there is an evolution from contractual research models to the open innovation model of cooperation.

Potential advantages of using open innovation initiatives in this context are: the distribution of risks in the study of immature technologies, the university as a partner can explore the applicability of some prospective studies with the help of young researchers, the industry can delegate this responsibility to partner universities, conduct experimental experiments under the control of partners.

The university can interact with industry both in the development direction of cooperation within the framework of the research project, and in the framework of learning or technology transfer; and also the activity connected with knowledge for the decision of the future problems.

To ensure the sustainability of these university ecosystems, two elements become decisive:

- The existence of strong internal positioning in relation to supporting innovation;
- The need to establish broad links with the external environment.

4   Conclusions

World trends shaping education are a necessity that should be taken into account when forming the development pathway of higher education in Russia. Adherence to trends shaping education helps to form a stable education system, flexibly reacting to global and local changes. Entrepreneurial universities represent an effective form of modern university, followed by world trends, satisfying the interests of all stakeholders, stimulate innovative activity in all areas of the university's functioning, which is the potential for the formation of the innovative Russian economy.

Acknowledgements. The research was performed within the state task of the Ministry of Education and Science of the RF, Project No. 26.940.2017/4.6 "Management of changes in the system of higher education on the basis of the concept of sustainable development and coordination of interests".

References

Dabić, M., Švarc, J.: About the concept of an entrepreneurial university: is there an alternative? [Zum konzept der unternehmer-universität: Gibt es denn eine alternative?] Drustvena Istrazivanja **20**(4), 991–1013 (2012). https://www.scopus.com/inward/record.uri?eid=2-s2.0-84855769240&doi=10.5559%2fdi.20.4.04&partnerID=40&md5=59ab6de209bf048c6b559e9bf8ad7c41. https://doi.org/10.5559/di.20.4.04

Etzkowitz, H.: The evolution of an Entrepreneurial University. Int. J. Technol. Global. **1**(1), 64–77 (2004). https://www.scopus.com/inward/record.uri?eid=2-s2.0-33745696205&partnerID=40&md5=d4fa9f04c81eb5f747e2902149cbe2cd

Guerrero, M., Urbano, D.: The development of an Entrepreneurial University. J. Technol. Transf. **37**(1), 43–74 (2012). https://www.scopus.com/inward/record.uri?eid=2-s2.0-84856269648&doi=10.1007%2fs10961-010-9171-x&partnerID=40&md5=15bf3bd0eea25ea29164c19e1036f4db. https://doi.org/10.1007/s10961-010-9171-x

de Wit, H., Hunter, F.: The future of the process of internationalization of higher education in Europe. International Higher Education, vol. 87, p. 40. Russian Version of the Newsletter of International Higher Education, Boston College, USA (2016)

Konstantinov, G.N., Filonovich, S.R.: What is an Entrepreneurial University? Educ. Issues, **1** (2007). http://cyberleninka.ru/article/n/chto-takoe-predprinimatelskiy-universitet

OECD: Trends Shaping Education 2016. OECD Publishing, Paris (2016). http://dx.doi.org/10.1787/trends_edu-2016-en

Methodology for Calculating Indicators to Monitor the Effectiveness of Higher Education Institutions of April 3, 2014 N AK 39/05vn. http://unecon.ru/sites/default/files/disshleb nikovkv.pdf

Philip, G., Altbach, J.K.: The internationalization of higher education: motivations and realities. J. Stud. Int. Educ. (2007). http://jsi.sagepub.com/content/11/3-4/290.short?rss=1&ssource=mfc. Accessed 15 Apr 2016

Shchedrovitsky, P.: The future has already come. Internet-newspaper (2016). www.vyatsu.ru, https://www.vyatsu.ru/internet-gazeta/buduschee-uzhe-nastupilo-petr-schedrovitskiy-o-per. html?print=Y

Wong, P.-K., Ho, Y.-P., Singh, A.: Towards an "Entrepreneurial University" model to support knowledge-based economic development: the case of the National University of Singapore. World Dev. 35(6), 941–958 (2007). https://www.scopus.com/inward/record.uri?eid=2-s2.0-34249043649&doi=10.1016%2fj.worlddev.2006.05.007&partnerID=40&md5=95830495dd da8cbaeafc901db56a47f2. https://doi.org/10.1016/j.worlddev.2006.05.007

Perspective Trends of Development of Professional Pedagogics as a Science

Svetlana M. Markova$^{(\boxtimes)}$, Ekaterina P. Sedykh,
Svetlana A. Tsyplakova, and Vadim Y. Polunin

Nizhny Novgorod State Pedagogical University Named After Kozma Minin
(Minin University), Nizhny Novgorod, Russian Federation
ngpu.profped@yandex.ru

Abstract. The article deals with the relationship between the development of science and changes in the content and structure of scientific and pedagogical activity. The role of science in the organization of industrial production, in the development of man, his productive forces is disclosed. The role of the integration of science as the main factor in raising the efficiency of social production, accelerating scientific and technological progress is shown. The main directions and ways of development of professional pedagogy as sciences are allocated: the strategy for the development of professional education, the cooperation of education and production, science, production and the academic subject, the laws of pedagogy and the laws of production, teaching, theoretical and educational activities; justification of the new principles of education and upbringing; interrelation of general, polytechnic and professional education, interrelation of production and social systems; integration processes (the formation of general sociological laws of the development of science, the formation of general scientific laws of the development of science, the formation of complex sciences, the intensification and industrialization of new researches). Factors influencing the development of pedagogical science are singled out: development of social, scientific and technical progress; interaction of general education, professional school; development of methodological, theoretical problems and empirical basis of pedagogical science; the combination of training with productive work in different levels of education; development of organizational forms, methods, means of education and upbringing.

Keywords: Professional pedagogy · Professional education
Socio-economic development · Scientific and technical progress
Development prospects

1 Introduction

At the present stage of social and economic development, the level of scientific knowledge in the system of professional education has significantly increased the need for creative development of the theory and practice of professional education based on the interaction of various sciences is increasing.

The use of scientific discoveries and the transition to scientific and industrial production, conditioned by integration processes in the development of science,

© Springer International Publishing AG, part of Springer Nature 2018
E. G. Popkova (Ed.): HOSMC 2017, AISC 622, pp. 129–135, 2018.
https://doi.org/10.1007/978-3-319-75383-6_17

evoke in modern conditions changes in the structure and content of scientific and pedagogical activity, in the interaction of society, science and technology, in the dynamics of economic and social processes.

2 Theoretical Basis of the Research

The work of many authors is devoted to study the prospects of the development of pedagogical science and professional education. They resolved issues of interaction between the development of society and the development of education (Bestuzhev-Lada I.V., Gershunsky B.S.); socio-economic and scientific and technical forecasting of the development of society and professional education (Batyshev S.Y., Belyaeva A.P., Gershunsky B.S.); trends in the development of professional education (Belyaeva A.P., Rabitsky A.I., Tkachenko E.V.) [9, 10].

3 Research Methodology

This study is aimed to determine the most probable ways of developing professional education in a broad scientific and pedagogical context. The integrative approach, scientific and technical progress, the idea of the unity of scientific knowledge, the unity of fundamental and applied knowledge, the interdependence of society, science and production are the methodological foundations.

Scientific and technological progress is the development of the productive forces of society, when it opens the possibility to automate production work, production management, and, thanks to information and communication technologies, increase the efficiency of many intellectual processes [2, 3, 6]. In this article, we are talking about the integration processes of professional education.

4 Analysis of Research Results

Summing up the various definitions of integration, we can conclude that in modern conditions integration is the inter-penetration and mutual enrichment of all the main spheres of labor and social activity on the basis of socio-economic development and has the properties of generalization, complexity, compactness and organization. With the acceleration of scientific and technological progress, it leads to a change in the nature and content of labor, the generalization and combining of professions, the emergence of new and universal professions [8, 10].

The successful solution of professional and pedagogical problems in the conditions of the modernization of the professional school largely depends on the integration processes taking place in science, technology, production, in the social and public spheres [9].

The study of integration processes in science is conditioned by existing contradictions in its development:

- the processes of integration and differentiation of science;
- methods of scientific research activities and means of technological support of complex scientific researches;
- the growing volume of scientific information and the means of its use;
- a new conceptual apparatus that appears in the new branches of scientific knowledge, and existing general scientific concepts.

At the same time, the leading law of the development of science is the integration of sciences and scientific knowledge.

Integration changes the content and structure of modern scientific knowledge and becomes a means of achieving the unity of knowledge in the content, structural, scientific, organizational and professional aspects.

To determine the main directions of the development of professional pedagogy as a science, it is necessary to single out the main integration processes in science that affect its development.

1. Formation of general sociological laws of the development of science that determine the direction of development of science, the basis of which is material production. In the conditions of accelerating scientific and technological progress, science opens the prospects for the development of production and technology. Its social role in the development of society is constantly growing [11]. The following regularities can be distinguished:

   - the dependence of science on production;
   - the dependence of science on production, manifested in different ways in different historical periods and at various stages in the development of production and science;
   - the dependence of the development of representations, categories, concepts, hypotheses, theories, means and methods of cognition from the reflection of reality in the public consciousness, the cognitive activity of human consciousness;
   - dependence of socio-economic and technical development of society on the development of science.

2. Formation of general scientific laws of the development of science. These regularities include the development of scientific knowledge, the interrelation of quantitative and qualitative aspects of the development of scientific knowledge, the movement of knowledge to absolute truth, the interrelation of integration and differentiation.

3. Formation of complex sciences, considered as a form of integration of knowledge on the subject and object. One object can serve as an object of research of a number of scientific branches, just one subject can become an object of study of many scientific disciplines. A comprehensive holistic study of a particular object makes it possible to synthesize knowledge of all its aspects within the framework of a single science and make such a knowledge system a constructive guide for practical knowledge.

4. Intensification and industrialization of new researches. The industrialization of science is carried out under the influence of scientific and technological progress using the methods of large-scale industrial production. The intensification of science depends on the means and methods of research, the processes of integration and differentiation of scientific activity, from forms of creative research cooperation. These processes affect the effectiveness of pedagogical research, which has made it possible to create a new form of intensification of research - educational, research, and production complexes.

Professional pedagogy can be considered as a complex science, covering theoretical areas, objects, subjects and empirical areas of the entire professional education system [7, 8].

The main tasks that determined the need for the development of pedagogical science and teacher education are the following:

– wide introduction of new ideas, solutions, scientific developments;
– integration of science, production practice;
– creation of new organizational forms of integration ensuring the rapid passage of scientific ideas from their inception to widespread application in practice;
– technical focus of research activities;
– association of fundamental, pedagogical and branch science for the purpose of interaction of various fields of knowledge;
– creation of organizational structures: interdisciplinary research centers, providing interdisciplinary research of actual problems;
– the strengthening of the human factor in scientific and social activities.

On the basis of the foregoing, the main directions of the development of professional pedagogy are considered.

The emergence of professional pedagogy is based on the practical needs of society in the training of workers and specialists.

Professional pedagogy studies laws, patterns of learning, education, development and education of students in professional schools, develops a scientifically grounded system of events and conditions in accordance with the goals and objectives of the formation of workers in the educational, training and production processes.

Professional pedagogy as a branch of pedagogical science is formed with the development of professional education [9]. It involves the development of a complex set of activities for the development of management bodies and types of educational institutions, improving the efficiency of training workers and specialists because of modern scientific and production achievements, scientific substantiation of forms, methods and means of training, improving skills and improving the training of professional education teachers.

Interest in professional pedagogy is determined by the following circumstances:

– accelerated development of professional pedagogy;
– the growth of scientific information about the theory and practice of education, training and education of workers and specialists;
– development of methods of scientific knowledge, which determine the specificity of professional pedagogy as a science;

– the complication of links between the pedagogy of professional education and other sciences;
– growing integration of science.

Professional pedagogy solves the specific tasks of preparing a special type of worker for the sphere of material production and social life.

Actual problems of research in professional education are: the strategy for the development of professional education, the cooperation of education and production, science, production and the academic subject, the laws of pedagogy and the laws of production, teaching, theoretical and production activities; justification of the new principles of education and upbringing; the relationship of general, polytechnic and professional education, the interrelationship of production and social systems [1, 4, 10].

In professional pedagogy, the most important are the studies conducted in the following areas:

– the development of a prognostic model of workers and specialists; modeling and designing of teaching, educational and production process;
– determination of the optimal relationship between the types of training, the terms of training, depending on the professional group;
– definition of perspective and effective systems of industrial training;
– professional education of workers and specialists.

Theoretically, problems related to the content of professional education become topical. The content of professional education should take into account the correlation of social, economic and professional factors [12]. At the same time, it should be noted the importance of the problems associated with the upbringing of the younger generation. It is necessary to develop a scientifically based strategy for professional education. Improvement of its content, forms and methods at all levels of education.

For the professionalism of the personality, the most important reference point are three types of socio-technological organization of production: pre-industrial, mass-industrial, scientific-industrial type of production.

Under the conditions of the scientific and industrial type of production, the requirements for the professional training of future workers and specialists are increasing, which calls for the improvement of the content of professional education in the organizational, structural and procedural terms in order to achieve a new quality level of education and training of students. Levels of professionalism allow us to consider the model of training skilled workers and specialists in terms of organizational forms; logical and procedural relations focused on socio-technological types of production.

For the pre-industrial type of production, the main qualities of the individual are initiative and enterprise.

For the preparation of industrial-type workers, the leading professional and social qualities of the individual are culture, morality, discipline, accuracy, etc.

To prepare workers in the scientific and industrial type of production, it is not only the process of professionalism that becomes important, but the process of socialization that is carried out at all levels of education.

The main characteristics of workers and specialists are the general educational level; professional level, general culture; political culture, ecological culture, technical culture, economic culture, the quality of the individual, the terms of training, and advanced training [4, 5].

The greatest influence on the development of pedagogical science is rendered by:

- development of social and scientific and technical progress;
- new requirements of science, technology and production to man, to the level of his education, development and upbringing;
- transition to a system of continuing education for young people;
- development of methodological, theoretical problems and empirical basis of pedagogical science;
- interaction of general education, professional school;
- the combination of training with productive work in different levels of education;
- development of organizational forms, methods, means of education and upbringing.

The goals of the development of pedagogy for the future are determined on the basis of an analysis of the contradictions between the social order and the possibilities for its realization. The resolution of contradictions is regarded as the goal for which scientific and pedagogical activity is directed.

Professional pedagogy integrates in the justification of pedagogical categories the knowledge of other social, natural- scientific, technical sciences.

The study of pedagogical phenomena requires initial data on the professional and skill structures of the working class, on changes in the content of labor of workers, on the trends and requirements of scientific and technological progress for the training of workers and specialists.

In this regard, there is a need for socio-economic research, which are the main provisions for the development of a system of education and training of future workers and specialists. Thus, pedagogy determines the level of education and qualifications of trainees, forms of organized, purposeful management of education in accordance with the goals set by society [11].

The essence of professional education, its social significance determines professional, pedagogical, psychological, technical interdependencies.

At the same time, the logic of the development of pedagogy science and its relationship with pedagogical practice, technology, and production becomes important. The development of professional pedagogy is influenced by objective and subjective factors. Subjective factors include: needs and scale of development of practice, high rates of development of the system of professional education, etc. Subjective factors include: difficulties in mastering related specialties, the use of methods and means of various sciences, etc.

Professional pedagogy studies and develops concepts, ideas, principles, patterns, theories of teaching and educating professional educational organizations. At the same time, the mutual influence of various sciences, the creative use of new opportunities change the methods of pedagogical science, expand factors and phenomena that need to be investigated in the present and the future.

Integration of knowledge of various sciences leads to the emergence of new facts, methods, concepts that ensure the change in the qualitative and quantitative characteristics of the object [6].

Pedagogical research is carried out at various levels of scientific integration (inside-disciplinary, interdisciplinary, general scientific and methodological).

In connection with the complexity, multidimensionality of the problems of professional education, the study has an integrative character. For example, the use of theories, principles and methods of various branches of knowledge, which allows to study the pedagogical process comprehensively and in integrity. Mutual penetration of scientific ideas and research methods is an objective source of improving the process of forming professional competences of the individual workers and professionals.

References

1. Markova, S., Depsames, L., Tsyplakova, S., Yakovleva, S., Shherbakova, E.: Principles of the building of an objective-spatial environment in an educational organization. IEJME: Math. Educ. **11**(10), 3457–3462 (2016)
2. Markova, S.M., Sedhyh, E.P., Tsyplakova, S.A.: Upcoming trends of educational systems development in present-day conditions. Life Sci. J. **11**(11s), 489–493 (2014)
3. Khizhnaya, A.V., Kutepov, M.M., Gladkova, M.N., Gladkov, A.V., Dvornikova, E.I.: Information technologies in the system of military engineer training of cadets. Int. J. Environ. Sci. Educ. **11**(13), 6238–6245 (2016)
4. Barber, M., Donneliy, K., Rizvi, S.: An avalanche is coming. Higher education and the revolution ahead. Institute for Public Policy Research (2013)
5. Hanushek, E.A.: The economic value of higher teacher quality. Working Paper No. 56, National Center for the Analysis of Longitudinal Data in Education Research (2010). http://www.urban.org/UploadedPDF/1001507-Higher-Teacher-Quality.pdf. Date Views 03 April 2017
6. Aleksandrova, N.M., Markova, S.M.: Problems of development of professional pedagogical education. Bull. Univ. Minin **1**(9), 11 (2015)
7. Vaganova, O.I.: General and professional pedagogy. Chronicles of the United Fund of Electronic Resources Science and Education **7**(62), 69 (2014)
8. Markova, S.M., Polunin, V.Y.: Theory and methods of professional education: theoretical basis. Bull. Moscow State Univ. Humanit. M.A. Sholokhova. Pedagogy Psychol. **4**, 40–44 (2013)
9. Tkachenko, E.V.: Continuing professional education in Russia: problems and prospects. Domestic Foreign Pedagogy **3**, 11–22 (2015)
10. Tkachenko, E.V., Smirnov, I.P.: Conceptual idea of the strategy of professional education. Prof. Educ. Russia Abroad **1**(13), 6–10 (2014)
11. Markova S.M., Tsyplakova S.A.: Management of the pedagogical process as a system. School of the Future **4**, 138–144 (2016)
12. Markova, S.M., Tsyplakova, S.A.: Interaction of basic and variable parts of the structure of curricula of professional education. Int. J. Appl. Fundam. Res. **5-1**, 115–117 (2016)

Didactic Foundations of Designing the Process of Training in Professional Educational Institutions

Natalia V. Bystrova[1](✉), Elena A. Konyaeva[2], Julia M. Tsarapkina[3], Irina M. Morozova[4], and Anna S. Krivonogova[5]

[1] Nizhny Novgorod State Pedagogical University Named After Kozma Minin (Minin University), Nizhny Novgorod, Russian Federation
bystrova_nv@mail.ru
[2] South Ural State Humanitarian-Pedagogical University, Chelyabinsk, Russian Federation
elen-konyaev@yandex.ru
[3] Moscow Agricultural Academy Named After K.A. Timiryazev, Moscow, Russian Federation
julia_carapkina@mail.ru
[4] Penza State Technological University, Penza, Russian Federation
89063981816@mail.ru
[5] Russian State Vocational and Pedagogical University, Ekaterinburg, Russian Federation
as.krivonogova@mail.ru

Abstract. The article is devoted to the development of didactic bases for designing the learning process in professional educational institutions. From this standpoint, the authors propose the use of the prognostic approach, the didactic forecasting. The future state of the pedagogical process is determined on the basis of socioeconomic, scientific, technical, cultural, technological forecasting. A characteristic feature of the didactic design of the learning process in a professional educational institution is the presence of two components of the learning process: theoretical and industrial training, which necessitates the design of theoretical and production training. Designing theoretical training is associated with the development of modular programs of general educational, general technical and special subjects, pedagogical technologies that ensure the implementation of innovative approaches to the development of educational and cognitive activities. The design of production training is associated with the design of the production process, the material and technical and socio-technical environment. The design of the material and technical environment is associated with the equipment of study rooms, training workshops, with the definition of quantitative and qualitative characteristics of the raw materials, with technical and technological capabilities for manufacturing products. The design of the social and industrial sphere is associated with production, economic relations, and professional communication. Also, the article discusses the features of designing the learning process in a professional school related to establishing links between theoretical and production training, pedagogical and production tasks, between professional knowledge and production activities.

© Springer International Publishing AG, part of Springer Nature 2018
E. G. Popkova (Ed.): HOSMC 2017, AISC 622, pp. 136–142, 2018.
https://doi.org/10.1007/978-3-319-75383-6_18

Keywords: Didactic forecasting · Didactic designing · Theoretical training
Industrial training · Professional educational institution

1 Introduction

Designing the pedagogical process in a professional school is the most important organizational and pedagogical condition for increasing the effectiveness of professional training in the system of continuing education.

Thus, in the most general terms, design is understood as the development of curricula and programs, projects of future composite objects and systems based on their development tendencies, the factors of the possibility of fulfilling long-term goals, the analysis of objective reality, the establishment of links between subjects in general education, general technical and special cycles, definition of methods and forms of organization of the pedagogical process, the design of professional activity, the system that provides the formation of professional knowledge, skills and certain types of training and production work. In this study, the design of the pedagogical process is considered from the point of view of the prognostic approach.

2 Theoretical Basis of Research

The research of many authors was devoted to the design of the pedagogical process: Batysheva S.Y., Belyaeva A.P., Bezrukova V.S., Vazina K.Y., Dumchenko I.I., Kathanova K.N, Makhmutova M.M, Morozova E.P, Naina A.I., et al. They solved questions of methodological and theoretical bases for designing the pedagogical process in the conditions of continuous multilevel professional education (Batyshev S.Y., Belyaeva A.P.), content and structure of professional activity (V.Y. Vazina), didactic bases of industrial training (Bezrukova V. S., Dumchenko I.I., Melnikova A.P., Kathanov K.N., Nain A.Y., et al.), the use of problem training in the lessons of industrial training (Makhmutov M.M., Vlasov V.G.), organizational and pedagogical bases of planning of educational and production activities (Morozov E.P., Kathanov K.N., et al.).

The development of designing the pedagogical process should be carried out on the basis of scientific forecasting, and considered in various sciences. Scientific didactic prediction can be defined as the construction of an ideal model of learning and its results from possible components based on the disclosure of objective laws of motion and the development of the learning process. The content of such forecasting is the process of learning and education with all its attributes and laws of development.

As a **research** methodology, a prognostic approach is distinguished. Didactic forecasting is designed to solve a number of important problems related to the study of the future development of didactic phenomena. There are two main directions for the development of didactic forecasts:

– forecasting of the system, content and methods of teaching and education at all levels;
– predicting the results of learning and education, depending on various conditions and different terms of training.

In recent years, the development of the first direction - the forecasting of the system and the content of learning and education - has begun.

The second direction of didactic forecasting, connected with the forecasting of learning outcomes, includes the solution of the following tasks, which are very important and complex:

- the definition of the components of the learning process, factors affecting the formation of the product of training;
- forecasting the development of individual components of the learning process: perception of knowledge, their assimilation, application in practice;
- forecasting the development and outcome of learning for an individual student.

The general task of didactic forecasting is to get an idea of the future result of didactic activity on the basis of scientific research, depending on the conditions and the selected teaching aids and to compile a scientifically based management plan for the learning process based on forecast data.

The forecast makes it possible to form a reasonable opinion about the possible outcome of the learning process; the use of forecasting data makes it possible to plan and implement the learning process purposefully - in such a way as to achieve the maximum possible effect under the existing conditions; The forecasting data allow to correct the learning process in order to achieve the desired result.

At this level, we are talking about the actual pedagogical and didactic design: on the prognostic justification of the goals, content, methods, means and organizational forms of education, upbringing and development of students at different levels of education. These components act as the main design objects. In their totality, these components form "pedagogical, co-oriented educational systems that are substantially different in their specific content depending on the level and profile of education." These systems must be objects of design. The results of design are reflected in the professional qualifications and models of graduates of educational institutions of various types, in the content of curricula and plans, textbooks, in the developed tools, methods, organizational forms of the upcoming pedagogical activity.

The methodological and theoretical foundations of designing activity in professional educational institutions are social, economic, production, scientific and technical, psychological-pedagogical, didactic laws; the main trends in the development of professional education (humanization, democratization, integration, differentiation, intensification, and cooperation), the main functions of the theory of scientific management.

The beginning of the organization in the design activity of the teacher is the Federal State Educational Standard and the professional standard of the teacher of professional training. The professional activity of the students is a system-forming factor, the qualification characteristic of the future specialist, who owns the educational, design, professional, research, organizational and technological activities

3   Analysis of the Research Results

As shown by the analysis of the pedagogical process, the educational and cognitive activity of the student is the theoretical basis for the implementation of educational and production activities. During the study (cognition), students learn objective laws, facts, concepts, phenomena.

Educational and production activities in the professional school are the leading, it is in the professional training of future workers and specialists that the unity of general educational, polytechnic and special knowledge is manifested.

At the same time, the integration of the content of professional training, the regulation and harmonization of activities in various learning spaces, the establishment of a system of relations between the subjects of the pedagogical process, and the complex interrelationships of pedagogical and production processes are formed.

Investigating the peculiarities of the pedagogical process in the new socioeconomic conditions, one cannot fail to mention such activities as scientific-methodical, commercial, research, and organizational. Their main task is to ensure the cooperation of science, education and production, which is a prerequisite for the existence of a new type of educational institution.

Social and economic activities are aimed at the development of professionally significant personal qualities, the formation of professional knowledge, skills, development of the social needs of man, the processes of professional self-determination and self-awareness.

Material, technical, financial and economic activities are aimed at creating the material and technical base of the educational institution and financial and economic support for the pedagogical process.

The activities we have identified are united in the content of the pedagogical process, are in a complex interconnection and coordinated by management activities. The allocated system of actions underlies definition of designing activity of the teacher in professional school.

In the system of design activity, it is possible to single out various subsystems that are part of a single complex of the pedagogical process organized in a professional school. Design subsystems can be viewed as relatively independent, consistent and interconnected units and as components of a single cycle of the pedagogical process.

The main subsystems are design of theoretical training and design of production training. When considering the pedagogical process as a social system, it is valuable for our research to be able to design a system of relations in the pedagogical process, the trajectory of personality development, various types of activities, as well as forms and types of education.

This kind of social projection, in our opinion, is essential for this concept, because both logically and historically it is connected with the essential characteristic of a person as a subject of history, an individual creating the conditions for his self-activity.

It is necessary to indicate the differences in the design of theoretical training in conditions of continuous multi-level education from the traditional one. Traditional design is based mainly on the chosen professional guidance of students; it does not fully reflect the increasing qualitative changes in the content of labor, characterized by processes of socioeconomic, technical and technological integration and differentiation, the inter-penetration of material and spiritual factors. Designing from new educational positions, as A.P. Belyaeva notes, involves the development of a generalized system of theoretical knowledge at the following levels of generalization:

(1) general production (professional training inter-sectoral in cross-cutting occupations);
(2) general industry (training for occupations of a broad profile common to the entire industry);
(3) general professional (training in the professions of a wide profile and related professions of individual industries within the industry);
(4) private-professional (training in groups and individual occupations of a narrow profile).

Designing theoretical training is associated with the development of modular programs of general educational, general technical and special subjects, pedagogical technologies that ensure the implementation of innovative approaches to the development of educational and cognitive activities.

In the context of this article, the design of industrial training is of particular interest. In our opinion, the design of production training should be viewed as an integrative phenomenon, expressed in the unity of designing the production process, the material, technical and socio-technical environment.

To implement the technological process, it is necessary to determine the elements of the system of professional activity: movement, action, reception, operation, a complex of labor operations, works, a complex of educational and production works.

The design of the material and technical environment is associated with the equipment of study rooms, training workshops, with the definition of quantitative and qualitative characteristics of the raw materials, with technical and technological capabilities for manufacturing products.

The design of the social and industrial sphere is associated with production, economic relations, professional communication. As studies show, this aspect in professional education is almost not represented. This requires the use of innovative approaches to the organization of the process of industrial training, active methods and forms of training, and professional-pedagogical technologies.

The essence of the design activity is to establish the conformity of pedagogical and production technology, the unity of theoretical and industrial training, to establish the interrelationship of the types of activity of the pedagogical process, to ensure the interrelationships of science, production and education.

Designing of professional training is carried out on a technological basis. The essence of the procedural component is the technological procedures of professional training, coordinated with the production processes, operations, and actions. The main phenomena of this approach are designing forms and methods of the pedagogical process, designing the content of training, designing a technological approach to learning.

The introduction of the design of the pedagogical process made it possible to ensure a purposeful interaction of the educational and production activities of the students, to realize the potential of the teachers in the process of solving creative tasks for mastering the integrated professions, to develop the student's abilities and needs for self-education and self-development.

Thus, based on the identified factors that provide design activity in a professional school, we can note the following main features of it.

The boundaries of the application of design activities in professional schools are widening. The pedagogical process unites all the necessary components of professional training: general educational, general technical, special and production. In this regard, the design activity is carried out in training and in production conditions. It ensures the establishment of links between theoretical and industrial training; between pedagogical and production tasks, between professional knowledge and production activities.

4 Conclusions

The design activity has a complex and multidimensional character. All engineering and pedagogical workers, specialists of various branches of knowledge are involved in the design activity:

1. In its composition, it has common components regardless of the subject of the activity; obeys the general laws, regardless of the specifics of the content of general education, general technical and professional training. Design activities include elements of differentiation in connection with the specific content of training workers in various professions.
2. In terms of production training, the design activity acquires a specific character. This is related to systems, forms of production training and the place of its conduct. In this regard, special attention is paid to the definition of a system of means and to the identification of the conditions for the educational and production activities of students, and also to quantitative and qualitative characteristics.

Thus, the system of the pedagogical process in the professional school determines the appropriate design activity, which, in turn, improves and develops the pedagogical process.

References

1. Khizhnaya, A.V., Kutepov, M.M., Gladkova, M.N., Gladkov, A.V., Dvornikova, E.I.: Information technologies in the system of military engineer training of cadets. Int. J Environ. Sci. Educ. **11**(13), 6238–6245 (2016)
2. Markova, S., Depsames, L., Tsyplakova, S., Yakovleva, S., Shherbakova, E.: Principles of the building of an objective-spatial environment in an educational organization. IEJME Math. Educ. **11**(10), 3457–3462 (2016)
3. Markova, S.M., Sedhyh, E.P., Tsyplakova, S.A.: Upcoming trends of educational systems development in present-day conditions. Life Sci. J. **11**(11s), 489–493 (2014)
4. Smirnova, Z., Vaganova, O., Shevchenko, S., Khizhnaya, A., Ogorodova, M., Gladkova, M.: Estimation of the educational results of the bachelor's program students. IEJME Math. Educ. **11**(10), 3469–3475 (2016)

5. Aleksandrova, N.M., Markova, S.M.: Problems of development of professional pedagogical education. Bull. Univ. Min. **3**(11), 13 (2015)
6. Koldina, M.I.: Designing of evaluation tools for the implementation of the basic educational program of higher professional education of a new generation. Sci. Methodical Electron. J. Concept **5**, 41–45 (2015)
7. Krivonogova, A.S., Tsyplakova, S.A.: Technology of realization of the project activity of the teacher of professional training. Bull. Univ. Min. **1**(9), 16 (2015)
8. Markova, S.M.: Patterns of pedagogical designing in conditions of professional education. Theory Pract. Soc. Dev. **19**, 206–208 (2014)
9. Markova, S.M.: Designing of pedagogical systems and their implementation in the conditions of the regional system of professional education (on the example of the University of Minin): monograph, 168 p. The University of Minin, Nizhny Novgorod, Flint (2017)
10. Markova, S.M.: Theory and methods of professional education: Textbook/N, p. 86. NGPU named after K. Minina, Novgorod (2013)
11. Markova, S.M., Polunin, V.Y.: Theory and methods of professional education: theoretical basis, vol. 4, pp. 40–44. Bulletin of the Moscow State University for the Humanities named after M.A. Sholokhova. Pedagogy and Psychology (2013)
12. Markova, S.M., Tsyplakova, S.A.: Management of the pedagogical process as a system. Sch. Future **4**, 138–144 (2016)
13. Sedykh, E.P.: Modeling the management of the pedagogical process in a professional school. World Sci. Discov. **5**, 170–180 (2012)
14. Tsyplakova, S.A.: Axiological bases of pedagogical designing. Kazan Sci. **4**, 221–224 (2013)
15. Tsyplakova, S.A.: A systematic approach to the design of the pedagogical process as a social system. World Sci. Discov. **3.1**(39), 154–164 (2013)
16. Tsyplakova, S.A, Barinova, A.N, Grishanova, M.N., Tsareva, I.A.: Formation of the project competence of bachelors of professional training within the framework of educational and project activity. Azimuth Sci. Res. Pedagogy Psychol. T.5. **4**(17), 286–288 (2016)
17. Tsyplakova, S.A, Grishanova, M.N, Korovina, E.A, Somova, N.M.: Theoretical bases of designing of educational systems. Azimuth Sci. Res. Pedagogy Psychol. T.5. **1**(14), 131–133 (2016)
18. Barber, M., Donneliy, K., Rizvi, S.: An avalanche is coming. Higher education and the revolution ahead. Institute for Public Policy Research (2013)
19. Hanushek, E.A.: The economic value of higher teacher quality. Working Paper No. 56, National Center for the Analysis of Longitudinal Data in Education Research (2010). http://www.urban.org/UploadedPDF/1001507-Higher-Teacher-Quality.pdf. Accessed 03 Apr 2017

Post-graduate Information Support
for Graduates of Pedagogical Universities

Marina L. Gruzdeva$^{(\boxtimes)}$, Olga N. Prokhorova, Anna V. Chanchina,
Elena A. Chelnokova, and Elena V. Khanzhina

Nizhny Novgorod State Pedagogical University Named After Kozma Minin
(Minin University), Nizhny Novgorod, Russian Federation
grul234@yandex.ru, prohorova_olga@mail.ru,
avchl@list.ru, chelnelena@gmail.com,
e.hanzhina@rambler.ru

Abstract. The article describes the information service of graduate support of graduates of a pedagogical university. The authors disclose the concept of "postgraduate support" as a set of purposeful complex measures involving the cooperation of a teacher and mentor with a young specialist with the aim of ensuring successful entry into the profession, effective implementation of pedagogical activity, adaptation to a professional environment and overcoming crises and barriers arising in the process of implementing a professional activities. The mechanism of postgraduate support, which is based on the institute of double mentoring, is described.

The article describes the electronic postgraduate support service that has been developed and implemented at the Nizhny Novgorod State Pedagogical University named after Kozma Minin for postmodern support of graduates of the university, which provides online support for a young specialist in professional activities. Electronic service is a specially organized Internet portal on which methodical services and Internet resources are concentrated.

The article contains the content of the electronic service, which includes: Diary reflections, a resource for conducting mentoring sessions, a personal chat, a bank of methodical ideas, an electronic portfolio, a tool for conducting webinars and thematic forums, etc.

The article contains some survey results that were conducted among graduates using the electronic support service.

Keywords: Graduate support of graduates · Pedagogical education
Electronic service · Information support

1 Introduction

One of the most important tasks of the teacher's training is postgraduate psychological and pedagogical support of the graduate on the path to the realization of professional activity (overcoming barriers and difficulties of the subject, strengthening confidence in oneself and their capabilities). The Mininsky University is working on the adaptation of a young specialist in professional development with a view to successfully "entering" and adapting to professions. The purpose of this work is to optimize the process of

E. G. Popkova (Ed.): HOSMC 2017, AISC 622, pp. 143–151, 2018.
https://doi.org/10.1007/978-3-319-75383-6_19

professional development of teacher, contributing to the development of his abilities to perform his duties in his own capacity and qualitatively.

2   Theoretical Basis of the Research

Postgraduate support is a set of purposeful comprehensive measures that involve the cooperation of a teacher and mentor with a young specialist in order to ensure successful entry into the profession, effective implementation of pedagogical activity, adaptation to the professional environment and overcoming crises and barriers arising in the process of professional activity.

The graduate's graduate support mechanism is based on the double mentoring institute. A mentor from a higher educational institution and a mentor from an employer organization are attached to the graduate who facilitate the implementation of information and methodological support, psychological and pedagogical support, and also contribute to the professional socialization of the young teacher (Fig. 1). One of the means of escorting in the mechanism of postgraduate support of a young teacher is electronic service.

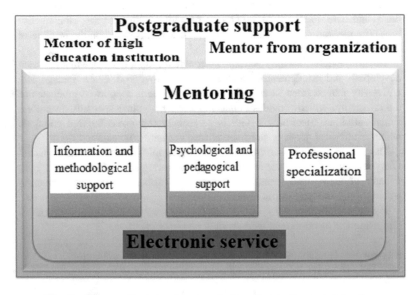

Fig. 1. The mechanism of postgraduate support of a young teacher

Accompaniment includes several interrelated stages aimed at successfully entering the profession and adapting the graduate to a professional environment: The stage of primary entry into professional activity (first half); Stage of the primary adaptation of the graduate (second half of the first year), the motivational-value stage (first half of the second year), the stage of professional growth (second half of the second year) (Fig. 2).

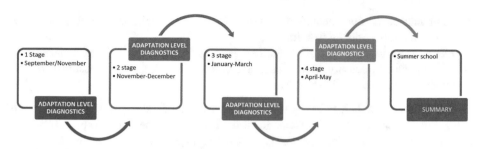

Fig. 2. The main stages of postgraduate support for the first year

Many scientists believe that accompaniment is a very versatile movement. "Changing the focus of support, you can get different types of it: Pedagogical, methodical, information, scientific-methodical, information-methodical, tutorial, consulting, etc." [5, 7]. Postgraduate support of a young teacher can be displayed with the help of structural, procedural and activity characteristics.

3 Research Methodology

During the research, we used theoretical and practical methods: Analysis of literature on the topic of research, modeling, pedagogical experiment, testing, questioning, observation, methods of mathematical statistics.

The purpose of pedagogical support is the purposeful development of the person's personality, carried out through special pedagogical systems. A young teacher often has enough theoretical training, which he received at the university, but lacks his own pedagogical experience and it is difficult for him to apply the learned methods and methods of teaching in practice.

Methodical support is considered as a complex method of methodical work, consisting of diagnosing the emerging professional pedagogical problems; Informing about the ways of their solution and assistance at the implementation stage. "This method involves the interaction of an escorted person and an accompanying person, aimed at solving the problems of professional activity that are actual for the teacher, carried out in the processes of actualization and diagnosis of the problem's essence, information search of a possible way of solving the problem, consultations at the stage of choosing the path, constructing an action plan and primary implementation of the plan" [5, 7].

Information support is the provision for young teachers of access to information stored in various databases and computer archives, in printed and electronic information, etc. Information support for the work of young teachers is the pedagogically and technically organized interaction of the subjects of the educational process.

Postgraduate support of graduates of higher educational institutions can be organized in the following forms:

- Participation of graduates in production meetings, pedagogical councils, methodological associations, professional associations and other events of the educational organization;

- Maintenance of documentation in terms of practical training of young teachers (schedule, report documentation);
- Compulsory work in scientific seminars held at graduate-supervising departments;
- The graduate's activity in the use of electronic service, which assumes online support in professional activities.

E.I. Kazakova and A.P. Tryapitsyna consider pedagogical support as a help to the subject in the formation of the orientational field of development, the responsibility for actions in which the subject himself is responsible [4].

E.A. Aleksandrova considers pedagogical support as a type of pedagogical activity, the essence of which is in the preventive process of teaching the subject to plan independently his own way of life and of the individual professional route, organize life activities, resolve problem situations, and in the permanent readiness to respond adequately to the situation of his emotional discomfort.

Foreign authors also believe that the professional formation of teachers should begin after the "exit" from the educational institution under the guidance of experienced teachers and cannot stop until the end of their professional activities [6, 8]. Professional development is a joint activity, the authors of the article "European teacher education policy: Recommendations and indicators. "The most effective way is" with conscious interaction, and not only teachers with teachers, but also teachers with the administration, parents and other members of the educational community" [9].

Prospects of people interaction through the Internet, which removes the problem of distance between people, the possibility of using cognitive technologies, require their implementation, set in motion an information educational space. In the educational process, various forms of distance learning are used. L.N. Bakhtiyarova notes that intensive development of information and communication technologies contributes to their wide penetration into all spheres of human life and, first of all, to education, transforming both the models of the educational process and the role of students and teachers [3]. Virtual educational environments have become not only a means of learning, but also an integral part of the educational process. An example can serve both distance learning systems, and various educational environments, for example, profile training.

It should be noted that the information educational space can be represented as a set of information educational environments actively interacting with each other, these are information educational environments of individual educational institutions and entire regions, these are information educational environments of small units, for example, the department of the university. Each of these environments develops and changes. S.M. Markova notes that the department is a new educational system that develops because of the developed educational standards [6, 8].

New storage tools, transformation and transmission of information affect the methodological support of the learning process. Lately there has been much talk about interactive technologies in education as a form of information exchange between students and the surrounding information environment. Interactive methods of teaching allow to intensify the understanding process due to more active inclusion of students in the process, form the ability to think extraordinarily, and initialize the disclosure of new opportunities for learners. Components of the information educational environment are constantly being transformed, including this refers to the social component. Currently, the European educational strategy sets the task of moving from education of knowledge and skills to education, ensuring the readiness of students to act and live in a rapidly in changing environment [4].

Information support for graduates of the University, the Minin University has developed an *electronic postgraduate support service* - an interactive multifunctional service that allows online support of a young specialist in professional activities. Electronic service is a specially organized Internet portal on which methodical services and Internet resources are concentrated.

The structure of the electronic service of postgraduate support presupposes the availability of interaction opportunities for subjects of postgraduate support in three main modes: Frontal, group and individual. The last mode of interaction requires the availability of work of subjects of postgraduate support in private offices (Fig. 3).

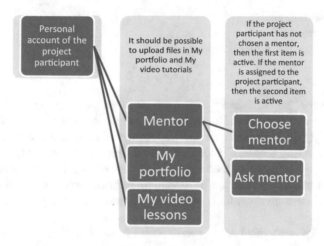

Fig. 3. Structure of the site page of the university for postgraduate support of project participants

Functional requirements for the electronic service and assume the availability of opportunities for:

- Conducting webinars
- Discussion (forum)
- Presentation of video tutorials
- Work with a diary of reflection
- On-line mentoring
- Psychological and pedagogical support
- Providing information and methodological support.

Contents of the electronic service:

1. Diary of Reflection.
2. Resource for conducting mentoring sessions.
3. Personal chat.
4. Bank of methodical ideas.
5. Electronic portfolio.
6. Webinars and thematic forums.

The diary reflects the assessment of the subject's assessment of the processes of adaptation to the professional environment, the difficulties, and crisis points in professional activity, the degree of satisfaction of the subject with the process and the results of activities, and the success of professional activity.

The diary of reflection is filled by the subject as needed, but not less than once every two weeks, and contains several sections:

- My professional goals
- My achievements
- Reflective sheet of professional activity
- Feedback
- My calendar

The resource for conducting mentoring sessions presupposes the possibility of downloading video lessons and materials that reflect the real process of professional activity of a young specialist with the possibility of conducting mentoring sessions on-line. With young teachers in on-line mode, individual consultations are held to compile thematic plans or plans for educational work, by choosing and optimizing methods and means of instruction for organizing certain types of lessons.

A *personal chat* involves individual online counseling of a young specialist by a teacher-mentor.

The bank of methodical ideas presupposes methodological content, which contributes to raising the level of competence of a young specialist in solving professional problems of different levels. This tool contains specific methodological recommendations for solving problems, analytical material, video presentations, materials on the teacher's self-presentation, a bank of solutions to pedagogical situations, literature, etc.

Methodical bank

- Video lessons
- Developing lessons
- Methodical findings

The electronic portfolio assumes the fixation of the personal success of the young specialist in professional activity. Has the opportunity to showcase the portfolio to the professional community. Portfolio is compiled in electronic form and includes: Information on academic performance for the period of study at the university, an autobiographical certificate, copies of documents confirming the student's achievements over the past 2 years, recommendations from educational organizations or from the university (in the case of participation in the competition for academic baccalaureate), additional information.

Thematic forums involve a group discussion of a variety of professional topics, tasks and situations. At the same time, the teacher-instructor acts as a moderator of the forum and a consultant on these topics (Table 1).

Forum	*Themes are asked from the postgraduate support center*
•How do you like the first teacher council?	
•Planning time in a new life after a carefree student's years	
•Score for a quarter ... Are there any doubts?	
•Do I need classroom leadership?	
•«I "through the eyes of my pupil	
•"Favorites" in the classroom: the overcoming of the subjective perception of students	
•...	

Forums are preceded by webinars of teacher-instructors

Table 1. Example topics of webinars and discussions

Webinar	Forum
What is a professional category? Why is it needed? How to properly collect a portfolio from the first working days	How do you like the first teacher council?
Participation in the contest "Pedagogical Hope"	Planning time in a new life after a carefree student's years
Diary.ru. How to work with him	Score for a quarter... Are there any doubts?
How to keep filling a paper journal	Do I need classroom leadership?
Participation in professional Olympiads	The basics of successful preparation of pupils for the Olympiad
How to keep discipline	"I" through the eyes of my student.
Objectivity when grading	"Favorite student" in the class: Overcoming the subjective perception of students
Network projects with class	We share the experience of project development
Integrated lessons	
What is a social project? How to implement it if there is no classroom management?	Opportunities and benefits of implementing social projects in the classroom
Project method. Actual projects.	Your Social Design Experience
If nevertheless gave a classy leadership: How to hold the first parent meeting	Parents of students: How to find contact and organize effective cooperation
About the creation of work programs	We share the experience of working programs
Where to give more time? Papers or children? Prioritization	We share the experience of effective time management - we are looking for the answer to the question: "How to make it all?"

Graduates of the university participating in the testing of the electronic service actively used it in their professional activities, especially personal chat and thematic forums turned out to be in demand.

4   Analysis of Research Results

After the introduction of the electronic service among the graduates of the bachelor's and magistracy, a survey was conducted, which, among other things, aimed to find out: Whether the electronic service of postgraduate support helps the professional development of young teachers.

For example, the question "Did you need to ask for help from more experienced colleagues in the process of work?" 70% of the respondents answered positively, and the clarifying question "Who would you rather ask for help with?" Was distributed as follows (Fig. 4):

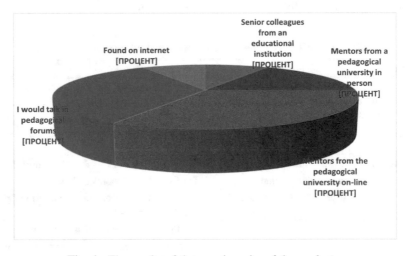

Fig. 4. The results of the questionnaire of the graduates

Thus, the carried out approbation of the electronic service of postgraduate support of graduates of pedagogical universities showed not only its viability, but also the theoretical validity and practical significance that manifests itself both in the growing professionalism of young teachers and in the growth of satisfaction with their chosen profession.

The electronic service of postgraduate support of the graduates of the Minin University is being realized at the current level at the level of graduates of the bachelor's and master's degrees. Currently, an escort service is being developed for students of advanced training courses and retraining of teaching staff.

Competent organization of postgraduate support of graduates of pedagogical universities in the period of adaptation and the initial stage of professional activity will ensure an increase in the professional competencies of young teachers and create the necessary conditions for improving the quality of domestic education.

References

1. Agapova, E.N., Agapova, E.H., Trapitsyn, S.Y.: Accompanying the process of adapting the initial stage of the teacher's professional activity. Universum: Bulletin of the Herzen University, no. 3, 36–46 (2013)
2. Alexandrova, E.A.: Pedagogical support of self-determination of senior schoolchildren. Research Institute of School Technology, p. 336 (2010)
3. Gruzdeva, M.L., Bakhtiyarova, L.N.: Pedagogical methods and methods of work of university teachers in conditions of information educational environment. In: Theory and Practice of Social Development, no. 1, pp. 166–169 (2014)
4. Kazakova, E.I., Herzen, A.I.: Technology of designing personal achievements. SPb.: Publishing House of the Russian State Pedagogical University. p. 195 (1994)
5. Nikolaeva, T.P., Bicheva, I.B.: Education as a prerequisite for achieving sustainable development and the basis for future progress. Bulletin of the University of Minin, no. 4 (2014). http://www.mininuniver.ru/scientific/scientific_activities/vestnik/archive/no8
6. Markova, S.M.: Innovative development of the department in the structure of the university. Bulletin of the University of Minin, no. 1 (2014). http://www.mininuniver.ru/scientific/scientific_activities/vestnik/archive/no5
7. Starodubtseva, E.V.: Pedagogical support of students in the system of postgraduate education. Pedagogy: Traditions and innovations: Materials of the international. Sci. Conf. (Chelyabinsk, October 2011). II. - Chelyabinsk: Two Komsomol members. pp. 121–123 (2011)
8. International Perspective Teacher Preparation - New Paradigm in Teacher Education, What Do Teachers Need to Know? [Electronic resource]. http://education.stateuniversity.com/pages/2484/Teacher-Preparation-International-Perspective.html. Accessed 24 May 2014)
9. Fredriksson, U.: European teacher education policy: recommendations and indicators. EuropeanCommission, Joint Research Centre, Institute for the Protection and Security of the Citizen, Centre for Research on Lifelong Learning [Electronic resource]. http://www.pef.uni-lj.si/atee/978-961-6637-06-0/715-723.pdf. Accessed 24 May 2014
10. Hans, A., Akhter, Sh.: Emerging Trends in Teacher's Education. The Macrotheme Review 2 (2), Spring (2013). Special Issue on Education and Training. http://macrotheme.com/yahoo_site_admin/assets/docs/3HansMR22SI.40131741.pdf. Accessed 24 May 2014

Accounting for Costs and Expenses: Problems of Theory and Practice

Igor E. Mizikovsky[1(✉)], Tatyana Y. Druzhilovskaya[1],
Emilia S. Druzhilovskaya[1], Ekaterina P. Garina[2],
and Elena V. Romanovskaya[2]

[1] N.I. Lobachevsky Nizhny Novgorod National Research University,
Nizhny Novgorod, Russian Federation
core090913@gmail.com,
drujilovskayatu@yandex.ru, druzhilovskaya@yandex.ru
[2] Nizhny Novgorod State Pedagogical University Named After Kozma Minin
(Minin University), Nizhny Novgorod, Russian Federation
e.p.garina@mail.ru, alenarom@list.ru

Abstract. The organization and maintenance of accounting require accurate methods and the most accurate results of their application, which entails the need to use generally accepted economic categories. Inaccurate or incorrect use of the categories "expenses" and "costs" leads to a distortion of a number of key performance indicators of the enterprise, primarily the financial result. In certain accounting standards there is inconsistency in the definition of the concept of "expenses", its economic meaning, there is no definition of the category "expenses". The purpose of the article is to determine the relationship between the concepts of "expenses" and "costs", clarify their wording, identify differences in regulations on cost accounting systems RAS and IFRS and give them an assessment, analyze the inclusion models in the cost of indirect expenses. System analysis of the relevant segments of theory and practice of accounting led to the conclusion that it is necessary to bring all documents, regulatory and regulatory costs and expenses in line, to improve the methodology for recording expenses in the cost of finished goods and work in process, the inclusion in the accounting standards of the cost accounting in the cost of finished goods, products and work in progress.

Keywords: Costs · Expenses · Work in progress · Finished products
Accounting regulations · International accounting standards
Accounting and calculation of production costs

1 Introduction

One of the most important objects of accounting organizations of any field of activity are costs and expenses. At the same time, it is these accounting objects that are among the most problematic, both from the point of view of accounting theory and from the point of view of practical accounting.

In systems of both national Russian accounting standards (RAS) and international financial reporting standards (IFRS) there is an unambiguous definition of the concept

E. G. Popkova (Ed.): HOSMC 2017, AISC 622, pp. 152–162, 2018.
https://doi.org/10.1007/978-3-319-75383-6_20

of "expenses". According to Russian PBU 10/99, expenditures are understood as "the reduction of economic benefits resulting from the retirement of assets (cash, other property) and/or liabilities, resulting in a decrease in the capital of this organization, except for the reduction of contributions by decision of the participants (property owners)". Almost completely coincides with the above definition of the interpretation of consumption in the IFRS system. According to chapter 4 of the Conceptual Framework for Financial Reporting, "expenditures represent a reduction in economic benefits during the reporting period in the form of retirement or" depletion "of assets or an increase in liabilities that result in a decrease in equity that is not related to its distribution between participants of the capital". In both definitions, costs are understood as a reduction in economic benefits, which is expressed in the fact that the value of expenditures decreases the organization's own capital, namely, its the financial result. We give examples of situations in which an organization should recognize the flow according to the definitions given (Table 1).

Table 1. Examples of situations leading to the formation of a flow in accordance with the definitions of this concept in PBU 10/99 and the conceptual basis for the financial statements of IFRS.

Economic operation	Impact on the amount of economic benefits	Impact on the amount of financial result and equity
Write-off of materials unsuitable for use	Reduction of economic benefits due to write-off of materials	The value of written-off materials decreases the financial result and equity
Impairment of non-current assets	Reduction of economic benefits due to a decrease in the carrying amount of non-current assets	The value of the decrease in the carrying amount of non-current assets decreases the financial result and equity
Writing off the cost of sold finished products	Reduction of economic benefits due to write-off of cost of sold finished products	The value of the written off cost of sold finished products decreases the financial result and equity
Writing off the cost of goods sold	Decrease in economic benefits due to write-off of cost of goods sold	The amount of the written off cost of goods that was sold decreases the financial result and equity
Interest accrual on loans and borrowings received for economic needs	The decrease in economic benefits due to the increase in the loan and borrowings liabilities by the amount of these percentages	The amount of interest on loans and borrowings reduces the financial result and equity
Fee accrual due to non-fulfillment of the contract with the supplier	Reduction of economic benefits due to an increase in liabilities to the supplier by the amount of the accrued fine	The amount of the accrued fine decreases the financial result and equity

A list of situations involving the formation of a flow could be continued further. The common in all these situations is that they determine either a reduction in the asset or an increase in the liabilities of the organization.

It would seem that thanks to the introduction in PBU 10/99 and in the Conceptual Basis of IFRS financial statements, a clear definition of the concept of "expenditure" in any situation can unequivocally solve the question of whether the expenditure is formed in it or not. However, the analysis of various situations leads to the conclusion that there is no such uniqueness. So, for example, should the expense of the cost of materials written off into production be recognized as a cost, the amount of cash outflows aimed at paying off accounts payable? After all, these situations attract a decrease in assets: materials and money. Should one recognize the expense of calculating the arrears of wages: because this increases the obligations of the organization?

2   Theoretical Basis of the Research

And the analysis of these situations shows that in the accounting, the reduction of economic benefits, financial results and equity in such situations is not recognized. The reason here is that the disposal of materials expended in production is fully offset by an increase in the same value of the finished product or work in process. That is, one asset (materials) is transferred to another asset (finished goods or work in progress). Withdrawal of cash on settlement of accounts payable leads to a decrease in the same amount of accounts payable. The accrual of arrears in wages leads to an increase in the amount of this debt of the cost of finished goods or incomplete production produced by employees. That is, the amount of increase in liabilities to employees increases the asset - finished goods or work in progress.

If in Russian PBU 10/99 and the Conceptual Basis of IFRS financial statements the interpretation of costs as a reduction in economic benefits and the amount of equity is presented, then this approach to characterizing the economic nature of the expense is not always supported by other accounting standards. If you refer to the texts of various Russian standards, you can find in them the use of the term "expense" in cases where the reduction of economic benefits and equity of the organization does not occur. For example, PBU 5/01 specifies that the actual cost of acquired inventories includes "the costs of procuring and delivering inventories to their place of use, including insurance costs." But the value, which in PBU 5/01 is called "insurance costs" increases the value of the asset and does not in any way lead to a reduction in economic benefits and capital. In another standard - PBU 14/2007 states that "the costs of acquiring an intangible asset are ..." (hereinafter, the list of "costs" included in the original value of intangible assets"). Thus, in this standard, too, the value of the increase in the value of an asset (rather than a decrease in capital) is called the term "expenses". In PBU 17/02, the discrepancy with the definition of the concept of "expenditure" already exists in the name of this standard: "Accounting for expenditures on research, development and technological work". According to this provision, expenses for research, development and technological work, if there are certain conditions (listed in the provision), are recognized in the balance sheet assets. Thus, the asset itself, formed in the process of research, development and technological works, is called an asset "costs". The list of situations where the term "expenses" is used in a meaning that contradicts its definition in PBU 10/99 and the Conceptual Basis for IFRS financial statements can be continued.

We believe that the reason for the discrepancy between the definition of the concept of "expenses" in PBU 10/99 and the meaning in which this term is applied in various accounting standards is largely the fact that this concept is used in a significant number of cases in the meaning of the concept of "costs". It is expenses, and not expenses, that form the cost of assets acquired or created by own forces. The reason for the identification of the notions "expenses" and "costs" in a number of normative documents on accounting is the inconsistency in the RAS system of questions of definition and interpretation of the economic content of the concept "reserves".

It should be noted that to date, in the domestic accounting science there is no single approach to the definition of such an important concept as "costs for ordinary activities". Surveys carried out by the authors of the chief accountants of the leading enterprises of the manufacturing industries of the Nizhny Novgorod industrial cluster made it possible to conclude that in most accounting practices, the terminological design in question is not essentially objectified in the system of local acts and, consequently, in the conceptual space of enterprises. This situation is explained by the lack of definition of the concept of costs in the legal framework of accounting. There is a paradox: the concept, in essence, which does not have normative legal and generally accepted scientific status, is actively used as a key in many sectoral instructions and recommendations on cost accounting and costing of products (works, services, goods).

A common approach to the terminological identification of expenses is their semantic identification with the concept clearly defined in the normative-legal space "costs" [1]. Such a "simple" solution is held by L.A. Malenkova and V.V. Prudnikov, defining the concept of costs, as "the totality of the organization's expenses for the production of products and its realization." The definition of costs, which are semantically identical to expenditures, is given by G.S. Klychova and R.R. Khairullin, who argue that "costs are the costs of enterprises, entrepreneurs, private producers for the production, circulation, sale of products that are expressed in cash."

It is obvious that the costs are immanently oriented to the generation of value added, in other words their implementation is subordinated to a specific goal. N. P. Kondrakov, M.A. Ivanova defines the concept of "expenses" as "the cost of resources used for specific purposes." In the context of this work, the connotation of the goal with economic activity is obvious. It should also be noted that the authors keep the approach of semantic identification of expenses as the equivalent of the cost of resources.

The above scientific judgments can be grouped according to the semantic criteria into three types: determined the essence of expenses on the basis of identification with costs, through resources and "repelled" from their purpose (Table 2).

It is obvious that the targeted approach is focused, first of all, on the implementation of key needs of management accounting of the business entity. The approach aimed at the disclosure of the concept of expenses as an equivalent of the cost of the enterprise resources is designed to reflect information about them in the information registers of accounting and the implementation of the control function. It should be noted that this definition includes the mandatory properties of costs - systematic and

Table 2. Approaches to the semantic identification of the concept of "expenses" [4, 5]

Content of the approach (keywords)	The authors who apply it
1. Identification with costs	L.A. Malenkova, V.V. Prudnikov, E.V. Bekhtereva G.S. Klychova, R.R. Khayrullin G.I. Prosvetov, G.Yu. Kasyanova, A.N. Azrielyan A.S. Bakayev
2. Cost of resources	A.M. Gershun, A.V. Molvinsky, L. S. Vasilyeva authors of the project of the Federal Standard of Accounting "Stocks" V.K. Sklyarenko, M.A. Vakhrushin, O.E. Nikolaeva, T.E. Shishkova, A.S. Tsarenko, N.P. Kondrakov, M.A. Ivanova
3. Business Objective	N.P. Kondrakov, M.A. Ivanova, Y.V. Sokolova, T.V. Kozyreva, I.E. Mizikovskiy, V.E. Kerimov, T.A. Korneeva, T.A. Karpova, S.F. Sharafutina, M.M. Gazaliev, V.A. Osipov, V.B. Ivashkevich

measurable. At the same time, there is a need to supplement the definition of the concept of costs from the position of accounting financial accounting, first of all, proceeding from the process of generating reports.

3 Analysis of the Results of the Study

Comparing various scientific points of view on the application of the concept of "expenses" and interpretation of its economic essence, we believe it expedient to clearly distinguish between the concepts "expenses" and "costs". The expense, as was shown when considering the definition of this concept in PBU 10/99 and the Conceptual Basis of IFRS financial statements, represents a decrease in the economic benefits and the size of the organization's equity. In contrast, costs positioned as actual or expected consumption of resources motivated by the receipt of planned financial results may lead to an increase in the valuation of assets, which, as follows from the definition of assets in domestic and international standards, are not expected to decrease but increase economic benefits. Let us compare two situations [3]. In the first situation, the materials were consumed in production, in the second, they were written off as a result of their spoilage and unfitness to use. In the first situation, the cost of production increased by the amount of consumed materials, which led to an increase in either the cost of the work in progress or the value of the finished product. In the second situation, there was a decrease in assets, which led to a reduction in economic benefits and capital, which led to the need to recognize the expense. We believe that it is the term "expenses" that should be used when it comes to the formation of the cost of assets acquired or produced by own forces. In the first case, we are talking about the amount of costs included in the cost of purchased materials, fixed assets, intangible assets, financial investments and other assets. In the second case, the expenses of production is determined by the cost of finished goods, work in progress, self-created real estate and other assets. The author's characteristic of the content and correlation of the concepts "expenses" and "costs" is proposed by T. Yu. Druzhilovskaya in the work "Characteristics of" expenses "and" costs "for accounting financial accounting.

The most important issue of cost accounting is the methodology of their inclusion in the cost of assets. In this case, two situations should be divided: the formation of the cost of assets purchased by the organization for a fee and the formation of the cost of assets created by the organizations themselves. In the first case, both national Russian accounting standards and international financial reporting standards contain clear lists of those costs that should be included or not included in the cost of the relevant asset (although these lists in the RAS and IFRS themselves differ slightly for those or other assets). In the second case, the approaches of Russian and international standards to the regulation of methods of accounting for costs in the cost of assets created by own forces differ significantly.

Another example of the difference between RAS and IFRS is the regulation of expenses accounting contained in these standards in the cost of finished goods and work in process. In addition, a significant difference between Russian PBU 5/01 and IAS 2 is the requirements for disclosure in reporting of information on the methodology for recording expenses in the cost of finished goods and work in process.

We believe that the useful information provided in IAS 2 and absent in PBU 5/01 is a characteristic of the accounting policies adopted for the valuation of inventories that include the formula used to calculate the cost price. The lack of this information does not provide an opportunity to assess the methodology for recording expenses in the cost of finished products and work in progress and used in the practical activities of organizations.

Speaking of practice, it should be recognized that expenses are the most important object of accounting for any organization. We have studied the practice of expenses accounting and approaches to the formation of the cost of finished products and work in progress in the Russian organizations of various industries. Table 3 shows excerpts from the accounting policy and explanatory notes to the financial statements of a number of the largest Russian organizations that characterize the approaches to expenses accounting used by these organizations.

As can be seen from Table 3, at the present time there is no uniformity of methods applied about cost accounting. For example, some organizations prefer to estimate work in progress on direct normative (planned) cost items, others - on the shop cost price on the basis of the standard method of accounting using a non-manufacturing option for recording production costs, a number of organizations estimate the work in progress at the actual production cost. Ready-made products are considered by some organizations at actual costs, others - at the planned full production cost. Some organizations use the normative method of expenses accounting, a number of organizations use the shown method.

At the same time, according to the data of Table 3, many large Russian organizations prefer to take into account general economic expenses as expenses of the reporting period, not including them in the cost of production. The distribution of general production costs, most organizations produce in proportion to the basic wages of production workers or the wages fund of specialists directly engaged in the performance of work.

Table 3. The Russian practice of expenses accounting [2, 6]

Organizations and their activities	Applied methods for expenses accounting	Percentage composition of expenses
PJSC "KAMAZ". Manufacture of trucks	With the ordering method of accounting for production expenses, an estimate of work in progress is carried out at a limited actual cost price. The evaluation of work in progress is carried out according to direct normative (planned) expenses items. Finished products are accounted for by actual expenses associated with its manufacture (at a limited actual production cost without taking into account general economic expenses). Analytical accounting of finished products in the warehouse is conducted in the context of the nomenclature at discount prices. As a record price, a limited normative cost is applied (without taking into account general economic expenses)	Material costs - 65.82% Labor costs are 11.91% Deductions for social needs - 3,32% Depreciation of fixed assets - 1.38% Utilization charge-8.26% Other costs - 9.31%
JSC AVTOVAZ. Manufacture of cars	Incomplete production is estimated in the volume of the workshop cost based on the standard method of accounting with the use of a semi-finished option for recording production costs without taking into account general economic expenses. General production expenses are distributed in proportion to the actual amounts of direct labor costs. Finished goods are valued at full actual cost, excluding sales and general operating expenses	Material costs - 69.77% Labor costs - 9.46% Depreciation - 5.69% Utilization fee - 3.83% Deductions for social needs - 2.73% Other expenses - 8.52%
OJSC "GAZ". Manufacture of spare parts and automotive components	In the main production, the method of accounting for production expenses and calculating the cost of production at actual costs is applied. Accounting for the costs of auxiliary and servicing facilities is made according to the actual expenses in the context of orders and objects of servicing industries and farms. The expenses collected during the month on account 25 are to be distributed to individual types of products in proportion to the basic wages of production workers or	Material costs - 61.2% Labor costs are 16.9% Depreciation - 3% Deductions for social needs - 5,3% Other costs - 13.6%

(*continued*)

Table 3. (*continued*)

Organizations and their activities	Applied methods for expenses accounting	Percentage composition of expenses
	to the salary fund of specialists directly engaged in the performance of work. Finished goods are reflected in the accounting at the actual production cost. General economic expenses are fully written off into the cost of sales (works, services), (except for situations when they are directly related to the creation or acquisition of an asset)	
OJSC U-UAZ. Production of helicopters and other aircrafts	Calculation of the cost of a unit of finished products, the cost of work performed, services rendered is carried out by the order method At the end of the reporting period (month), the total amount of expenses is deducted from the account 25 in proportion to wages for orders of the main production and for orders of auxiliary production	Material costs - 58.07% Labor costs are 8.11% Deductions for social needs - 5,3% Depreciation - 1,75% Other costs - 29.73%
PJSC MMC Norilsk Nickel. Non-ferrous metallurgy	Finished products are estimated at the planned full production cost (that is, without taking into account managerial and commercial expenses) Incomplete production is estimated at the actual production cost The evaluation of semi-finished products is carried out at the accounting value	Materials, raw materials and semi-finished products - 16.90%. The cost of acquiring energy resources is 6.54%. Labor costs are 16.9%. Deductions for social needs - 5,3%. Expenses for payment of services of industrial character of the foreign organizations - 21,36%. Taxes - 4.92%. Depreciation - 10,60% Export customs duties - 2.9%. Other costs - 10.38%
PJSC GAZPROM. Oil production, production of petroleum products and generation of electrical and heat energy	Incomplete production is estimated in the amount of actual production costs without taking into account general economic expenses Finished goods are estimated at the reduced actual production cost without	Materials - 5.85% Labor remuneration and deductions for social needs - 11.06%

(*continued*)

Table 3. (*continued*)

Organizations and their activities	Applied methods for expenses accounting	Percentage composition of expenses
	taking into account general economic expenses	Depreciation of fixed assets and intangible assets - 17.66% Other - 65.43%
PJSC "Polyus Gold". Extraction of precious metal ores and sands	The account 26 at the end of the reporting period is written off to 90 accounts	Material costs - 0% Labor costs - 29.75% Deductions for social needs - 6.06% Amortization - 0,03% Other costs - 64.16%
PJSC RusHydro. Power generation by hydroelectric power plants	Inventories are taken into account in the amount of actual costs of acquisition (production)	Material costs - 29.79% Labor costs - 12.85% Deductions for social needs - 2.96% Depreciation - 22.29% Other costs - 32.11%
OJSC Rosneft Oil Company. Industry	Information is not disclosed	Material costs - 82% Labor costs 0.4% Deductions for social needs - 0.08% Depreciation - 3.1% Expenses related to the exploration and evaluation of oil and gas reserves - 0.46% Other costs - 13.87%
PJSC "Vimpel-Communications". Provision of communication services	Management expenses are charged to the expenses of the reporting period to account 90	Material costs - 51.75% Labor costs - 8,333% Deductions for social needs - 6.06% Depreciation - 13.16% Other costs - 24.12%
The group of companies "MCC" EuroChem ". Production and distribution of mineral fertilizers and extraction of minerals	Information not disclosed	Raw materials and materials - 35.12% Other materials - 10.39% Staff costs, including contributions to social funds - 13.54% Depreciation of fixed assets and intangible assets - 9.87% Other costs - 31.08%

(*continued*)

Table 3. (*continued*)

Organizations and their activities	Applied methods for expenses accounting	Percentage composition of expenses
PJSC Severstal. Extraction and production of metal products	Information not disclosed	Information not disclosed

It should be noted that some organizations do not disclose information on the methods used to account for production costs, indirect costs distribution bases, methods for estimating work in progress and finished products. The reason for this is the absence in the system of Russian PBU requirements for the disclosure of this information. The percentage composition of expenditures among various Russian organizations is also significantly different. The data in Table 3 allow us to conclude that for industrial organizations the greatest relative weight is possessed by material costs (more than 50% for the organizations studied). The least specific weight belongs to such elements of expenses of the specified organizations as depreciation and deductions for social needs. Another is the percentage composition of costs for extracting organizations: The labor costs and depreciation have a fairly large share. We also draw attention to the fact that some organizations do not disclose information on the percentage composition of costs in the financial statements prepared in accordance with IFRS.

4 Conclusions

The conducted researches allow drawing the following conclusions concerning the current state of accounting of cost and expenses. Both in the system of national Russian accounting standards, and in the IFRS system, there are provisions containing the theoretical bases of cost accounting: the definition of this concept is formulated, the classification of expenses is presented, and the criteria for recognizing expenses in accounting and reporting are proposed. At the same time, the formulated theoretical propositions are not applied consistently in the totality of both Russian and international standards, and the term "expenses" in many cases is used in a meaning that contradicts the definition of this concept. In a significant number of situations, there is an identification of the concept of "costs" with the concept of "expenses". At the same time, the theoretical bases of accounting for costs are not set out both in the RAS system and in the IFRS system. Not fully described in the systems of RAS and IFRS and the methodological basis of accounting for costs. The inadequacy of the theoretical and methodological bases for recording costs and expenses in accounting standard systems causes different interpretations in the interpretation of these concepts in the scientific literature, as well as the existence of problematic aspects of accounting for costs and expenses in the practical activities of organizations.

We believe that a number of the following measures are needed to solve the problematic aspects of cost and expenses accounting. In order to sequencing approaches to the accounting of costs in the system of Russian accounting standards, a clear and unambiguous formulation of the definition of the concept of "expenses" and "costs" in all regulatory documents on accounting in accordance with their definitions should be given. It is necessary to introduce in the system of accounting standards the regulation of general approaches to the methodology of cost accounting in the cost of finished goods and work in progress. It is advisable to include in the accounting standards regulating the accounting of inventories, the requirement on the need to disclose the accounting methodology of expenses in the cost of finished products and work in process. We also believe that the system of IFRS, which in many ways is the basis for reforming Russian accounting, requires further improvement. At the same time, the fact that recently international standards have been developing quite intensively allows us to hope that the problematic issues of accounting for costs and expenses in this system will be solved in the future.

References

1. Mizikovsky, I.E.: Accounting Management Accounting, 2nd edition of the revision and add-M. Magister, SIC INFRA-M, 144 p. (2016)
2. Mizikovsky, I.E., Maslova, T.S., Druzhilovsky, T.Y., Druzhilovsky E.S., Bazhenov, A.A.: Formation of the accounting and control space of organizations of the state (municipal) sector of the economy: theoretical and methodological aspect. Monograph/Nizhny Novgorod, 236 p. (2016)
3. Druzhilovskaya, T.Y., Druzhilovskaya, E.S.: Problems and prospects for the development of accounting at the present stage. In the book: The International Scientific and Practical Congress of Economists and Lawyers "Science engineering and economic paradigm of modern society", pp. 154–158. ISAE, Consilium (2014)
4. Druzhilovskaya, T.Y.: Characteristics of "expenses" and "costs" for financial accounting. Int. Acc. 2(344), 31–45 (2015)
5. Druzhilovskaya, E.S.: Methodology for estimating inventory in accounting. Int. Acc. 7, 16–25 (2012)
6. Kuznetsov, V.P., Garina, E.P., Semakhin, E.A., Garin, A.P., Klychova, G.S.: Special aspects of modern production systems organization. Int. Bus. Manag. 10(21), 5125–5129 (2016)

Man as the Subject of Possible/Impossible in the Russian Nominations of the Feature of the Subject

Natalia E. Petrova[✉], Natalia M. Ilchenko, Olga A. Patsyukova,
Galina S. Samoylova, and Anastasia N. Moreva

Nizhny Novgorod State Pedagogical University Named After Kozma Minin
(Minin University), Nizhny Novgorod, Russian Federation
petrova_ngpu@mail.ru, ilchenko2005@mail.ru,
olalpa@mail.ru, galasam2010@yandex.ru,
linguanastya@yandex.ru

Abstract. In the article, an anthropocentric approach to the analysis of factual material is realized, allowing seeing behind the units of language the work of human consciousness, view of the world, assessments and values. The object of the study is verbal adjectives formed from verbs with the help of the suffix -em-/-im-. The choice of material is because the modal component is consistently present in the semantics of these adjectives - the meaning of the possibility or impossibility of action, the subject of which is a person. On the basis of the nature of the motivating action, the semantic groups of these adjectives are distinguished, representing the spheres of interaction of a person with the world that is important for the Russian consciousness: Cognitive and speech-intellectual activity, sensory perception of the object, the physical and mental state associated with the object, ethical relations and the evaluation of the object, the forms of behavior associated with the object, the relationship of possession, the creative and goal-oriented activity of man in relation to the objects of the external world and himself. The authors consider varieties of modal semantics, characteristic of the adjectives analyzed. It is noted that the possibility/impossibility of the key subject for an action can be conditioned, firstly, by the internal potential of a person - his intellectual, physical, psychological qualities and abilities; Secondly, the prescription that is allowed and what is prohibited; Thirdly, the axiological factor-the estimated relation to the determined subject in terms of "good/bad", "useful, appropriate/not useful, inexpedient". In view of this factor, the adjectives of various semantic groups are related to the categories of norm and evaluation. In summary, a conclusion is made about pronounced anthropocentricity of the deverbativs with suffixes -em-/-im-, indicating that the possibility or impossibility of human interaction with the objective world is an important means of understanding objects and giving them a feature.

Keywords: Adjective · Modality · Possible · Action · Anthropocentrism
Norm · Subject · Nomination · Attribute

© Springer International Publishing AG, part of Springer Nature 2018
E. G. Popkova (Ed.): HOSMC 2017, AISC 622, pp. 163–169, 2018.
https://doi.org/10.1007/978-3-319-75383-6_21

1   Introduction

In the process of scientific knowledge of the world, the human factor is manifested in the fact that the study of any object is carried out not only from the point of view of man and in the interests of man, but also in order to comprehend the nature and essence of man himself [7, p. 19]. The object of linguistics is the natural human language, "in the construction and application" of which reflects not the objective world surrounding the person, but the subjective image of this world, created "in our soul" [2, p. 80]. Thus, not only the established paradigm of humanitarian knowledge, but also the very nature of language, makes the anthropocentric approach to this study urgent [9, p. 90], which prompts the researcher to switch attention from the language system to carrier. An attempt to implement such an approach is presented in this article on the basis of the   Russian verbal adjectives with the suffixes -em-/-im-: *Explainable and unrepresentable.*

2   Theoretical and Methodological Basis of the Research

In our understanding of the anthropocentric approach to language, we rely on the work of domestic and foreign scientists Potebni, Yu, Apresyan, Arutyunova [1], Karaulova [7], Cubreacova, Shmeleva, Cherneyko, Humboldt [2], Verzbicka, Muracosy et al. The analysis of the actual material takes into account the research of adjectives with suffixes -em/-im-, presented in the works of Bulakhovsky, Ivanova, Zaneginoy [5], Ivanchi [6], Krysko, Kunavina, Kim et al. In the interpretation of the concepts of "evaluation", "modality," "norm" expressed by the means of language, we are guided by the work of Arutyunova [1], Wolf, Ivina, Rudnev [11], Beller, Thomasson, and von Wright [12].

   Deverbatives with suffixes -em-/-im- denote the feature of the object through the possibility or impossibility of a particular action that refers to or is directed from the outside to the object being determined. The subjective characteristic of the action reflects a different way of understanding the trait by a person and is important for the conceptual content and use of these adjectives. This can be shown on the example of adjectives, *fathomless* and *ineradicable.* In the first case, the sign "preserving the freshness that does not lose its importance with time, [10, p. 487] is expressed through an appeal to the action to *wither*, the subject of which is the object itself: *Unfading glory is* glory that can *not fade.* In the second case, a relatively close sign, "one that is difficult to exterminate, destroy" [10, p. 449], i.e. constantly, for a long time persisting, is expressed through an appeal to an action aimed at an object from outside: *Indestructible Smell is* a smell that cannot be *exterminated.* Thus, through the adjective *unfading* attribute, "permanent, persistent with time," is interpreted as an intrinsic property of an object irrespective of the surrounding world and man, whereas through an adjective an *ineradicable* similar feature is interpreted because of resistance to external influences. As a result, the adjective *unfading* carries information only about the object being determined, and the adjective is *indestructible* - about the subject and the subject of the impact on this object.

   The way in which the sign was expressed also caused a different connection between these adjectives and axiological semantics. The sign of the *unfading* is inseparably linked with the idea of what constitutes a value to human *Love, beauty, fame, talent, youth*. The *indestructible* attribute can be connected both with what is valuable (an *ineradicable desire for freedom, an ineradicable interest in the book*), and with antivalues (an *ineradicable desire for violence, an indestructible filth*), since a person is inclined to "exterminate" everywhere, including in to itself, something bad, inexpedient.

All the above leads to the conclusion that adjectives with suffixes -em-/-im- are characterized by a particularly pronounced anthropocentricity in the case when they are formed from verbs denoting the action of a person. Interpretation of their meaning appeals to the potential of human possibilities; therefore, denoting the feature of the object, they simultaneously create a kind of "portrait" of a person interacting with the surrounding world and at the same time making himself a "measure of all things" they know "[3, p. 33].

Proceeding from this, for the present research these adjectives were selected. Further, because of the nature of the motivating action, the semantic groups of the adjectives will be singled out, which will make it possible to conclude that the spheres of influence on the subject important for the Russian consciousness. In addition, we will trace the connection of adjectives of various semantic groups with categories of norm and evaluation. For illustration, examples from the database of the National Corpus of the Russian language (main and newspaper corps) will be used.

3   Analysis of the Results of the Study

Most of the deverbativs on -em-/-im- (more than 190 words out of just over 270, recorded in [4, pp. 356–357]) characterize the subject from the point of view of human actions. The verbs that motivate these adjectives represent different spheres of human life and activity. It can be:

- Cognitive activity: *Learn* → *(not) knowable, explain* → *(not) understandable, compare* → *(not) comparable, calculate* → *(not) computable, disprove* → *irrefutable*, etc.;
- Speech activity: *Describe* → *indescribable, pronounce* → *unpronounceable, translate* → *(not) translate, transmit* → *non-transferable*, etc.;
- sensory perception of the object: *See* → *(not) visible, hear* → *(not) audible, touch* → *(not) tangible, feel* → *(not) palpable, view* → *(not) foreseeable, recognize* → *(not) recognizable*, etc.;
- The physical or mental state associated with the object: *Tolerate* → *(not) tolerant, transfer* → *(not) tolerable, endure* → *(not) tolerable*, etc.;
- Psychological impact on the object: *Frighten* → *intrepid, shake* → *unshakable, beg* → *unforgiving, tame* → *indomitable*, etc.;
- ethical relations, object evaluation: *Punish* → *(not) punishable, allow* → *(not) permissible, reward* → *unrewardable, value* → *invaluable, atoneate* → *unrequited*, etc.;

- activities related to the modification, including destruction, of the object: *Decompose* → *(not) decomposable, fix* → *(not) correctable, change* → *irreplaceable, restore* → *(not) recoverable, heal* → *(not) curable, remove* → *(not) removable*, etc.;
- goal-setting in relation to an object: *Implement* → *(not) feasible, reach* → *(not) attainable, overcome* → *(not) surmountable, win* → *invincible*, etc.;
- The behavior associated with the object: *Hide* (something) → *unconcealed, imitate* (someone) → *inimitable, bicker* (with someone) → *indisputable*, etc.;
- Ownership relations and contractual relations: *Take away* → *inalienable, alienate* → *inalienable, terminate* → *(not) dissolvable*, etc.

In all these groups, the structure of adjectives simulates a situation in which a person can or cannot perform a certain action with respect to an object, which becomes a sign of the feature of this subject. In this case, the modality contained in the semantics of such adjectives has a different nature.

First, the possibility/impossibility of the key to the feature of the action can be related to the internal potential of man - his intellectual, physical, psychological qualities and abilities. This is typical of the adjectives of most of the groups listed above, such as, for example, *(not) explainable, indescribable, (not) countable, (not) visible, (not) tangible, unbearable, (not) carried, inexorable, (not) (Not) attainable, indestructible, irresistible, incurable, irreparable* and similar: *The fire goes **inexplicable*** [no one can explain - aut.] *Way On Great Saturday in the Church of the Holy Sepulcher in Jerusalem* (Komsomolskaya Pravda, 2014.04.09); *As a result, already in the course of the concert, the pain became simply **unbearable*** [the artist and anyone else in his place can not stand the *test*] and after the fourth song the show still had to be interrupted (Komsomolskaya Pravda, 2013.03.13).

Secondly, the possibility/impossibility of the key to the feature of the action can be associated with a specific prescription: *Inherent, (not) alienable, indissoluble, (not) punishable, and (not) permissible* and under. Such adjectives express the modality of norms [13], which "prescribe what should, what is allowed, what is forbidden" [10, p. 175]. According to Beller, deontic statements sometimes signal that people can go beyond the conventions, referring to the alternative or additional conditions necessary for the settlement of the deontic situation [12, p. 308]: *A small amount of wine is **allowed** only at an official reception, for which the bar is needed* (Izvestia, 2014.07.09).

As von Wright rightly points out, the very fact of allowing or banning anything in the society is teleologically connected with the evaluation [14, p. 353], therefore the adjectives of this group often express a concomitant value: *The fact that we, men, sometimes allow ourselves to women - this is **punishable*** [badly, not allowed, must be punished - author] (Komsomolskaya Pravda, 2013.10.14); *Nechayev was one of those who proclaimed this "right" **inalienable*** [valuable, good, can not be taken away - auth.] The *right of a revolutionary* (Culture, 2002.04.08).

Finally, the possibility or impossibility of an action can be caused by an axiological factor-an estimated relation to a particular subject in terms of "good/bad", "useful, appropriate/not useful, inexpedient". This is typical of the semantics of such adjectives as *(not) acceptable, invaluable, (not) tolerant, (not) apologetic, (not) permissible* and under. There is a certain pattern: In the adjective with the prefix, *not* the inability to perform an action by virtue of certain axiological settings forms an unambiguous value, for example: *The first time was a little sad, but then I realized that this is a good option. I can get **invaluable*** [good, useful - aut.] *Coaching experience* (Izvestia, 2014.05.15); *this state of things is **intolerable*** [bad, harmful - auth.], *But it can only be corrected by differently directed systemic methods* (Komsomolskaya Pravda, 2013.09.12).

If the possibility of action is not denied, then the estimate only tends to one pole or another, so adjectives are often *not* used with lexical and grammatical indicators of the relative sign: *His voice - **in the best cases, tolerable** (and at worst it looks like the sound of a tearing foam plastic)* (Komsomolskaya Pravda, 2013.05.01); *This experi̇ence does not yet have the Russian Academy of Sciences, but this is a **more acceptable** path, which, along with the state, requires* (Labor-7. 2007.04.17).

Our language material allows, through the prism of a possible/impossible for a person, to examine in more detail the relationship between the nomination of a feature and the concept of a norm. Above we have identified a small group of adjectives expressing the significance of deontic modalities. However, a much larger number of adjectives considered relates to the "generic concept" of the norm, which unites "all kinds and forms of order" [1, p. 6]. Defining the content of this broad notion of the norm, Arutyunova enumerates particular groups of concepts whose invariant sign is one or another manifestation of the norm. Among these features, for us, the one on the basis of which the concepts "model", "sample", "stereotype", - standardization are united is important. "To conform to the norm and to observe order means to be" like everyone else "and" as always "<...>". The normative field borders on the concepts of ordinary, ordinary, predictable, habitual... "[1, p. 7]". It is this interpretation of the norm that is reflected in the meaning of those nominations of the trait, which are based on an idea of a person's capabilities.

Most of them are paired formations, differing only in the presence or absence of a prefix of *not - Imaginable - unimaginable, replaceable - irreplaceable, comparable - incomparable* and under. Comparison of the values of paired adjectives allows us to reveal a certain regularity. Adjectives without a prefix do *not*, as a rule, designate a feature that is rated by the speaker as normative in the sense that it corresponds to generally accepted concepts and the routine of things. The adjective with *not*, on the contrary, interprets the feature of the subject as anomalous, unusual from the point of view of the human perception of the world, because it goes beyond the bounds of human possibilities. This different attitude of adjectives to the norm is often empha-sized by the context:

*Thus, experts note, the desire to invest in durable goods is quite **natural** and **understandable**...* (Ogonyok, 2014); *and yet this **mystery** turned out to be **solvable*** (Knowledge is Power 2014); *this is nothing **complicated No**, and if you set yourself this goal of life, then it is quite **achievable*** [Siberian lights. 2013].

*This amazing feeling that arose in the night battle, where you can not distinguish in three steps who it is - a friend or an enemy ready to kill you, was associated with a second, no less **surprising and inexplicable** feeling of the general course of the battle* (V. Grossman. Life and destiny. Part 1. 1960); *All of us in one way or another agree that today in the world there are accumulating reasons for the crisis of the structural one, i.e. **Insoluble** within the **standard** political and investment decisions of our days*

(Expert, 2015); *He likes to start the film with a landscape of dizzying beauty and peace, and it will be like an ideal **unattainable** for **vain** life* (Izvestia. 2002.04.26).

In the above examples, we have identified those means of the context that form a semantic opposition to the adjectives (the *riddle of* vs. *is still solvable, the unsolvable* vs. *standard, unreachable* vs. *vain*) or, on the contrary, approaching them (naturally - *understandable, difficult not - achievable, surprising - inexplicable*). In any case, the gravitation of adjectives is emphasized without *not* any notion of norms: "Order", "pattern", "ordinary", "natural", and non-adjectives - to the concepts of the antinorm: "Paradox", "exception", "uniqueness". In quantitative terms, *not* adjectives clearly prevail: The adjective without a prefix does *not* always have a paired variant with a prefix, whereas the *not-* adjective often functions outside the pair. This pattern is explained by the desire of a person to fix and designate by language means primarily anomalous [1, p. 4], and in our case - beyond the possible.

The connection with the concept of the antinorm has led to a wide use of the adjective type *unattainable* in the role of intensities, i.e. Indicators of the highest degree of manifestation of another feature: *Inexpressible joy, unimaginable self-confidence, unthinkable happiness, indescribable villainy, incorrigible slander, irresistible aversion*, etc. see: [5, 8, p. 328]. In itself, this function is inextricably linked with the subjective interpretation of the world, so that all "adjective-intensifiers are indicators of the stressed anthropo-orientation of the utterance" [6, p. 31]. In the case of adjectives with suffixes -em-/-im- this is especially obvious, because measure of the intensity of the trait is the limits of what is possible for man.

4   Conclusion

Adjectives with suffixes -em-/-imitate such a fragment of the linguistic picture of the world in which the objects-carriers of the feature appear as objects of the cognitive, speech-thinking, sensual, creative, goal-setting activity of human. At the same time, the possibility or impossibility to perform certain actions with respect to the subject turns out to be an important means of understanding this subject and giving it a feature for the Russian national consciousness.

References

1. Arutyunova, N.D.: Anomalies and language (to the problem of the language "picture of the world"). Questions Linguist. **3**, 3–19 (1987)
2. Humboldt, W.: On the difference in the structure of human languages and its influence on the spiritual development of humanity. Selected Works on Linguistics, 400 p. (1984)
3. Gurevich, V.V.: About the "subjective" component of language semantics. Questions Linguist. **1** (1998)
4. Zaliznyak, A.A.: Grammatical dictionary of the Russian language: Change of words. OK. 100 000 words, 2nd edn. The stereotype. M.: Russian language, 880 p. (1980)
5. Zanegina, N.N.: Adjectives with the intensity value, the measure of the manifestation of the sign. http://lexrus.ru/default.aspx?s=0&p=2917. Circulation date is 12 May 2017

6. Ivancha, A.V.: Anthropo-oriented adjectives with the meaning of intensity in the Russian language. Izv. Saratov Univ. New Ser. Philology J. **12**(1), 30–33 (2012)
7. Karaulov Yu, N.: Russian language and language, 7th edn., 264 p. Publishing house LCI (2010)
8. Petrova, N.E.: On the factors of expressiveness of verbal adjectives with suffixes -em-/-im- rational and emotional in language and speech: means of artistic imagery and their stylistic use in the text: Interuniversity. Sat. Scientific works, dedicated to the 85th anniversary of Professor A.N. Kozhena: MGOU, pp. 326–331 (2004)
9. Radbil, T.B.: Basics of studying the language mentality: Training tutorial. Flint, 328 p. (2010)
10. Dictionary of the Russian language: In the 4th volume/USSR Academy of Sciences, Institute of Russian language. Evgenieva, A.P. (ed.) 3rd edn. Stereotype. T. II. M.: Russian language (1986)
11. Rudnev, V.P.: Dictionary of culture of the XX century. M.: Agraf, 384 p. (1998)
12. Beller, S.: Deontic norms, deontic reasoning, and deontic conditionals. Think. Reason. **14**(4), 305–341 (2008)
13. Thomasson, A.L.: Norms and necessity. South. J. Philos. **51**(2), 143–160 (2013)
14. von Wright, G.H.: Valuations – or how to say the unsayable. Ratio Juris. **13**(4), 347–357 (2000)

Determining the Value of Own Investment Capital of Industrial Enterprises

Yaroslav S. Potashnik[✉], Ekaterina P. Garina,
Elena V. Romanovskaya, Alexander P. Garin,
and Sergey D. Tsymbalov

Nizhny Novgorod State Pedagogical University Named After Kozma Minin
(Minin University), Nizhny Novgorod, Russian Federation
yaroslav.sandy@mail.ru, e.p.garina@mail.ru,
alenarom@list.ru, rp_nn@mail.ru, keo.vgipu@mail.ru

Abstract. The article is devoted to the determination of the value of own investment capital of industrial enterprises while preparing the economic justification for investment projects of industrial enterprises. The cost of own investment capital is one of the main factors that form the requirements for the profitability of projects. The accuracy of determining the value of own investment capital significantly influences the reliability of the conclusions about the economic attractiveness of projects. The article presents in a formalized form the conditions for the acceptability of investments, taking into account the value of investment capital (individual elements) in the context of the organizational and legal forms used. The approaches to determining the value of own investment capital are specified. Methods for adjusting the cost of own investment capital of the alternative are suggested, if the investment project under consideration differs in terms of the level of financial risk. The first method can be used by public industrial enterprises. It is based on the capital asset pricing model (Capital Asset Pricing Model) and Hamada's equation. The second method assumes that the industrial enterprise has the possibility of applying for credit resources to the relevant institutions. It is based on assessments of the level of financial risk associated with the enterprise with an investment decision. The article describes the main stages of implementation of methods. Examples are given that reflect the necessary calculations.

Keywords: Investment · Equity · Financial risk · Cost of capital
Industrial enterprise

1 Introduction

Determining the value of the invested capital is one of the most important tasks that need to be addressed when preparing an economic justification for investment projects of industrial enterprises. The cost of investment forms the requirements for the profitability of projects. If at the planning stage the value of the invested capital is understated, then a situation is possible, in which the implementation of an attractive investment project according to preliminary calculations will reduce the cost of the enterprise. If the cost of capital is overstated, then an attractive project may not be

© Springer International Publishing AG, part of Springer Nature 2018
E. G. Popkova (Ed.): HOSMC 2017, AISC 622, pp. 170–178, 2018.
https://doi.org/10.1007/978-3-319-75383-6_22

accepted for implementation, advantageous opportunities will be missed. Various studies of the value of investment capital are devoted to the works of many scientists, including L. Abalkin, D. Lvov, P. Vilensky, A. Demodaran, U. Sharp et al. However, a number of issues, in our opinion, need further elaboration. In particular, methods for estimating the cost of equity capital of Russian industrial enterprises need to be improved taking into account the financial risk of the investment project. In this regard, the purpose of this work is to clarify and develop approaches to assess the value of the capital of an industrial enterprise, allowing taking into account the level of financial risk of an investment project.

2 Theoretical Basis of the Research

Investments are money, securities, other property, including property rights, other rights having monetary value, invested in objects of entrepreneurial and (or) other activity for profit and achieve a different useful effect [11, p. 2]. An investment project is an organizational form of implementing investment by enterprises. An investment project is a set of actions (works, services, acquisitions, management operations and decisions) aimed at achieving the stated goal and requiring investments for implementation [7, p. 10].

The formulation of the conclusion on the economic feasibility of the implementation by an industrial enterprise of a particular commercial investment project (hereinafter referred to as the project) is based on an appropriate economic justification, during which the volumes and terms of the proposed investments are determined, as well as the possible economic effect. If the possible economic effect corresponds to the targeted economic effect (is equal to or exceeds it), the project implementation can be considered economically viable, if not, inappropriate.

To date, a wide range of indicators have been developed, characterizing from one side or another the economic effect of investment (for example, profit, net cash flow, profitability index of costs, etc.). However, many researchers (for example, Brayley R., Myers S., Van Horne D., Wachovich J., etc.) agree that the main indicator is the profitability index (return) of investments, reflecting the share of the net cash flow generated by the investment project in the total volume of project investments (investment capital). Van Horn D. and Vakhovich G. point out that the project can be considered acceptable if the return on investment provided by it is greater than or equal to the cost of investment capital, which is understood as the minimum profitability expected by the owners (suppliers) of capital [3, p. 661]. This condition can be displayed in the form of the following expression:

$$Rp \geq \left[\sum_{i=1}^{n} WiRi \right] \quad (1)$$

Where Rp is the project investment profitability; i - number of the source of investment capital; n - number of sources of investment capital; Wi - share of financing from source i in total investment capital; Ri - the value of the share of investment capital formed from funds from the source i (the minimum profitability expected by the capital providers from the source i).

The main sources of investment capital of industrial enterprises are their own capital, including net cash flow generated by projects, borrowed funds for the use of which the payment of interest (obtained as a result of borrowing, issuance of bonds and promissory notes of enterprises, implementation of other loans, Leasing, etc.), as well as borrowed funds, for the use of which usually does not provide for the payment of interest (received in the form of subsidies, as well as accounts payable). Given that in most cases, the structure of the investment capital of the project is not homogeneous, and the fact that the value of funds for which use is not provided for the payment of "explicit" interest is zero, condition (1) can be presented in the following expanded form:

– For public industrial enterprises:

$$Rp \geq [WdRd + WpsRps + WcsRcs] \tag{2}$$

Where Wd - the share of investment capital formed at the expense of borrowed funds, for the use of which the payment of "explicit" interest is provided; Rd - the cost of investment capital formed at the expense of borrowed funds, for the use of which the payment of "explicit" interest is provided; Wps - the share of investment capital formed at the expense of funds owned by the owners of the company's preferred shares; Rps - the value of the share of investment capital formed at the expense of the owners of preferred shares of the enterprise; Wcs - the share of investment capital formed at the expense of funds owned by the owners of the company's ordinary shares; Rcs - the value of the share of investment capital formed at the expense of the owners of ordinary shares of the enterprise (hereinafter - the value of own investment capital);

– For non-public industrial enterprises:

$$Rp \geq [WdRd + WeRe] \tag{3}$$

Where We - the share of investment capital formed at the expense of the company's own funds; Re - the value of the share of investment capital formed at the expense of the company's own funds (hereinafter - the value of own investment capital).

3   Research Methodology

Most scientists and practitioners agree that the value of the company's own investment capital should be determined on the basis of the alternative cost of funds, i.e. The profitability that owners of capital can receive in case of refusal to implement this project (hereinafter - the basic project) with the subsequent investment in the best available alternative (hereinafter - alternative project). When determining the value, it is necessary to take into account the inherent financial risk, which is understood as the variability of the return on investments in relation to its expected value [3, p. 191].

If the basic and alternative projects have approximately the same level of financial risk, the value of their own investment capital for the basic project can be assumed equal to the profitability (profitability) of its own investment capital in the implementation of the alternative project, i.e.:

$$\begin{cases} FRLb = FRLa \\ Veb = \operatorname{Re} a \end{cases} \tag{4}$$

Where *FRLb* is the level of financial risk of the basic project; *FRLa* - the level of financial risk of an alternative project; *Veb* - the cost of own investment capital for a basic project; $\operatorname{Re} a$ - profitability (profitability) of own investment capital in the implementation of an alternative project.

If the basic and alternative projects differ in terms of financial risk (for example, the goal of the basic project is to diversify the business portfolio of an enterprise, and alternative - to continue the current operating activity), then to determine the value of its own investment capital in relation to the basic project, the return on its investment Capital on the alternative project to adjust by an amount reflecting the difference in the level of financial risk of projects, i.e.:

$$\begin{cases} FRLb > (<)FRLa \\ Veb = \operatorname{Re} a + (-)RA \end{cases} \tag{5}$$

Where *RA* is the discount (discount), taking into account (i) the difference in the level of financial risk of projects.

Scientists and practitioners have developed a significant number of approaches to the definition of $\operatorname{Re} a$. Briefly consider the features of those of them that are most often used in the practical activities of industrial enterprises of the Nizhny Novgorod region.

The model of discounting dividends (model the dividend discount) is applied when there is an opportunity with a high degree of reliability to determine the flow of future dividends from the project. The cost of equity is found by the iteration method from the formula:

$$Peo = \sum_{t=1}^{T} \frac{Dt}{(1 + \operatorname{Re} a)^t} \tag{6}$$

Where *Peo* - the amount of equity invested in the project at time 0; *T* is the period of project implementation; *T* - number of project implementation periods; *Dt* is the amount of dividends per unit of own investment capital, payment of which is expected at the end of the period *t*.

The method of cumulative construction (build-up method) is based on the use of the formula:

$$\operatorname{Re} a = Rf + \sum_{y=1}^{Y} IRy \tag{7}$$

Where Rf is the risk-free rate of return in the period under review; Y - number of the considered risk factor inherent in the project implementation; Y - number of considered risk factors; IRy is the risk premium due to the action of factor k.

A risk-free rate is the rate that reflects the profitability of investments not associated with risk, i.e. Such, whose profitability is known in advance and the probability of non-receipt of which is minimal [2, p. 205]. We agree with the opinion that at present it is not possible to talk about the presence of zero-risk investment within the Russian economy [4, p. 416]. However, as risk-free one can consider the rate of return on state securities (for example, OFZ) with the corresponding period of the project's implementation with a maturity.

The number of risk factors and the corresponding amount of surcharges are determined by the method of expert assessments.

To determine the value of RA (see FP-5), most experts recommend using the method of expert assessments with a preliminary calculation of indicators characterizing the volatility of project profitability (variance, standard deviation, coefficient of variation). The merits of the method of expert assessments include universality and convenience for enterprises. However, it has a number of shortcomings (subjectivity of expert assessments, the complexity of selecting participants with the required qualifications), which reduces the reliability of the results. In this regard, we propose methods, the use of which will, in our opinion, increase the reliability of determining the value of RA. The essence and examples of using these methods are presented below.

4 Analysis of Research Results

The first method can be used by public industrial enterprises. It is based on the capital-asset pricing model (CAPM), which assumes the following assumptions [3, c. 202]: Investors investing in securities are well informed; Operational costs of investors are small; Restrictions on investment can be neglected; none of the investors has sufficient funds to influence the market price of shares of enterprises. The amount of the discount (discount), taking into account difference in the level of financial risk of projects is determined from the expression:

$$RA = Rcspr - Rcsa = Rf + (\overline{Rm} - Rf)\beta pr - Rf - (\overline{Rm} - Rf)\beta a$$
$$= (\overline{Rm} - Rf)(\beta pr - \beta a) \tag{8}$$

Where $\overline{Rm}$ - the expected return on the market portfolio in the period under review; βpr - value of the coefficient "beta" for the company's ordinary shares in this period in the case of the implementation of the basic project; βa - the value of the coefficient "beta" for the company's ordinary shares in this period in the case of an alternative project.

The expected return on the market portfolio is determined by the method of expert assessments [13, p.]. At the same time, the forecast of the profitability of not only the market portfolio, but representative, as one of the indexes of the Moscow stock exchange (for example, the broad market index, one of the industry indexes, etc.) can be used in Russia.

The coefficient "beta" characterizes the riskiness of a security. It reflects the sensitivity of its profitability to changes in the yield of the stock index. The higher the value of the "beta" coefficient, the higher the level of financial risk associated with investing in this security and the higher the profitability required by investors. To calculate the beta coefficient, we can use the following formula [1, p. 130]:

$$\beta = \frac{K \cdot Gcs}{Gi} \tag{9}$$

Where K - the degree of correlation between the level of profitability of the company's ordinary shares and the average level of return of the stock index in the period under review; Gcs - standard deviation of the profitability of the company's ordinary shares in this period; Gi is the standard deviation of the yield of the stock index in a given period.

We give an example of calculations using formula (8). If in the period under consideration $Rf = 7\%$, $\overline{Rm} = 12\%$, $\beta pr = 1.6$, $\beta a = 1.3$, then $RA = (12\% - 7\%)$ $(1.6 - 1.3) = 1.5\%$. Thus, the financial risk of the basic project is higher than that of the alternative project, and the cost of the investment capital of the basic project is 1.5% points. Higher than an alternative, project.

When using formula (8), it is necessary that the values of β projects correspond to the level of financial leverage planned in the course of their implementation.

If the data for calculation βpr and (or) is βa not enough, then the values of the coefficients can be determined using the following procedure:

(1) Enterprises are selected (about 10), whose activities are similar to the envisaged project and for which common shares are known (can be calculated) in the period under consideration the values of the coefficients "beta";
(2) For each enterprise, the values of the beta coefficient and the corresponding level of financial leverage are calculated (calculated);
(3) For each enterprise on the basis of R. Hamada's formula, "non-leveraged beta" (βuan) is defined:

$$\beta uan = \frac{\beta an}{\left(1 + \frac{Kdat}{E}\right)} \tag{10}$$

Where βan - the initial value of the coefficient "beta" of the enterprise - an analog (with a financial lever); $Kdat$ - the value of the borrowed capital used by the enterprise after paying corporate income tax (it is taken into account that the interest accrued for the use of borrowed funds in the amount established by Article 269 of the Tax Code of the Russian Federation can be attributed to expenses, which reduces the tax base when calculating the profit tax of organizations); E - cost of equity of the enterprise. In the absence of data on the cost of borrowed and equity capital, we can use approximate values of the ratio between them;

(4) The arithmetic mean calculated for the enterprises of "leverless beta" is calculated, which is accepted as the "leverless beta" of the project (βup);
(5) The meaning of the "leverless" of the project is adjusted in accordance with the planned level of financial leverage according to the formula:

$$\beta p = \beta up \left(1 + \frac{Kdat}{E} \right) \tag{11}$$

We give an example corresponding to the procedure described above. The initial data and the results of the main calculations are presented in Table 1.

Table 1. Determination of the coefficient βp

Company - Analogue	Meaning of indicators		
	$Kdat/E$	βan	βuan
1	0.59	1.65	1.04
2	0.32	1.48	1.12
3	0.72	1.75	1.02
4	0.38	1.52	1.10
5	0.79	1.82	1.02
6	0.37	1.51	1.10
7	0.91	2.02	1.06
8	0.47	1.54	1.05
9	0.81	1.89	1.04
10	0.64	1.67	1.02

$\beta up = (\sum \beta uan)$: 10 = 1,06
If the design ratio $Kdat/E$ is 0.25, then =
$1.06 \times 1.25 = 1.33$

Thus, the value of the beta coefficient for the project in question can be taken as 1.33.

The second method can be used if the industrial enterprise has the possibility of applying for credit resources to banks (at least three) having considerable experience in selecting investment projects for financing. The method involves collecting information on effective annual interest rates (EAIR), under which banks are ready to provide the enterprise with a loan for the implementation of basic and alternative projects. The term and volume of loans should correspond to the parameters of the basic project. For both variants, the average effective annual interest rates (AEIR) are calculated and the difference between them is determined, which will correspond to the RA value. A methodological example of using the proposed method is presented in Table 2.

According to Table 2, the implementation of the basic project increases the level of financial risk associated with the enterprise, $RA = 6.96\%$. If the return on own

Table 2. Determination of *RA*

Bank	The value of EAIR, %		RA, %
	Base project	Alternative project	
1	25.50	18.00	7.50
2	24.00	16.05	7.95
3	22.00	16.00	6.00
4	24.50	17.50	7.00
5	23.00	16.50	6.50
6	24.00	16.75	7.25
7	23.50	17.00	6.50
Average	23.78	16.82	6.96

investment capital when implementing an alternative project is, for example, 23%, then the value of own investment capital in relation to the basic project will be: $Veb = \mathrm{Re}\, a + RA = 23\% + 6,96\% = 29,96\%$.

The methods presented above have been tested in OAO GAZ. It showed their high practical importance. According to an expert survey conducted among specialists of the enterprise, the application of methods allowed increasing the reliability of determining the value of the company's own investment capital by an average of 18%.

5 Conclusions

The approaches to the estimation of the value of the company's own investment capital have been clarified, taking into account the financial risk of the project; methods for determining the amount of adjustment of the alternative have been proposed. The first method is based on the pricing model for capital assets, the second - on the quantitative assessments by credit institutions of the level of financial risk associated with the enterprise in this or that investment decision. Approbation showed that their application allows increasing the reliability of determining the value of own investment capital and the economic feasibility of implementing investment projects.

References

1. Blank, I.A.: Financial risk management. Nika-Center, Kyiv (2005)
2. Brayley, R., Myers, S.: Principles of Corporate Finance/Trans. with English. N. Baryshnikova. Olymp-Business, Moscow (2008)
3. Van Horne, D., Wachovich, J.: Fundamentals of Financial Management, 12th edn. Translated from English. LLC I.D. Williams, Moscow (2006)
4. Vilensky, P.L., Livshits, V.N., Smolyak, S.A.: Evaluation of the Effectiveness of Investment Projects. Theory and Practice: Textbook. Allowance - 2 edn., Pererab. and additional. The Case, Moscow (2008)

5. Garina, E.P.: Business solutions for the creation of a product in the industry. Bull. Univ. Minin **1** (2014). http://vestnik.mininuniver.ru/reader/search/biznes-resheniya-po-voprosu-sozdaniya-produkta-vp/. Accessed 10 June 2017
6. Egorova, A.O., Kuznetsov, V.P., Zokirova, N.K.: Features of the influence of risk factors on the activities of engineering enterprises. Bull. Univ. Minin **4** (2016). http://vestnik.mininuniver.ru/reader/search/osobennosti-vliyaniya-faktorov-riska-na-deyatelnos/. Accessed 10 June 2017
7. Methodological recommendations for the evaluation of investment projects: manual for economists. RAS, Moscow (2004)
8. Potashnik, Y.S.: Estimation of the cost of the company's own capital taking into account the financial risk of the investment project. J. Actual Prob. Econ. Law **3**(31), 90–94 (2014)
9. Raizberg, B.A., Lozovsky, L.S., Starodubtseva, E.B.: Modern economic dictionary, 6 edn., Pererab. and additional. INFRA-M, Moscow (2008)
10. Rimer, M.I., Kasatov, A.D., Matienko, N.N.: Economic Evaluation of Investment. Ed. M. Rimera. St. Petersburg, Peter (2008)
11. Federal Law No. 39-FZ of 25.02.1999: On Investment Activities in the Form of Capital Investments
12. Hamada, R.S.: Portfolio analysis, market equilibrium and corporation finance. J. Financ. **24**, 19–30 (1969)
13. Sharpe, W.: Capital asset prices: a theory of market equilibrium under conditions of risk. J. Financ. **19**, 425–442 (1964)
14. Kuznetsov, V.P., Romanovskaya, E.V., Vazyansky, A.M., Klychova, G.S.: Internal enterprise development strategy. Mediter. J. Soc. Sci. **6**(1) S3, 444–447 (2015)
15. Kuznetsova, S.N., Garina, E.P., Kuznetsov, V.P., Romanovskaya, E.V., Andryashina, N.S.: Industrial parks formation as a tool for development of long-range manufacturing sectors. J. Appl. Econ. Sci. Volume XII **2**(48), 391–402 (2017)
16. Garina, E.P., Kuznetsov, V.P., Egorova, A.O., Garin, A.P., Yashin, S.N.: Formation of the system of business processes at machine building enterprises. Eur. Res. Stud. J. **19**(2 Special Issue), 55–63 (2016)

Information Technologies as a Factor in the Formation of the Educational Environment of a University

Elvira K. Semarkhanova(✉), Lyudmila N. Bakhtiyarova,
Elena P. Krupoderova, Klimentina R. Krupoderova,
and Alexander V. Ponachugin

Nizhny Novgorod State Pedagogical University Named After Kozma Minin
(Minin University), Nizhny Novgorod, Russian Federation
samerkhanovaek@gmail.com, l_bach@rambler.ru,
krupoderova_ep@mininuniver.ru, kklimentina@gmail.com,
sasha3@bk.ru

Abstract. **The goal is to** analyze the existing problems arising in the process of forming the educational environment of the university.

Project/Methodology/Approach - General scientific methods are used: analysis of domestic and foreign literature, comparisons, as well as special-pedagogical experiment and the method of expert assessments for the systematization of knowledge.

Main results of the research - Different approaches to the definition of the information educational environment of a modern university are considered and a complex of problems of formation at various stages is revealed. Defined didactic principles and requirements to the elements of the educational environment of the university. The key issues of creation of the content of the electronic educational resource are analyzed. The article pays attention to the development and introduction of electronic educational and methodological complexes of educational disciplines in the educational process of higher educational institutions. The study addresses the importance of creating an information educational environment of the university, and the role of interactive forms of learning in the teaching of individual academic disciplines. The key role of the use of multimedia technologies in the modern learning process is noted.

Practical significance - The results of the formation of the electronic educational environment of the University of Kozma Minin are analyzed. The issues of modernization of the corporate local-computer network of the university are considered and its importance as an infrastructural element of the complex educational environment is singled out. The role of the university department in the formation of the content of the educational environment is defined. The prospects of development and improvement of the educational environment of the university are revealed.

The received data - The basic principles of creation and key qualitative characteristics of electronic educational-methodical complexes are defined. The structure and components of the educational environment of the university are presented.

© Springer International Publishing AG, part of Springer Nature 2018
E. G. Popkova (Ed.): HOSMC 2017, AISC 622, pp. 179–186, 2018.
https://doi.org/10.1007/978-3-319-75383-6_23

Keywords: Website · Information and educational environment
Information resources · Electronic educational and methodological complex
Educational process · Electronic educational resources

1 Introduction

Today, the rapid development of information technology leads to the fact that modern information technologies are actively introduced into all spheres of human activities, including the system of higher education. Taking into account the growing importance of information technologies in solving scientific, research, pedagogical, methodological problems, the question arises of improving the teaching methodology. The creation in the higher educational institution of the information and educational environment (IEE) is the most important factor in the introduction of the federal state educational standard of the new generation [6].

The solution to this problem can be a fundamentally new approach to creating educational content and presenting it to consumers of educational services. An example is the use of multimedia and interactive elements in the course of training or multimedia training and the formation on their basis of a comprehensive educational environment of the university (university EE).

2 Theoretical Basis of the Research

According to Professor B.V. Palchevsky: "… delaying the development and implementation of next radical innovations is fraught with stagnation of the entire education system. Therefore, the basic innovations have a special social value, they need more experiments than others and should be the subject of research".

The formation of the educational environment of the university is considered in the works of A. Andreeva, G. Belyaeva, V. Bykova, V. Guzhov, M. Zhaldak, and Yu. Zhukova, I. Zakharova, O. Kazanskaya, T. Krasnoperova, E. Mashbitsa, Yu. Petrov, V. Yasvina et al.

According to scientific reference literature, the term "environment" is understood as the totality of economic, political, social, every day, spiritual, territorial, natural and other conditions for the existence of human and society.

The concept of "single information space" includes a set of databases and data banks, technologies for their maintenance and use, information and telecommunication systems and networks operating on the basis of common principles and by common rules - this ensures the information interaction of organizations and people, and satisfies their information needs [9].

In the scientific literature, there is no unambiguous, definition of the term "information educational environment".

Some scientists under the information EE of the university understand:

- The pedagogical system and software, which includes, for example, financial, economic, material and technical, regulatory legal and marketing subsystems, and a management subsystem (A.A. Andreev);

- Organizational and methodical tools, a set of hardware and software that provide quick access to information and carry out educational scientific communications (O.O. Sokolova);
- A system in which all participants in the educational process are involved and connected at the information level: the administration of the educational institution teachers - students - parents (O.O. Kravchina);
- A complex system that allows the intellectual and cultural potential of the university, the content and activity components of the teachers and students themselves to accumulate along with the program-methodological, organizational and technical resources, while the management of this system is based on the objectives of the society, students and teachers (I.I. Zakharova) and others.

3   Research Methodology

In the article, an analysis was made of the domestic and foreign scientific literature on the subject of the study, because of which the conclusion was made on the need for active participation of students in the learning process. Different models of network interaction between teachers and students were compared in the framework of the university's information system. As a result of the pedagogical experiment, the necessity of using interactive teaching methods using electronic educational systems, for example, a modular object-oriented dynamic learning environment - MOODLE, is substantiated.

The university includes a wide range of information resources and a wide range of users. Monitoring of general information resources shows the need for their development in several areas, reflecting the specifics of their functioning.

Intra-university information resources should be divided into the following: colleges, departmental, cathedral, library, additional units of the university and individual [7].

Monitoring of the processes of active preparation and use of various kinds of information in the university shows that the most extensive in content and demanded teachers and students are information resources, formed and provided at the department level.

These information resources can account for up to 75% of all intra-university information, excluding library information. The volume of the departmental information is the basis for the traffic of the intra-university network.

In some universities, the departments have their own websites, on which educational resources are established. It is obvious that it is advisable to create an ITS of the university in the form of an educational portal, where educational websites of different departments and other units participating in the educational activity of the university will be merged. The creation of such a portal requires some organizational, administrative and technical measures [7].

The analysis of the educational websites of the university showed that there is a need to move from individual developments to the system solution of this issue - to create an information EE of the university.

4 Analysis of Research Results

The development of pedagogical education requires the formation of a new environment characterized by innovative content, forms, methods and tools aimed at addressing not only actual, but also prognostic tasks in the activity of the teacher [8].

It can be noted that part of the problems of universities in Russia with a certain degree of specific features in general coincides with universal challenges to higher education. However, Russian specific features have a significant impact on the role of universities in the integration and internationalization of the educational space. With a competent approach to solving problems, specific features can be turned into factors that, by their uniqueness and effective mechanisms of functioning, will help strengthen the position of Russian universities in the world educational process.

Another urgent problem of Russian universities is the task of improving the quality of the material and technical base of the educational process. Not all Russian universities have modern libraries with an electronic indexing system and various search options, the possibility of free access to scientific publications of leading international educational institutions and electronic libraries [4].

The creation of innovative educational technologies is directly connected with the development and use in the university of high-quality electronic educational resources (EER) within the ITS of the university. It is important to take into account the availability of qualitative EERs simultaneously with their availability within the ITS [11]. In general, the problem of the formation of an ITS of a university is reduced to an operational solution of the following tasks:

- The formation of a software and hardware platform that provides storage, retrieval, selection, delivery and presentation of the educational resource to end users and participants in the formation;
- Implementation taking into account the capabilities of the educational content provided by the platform, allowing the learner to effectively perceive and assimilate the training material presented.

The complex approach and successful solution of these problems leads to the formation of an open EE university, which allows the high-quality implementation of educational services, both in traditional contact learning technologies and using distance education technologies [1]. The main components of the modern educational environment are shown in Fig. 1.

At the initial stage, the design of the very concept of the university EE assumes the presence of a developed pedagogical reflection of developers. The paradigm of modern EOS is a complex pedagogical system, including the following components:

- Quality content;
- Tools that provide interactivity;
- Psychological and pedagogical support of the student at all stages of his education.

It can be concluded that ESM, as a necessary element of the university EE, suitable for using it as an alternative to a printed textbook or electronic counterpart, must correspond to three aggregated indicators:

– A good study in terms of methodology and methodology of training;
– Application of tools and technologies providing a qualitative visual series;
– Access to selected reference materials on related terms, which can be located on both a local and remote resource.

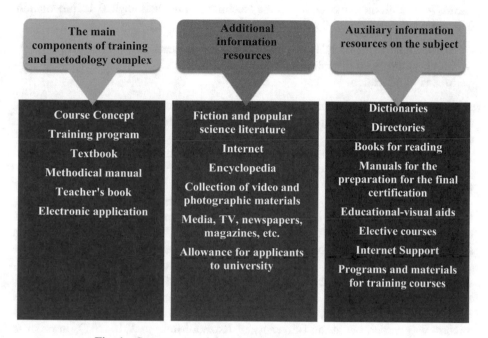

Fig. 1. Components of the modern educational environment.

The problems of making interactivity in the ESM should be considered from the position of modeling the student's learning activity. The basis for designing the educational material is a person-centered approach, implying that the motivation for student interaction with content in the ESM is the student's personal cultural and educational needs. In the context of the changing paradigm of modern education and the acceleration of technological processes, the main difficulty in designing is the knowledge of the student's possible requests and ensuring their satisfaction within the framework of ESM or the EE in general [6].

Scenarios of interaction of students with EE mean:

– The choice of the proposed content and form of presentation of the material;
– The choice of the level of complexity of the material from several proposed;
– Selection of the necessary time intervals within the calendar schedule when studying the material;
– Realization of the creative approach in the study of the material;
– Formation and development of the creative and personal abilities of the learner in process of interaction with the university EE.

From this, follows that the rational structure of ESM should be based on a well-thought-out pedagogical concept. In general, the ESM should include a textbook, workshop and test materials.

For the systematization of local regulatory acts of higher education institutions, a broader concept has emerged - the electronic educational and methodological complex (EEMC). In addition to the totality of the teaching and methodological documentation, it is usual to include the following elements [5]:

- The technology of studying the discipline - for the teacher;
- Block-summary of discipline - for the student;
- Means of education.

As part of the EEMC, multimedia documents and some elements of interactivity are necessarily present. It is not just the digitizing of a SMC from a paper equivalent into a digital format doc or pdf.

To a significant extent, electronic educational and methodical complexes developed and presented in the educational environment of the University of Kozma Minin [10] meet these requirements.

Another important component of the information EE of the university is its corporate LAN. The level of network equipment and hardware, logical and physical topology, the use of modern telecommunication services, the process of authentication and user identification, and the complex level of network security determine the degree of implementation and ease of use of educational resources in it. Many regional universities in the country use the network infrastructure laid back in the mid-90's, which adversely affects the ability to deploy modern services, as well as the overall LAN performance. Realizing this, a comprehensive modernization of the network infrastructure was carried out at the University of Kozma Minin in 2016–2017, in order to increase the availability of the network and educational resources functioning in it.

At the initial stages of the formation in the university of a modern educational environment, the Department of Applied Informatics and Information Technologies in Education took part in the creation and development of the system, in particular on educational segment. Organizational and technological elements supporting the educational process were proposed and implemented. Work is underway to create a file-server system for the organization of normative and methodological materials. The intensive dissemination of the Internet and related Web technologies as a technological basis for training determines the task of introducing and using these technologies in the university. The further solution of this problem at the department is moving along three interrelated directions:

- Creation and operational support of the information environment of the department in the form of a Web-site in the intranet of the university network, with the possibility of remote access to it;
- Further improvement of the electronic educational environment content of the university, and presentation of a wide range of training directions and educational materials on it.
- Participation in the project on creation of mass open online courses on the directions of preparation of the department.

The knowledge and experience gained in the creation and filling of the information environment of the department made it possible to outline the ways and give recommendations on the creation of the ITS of the university.

5 Conclusions

The formation of the information society in Russia initiated the process of informatization of higher education. In general, the effectiveness of informatization of the educational process is determined by the quality of management activity of the university leadership, the formation and continuous development of the educational environment.

The modern educational environment of a higher educational institution cannot be imagined without the comprehensive use of information technology.

Electronic educational environment may include the following elements:

- Management of educational process;
- Educational work;
- Provision of communications;
- Resource management;
- Management of the contingent of students; personnel management of the educational institution.

When forming and supporting electronic educational environment, the following requirements should be taken into account:

- Technical (computer technology, network availability, use of Wi-Fi technology in university campuses);
- Program (information security, integration, interaction);
- Academic (methodological content, compliance of educational content with the requirements of state standards);
- Social (ethical, cultural, regulatory and legal aspects);
- Human resources (ICT - literacy, psychological readiness, availability of specialists).

From this it follows that in order to solve problems related to the informatization of the educational process of higher educational institutions, it is necessary to ensure the high quality of the educational environment. The information EE of the university can be considered qualitative in accordance with the following indicators:

- The presence of a clear organizational structure in which electronic information resources are accumulated and stored and information services are provided;
- Availability of a modern material base necessary to create an information educational environment, use of advanced information technologies (electronic catalogs, access to the Internet, etc.) and licensed software;
- Information literacy of end-users and personnel involved in the creation of this environment.

In conclusion, we can conclude that an important factor in the quality of the educational process is information technology and the infrastructure of the educational institution created on their basis. The modern educational environment of the university is the key to successful implementation of information technologies in education at all levels, which makes it possible to automate and simplify the educational, managerial and any other activity of the educational institution.

References

1. Bazilevich, S.V.: The use of innovative and interactive teaching methods for lecture and seminar classes. Sci. Krasnoyarsk **4**(04), 103–112 (2012). S.V. Bazilevich, T.B. Brylov, V.R. Glukhikh, G.G. Levkin. http://www.informio.ru/files/main/documents/2016/06/Bazilevich_Brylova_Gluhih_Levkin.pdf
2. Deeva, E.M.: The application of modern interactive teaching methods in the university: workshop. E.M. Deeva. - Ulyanovsk: UlSTU 2015, p. 116 (2015). http://venec.ulstu.ru/lib/disk/2016/5.pdf
3. Kazanskaya, O.V., Guzhov, V.I.: Formation of the information educational environment of the technical university. http://www.ict.edu.ru/ft/004064/kaz_guzh.pdf
4. Красноперова Т.В.: Creation of an information and educational environment on the territory of the municipality of Bratsk. Crede Experto: Transp. Soc. Educ. Lang. Int. Inf. Anal. J. **2**(09) (2014). http://ce.if-mstuca.ru
5. Labuz, L.S., Mazaeva, L.N.: Information technologies in higher vocational education: problems and prospects. Sci. Methodical Electron. J. "Concept", **37**, 90–95 (2016). http://e-koncept.ru/2016/56791.htm
6. Novikov, V.M.: Problems and prospects of Russian higher education. http://www.vestnik.mgimo.ru/sites/default/files/pdf/novikova.pdf
7. Petrov, Y.I.: Formation of the information and educational environment of the university. http://berestneva.am.tpu.ru
8. Sedykh, E.P., Fedorov, A.A.: The simulator of the "additional pedagogical reality": theoretical substantiation. Bull. Univ. Minin **2** (2017). https://elibrary.ru/download/elibrary_29842373_50013253.pdf
9. Shakhina, I.Y.: Informational educational environment of the educational institution. http://elib.bsu.by/bitstream/123456789/105333/1/Шахина-427.pdf
10. Samerkhanova, E.K., Krupoderova, E.P., Krupoderova, K.R., Bahtiyarova, L.N., Ponachugin, A.V.: Networking of lecturers and students in the information learning environment of higher school by means of cloud computing. IEJME Math. Edu. **11**(10), 3551–3559 (2016)
11. Sobon, A.: Innovative computer software supporting internationalization of university, AON, Warsaw (2015). http://www.tsi.lv/sites/default/files/editor/science/Conferences/MIP2016/mip_2016_2v_1.pdf

Organization of the Research Activities
of Service Majors Trainees

Zhanna V. Smirnova[✉], Maria V. Mukhina, Lyubov I. Kutepova,
Maxim M. Kutepov, and Olga I. Vaganova

Nizhny Novgorod State Pedagogical University named after Kozma Minin
(Minin University), Nizhny Novgorod, Russian Federation
z.v.smirnova@mininuniver.ru, mariyamuhina@yandex.ru,
lubovkutepova@mail.ru, kmm-asb@mail.ru,
vaganova_o@rambler.ru

Abstract. The theoretical foundations of the organization of research activities of the trainees in the field of service areas are considered, taking into account the requirements of the system approach. The main tasks of students' research activity are identified and their content characteristics are determined. The basic directions of the organization of scientific research activity of students of service majors are determined. The experience of organization of research activities of students in the studied field is presented on the example of major 43.03.01 Service, ensuring the continuous participation of students in scientific work during the entire period of study. This system is distinguished by the continuity of methods and forms from course to course; from one academic discipline to another; from one type of study to another, which is one of the most important principles of this system work. Research activity is an essential part of the system of training of a highly qualified, demanded by the labor market, initiative specialist who is able to think critically and take innovative methods and technologies to achieve high results. The result of educational and research activities in the educational institution is the formation of general cultural and professional competencies, personal qualities of the student, his motivation, reflection and self-esteem. Conclusions and experience in organizing research activities of students can be used in vocational education institutions.

Keywords: Research activity of students · Scientific research
Model of professional development of the future specialist · Student publication
Research competence · Research project

1 Introduction

One of the priorities of the modern higher school is the training of highly qualified specialists who are able to continually replenish and deepen their knowledge and experience, raise theoretical, practical and professional level as well as approach any task creatively. With the increasing role of human factor affecting various spheres of life and the activities of society, the requirements imposed on the professionalism of future graduates of vocational education have become more complicated [4].

© Springer International Publishing AG, part of Springer Nature 2018
E. G. Popkova (Ed.): HOSMC 2017, AISC 622, pp. 187–193, 2018.
https://doi.org/10.1007/978-3-319-75383-6_24

Nowadays, there is a demand for highly qualified, competitive specialists that can come up with new creative ideas and able to solve the set tasks skillfully, predict and model the results of their own professional activity, adapt quickly and successfully in a developing society [1].

Involvement of students in scientific activity is an integrative indicator of the quality of education, namely, the achievement of the required level of education.

In the context of the implementation of the GEF of higher professional education, the model of professional development of the future specialist, which is oriented to forecast and record all future changes, is becoming increasingly important. The task of this model is to prepare a specialist for an "exit" beyond the framework of daily practice; to teach him evaluate and solve various problems competently being able to resolve them on the basis of their own value focus; to perceive any difficulty as an incentive for further development. At the moment, attention is focused on the intellectual development of the individual, encouraging creativity and self-reliance [2].

The educational process is gradually evolving into the process of scientific research. Reproductive activity of the teacher is gradually disappearing into the past, and its place is occupied by the partially search and actual research. In the research activity of students, the function of vocational training is considered being aimed at the needs of practice.

2   Theoretical Basis of the Research

The problem of research activities of students is not new. Many Russian scientists and researchers reflected the use of practical experience of universities in the conduct of research work of students in their work. Among them are S.I. Archangel, G.I. Zhiltsov, E.P. Elyutin, I.I. Ilyasov, V.V. Krayevsky, A.M. Novikov, V.A. Slastenin, M.G. Yaroshevsky, M. Barber [10, 11] and others.

After analyzing the scientific literature devoted to the problems of research, we revealed the main tasks of this type of activity and determined their content characteristics.

Among the main objectives of research activities of students should be noted the following: (1) to organize the support of the teacher of students to provide them with timely assistance in mastering profession; (2) develop creative thinking and initiative in solving practical problems; (3) develop interest and motivation to engage in research and development of research skills; (4) it is necessary to expand the theoretical horizon and scientific erudition; (5) master methods of scientific knowledge; (6) to form the skills of dealing with scientific literature; (7) development of the skills of literate use of Internet resources.

The research work of students is one of the most important forms of the educational process, since the knowledge gained in the process of research will help to form a competent future specialist. The presence of scientific laboratories and clubs, student scientific societies and conferences promotes the development of a full-fledged scientific activity in the university, motivates the student to take an active part in it as they can come across like-minded people with whom they can share the result of their research.

Each university student, to a greater or lesser degree, is engaged in scientific work. During the training period he is forced to write abstracts, course papers, theses, and such activities are impossible without research. Only the student himself depends on the results of his work, choose what the topic of his research will be, the timing of the work and whether the work is ready at all. In the process of performing the work there is a check of the creative qualities of the student, his responsibility, whether he is able to prove his position reasonably.

An important goal is pursued by student publications under the scientific guidance of teachers: revealing the results of their work the student has a motivated desire to work even harder and as a result receive serious scientific results for this, which could not go unnoticed by the public.

Training of mobile professionals is based on two types of activities - scientific and educational. It is no accidentally that the research and development competencies that underlie the knowledge of the surrounding world, the research of its objects, phenomena and processes are included, in accordance with the GEF, into the general competences that are particularly relevant to the situation of multiple choice, the dynamics of change, the numerous problems inherent in modern reality [5].

They contain the most important human abilities to self-knowledge, ability to solve problems, to make the best choice of strategies of behavior and activity.

Forming students readiness for professional mobility, one should be guided by the following principles of organization of research activities: to develop the need for creative self-realization in the framework of professional activity; repeat and consolidate the acquired theoretical knowledge; form the ability to process information; form the foundations of research activities as a component of common competencies: develop a propensity to systematize incoming information; quickly and accurately identify the problem; plan stages of research work; carry out investigations; analyze and summarize the results; to develop communicative and corporate competences with the participation of several people in research work, that is, develop the ability to work in a team [7].

Interest in scientific activity can be formed in various ways with different approaches. We select the invariant basis, the components of which are the following skills:

- stating the research problem;
- setting goals and objectives of the study;
- defining the object and the subject of the study;
- promoting the hypothesis of research and proposing a way to verify it, the ability to distinguish hypotheses from scientific theories;
- selecting and implementing research methods;
- working with information (find information and critically approach its evaluation, systematization, be able to analyze and generalize unordered information, to distinguish between information facts and opinions, descriptions and explanations, hypotheses and theories, arguments and conclusions);
- performing observations, measurements, descriptions, experiments, analysis of phenomena;

– drawing conclusions based on the experimental data;
– holding discussions and defending their point of view [6].

For successful formation of research and development competencies it is necessary to:

– involve as many participants as possible in active scientific work at the initial stages of training;
– ensure participation of junior students in the work of scientific seminars, conferences of teachers, as this will facilitate immersion of students in research activities from the first days of training;
– give students the opportunity to choose the topic of scientific research, not only in terms of their profile, but also from other professional fields.

Involvement of students in research work is one of the primary tasks of Minin University. The implementation of research projects is carried out both at the university level and outside the university.

As a rule, topics of research activity of students who study at Minin University are related to: in-depth study of separate sections of lecture material; the development of computer programs and their use in the educational process to improve the teaching of general professional and special disciplines; the theme of the department's work, reflect the student's personal interests.

As the basic directions of the organization of research activity of students of the Minin University it is possible to formulate the following:

– joint participation of students and teachers in the performance of various types of research work;
– active participation of students in research;
– development of students' abilities for independent judgments and conclusions;
– increasing the effectiveness of research activities of students;
– increased participation of teaching staff in the organization and management of research activities of students.

3   Research Methodology

As a methodological basis for the organization of research activities of students a systematic approach is chosen. The essence of the system approach is to determine the goals and objectives of scientific work. Next, we should create a concept for the preparation of future university graduates for research activities, that is, identify the main directions, the strategy for their implementation; choose the program and methodology. It is necessary to determine the structural components of this system, to establish the nature of relationship between them, to identify levels and criteria for assessing the effectiveness of scientific work, to choose the forms, methods, means of implementing this program, to determine what will be reflected, diagnosed and corrected in the activity of students in the field of scientific research.

The system approach is realized through the step-by-step solutions of tasks: to develop the basic concept, to determine its provisions; to create a material and scientific-methodological base; to develop a schedule of scientific work in accordance with the specifics of the faculty and the boundaries of the educational process; to provide competent scientific guidance to students; include them in more research activities, taking into account their competence and experience in the research field [1].

4 Analysis of the Research Results

We will describe the experience of organizing research activities of students in the field of training of students with major 43.03.01 Service. The uniqueness of this training area is that the training program is based on universal bachelor's degree and the student has the opportunity to choose a 3-year profile for training. The modular system of education, the cumulative rating system (10-point), the availability of disciplines in English in the curriculum, various electronic resources, the design and scientific activity, the invitation of practitioners, the availability of basic chairs, the organization of practices at the base enterprises belong to the specific features of training.

System of research activities of students whose major is 43.03.01 The service provides continuous participation of students in scientific work during the entire period of study [9]. This system is distinguished by the continuity of methods and forms from course to course, from one academic discipline to another, from one type of study to another, which is one of the most important principles of the work of this system. It should be noted that there is gradual increase in the complexity and volume of knowledge and skills that students acquire in the course of their research activities [8].

For example, at the 1st and 2nd year of study, students write essays, conduct research in laboratories, do term papers, that is, they learn the elements of research work, and they are trained to get skills of this work.

Research works are presented in the following main forms: term papers; Educational and research projects; Reports; Messages on the topic; Diaries of pedagogical observations; Algorithm for solving a specific problem; The construction of didactic means; Annotated bibliography; Terminological dictionaries; Abstracts; Annotations; Plan to solve the problem (simple or complex).

In the third year students must take part in university conferences and scientific works competitions. The volume of work is increasing; the tasks and forms of scientific research are becoming more complex. The student is increasingly using his creative approach.

The 4th year for the trainee carries complex course and bachelor's projects.

At this stage, the student can submit his research papers in the form of: a research project, a scientific report, a program, a reference book, a report, an article, a speech, a final qualifying work, a methodological recommendation on various activities.

The best student works are directed to regional, national and all-Russian competitions. For example, students in the field of training major 43.03.01 Service took an active part in the project "Improvement of urban areas", which includes a number of measures to improve the urban area of Nizhny Novgorod. During the development of this project, a number of studies were involved into the research of the problem of improvement in Nizhny Novgorod. In the process of collecting information and working with the basic standard documents on housing and communal services, students get acquainted with basic information on housing and communal services, study basic laws on housing and communal services, a number of projects on landscaping of Nizhny Novgorod.

Students training in the field of major 43.03.01 The service takes an active part in the regional scientific and practical conference "Integration of information technologies into the system of professional and additional education", the scientific and practical conference "Traditions and Innovations in Contemporary Design", the XIV International Scientific and Practical Conference of Students, Graduate students and young scientists "Student genius - 2016", III All-Russian scientific-practical conference "My professional START".

Development of research competence is facilitated by participation in the excursion-research "3D technologies in computer graphics", a lesson-seminar "Modern computer technologies", a lesson-seminar "150 ideas in service activities", excursions connected with research activities in the housing complex "Water World".

The received diplomas at the All-Russian competition of socially significant projects "My initiative in education-2016", the All-Russian competition "My profession is my future", the III All-Russian competition of draft educational publications, the Regional competition for the best scientific work of students testify to the high level of organization of research activity of students.

5   Conclusions

The result of the students' learning, research and research activities depends on the competent implementation of a complex and interrelated process consisting of multiple links.

On the basis of all stated above, it can be concluded that research activities are a complex, multifaceted part of the structure of educational work, which is built on the motivation of the student, the teacher being responsible for that; the methods and forms of scientific knowledge that are important for a full research process. This is an essential part of the training system of a highly qualified, demanded, initiative specialist who is able to think critically and take innovative methods and technologies to achieve high results. Therefore, the educational institution should form an educational environment aimed at developing cognitive interest of students as well as their independence.

All activities for organizing the scientific work of students should be systemic in nature and be solved on the basis of a systematic approach. The final result of educational research and research activities in an educational institution is the formation of general cultural and professional competencies, personal qualities of the student, his motivation, reflection and self-esteem.

References

1. Vaganova, O.I., Ermakova, O.E.: System-activity approach in the development of professional and pedagogical education. Bull. Univ. Minin. **4**(6) (2014)
2. Vaganova, O.I., Khizhnaya, A.V., Trutanova, A.V., Gladkova, M.N., Luneva, Yu.B.: Sillabus as a means of organizing independent work of students. Int. J. Appl. Fundam. Res. **11-5**, 968–970 (2016)
3. Kostylev, D.S., Saliaeva, E.Y., Vaganova, O.I., Kutepova, L.I.: Realization of the requirements of the federal state educational standard for the functioning of the electronic information and educational environment of the Institute. Azimuth Sci. Res. Pedagogy Psychol. **5**(2(15)), 80–82 (2016)
4. Kutepov, M.M., Korovina, E.A.: The role of interactive technologies in the formation of professional competencies of a university student. The Successes Modern Sci. **1**(10), 72–73 (2016)
5. Kutepova, L.I., Nikishina, O.A., Aleshugina, E.A., Loshkareva, D.A., Kostylev, D.S.: The organization of independent work of students in conditions of the information-educational environment of the university. Azimuth Sci. Res. Pedagogics Psychol. **5**(3(16)), 68–71 (2016)
6. Mukhina, M.V.: Perfection of the process of training specialists in service directions at the Kozma Minin NGPU. In: Science and Education: Problems and Prospects: A Collection of Articles of the International Scientific and Practical Conference, 13 March 2014, Ufa: at 2 pm Part 2/off. A.A. Sukiasyan. - Ufa: RIC BashGu, pp. 111–113 (2014)
7. Mukhina, M.V., Smirnova, Zh.V., Suhareva, E.V.: Perfection of the process of forming professional competencies of specialists in the service sector in the NGPU named after Kozma Minin/M.V. Mukhina. Privolzhsky Sci. Bull. **6**(34), 121–124
8. Prokhorova, M.P., Vaganova, O.I., Gladkova, M.P., Gladkov, A.V.: Dvornikova, E.I.: Independent work of students in conditions of implementation of educational standards of higher education. Successes Modern Sci. **1**(10), 119–123 (2016)
9. Shevchenko, S.M., Mukhina, M.V., Kutepova, L.I., Smirnova, Zh.V.: Designing of the basic educational program of preparation of the bachelor in a direction "Service": the competence approach. The Internet J. Sci. **5**(24), 182 (2014)
10. Barber, M., Donneliy, K., Rizvi, S.: An avalanche is coming. Higher education and the revolution ahead. Institute for Public Policy Research (2013)
11. Barber, M., Donnelly, K., Rizvi, S.: Oceans of Innovation: The Atlantic, The Pacific, Global Leadership and the Future of Education. IPPR, London (2012). http://www.ippr.org/publication/55/9543/oceans-of-innovation-the-atlantic-thepacific-globalleadership-and-the-future-of-education. Date Views 04 March 2017

Electronic Testing as a Tool for Optimizing the Process of Control over the Results of Educational Training Activities

Zhanna V. Chaikina(✉), Sofya M. Shevchenko, Maria V. Mukhina,
Olga V. Katkova, and Lyubov I. Kutepova

Nizhny Novgorod State Pedagogical University Named After Kozma Minin
(Minin University), Nizhny Novgorod, Russian Federation
jannachaykina@mail.ru, shevchenko.sm@mail.ru,
lubovkutepova@mail.ru, mariyamuhina@yandex.ru,
katkova.ov@yandex.ru

Abstract. The article deals with the problem of ensuring the objectivity of the pedagogical control of results of the academic activity of students. One of the leading modern means of pedagogical control in educational institutions of higher education is pedagogical testing, carried out with the help of information and communication technologies. In K. Minin Nizhny Novgorod State Pedagogical University for the implementation of effective educational activities in the main areas of implementation and training profiles for students, the electronic educational environment LMS Moodl has been used. The authors of the article consider the electronic educational environment of LMS Moodl to be an effective tool for creating control and measuring materials for assessing the results of the learning activity of students. The article reveals the possibilities of this electronic environment for the creation by teachers of the university of electronic educational materials, means of control, including pedagogical tests. The article describes the algorithm for creating parallel forms of electronic tests based on the tool capabilities of the electronic resource of the LMS Moodl environment. The authors of the article suggest using "random questions" to create parallel (multi-variant) forms of tests. "Random questions" are added to the test directly at the design stage and regulate the inclusion of a specific task from the e-course issues bank in this question. In addition, the authors give the results of an experimental study of the effectiveness of the use of electronic tests created in the electronic environment of LMS Moodl based on parallel forms of the test.

Keywords: Pedagogical testing · Pedagogical control
Electronic educational environment · Electronic tests
Parallel forms of tests · Test tasks

1 Introduction

At present, large-scale optimization of the work of higher educational institutions of Russia is being carried out. A well-developed information and communication technology's (ICT) infrastructure is being formed, and a massive increase in the

© Springer International Publishing AG, part of Springer Nature 2018
E. G. Popkova (Ed.): HOSMC 2017, AISC 622, pp. 194–200, 2018.
https://doi.org/10.1007/978-3-319-75383-6_25

qualification of higher school employees is in parallel. Teachers of higher educational institutions more and more successfully use electronic educational resources in the educational process.

According to the latest regulatory documents of the education system, computer (electronic) testing is defined as pedagogical testing using a computer running a special program that is designed to provide the required presentation of test questions and the necessary processing of test results.

The importance of electronic testing in the learning process can not be overestimated. The main advantages of electronic testing include: The ability to simulate test tasks; Efficiency of evaluation; Objectivity of evaluations; Possibility of reflection, feedback from students; Ease of use [3, 4].

2 Theoretical Basis of the Research

The problem of designing pedagogical tests is considered in numerous studies of domestic [1, 9] and foreign authors [2, 4, 10]. The electronic educational environment of LMS Moodle includes a rich arsenal of various tools for the creation of test, control and measurement materials for evaluating the learning outcomes of students. The main element in LMS Moodle for testing knowledge of students is a test that allows the teacher to ask different question formats and ways of selecting answers.

The environment of LMS Moodle has a great potential for creating test tasks of different types, types and forms of presentation. This electronic environment makes it possible within the course to create a test and to set up a system for evaluating the answers of students for certain learning tasks. In order to fully realize these opportunities, the teacher needs to devote a lot of time to the process of designing assignments.

On the available tools of the LMS Moodle environment, the teacher can prepare a bank of test tasks according to his electronic course [3, 8].

Using these tools, the teacher - the author of the electronic course can make a test task of any complexity, focusing on the level of student preparation and specific training tasks: From simple tests containing popular and original types of answers, to complex test tasks that combine several questions.

When creating test tasks in the LMS Moodle environment, it is required to distinguish the functions of such elements of the electronic environment as the "Test Tasks Bank" and "Test". Function the questions bank cumulates all the test tasks of this course, allows the teacher to systematize and manage certain volumes of test questions of the electronic course on discipline, and provides access to test tasks from published categories of questions of other author's courses.

The test is an element of the electronic course, with which the learner directly works, he contains a certain set of test tasks collected in a certain order.

In modern studies, the term "Pedagogical test" is considered as a representative pedagogical system of test tasks of increasing difficulty, a specific form, which allows qualitatively evaluate the structure and measure the level of preparedness of subjects [5].

Tests used to evaluate the results of training activities should have several options [1]. Since, this makes it difficult to write-off and guess during the control event, which makes it easier to conduct simultaneous testing of large groups of students. In addition, one of the test quality indicators is the reliability of test results, which in practice is usually considered as a test characteristic reflecting accuracy of test measurements, and the stability of test results to the action of random factors [9].

The problem of determining the reliability of tests and test tasks is considered in the studies of Russian [1, 9] and foreign authors [4]. Reliability is often defined as a measure of the correlation between two parallel test cases. Therefore, ideally, we should try to make several variants of the test, and for this, different versions of the same task must be developed.

When developing parallel forms of tests, experts recommend using a facet, the so-called special form of recording several variants of the content of the same test task. This principle is fundamental when creating parallel tests.

Theoretically, before control activities with the help of the developed test, it is necessary to test for a separate sample of students with a minimum of 100 people in order to reject poorly performing tasks. This stage (it can be termed as a "test approbation") is mandatory in the preparation of test tasks and the construction of tests. However, in the conditions of the educational process, most often the empirical testing and testing of the test is carried out during the training on the first group of students.

The correctness of each pre-compiled test task (question) can definitely be indicated only by the results of an empirical study. Only after carrying out an experimental check of the test results and their statistical processing is the conclusion about the validity and reliability of the test results. Domestic experts note that in this case it is necessary to talk about the validity and reliability of the test results, and not the test itself [1]. Transformation of tasks in the test form into test tasks begins with the moment of statistical check of each task for the presence of test-forming properties. Each task, before becoming a test, must go through a stage of statistical and test case study.

The electronic educational environment LMS Moodle has a certain set of tools for assessing the quality of developed tests. Element of the course "Statistics" allows the teacher to easily determine the effectiveness of the created test tasks. This procedure becomes possible only after the trial testing of students using the created test. The process of empirical evaluation of test quality and test tasks is described in sufficient detail in the domestic scientific literature [5].

The main indicators of the effectiveness of the developed tests in the electronic environment of LMS Moodle are standard deviation, the index of differentiation and the coefficient of differentiation.

(1) Standard deviation: The indicator reflecting the scatter of points that the students received when answering a certain test question. In the event that all students respond to the assignment in the same way, the scores will be zero. This result indicates that this task is not a test task and, therefore, should be excluded from the test.

(2) The index of differentiation: Is a coefficient of discrimination, that is, an indicator of the ability of each specific task to separate the best students from the worst. This index takes values between +1 and −1. In the event that a differentiation index with a negative value is obtained, it means that the subjects from the strong group answer this question better than the strong ones. Such test tasks should be discarded, since they actually reduce the accuracy of the entire testing procedure.

3) The coefficient of differentiation: The second way to measure the ability of a particular test task is to differentiate between strong and weak subjects. This parameter can also take values between +1 and −1. This coefficient shows the relationship between the results of the students performing the test as a whole and the answers received by the subjects in the performance of a specific task.

In the event that the values of this coefficient have positive values, it means that the used test tasks do differentiate the trainees with high and low level of preparation. When the coefficient takes negative values, this circumstance indicates that poorly trained students respond to these test tasks better than well-prepared ones. Similar tasks with a negative value of the differentiation coefficient can not be called test, since they do not meet the basic requirements of testing tasks related to the assessment of the level of student preparation. Such tasks should be excluded from the test.

The coefficient of differentiation is more sensitive for detecting the effectiveness of the measuring ability of test questions in comparison with the index of differentiation. The advantage is that the former uses the data of the whole set of learners, and not just the results of more successful and less successful students.

The pedagogical test is a system of tasks assembled in a certain way, each of which is necessary for performing a test of its function. Removing at least one job from the test system results in a space in the list of tested knowledge, which reduces the quality of measurements. To increase the accuracy of the measurement, the test uses the location of the test tasks (questions) in order of increasing complexity. The test includes the minimum number of tasks necessary to obtain accurate results.

3 Methodology of the Study

The purpose of the study was to theoretically substantiate and develop a technology for assessing the quality of training of university students using multivariate (parallel) forms of tests and experimentally test the effectiveness of implementation.

Two electronic courses developed in the electronic environment of Moodl, existing on the basis of the University of Minnesota, participated in pilot test: "Cutting of materials", "Fundamentals of technical creativity". Electronic courses in these disciplines have been used in e-learning practice for more than two years. In addition, test tasks from the banks of questions of these courses were preliminarily tested and adjusted.

For the experiment, two parallel tests were developed for each course: Experimental and control on certain topics of the course with the same number of tasks and the same difficulty for the intermediate control of students' knowledge. The tasks in the test were selected open and closed: The question with a short answer, the question of multiple choice, the question of correspondence, the question of restoring the sequence, the tasks of the experimental and control tests were not duplicated, but in form of presentation, content and difficulty were equivalent. The questions in the tests were arranged according to the degree of difficulty increase, the function of mixing questions in the test was turned off.

These tests contained 30 questions. The time for electronic testing was laid in the program and was 30 min.

Each question was rated at 1 point; therefore, the maximum possible score is 30, which corresponds to 100% of the performance.

At the end of the time provided for testing, LMS Moodl automatically completes the testing process and displays the number of points and the final result in the form of % completion of the submitted tasks on the monitor screen. The student had the right to finish testing before the expiration of the given time.

The control tests for each subject were the same and had no options. The construction of experimental tests for each discipline had goal creation of a multivariate means of control, which ensures the objectivity and reliability of the results of the evaluation.

In order to prepare multivariate (parallel) tests with several presentation options for the test subjects, the LMS Moodl program was used to develop the tests, namely, creating a test task in the "body" of the test in the form of a random question. These random questions do not have their own content, they do not contain the text of the assignment and the answers. Random questions added to the test directly at the stage of constructing the test, perform a "programming" function that regulates the inclusion of a specific task from the e-course issues bank in this question. To ensure that the function "add a random question" is included in the content of the test. Questions are created in the questions bank in which categories the questions of a certain content and form are cumulated, to which the program applies while creating a random question.

In the course of the research, an algorithm for creating a multivariate (parallel) electronic test was developed, which includes the following steps:

(1) The allocation of a module containing the subjects of the discipline subject to control;

(2) The content of each topic is divided into separate educational elements - didactic units. Didactic units are the minimum units of discipline content, which are usually oriented to one or another educational task. This educational problem can be connected with science (phenomenon, information, hypothesis, proof, theorem, axiom, experiment, fact, knowledge, theory, principle, method of research, etc.) or practice (skill and an indicative basis of activity, etc.);

(3) The specification of the test is developed, the number of test tasks for each training element and the test as a whole (the length of the test) is determined;

(4) Each test task (initial), according to the principle of facetedness, is created in several variants;

(5) In the questions bank, categories are created according to the number of test tasks (initial), these categories can be combined into larger categories according to the developed educational elements of the module topics;

(6) Test tasks created by the principle of facetedness are stored in categories (subcategories);

(7) The test is constructed by the way of adding a random question to the "body" of the test; when creating the question, a subcategory with a set of facet test tasks is indicated;

(8) The test settings for electronic testing are carried out.

The values of the standard deviation, index and coefficient of differentiation were taken into account when assessing the effectiveness of test tasks (questions) in the trial testing. In the trial (preliminary) testing, mainly students studying in the field of training "Pedagogical education" and "Operation of transport-technological machines and complexes", who work in the considered electronic courses of disciplines, took part.

In the experimental and control tests, only those test tasks were included, according to which the index values and the differentiation coefficient were not less than 0.3. Test tasks with lower or negative values of the index and differentiation coefficient, as well as assignments with the value of the mean square deviation close to zero, were rejected and not included in final versions of the tests.

4   Analysis of the Results of the Study

When processing the test results for control and experimental tests of e-courses, the indicators were used, which were determined using the electronic system LMS Moodl, namely the average score and the median estimate of the milestones (see Table 1).

Table 1. Results of testing based on the use of control and experimental tests

Criteria	OTT Course		Course "RM"	
	CT	ET	CT	ET
Average rating from all attempts (%)	72.38	55.07	76.42	59.31
Median score for all attempts (%)	76.67	58.00	79.05	60.87
Scale of points	18.3	23.1	16.2	20.8

The study showed that the distribution of the results is not symmetrical for both control and experimental tests, the average value of individual scores deviates to the right from the middle of the scores, counting from 1 to 30, which indicates a certain ease of the tasks of the tests being analyzed. In addition, the test tasks of the control tests were easier for the students.

At the same time, the average score based on testing results using experimental tests of all e-learning courses is closer to 50% in value than in control tests. This circumstance testifies to the reliability of the test measurements [9].

The difference between the maximum value of the primary score and minimum value is called the scores of the subjects. Other things being equal, the best test is the one whose scores are higher. In this connection, the results of testing with the help of experimental tests are more reliable than the results of control tests compiled in different electronic courses of disciplines.

Thus, the study showed that in the process of monitoring the results of training activities of students it is more effective to use electronic tests created on the basis of random questions, which increases the objectivity of control and reduces the probability of copying. The electronic educational environment LMS Moodl has sufficient capabilities for the development of multivariate (parallel) forms of tests that facilitate the implementation of control measures to assess the learning achievements of students.

References

1. Avanesov, B.C.: Composition of Test Tasks. Training Book, 3 edn., 240 pp. Testing Center (2002)
2. Greaney, V., Kellagan, T.: Assessment of Educational Achievements at the National Level, 108 pp. T.N. Leonova, Scientific, Logos (2011). Chelyshkova, M.B. (ed.)
3. Gruzdeva, M.L.: The method of information modeling as a means of learning and an instrument of knowledge of reality. Bull. Univ. Minin 2(10), 13 (2015)
4. Crocker, L., Algina, J., Naidenova, N.N., Simkina, V.N., Chelyshkova, M.B.: Introduction to the classical and modern theory of tests. In: Zvonnikova, V.I., Chelyshkova, M.B. (eds.) Under the Society, 668 pp. Logos (2010)
5. Krasnov, Y.E.: Guidelines for the development of test tasks and the construction of pedagogical tests. http://charko.narod.ru/tekst/metodiki/krasnov.pdf. Accessed 20 Mar 2017
6. Terekhina, O.S., Chaikina, Z.V.: Informational educational environment in the organization of independent work in the university. In: The Collection: Integration of Information Technologies into the System of Additional Education in the Field of Technical Creativity, A Collection of Articles on the Materials of a Regional Scientific and Practical Conference, pp. 37–40. Nizhny Novgorod State Pedagogical University. K. Minin (2016)
7. Terekhina, O.S., Chaikina, Z.V.: Information and communication technologies as a means of activating the independent activity of students - future technology teachers. World of Science. vol. 4(3), p. 22 (2016)
8. Chaikina, Z.V., Smirnova, Z.V.: Use of information and communication technologies in pedagogical testing. Mod. High Technol. 11-2, 392–396 (2016)
9. Chelyshkova, M.B.: Theory and Practice of Designing Pedagogical Tests, p. 432. Logos, Moscow (2002)
10. Roid, G.H., Haladyna, T.M.: Technology for Test-Item Writing. Academic Press, New York (2008)

"Reformer" Before the Reformation: Regarding the Issue of Proto-Protestant Views of John Wycliffe

Tatiana G. Chugunova[1]([✉]), Lydia V. Sofronova[1], Anna V. Khazina[1],
Elena S. Balashova[1], Vladimir M. Tyulenev[2],
and Vusala S. Khasanova[1]

[1] Nizhny Novgorod State Pedagogical University Named After Kozma Minin
(Minin University), Nizhny Novgorod, Russian Federation
tat-chugunova@yandex.ru, lidiasof@yandex.ru,
annhl@yandex.ru, balashova.I.s@gmail.com,
vhasanova@yahoo.com
[2] Ivanovo State University, Ivanovo, Russian Federation
tulenev31@bk.ru

Abstract. The purpose of the article is to analyze the teachings of the English Christian thinker of the 14th century John Wycliffe, in particular those of his provisions, which were later developed by the Protestant reformers of the sixteenth century (on the supremacy of Holy Scripture, the abolition of the papacy, the use of preaching, etc.). Using the method of text analysis, the authors of the article point to the same interpretation of thinkers belonging to different historical epochs, the most important religious and political doctrines. The authors also pay attention to the fact that all the works of John Wycliffe were devoted to the development of the correct approach to the study and interpretation of the Bible. The English theologian claimed that Holy Scripture is the main guide in all matters of man's spiritual life, and that only in it alone is revealed the true meaning of faith. In addition to translating the Holy Scripture into the mother's language, J. Wycliffe concertize his views on religious organization, secular power and the relationship of these two spheres. The works of the Evangelical Doctor were devoted to the active apologetic of monarchical power, the creation of a strong secular state led by the king, which was subsequently reflected in the works of Anglican and then Puritan theologians: U. Tindela, T. Cranmer, J. Rogers, H. Latimer and others.

Keywords: Reformer · Protestantism · The reformation · The church
The holy scriptures · The papacy · The antichrist · The sermons

1 Introduction

English Christian thinker John Wycliffe (1320-1384) is rightfully considered the forerunner of the European reformers of the 16th century. In his many works, he put forward those ideas that the Protestant theologians subsequently developed. The speeches of the Oxford theologian laid the foundation for that struggle, which, according to Ellen White's just remark, "led to the liberation of not only individuals,

© Springer International Publishing AG, part of Springer Nature 2018
E. G. Popkova (Ed.): HOSMC 2017, AISC 622, pp. 201–207, 2018.
https://doi.org/10.1007/978-3-319-75383-6_26

but also churches and entire nations" [5, p.72]. John Wycliffe stated in his sermons that the truth should be sought in the Holy Scripture, and not in the traditions of the apologists of the Roman Catholic Church. It was he who sowed the interest in the Bible, completely translating it into English from the Vulgate - the Latin text of Jerome, approved by the Catholic Church. The problems of the formation of national languages based on the text of the Bible, the creation of a strong secular state, touched upon in the writings of the evangelical doctor (so called Wycliffe his contemporaries) were later reflected in the works of M. Luther, W. Tindel, W. Zwingli, J. Calvin and many others. In this article, an attempt is made to analyze the main provisions of the religious and political doctrines of J. Wycliffe, which became the ideological basis for the formation of views of European religious reformers of the XVI century.

2   Research Methodology

The problematic field of this work is at the intersection of several research areas: intellectual and personal history, the history of religion, biblical studies and hermeneutics. The historical-genetic method used in this study makes it possible to identify the specific features of John Wycliffe's religious and political doctrines and to trace the way in which his outlook is formed. The method of hermeneutical analysis provides an opportunity for a correct interpretation of the texts of the English theologian both from the point of view of belonging to the genre, the era, and as evidence of his inner peace and spiritual life. With the help of the historical comparative (comparative) method, it becomes possible to correlate the views of J. Wycliffe with the intellectual and spiritual quest of the era of Reformation.

3   Analysis of Research Results

The Oxford theologian anticipated many of the ideas of future reformers by opposing the basic tenets of Catholicism, rejecting indulgences and veneration of saints, as well as the doctrine of transubstantiation. Long before the speeches of Protestant theologians, John Wycliffe questioned the necessity of the existence of the institution of the papacy and called the Roman high priest "the godless monster," "the most terrible Antichrist," and his prelates - servants of the Antichrist. [15, pp. 423-427]. In his opinion, "these" shepherds of sheep "loved the worldly more than Christ, and, grazing their sheep, pour in them the mortal poison of the Antichrist" [15, p. 425]. Here is what the evangelical doctor wrote about the head of the holy throne in his main Latin-speaking theological composition, Trialogue: "The Avignon Sister, whom many consider the Pope, and some as the Vicar of Christ on earth, is the source of all ungodliness in a militant church, and if it is possible to say so, then the approaching Antichrist" [15, p. 423]. He, according to the Oxford theologian, leads a life different from the life of Christ, the Roman pontiff personally and in the person of his cardinals blaspheme the Savior, for Christ was poor, humble and obedient to the God-Father, and the papal curia lives differently [15, p. 423]. "Antichrist thinks of himself as God on earth and tries to take the whole church under his authority," concludes the thinker

[15, p. 423]. John Wycliffe imputes to the pope the most difficult sins, alienating him from God and plunging into the Devil's slavery. Similar expressions will also be found in the works of his compatriot William Tindel (1494-1536), interpreter of the Bible in English, who declared that the Pope had distorted the order of the world with lies and distortion, overthrew the Kingdom of Christ and established the Kingdom of the Devil, becoming his viceroy [11].

As for obedience to the law of God, the Roman pontiff, according to Wycliffe, distorts him, contrasting his law, built on "universal domination, mud, blasphemous pride" to the law of Christ [15, p. 423]. The evangelical doctor believes that a true Christian must live in accordance with the law of God, create the Covenant of Christ, fulfilling his holy life and study, and not believing instead of Christ in false prelates and sinful priests who are incapable of another life, except in pride and lust, laziness and gluttony [12, p. 259].

The evangelical doctor, as later the reformers of the 16th century, pays special attention to money-grubbing and an irrepressible thirst for territorial enrichment of the papal curia: "With one hand the pope collects treasures, the other - troops to settle conflicts by armed means" [15, p. 425]. According to Wycliffe, the Roman Catholic Church "became famous" for disbelief, homicide and the acquisition of untold riches. The Oxford theologian condemns the Catholic episcopate for illegally collecting tithes, taxes, posthumous gifts, donations in favor of the church, insinuations with wills, outright lasciviousness and much more, arguing that Catholicism is "a religion founded by sinful people."

The abuses of the Catholic clergy, promulgated in the writings of J. Wycliffe, became at the beginning of the sixteenth century the subject of sharp criticism from his compatriot, the English humanist theologian, the dean of the Cathedral of St. Paul J. Colet (1466-1519) in his famous "Cathedral Sermon", in which the theologian called for the return of the church to the lost ascetic and apostolic ideal. Colet underwent severe criticism of the way of life of the spiritual estate, consisting of four main vices: "devilish pride, carnal voluptuousness, greed for earthly goods and worldly pursuits." With great regret, the English humanist said, "the canons are captured by the love of the world, and not of God" [4].

The problem of the secularization of the Catholic clergy will be examined in sufficient detail in the first third of the 16th century English theologian William Tyndel. He will accuse the papacy of many troubles that have fallen to the lot of Europeans, and among them: The great migration of peoples, the crusades, and the disintegration of the Christian church, numerous wars and international conflicts. In his writings, W. Tindel will often refer to John Wycliffe, one of the first, according to him, to openly condemned the actions of the head of the holy throne and his wards. Based on Matt. 6, W. Tindel will remind the clergy mans that it is not possible to serve two masters: God and mammon; i.e. riches, covetousness, lust and vanities of this world [11].

The evangelical doctor's statements were not limited to reproaches and censures against the papal Rome, he urged secular authorities not to obey the Roman pontiff: "The princes and lords must understand that we are not needed such a pope, not such a clergy for the study of Scripture, but humble laymen" [15, p. 427]. The secular princes, in the opinion of the theologian, must defend their kingdoms from the assassination of this Devil, drive out the envoys of the Pope-Antichrist everywhere and in no case obey his laws [15, p. 427]. Such bold statements by the Oxford theologian against the papal curia would greatly undermine the already shaken authority of the Roman pontiff, and in the 16th century, religious reformers without any fear will call the head of the holy throne the Antichrist, leading the papacy to the most terrible of the accusatory concepts of the Gospel - the concept of the anti-Christ establishment. The initiator of the European reform movement, Martin Luther (1483–1546), in his famous treatise "To the Christian Nobility of the German Nation on the Correction of Christianity" will give the Roman high priest an unflattering description: "The pope, rather, should be considered the enemy of Christ, called in the Scriptures" the Antichrist. « After all, his whole essence, all his actions and undertakings contradict Christ and only destroy the meaning of the deeds of Christ [3, p. 42].

Not finding excuses for the actions of the spiritual authorities and not hoping for their correction, Wycliffe declares the complete subordination of spiritual power to secular one. In the treatise "On the Royal Service," the evangelical doctor insists that the spiritual power be under the control of the secular, since the last one is perfect [14, p. 13]. According to the Russian researcher Kuznetsova, an evangelical doctor was "a consistent supporter of Supremacy over the Church" [2, p. 134]. It is this provision that will become one of the leading in the political program of reformers.

By denying the pope supremacy over the whole Christian world, Wycliffe called for believing in his true head - Jesus Christ and come to him through the study of the Word of God, as set out in the Holy Scripture. Wycliffe proclaimed the Holy Scripture the only source of faith, an absolute measure in all matters of Christian dogmatism, and therefore he considered acquaintance with this book as an important step for all believers. Domestic researcher of life and creativity of Wycliffe and Kuznetsov notes that, according to the evangelical doctor, "Holy Scripture is sufficient for the righteous life of people without a Catholic church and its rituals" [2, p. 119]. As already mentioned above, Wycliffe translated the Bible into English, thus giving the opportunity to read it not only to learned men who know the languages of the original (Hebrew and Ancient Greek) and study the Holy Scripture in the original or Latin translation, but also to ordinary people in their mother language. Later the reformers will translate the Bible into vernaculars, but as a rule, not from the Latin translation, as Wycliffe did, but from the original sources - Hebrew and Greek texts. However, as W. Cooper observes, the Wycliffe's translation of the New Testament was very close in spirit to the ancient Greek original [10, p. 36].

The study of Holy Scripture has become one of the priorities of the reform program of the Oxford theologian. Wycliffe developed his concept of the Bible as the law of God, therefore it is no accident that chapter 31 of the third book of the "Trialogue" is called: "Christ's law, i.e. Sacred Scripture, infinitely exceeds other laws " [15, p. 238]. In this chapter, the author explores the semantic meaning of the term and gives him the following interpretation: "First of all, the words" Holy Scripture "denote Jesus Christ and the book of life in which all truth is inscribed, and according to John 10," the Scripture can not pass away, which the Father sanctified and let into the world [15, pp. 238-239]. The Evangelical Doctor speaks of the superiority of the Holy Scripture to other religious books [7]. "Sacred Scripture is much more certain than any other books ... Because Jesus Christ is much higher than any ordinary person, so is His book or the Scripture containing His law much more valuable than any other book," concludes the

Oxford theologian [15, p. 239]. According to Wycliffe, "Scripture is authentic in any part of it, because it contains the sayings of the Lord Jesus, and He can not lie and be mistaken, and mislead anyone" [15, p. 239].

Wycliffe insisted on studying the Scriptures by every believer, which was a bold step for that time, as the church had a monopoly on reading and interpreting the Bible. Only the Reformation was able to leave the believer alone with the Bible and give him the opportunity to judge God himself. Oxford professor anticipated this important requirement of the European reformers, he recognized for every believer the right to be guided in matters of faith by his own interpretation of the Bible, but in his works Wycliffe gave some recommendations for the correct study of this book. The evangelical doctor insists that every believer try to find in the Scripture a literal meaning and beware of interpreting this complex source without having the Holy Spirit in his heart, for "such an interpreter, according to Jerome, is a heretic, and he is even worse than the one who blasphemers, Pretending that he gives the Holy Scripture a meaning that he himself considers unknowable for himself" [15, p. 243].

An evangelical doctor criticizes scholastic scholars who, in pursuit of meaning, are addicted to reasoning and forget that many things should simply be believed. According to Wycliffe, there is nothing in the Holy Scripture about which "the heretic himself understood that this is ominous." And then he, according to the Oxford theologian, "does not submit anything to his judgment, for the greatest mistake in his understanding comes from pride and stupid prejudice in the superiority of one's own logic, while the logic of Scripture is the most correct, most accurate and most used" [15, pp. 241–242]. As rightly notes Shchelokova, "Wycliffe attributes an infinite authority to Holy Scripture and fundamentally distinguishes the Word of God and its human interpretation" [8, p. 188]. Subsequently, religious reformers will also criticize scholasticism based on the use of logic, reasoning, references to church authorities and a fourfold interpretation of biblical concepts.

Protestant churches are traditionally called evangelical, because they recognize the primacy of Scripture before the Tradition, the Words - before the church. John Wycliffe, as mentioned above, was also named by his contemporaries as an evangelical doctor, because that is how he prioritized, asserting that the Gospels are more authoritative than the church [12, p. 255]. Understanding Wycliffe of the Bible as the Word of God is consistent with the Protestant outlook. It is no accident that many evangelical doctor interpreted many key concepts of the Holy Scripture not in the traditional but in the Protestant manner, for example, the term clergyman {*priest*}, important for the Catholic church, he replaced with the elder in translating a number of passages of the New Testament (Matthew, 16:21; Mk., 7: 5, 8:31) [9]. The English translator U. Tindel did the same, enlisting the authority of his senior colleague [6, p. 90].

Great place in his reform program Wycliffe preached sermons as an important component of the spiritual life of a Christian. The concept of preaching is one of the main themes of his numerous treatises. According to Wycliffe, Christ Himself more determined the clergy to preach than to serve the Mass, for nowhere in the Gospel is the open text of the morning and evening (but only of the sacraments) spoken of, and the need for preaching is said everywhere [1, p. 374; 13, p. 112]. Thus, the Oxford theologian unambiguously hints at replacing the Roman clergy with simple preachers who could bring the true teachings of Christ to the people. English reformer William Tindel, who insisted on the destruction of the institution of the priesthood, will later say: "The preaching of the Word of God is hateful to them, for it is impossible to preach Christ and not preach against the Antichrist, i.e., all those who by swords and false doctrines are trying to use the true teaching of Christ " [11, p. 114].

4   Conclusion

Therefore, as the analysis shows, John Wycliffe anticipated many of the demands of the European reformers of the sixteenth century and formulated the basic principles of the future Protestant doctrine. His philosophical-theological and political ideas became extremely relevant during the autumn of the middle Ages and in many ways contributed to the emergence of a new Christian religion - Protestantism. Like future reformers, Wycliffe wanted the "true Word of God" to become available to every Christian. The emotional mood of the Evangelical Doctor in exposing the vices of the Roman Catholic Church was in many respects similar to that of the Protestant theologians of the 16th century. However, unlike future reformers, Wycliffe did not have a clear plan for creating a new church, reformed on the basis of Holy Scripture. His ideas are more enlightening in nature and are aimed primarily at eradicating the vices of the existing church, rather than at its elimination.

References

1. Wycliffe, J.: Mirror of the Antichrist. How the Antichrist and his servants take the true priests from the preaching of the Gospel of Christ by four fouls. (Translation and introduction by T.G. Chugunovoy). Alm. Intellect. Hist. (43), 368–375 (2013). Dialog with time
2. Kuznetsov, E.V.: The Lollards movement in England (late 14th–15th centuries). Studies on the history of popular movements in Western European countries XIII-XV centuries. Sch. Notes (95), 25–285 (1971). Historical series. Gorky State University, Gorky
3. Luther, M.: To the Christian Nobility of the German Nation on the Correction of Christianity, pp. 11–94. Oko, Kharkov (1994). The time of silence has passed. Selected works of 1520-1526 years. Translation from German. U.A. Golubkina
4. Sofronova, L.V., Erokhin, V.N.: J. Colet and the statute of the Cathedral of St. Paul: Attempt to implement the church reform, vol. 1. Bulletin of the University of Minin (2015). http://www.mininuniver.ru/scientific/scientific_activities/vestnik/archive/1-9
5. White, E.: The Great Struggle. Translation from English, 704 p. Source of life, Zaoksky (2011)
6. Chugunova, T.G.: Interpretation of key biblical terms in the English translation of the Bible by U. Tindela. Alm. Intellect. Hist. (2012). (38), 82–99 (2012). Dialog with the times
7. Chugunova, T.G.: Sacra Scriptura: the problem of understanding the term in the works of J. Wycliffe. Bull. UNN 1(2), 263–267 (2012). Series: Philology, Journalism: Foreign Lang. N. Novgorod
8. Shchelokova, N.V.: Philosophical and Theological Views of John Wycliffe, pp. 186–190. Gorky State University, N.I. Lobachevsky, C. I. N. Novgorod (2001). XII Reading the Memory of Professor S.I. Arkhangelsky. Materials of the International Conference

9. Evangelia, B.N.T.: The Gothic and Anglo-Saxon Gospels in Parallel Columns with the Versions of Wycliffe and Tyndale, 3rd edn. Reeves and Turner, London (1888). With Preference and notes by J. Bosworth, assisted by G. Waring. XXXVI, 584 p

10. Cooper, W.R.: The Wycliffe New Testament. Tyndale Soc. J. (10), 6–9 (1998)

11. Tyndale, W.: The Obedience of a Christian Man and How Christian Rulers Ought to Governe, pp. 97–183. Yale University Press, London (1573). Whole works of W. Tyndall, John Frith and Doct. Barnes, three worthy Martyrs and principall teachers of this Church of England edited by J. Foxe; Printed by J. Daye. London

12. Wycliffe, J.: How Antichrist and his clerks travail to destroy Holy Writ. In: Mattew, F.D. (ed.) The English Works of Wycliffe Hitherto Unprinted, pp. 254–262. Trubner, London (1880)

13. Wycliffe, J.: Speculum de Antichristo. In: Mattew, F.D. (ed.) The English Works of Wycliffe Hitherto Unprinted, pp. 108–113. Trubner, London (1880)

14. Wycliffe, J.: Tractatus de Officio Regis. London (1887). Edited by A.W. Pollard and C. Saule

15. Wycliffe, J.: Trialogus cum supplemento trialogi. Typographeo Clarendonia, Oxonii (1869). Edited by G.V. Lechler

Evaluation of the Effect from Organizational Innovations of a Company with the Use of Differential Cash Flow

Sergey N. Yashin[1]([✉]) [iD], Yury V. Trifonov[1], Egor V. Koshelev[1] [iD],
Ekaterina P. Garina[2], and Viktor P. Kuznetsov[2]

[1] Nizhny Novgorod State University Named After N.I. Lobachevsky,
Nizhny Novgorod, Russian Federation
jashinsn@yandex.ru, ekoshelev@yandex.ru,
decanat@ef.unn.ru
[2] Novgorod State Pedagogical University,
Nizhny Novgorod, Russian Federation
keo.vgipu@mail.ru, kuzneczov-vp@mail.ru

Abstract. At present, interest in the problems of evaluating the profitability of organizational innovations arises in Russia. Despite their importance, due primarily to the production need, it is often difficult to assess their effectiveness, because such measures do not have a separate commercial result. The creation of an appropriate methodology to solve this problem is necessary in order to choose the most advantageous option of such managerial solutions from the set of available alternatives. In this case, you can use a technique based on the application of differential cash flow, which is the difference between an alternative and a basic version of the organizational solution. After calculating such a flow, the expected value of the net present value is calculated according to the available set of scenarios. However, the solution of this practical problem is not limited to solving the presented direct problem. It is also important to solve the inverse problem, the essence of which is to calculate the optimal value of the parameter characterizing the idea of organizational innovation. So, in the presented example, as such a parameter, the size of the trade discount was considered for the buyers of the products, which pay for it immediately. To solve the direct and inverse tasks of assessing the effectiveness of organizational innovation, we presented the necessary computational methods, as well as computer simulation methods.

Keywords: Innovative project · Organizational innovations
Differential cash flow · Assessment of the effect of innovation

JEL Classification: C88 · G32 · O22

1 Introduction

At present, interest in the problems of evaluating the profitability of technical, technological and organizational innovations arises in Russia. Organizational innovations of the company, of course, are also important, both technical and technological.

However, it is not possible to directly evaluate the effectiveness of such measures, since they do not have a separate commercial effect expressed in money. In addition, to refuse organizational innovations for this reason is impractical, since most of them are dictated by production necessity (Galor and Tsiddon 1997, Klychova et al. 2016a). In addition, organizational and managerial innovations play an important role in the system of factors of increasing the competitiveness of the company and the successful achievement of its strategic goals. In this case, a transition from a traditional functional approach in management to flexible adaptive management models is necessary.

For these reasons, an appropriate methodology is needed to assess the effect or effectiveness of organizational innovation of the company. Such a methodology justifies itself only in conditions when the goals of the company or investor are clearly formulated and quantitative guidelines for achieving these goals are presented. Then such management decisions, in fact, will bring real benefits to the development of the firm.

Thus, in Lazarev's work (2013), the problem of evaluating organizational innovations is proposed to be solved using matrices that allow establishing infra-firm qualitative and quantitative links of management, self-management, progress and regress in the process of organizational development of the firm.

In Krylov's work (2015) with the purpose of project management of innovative activity of the enterprise, it is offered to evaluate efficiency of organizational modernization among other directions. For this purpose, the organizational effect of the number of realized projects is calculated.

There are also strictly formalized approaches to the evaluation of organizational innovations, based on the creation of appropriate mathematical models. Therefore, Kolbachev and Peredery (2015) propose to use the entropy approach, based on the evaluation of the information characteristics of the organizational structure.

Also of practical interest are the methods for modeling factor systems for assessing the effectiveness of organizational and technological solutions. For example, Lapidus (2014) proposes to explore the parameters of the integrated potential of such solutions, which allow the system to become flexible, allowing it to adapt to the changes that inevitably occur, and at the same time, strive to optimize organizational, technological and management decisions.

Despite the fact that many important scientific results have been obtained in the evaluation of organizational innovations, approaches based on methods of comparing already existing basic and new more progressive versions of organizational innovation projects have not yet been developed. Here we are just talking about the use of differential (incremental) cash flows in assessing the effectiveness of relevant management decisions.

In practice, there are always several rational options for achieving the set goals of the company. That is, there are alternative solutions from which to choose the optimal. First of all, choose the option that is the most economical in comparison with others.

2 Theoretical Basis of the Research

The least expensive option of the decision is taken as the basic one. The remaining alternatives are compared with the basic one. Cash flows are then defined as the difference between the respective flows of the analyzed and basic alternative. The money flow defined in this way ( ) is called *differential (incremental)* (Brigham and Gapenski 1993; Limitovsky 2008):

$$\Delta CF = CF_a - CF_0$$

Where CF_0 - cash flow in the basic option in the period of time t (rub.);

CF_a - the same for the considered alternative (rub.).

In each period of the planning horizon, the differential cash flow is caused by a change in the amount of investment at the beginning of the period of the organizational innovation project. The return from such events will be stretched in time. For this reason, it is necessary to bring the differential cash flows to the point of project evaluation using the appropriate discount rate. As such a rate, it is reasonable to take the weighted average cost of capital (WACC) for the firm carrying out the project.

Table 1 shows the methodological approaches to the estimation of differential cash flow for various types of company innovations (Limitovsky 2008).

Table 1. Organizational innovations of the company, the effect of which is calculated using differential cash flow.

Event	Cash in flow	Cash out flow
Change in collection policy (relationship with debtors)	Reduction of non-payments, disinvestment with reduction of the term of repayment of receivables	The costs of implementing a new collection policy
Change in payment policy (relationship with creditors)	The emergence of a source of temporarily free cash	Instability of supply, risk of loss of suppliers
Change in the organization of work, inventory management	Disinvestment of related current assets (decrease in demand for inventories, etc.)	Expenses for the implementation of the event
Introduction of new productive machinery	Additional cash flow as a result of increased productivity	Investments in the purchase of equipment, development of new equipment, change in risk
Introduction of new unproductive equipment	Cost reduction compared to the basic option	Investments in the purchase of equipment, etc.
Staff development (replacement of specialists with more qualified ones)	Higher productivity in the future, profit from better work performance	Higher pay

(*continued*)

Table 1. (*continued*)

Event	Cash in flow	Cash out flow
Sale of assets	Receiving income from sales	Loss of current income from assets
Use of commercial intermediaries	Inflow from the acceleration of inventory turnover of finished goods, increase in sales volumes	Remuneration of intermediaries
Reorientation to new sources of raw materials, materials	Higher productivity, disinvestment of stocks, reducing transportation costs, lowering prices, etc.	Change in technological risk, costs associated with the development of new sources

Finally, the use of differential cash flow can also have practical application for assessing the effectiveness of social projects that are implemented by the state. In this case, for the basic version of the project, one can take one that was implemented earlier by the corresponding state structure. Change of any parameter of the new social project entails the appearance of a differential cash flow. By discounting this flow at the time of assessment, you can calculate the net present value of the new social project in relation to the basic option ( ). If it is positive, then the social project in these conditions is beneficial. It is also possible to investigate the effect of several parameters on the effect of the project, both individually and in a complex (Klychova et al. 2016b).

3   Research Methodology

With the help of differential cash flow, various organizational solutions can be justified, for example, changing the company's credit policy with regard to debtors, its payment policy towards creditors, changing approaches to managing inventories, and so on. Such management decisions can include the situation of justification of a flexible price scale, depending on the timing of payment of invoices by customers. In this case, the value d *of the* trade discount is set, and then it is necessary to find out whether it is profitable to sell at a discount d.

However, the task can be put in a different way. Say, we need to calculate the maximum discount size d, at which the corporation will not suffer losses. To solve similar problems, you can also use differential cash flow. Then the expected value of its net present value ( ) is used as a criterion for making a decision about the discount. For this purpose, the linear interpolation formula can be used in the calculations (Kruschwitz 1999), in this case will be of the form

$$d = d_1 + \frac{E[\Delta NPV]_1}{E[\Delta NPV]_1 - E[\Delta NPV]_2} (d_2 - d_1), \tag{1}$$

as well as the corresponding methods of computer modeling, for example, in the packages *Maple* and *Matlab*.

We will illustrate the methodology for assessing the effect of organizational innovation of the company using differential cash flow on a specific example (Limitovsky 2008).

Analysis of research results

The firm is engaged in the production of Russian national toys. The average maturity of receivables is 3.32 months. The company's management believes that using a discount of 7% of the price of the goods for immediate payment of the product can lead to the following results:

(a) in an optimistic case, 40% of all products will be sold at a discount (the probability of this event is 0.6);
(b) in the pessimistic case, 15% of the total goods will go at a reduced price (the probability of this event is 0.4).

Is it advantageous for the company to introduce a discount if:

- the average annual sales volume is 17 328 thousand USD. USA;
- the current costs are 63.6% of the sales volume and will not change with the introduction of a discount;
- the income tax rate is 20%.

It is assumed that this organizational event will last 10 years, the average cost of the company's capital at the moment is 21% per year in USD.

The maturity of receivables in months can be found as

$$t = \frac{\text{receivable s}}{\text{annual sales volume}} \times 12$$

With a maturity of accounts receivable of 3,32 months and the annual revenue of $17,328 the firm's investments in receivables are

$$\frac{3,32}{12.} \times 17,328\,\$ = 4794,1\,\$$$

In an optimistic scenario, from the moment the discount is introduced, 40% of the total sales volume will go at a discount price, then the amount of investment (I) in accounts receivable will also decrease by 40%:

$$\Delta I = -0,4 \cdot 4794,1 = -1917,6 \,(\text{Thousand USD}).$$

In a pessimistic scenario, the amount of investment in receivables will be reduced by 15%, i.e.,

$$\Delta I = -0,15 \cdot 4794,1 = -719,1 \,(\text{Thousand USD}).$$

At the moment, the company's annual net profit is

$$CF_{base} = (17328 - 0,636 \cdot 17328)(1 - 0,2) = 5045,9 \text{ (Thousand USD)}.$$

After the introduction of the discount, the profit will change either to the value

$$CF_{opt} = (17328 \cdot 0,6 + 17328 \cdot 0,4(1 - 0,07) - 0,636 \cdot 17328)(1 - 0,2)$$
$$= 4657,8 \text{ (Thousand USD)},$$

Or before

$$CF_{pes} = (17328 \cdot 0,85 + 17328 \cdot 0,15(1 - 0,07) - 0,636 \cdot 17328)(1 - 0,2)$$
$$= 4900,4 \text{ (Thousand USD)}.$$

In the optimistic case, the change in net profit will be

$$\Delta CF_{opt} = 4657,8 - 5045,9 = -388,1 \text{ (Thousand USD a year)},$$

But in the pessimistic –

$$\Delta CF_{pes} = 4900,4 - 5045,9 = -145,5 \text{ (Thousand USD a year)}.$$

The diagram of differential cash flows under the conditions considered is shown in Fig. 1.

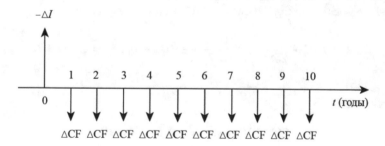

Fig. 1. The diagram of differential cash flows.

Net present value in an optimistic scenario is

$$\Delta NPV_{opt} = 1917,6 - 388,1 \cdot a_{10;21\%} = 1917,6 - 388,1 \frac{1 - 1,21^{-10}}{0,21}$$
$$= 344,2 \text{ (Thousand USD)},$$

but in the pessimistic –

$$\Delta NPV_{pes} = 719, 1 - 145, 5 \cdot a_{10;21\%} = 719, 1 - 145, 5 \frac{1 - 1,21^{-10}}{0,21}$$
$$= 129, 2 \,(\text{Thousand USD}).$$

Thus, even in the worst case, the use of discount is beneficial for the firm. The expected net present value is

$$E[\Delta NPV] = \Delta NPV_{opt} \cdot 0, 6 + \Delta NPV_{pes} \cdot 0, 4 = 344, 2 \cdot 0, 6 + 129, 2 \cdot 0, 4$$
$$= 258, 2 \,(\text{Thousand USD}).$$

Now we set the inverse task: *What is the maximum size of the discount the firm can use, given the characteristics of the volume of sales of products, the average maturity of receivables and current production costs?*
The expected net present value function, depending on the discount d, will look like

$$E[\Delta NPV] =$$
$$= \big\{ 1917, 6 + [17328(0, 6 + 0, 4(1 - d) - 0, 636)0, 8 - 5045, 9]a_{10;21\%} \big\}0, 6 + .$$
$$+ \big\{ 719, 1 + [17328(0, 85 + 0, 15(1 - d) - 0, 636)0, 8 - 5045, 9]a_{10;21\%} \big\}0, 4$$

The discount size d can be found from this expression, for example, by the method of linear interpolation (Kruschwitz 1999) using formula (1). In this case, we already know. Its exact value is a thousand USD. Then using the expression for the function of the expected net present value, depending on d, it is not difficult to find, for example, thousand USD. Substituting the data into formula (1), we obtain

$$d = 0, 07 + \frac{258, 071}{258, 071 + 247, 722}(0, 1 - 0, 07) = 0, 085307,$$

i.e. d = 8,5307%.
Check results

$$E[\Delta NPV]_{8,5307\%} = -0, 001699 \approx 0.$$

To avoid these rather cumbersome calculations on the calculator, and also to find the exact value of the rate d, if necessary, you can calculate it, for example, in a package *Maple*. To do this, type the following text in the program:

```
> solve((1917.6+(17328*(.6+.4*(1-x)-.636)*.8-5045.9)
*(1-1.21^(-10))/(.21))*.6+(719.1+(17328*(.85+.15
*(1-x)-.636)*.8-5045.9)*(1-1.21^(-10))/(.21))*.4=0);
```

The result is obtained.
In addition, to obtain a visual representation of the dependence on d, you can use, for example, the *Matlab* package. To do this, type the following text in the program:

```
>> x=0:0.001:0.16
>> y=(1917.6+(17328*(0.6+0.4*(1-x)-0.636).*0.8-5045.9)
.*(1-1.21^(-10))/0.21).*0.6+(719.1+(17328*(0.85+0.15
*(1-x)-0.636).*0.8-5045.9).*(1-1.21^(-10))/0.21).*0.4
>> plot(x,y)
>> grid on
>> xlabel('{\itd}')
>> ylabel('E[\DeltaNPV] (`000 USD)')
```

In Fig. 2 shows the result of the program.

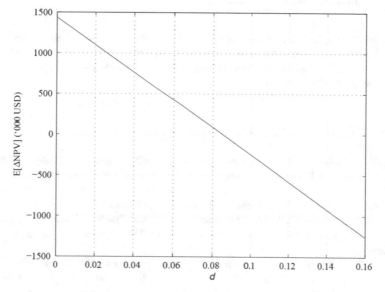

Fig. 2. Dependence of expected net present value on discount *d*

4 Conclusions

1. Despite the importance of organizational innovation of the company, due primarily to the production need, it is often difficult to assess their effectiveness, because such measures do not have a separate commercial result. The creation of an appropriate methodology to solve this problem is necessary in order to choose the most advantageous option of such managerial solutions from the set of available alternatives.
2. In this case, you can use a technique based on the application of differential cash flow, which is the difference between an alternative and a basic option of the organizational solution. After calculating such a flow, the expected value of the net present value is calculated according to the available set of scenarios.

3. However, the solution of this practical problem is not limited by the solution of the presented direct problem. It is also important to solve the inverse problem, the essence of which is to calculate the optimal value of the parameter characterizing the idea of organizational innovation. So, in the presented example, as such a parameter, the size of the trade discount was considered for the buyers of the products, which pay for it immediately.
4. To solve the direct and inverse tasks of assessing the effectiveness of organizational innovation, we presented the necessary computational methods, as well as computer simulation methods.
5. The use of differential cash flow can have practical application for assessing the effectiveness of various projects that do not have a separate commercial result, that is, both projects of organizational innovations of the company and social projects that are implemented by the state.

Acknowledgments. The article was prepared with the financial support of the Russian State National Scientific and Technical Foundation. Grant No. 15-02-00102 "Formation of a mechanism for managing the innovative development of the industrial region (exemplified by the Nizhny Novgorod region)."

References

Lazarev, V.N.: Problems of the evaluation of organizational innovation. Bulletin of the Perm National Research Polytechnic University. Socio-Economic Sciences, No. 20, pp. 98–105 (2013)

Krylov, A.G.: A model for assessing the economic efficiency of project management of enterprise's innovation activity. Electronic Scientific and Practical Journal "Economics and Management of Innovative Technologies", No. 2 (2015). http://ekonomika.snauka.ru/2015/02/7364

Kolbachev, E.B., Peredery, M.V.: Planning the development of organizational structures and business processes as an innovative task. Vestnik YURSTU (NPI), No. 1, pp. 4–10 (2015)

Lapidus, A.A.: Potential efficiency of organizational and technological solutions of the construction site. Vestnik MSSU, No. 1, pp. 175–180 (2014). https://doi.org/10.22227/1997-0935.2014.1.175-180

Limitovsky, M.A.: Investment projects and real options in emerging markets: training manual, 4th edn., revised and supplemented. M.: Yurayt (2008)

Brigham, E.F., Gapenski, L.C.: Intermediate Financial Management, 4th edn. The Dryden Press, Orlando (1993)

Galor, O., Tsiddon, D.: Tecnological Progress, Mobility, and Growth. Am. Econ. Rev. **87**, 363–382 (1997)

Klychova, G.S., Kuznetsov, V.P., Trifonov, Y.V., Yashin, S.N., Koshelev, E.V.: Upgrading corporate equipment as an Asian real option. Int. Bus. Manage. **10**(21), 5130–5137 (2016a). https://doi.org/10.3923/ibm.2016.5130.5137

Klychova, G.S., Kuznetsov, V.P., Yashin, S.N., Koshelev, E.V.: Concept of Integrated management of the financial flows of the investing region. Acad. Strateg. Manage. J. **15** (Special Issue 1), 198–209 (2016b)

Kruschwitz, L.: Finanzierung und Investition, Munchen, Wien: R. Oldenbourg Verlag (1999)

Evaluation of Technological Innovations of a Company by the Methods of Chain Repeat and Equivalent Annuity

Sergey N. Yashin$^{(\boxtimes)}$ (ID), Yury V. Trifonov, Egor V. Koshelev (ID),
Julia A. Grinevich, and Sergey L. Ivankovsky

N.I. Lobachevsky Nizhny Novgorod State University, Nizhny Novgorod,
Russian Federation
jashinsn@yandex.ru, decanat@ef.unn.ru,
ekoshelev@yandex.ru, julia-grinevich@mail.ru

Abstract. When assessing technological innovation, one cannot rely on any one method of analysis. Based on the investor's goals and the planning horizon, each specific method carries some additional information about the effectiveness of investment chains. In order to make the most optimal decision about replicating technological innovation projects, it is necessary to take into account the specific features of operation of a particular new equipment, in particular concerning the planned lifetime. Among the many approaches used to solve this problem, the chain repeat method and the method of equivalent annuities are the most popular among financiers. In this case, often a preference is given to the method of equivalent annuities as the most simple for calculations. Despite the fundamental differences between the method of chain repeat and the method of equivalent annuities, they allow one to obtain the same result for investment chains that have terms that are multiples of the investor-planning horizon. If this condition is violated, then the problem of choosing the most profitable chain of technological innovations is considerably complicated. This factor can significantly affect the correctness of the final decision of the company. Therefore, in the example considered, if we were guided only by the results of the equivalent annuity method, we would get an incorrect conclusion about the optimal chain of technological innovations of the company.

Keywords: Technological innovation · Investment chains
Chain repeat method · Equivalent annuity method

JEL Classification: D92 · G31 · O22

1 Introduction

At present, interest to the problems of assessing the profitability of technical, technological and organizational measures is emerging in Russia. Refusal of such measures often leads to a decrease in the competitiveness of enterprises, loss of clients, strategic positions in the market. It is important to understand that many actions of the company's management are often dictated not by considerations of economic benefit, but by production necessity (Galor and Tsiddon 1997; Damodaran 2002; Klychova et al. 2016).

In the field of assessing the effectiveness of technological innovation of companies, many different methodological approaches have been developed. Therefore, for example, Kondrashov (2014) for assessing technological innovation in the agronomic-industrial complex suggests taking into account indicators of a general economic and financial nature that allow to assess the enterprise's ability to develop because of introducing innovations in the future. For this, Kondrashov (2014) considers the qualitative characteristics of the innovative attractiveness of enterprises, which offers to be evaluated using the expert method. These characteristics make it possible to take into account the opinion of reputable specialists in the field of investment and innovation management.

However, such an approach, in our opinion, is not strict enough in the scientific sense, since the opinions of experts are always highly individual. This irreversibly affects the nature and quality of the assessments they give.

Bakaev (2014), on the contrary, proposes an evaluation technique whose main advantages are: the possibility of assessing the level of innovative development of the company in dynamics; integrated research of the level of innovative development; the possibility of comparing the actual values of indicators with ones that, evaluated by experts. A feature of the proposed methodology is not only the possibility of quantitative assessment of the level of innovation development in general, but also the diagnosis of individual factors. This approach allows us to determine what factors need to be intensified to bring the company to a higher level of innovative development.

On the other hand, Vlaskin and Lisin (2015) apply the principle of template business modeling to evaluate technological innovations. Particular attention is paid to the problem of assessing the effectiveness of the chosen business model of innovation and technological activities. To solve this task using the process approach, structural and logical links between blocks of the template business model of the technological start-up are established and key performance indicators for each main business process are proposed. Based on the developed system of balanced indicators, a method for assessing the effectiveness of business processes of an innovative enterprise was developed.

However, considering the technological innovations of companies, the analyst faces the problem of evaluating multiple investments. Such a situation can arise in practice, when it is assumed that technological innovations in the future will be repeated periodically. At the same time, the evaluation of such innovations is based on the fact that any rational investor takes into account only the possibilities of acquiring or designing such equipment or entire technological lines that will at least be as good as those already available. Such an approach, of course, does not exclude the possibility that the equipment will be better than that already used. In this case, a monetary valuation of the worst possible future equipment alternatives assumes a calculation according to a pessimistic scenario. This estimate will then be the lowest threshold for making a final decision on the chain of investment in technological innovation.

Among the many approaches used to solve this problem, the chain repeat method is the most popular among financiers (Brigham and Gapenski 1993; Kruschwitz 1999; Yashin et al. 2014) and the method of equivalent annuities (Brigham and Gapenski

1993; Limitovsky 2008 Yashin, Koshelev and Podshibyakin 2014). In this case, often a preference is given to the method of equivalent annuities as the most simple for calculations. (Brigham and Gapenski 1993; Limitovsky 2008). However, this approach does not always lead to the correct solution of the problem of estimating the investment chain.

In order to make the most optimal decision about replicating technological innovation projects, it is necessary to take into account the specific features of operation of a particular new equipment, in particular concerning the planned lifetime. This factor can significantly affect the correctness of the final decision of the company.

2 Theoretical Basis of the Research

In the case of a one-time investment, it is assumed that the investor, during its planned period after the expiration of the optimal period of operation, as it were, "takes a well-deserved rest" and if it implements any investments, it is only complementary. This case, of course, is far from reality. As a rule, after the expiration of the economic life of the first project, the investor starts the second (third, fourth …) project.

We are here to consider the sequence of investment or, as they say, the investment chain. In this regard, it is necessary to distinguish between identical and non-identical investment chains (Kruschwitz 1999).

1. The investment chain is called identical when all individual projects included in the chain have the same net reduced income (NPV). This does not necessarily mean the premise that all projects have identical cash flows. However, for the sake of simplicity, we further discuss always a special case in which all projects generate a chain of identical cash flows. Such identical chains have the property that NPV of individual projects coincide at any discount rates.
2. Non-identical investment chains are spoken of when NPV chain projects differ from each other. In the event that the cash flows of individual projects do not coincide with each other, the probability of a difference in NPV is very high.

In addition to these two kinds of investment chains, we would like to distinguish between the final planned time period (temporary enterprise) and the endless planning period (permanent enterprise). If we combine the term of the planning period (finite or infinite) and types of investment chains (identical or non-identical) with each other, we get 4 possible planning situations in Table 1 (Kruschwitz 1999).

Table 1. Planned situations with multiple investments.

		Chain of investments
		Identical
Planning period	*Finite*	chain effect
	Endless	infinitely identical chain investments

This approach needs clarification. In a temporary enterprise, the use of identical investment chains leads to the so-called "chain effect" or "general law of replacement". After all, in the final chain, the optimal lifetime of a given project is always longer than its "predecessor" (and shorter than the term of its "successor"). This is not immediately clear-the phenomenon was discussed intensely and with passion in publications on the theory of investment without clarifying the practical significance of this circumstance (Kruschwitz 1999). Identical investment chains, apparently, are really rare. Therefore, just in the case of a temporary enterprise that can "forecast" its planning horizon, it would be better to limit ourselves to analyzing non-identical investment chains.

The situation is different in the case of a permanent enterprise, that is, in the case of an investor with an infinite horizon of planning. Here it seems quite unreasonable to believe that an investor can somehow reliably predict the cash flows of the 10th, 20th … and even more so of the 1000th project. In this situation, it is more reasonable to assume identical cash flows (identical investment chains).

Thus, further we will consider infinitely identical chains of investments in technological innovations, then moving to comparison of different variants of termination of these chains in time. In this case, the planning horizons will be as short as possible, which will avoid the previously mentioned "chain effect".

3   Research Methodology

Comparing the method of chain repeat and the method of equivalent annuities, we give them a brief description.

The method of chain repeat. The essence of this method is the periodic repetition of a particular project. For this, the net present project income (NPV) is calculated, which the firm expects to have as many times in time as the project Brigham and Gapenski 1993 is planned to repeat; Kruschwitz 1999) (Fig. 1).

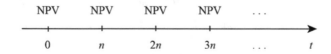

Fig. 1. Chain of investments.

After this, all NPV of the repeating project are discounted to zero and add up. Choosing the chain of investment in technological innovation, which will bring the greatest effect in monetary terms.

Method of equivalent annuities. The choice of the most profitable investment chain in this case is made by comparing the equivalent annuities of the chains. That is, it is calculated how much the annual annuity (EAA) can generate a particular project. Equating the present value of such an annuity to NPV (Brigham and Gapenski 1993; Limitovsky 2008), we find that

$$NPV = EAA \cdot a_{n;i},$$

Where is the discount multiplier for an annuity of n year's duration with a rate of i percent for the year, calculated as

$$a_{n;i} = \frac{1 - (1+i)^{-n}}{i}.$$

From here, you can find the value of the equivalent annuity for a particular project:

$$EAA = \frac{NPV}{a_{n;i}}. \tag{1}$$

In this case, the investment chain that generates the largest annual equivalent annuity is selected.

Despite the fundamental differences between the method of chain repeat and the method of equivalent annuities, they allow one to obtain the same result for investment chains that have terms that are multiples of the investor planning horizon. If this condition is violated, then the problem of choosing the most profitable chain of technological innovations is considerably complicated. Then one can not uniquely apply any one of the considered methods, for example, the method of equivalent annuities as the most simple to understand and calculate. As practice shows, often the method of chain repeat is truer for projects that differ significantly in terms of time.

However, it would be wrong to take a decision in this case, relying only on one of the methods. It is necessary to thoroughly investigate the problem of evaluation, starting with trivial methods, such as, for example, calculating the NPV of each project from different investment chains.

Let us illustrate these arguments on a concrete example.

4   Analysis of Research Results

Corporation "SIA International" (c. Moscow) is engaged in the production and wholesale of medicines. This company considers the expediency of acquiring a new technological line. In the market, there are two models with the parameters suitable for the firm, the data on which are presented in Table 2.

Table 2. Parameters of technological lines.

	Line A	Line B
Price (thousand rubles)	9 500	13 000
Generated annual income (thousand rubles)	2 100	2 250
Service life (years)	8	12
Liquidation value (thousand rubles)	500	800
WACC firms (%)	11	11

It is necessary to justify the expediency of acquiring a particular technological line. Acquisitions of line A will be called project A, and acquisition of line B - project B. Then we estimate NPV projects:

$$\text{NPV}_A = -9\,500 + 2\,100 \cdot \frac{1 - 1,11^{-8}}{0,11} + \frac{500}{1,11^8} = 1\,523,821 \text{ (thousand rubles.)},$$

$$\text{NPV}_B = -13\,000 + 2\,250 \cdot \frac{1 - 1,11^{-12}}{0,11} + \frac{800}{1,11^{12}} = 1\,836,474 \text{ (thousand rubles.)},$$

therefore. However, projects vary considerably in terms of time. Therefore, for the analysis of projects, it is necessary to apply the chain repeat method. Project A can be repeated every 8 years, and project B - every 12 years. Then compare them for the smallest multiple period, that is, 24 years.

In Fig. 2 we show the flows of NPV projects: Above the time axis for project A, and below for project B.

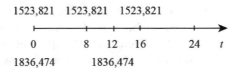

Fig. 2. NPV project flows A and B

NPV of a recurring project A for 24 years will be

$$\text{NPV}_{\Sigma A} = 1\,523,821 + \frac{1\,523,821}{1,11^8} + \frac{1\,523,821}{1,11^{16}} = 2\,471,971 \text{ (thousand rubles.)}.$$

NPV of a recurring project A for 24 years will be

$$\text{NPV}_{\Sigma B} = 1\,836,474 + \frac{1\,836,474}{1,11^{12}} = 2\,361,413 \text{ (thousand rubles.)}.$$

$\text{NPV}_{\Sigma A} > \text{NPV}_{\Sigma B}$, so now $A \succ B$.

The same result can be obtained using the method of equivalent annuities. So by the formula (1) we get that

$$\text{EAA}_A = \frac{\text{NPV}_A}{a_{8;11\%}} = \frac{1\,523,821}{5,146123} = 296,11 \text{ (thousand rubles.)}.$$

$$\text{EAA}_B = \frac{\text{NPV}_B}{a_{12;11\%}} = \frac{1\,836,474}{6,492356} = 282,867 \text{ (thousand rubles.)}.$$

$\text{EAA}_A > \text{EAA}_B$, therefore $A \succ B$.

However, this result should also be seriously challenged. The fact is that 24 years is too long to periodically repeat the acquisition of the same technological line, even if only two times in the case of line *B*. For 12 years, many parameters of the technological line will change. This is primarily (1) the price, (2) the liquidation value and (3) the WACC or the firm. In addition, and this is perhaps the most important, the technological line for 12 years will become morally obsolete. Therefore, acquiring old technology will no longer make sense.

You can solve this problem by assuming the divisibility of projects in time. Namely, suppose the divisibility of project *A* in half. This will make it possible to compare different projects for the period of the greatest of them, that is, Project *B*.

Then we calculate the NPV of the dividend project *A* for 12 years:

$$\text{NPV}_{\Sigma A} = 1\,523,821 - \frac{9\,500}{1,11^8} + 2\,100 \cdot \frac{1 - 1,11^{-4}}{0,11} \cdot \frac{1}{1,11^8}$$
$$= 228,609 \text{ (thousand rubles.)}.$$

This is less than a $\text{NPV}_B = 1\,836,474$ thousand rubles, so the final conclusion will be: $B \succ A$. This means that the company should purchase the 2nd model of the production line, that is, the line *B*.

5 Conclusions

When assessing technological innovation, one cannot rely on any one method of analysis. Based on the investor's goals and the planning horizon, each specific method carries some additional information about the effectiveness of investment chains. So in the considered example, if we were guided by the results of the method of equivalent annuity most popular among financiers, we would get an incorrect conclusion. We would have chosen the repetition of project *A*, although at the most reasonable minimum planning horizon - 12 years - this decision would not be optimal. In this case, the optimal investment chain *B* allows us to identify the chain repeat method. However, if we considered another horizon of planning - 24 years - then the result would be directly opposite. We would choose the investment chain *A*.

These results can be useful for financial analysts and project managers when choosing the most profitable option for technological innovation of the company.

Acknowledgments. The article was prepared with the financial support of the Russian State National Scientific and Technical Foundation. Grant No. 15-02-00102 "Formation of a mechanism for managing the innovative development of the industrial region (exemplified by the Nizhny Novgorod region)."

References

Bakaev, D.N.: Methodology for assessing the innovative activity of meat industry enterprises. Vestnik VGUIT, No. 3, pp. 163–167 (2014)

Vlaskin, A.A., Lisin, E.M.: Development of a model for assessing the effectiveness of innovative entrepreneurship based on the principle of template business modeling. Int. Sci. Res. J. 6(37), Part 3, 23–27 (2015).

Kondrashov, K.A.: Methodology for assessing the innovative attractiveness of enterprises of the agronomic industrial complex. Discussion, No. 7(48) (2014). http://www.journal-discussion.ru/publication.php?id=1143

Limitovsky, M.A.: Investment projects and real options in emerging markets: training manual, 4th edn. M.: Yurayt (2008). Revised and supplemented

Yashin, S.N., Koshelev, E.V., Podshibyakin, D.V.: Index method of forming the investment program of the corporation with a limited budget for financing. Finances and credit, No. 24 (600), pp. 2–8 (2014)

Brigham, E.F., Gapenski, L.C.: Intermediate Financial Management, 4th edn. The Dryden Press, Orlando (1993)

Damodaran, A.: Investment Valuation: Tools and Techniques for Determining the Value of Any Asset. Wiley, New York (2002)

Galor, O., Tsiddon, D.: Tecnological Progress. Mobility, and Growth, American Economic Review 87, 363–382 (1997)

Klychova, G.S., Kuznetsov, V.P., Trifonov, Y.V., Yashin, S.N., Koshelev, E.V.: Upgrading corporate equipment as an Asian real option. Int. Bus. Manage. 10(21), 5130–5137 (2016). https://doi.org/10.3923/ibm.2016.5130.5137

Kruschwitz, L.: Finanzierung und Investition, Munchen, Wien: R. Oldenbourg Verlag (1999)

Legal Foundations of Human Society

The Russian Information Systems of the Housing and Utilities Sector: Peculiarities of Legal Regulation and Application

Andrey A. Inyushkin[✉], Elena S. Kryukova, Yury S. Povarov,
Evgenia V. Ruzanova, Valentina D. Ruzanova,
and Nikolay G. Frolovskiy

S.P. Korolev Samara National Research University, Samara, Russia
cl-su@mail.ru

Abstract. The article studies the issues of application of information systems in the housing and utilities sector in the Russian Federation. Topicality of the studied issues is predetermined by the necessity for improving the law for the purpose of overcoming the difficulties that appear in practice due to the Russia's active implementation of state and private information systems in the housing and utilities sector. The purpose of the article is to determine the main directions of improvement of the legal mechanism that determines the forms and means of using the information systems in the housing and utilities sector on the basis of determination and analysis of this mechanism's drawbacks. The main approaches to studying this problem are the dialectic, system, intersectorial, etc., which allows for the complex consideration of the mechanism of the information systems usage in view of the needs of the housing and utilities sector. The results are as follows: the current state of legal regulation in the sphere of the Russian information systems is determined, and the peculiarities of application of the information systems in the housing and utilities sector are determined, in particular, during general meeting of the residents of an apartment block via postal vote; the procedure of using the state and private information systems the housing and utilities sector is studied; the directions for improvement of the law in this sphere are given. The article poses practical value for specialists of various profiles that work in the housing and utilities sector, scholars who conduct analysis of the legal, information, and economic aspects of functioning of the housing and utilities sector, and persons who deal with the legislative activities.

Keywords: Information systems · Data bases · Housing and utilities sector
Intellectual property · Premises' owners voting

1 Introduction

1.1 Establishing a Context

In the recent decades, the information and communication technologies have been developing in Russia, which reflected the legal system of the country as well. Social

© Springer International Publishing AG, part of Springer Nature 2018
E. G. Popkova (Ed.): HOSMC 2017, AISC 622, pp. 227–233, 2018.
https://doi.org/10.1007/978-3-319-75383-6_29

transformation in Russia, democratization of public life, and informatization of activities of the public authorities bodies require deep change of the structure and character of the information provision (Khusyaynova 2016). The legislator has been showing peculiar interest in regulation of public relations with the legal and technological methods (Amelin 2016). Article 23 of the Federal Law No. 149-FZ dated July 27, 2006 "On information, information technologies, and protection of information" (hereinafter – the Federal law on information) establishes the definition of the information systems – totality of the information containing in data bases and ensuring its processing with information technologies and technical means. At that information is the data (messages, etc.) regardless of the form of its presentation, and data bases are the objects of intellectual property (Inyushkin 2016). In March 2014, the Federal law dated March 12, 2014 No. 35-FZ introduced a new object of intellectual property – Internet site, which has a legal regime of a complex item. Until this law, the relations in the sphere of information turnover with the help of the information and telecommunication network Internet had been regulated in the form of the legal regime of data bases. In particular, higher judicial bodies treated Internet site as a data base, which led to allied rights of the creator of the data base for the information containing on the Internet (decision of the Court on intellectual rights dated March 7, 2014 No. C01-114/2013 for the case No. A56-58781/2012). Definition of web-site on the Internet is found in Article 2 of the Federal law on information, but it does not contain the interconnection with data bases. At that, all information systems use the Internet – so they are Internet sites. In literature, appearance of an electronic information system is related to connecting the information resources of a certain subject of law to the program product on the basis of a certain data base which is included into a specific personal computer or a system of computers (Bachilo 2009). Diversity of the existing types of information is predetermined by diversity of human activities (Shaikh and Londhe 2016); at that, the law provides the classification of information into state, municipal, and others (Article 13 of the Federal law on information); information system became were widespread in the housing and utilities sector.

1.2 Literature Review

The general problems of the issues of regulating the information systems were reflected in the works of Bachilo (2009); Eremenko (2012); Voynakanis (2013). Interconnection between data bases and the information on the Internet was studied by Manap et al. (2013); Inyushkin (2015, 2016). The issues of protection of information in various information systems were analyzed by Tarasov (2014); Amelin (2016); Shaikh and Londhe (2016). The peculiarities of application of specialized information systems in the housing and utilities sector were studied by Pichugin (2011); Khusyaynova (2016).

1.3 Establishing a Research Gap

The previous studies, information systems were viewed without consideration of the specifics of the sphere of their usage, which did not allow determining the factors influencing the effectiveness of such usage in the housing and utilities sector and determining the ways of overcoming the obstacles for their full implementation.

The research determines a range of contradictory provision of the law regarding the conditions and the order of applying the state and private information systems in the housing and utilities sector and determines the directions of improving the mechanism of functioning of these systems for the purpose of the information provision of realization of housing rights of the citizens.

1.4   Aim of the Study

The purpose of the study is to determine the main directions of improvement of the legal mechanism that determines the forms and methods of using the information systems in the housing and utilities sector on the basis of determining and analysis of this mechanism's drawbacks.

2   Methodological Framework

2.1   Research Methods

General and private methods of cognition were used: dialectic, historical, system, formal and logical, rather-legal, intersectorial, etc. Their complex application allowed determining the peculiarities of functioning of the systems of the housing and utilities sector, the circle of subjects of relations for usage of the corresponding information systems, and the hierarchy of legal norms in this sphere.

2.2   Research Basis

The research basis includes the works of the Russian and foreign scholars who studied various aspects of functioning of information systems, including in the housing and utilities sphere, the law in the sphere of information systems, and practice of its application.

2.3   Research Stages

The problem was studied in two stages:

- 1st stage: selection and analysis of the existing scientific literature on the topic of the research, as well as the law in the sphere of information systems and practice of its application in the housing and utilities sector;
- 2nd stage: formulation of the conclusions received in the course of analysis of the scientific literature, the law, and the practice of its application in the housing and utilities sector, preparation of materials for publication.

3   Results and Discussions

3.1   Legal Regulation in the Sphere of Information in the Housing and Utilities Sector

The state information system of the housing and utilities sector is one of the varieties of state information systems (hereinafter – SIS HUS), which – due to the large number of its users and social significance – is related to those for whom information technologies play one of the most important roles. Information systems allow for participation of the maximum number of citizens in voting for various issues that emerge during management of an apartment block, which means realization of housing rights. The literature notes that SIS HUS is very important during solving the inventory tasks by means of accumulation and systematization of a large massive of information (Pichugin 2011). The Order of the Ministry of Communications of the RF No. 504, and the Order of the Ministry of Construction of the RF No. 934/pr dated December 12, 2014 made www.dom.gosuslugi.ru the official web-site of SIS HUS. Therefore, this address on the Internet is the web-site in which SIS HUS is located (Inyushkin 2015). Practical usage of SIS HUS causes a lot of argues. The problem of usage the information system in the housing and utilities sphere appeared during the general meeting of the owners of an apartment block by voting, which was provided to the citizens by Article 47.1 of the Housing Code of the RF.

3.2   The Procedure of Usage the State Information System of the Housing and Utilities Sector

The key role during usage of the information system belongs to the administrator of the general meeting, who, according to the provisions of the housing law, places the announcement of the general meeting of premises' owners in the apartment block, sends the notification to each owner in the apartment block via the system, accepts the written decisions of the owners in the apartment block on the issues that were voted (in case they have not voted in the electronic form), specifies the data on the voter, and post the electronic form of the owner's decision in the system. At that, voting on the issues of the agenda of the general meeting of premises' owners in the apartment block with the usage of the system is performed in the electronic form. According to the Federal law dated April 6, 2011, No. 63-FZ "Regarding electronic signature", it is possible to use a simple electronic signature for legally significant actions, which can be combined with the certificate of the verification key of the electronic signature and the qualified certificate of the verification key of the electronic signature. Electronic signature is the guarantee for identification of information and its sender (Tarasov 2014). Article 47.1 of the Housing Code of the RF does not envisage the corresponding detalization and does not specify the implementation of this norm regarding the owners' voting. Probably, for the purpose of observing the balance of interests of owners and initiators of the general meeting, it is expedient to use electronic signature with a usual certificate of the verification key. In this situation, the persons participating in voting will be reflected in the system, and falsification of votes and decisions of the meetings is improbable.

3.3   Application of Non-Government Information Systems of the Housing and Utilities Sector, Including Data Bases in the Housing and Utilities Sector

The issue of applying other information systems during a general meeting of owners in the apartment block in the form of postal vote, which is allowed by the legislator with observation of the similar order and terms, is rather complex. Probably, for the purpose of application of Article 47.1 of the Civil Code of the RF after January 1, 2018, it will be necessary to determine the circle of subjects of these relations – similarly to the subjects of relations during usage of SIS HUS. In this case, the list of participants of these relations expands, which allows for their maximum regulation. During usage of other information systems, the Internet will be used in any case – therefore, the mandatory subjects that participate in the relations of postal vote of the owners will also include the owner of the Internet site. He shall determine the order of using the web-site, including the order of posting the information on it. Besides, during the usage of other information systems, new providers of communications will appear who provide the access to the Internet and, from the point of view of the Civil Code of the RF, are information intermediaries. It should be noted that the system treatment of the law allows using the legal regime of data bases with other information systems used in the housing and utilities sector. As is provided in the literature, only the work of actors, performers, conductors, and directors can be deemed the result of intellectual activities in the list of joint rights objects that conform to the requirements of Article 1228 of the CC of the RF on creation of such results with creative work of their authors (Eremenko 2012; Voynikanis 2013). The activities of manufacturers of phonograms, data bases, and information organizers and publishers could be qualified as organizational and technical. Thus, allied right of the manufacturer of data bases conforms to the doctrine "Sweat of the brow", which was described in the scientific literature on many occasions (Manap et al. 2013). In the conditions of straight "direction" of the analyzed non-government information systems at the housing and utilities sector, application of the civil and legal norms on data bases will be subject to the special norms of the housing law.

4   Conclusions

Legal regulation in the sphere of information systems in the RF is development very quickly. In the conditions of wide implementation of informatization in the housing and utilities sector, we offer to implement the following system legal corrections:

– establish the full list of subjects in relations for usage of information systems in the housing and utilities sector, as it is not possible to determine them due to lack of certainty in the ratio of general and special legal norms on information systems;
– specify the legal regime of data bases that perform the role of other information systems, in particular – distinguish specialized data bases for solving the national tasks.

– in order to avoid falsification of the decisions of general meetings of premises'
owners in the apartment block and for the purpose of keeping the balance of the
owners and initiators of the general meeting, it is expedient to use the electronic
signature with the usual certificate of verification key.

References

Manap, N.A., Hambali, S.N., Tehrani, P.M.: Intellectual creation in database: a superfluous test?
J. Intell. Property Rights **18**(4), 369–376 (2013)

Shaikh, S.-A., Londhe, B.R.: Intricacies of software protection: a techno-legal review. J. Intell.
Property Rights **21**(3), 157–165 (2016)

Amelin, P.B.: Federal law or information system: invitation to discussion. In: Administrative
Law and Procedure, No. 4, pp. 64–67 (2016)

Bachilo, I.L.: Information law: study guide for universities. - M.: Higher education, Yurait-Izdat,
p. 238 (2009)

Voynikanis, E.A.: Right of intellectual property in the digital age: the paradigm of balance and
flexibility. M.: Yurisprudentsiya, p. 208 (2013)

Eremenko, V.I.: Regarding the legal protection of allied rights in the Russian Federation. In: Law
and Economics, No. 2, pp. 30–55 (2012)

Inyushkin, A.A.: Information in the system of objects of civil rights and its interconnection with
intellectual property by the example of data bases. In: Information Law, No. 4, pp. 4–7 (2016)

Inyushkin, A.A.: The ratio of web-site on the Internet and data base. volume: Problems and
perspectives of socio-economic reformation of modern state and society. Materials of the 20th
international scientific and practical conference. The scientific publishing center "Institute of
strategic research". - Moscow, pp. 56–58 (2015)

Pichugin, I.L.: Application of GIS technologies – the effective methods of monitoring of the
objects of the housing and utilities sector. Bulletin of Orel State Agrarian University, vol. 31,
No. 4, pp. 76–79 (2011)

Tarasov, A.M.: Cryptography and electronic digital signature. In: Russian Investigator, No. 1,
pp. 50–54 (2014)

Khusyaynova, S.G.: Imperfection of the law on disclosure of information in the housing and
utilities sector: a prosecutor's opinion. In: Russian Justice, No. 11, pp. 16–18 (2016)

The Housing Code of the RF dated 29 December 2004 No. 188-FZ. Collection of laws of the RF.
No. 1 (part 1), Article 14, 03.01.2005

Federal law dated 3 July 2016 No. 267-FZ "Regarding the changes in the Housing Code of the
RF". Collection of the laws of the RF. No. 27 (Part I). Article 4200. 04.07.2016

Federal law dated 12 March 2014 No. 35-FZ "Regarding corrections in part 1, 2, and 4 of the
Civil Code of the RF and certain laws of the RF". Russian Gazette. No. 59. 14.03.2014

Federal law dated 27.07.2006 No. 149-FZ "Regarding information, information technologies,
and protection of information". Collection of the laws of the RF. No. 31 (Part 1). Article 3448
31.07.2006

Federal law "Regarding the state information system in the housing and utilities sector".
Collection of the laws of the RF. No. 30 (Part I). Article 4210 28.07.2014

Federal law dated 06.04.2011 No. 63-FZ "Regarding electronic signature". Collection of the laws
of the RF. No. 15. Article 2036, 11.04.2011

The Order of the Ministry of Communications of the RF No. 504, and the Order of the Ministry of Construction of the RF No. 934/pr dated 30 December 2014 "Regarding the official web-site of the state information system of the housing and utilities sector in the information and telecommunication network 'Internet'". Russian newspaper, vol. 42, 02.03.2015

Decision of the Court on intellectual rights dated 7 March 2014 No. C01-114/2013 on the Case No. A56-58781/2012

Migration Processes in the Legal Life of Society

Vitaly A. Ponomarenkov[1]([⊠]), Tatyana S. Cherevichenko[1],
Elena A. Efremova[1], Anna G. Bordakova[1], and Alexey V. Azarkhin[2]

[1] Samara State University of Economics, Samara, Russia
2770402@mail.ru, cherev_777@mail.ru,
lenoksamara97@gmail.com, ansan12263@mail.ru
[2] Samara Law Institute of the Federal Penitentiary Service of Russia,
Samara, Russia
aazarhin@mail.ru

Abstract. Topicality. Topicality of the studied problem is predetermined by the fact that the global migration processes take place in the world, and they bring positive and negative consequences for the society.

The purpose of the article is to find the reasons and mechanisms of migration, as well as their social consequences for the modern society.

The methodology of this work includes the cognitive and systemic & determining approach which required using a complex of general scientific, specific scientific, and special methods of cognition.

The article opens the essence of migration processes and shows various regulatory mechanisms of influencing the migration processes.

Practical significance. The article poses practical value for researchers who study migration processes and conduct socio-economic forecasting.

Keywords: Migration · Migration policy · Labor potential · Geopolitical factor

JEL Code: K 100 · K 150 · G 290

1 Introduction

Over the long history of mankind, people have constantly moved on the territory of the planet. Article 13 of the Universal Declaration of Human Rights says, "Everyone has the right to freedom of movement and residence within the borders of each state. Everyone has the right to leave any country, including his own, and to return to his country" (Universal Declaration of Human Rights 1948).

In the modern world, such movement of people (migration of people) has different consequences.

Migration (under certain conditions) supports the process of global economic growth, stimulates the development of countries and societies, enriches difference cultures and civilization, and plays an important role in solving national, regional, and global problems. However, in certain cases, it becomes a serious problem that can influence economic development of the countries and determine the level of social tension, political stability, demographic situation, ethnic relations, etc.

E. G. Popkova (Ed.): HOSMC 2017, AISC 622, pp. 234–240, 2018.
https://doi.org/10.1007/978-3-319-75383-6_30

In the process of globalization, the Russian Federation opened its borders for migrants, which led to growth of unemployment among the indigenous population and aggravation of social and legal conflicts, which led to increase of dissatisfaction with the state policy, etc.

Wise immigration policy and correct legal regulation in the regions of the RF should influence the social processes of the life of the society and provide a positive potential for development of the society's legal life.

2 Materials and Methods

2.1 Research Methods

The following methods were used in the process of the research: systemic & structural, comparative, and sociological.

2.2 Experimental Basis of the Research

The experimental basis of the research is Samara State University of Economics.

2.3 Stages of the Research

The problem was studied in three stages:

- first stage: theoretical analysis of existing methodological approaches given in the scientific literature and dissertation works on the problem, as well as theory and methodology of socio-legal studies; distinguishing the problem, purpose, and methods of the research, compilation of the plan.
- second stage: formation and substantiation of the complex of social conditions of the research; conduct of field survey, using the sociological method.
- third stage: analysis of conclusions received in the course of the research: specification of theoretical and practical conclusions, generalization and systematization of the received results.

3 Results

523 respondents were surveyed; the analysis of the regime of living activities and labor of migrants in Samara Oblast was conducted.

4 Discussions

A lot of unskilled labor force enters Russia; also, a negative tendency is emigration of highly-qualified personnel, especially representatives of science and higher education, employees of the healthcare sphere, which services have an important social and technological value for development of the country's potential.

The reasons for external migration include economic, social, family, religious, cultural, political, military, and existential (voluntary migration, people who want to live in another country) (Madison 2006) factors of society's life.

In its turn, the reasons for internal migration are search for a job, improvement of housing conditions, increase of living standards, change of way of life, etc.

Internal migration is popular in the countries with large territory and various climate and economic conditions – e.g., Russia. At that, for Russia internal migration is an important component of demographic and socio-economic development of the country on the whole and separate regions in particular.

In the countries with wide territory, a large role belongs to internal seasonal migration of work force, i.e., temporary movement of employees into rural territory for agricultural works, and from rural territory to cities.

The cause of international migration is economic difference in the level of wages, which could be received for the same work in different countries.

Lack of specialists in a certain profession or in certain country raises the wages for representative of this profession and stimulates the inflow of highly-qualified migrants, so external migration of work force is peculiar for increase of the share of highly-qualified specialists in it.

Migration is partially caused by such reasons as war (military action) – e.g., migration from Syria under attacks of the "Islamic state" to Europe; natural catastrophes and political, ethnic, and confessional conflicts. At that, forced migration can be a means of social control of authoritarian and totalitarian regimes, while voluntary migration is a means of social adaptation and a reason for growth of urban population.

Modern migration processes are a many-sided phenomenon that influences all aspects of development of society and the state.

Most of the modern theories of migration agree that migration is profitable for accepting countries and for the countries of departure (under the condition of control over the flows). At that, according to the scholars, migration does not negatively influence the level of unemployment and the level of wages in accepting countries (Haisken-DeNew and Zimmerman 1995; Brücker 2002).

Russia, with its low birth rate, high death rate, and large territory, needs a high-quality migration structure. That's why the task to "make our country attractive for immigrants" requires real actions from Russia, aimed at development of well-balanced migration policy, including the policy of acceptance of migrants, legal and institutional provision of their stay and employment (in case of temporary labor migration), and integration and naturalization (in case of permanent migration).

Of course, labor potential of migrants is different, so there should be measures of professional training and additional training of migrants, as well as socio-cultural adaptation in society.

Migrants (with low level of education, from remote and rural regions) are less socio-adaptable, so they are less inclined to use the existing social institutes and services – legal, educational, medical, national & cultural, etc. Most of social contracts are conducted by them through informal connections, primarily through relatives and friends, as well as the existing shadow institute of brokers in the sphere of organization of migration and employment of migrants. All this increases "migration risks" and decreases protection of migrants. On the other hand, such migrants possess underdeveloped legal awareness and prefer not to protect their rights or to do this through their informal (simply speaking, criminal) brokers.

From the beginning of mass labor migration, large and flexible networks have appeared, which are used by new generations of migrants for organization of trips and for emigration to Russia for permanent residency. These networks are not well-developed. Unlike traditional expat communities, they are usually informal and weakly institutionalized. However, they fill in the vacuum that appears due to lack of official services but are often more effective than official structures.

Most (up to 80%) migrants find the job through their relatives and acquaintances, i.e., with the help of existing informal migrants' networks. On this basis, the "institute" of private brokers is formed – these brokers work with most of the flow of labor migrants, which leads to a substantial socio-economic imbalance – non-payment of taxes, outflow of money from the country, distribution of infection diseases, growth of the level of illegal migration, etc.

Thus, the reasons for negative tendencies of the migration processes are the following:

1. Ineffectiveness of state migration policy, i.e., lack of clear systematized mechanisms for solving the problems related to stay of foreign citizens on the territory of the Russian Federation (unequal distribution of migrants on the territory of the country), which requires changing the laws that envisage the possibility of redistribution of migrants for the subjects of the RF.
2. Mismatch between the migration and labor laws, which are related to organization and functioning of the modern labor market, which leads to formation of shadow sectors of economy that support high demand for unskilled and low-paid work force.
3. Lack of the effective system of social protection and support for migrants.
4. Lack of effective system of social adaptation and integration of immigrants into the Russian society (e.g., the possibility of serving religious cults, creating conditions for implementation of socio-cultural needs, a chance for socio-psychological adaptation in society, etc.)
5. Formation of the conditions of tolerant attitude of the country's population and migrants, as well as other conditions necessary for development of normal living (Simanovich 2008).

The modern problems of illegal migration in Russia are caused by the fact that the legal field of migration is very narrow due to strict laws, and, thus, illegal migrants cannot legalize their position due to lack of procedures and the corresponding legal basis (Balashov 2009).

A large share of illegal migrants are socially vulnerable, as they are isolated from access to social institutes of the modern society and do not have a possibility to implement the universal rights and freedoms of human – so they are obliged to go to the sphere of shadow economy and criminal business.

Criminal migration is one of the negative aspects of the modern legal life of society, which negative consequences are typical for Russia, but its scale and character make it a threat to national security.

It is largely caused by the role of the criminal component of illegal migration. During criminal migration the persons understand that they violate criminal laws, which makes criminal migration a socially dangerous phenomenon. At that, criminal migration is characterized by relative mass, as we speak of large numbers of people who move into different countries with the purpose of performing illegal activities (performing crimes).

Criminal migration is a latent phenomenon, which is manifested in hidden character of territorial movements for the purpose of performing crimes. At that, criminal migration is a part of the context of various phenomena: socio-economic, political, national & ethnic, demographic, etc.

From the socio-legal aspect, criminal migration is a part of the developed infrastructure of criminal business; in view of the level of criminal penetration into legal business and high mobility of offenders, it is possible to speak of a potential threat to the state of the whole system of economic relations in a region or the country.

From the national & ethnic point of view, criminal migration is manifested in territorial movement of the nations and ethnic groups that are peculiar for high mobility in the criminal aspect. From the point of view of socio-demographics, criminal migration is "mixed" with natural migration.

As of now, criminal migration is a certain means of provision of sordid wishes of certain groups and the states which are used for achieving certain sordid purposes of illegal enrichment.

Solving the problems of illegal migration requires a lot of effort from the state and population; thus is true also for increasing the level of legal awareness among the population, development of a new common goal and socio-legal standards.

These problems could be solved with the following:

(1) improvement of the state social and migration policy;
(2) changing the existing migration law;
(3) effective opposition to illegal immigration;
(4) attraction of highly-qualified foreign specialists into the country;
(5) integration and adaptation of immigrants into the Russian society, as well as development of measures for formation of tolerant attitude to them with the population of the RF.

Migration in the modern world is a factor that can influence the economic development of countries and determines the level of social tension and political stability, demographic situation, and ethnic relations. Migration is considered to be an independent direction of the state policy not in all countries.

Wise policy and planned legal regulation are the tools for the state to influence these processes. It is possible to view migration as a solution to the demographic problem: inflow of able-bodied population with high level of education and professional qualification, as well as children. On the contrary, in case of lack of wise political and legal approaches, migration could turn into a negative factor that increases social tension, provokes ethnic conflicts, and aggravates criminal situation (Nemytina 2001).

It is necessary to state that against the background of low population density of Russia, the controlled and regulated migration could save the country, its nationhood, etc.

State, as a central element of the political system, influences all spheres of society's living activities. According to K.S. Gadzhiev, "state is an entity in which ethnic & national, socio-cultural, property, and civil interests of people are combined in various forms and levels" (Gadzhiev 1999).

Being a carrier of sovereignty (Article 4 of the Constitution of the RF), the state uses the corresponding institutes to enter the relations with social subjects and performs the main functions for managing the activities of the society.

The state's activities for managing the economic, social, national, migration, and other processes acquire the character of state policy (Zametina 2001). That's why the state has to develop and implement migration policy.

During development and implementation of state and regional migration policy, it is necessary to create:

- a single data base of migrants on the territory of the country;
- develop rules for quotas of migrants on the basis of the principle of proportional settlement, avoiding creation of ethnic "enclaves" of migrants;
- during adaptation of migrants, it is necessary to bear in mind that their integration into the RF does not mean that they have to refuse their ethnic features;
- controlled migration is an important factor of provision of economic development and the strategy of national security, constitutes an important resource and condition for the state's development.

The influence of the processes of globalization is unequal, which leads to disproportions of increase of the scales and volumes of migration.

5 Conclusion

It should be concluded that migration is one of the regularities of developing processes of socio-economic life of society in the conditions of globalization.

Due to Russia's lacking mobile work force and due to refusal from low-paid and "not prestigious" work, as well as destruction of the system of professional technical education, demographic decline, and other negative phenomena of the modern life of the society that led to low demand for free jobs, especially in the sphere of real production, labor migration would perform an important and positive function of industrial production, as well as development of non-production sphere (education, science, medicine).

This requires wise migration policy and solution of the following tasks that appear in the life of the modern society:

1. Differentiated approach to limitation of labor migrants as to the levels:

- provision of permanent residence to highly-qualified specialists with a perspective of obtaining citizenship of the RF (after the trial period);
- provision of work permit for several years to workers of industrial production and agriculture with a perspective of extension of the period (under condition of preservation of legally set requirements);
- training representatives of labor migration for the specialties that are necessary for the Russian economy;

- creating conditions that ensure adaptation of migrants (teaching the basics of life activities in the country, determining the temporary residence place, establishing a system of social control over migrants and activities of employers, coverage of the Russian labor law for migrants, etc.);
- creating conditions that favor the controlled labor migration into East Siberia and the Far East.

All this will allow creating the conditions for solving the migration problem of the legal life of society in the conditions of globalization and ensure observation of rights, freedoms, and interests of migrants and the population of accepting countries.

References

Universal Declaration of Human Rights. Resolution 217 A (III) of the UN General Assembly dated 10 December 1948

Madison, G.: Existential migration. Existent. Anal. **17**(2), 238–260 (2006)

Haisken-DeNew, J.P., Zimmerman, K.F.: Wage and mobility effects of trade and migration, CEPR Discussion Paper 1318, London (1995)

Brücker, H.: The Employment Impact of Immigration: a Survey of European Studies (2002)

Simanovich, L.N.: Actual problems of immigration in the Russian Federation. Migration picture of modern Russia. Migr. Law **4**, 24–27 (2008)

Balashov, Z.V.: Problems of criminal migration in modern Russia. Sci. Works **3**(9), 393–396 (2009)

Nemytina, M.V.: Regulation of migration processes in Russia: ratio of politics and law. Migration in Russia: Problems of legal provision. Saratov, p. 37 (2001)

Gadzhiev, K.S.: Political philosophy, p. 278 (1999)

Zametina, T.V.: Certain problems of correlation of the national and migration policy in the Russian Federation. Migration in Russia. Problems of legal provision. Saratov, p. 36 (2001)

Tools for Sustainability Management of Socio-ecological Systems in the Globalizing World

Aleksey F. Rogachev[1]([⊠]) [iD], Viktoria N. Ostrovskaya[2],
Alexandr S. Natsubidze[3], Tatiana N. Litvinova[1],
and Elena A. Yakovleva[4]

[1] Volgograd State Agrarian University, Volgograd, Russia
rafr@mail.ru, litvinoval358@yandex.ru
[2] North-Caucasus Federal University, Stavropol, Russia
ostrovskayav@mail.ru
[3] Moscow Institute of State and Corporate Governance, Moscow, Russia
[4] Voronezh State Forestry Engineering University Named After G. F. Morozov,
Voronezh, Russian Federation
elena-12-27@mail.ru

Abstract. The article is devoted to solving the following problem: further development of human society with preservation of current rate of production volumes growth and increase of the number of population may lead to the global catastrophe. The purpose of the work is to develop effective tools of sustainability management of socio-ecological systems in the conditions of globalization. The methods of the research include modeling and multi-criterial optimization, with the help of which the authors formulate the task of provision of socio-ecological systems sustainability in the conditions of globalization and offer a managerial model for solving it. The basis of the sustainability management tools of socio-ecological systems in the globalizing world should be economic & mathematical model of multi-criterial optimization of sustainability of economic and ecological development of modern global economic system. The authors denote three directions of solving this optimization task and develop the corresponding instrumentarium and the proprietary model of sustainability management of socio-ecological systems in the globalizing world.

Keywords: Management · Socio-ecological systems
Sustainable development · Globalization

1 Introduction

Striving for satisfying various interests of various economic agents under the conditions of limitation of resources and possibilities of the environment over the whole history of humankind led to aggravation of the ecological situation in the world. The key problem consists in the fact that further development of humanity with preservation of the current rate of increase of production volumes and increase of population might lead to the global catastrophe.

© Springer International Publishing AG, part of Springer Nature 2018
E. G. Popkova (Ed.): HOSMC 2017, AISC 622, pp. 241–247, 2018.
https://doi.org/10.1007/978-3-319-75383-6_31

Modern culture of consumption dictates need for increase of volumes of GDP. Quick growth of population in combination with the idea of humanity also causes growth of total needs of humankind. All this does not allow refusing from the plans of economic development. However, it is impossible to ignore the ecological consequences of intensive development of economy.

Over the recent decades, developed countries actively use such tool of sustainability management of socio-ecological systems as transfer of hazardous production on the territory of other countries. It does not allow solving ecological problems in the global scale but only improves the position of certain countries as compared to others. At that, development of the global socio-ecological system is still unsustainable.

In the conditions of globalization, all countries of the world are in close interdependence, and the idea that lies in the basis of modern approach to sustainability management of socio-ecological systems, which consists in striving for solving the ecological проблем by means of other countries without significant limitations from their side, is false. This is reflected by ineffectiveness of existing and necessity for development of a new instrumentarium of sustainability management of socio-ecological systems in the conditions of globalization – which is viewed in this article.

2   Methods

Sustainable development supposes combination of interests of development of production – primarily, industrial (Grincheva 2016) – and environment protection (Hák et al. 2016). Unsustainable development of socio-economic and ecological systems leads to crises (Kopnina 2016), which could be manifested in the sphere of economy and in the sphere of ecology (Aznar-Márquez and Ruiz-Tamarit 2016).

Management of socio-ecological and economic systems supposes state interference with market processes (Moussiopoulos et al. 2010) for limiting the development of economy for the purpose of minimizing the damage to the environment (Ferrão et al. 2014; Frolov et al. 2015; Popkova et al. 2015; Nadtochey 2010).

In the conditions of globalization, the process of managing the socio-economic and ecological systems becomes complicated (Ren 2012), as it is influenced by more factors (Cheng et al. 2015), and it is necessary to take into account interests of a large number of interested parties (Sturn 2013; Marsh 2012).

Based on the performed literature overview on the topic of the research, it is possible to conclude that despite a high level of elaboration of certain aspects of the solved problem, modern authors study it primarily at the theoretical level.

At that, practical tools of sustainability management of socio-ecological systems in the conditions of globalization remain beyond the limits of the performed studies. It causes the necessity for development of empirical and methodological basis of management sustainability.

The research methods include modeling and multi-criterial optimization, with the help of which the authors formulated the task of provision of sustainability of socio-ecological systems in the conditions of globalization and offer a managerial model for its solution.

3   Results

The authors offer that the tools of sustainability management of socio-ecological systems in the conditions of globalization be based on the proprietary economic & mathematical model of multi-criterial optimization:

$$\begin{cases} S = (GDP/NP)/DE \rightarrow \max; \\ GDP \rightarrow \max; \\ NP \rightarrow \max; \\ DE \rightarrow \min. \end{cases} \tag{1}$$

where S – sustainability of socio-ecological system;

GDP – volume of GDP within system;

NP – number of system's population;

DE – damage to ecology of system;

In the formula (1), the sought variables are volume of GDP, number of population, and damage to the ecology of environment. The targeted function supposes that sustainability of socio-ecological system, determined as ration of GDP per capita to the damage to environment, should strive to maximum with the following limitations: GDP and number of population should strive to maximum, and loss to environment – to the minimum.

These limitations are dictated by the very idea of sustainable development of the socio-ecological system, according to which the number of population constantly grows, which leads to increase of needs and necessity for satisfying them, by growth of GDP. At that, provision of this level and living standards of the population requires preservation of favorable environment and minimization of the damage to ecology from production activities.

Obviously, the limitations set in the optimization task contradict each other – increase of GDP inevitable leads to increase of damage to the environment. This task could be solved in the global, national, and regional scales. With the help of modern software (for example, MathCad program), it is possible to obtain Pareto limit for non-dominating solutions (series of optimal solutions each of which corresponds to the given limitations).

While selecting the best solution, depending on the socio-economic, cultural, and political peculiarities of the viewed system, it is necessary to determine priority of each of the limitations. Thus, interests of the growth of population might prevail over the purposes of provision of favorable conditions of life. Then the number of population and GDP will increase, and the state of ecology will aggravate.

Or, quite on the contrary, the system might be interested in provision of maximally comfortable conditions for life. Then, either GDP per capita will decrease, or the number of population will reduce. In either case, there are three directions for solving this optimization task within sustainability management of socio-ecological systems in the conditions of globalization.

The first direction supposes reduction of each human's needs within change of GDP/NP ratio. At the first glance, this might seem unreal, as it is considered that needs are unlimited and the possibilities of their satisfaction with the help of GDP growth are limited by existing resources and production capacities.

However, with more detailed analysis of this problem, it is possible to see that with improvement of technologies of production, the very product changes – and it can satisfy more needs, i.e., there is increase of feedback from production in the sphere of satisfaction of needs. Thus, while preserving the volume of production, it is possible to increase the level of satisfaction of society's needs. Within this direction, it is possible to offer the following tools:

- Modernization of technologies and equipment. This will allow increasing feedback from production and labor efficiency;
- Application of marketing for management of needs. With the help of marketing tools it is possible to create new needs and reduce the old ones;
- Implementation of innovations and creation of new types of products. Development of new products, which can satisfy more needs, will allow reducing the production volume with preservation of the level of need satisfaction.

The second direction is oriented at adaptation to new, less favorable conditions of the environment within the change of NP/DE ratio. This could be conducted within the evolution. As is known, human society differs from animals by the fact that it does not adapt to the environment but changes the environment. However, human can also adapt to the new ecology.

This supposes artificial creation of special closed areas with favorable environment for special groups of population, improvement of medical and biotechnologies for curing new types of diseases that emerge in the conditions of life in unfavorable environment, conquering space for relocation of part of the population, etc. The following tools could be offered within this direction:

- Development of innovations. Development of new technologies will allow creating and restoring favorable environment;
- Reduction of requirements to the environment. Within adaptation to new ecological conditions, it is possible to change a human's body with the newest technologies;
- Refusal from hazardous production. This tool supposes import of products manufacture of which supposes serious damage to the environment, or full refusal from consumption of such products.

The third direction is based on minimization of the damage done to the environment in the process of production activities within changing the GDP/DE ratio. This direction is the most realistic and perspective. It supposes development of innovational technologies for reduction of production wastes with preservation of its scales. The following tools could be offered within this direction:

- Attraction of investments. Implementation of the latest production technologies requires significant financial resources;
- Toughening of ecological standards. This measure is aimed at state stimulation of the process of business ecologization;
- Growth of demand for ecological responsibility of business. This allows the whole society to motivate business for increase of ecological compatibility of production.

In practice, it is possible to realize all the given directions simultaneously or some of them that fit the socio-economic system. Together, they form the model of sustainability management of socio-ecological systems in the conditions of globalization (Fig. 1).

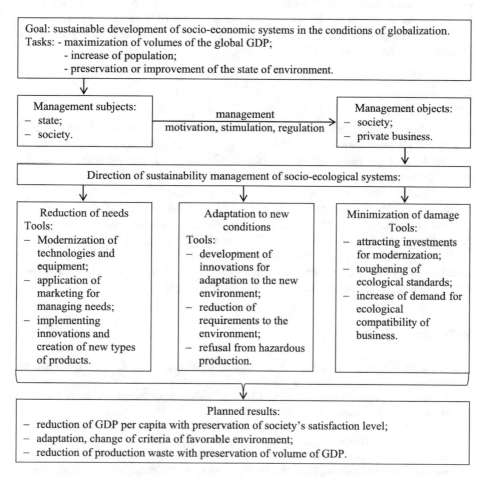

Fig. 1. Model of sustainability management of socio-ecological systems in the conditions of globalization

As is seen from Fig. 1, subjects of management in the offered model are state, which can regulate and stimulate society and business for increase of sustainability, and society, which can motivate private business for realization of the sustainability practice. As a result of realization of the model, GDP per capita should be reduced with preservation of society's satisfaction level, criteria of favorable environment should be adapted and changed, and volume of hazardous waste of production should be reduced with preservation of the GDP volume.

4 Discussion

Realization of the determined directions and the offered tools of sustainability management of socio-ecological systems will allow regulating the contradiction between socio-economic and ecological interests in the modern global economic system. At the national level – depending on the unique peculiarities of a specific country – it is possible to select and realize the best fitting direction of sustainability management depending on the set criteria of optimization.

5 Conclusion

It should be concluded that sustainable development of socio-ecological systems could be ensured by joint efforts of all economic agents. In the conditions of globalization, the problem of provision of sustainable development of socio-ecological systems should be solved by all members of international economic relations.

To a certain extent, a limitation of the results of this research is the full character of the offered tools and the model of sustainability management of socio-ecological systems in the conditions of globalization, as well as non-domination of solutions on Pareto limit for the developed model of multi-criterial optimization.

In the process of further research in the sphere of sustainable development, the attention should be paid to development of the mechanism of prioritization of optimality criteria for various socio-economic and ecological systems.

References

Hák, T., Janoušková, S., Moldan, B.: Sustainable development goals: a need for relevant indicators. Ecol. Ind. **60**, 565–573 (2016)

Aznar-Márquez, J., Ruiz-Tamarit, J.R.: Environmental pollution, sustained growth, and sufficient conditions for sustainable development. Econ. Model. **54**, 439–449 (2016)

Cheng, Y., Johansen, J., Hu, H.: Exploring the interaction between R&D and production in their globalization. Int. J. Oper. Prod. Manag. **35**(5), 782–816 (2015)

Ferrão, P., Ribeiro, P., Rodrigues, J., Lopes, A., Costa, E.L.: Environmental, economic and social costs and benefits of a packaging waste management system: a Portuguese case study. Resour. Conserv. Recycl. **85**, 67–78 (2014)

Frolov, D.P., Popkova, E.G., Stratulat, I.V., Shulimova, A.A.: Complex management of promising technologies: a case study of Russian nano-industry. Nanotechnol. Law Bus. **12**(2), 148–160 (2015)

Grincheva, N.: Sustainable development in cultural projects: mistakes and challenges. Dev. Pract. **26**(2), 236–250 (2016)

Kopnina, H.: The victims of unsustainability: a challenge to sustainable development goals. Int. J. Sustain. Dev. World Ecol. **23**(2), 113–121 (2016)

Marsh, P.: The New Industrial Revolution: Consumers, Globalization and the End of Mass Production, pp. 1–311 (2012)

Moussiopoulos, N., Achillas, C., Vlachokostas, C., Spyridi, D., Nikolaou, K.: Environmental, social and economic information management for the evaluation of sustainability in urban areas: a system of indicators for Thessaloniki, Greece. Cities **27**(5), 377–384 (2010)

Popkova, E.G., Abramov, S.A., Ermolina, L.V., Gandin, E.V.: Strategic effectiveness evaluation as integral part of the modern enterprise management. Asian Soc. Sci. **11**(20), 16–21 (2015)

Ren, X.: Building globalization: transnational architecture production in urban China. Urban Morphol. **16**(2), 165–166 (2012)

Sturn, R.: Economic citizenship rights as barriers to trade? Production-related local justice and business-driven globalisation. In: Merle, J.C. (ed.) Spheres of Global Justice, pp. 663–679. Springer, Dordrecht (2013). https://doi.org/10.1007/978-94-007-5998-5_53

Nadtochey, Y.I.: The phenomenon of one-sidedness in the policy of the USA and Russia. Int. Process. **8**(24), 85–95 (2010)

Analysis and Problems of Development of Regional Consumer Complex

Galina V. Golikova$^{(\boxtimes)}$, Galina N. Franovskaya,
and Olga B. Dzyubenko

Voronezh State University, Voronezh, Russia
ggalina123@yandex.ru, fgnvrn@yandex.ru,
shiyanovaolga1993@mail.ru

Abstract. The article is devoted to the issues of development of the consumer sphere of goods and services in regional economy and to the possibilities of increase of the consumer market's competitiveness in the conditions of economy's globalization and growth of territorial competition. It is considered that the consumer sphere of goods and services in middle-sized regions of Russia develops with insufficient intensity. Thus, it is necessary to develop the measures of state stimulation and regulation of the consumer market's development. As a result of research, the problems of development of the regional consumer complex are determined and the system of measures of state support for development of the consumer sphere of goods and services in a region is offered.

Keywords: Regional consumer complex · Consumer market
Development of the consumer sphere of goods and services

GEL Classification Codes: R 11 – General Spatial Economics:
Analysis of Growth, Development, and Changes

1 Introduction

In the conditions of economy's globalization and growth of territorial competition, development of a region depends on the capabilities to create and develop new structures that allow developing according to the global standards. Dynamic development of the consumer sphere of goods and services is one of the directions of increase of regional economy's competitiveness which allows solving a wide specter of socio-economic tasks. They include increase of living standards, increase of well-being and employment rate of the population, increase of region's competitiveness, and increase of innovative activity of socio-economic systems.

Characteristics and qualitative parameters of the regional consumer complex in Russia differ from the Western ones. It could be explained by the fact that Russia has more traditional spheres which ensure transportation and sales of the manufactured products. Besides, differentiation of the Russian regions as to various indicators of socio-economic development is very high [1].

The Central Federal District of the RF is the mode developed and the widest in the sphere of consumption of goods and services. A special role belongs to the transport

© Springer International Publishing AG, part of Springer Nature 2018
E. G. Popkova (Ed.): HOSMC 2017, AISC 622, pp. 248–255, 2018.
https://doi.org/10.1007/978-3-319-75383-6_32

and logistics, telecommunication, and financial services, which are the basis of the whole economy [2]. Still, they require constant improvement.

Consumer preferences and institutional conditions of the regions are very important for various economic sub-systems of Russia [3].

The influence of public authorities of various levels in the Russia's economic processes is very high [4].

Based on the documents of strategic planning of the RF and the Central Federal District (CFD), the key external factors of the consumer complex development in regions are the following [7, 8]:

- high dependence on import of food, raw materials, and components, machines and equipment, and everyday usage products;
- high level of cooperation between the region's economy and other subjects of the RF;

The key internal factors of development of the consumer sphere of goods and services in the CFD are the following:

- the dominating position in the Moscow region (80% of gross regional product of the region);
- substantial disproportions in the socio-economic development of the District's region;
- rates of the consumer market development of Moscow (50% of GRP is created in trade);
- tourism.

2 Analysis of Development of the Consumer Complex in the CFD of the RF

Let us characterize the structure of the consumer complex in the CFD. At present, the Central Federal District is peculiar for growth of the volume of sold goods and services on the whole and per capita. Still, the indicator "turnover of goods and services per capita" does not fully reflect the real socio-economic phenomena and processes, as the level of population's income is different for each region.

As of now, the consumer market is peculiar for the large volume of various goods and services. The consumer market of the CFD is determined by the indicators given in Table 1.

These data show that dynamics of turnover of retail of the CFD is positive. Its structure is in constant development and improvement. This is explained by growing indicators of the share of paid services in the total volume of retail, public catering, and paid services for the population.

Dynamics of development of the consumer complex in the regions of the CFD, which formed in 2011–2015, is peculiar for the positive tendencies (Table 2).

According to the statistical data, factual consumption of households grew by 169.3% in 2015; the average money income per capita grew by 165.5%; the average

Table 1. Turnover of retail, public catering, and paid services for the population of the CFD [9]

Indicators	2014	2015	2016
Total volume of retail, public catering, and paid services for population, RUB million	129,869.4	166,130.1	226,576.9
In % including:	100	100	100
Turnover of retail, RUB million	96,969.8	121,736.6	161,441.0
In % to total	74.7	73.3	71.3
Turnover of public catering, RUB million	6,549.4	7,837.6	10,543.5
In % to total	5.0	4.7	4.6
Paid services to population, RUB million	26,351.2	36,555.9	54,532.4
In % to total	20.3	22.0	24.1

Table 2. Dynamics of development of the consumer sphere of goods and services in the regions of the CFD in 2011–2015.

Indicators	Years					2011 to 2015
	2011	2012	2013	2014	2015	
Factual final consumption of households, in current prices, RUB billion	18,928	23,684	25,039	27,660	32,053	169.3%
Per capita, RUB	132,542	165,920	175,361	193,630	224,206	169.2%
In % as compared to the previous year, in compatible prices	112.5	109.4	95.5	104.0	105.8	−6.7 pct
Money income of the population per capita, monthly, RUB	12,540.2	14,863.6	16,895.0	18,950.8	20,754.9	165.5%
Real disposable money income of the population, in % as compared to the previous year	112.1	102.4	103.0	105.9	100.4	−11.7 pct
Average monthly nominal accrued wages of the people employed in economy, RUB	13,593.4	17,290.1	18,637.5	10,952.2	23,369.2	171.9%
Real accrued wages, in % as compared to the previous year	117.2	111.5	96.5	105.2	102.8	−14.4 pct
Average pension, RUB	3115,5	4198,6	5191,1	7476,3	8202,9	263,3%
Real volume of pensions, in % as compared to the previous year	104.8	118.1	110.7	134.8	101.2	−3.6 pct

(*continued*)

Table 2. (*continued*)

Indicators	Years					2011 to 2015
	2011	2012	2013	2014	2015	
Minimum subsistence level, per capita RUB per month	3,847	4,593	5,153	5,688	6,369	165.6%
In % as compared to the previous year	112.4	119.4	112.2	110.4	112.0	−0.4 pct
Population with the money income below the minimum subsistence level; million people	18.8	19.0	18.4	17.7	18.0	95.7%
In % of the total number of population	13.3	13.4	13.0	12.5	12.7	−0.6 pct
In % as compared to the previous year	87.0	101.1	96.8	96.2	101.7	14.7 pct
Ratio of the value of the minimum subsistence level to the monthly average accrued wages	327	348	334	341	340	13 pct

accrued wages of the people employed in economy grew by 171.9%; the average volume of pensions grew by 263.3%.

Besides the number of the population below the poverty line (people with money income below the minimum subsistence level) decreased by 4.3% by 2015; as to the share of the total number of the Russian population – by 0.6 pct.

At that, in "real" value, i.e., in view of inflation, all indicators of the money income of the population reduced over the studied period: for income – by 11.7 pct, for accrued wages – by 14.4 pct, for pensions – by 3.6 pct. This influenced the dynamics of the indicators of formation of the factual final consumption, which sources, according to the official statistics, are expenditures of household and social transfers in the natural form.

Dynamics of the structure of consumption for the directions of consumer expenditures is shown in Table 3.

Quicker growth of the population's expenditures for consumption of food products is confirmed not only by structural but also by "per unit" indicators.

It should be noted that the indicators of dynamics of development of retail and turnover per capita slowed down as calculated by compatible prices.

It is known that due to large difference in socio-economic development of the regions of the CFD, retail turnover is differentiated. This is true for the "per unit" indicator of turnover, calculated per capita for the corresponding territory, which varies from RUB 68,416 in 2015 for a resident of Orel Oblast (per year) to RUB 286,952 for a Moscow resident. In view of this, we studied the changes in distribution of the regions of the CFD as to retail turnover per capita for 2011–2015 in Table 4.

Table 3. Dynamics of the structure of consumption for the directions of consumer expenditures in the CFD

Directions of consumer expenditures for the purpose of consumption	Years					2011 and 2015, (+, −)
	2011	2012	2013	2014	2015	
Consumer expenditures – total	100.0	100.0	100.0	100.0	100.0	
Including: Expenditures for food products and non-alcoholic beverages	28.4	29.1	30.5	29.6	29.5	1.1
Alcoholic beverages, tobacco products	2.4	2.3	2.4	2.4	2.5	0.1
Clothing and shoes	10.4	10.4	10.4	10.8	10.1	−0.3
Household services, water, electric energy, gas, and other types of fuel	11.6	10.4	10.8	11.3	11.4	−0.2
Household articles, household appliances, and house care	7.3	7.5	7.0	6.2	6.5	−0.8
Healthcare	3.1	2.9	3.1	3.3	3.5	0.4
Transport	16.6	15.5	13.4	14.9	15.9	−0.7
Communications	3.8	3.7	3.8	3.8	3.7	−0.1
Leisure and culture events	6.4	3.7	7.3	6.8	6.8	0.4
Education	1.8	1.6	1.5	1.3	1.2	−0.6
Hotels, café, and restaurants	3.0	3.0	3.4	3.4	3.2	0.2
Other goods and services	5.2	5.9	6.4	6.2	6.0	0.8

The grouping showed that over the five-year period there took place a substantial shift in distribution of regions in the direction of growth of the retail turnover per capita.

This allows for the conclusion on the tendency of leveling of the realized consumer demand, which is reflected by the indicator of turnover of retail per capita. This is a positive aspect – in view of its influence on the formation of consumer behavior.

In the regional distribution of retail turnover per capita, the positive changes in the direction of regions' shift into the groups with higher intervals of the indicator and leveling of the realized consumer demand took place, confirmed by equal distribution of regions at the end of the research period.

The consumer complex of the CFD is developing very quickly. Wholesale and retail trade accounts for a third part of the gross regional products of the District. In 2011, the crisis phenomena in the economy led to reduction of retail turnover in the CFD by 4.8%. The trade sphere is peculiar for domination of the Moscow region. Retail turnover for the trade networks of the District constitutes 16.8% of the total volume, which shows a significant potential of development of trade networks.

One of the most urgent problems that hinder the development of the consumer complex in the regions of the CFD is insufficiently developed trade infrastructure and low rates and volumes of construction of new trade areas.

Table 4. Dynamics of retail turnover per capita in the CFD for 2011–2015

Indicators	Years					2011–2015 (+, −)
	2011	2012	2013	2014	2015	
RUB, in factual prices						
Turnover of retail, including:	171,320	188,757	207,394	231,873	237,737	66,417
Food products, including beverages and tobacco	82,919	90,415	101,623	113,386	119,582	36,663
Non-food products	88,401	98,342	105,771	118,487	118,155	29,754
In % as compared to the previous year, in compatible prices						
Turnover of retail, including:	116.5	110.2	109.9	111.8	102.5	−14.0
Food products, including beverages and tobacco	113.7	109.0	112.4	111.6	105.5	−8.2
Non-food products	119.3	111.2	107.6	112.0	99.7	−19.6

The measures of state policy in the sphere of development of the consumer complex should be oriented at increase of accessibility of areas and lands for development of trade and logistical organizations, quick increase of communal generating and distributing capacities, and development of the sphere of telecommunications and electronic payments.

The main mechanisms of development of the consumer complex infrastructure are construction of infrastructural objects, usage of the mechanisms of public-private partnership, provision of various subsidies, and creation of the necessary administrative conditions for the purpose of attracting private capital for development of infrastructural objects.

An important task of development of the consumer sphere of goods and services is reduction of disproportions in development of trade in industrially developed centers of the macro-region and remote communities, where the material and technical basis of trade companies is ineffective and old – both morally and physically.

At the regional level, for the purpose of development of the consumer sector, it is possible to use the mechanism of formation of sectorial and inter-sectorial clusters [5].

Analysis of the state and problems of the consumer complex allows determining a range of top-priority directions for further development of the consumer sphere of goods and services at the meso-level. They include:

(1) scientific substantiation of the territorial organizations of the consumer complex in view of peculiarities of settlement and social structure of population;
(2) development of the marketing studies aimed at the issue of the population's demand for goods and services;
(3) improvement of the structure of services by means of increase of the share of educational, medical, financial, information, service, and tourist & recreational services;

(4) development in the production sphere of consumer goods and services of small entrepreneurship and medium business, including attraction of foreign capital (creation of joint companies);

(5) active search for source of financing and development of the economic mechanisms for stimulating entrepreneurial activities, aimed at fuller satisfaction of the population's demand for goods and services;

(6) strengthening of the material and technical basis of industrial companies by means of restoration of the main funds and implementation of the world-level progressive technologies, which will allow diversifying the assortment and improving the quality of their products;

(7) implementing the achievement of the scientific and technical revolution into the consumer sphere, including communications means and information technologies – which will influence positively the development of the consumer complex and increase of the quality of goods and services for the population.

3   Conclusions

The Russian and foreign experience of functioning of the regional consumer complexes confirms the necessity for interaction between the public and local authorities and the key participants of the consumer sphere of goods and services. This aspect positively influences the processes of consolidation of the efforts of the state and business structures for increase of the region's competitiveness in the consumer sphere.

For development of the regional consumer complex, it is necessary to take into account advantages of the specific region, which will allow attracting investments. Besides, active development of the consumer sphere of goods and services requires state regulation and support in the form of provision of subsidies and state guarantees. Implementation of cluster project in the consumer complex may become a multiplier for development of the region's economy.

Bases on the monitoring of development of the consumer sphere of goods and services, it is possible to conclude that the situation in the consumer market is viewed as positive, and the forecast for its development is positive as well. However, public authorities and the entrepreneurial community have to show clear actions for solving the most topical problems of the consumer complex.

References

1. Ibragimova, N.A., Parakhina, V.N., Pankova, L.N., Kalyugina, S.N., Khanaliev, G.I.: Differentiation of regions of the Russian Federation as to level of budget revenue potential of municipal entities. J. Appl. Econ. Sci. 11(5(43)), 940–954 (2016)
2. Makarov, E.I., Nikolaeva, Y.R., Shubina, E.A., Golikova, G.V.: Impact of risks on stable and safe functioning of transport and logistics cluster of the transit region. In: Popkova, E.G. (ed.) Russia and the European Union, Development and Perspectives. Contributions to Economics, pp. 321–326. Springer International Publishing AG (2017)

3. Treshchevsky, Y., Nikitina, L., Litovkin, M., Mayorova, V.: Results of innovational activities of russian regions in view of the types of economic culture. In: Russia and the European Union, Development and Perspectives. Contributions to Economics, pp. 47–53 (2017). ISBN: 978-3-319-55256-9 (Print) 978-3-319-55257-6 (Online)
4. Risin, I.E., Treshchevsky, Y.I., Tabachnikova, M.B., Franovskaya, G.N.: Public authorities and business on the possibilities of region's development. In: Popkova, E. (ed.) Overcoming Uncertainty of Institutional Environment as a Tool of Global Crisis Management. Contributions to Economics. Springer, Cham (2017). https://doi.org/10.1007/978-3-319-60696-5_8, ISBN 978-3-319-60695-8 (Print), ISBN 978-3-319-60696-5 (Online)
5. Vertakova, Y., Risin, I., Treshchevsky, Y.: The methodical approach to the evaluation and development of clustering conditions of socio-economic space. In: Proceeding of the 27th International Business Information Management Association Conference – Innovation Management and Education Excellence Vision 2020: from Regional Development Sustainability to Global Economic Growth, IBIMA (2016)
6. Vertakova, Y.V.: Vector analysis of cluster initiatives of the region. In: Vertakova, Y.V., Polozhentseva, Y.S., Klevtsova, M.G. (eds.) Scientific and Technical Bulletin of St. Petersburg State Polytechnic Universitya. Economic Sciences, no. 1(211), pp. 43–50 (2015)
7. The Concept of long-term socio-economic development of the RF until 2020. http://www.consultant.ru/document/cons_doc_LAW_82134/28c7f9e359e8af09d7244d8033c66928fa27e527/
8. Strategy of socio-economic development of the Central Federal District until 2020. http://economy.gov.ru/wps/wcm/connect/16ee8ae4-4bdb-4616-9375
9. Economy of the Central Federal District. http://newsruss.ru/doc/index.php. Accessed 25 Apr 2017

Information Provision of Planning the Balance of the Innovational and Investment Spheres of Activities

Olga A. Boris[1](✉), Pavel N. Timoshenko[2],
and Valentina N. Parakhina[1]

[1] North Caucasus Federal University, Stavropol, Russia
{boris, vparakhina}@ncfu.ru
[2] Nevinnomysk State Humanitarian and Technical Institute,
Nevinnomyssk, Russia
timpol@bk.ru

Abstract. The issues studied in this article are very topical, as all specialists are confident in the necessity for provision of well-balanced innovational and investment spheres of activities. For management of these processes, it is important to have correct information on their dynamics, as well as methodologies and tools of its processing for assessment of balance.

For solving the set problem, the essence and indicators of balance of the innovational and investment spheres of activities were analyzed. It was determined that the notion "balance" has the financial, economic, ecological, organizational, and technical & technological senses. The system of indicators that reflects them has to conform to certain requirements (accessibility, authenticity, fullness, and compatibility) and include the indicators of balance of the required resources and their sources, as well as coordination of the required and existing (planned, forecasted) rates of change of the results of their usage.

It is especially important to determine and prevent the critical level of imbalances in the state system of management, including the sphere of innovational development and its investment provision. One of the key aspects in this process is formation of the comprehensive innovational & investment policy at the state level as a primary precondition for the balance of the innovational and investment spheres of activities and the basis for its planning.

The authors identify the sources of the information necessary for conduct of evaluation of balance and methods of assessment of its indicators and distinguish the most acceptable of them, which, due to high information content and relative simplicity, include the following: the matrix (for evaluation of the resources' balance) and correlation methods (for evaluation of balance of dynamics of the basic indicators of development).

The principles (of normative & legal provision, strategic orientation, coordination of interests, and constant monitoring) and organizational & methodological approaches to information provision of the process of planning of balance are determined.

The article provides the results of evaluation of balance at the level of the sphere (industry). The determined limitations in development show that it is necessary to reorganize management of regional and sectorial systems from the positions of provision of balance. For this, the mechanism of information

© Springer International Publishing AG, part of Springer Nature 2018
E. G. Popkova (Ed.): HOSMC 2017, AISC 622, pp. 256–268, 2018.
https://doi.org/10.1007/978-3-319-75383-6_33

provision of planning of balance of the innovational and investment spheres of activities is offered.

Keywords: Information provision · Planning · Well-balanced development Innovative activities · Investment sphere · Industry Methods of evaluation of balance

1   Introduction

The value of indicators of balance development is that they provide factual basis for determining:

strategic feedback that shows to the persons who make decisions the current status of the organizations from several perspectives;
diagnostic feedback from various processes for management of changes;
temporary tendencies of the change of effectiveness of work with control over indicators;
feedback between the methods of measuring and selection of controlled indicators;
quantitative incoming parameters for the methods of forecasting and modeling for the systems of decision support.

Topicality of the problem of information provision of planning of balance development of industry is caused by a lot of problems of the methodological, organizational, technological, and economic character.

According to the works (Niven 2004; Sirotkina and Vorontsova 2014; Khamidulina and Suldina 2015, Goncharov and Sitorkina 2015a, 2015b; Kruglyakova et al. 2017), as well as Parakhina and Khanaliev 2011, Boris 2014, Timoshenko and Gorbenko 2016, there is no methodology for balance evaluation that is acceptable for various spheres. Therefore, it is necessary to compare different approaches that allow evaluating the level of "imbalances" and planning their reduction.

Organizational problems are manifested in the fact that the existing system of planning supposes evaluation of balances of the material and financial characters (which is reflected in the works on planning of regions and separate spheres – (Basovsky 2002; Maximenko 2009; Tadtaev and Parakhina 2016) and does not deal with the issues of collecting information for coordinating the tendencies of innovational & investment development.

The technological problems are related to weak elaboration of the technologies of receipt and processing of interconnected information of the innovational and investment spheres of activities – as is seen in the works (Anshina 2006; Arsenyeva et al. 2016; Bergman and Charles 2003; Goncharenko 2014; Doloro and Parto 2003), etc.

The economic problems are determines by increasing negative influence of imbalance and uncertainty of the forecast situation in the sphere of the innovational & investment spheres of activities on the results of development of the spheres of the Russian national economy – which is shown by the works (Glazyev 2013; Parakhina et al. 2015; Treshchevsky et al. 2017).

2 Methods

The following methods were used:

- monographic, for studying the essence, principles, and approaches to creation of the system of information provision planning of balance of the innovational and investment spheres of activities and formation of the investment policy at the state level;
- economic & statistical – for receipt, selection, and evaluation of information on the indicator of balance of the innovational & investment spheres of activities;
- index, matrix, economic & mathematic modeling – for complex evaluation of the level of the system's development balance;
- induction and deduction for the formation of conclusions for the results of calculations of the indicators of balance of the innovational & investment spheres of activities in the analyzed systems.

3 Main Part

1. *Studying the essence and the indicators of balance of the innovational & investment spheres of activities*

For the purposes of monitoring and diagnostics of the influence of various conditions on well-balanced innovational development, for adequate managerial interaction at the macro-economic level, it is necessary to apply special research methods, developed in view of peculiarities of the component structure of balance. This condition predetermines the topicality of substantiation the indicators of well-balanced development and further development of methodological approaches to conduct of monitoring of the socio-economic position and interpretation of the received results from the positions of well-balanced development. The system of indicators of socio-economic development should conform to the requirements of completeness, accessibility, and correctness of the information; possibility of expansion of spatial and time limits; compatibility; information content of results.

That's why the indicators should be developed on the basis of priorities of the strategic plan which contains the key factors of development of business and criteria of selecting the indicators that are most interesting for the managers. Then, the processes of information collection are planned and bringing them into the numerical from for storage, reflection, and analysis.

During determining the criterion of balance, it is necessary to consider several aspects of this category, providing each indicator in two "views" (the present and the future) and as to a certain level of the production complex (a company or the sphere on the whole, a region or the country on the whole). Integrating various approaches to the forms of balance (Timoshenko and Gorbenko 2016), we offer – for the purpose of similarity of evaluation of the internal and external balance – to study the following list of forms of balance: financial; economic; material; targeted (large industrial systems – sphere, region, country)/technical & technological (companies and organizations); innovational; labor; social, ecological.

Let us view the essence of balance of industrial complex by its forms and select one of the indicators as the main criterion of its assessment.

Firstly, financial balance is the balance of needs and sources of financial resources, which could be characterized by profitability or profitability of capital in industry, as only a certain level of profitability allows accumulating own assets for development (at least, profitability should be above the inflation level) and attracting borrowed assets (profitability should be above the interest rate for the credits).

Secondly, economic balance, as ratio between industrial production (its results) and needs for them (demand for industrial production). This ratio cannot be reflected by enlarged indicator, though it may be characterized by the level of business activity and turnover of capital in the industry on the whole and for the spheres and companies: if the products does not satisfy the demand, the company acquired more products, and the stock of final products and unfinished goods are increased, the capital becomes "dead", and its turnover reduces.

Thirdly, labor balance, as proportional distribution of public labor (resources) between various spheres of industry and balance between the required and existing human capital (quantity and quality), which could be expressed by the indicators of effectiveness of human resources use, as the lack (or "excess") of labor force leads to its "overload", loss of the potential of labor activity (or "underutilization") – as a result, labor efficiency decreases.

Fourthly, targeted and technical & technological balance as the balance of existing and required (according to the priorities of development) production capacities and correspondence of the performed changes to the set strategic goals of development of industry. At a company, this comparison is conducted for each area of product creation (in the quantitative and qualitative aspects). We think that this is reflected in the level of return on assets. Presence of "narrow" spots in the technological chain leads to underutilization of other areas of the production chain and lower return of production funds on the whole.

Fifthly, material balance – as the balance of existing and required material resources (raw materials, spare parts, etc.). Lack of one resource leads to its urgent and more expensive purchase, use of replacements, reduction of quality and/or growth of the cost of products cost in the part of material expenditures. As a result, the share of material expenditures in the products' cost grows.

Sixthly, balance of innovational processes, as the ratio between the existing and used innovational potential. Underuse of the innovational potential leads to overrun of the rates of growth of expenditures for innovations and development over the rates of growth of the innovational product.

Seventhly, social balance as a reflection of the level of social provision of employees, expressed by the level of wages and the volume of expenditures for the social programs in the industrial complex and their growth in view of inflation.

Eighthly, ecological balance as a ratio of growth rates of production (reflects the usage of natural resources) and the costs of reproduction of natural resources (assets allocated for ecological programs).

This structure of the indicators of balance is applicable to the industrial sphere of the country on the whole and reflects the balance of its development.

2. *Necessity for formation of the integrated innovational & investment policy and the system of collection of information on its implementation at the state level*

For evaluation of balance of production complexes, it is necessary to collect the required information on the state of their innovational and investment spheres and to observe certain priorities that ensure coordination of their development. These priorities are set by the public authorities bodies within the corresponding policies, strategies, or concepts of development of the spheres, regions, and the country on the whole.

An important element of the state economic policy is innovational policy, which supposes selection of the top-priority directions in development of science and technology and full support from the state in their development. As a rule, when developed independently, investment policy creates conditions for implementing the innovational programs and projects – but analysis of the goals, mechanisms, and tools of state regulation of investment policy that are used in practice show that they are ineffective as to the innovational sphere.

Thus, it is expedient to use the complex and systemic approaches during formation of the innovational & investment policy at the state level, which allows viewing it as a union of two spheres: innovational and investment; as well as interconnected components: goal, limitations, the processes of entering and exiting (Fig. 1).

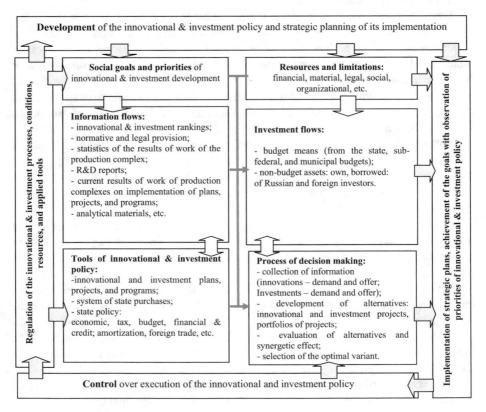

Fig. 1. The system approach during formation of the state innovational & investment policy

During formation and implementation of the state integrated innovational & investment policy, it is necessary to pay attention to such constant negative factors as inequality of distribution of the investment potential among the country's regions and deep differences in the regional investment risks.

3. Identification of the methods of evaluation of the balance indicators

The main idea of the increase of the level of balance of functioning supposes creation and usage of its indicators. The indicators of balance are measured characteristics of products, services, processes, and operations that are used by the company for tracking and increasing the balance.

Analysis of the methodological approaches to evaluation of balance of innovational & investment development of the production complexes allowed determining the following.

Most often the researchers use the system of well-balanced indicators offered by Kaplan and Norton (2003), Neeley et al. (2003), Niven (2004). It includes the following groups: finances, satisfaction of customers, internal business processes, and the company's capability to develop and grow. However, this concept does not draw a difference between achievement of internal and external balance – i.e., this system could be applied for the companies of various levels, but not for their aggregated totality in the form of sectorial complexes, regional and national economies.

The opinion of Goncharov and Sirotkina (2015a, 2015b) on the issue of well-balanced regional development on the basis of implementation of the stakeholder and Social Darwinism approaches, which reflect the external and internal balance of the socio-economic system, is especially interesting.

However, the works of various authors do not set the criteria of balance that would ensure the compatibility of evaluations of organizations and complexes of different levels. It seems that the indicators of this evaluation could be formed on the basis of the selected methods of provision of correspondence and become the calculations of:

(1) index of correspondence to the criterial indicator of balance;
(2) indicator of observation of normative proportions and limitations;
(3) level of implementation of the necessary functions and operations;
(4) coefficient of correlation of the rate of change of various indicators and the set ranks of priority.

On the basis of private indicators of balance of innovational & investment development for the formed groups $Ki(i = 1, 2, \ldots, n)$, the complex indicator is calculated (1):

$$Kp = fp(K1, K2, \ldots, Kn), \tag{1}$$

where $fp(K1, K2, \ldots, Kn)$ – function of variables $(K1, K2, \ldots, Kn)$.

The methodological basis for determining this indicator could be the multi-factor model. The model of multi-factor analysis is of the universal character, as it allows studying various phenomena and processes. According to this model, balance should be evaluated at all levels of analysis: federal, regional, sectorial, and entrepreneurial. The system of criteria of balance of innovational & investment development forms on the basis of analyzing the general strategic goals of the economic system, problem spheres for innovational activities, and investment possibilities of the economic system.

The drawbacks of the traditionally used economic & mathematical and statistical methods are the necessity for quantitative determination of the evaluation's indicators. For overcoming the drawbacks of the traditional methods, it is possible to use the theory of fuzzy logic, theory of graphs and scenario approach, which are applied in the works of Prigozhin (2003), Nedosekin (2003), Katkova et al. (2013), etc.

In the process of analysis, the most acceptable methods of evaluating the balance of innovational & investment development of the production complex were determined; the following ones were assigned to them due to high information content and relative simplicity of usage: matrix (for evaluation of the resources' balance) and correlation methods (for evaluation of balance of dynamics and the basic indicators of development) (Timoshenko and Parakhina 2017). Determination of the methodological approaches to evaluation of balance and conditions that determine it is the initial position for development of the recommendations aimed at effective management of economy for the purpose of provision of its well-balanced development.

4. *Conduct of evaluation of balance in the sectorial and regional levels*

For evaluation of balance of the industry development, the growth rates of each indicator is assigned with factual rank; based on coordination of normative (Table 1) and factual ranks, the level of balance of development of industry with the usage of the correlation coefficients of the ranks of Spearman (Kc) and Kendall ($Kк$) is used:

$$Kc = 1 - \frac{6\Sigma di}{n \times (n2 - 1)}, \tag{2}$$

$$Kk = \frac{(Sp - Sn)}{\frac{1}{2} \times n \times (n - 1)}, \tag{3}$$

where
di – second degree of difference between the normative and factual ranks;
n – number of the observed features (indicators);
Sp – number of the rank ratios that were observed;
Sn – number of violated rank ratios

The results of evaluation of balance, performed with the usage of the offered totality of indicators, show that the level of balance of development of the Russia's industrial complex was below average in the pre-crisis period of 2011–2012: less than half of the required normative ratios were observed.

In the period of quick decrease of oil prices and implementation of economic sanctions, the level of balance drops down. However, apart from the export-import operations (due to the policy of import substitution), it restores rather quickly: even at the higher level than in 2011–2012. However, export-import operations are the top-priority direction, so the level of balance is lower than before the sanctions (Fig. 2).

Table 1. The ranked basic list of indicators for evaluation of balance that takes into account the priorities of modern development of industry (innovational and social orientation of development)

	Rank of norms	Indexes of growth of indicators *				
		2011	2012	2013	2014	2015
Index of growth of balanced financial result	1	117.77	102.16	77.58	105.48	136.31
Index of growth of gross added value of industrial production	2	119.26	111.65	106.26	108.29	111.59
Index of investments into fixed capital	3	120.64	118.73	108.99	105.43	110.24
Index of growth of fixed funds in industrial production	4	117.35	113.89	113.51	113.70	113.22
Index of growth of the volume of innovational industrial goods	5	158.47	135.85	122.43	98.86	107.27
Index of the volume of export of industrial goods	6	130.13	101.55	100.24	94.56	69.07
Index of industrial production for the types of economic activities - processing production	7	108	105.1	100.5	102.1	94.6
Index of industrial production, total	8	105	103.4	100.4	101.7	96.6
Index of the volume of import of industrial goods	9	133.57	103.76	99.38	91.04	63.65
Index of growth of the expenditures for scientific and technical works	10	168.42	116.73	123.65	113.05	125.81
Index of growth of internal expenditures for R&D	11	125.62	121,09	118.81	102.72	122.52
Index of growth of the number of industrial companies that perform scientific and technical works	12	117.65	97.86	97.08	103.38	134.91
Index of growth of expenditures for payment of labor for production of industrial goods	13	114.88	109.56	109.44	106.71	107.51
Index of growth of investments into fixed capital, aimed at protection of environment	14	105.92	110.02	123.57	124.49	107.83
Index of growth of current expenditures for protection of environment	15	100	100	106.69	103.35	109.97
Index of growth of material expenditures for production of industrial goods	16	123.03	108.82	106,07	108.64	110.79
Index of growth of expenditures for production of industrial goods	17	122.66	110.10	111.19	107.80	106,41
Index of growth of average annual number of companies' employees	18	99.90	98.61	99.10	97.66	98.41
Index of growth of average annual number of employees in the industrial production	19	100	99.25	99.24	97.71	100

*Based on the data of the Federal State Statistics Service. http://www.gks.ru/wps/wcm/connect/rosstat_main/rosstat/ru/statistics/enterprise/industrial/

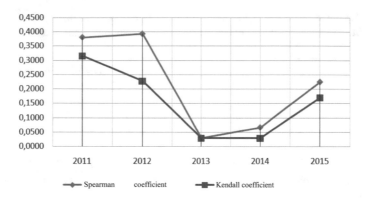

Fig. 2. Evaluation of balance of the Russian industry development (in view of export-import operations)

It is possible to say that social priorities are damaged. Without considering the social priorities, balance of industry is higher and grows in a stable way (Fig. 3).

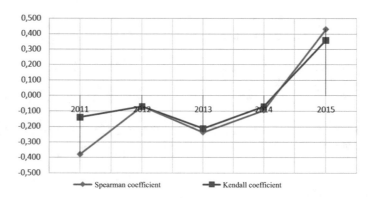

Fig. 3. Evaluation of the production and financial balance of the industrial complex of the RF

On the contrary, balance of the industrial complex reduced in 2014 and reached zero level in 2015. The obtained data show that the financial and economic problems cannot be solved by means of social results and expenditures, though attention is not paid to the social problems.

In the course of the complex diagnostics of well-balanced development, the key problems were determined; a conclusion was made that the available reserves of the previous stage of development are depleted, and there's necessity for the innovational approach, which supposes re-industrialization and development of the powerful "innovational sector" in the reproduction system.

5. *Organizational and methodological approaches to provision of the process of planning of balance with the necessary information*

Overview of the existing approaches to formation of the information and analytical sub-system allowed distinguishing the basic functional tasks and methods of their solution (Fig. 4).

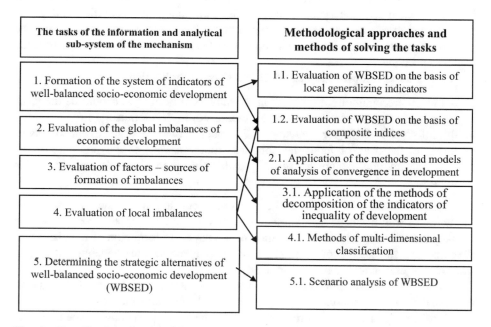

Fig. 4. Classification of tasks of the information and analytical sub-system and methodological approaches to their solution

Management of innovations at the company level and at the level of industrial complex is the criterial sign of the possibility for well-balanced management of changes in the strategic aspect. The innovational sub-system determines the necessity and possibility for well-balanced changes in other components of an industrial company management.

Developing the organizational and methodological approaches to information provision of the process of planning of balance, it is possible to distinguish the following principles of implementation of this process:

- normative and legal provision, which supposes coordinated improvement of the normative and legal provision of investment & innovational activities;
- strategic orientation which shows concentration of efforts on the strategic directions of balance;
- coordination of interests, including organization of interaction of public authorities and business for the purpose of mobilization of investments at the top-priority directions of innovational policy;
- constant monitoring, includes constant control and analysis of positive and negative aspects of the innovational & investment development.

4 Conclusions

The necessity for managing the balance of balance of development of socio-economic systems, including the industrial complexes, is caused by the fact that violations of balance lead to quick reduction of effectiveness of the companies' activities, emergence of crises and threats to national economic security. The mechanisms of managing the balance of industry's development should ensure the prevention of emergence of these negative phenomena.

Their information provision could be based on the following:

– the initial precondition for developing the models of evaluation should be the ideas of the main reasons for emergence of imbalances in development;
– the information basis for evaluation of balance should be organization of monitoring of the state of industrial complex, timely setting of the problems of its development, and diagnostics of the possible threats to economic security of the region.

From the position of the systemic and cybernetic approach, the mechanism of formation of well-balanced development could be treated as a system for conducting the targeted activities related to evaluation of the threats to formation of imbalances of development, forecasting of their negative consequences, and development of the strategies of prevention or their neutralization.

Within the information and analytical sub-system, the level of threats to well-balanced socio-economic development is evaluated, and the possible tools and levers of management are determined. One of the basic providing elements is the information and analytical block, which forms the foundation of the strategy and policy of well-balanced changes in regional economic systems. The central role in these documents belongs to the integrated innovational & investment policy aimed for creation of the favorable investment climate and the organizational and economic basis of positive structural changes in the innovations-oriented economy of the region.

References

Arsenyev, V.A., Litvinova, S.A., Parakhina, V.N., Kozenko, Z.N., Denisov, M.Y.: Mechanisms of innovational development of countries with transitional economy. Contemp. Econ. **10**(4), 373–380 (2016)

Bergman, E.M., Charles, D.: Innovative Clusters: Drivers of National Innovation Systems. Organization for Economic Cooperation and Development, San Antonio (2003)

Doloreux, D., Parto, S.: Regional innovation systems: a critical review. Int. J. Innov. Manag. **7** (2003)

Goncharov, A.Y., Sirotkina, N.V.: The mechanism for managing balanced development of regions with dominant economic activities. News High. Educ. Establ. Technol. Text. Ind. **4** (358), 35–43 (2015a)

Kaplan, R., Norton, D.P.: Alignment: Using the Balanced Scorecard to Create Corporate Synergies. Harvard Business School Press, Boston (2003)

Kruglyakova, V.M., Treshchevsky, Y.I., Bredikhin, V.V.: The development of the textile industry in the context of harmonizing national, sectoral and regional strategies. News High. Educ. Establ. Technol. Text. Ind. **1**(367), 60–67 (2017)

Parakhina, V.N., Timoshenko, P.N.: Internal balance of the industrial complex: the fundamental necessity and methodology of assessment. Bull. North-Caucasian Fed. Univ. **3**(60), 124–131 (2017)

Parakhina, V., Boris, O., Midler, E.: Evaluation of innovative regional development Russia. Asian Soc. Sci. **11**(5), 201–208 (2015)

Treshchevsky, Y., Nikitina, L., Litovkin, M., Mayorova, V.: Results of innovational activities of Russian regions in view of the types of economic culture. In: Popkova, E. (ed.) Russia and the European Union Development and Perspectives Part (Series Contributions to Economics), Contributions to Economics, pp. 47–53. Springer, Cham (2017)

Anshina, V.N.: Innovational Management: Concepts, Multi-level Strategies, and Mechanisms of Innovational Development (2006). Delo, M

Basovsky, L.E.: Forecasting and planning in the market conditions. In: INFRA-M (2002)

Boris, O.A.: The Socially Oriented Innovational Company: Theory and Practice of the Holistic Development. The Publishing and Information Center "Fabula", Stavropol (2014)

Glazyev, S.Y.: Regarding the purposes, problems, and measures of state policy of development and integration, Moscow, 29 January 2013. http://www.glazev.ru/econom_polit/305/

Goncharenko, L.P.: Management of investments and innovations. In: KnoRus, 160 p. (2014)

Goncharov, A.Y., Sirotkina, N.V.: Well-balanced regional development: the stakeholder and social darwinism approaches. Reg. Syst. Econ. Manag. **3**(30), 10–17 (2015b)

Kaplan, R., Norton, D.: Well-balanced System of Indicators. Olimp Business CJSC, From strategy to action. M. (2005)

Konovalova, M.E.: Structural Balance of Public Reproduction: Issues of Theory and Methodology, 184 p. Samara State University of Economics Publ., Samara (2009)

Maximenko, L.S.: Development of the Theory and Methodology of Planning: Structural and Logical Approach. North Caucasus State Technical University, Stavropol (2009)

Niven, P.: Well-balanced system of indicators step-by-step. Maximum increase of effectiveness and establishment of the received results. Balance Business Books, 328 p. (2004)

Neeley, E., Adams, K., Kennerly, M.: The prism of effectiveness: map of well-balanced indicators for measuring success in business and management. Balance Club, Dnipro (2003). 400 p

Parakhina, V.N., Khanaliev, G.I.: The conceptual issues of managing the well-balanced development of regions. Bulletin of North Caucasus Federal University **3**, 244–249 (2011)

Sirotkina, N.V., Vorontsova, I.N.: The notion and essence of well-balanced development of region, competitiveness, innovations, finances, no. 1, pp. 55–59. Institute of management, marketing, and finance, Voronezh (2014)

Tadtaev, D.M., Parakhina, V.N.: Strategic planning of innovational development of the Republic of South Ossetia. Retrospective analysis and the modern model. Bull. North Caucasus Fed. Univ. **4**(55), 143–147 (2016)

Timoshenko, P.N., Gorbenko, L.I.: Forms of balance of the innovational development of organizations. Bull. North Caucasus Fed. Univ. **6**, 151–157 (2016)

Khamidulina, A.M., Suldina, G.A.: Evaluation of balance of socio-economic development of municipal entities in the region. Bull. Kazan (Volga) Fed. Univ. **3**(17), 55–59 (2015)

Katkov, E.V., Borodin, A.I., Streltsova, E.D.: Fuzzy logic in evaluation of the investment attractiveness of projects. Appl. Inf. **46**(4), 19–24 (2013)

Administration of Oslo. Recommendations for collection and analysis of the data on innovations/Joint publication of the OECD and Eurostat. 3rd edn., 107 p. CRSS (2010)

Nedosekin, A.O.: Strategic planning with the use of fuzzy descriptions. Audit Financ. Anal. **2**, 38–42 (2003)

Prigozhin, A.I.: Methods of organizations' development. In: ICFED, 863 p. (2003)

Complex Development of Competence-Based Potential as the Innovational Task of Industrial Companies Management

Olga A. Boris[✉], Irina I. Kuzmenko, Elena N. Lepyakhova,
and Valentina N. Parakhina

North Caucasus Federal University, Stavropol, Russia
{boris, vparakhina}@ncfu.ru, irina-stav26@mail.ru,
lika64@list.ru

Abstract. The purpose of the work is to determine and overcome the contradictions in provision of the complex development of a company's competences.

The methodological basis of the research includes dialectics and the systemic approach to changes of the component structure of a company's competence-based potential.

The differences between the managerial categories – competence and competency – are analyzed. It is shown that they are interconnected as a whole and in parts. Competency consists of separate competences and has relative evaluation (level of competency) which is assesses quantitatively. Competences are specifics, for the elements of competency have their own qualitative characteristics in a certain sphere.

The authors offer to treat the totality of competences of personnel the competence-based potential of the company, as the skills of employees constitutes the core of the key competences of the company.

Development of competence-based potential is predetermined by development of competences of personnel – primarily, managerial personnel. Thus, it is necessary to develop the system of personal management of managerial personnel, including formation of its modern structure, aimed at development and effective usage of six basic resources of a person, six main potential of the manager and employees of the organization, which include: potential of time management; potential of education level; potential of paying capacity; potential of activity and working efficiency; charismatic potential (influences on people) (public authorities); potential of psychological sustainability. Development of personal potential is ensured by the existing system of coaching.

Keywords: Competences · Competency · Personal management
Coaching · Development of personnel

JEL codes: M 54 · M 53 · J 24

© Springer International Publishing AG, part of Springer Nature 2018
E. G. Popkova (Ed.): HOSMC 2017, AISC 622, pp. 269–276, 2018.
https://doi.org/10.1007/978-3-319-75383-6_34

1 Introduction

Topicality of these issues is related to the importance and popularity of development of company's competences as a reflection of its capabilities to develop, occupy its positions in the competitive market, and oppose the aggressive behavior of the environment (Lovkova et al. 2016; Arsenyev et al. 2016).

There are differences between the managerial categories – competency and competence – regarding organization and its employees, though they are often used as similar, especially as to organizations. Company's competences reflect (Hamel and Prahalad 1990) its skills and technologies which allow providing certain values to the customers. Employee's competences are his knowledge and skills which allow him to work effectively.

As knowledge, skills, and technologies have a tendency for aging, the most important component of development of organization's competency is constant work on restoration of competences of its personnel, provisionи of the potential of their improvement and effective usage (Parakhina et al. 2017). This task could be solved on the basis of creation and improvement of the system of personal management in organization.

2 Methods and Their Initial Provisions

According to the results of the research, competency consists of separate competences and has relative evaluation (level of competency) which is assessed quantitatively. Competences are specifics, for the elements of competency have their qualitative characteristics in a certain sphere.

Competency and competence are interconnected as a whole and in part; employees' competence is the basis of organization's competency, which actually integrates subjective and objective components into one complex, the attributes of which are brought down to their sum:

$$Ko = f(koi, ksj),$$

where
Ko – competence of organization;
koi – competences determined by i objective factors (used by technologies, possibilities of their development in the form of the research basis);
ksj – competences determined by j subjective factors (skills of the personnel that allow them to work effectively and to use and develop the leading technologies).

We offer to treat the totality of competences of personnel as the competence-based potential of the company, for the skills of the employees constitute the core of the key competences of the company.

Evaluation of the level of competence-based potential of a company is conducted in comparison to other companies in such spheres as a capability to work and achieve success at various markets (commodity, capital, and resources); capability for changes, development, and innovations; capability for social responsibility; capability to pre-serve its potential in negative external conditions.

Development of competence-based potential is predetermined by development of competences of personnel – primarily, managerial. Thus, it is offered to develop the system of personal management of the managerial personnel.

Thus, it is offered to rank the components of personnel competency, distinguish the key ones, and offer the development of the system of personal management of managerial personnel, including formation of its modern structure.

3 Results

3.1 Structure of Personal Management as a Reflection of the Systemic Approach to Development of Competence-Based Potential of Industrial Company

Totality of an industrial company's competences, determined by the personnel's skills which allow them to work effectively and use and develop the leading industrial technologies, constitutes the core of the organization's competency. The essence of its competence-based growth consists in complex development of the manager's potential and the potential of the whole personnel of the company.

Each organization's manager has six types of resources and personal qualities, as well as potentials, the effective usage of which will provide the industrial company with long-term growth by application of motivators of the personnel and with development of competence-based potential of the company (Parakhina et al. 2017).

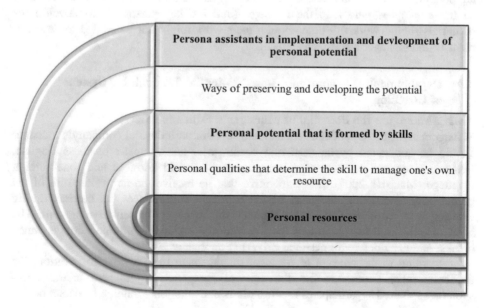

Fig. 1. Components of personal management of an industrial company

Basic resources of a personality include time, knowledge, money, physical and mental health, and power.

Personal qualities: organized nature, competency (education, experience), paying capacity, work efficiency (activity), resilience, and influence (charisma).

Figure 1 shows the concept of the structure of personal management of an industrial company.

Accordingly, the main potentials of the manager of an industrial company include: potential of time management; potential of education; potential of paying capacity; potential of activity and working efficiency; charismatic potential (influence on people) (power); potential of resilience (psychological stability), the effective usage of which will provide the company with long-term growth by application of the motivators of personnel and, therefore, development of the company's human resources.

3.2   Peculiarities of Formation of Competence-Based Potential of an Industrial Company's Manager

The manager of an industrial company, due to his busy schedule, cannot pay proper attention to managing his potential of education, as he is not very well aware of the technological, marketing, and other innovations. So there's a necessity for an assistant (Fig. 2).

The manager who knows what to study and how to study can achieve the set goal. If he can assess where to receive knowledge and where to obtain the necessary information for solving certain problems, he will be able to increase his professional level and acquire any necessary additional knowledge (Novikov 2016).

According to the recent research, the main potentials of the ones presented in Fig. 2 are potential of education level, as well as of organized nature and time management – so the employees who help the manager (and the key personnel) to develop his potentials should work full-time, though their work might not be related to development of the manager's self-management.

3.3   Development of Competence-Based Potential Through the System of Coaching

3.3.1   Managing the Potential of Manager's Education Level

Management of the potential of educational level of an industrial company's manager should be conducted by an academic coach, who has to speak business English. Besides, a modern manager of a large company should know at least one foreign language (Markova 2004). It is imperative that he receive the news literature of the global publishers in the language of the original. This reduces the barriers during entering new world markets and stimulates the manager for self-development. The most optimal time for presenting the innovational material – morning coffee, lunch breaks, and car rides (with personal driver) (Ponomarev 2003).

According to the concept of personal management of industrial corporation, the academic coach determines the directions of personnel development in the corresponding sphere. He participates in attestations and additional training of personnel, as

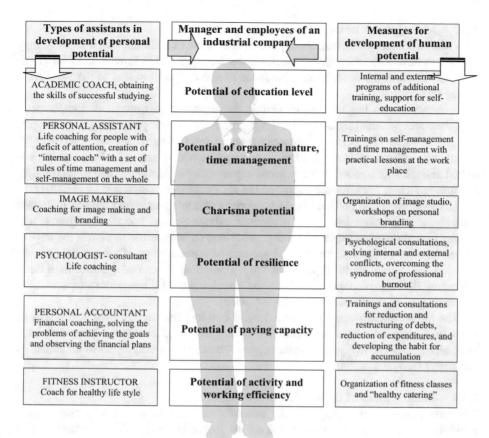

Types of assistants in development of personal potential	Manager and employees of an industrial company	Measures for development of human potential
ACADEMIC COACH, obtaining the skills of successful studying.	Potential of education level	Internal and external programs of additional training, support for self-education
PERSONAL ASSISTANT Life coaching for people with deficit of attention, creation of "internal coach" with a set of rules of time management and self-management on the whole	Potential of organized nature, time management	Trainings on self-management and time management with practical lessons at the work place
IMAGE MAKER Coaching for image making and branding	Charisma potential	Organization of image studio, workshops on personal branding
PSYCHOLOGIST- consultant Life coaching	Potential of resilience	Psychological consultations, solving internal and external conflicts, overcoming the syndrome of professional burnout
PERSONAL ACCOUNTANT Financial coaching, solving the problems of achieving the goals and observing the financial plans	Potential of paying capacity	Trainings and consultations for reduction and restructuring of debts, reduction of expenditures, and developing the habit for accumulation
FITNESS INSTRUCTOR Coach for healthy life style	Potential of activity and working efficiency	Organization of fitness classes and "healthy catering"

Fig. 2. The concept of formation of the structure of the company's personal management

well as periodic seminars with personnel on importance and need for self-education for development of certain knowledge.

3.3.2 Time Resource Management

The next potential that lies in the basis of the concept of personal management of modern industrial companies is management of the time resource. Over the recent 2–3 decades, each manager of a large company has had a feeling of scarce time for different things (Zayvert 1995).

Time management (or time potential management) of a corporate's manager should be conducted by the secretary or personal assistant. Personal assistant or life coach manager have to organize seminars on time management with practical lessons at the work place. It is important to understand that individual time management is very difficult – it should be a corporate system that connects the tasks of all employees. The complex approach to implementation of such system is much more convenient with the usage of special programs or online services.

3.3.3   Management of the Paying Capacity Resource

As a rule, a company's manager has several directions of activities and a lot of money assets that have to be managed. Therefore, the manager has to have a financial coach who will conduct the accounting of finances of the corporation and of the manager's family (Savenok 2006).

Financial coach has to conduct certain events for improvement of the financial discipline and financial literacy of the large company's employees. Thus, for example, it could be courses on personal finances management, during which the employees are taught the methods of family budget management, which has to be thoroughly controlled and analyzed – for it helps to avoid unnecessary expenditures and to accumulate the sum sufficient for implementation of the set goals (Shevtsova and Gorb 2007).

At the seminars on financial literacy the corporation's employees have to be taught their rights and obligations, shown obvious advantages of financial programs that consist in authomatization of all routine processes (it is possible to divide the expenditures into the categories, compile the scheme of accumulation of money for summer vacations, receive plans and reports), and given practical lessons on application of these programs. Besides, modern banks "connect" bank cards to the programs of automatic accounting of all transactions (Kosolapov 2016).

The higher effectiveness, it is necessary to conduct the courses on development of financial literacy on a mandatory basis for all employees of the organization; they have to be repeated every two years (depending on the number of employees at an industrial company) (Stoyanova 2003).

3.3.4   Management of the Resilience Potential

The next potential of manager is the potential of resilience. Stress in professional activities is an inseparable part of life of an industrial company's manager, which makes the issue of stress management very urgent. High changeability and complexity of the world make stresses constant companions of a modern manager (Sergeeva and Tokareva 2012).

Phenomena of the global economic crisis influence not only industrial companies and the state but also every human. The Russian industry works in the conditions of permanent crisis, but it does not use the means of psychological protection of its employees. The main goal of the employees is to survive, which is peculiar for industrial companies and for each human in the conditions of partial employment, which leads to constant tensions at work, lack of confidence in future and perspectives of work in the company. A psychologist has to provide help in management of stress for the personnel of a large organization.

3.3.5   Development of Potential Leadership

A good manager of corporation should be a leader, and leader has to have charisma. The main role of a manager in the management of a large industrial company is to show an example and lead the people. A charismatic manager has authority and looks like a confident person. He has to show approval and gratitude with his employees, possess communication capabilities for explaining his ideas to the people who surround him, and be able to motivate himself and other people for achieving success. However, not all top managers of large companies have charisma, which negatively influences the development of organization. Charisma could and should be developed by development and activization of internal potential. The obstacles and barriers that hinder manifestation of charisma are hidden in the manager himself. Their manifestation reduces the work efficiency of the manager and the whole industrial company, so the manager needs help in order to overcome them. According to the offered concept of personal management, this help could be provided by an image maker together with a psychologist (Robert and Tilman 1988).

Image maker is a specialist dealing with creation of an attractive highly-professional image of an industrial company's manager (Rogacheva 2004).

According to the concept of personal management of a company, the assistant for development of charisma potential deals with creation of internal basis for the formation of public opinion – the so called basis, i.e., the looks and capability to present oneself. Thus, the profession of coach for image making and/or branding includes the skills of stylist and specialist on the oratory skill. As to the manager, the described competences are supplemented by selection and purchase of clothing and footwear.

3.3.6   Management of the Working Efficiency Potential

The most important potential of a manager is the potential of activity and working efficiency.

Over the course of history of humankind, movement has been the most important condition of its survival in the process of evolution, which led to establishment of people. In the modern computerized conditions, managers and other managerial staff of corporations deal with small volumes of physical movements; the volume of muscle work decreased, but nervous tension increased. In the situations of nervous overloads the hormones are directed into blood, and their excess negatively influences the nervous system of managers, bereaves them of sleep and creates restlessness (Lukyano 2002). The movement activity stimulates restoration of hood mood, returns lost calmness, and allows working for a long time without any harm to health (Fomin and Vavilov 2004; Dyadichkin 1990). Thus, management of the potential of activity and working efficiency of a corporation's manager should be conducted by fitness instructor and coach responsible for healthy way of living of the company's employees.

4   Conclusions

Thus, according to the concept of personal management of industrial corporation, depending on the size and needs of the manager and employees, a department could be formed for each direction of development of competences – it could be one department of personal management with coaches on development of separate components of self-management (human resources). Assistants or coaches could be full-time employees of the corporation or be invited on the conditions of outsource. This will reduce the expenditures, but will also reduce the effect of influence on all employees. The offered concept unifies self-management and HR management and allows the manager to distribute work time in the optimal way by delegating a part of work to another person.

The system of personal management in the offered structure stimulates the development of professional qualities of managers and employees of the organization that are required for execution of work responsibilities, which ensures development of competence-based potential of an industrial company.

References

Prahalad, S.K., Hamel, G.: The Core Competence of the Corporation. Harvard Bus. Rev. **68**(3), 79–91 (1990)

Novikov, A.M.: The notion "education" in the post-industrial society (2016). http://www.anovikov.ru. Accessed 11 Nov 2016

Markova, O.Y.: Education in the system of values of modern human's culture. In: Materials of the International Symposium, Zugdidi, Georgia, 19–20 May 2004, pp. 177–187. SPb: St. Petersburg Philosophic Society. – no. 2 (2004)

Ponomarev, R.E.: Educational space as the basic notion of the theory of education. Pedagogical Educ. Sci. **1**, 29–31 (2003)

Zayvert, L.: Your time is in your hands. INFRA-M (1995)

Savenok, V.S.: Creating a personal financial plan. Path to financial independence. Piter, SPb (2006)

Shevtsova, S.G., Gorb, M.I.: 10 ways to become richer. Personal budget. Piter, SPb (2007)

Kosolapov, A.: Methods of saving personal money (2016). http://www.lkapital.ru/articles_review_1_52.htm. Accessed 21 Oct 2016

Stoyanova, E.S.: Financial management: theory and practice: study guide. Perspektiva (2003)

Sergeeva, V.B., Tokareva, A.A.: Leadership in management. Bull. NSIER. **9**, 81–102 (2012)

Parakhina, V.N., Perov, V.I., et al. (eds.): Self-management. MSU Publ, Lomonosov (2012)

Robert, M.-A., Tilman, F.: Psychology of individual and group. Progress (1988)

Rogacheva, M.I.: Image – component of manager's success and labor organization. HR Manage. **17**, 60–62 (2004)

Lukyano, V.S.: Regarding preservation of health and working efficiency. Medgiz (2002)

Fomin, N.A., Vavilov, Y.N.: Physiological foundations of moving activities. Physical training and sport (2004)

Dyadichkin, V.P.: Psychophysiological reserves of increase of working efficiency. HSE, Minsk (1990)

Lovkova, E.S., Starikova, T.V., Sirotkina, N.V.: Problems of activation of innovative activities in the textile industry. Bull. High. Educ. Establishments. Technol. Text. Ind. **5**(365), 22–25 (2016)

Arsenyev, V.A., Litvinova, S.A., Parakhina, V.N., Kozenko, Z.N., Denisov, M.Y.: Mechanisms of innovational development of countries with transitional economy. Contemp. Econ. **10**(4), 373–380 (2016)

Parakhina, V.N., Boris, O.A., Timoshenko P.N.: Integration of social and innovative activities into industrial organization. In: Popkova, E.G., et al. (ed.) Integration and Clustering for Sustainable Economic Growth, pp. 225–242. Springer (2017)

Parakhina, V.N., Ustaev, R.M., Boris, O.A., Maximenko, L.S., Belousov I.N.: Study of tendencies of formation and evaluation of HR innovational potential of the regions of the Russian federation. In: Popkova, E.G. (ed.) Overcoming Uncertainty of Institutional Environment as a Tool of Global Crisis Management, pp. 295–301. Springer (2017)

The Problems of Financing
of Entrepreneurship Infrastructure
in Developing Countries and Their Solutions

Irina M. Morozova[1]([✉]), Tatiana N. Litvinova[2],
Natalia V. Przhedetskaya[3], and Veronika V. Sheveleva[3]

[1] Volgograd State Technical University, Volgograd, Russian Federation
morozovaira@list.ru
[2] Volgograd State Agrarian University, Volgograd, Russian Federation
litvinoval358@yandex.ru
[3] Rostov State University of Economics, Rostov-on-Don, Russia
nvpr@bk.ru, beloveronika@yandex.ru

Abstract. The purpose of the research is to determine the problems of financing of entrepreneurship infrastructure in developing countries and their solutions. The authors use the method of logical analysis as to the level of national debt, volume of direct foreign investments, and cost of public-private projects in the infrastructure. The authors also use the method of correlation analysis for studying dependence of the level of national debt, volume of direct foreign investments, and cost of public-private projects in infrastructure on the index of development of transport and telecommunication infrastructure, as well as dependence of the values of the indices of development of transport and telecommunication infrastructure and the number of registered new companies. The information and analytical support for the research includes the materials of the World Bank, the International Telecommunication Union, and the World Economic Forum for 2016. The research objects are China, Brazil, Turkey, and Russia. The authors determined three key problems of financing of entrepreneurship infrastructure in developing countries: national debt which limits the state's capabilities for financing of infrastructural projects, unattractiveness of domestic infrastructural projects for foreign investors, and institutional barriers on the path of implementation of the mechanism of public-private partnership in the infrastructural sphere. For solving the above problems, the authors offer the framework strategy and practical recommendations: reorientation of the international financial support for developing countries at the infrastructural sphere, development of special economic areas in the infrastructural sphere for attracting direct foreign investments, and development of institutional provision of public-private partnership in the infrastructural sphere.

Keywords: Problems of financing · Entrepreneurship infrastructure
Developing countries

© Springer International Publishing AG, part of Springer Nature 2018
E. G. Popkova (Ed.): HOSMC 2017, AISC 622, pp. 277–283, 2018.
https://doi.org/10.1007/978-3-319-75383-6_35

1   Introduction

Modern global economy is very polarized. Countries are divided into conventional, but stable, categories; the largest categories are developed and developing countries. Differentiation of the level of their development is so high that a lot of projects of developed countries for overcoming the underrun of developing countries lead to a very small result. In the interests of elimination or at least leveling of structural disproportions in development of the global economy with simultaneous maximization of its growth, it is necessary to accelerate the rate of development of developing – not only in comparison with previous periods but in comparison with developed coutnries.

As the most important source of economic growth is entrepreneurship, it draws close attention of the modern global society. A lot of scientific studies in the sphere of entrepreneurship show that its development required the corresponding modern and sufficient infrastructural provision. The differences in entrepreneurship infrastructure occupy the central place in the system of limitation of the categories of developed and developing countries, so their research is topical for modern science and practice.

The working hypothesis of the research is that entrepreneurship infrastructure in developing countries is peculiar for low global competitiveness. This leads to low business activity in these countries, hindering their economic growth and increasing their underrun from developed countries. One of the origins of this phenomenon is the problems of financing of entrepreneurship infrastructure in developing countries. The purpose of this research is to determine the problems of financing of entrepreneurship infrastructure in developing countries and to determine their possible solutions.

2   Materials and Method

In this research, entrepreneurship is treated in the narrows sense. It includes only the material – transport and telecommunication - component. At that, the non-material component, which includes the institutional conditions, provision with financial and human resources, etc., goes beyond the limits of the selected research object. Verification of the working hypothesis supposes consecutive verification of the three following sub-hypotheses:

- hypothesis H_1: developing countries are peculiar for deficit of financial resources for development of entrepreneurship infrastructure. Its verification supposes the logical analysis of the national debt level, volume of direct foreign investments, and cost of public-private projects in the infrastructure;
- hypothesis H_2: there exists close connection between the volume of financial resources, accessible for developing стран, and the level of development of their entrepreneurship infrastructure. It is verified through finding the correlation between the level of national debt, volume of direct foreign investments, and cost of public-private projects in the infrastructure with the index of development of transport and telecommunication infrastructure;

– hypothesis H_3: business activity in developing countries is in close interconnection (predetermined) with the level of development of entrepreneurship infrastructure. In order to verify this hypothesis, it is necessary to assess the correlation of the indices of development of transport and telecommunication infrastructure and the number of registered new companies.

The information and analytical support for this research includes the materials of the World Bank, the International Telecommunication Union, and the World Economic Forum. According to the criterion of accessibility of the official statistical data, the objects of the research are such developing countries as China, Brazil, Turkey, and Russia. For the purpose of obtaining more authentic results, the study is performed with the 2016 data. The indictors of financial provision, development of entrepreneurship infrastructure, and business activity in the studied countries in 2016 are systematized and shown in Table 1.

Table 1. Indicators of financial provision, development of entrepreneurship infrastructure, and business activity in the studied countries in 2016

Indicators		Country			
		China	Brazil	Turkey	Russia
x_1	National debt, % of GDP	237.18	67.48	29.10	13.52
x_2	Volume of direct foreign investments, % of GDP	1.52	4.39	1.43	2.57
x_3	Public-private projects in the sphere of transport, % of GDP	0.06	0.05	0.09	0.01
x_4	Public-private projects in the sphere of telecommunications, % of GDP	0.20	0.70	0.22	0.43
y_1, x_5	Index of development of transport infrastructure, points from 1 to 6	4.70	4.00	4.40	4.90
y_2, x_6	Index of development of telecommunication infrastructure, points from 1 to 6	5.05	6.03	5.58	6.91
y_3	Number of registered new companies	167,280	73,614	57,760	427,388

Source: compiled by the authors on the basis of materials (The World Bank (2017a, b, c, d, e), (World Economic Forum 2017), (International Telecommunication Union 2017).

3 Discussion

The conceptual and applied issues of infrastructural provision of entrepreneurship in developing countries are studied in the works (Popkova et al. 2016a), (Ragulina et al. 2015), (Bogoviz et al. 2017), (Bogdanova et al. 2016), (Popova et al. 2016b), (Kuznetsov et al. 2016), (Kostikova et al. 2016), and (Simonova et al. 2017). At that, the financial component of development of infrastructural provision in developing countries is not sufficiently studied from the scientific point of view, which requires further scientific research in this sphere.

4 Results

The results of the correlation analysis of the indicators of financial provision, development of entrepreneurship infrastructure, and business activity in the studied countries in 2016 are shown in Table 2.

Table 2. Level of correlation of the indicators of financial provision, development of entrepreneurship infrastructure, and business activity in the studied countries in 2016

Variables	Correlation of variables		
	y_1	y_2	y_3
x_1	−0.83	−0.76	–
x_2	0.82	0.85	–
x_3	0.78	–	–
x_4	–	0.71	–
x_5	–	–	0.81
x_6	–	–	0.71

Source: compiled by the authors.

As is seen from Table 2, correlation of the values of the index of development of transport infrastructure in developing countries to national debt is strong and negative (−0.83), to volume of direct foreign investments – strong and positive (0.82), to number of public-private projects in the transport sphere – moderately expressed and positive (0.78).

Correlation of the values of the index of development of telecommunication infrastructure in developing countries to national debt is moderate and negative (−0.76), to volume of direct foreign investments – strong and positive (0.85), to number of public-private projects in the sphere of telecommunications – moderately expressed and positive (0.71). Correlation of the number of registered new companies to the index of development of telecommunication infrastructure is strongly expressed and positive (0.81), to the index of development of transport infrastructure – moderately expressed and positive (0.71).

The results of the performed analysis prove all three offered hypotheses and the main working hypothesis of the research. Thus, deficit of financial resources in developing countries hinders the formation of entrepreneurship infrastructure, which slows down business activity in these countries. We determined three main problems of financing of entrepreneurship infrastructure in developing countries, which are brought down to the following.

1st problem: deficit of financial resources with the state, which does not allow it to finance infrastructural projects in full. This problem could be solved only with the help of the mechanism of international support for development of infrastructure in developing countries. Infrastructure of entrepreneurship should become the initial targeted object for provision of international subsidies, credits, etc. for developing countries, as this allows achieving stronger effect in the sphere of growth of these countries' economies, thus ensuring high effectiveness of support.

2nd problem: low investment attractiveness of economy, which leads to weak investment flows. As infrastructural projects are peculiar for the longest period of investments return and high risk, they are not very attractive for investments. Together with deficit of investments resources in the economy on the whole, its influence is very large in the infrastructural sphere.

This problem could be solved by increase of investment attractiveness of infrastructural projects for foreign investors, fro it is very difficult to achieve the increase of investment attractiveness of economy on the whole in the mid-term and short-term. Attraction of direct foreign investments could be achieved with the help of developing special economic areas that specialize in the infrastructural sphere of economy.

3rd problem: insufficient development of projects in the form of public-private partnership (PPP) in the infrastructural sphere. Weakness of the institutional mechanisms of cooperation between the state and private business does not allow using this highly-effective means of expanding the financing of infrastructural projects, which is actively used in developed countries, with the same success and intensity. Creation of strong institutional provision of PPP in the infrastructural sphere allows solving this problem.

Based on the above, we offer the following framework strategy of solving the problems of financing of entrepreneurship infrastructure in developing countries (Fig. 1).

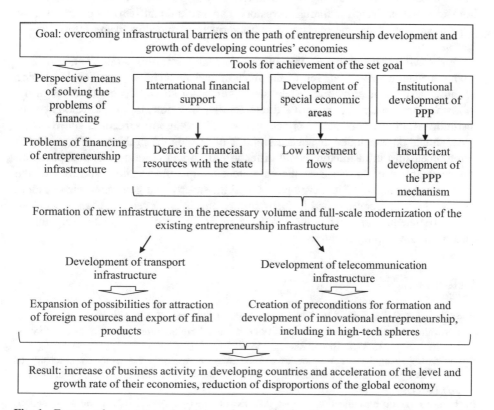

Fig. 1. Framework strategy of solving the problems of financing of entrepreneurship infrastructure in developing countries Source: compiled by the authors.

As is seen from Fig. 1, implementation of the offered practical recommendations for overcoming the determined problems of financing of entrepreneurship infrastructure in developing countries ensures formation of new infrastructure in the necessary volume and full-scale modernization of the existing entrepreneurship infrastructure. This stimulates the development of transport infrastructure and expansion of possibilities for attraction of foreign resources and export of final products, as well as development of the telecommunication infrastructure and creation of preconditions for formation and development of innovational entrepreneurship, including in the high-tech spheres.

As a result, the set goal is achieved – the infrastructural barriers on the path of development of entrepreneurship and of growth of developing countries' economies are overcome, business activity in developing countries is raised, the level and growth rate of their economy are increased, and disproportions of the global economy are reduced.

5 Conclusions

In the course of the research, we determined three key problems of financing of entrepreneurship infrastructure in developing countries: national debt which limits the state's capabilities for financing of infrastructural projects, unattractiveness of domestic infrastructural projects for foreign investors, and institutional barriers on the path of implementation of the mechanism of public-private partnership in the infrastructural sphere.

For solving these problems, the authors offer the framework strategy and practical recommendations: re-orientation of the international financial support for developing countries at the infrastructural sphere, development of special economic areas in the infrastructural sphere for attractive direct foreign investments, and development of institutional provision of public-private partnership in the infrastructural sphere.

It should be concluded that successfulness of solving the problems of financing of entrepreneurship infrastructure could differ in developing countries, as it depends on the peculiarities of the state of the entrepreneurship infrastructure, specifics of the problems of financing of its development, and the macro- and geo-economic situation. That's why it is recommended to focus further efforts on development of specialized national strategies of solving the problems of financing of entrepreneurship infrastructure in developing countries.

References

Bogdanova, S.V., Kozel, I.V., Ermolina, L.V., Litvinova, T.N.: Management of small innovational enterprise under the conditions of global competition: possibilities and threats. Eur. Res. Stud. J., **19**(2 Special Issue), 268–275 (2016)

Bogoviz, A.V., Ragulina, Y.V., Kutukova, E.S.: Ways to improve the economic efficiency of investment policy and their economic justification. Int. J. Appl. Bus. Econ. Res. **15**(11), 275–285 (2017)

International Telecommunication Union: ICT Development Index (2017). http://www.itu.int/. Accessed 6 Nov 2017)

Kostikova, A.V., Tereliansky, P.V., Shuvaev, A.V., Parakhina, V.N., Timoshenko, P.N.: Expert fuzzy modeling of dynamic properties of complex systems. ARPN J. Eng. Appl. Sci. **11**(17), 10601–10608 (2016)

Kuznetsov, S.Y., Tereliansky, P.V., Shuvaev, A.V., Natsubize, A.S., Vasilyev, I.A.: Analysis of innovate solutions based on combinatorial approaches. ARPN J. Eng. Appl. Sci. **11**(17), 10222–10230 (2016)

Popova, L.V., Popkova, E.G., Dubova, Y.I., Natsubidze, A.S., Litvinova, T.N.: Financial mechanisms of nanotechnology development in developing countriesJ. Appl. Econ. Sci. **11** (4), 584–590 (2016b)

Ragulina, Y.V., Stroiteleva, E.V., Miller, A.I.: Modeling of integration processes in the business structures. Mod. Appl. Sci. **9** (3), 145–158 (2015)

Simonova, E.V., Lyapina, I.R., Kovanova, E.S., Sibirskaya, E.V.: Characteristics of Interaction Between Small Innovational and Large Business for the Purpose of Increase of Their Competitiveness. Russia and the European Union Development and Perspectives, pp. 407–415 (2017)

The World Bank: Foreign direct investment (2017a). https://data.worldbank.org/indicator/BX. KLT.DINV.CD.WD. Accessed 6 Nov 2017

The World Bank: Current account balance (2017b). https://data.worldbank.org/indicator/BN. CAB.XOKA.CD. Accessed 6 Nov 2017

The World Bank: Investment in telecoms with private participation (2017c). https://data. worldbank.org/indicator/IE.PPI.TELE.CD. Accessed 6 Nov 2017

The World Bank: Investment in transport with private participation (2017d). https://data. worldbank.org/indicator/IE.PPI.TRAN.CD?view=chart. Accessed 6 Nov 2017

The World Bank: New businesses registered (2017e). https://data.worldbank.org/indicator/IC. BUS.NREG?view=chart. Accessed 6 Nov 2017

World Economic Forum: The Global Competitiveness Report 2016–2017 (2017). http://www3. weforum.org/docs/GCR2016-2017/05FullReport/TheGlobalCompetitivenessReport2016-2017_FINAL.pdf. Accessed 6 Nov 2017

Popkova, E.G., Chechina, O.S., Abramov, S.A.: Problem of the human capital quality reducing in conditions of educational unification. Mediterr. J. Soc. Sci. **6**(3), 95–100 (2016a)

Infrastructure as the Key to Domestic Companies' Entering the Global Markets

Tatiana N. Litvinova[(⊠)]

Volgograd State Agrarian University, Volgograd, Russian Federation
litvinova1358@yandex.ru

Abstract. The purpose of the research is to determine the infrastructural barriers for the companies' entering the global markets by the example of modern Russia and to develop the practical recommendations for overcoming them. The methodology of the work is based on systematizing and logical analysis of statistical data, which are formed from the Global Competitiveness Report for 2017–2018 (World Economic Forum 2017). The authors perform regression and correlation analysis, with the help of which the dependence of domestic companies' entering the global markets on the infrastructural provision is determined. The research objects are the four largest countries with developing economies: Russia, Brazil, China, and India. The main conclusion is that the counties with developing and forming market economies are peculiar for non-conformity of the infrastructural provision of entrepreneurship to the global standards that are set by the developed countries. This is an obstacle on the path of opening their export potential. The authors prove that modern Russia possesses wide possibilities in the sphere of entrepreneurship transnationalization. In order to accelerate this process, the practical recommendations for solving the determined problems in the sphere of infrastructural provision of entrepreneurship in Russia, which hinder its successful transnationzalization, are offered. In order to achieve the synergetic effect, it is recommended to use them in the systemic and complex way in modern Russia. For this, the algorithm of the domestic companies' entering the global markets through the prism of infrastructural provision development is offered.

Keywords: Infrastructure · Entering global markets · Domestic companies
Transnationalization of entrepreneurship

1 Introduction

In the modern global economic system, which is currently under the influence of the processes of economic integration and globalization, there are significant transformation changes of the national sectorial markets. In order to avoid international economic isolation, modern economic systems join the international trade association, the most vivid example of which is the World Trade Organization – which makes them terminate the policy of protectionism.

Erasure of customs barriers leads to appearance of foreign rivals on the previously closed national markets, which possess larger marketing and resources capabilities and experience, which helps them to oust the domestic companies. For the countries of the world

© Springer International Publishing AG, part of Springer Nature 2018
E. G. Popkova (Ed.): HOSMC 2017, AISC 622, pp. 284–290, 2018.
https://doi.org/10.1007/978-3-319-75383-6_36

to receive profits from the international trade and economic integration, they have to change their internal orientation and the domestic companies have to enter the global markets.

This research is built on the hypothetical and deductive principle and is based on the idea that most countries with developing and forming market economy are peculiar for non-conformity of infrastructural provision of entrepreneurship to the global standards, set by the countries with developed economy. This is an obstacle on the path of opening their export potential. The goal of this research is to determine the infrastructural barriers on the path of the domestic companies' entering the global markets by the example of modern Russia and to develop practical recommendations for overcoming them.

2 Materials and Method

The methodology of the research is based on systematization and logical analysis of statistical data that are formed from the materials of the Global Competitiveness Report for 2017–2018 (World Economic Forum 2017) (Table 1). The data are divided and grouped according to the components of entrepreneurship infrastructure, which is viewed in the wide sense and includes the transport, institutional, human, financial, technological, and innovational infrastructure.

Table 1. Indicators of infrastructural provision of entrepreneurship in Russia and the Russian companies' entering the global markets in 2017.

Indicator		Corresponding index	Values of the indices for the countries			
			Russia	Brazil	China	India
Business infrastructures	Transport	2nd pillar: Infrastructure, points, 1–6	4.9	4.1	4.7	4.2
	Institutional	1st pillar: Institutions, points, 1–6	3.6	3.4	4.4	4.4
	Human	5th pillar: Higher education and training, points, 1–6	5.1	4.2	4.8	4.3
		7th pillar: Labor market efficiency, points, 1–6	4.3	3.7	4.5	4.1
	Financial	8th pillar: Financial market development, points, 1–6	3.4	3.7	4.2	4.4
	Marketing	6th pillar: Goods market efficiency, points, 1–6	4.2	3.8	4.5	4.5
	Technological	9th pillar: Technological readiness, points, 1–6	4.5	4.6	4.2	3.1
	Innovational	12th pillar: Innovation, points, 1–6	3.5	3.2	4.1	4.1
Presence of domestic companies at the global market		10.04 Export, % of GDP, 1–100	25.9	12.1	20.6	18.8
		11.04 Nature of competitive advantage, points, 1–6	3.4	2.8	4.4	4.4

Source: compiled by the author on the basis of: World Economic Forum (2017).

Based on these data, the index of development of entrepreneurship infrastructure in a country was calculated (Iinfr), as a direct average of the values of indices of all the components of infrastructural provision of entrepreneurship (business infrastructure). Also, the index of global presence of domestic entrepreneurship (Ientr) was calculated, as direct average of the volume of export and the index of global competitive advantages.

For the sake of compatibility of the data, the absolute values of the indices (Abs.) were transformed into % (Rel.). Based on the calculated indices, the regression and correlation analysis was performed, and the dependence of domestic companies' entering the global markets on infrastructural provision was determined. The research objects are the four largest countries with developing economy: Russia, Brazil, China, and India.

3   Discussion

The conceptual and applied issues of establishment and development of the global entrepreneurship are viewed in the works (Popkova et al. 2016; Ragulina et al. 2015; Bogoviz et al. 2017; Bogdanova et al. 2016; Popova et al. 2016; Kuznetsov et al. 2016; Kostikova et al. 2016; Simonova et al. 2017). The essence and practical peculiarities of formation and development of entrepreneurship infrastructure are studied in the works (Mottaeva and Gritsuk 2017; Ray 2016; Tammela and Salminen 2016; Revoltella et al. 2016; Misbakhova et al. 2016).

The scientific literature overview on the selected topic showed insufficient elaboration of the interconnection between infrastructural provision and development of the global entrepreneurship. Therefore, the infrastructural aspect of domestic companies' entering the global markets requires further attention and study.

4   Results

The results of the calculations are shown in Table 2.

Table 2. Calculation of indicators for the regression and correlation analysis

Indicators		Values of the indicators for the countries							
		Russia		Brazil		China		India	
		Abs.	Rel.(%)	Abs.	Rel.(%)	Abs.	Rel.(%)	Abs.	Rel.(%)
Infrastructure	Transport	4.9	81.67	4.1	68.33	4.7	78.33	4.2	70.00
	Institutional	3.6	60.00	3.4	56.67	4.4	73.33	4.4	73.33
	Human	5.1	85.00	4.2	70.00	4.8	80.00	4.3	71.67
		4.3	71.67	3.7	61.67	4.5	75.00	4.1	68.33
	Financial	3.4	56.67	3.7	61.67	4.2	70.00	4.4	73.33
	Marketing	4.2	70.00	3.8	63.33	4.5	75.00	4.5	75.00
	Technological	4.5	75.00	4.6	76.67	4.2	70.00	3.1	51.67
	Innovational	3.5	58.33	3.2	53.33	4.1	68.33	4.1	68.33

(continued)

Table 2. (*continued*)

Indicators	Values of the indicators for the countries							
	Russia		Brazil		China		India	
	Abs.	Rel.(%)	Abs.	Rel.(%)	Abs.	Rel.(%)	Abs.	Rel.(%)
Iinfr	–	69.79	–	63.96	–	73.75	–	68.96
Share of export	25.9	25.90	12.1	12.10	20.6	20.60	18.8	18.80
Competitive advantages	3.4	56.67	2.8	46.67	4.4	73.33	4.4	73.33
Ientr	–	41.28	–	29.38	–	46.97	–	46.07

Source: compiled by the authors.

Based on the data from Table 2, the regression and correlation analysis is performed. The following model of the paired linear regression is obtained: $y = 1.79 + 1.82x$, which shows that increase of the index of development of entrepreneurship infrastructure in developing countries by 1 point leads to increase of the index of global presence of domestic entrepreneurship by 1.82 points. The correlation coefficient of these indicators constitutes 93%, which reflects their strong connection and statistical significance of the compiled regression model.

The results of the performed analysis by the example of the selected countries with developing economies prove the offered hypothesis and show that infrastructural provision influences the domestic companies' entering the global markets very strongly. For determining the perspectives of overcoming the infrastructural barriers on the path of transnationalization of domestic entrepreneurship, let us concentrate on one of the viewed countries – modern Russia.

As is seen from Table 2, the index of development of entrepreneurship infrastructure in the country (Iinfr) in Russia constitutes 67.79%, i.e., the potential of development of infrastructural provision of entrepreneurial activities is realized by 67.79%, which reflects large perspectives for bringing it in correspondence with the leading global standards.

The index of global presence of domestic entrepreneurship (Ientr) in Russia constitutes 41.28%. That is, the level of transnationalization of entrepreneurship in modern Russia is 41.28%. This shows large perspectives of further development of the process of transnationalization of Russian entrepreneurship. Based on deep study of the global competitiveness report of the Russian economy, the following key problems in the infrastructural provision of entrepreneurship in Russia, which hinder its successful transnationalization, were determined:

- Critical underrun from the leading globally oriented countries in the sphere of institutional entrepreneurship infrastructure (3.6 points out of 6 and the 116th position in the world). This causes high transaction expenditures of entrepreneurship and, accordingly, higher prices, as compared to foreign rivals;
- Strong deficit of financial entrepreneurship infrastructure (3.4 points out of 6 and the 107th position in the world). This complicates modernization of the production and distributive processes, thus reducing the qualitative (technical) characteristics of the Russian products as compared to foreign analogs;

– Weak development of the marketing entrepreneurship infrastructure (4.2 points out of 6 and the 80th position in the world). This hinders the formation and strengthening of the global Russian brands of products, hindering its promotion in the world markets;
– Insufficient development of technological entrepreneurship infrastructure (4.5 points out of 6 and the 57th position in the world). This does not allow for full automatization of business processes at the Russian companies, thus increasing their expenditures and hindering the receipt of benefits from the "scale effect".

Due to the above infrastructural barriers, the presence of the Russian companies in the global markets is very weak. Thus, the share of export constitutes 25.9% of GDP (94th position in the world). The character of competitive advantages is 3.4 points out of 6 (72nd position in the world). This is the sign of the initial stage of transnationalization of entrepreneurship in modern Russia.

At that, high level of development is peculiar for the transport (4.9 points out of 6, 35th position in the world), human (5.1 and 4.3 points out of 6 and 31st and 60th positions in the world), and innovational (3.5 points out of 6 and 49th position in the world) infrastructure. This provides large possibilities for Russia in the sphere of transnationalization of entrepreneurship. For accelerating this process, the following practical recommendations for solving the determined problems in the sphere of infrastructural provision entrepreneurship in Russia, which hinder its successful transnationalization, are offered.

Firstly, it is recommended to strengthen the institutional support for globally oriented entrepreneurship in Russia. Secondly, it is offered to increase the state support for globally oriented Russian companies. Thirdly, it is necessary to stimulate the development of transnational cluster processes with participation of the Russian companies, which stimulates the strengthening of their brands in the global markets. Fourthly, it is necessary to develop the technological infrastructure through attraction of private and foreign investors.

In order to achieve the synergetic effect, which increases the positive influence of the above recommendations, they should be applied in complex in modern Russia. For this, the algorithm of domestic companies' entering the global markets through the prism of development of infrastructural provision has been prepared – it is shown in Fig. 1.

As is seen from Fig. 1, development of the components of entrepreneurship infrastructure takes place consecutively; due to the systemic nature of this process, sustainable competitive advantages in all aspects of competitiveness are achieved: price, quality, and marketing. As a result, the growth of global competitiveness of domestic products and strengthening of the positions of domestic products in the global markets is achieved.

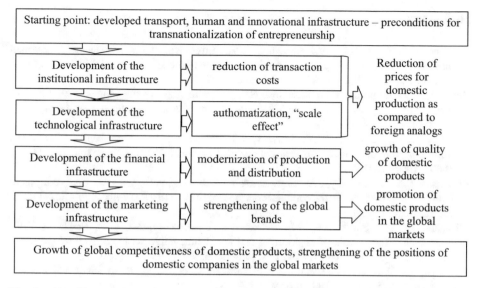

Fig. 1. Algorithm of domestic companies' entering the global markets through the prism of infrastructural provision development Source: compiled by the authors.

5   Conclusions

Summarizing the above, it is necessary to emphasize that infrastructure is a key to domestic companies' entering the global markets. That's why the infrastructural aspects of transnationalization of entrepreneurship should be paid a lot of attention in the process of development and implementation of the state foreign economic policy. According to the results of this research, countries with developing economy possess significant potential and perspectives in the sphere of improvement of infrastructural provision entrepreneurship, thus stimulating domestic companies' entering the global markets.

It should be noted that the results of this research are limited by the fact that the objects of studying the entrepreneurship infrastructure and domestic companies' entering the global markets are only four of the many countries with developing economy. This does not allow distributing the received conclusions for all countries with developing economy. During development of the practical recommendations, attention is paid to determination and overcoming of infrastructural barriers on the path of transnationalization of entrepreneurship in modern Russia. Research on the basis of other countries is a perspective direction for further scientific studies.

References

Bogdanova, S.V., Kozel, I.V., Ermolina, L.V., Litvinova, T.N.: Management of small innovational enterprise under the conditions of global competition: possibilities and threats. Eur. Res. Stud. J. **19**(2 Special Issue), 268–275 (2016)

Bogoviz, A.V., Ragulina, Y.V., Kutukova, E.S.: Ways to improve the economic efficiency of investment policy and their economic justification. Int. J. Appl. Bus. Econ. Res. **15**(11), 275–285 (2017)

Kostikova, A.V., Tereliansky, P.V., Shuvaev, A.V., Parakhina, V.N., Timoshenko, P.N.: Expert fuzzy modeling of dynamic properties of complex systems. ARPN J. Eng. Appl. Sci. **11**(17), 10601–10608 (2016)

Kuznetsov, S.Y., Tereliansky, P.V., Shuvaev, A.V., Natsubize, A.S., Vasilyev, I.A.: Analysis of innovate solutions based on combinatorial approaches. ARPN J. Eng. Appl. Sci. **11**(17), 10222–10230 (2016)

Misbakhova, C.A., Shinkevich, A.I., Belozerova, Y.M., Yusupova, G.F., Stakhova, L.V.: Innovation infrastructure of engineering and small innovative business in development of national innovation system. J. Adv. Res. Law Econ. **7**(2), 323–331 (2016)

Mottaeva, A., Gritsuk, N.: Development of infrastructure of support of small and medium business. MATEC Web Conf. **106**, 80–83 (2017)

Popova, L.V., Popkova, E.G., Dubova, Y.I., Natsubidze, A.S., Litvinova, T.N.: Financial mechanisms of nanotechnology development in developing countries. J. Appl. Econ. Sci. **11**(4), 584–590 (2016)

Ragulina, Y.V., Stroiteleva, E.V., Miller, A.I.: Modeling of integration processes in the business structures. Mod. Appl. Sci. **9**(3), 145–158 (2015)

Ray, N.: Business Infrastructure for Sustainability in Developing Economies, pp. 1–343 (2016)

Revoltella, D., Brutscher, P.-B., Tsiotras, A., Weiss, C.T.: Linking local business with global growth opportunities: the role of infrastructure. Oxf. Rev. Econ. Policy **32**(3), 410–430 (2016)

Simonova, E.V., Lyapina, I.R. Kovanova, E.S., Sibirskaya, E.V.: Characteristics of interaction between small innovational and large business for the purpose of increase of their competitiveness. In: Russia and the European Union Development and Perspectives, pp. 407–415 (2017)

Tammela, J., Salminen, V.: Modeling business innovation collaboration in open infrastructure. In: 2006 IEEE International Technology Management Conference, ICE 2006 (2016). https://doi.org/10.1109/ICE.2006.7477075

World Economic Forum: The Global Competitiveness Report 2017–2018 (2017). http://www3.weforum.org/docs/GCR2017-2018/05FullReport/TheGlobalCompetitivenessReport2017–2018.pdf. Accessed 8 Nov 2017

Popkova, E.G., Chechina, O.S., Abramov, S.A.: Problem of the human capital quality reducing in conditions of educational unification. Mediterr. J. Soc. Sci. **6**(3), 95–100 (2016)

The Mechanism of Optimization of the Tax Administration System with the Help of the New Information and Communication Technologies

Irina V. Gashenko[✉], Yulia S. Zima, V. A. Stroiteleva,
and N. M. Shiryaeva

Rostov State Economic University (RINH), Rostov-on-Don, Russia
gaforos@rambler.ru

Abstract. The purpose of the article is to develop the mechanism of optimization of the system of tax administration with the help of new information and communication technologies by the examples of modern Russia. For studying the connection between the system of tax administration and usage of new information and communication technologies, the authors use the method of correlation analysis. The authors calculate the correlation coefficients for time spent for tax accounting according to PricewaterhouseCoopers and the values of the E-government index according to the United Nations. For the sake of presentability of the data, the objects of the research are countries from various geographical regions of the world and with different values of studied indicators – Russia, the USA, China, the Maldives, the UAE, Ireland, and Singapore. The timeframe of the research is 2005–2016. With the help of this method, the authors also calculate the correlation coefficients for the values of the E-government index and the share of shadow economy according to the Association of Chartered Certified Accountants, as well as the share of tax revenues of state budgets of the countries according to the World Bank in 2016. The authors show that an important step on the path of overcoming the consequences of the global economic crisis is optimization of the tax administration system, which allows reducing the scale of shadow economy and ensuring more revenues for state budgets. A perspective tool for achieving this goal is implementing the new information and communication technologies into the process of tax administration. The authors analyze the problems that countries of the world face in the process of optimization of the system of tax administration with the help of new information and communication technologies by the example of modern Russia.

Keywords: Optimization · Tax system · System of tax administration
New information and communication technologies

1 Introduction

Viewing the national economic system from the positions of the dominating systemic approach, it is obvious that highly-effective and continuous functioning of this systems requires well-coordinates work of all sub-systems. Due to transition to the modern

© Springer International Publishing AG, part of Springer Nature 2018
E. G. Popkova (Ed.): HOSMC 2017, AISC 622, pp. 291–297, 2018.
https://doi.org/10.1007/978-3-319-75383-6_37

economic systems to the phase of restoration after the global recession, the attention of the global society was switched from tactical tasks to strategic tasks, of which the primary one is modernization of economy.

However, a serious obstacle on the path of implementing the new strategic course of development of countries of the world is deficit of financial resources that appeared as a result of the global crisis. This increased the importance of solving tactical tasks related to provision of replenishment of state budgets of countries of the world and drew attention to providing sub-systems, among which the central role belongs to the tax sub-system, with the help of which the state's financial resources are formed which are necessary for implementation of strategic tasks of economy's development.

In the tax sub-system, the consequences of the global recession are seen very clearly. The number and volume of profit of entrepreneurial structures, which are the main sources of revenues into the state budgets of the countries, reduced. Striving to cover the deficit of the state's budget, the governments raise the tax load for business, which often leads to growth of shadow economy and further reduction of the volume of tax revenues into the state budgets of the countries of the world. Due to this, the topicality of the issues related to optimization of the system of tax administration, aimed at fighting shadow economy and replenishment of state budgets, grew.

The working hypothesis of this research is that a perspective direction of optimization of the system of tax administration in the countries of the world is implementing the new information and communication technologies into the process of tax collection. However, low effectiveness of the process of informatization of modern Russia's economy does not allow for full implementation of the existing potential in the sphere of optimization of the system of tax administration. The authors verify the hypothesis and seek the goal of developing the mechanism of optimization of the system of tax administration with the help of the new information and communication technologies by the example of modern Russia.

2   Materials and Method

For studying the connection between the system of tax administration and usage of the new information and communication technologies, the method of correlation analysis is used. With the help of this method, the authors calculate the coefficients of correlation для time spent for tax accounts according to PricewaterhouseCoopers and the values of the E-government index according to the United Nations (Table 1).

For the sake of presentability of the data, the objects of the research are the countries from various geographic regions of the world and with different values of the studied indicators – Russia, the USA, China, the Maldives, the UAE, Ireland, and Singapore. The timeframe of the research is 2005–2016.

With the help of this method, the authors also calculate the coefficients of correlation for the values of the E-government index and share of shadow economy according to the Association of Chartered Certified Accountants, as well as share of tax revenues of state budgets according to the World Bank in 2016 (Table 2).

Table 1. Statistics of the time spent for tax accounts and the E-government index in 2005–2016

Country	Indicators	Values of the indicators for the years					
		2005	2008	2010	2012	2014	2016
Russia	Expenditures for tax accounts, hours/year	449	448	448	448	447	447
	E-government index	0.53	0.51	0.51	0.73	0.73	0.72
USA	Expenditures for tax accounts, hours/year	56	56	56	55	55	54
	E-government index	0.84	0.84	0.84	0.85	0.85	0.86
China	Expenditures for tax accounts, hours/year	80	80	80	80	80	80
	E-government index	0.6	0.6	0.6	0.6	0.6	0.6
Maldives	Expenditures for tax accounts, hours/year	0	0	0	0	0	0
	E-government index	0.43	0.43	0.43	0.43	0.43	0.43
UAE	Expenditures for tax accounts, hours/year	12	12	12	11	11	11
	E-government index	0.75	0.75	0.75	0.77	0.77	0.77
Ireland	Expenditures for tax accounts, hours/year	76	76	76	75	74	74
	E-government index	0.76	0.76	0.76	0.77	0.78	0.78
Singapore	Expenditures for tax accounts, hours/year	84	84	84	83	82	81
	E-government index	0.68	0.68	0.68	0.67	0.7	0.7

Source: compiled by the authors on the basis of: The United Nations (2017);
PricewaterhouseCoopers (2017).

Table 2. The country statistics of the share of shadow economy and share of tax revenues in 2016

Country	Share of shadow economy, % of GDP	Share of tax revenues, % of GDP
Russia	39.0	25.9
USA	7.8	11.3
China	10.2	22.6
Maldives	24.5	21.2
UAE	26.3	23.0
Ireland	22.1	21.8
Singapore	18.6	14.3

Source: compiled by the authors on the basis of: Association of Chartered Certified
Accountants (2017); The World Bank (2017).

3 Discussion

The authors use the materials of the works of modern authors in the sphere of tax administration and implementation of the new information and communication technologies, among which are Popkova et al. (2016); Ragulina et al. (2015); Bogoviz et al. (2017); Orudjev et al. (2016); Bogdanova et al. (2016); Popova, et al. (2016); Kuznetsov et al. (2016); Kostikova et al. (2016); Simonova et al. (2017).

4 Results

The results of the performed correlation analysis are shown in Tables 3 and 4.

Table 3. The level of correlation of the time spent for tax accounts and the E-government index in the selected countries

Country	Level of correlation	Character of connection (+, −)
Russia	46%	−
USA	99%	−
China	98%	−
Maldives	99%	−
UAE	98%	−
Ireland	99%	−
Singapore	99%	−

Source: calculated by the authors.

As is seen from Table 3, in all cases, except for Russia, there is strong positive correlation of time spent for tax accounts and the E-government index (98–99%). This shows that new information and communication technologies do allow for optimization of the systems of tax administration.

Introduction of the system of electronic document turnover in the system of corporate tax accounts allows for automatic formation of the accounts and their transfer to the tax bodies, which can conduct the administration automatically. Low correlation of these indicators in modern Russia (46%) shows that informatization of economy only slightly influences the tax sphere or is not effective at all.

Table 4. The level of correlation of the E-government index and the time spent for tax accounts, share of shadow economy, and share of tax revenues of government in the selected countries in 2016

	Level of correlation	Character of connection (+, −)
Time spent for tax accounts	98%	−
Share of shadow economy	91%	−
Share of tax revenues	95%	+

Source: calculated by the authors.

As is seen from Table 4, in 2016 there was high correlation of the E-government index and the time spent for tax accounts (98%), share of shadow economy (91%), and share of tax revenues of government (95%) in the selected countries. Accordingly, in the process of informatization of economy, time spent for tax accounts and the share of shadow economy reduce, while the share of tax revenues increases. This means that implementation of new information and communication technologies in the tax sphere allows achieving clear results in the sphere of provision of replenishment of state budgets.

Based on the above, it is possible to conclude that the potential of new information and communication technologies in the sphere of optimization of tax administration is very high. In modern Russia, it is not implemented sufficiently, which is due to the following reasons:

- alternative forms of corporate tax accounts: automatic tax accounts with the help of the new information and communication technologies is not mandatory in Russia, and high primary capital expenditures, required for starting it, make it a rare phenomenon in the Russian economic practice;
- corporate accounts of not all taxable business operations: low level of consumer consciousness and corporate responsibility are the reason that not all operations are reflected in corporate tax accounts, which - even in case of its authomatization – does not allow to reduction of shadow economy;
- a small share of electronic corporate transactions: lack of transparency of financial operations of business, including B2B and B2C transactions, which complicates the process of authomatization of corporate tax accounts.

For optimization of the system of tax administration with the help of new information and communication technologies in modern Russia, the following practical recommendations are offered:

- legislative establishment of mandatory authomatized accounting of economic operations that are taxable;
- increase of the level of consumer consciousness in corporate responsibility;
- legislative establishment of mandatory electronic transactions by entrepreneurial structures.

The mechanism of optimization of the system of tax administration with the help of new information and communication technologies on the basis of the offered recommendations is shown in Fig. 1.

As is seen from Fig. 1, the offered mechanism is aimed at achieving the goal of reduction of shadow economy and increase of the volume of tax revenues into state budgets of all levels of a country's budget system. The tools for achieving it include optimization of the system of tax administration with the help of new information and communication technologies. For this, the following algorithm is implemented: the first stage supposes realization of the offered recommendations for preparation and creation of the necessary conditions for optimization of tax administration.

The second stage supposes authomatization of the system of tax administration, for which it is necessary to expand the financing of modernization of equipment and technologies of tax administration, train specialists for tax bodies to use the new information and communication technologies, and implement new requirements and standards of tax administration. The third stage supposes monitoring and control over the work of the modernized system of tax administration – i.e., its debugging.

The result for the taxpayers includes the reduction of resources and time required for tax accounts, simplification of tax accounts, and ineffectiveness of shadow tax accounts. The result for the tax bodies includes reduction of resource intensity and simplification and acceleration of tax administration. The result for the state includes growth of tax revenues and replenishment of budgets in the necessary volume.

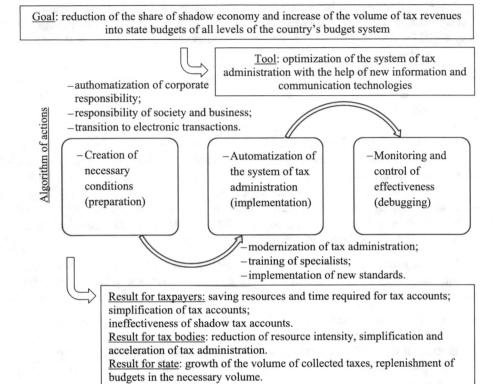

Fig. 1. The mechanism of optimization of the system of tax administration with the help of new information and communication technologies. Source: compiled by the authors

5 Conclusion

Thus, the working hypothesis has been proved. The authors show that an important step of the path of overcoming the consequences of the global economic crisis is optimization of the system of tax administration, which allows reducing the scale of shadow economy and ensures the replenishment of the state budgets in full volume. A perspective tools for achieving this goal is implementation of the new information and communication technologies into the process of tax administration.

The authors analyze the problems the countries of the world might face in the process of optimization of the system of tax administration with the help of the new information and communication technologies by the example of modern Russia. A certain limitation of the results of the performed research is emphasis on causal connections during developing the mechanism of optimization of the system of tax administration with the help of the new information and communication technologies with the generalized character of the implemented measures. Development of the strategies of optimization of the system of tax administration with the help of the new information and communication technologies in the countries of the world on the basis of the offered mechanism is a perspective direction for further scientific studies.

References

Association of Chartered Certified Accountants: Ranking of the countries with largest shadow economies (2017). http://www.accaglobal.com/russia/en.html. Accessed 8 Nov 2017

Bogdanova, S.V., Kozel, I.V., Ermolina, L.V., Litvinova, T.N.: Management of small innovational enterprise under the conditions of global competition: possibilities and threats. Eur. Res. Stud. J. **19**(2 Special Issue), 268–275 (2016)

Bogoviz, A.V., Ragulina, Y.V., Kutukova, E.S.: Ways to improve the economic efficiency of investment policy and their economic justification. Int. J. Appl. Bus. Econ. Res. **15**(11), 275–285 (2017)

Kostikova, A.V., Tereliansky, P.V., Shuvaev, A.V., Parakhina, V.N., Timoshenko, P.N.: Expert fuzzy modeling of dynamic properties of complex systems. ARPN J. Eng. Appl. Sci. **11**(17), 10601–10608 (2016)

Kuznetsov, S.Y., Tereliansky, P.V., Shuvaev, A.V., Natsubize, A.S., Vasilyev, I.A.: Analysis of innovate solutions based on combinatorial approaches. ARPN J. Eng. Appl. Sci. **11**(17), 10222–10230 (2016)

Popova, L.V., Popkova, E.G., Dubova, Y.I., Natsubidze, A.S., Litvinova, T.N.: Financial mechanisms of nanotechnology development in developing countries. J. Appl. Econ. Sci. **11** (4), 584–590 (2016)

PricewaterhouseCoopers: Russian FederationCorporate - Tax administration (2017). http://taxsummaries.pwc.com/ID/Russian-Federation-Corporate-Tax-administration. Accessed 8 Nov 2017

Ragulina, Y.V., Stroiteleva, E.V., Miller, A.I.: Modeling of integration processes in the business structures. Mod. Appl. Sci. **9**(3), 145–158 (2015)

Simonova, E.V., Lyapina, I.R. Kovanova, E.S., Sibirskaya, E.V.: Characteristics of interaction between small innovational and large business for the purpose of increase of their competitiveness. In: Russia and the European Union Development and Perspectives, pp. 407–415 (2017)

The United Nations: E-Government Survey 2005–2016 (2017). https://publicadministration.un.org/egovkb/Reports/UN-E-Government-Survey-2014. Accessed 8 Nov 2017

The World Bank: Tax revenue (2017). https://data.worldbank.org/indicator/GC.TAX.TOTL.GD.ZS?view=chart. Accessed 8 Nov 2017

Popkova, E.G., Chechina, O.S., Abramov, S.A.: Problem of the human capital quality reducing in conditions of educational unification. Mediterr. J. Soc. Sci. **6**(3), 95–100 (2016)

Highly-Effective Management of the Process of Innovations Commercialization as a Basis of Development of Modern Human Society

Aleksei V. Bogoviz[1]([⊠]) [iD], Svetlana V. Lobova[2] [iD],
Nelli A. Saveleva[3], Irina V. Lysak[4], and Sergei N. Makarenko[4]

[1] Federal State Budgetary Scientific Institution
"Federal Research Center of Agrarian Economy and Social Development
of Rural Areas – All Russian Research Institute of Agricultural Economics",
Moscow, Russia
aleksei.bogoviz@gmail.com
[2] Altai State University, Barnaul, Russia
barnaulhome@mail.ru
[3] Sochi State University, Sochi, Russia
savelevanelli@rambler.ru
[4] Southern Federal University, Rostov-on-Don, Russia
{ivlysak,snmakarenko}@sfedu.ru

Abstract. The purpose of the work is to study the social component of managing the process of commercialization of innovations and to develop the concept of highly-effective management of the process of commercialization of innovations as a basis of the modern human society development by the example of Russia. In order to verify the offered hypothesis, the work uses the method of statistical analysis – trend and correlation analysis, as well as the method of comparative analysis for calculating the relative indicators. The information basis includes statistical accounting of the Federal State Statistics Service for 2012–2016. The authors analyze and compare the indicators of the potential and efficiency of the country in the sphere of commercialization of innovations. As a result, it is concluded that insufficient attention to the social component in the process of managing the process of commercialization of innovations leads to emergence of a large lost socio-economic profit, caused by incomplete implementation of the accumulated potential of society's development. Highly-effective management of the process of commercialization of innovations as a basis of the modern human society's development supposes the shift of emphasis from management of companies' innovative activity at the final stage of the innovational process – implementation of the leading production technologies – to the initial stages of this process – creation and patenting of leading production technologies.

Keywords: Highly-effective management
Process of commercialization of innovations
Development of modern human society

E. G. Popkova (Ed.): HOSMC 2017, AISC 622, pp. 298–304, 2018.
https://doi.org/10.1007/978-3-319-75383-6_38

1 Introduction

The key role of innovations in achievement of economic growth is acknowledged by the modern academic society and state regulating bodies of the economic systems. Innovations are justly considered to be the sources of national entrepreneurship's competitiveness and, accordingly, of sustainable development of national economies, as well as the tool for overcoming the crises of economic systems and stabilization of the world economy. According to this, stimulation of innovative activity of entrepreneurial structures is paid a lot of attention in most of countries of the world.

At that, deep research of the essence of the innovational process shows that the central role and meaning in it belongs to human – research and innovator. Therefore, ignoring the social component of the innovational process violates its integrity. Emphasis on mostly economic goals of starting the innovational process does not allow taking into account the social component, thus presenting this process as a "black box", which does not allow achieving high effectiveness in its management. This creates a scientific and practical problem, which is not yet studied sufficiently and is not yet solved.

The initial point of this research is the offered hypothesis that insufficient attention to the social component in the process of managing the process of commercialization of innovations leads to emergence of large lost socio-economic profit, caused by incomplete realization of the accumulated potential of society's development. The purpose of this work is to study the social component of managing the process of commercialization of innovations and to develop the concept of highly-effective management of the process commercialization of innovations as a basis of the modern human society's development by the example of Russia.

2 Materials and Method

For verification of the offered hypothesis, the authors use the methods of statistical analysis – trend and correlation analysis, as well as the method of comparative analysis for calculating relative indicators. As the information basis, the authors use the materials of the Federal State Statistics Service for 2012–2016 (Table 1). The authors analyze and compare the indicators of potential and the indicators of the country's efficiency in the sphere of commercialization of innovations.

The indicator of potential of commercialization of innovations is total quantity of potentially created innovations in the country. It is calculated based on the supposition that each Ph.D. or doctor of science should create at least one leading technology.

Personnel involved with conduct of scientific research in R&D organizations could be unified into the groups of five people according to the international scientific standards (maximum number of authors of a scientific article in a peer-reviewed international journal cannot usually exceed five people). Each group should create at least one leading production technology per year.

The indicators of the result are the number of developed and used leading production technologies (own) and the number of patent applications. The level of

Table 1. Selected statistical information on R&D and innovative activities in Russia in 2016

Indicator	2012	2013	2014	2015	2016	2016/2012
Number of Ph.D.'s	33,082	35,162	34,733	28,273	25,826	0.78
Number of doctors of science	1,321	1,371	1,356	1,359	1,386	1.04
Number of R&D organizations	3,566	3,492	3,566	3,605	3,604	1.01
Number of personnel of R&D organizations	813,200	736,500	726,300	727,000	732,300	0.90
Average number of personnel in R&D organization	228.043	210.911	203.674	201.664	203.191	0.89
Number of R&D teams five people each	46	42	41	40	41	0.89
Number of potentially created innovations by R&D organizations	162,640	147,300	145,260	145,400	146,460	0.90
Number of developed leading production technologies	1,138	1,323	1,429	1,409	1,398	1.23
Number of patent applications	32,254	44,211	44,914	40,308	45,517	1.41
Number of used leading production technologies	191,650	191,372	193,830	204,546	218,018	1.14
Number of used leading production technologies of Russian origin	117,697	110,037	109,424	116,002	122,583	1.04

Source: compiled by the authors on the basis of: (Federal State Statistics Service 2016).

implementation of the existing potential is assessed by through comparing the indicators of the result with the indicator of potential. Also, their correlation and change over the five-year period are calculated.

3 Discussion

Specifics and the essence of managing the economic component of the process of commercialization of innovations is studied in detail in the works (Popkova et al. 2016; Ragulina et al. 2015; Bogoviz et al. 2017; Orudjev et al. 2016; Bogdanova et al. 2016; Popova et al. 2016; Kuznetsov et al. 2016; Kostikova et al. 2016; Simonova et al. 2017). However, despite the high level of elaboration of the economic component, the social component is not paid enough attention by modern researchers.

4 Results

The received results of statistical analysis of the indicators of R&D and innovative acitvities in Russia in 2016 are given in Table 1.

As is seen from Table 2, the potential of creation of leading production technologies in Russia reduced by 12% 2016, as compared to 2012, constituting 17.36 thousand. The level of its implementation is low. Thus, the share of developed leading

Table 2. The results of statistical analysis of indicators of R&D and innovative acitvities in Russia in 2016

Indicator	2012	2013	2014	2015	2016	2016 /2012	R^2
Total number of potentially created innovations in the country, thousand (x)	19.70	18.38	18.13	17.50	17.36	0.88	–
Share of developed leading production technologies, of x, %	0.58	0.72	0.79	0.80	0.80	1.39	84%
Share of the number of patent applications of x, %	16.37	24.05	24.77	23.03	26.21	1.60	61%
Share of the number of used leading production technologies of the Russian origin of x, %	59.73	59.86	60.34	66.27	70.58	1.18	2%

Source: compiled by the author.

production technologies constitutes 8% of the existing potential, the share of patent applications – 26.21%, the share of the used leading production technologies of the Russian origin – 70.58%.

The positive aspect is growth of the values of the indicators of the result by 39%, 60%, and 18% in 2016, as compared to 2012. Correlation of the number of developed leading production technologies and the potential of their creation is rather high, constituting 84%. Correlation of the number of patent applications and the potential is moderate, constituting 61%. Correlation of the number of used leading production technologies of the Russian origin and the potential is low, constituting 2%.

Thus, we determined the following problems at each stage of the innovational process:

– At the stage of R&D activities: low efficiency of R&D activities, caused by the fact that a lot of scientific studies are not finished or reach a negative results, which does not lead to creation of leading production technologies and creation of completely new leading production technologies. The share of completely new leading production technologies in Russia constituted 11% in 2016 (153 technologies);

– At the stage of registration of rights for the results of R&D activities – objects of intellectual property: not all created leading production technologies are registered as patents, which leads to absence of protection of intellectual property of the scholars;

– At the stage of implementation of leading production technologies into business processes of companies: not all domestic leading production technologies are implemented into production, as import of leading production technologies is very popular. A lot of technologies are implemented without any success, which does not lead to creation of innovational products. Thus, the share of innovational products in Russia constituted 6% in 2016 (RUB 3. 13 billion).

These problems show low effectiveness of managing the process of commercialization of innovations in Russia, the main reason of which is insufficient attention to the social component of this process. In order to increase the effectiveness of managing the process of commercialization of innovations in Russia, we offer the following practical recommendations:

–   Provision of increase of social responsibility of scholars for the performed scientific research and crated innovations. The scholars should be motivated for achieving high efficiency of the performed scientific research, which should end with creation of completely new leading production technologies and their successful implementation;
–   Simplification of the process of registration of rights for the objects of intellectual property through expansion of stimulation of R&D organizations;
–   development of cooperation between R&D organizations and innovations-active companies through creation of innovational clusters, technological parks, technological cities, etc. for increase of efficiency of the process of implementing innovations into the companies' economic practice.

Based in the above analysis, we offer the following concept of highly-effective management of the process of commercialization of innovations as the basis of the modern human society's development (Fig. 1).

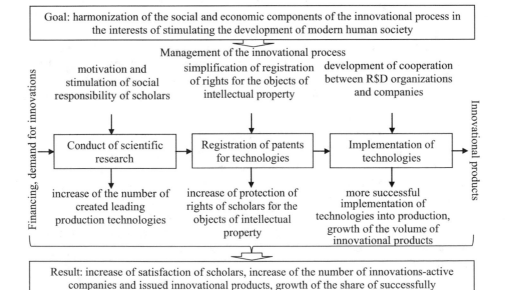

Fig. 1. The concept of highly-effective management of the process of commercialization of innovations as a basis of the modern human society's development. Source: compiled by the authors.

As is seen from Fig. 1, the offered concept is aimed at harmonization of the social and economic component of the innovational process in the interests of stimulating the modern human society's development. As a result of implementation of the offered practical recommendations, the number of created leading production technologies is increased, protection of scholars' rights for the intellectual property objects is raised, and more successful implementation of technologies into production and growth of the number of innovational products are achieved.

In total, these positive effects stimulate the increase of scholars' satisfaction, increase of the number of innovations-active companies and issued innovational products, growth of the share of successfully commercialized innovations, and increase of positive socio-economic effect from innovations.

5 Conclusions

Concluding the above, it should be noted that commercialization of innovations is an important component of the innovational process, as it fills this process with certain sense. For the purpose of maximization of efficiency of the innovational process and successful commercialization of innovations, it is necessary to set before R&D organizations not only economic (growth of the volume of sales and revenues, economic growth, etc.) but also social indicators (growth of the population's living standards, opening the innovational potential of scholars, etc.).

Highly-effective management of the process of commercialization of innovations as a basis of the modern human society's development supposes shifting the emphasis from managing the innovational activity of companies at the final stage of the innovational process – implementation of leading production technologies – to the starting stages of this process – creation and patenting of the leading production technologies. This is explained by the fact that successful commercialization of innovations requires all stages of the innovational process, as they are closely interconnected.

A certain limitation of the results of this research is the fact that the offered recommendations denoted the perspective directions of increase of effectiveness of managing the process of commercialization of innovations as a basis of the modern human society's development. Their more detailed elaboration and preparation for implementation in practice is a perspective vector for continuing the research presented in this article.

References

Bogdanova, S.V., Kozel, I.V., Ermolina, L.V., Litvinova, T.N.: Management of small innovational enterprise under the conditions of global competition: possibilities and threats. Eur. Res. Stud. J. 19(Special Issue 2), 268–275 (2016)

Bogoviz, A.V., Ragulina, Y.V., Kutukova, E.S.: Ways to improve the economic efficiency of investment policy and their economic justification. Int. J. Appl. Bus. Econ. Res. 15(11), 275–285 (2017)

Kostikova, A.V., Tereliansky, P.V., Shuvaev, A.V., Parakhina, V.N., Timoshenko, P.N.: Expert fuzzy modeling of dynamic properties of complex systems. ARPN J. Eng. Appl. Sci. **11**(17), 10601–10608 (2016)

Kuznetsov, S.Y., Tereliansky, P.V., Shuvaev, A.V., Natsubize, A.S., Vasilyev, I.A.: Analysis of innovate solutions based on combinatorial approaches. ARPN J. Eng. Appl. Sci. **11**(17), 10222–10230 (2016)

Popova, L.V., Popkova, E.G., Dubova, Y.I., Natsubidze, A.S., Litvinova, T.N.: Financial mechanisms of nanotechnology development in developing countries. J. Appl. Econ. Sci. **11**(4), 584–590 (2016)

Ragulina, Y.V., Stroiteleva, E.V., Miller, A.I.: Modeling of integration processes in the business structures. Mod. Appl. Sci. **9**(3), 145–158 (2015)

Simonova, E.V., Lyapina, I.R., Kovanova, E.S., Sibirskaya, E.V.: Characteristics of interaction between small innovational and large business for the purpose of increase of their competitiveness. In: Russia and the European Union Development and Perspectives, pp. 407–415 (2017)

Popkova, E.G., Chechina, O.S., Abramov, S.A.: Problem of the human capital quality reducing in conditions of educational unification. Mediterr. J. Soc. Sci. **6**(3), 95–100 (2016)

Federal State Statistics Service. Russia in numbers: short statistical collection. Federal State Statistics Service, Moscow (2016)

The Problems of Legal Regulation of the Development of Far Eastern Native Minorities During the Soviet State National Policy Formation Period (1920's–1930's)

Anna V. Akhmetova[✉] and Yana S. Ivashchenko

Komsomolsk-on-Amur State Technical University,
Komsomolsk-on-Amur, Russian Federation
wyrds@rambler.ru, iva_ya@mail.ru

Abstract. This paper addresses the formation of the legal status of aboriginal population of the Far East during the period of establishment of the new Soviet State government authorities between 1920's–1930's. The object of the research is the legal development of native small-numbered ethnic groups of the Far East. The study employs different methodological tools: historical-systematic, comparative law, historical-typological and other methods. The main problems resolved in this paper are handled in the light of modernization theory. The author defines basic laws and regulations reflective of the process of establishment of national statehood among the indigenous population of the Far East. The historic significance of the question of compliance of the objectives declared by the Soviet government with the actual situation in national districts of the Far East is shown. The conclusion on positive results of pursued policy is drawn; the problems relevant to the period in question are identified. The priority of political and ideological considerations over the economical interests in the national policy of the Soviet government can be considered as the most important positive factor for the indigenous peoples of the Far East. From the very beginning the approach of the Soviet government to the problems of the peoples of the North was based on political perspective rather than on profitability of investments. Originally the activity of the Soviet government had been focused on emphasizing the traditional peculiarities of aboriginal population. In the second half of 1930-s the orientation of ideological line was changed towards the forced integration of small-numbered ethnic groups into the Soviet society resulting in introduction of modernization activities related to national and territorial division of the national regions of the Far East.

Keywords: The Far East · Legal status · Native minorities · National district
National policy · Municipal authorities

1 Introduction

The problem of compliance of the legal system of the Russian Federation with international legislation on the status of traditional ethnic groups remains open. Unstable market-based economy of the Russian Federation coupled with social problems hinders

© Springer International Publishing AG, part of Springer Nature 2018
E. G. Popkova (Ed.): HOSMC 2017, AISC 622, pp. 305–314, 2018.
https://doi.org/10.1007/978-3-319-75383-6_39

the development of regulatory and legal framework governing life and activities of the native ethnic groups of the Russian North (Akhmetova 2012).

Governmental attention to this problem today is apparent as confirmed by the Decree of the President of the Russian Federation No. 1666 "On strategy of the state national policy of the Russian Federation until 2025" (Legislation Bulletin: 7477) dated 19.12.2012, that establishes preservation and development of cultures and languages of the nations of the Russian Federation, strengthening of their spiritual community and protection of the rights of ethnic minorities including the indigenous peoples as the key issues of the national policy of the Russian Federation calling for special attention from governmental and municipal authorities.

In order to give an adequate estimation of national policy methods formerly used or still employed by the government with regard to native minorities of the Russian Far East a clear understanding of mechanisms utilized in the relationships forming process between the government and the indigenous peoples at different historical stages is required.

2 Materials and Methods

The exploration of the topic was performed with involvement of the following data source base.

The State Archive of the Russian Federation has an archive fund P-3977 The Committee for development of the peoples of the North under the Presidium of All-Russian Central Executive Committee (The Committee for Affairs of the North). The documents of this archive fund include minutes of meetings of the Bureau of the Committee, the Committee for development of the peoples of the North, the Administrative and Legal Commission, the Commission for improvement of labor and living conditions of women under the Committee for Affairs of the North (1924–1930), minutes of meetings of the Committees for Affairs of the North (inclusive of the Far Eastern territory 1927-1934) and other documents reflecting the legal regulation of the Soviet policy towards the aboriginal population in 1920's–1930's.

The important information was found in the State Archive of Khabarovsk region П-35 (Khabarovsk regional committee) – the largest archive in terms of document volume and informational content. It possesses the data revealing the role of central and local organizations in economic, educational and cultural development of indigenous peoples of the Far East.

The main problems outlined in this paper are handled in the light of modernization theory. Despite the fact that many scientists consider the scientific and research potential of this theory to be exhausted, the modernization processes perfectly reflect transformations in life of the Far East aborigines during the Soviet period. Therefore the concept of modernization is renovated within the framework of the study of adaptive (overtaking) variant of modernization which is the case for the Soviet modernization – the attempt to reach qualitative and quantitative economic indicators of the leading western states at a swift rate. Socialist variant of modernization exercised in the USSR regardless of any ideological differences was a type of general modernization process, the adaptive response of non-western societies (Gavrov 2010). Socialist modernization model meshed with collectivist mindset of non-western societies including native minorities of the North. That's why the Soviet variant of transformation of ethnic and cultural development of traditional ethnic groups during the period under investigation became more successful than in western countries.

From this perspective the theory of A.I. Flier about the historical dynamics of social and cultural reality (Flier 2014), identifying different stages of this reality (preservation, destruction, development and renovation of the system) is of the utmost interest. Thus, Flier indicates different variants of system development dynamics. According to the author, transformation is the reinvention of essential or functional content of a tradition without changing the form of an action itself. Further Flier identifies the modernization or modification of a tradition, when the symbolically significant essence or the meaning of such a tradition is preserved in contrast to some form changes. In our opinion the Great October Socialist Revolution (1917) and the USSR modernization transformations in 1920's–1930's, when in two stages the old system was destroyed and the hundreds of traditions were left behind, give the perfect examples of overcoming the traditions. With regard to paucity of aboriginal societies their traditions were treated in a special way including transformation and modification approaches applied in natural sequence. The activity of the Committees for Affairs of the North was a typical example of transformation of the old generic system and imposition of new Soviet practices. The formation of integral cooperation is a manifestation of modernization approach. It was the formation of a command and administration system which resulted in completion of system renovation process. At first sight this process was suspended (from the technological point of view) by the Great Patriotic War. However, from the mental point of view it resulted in final alteration of traditions of the Far Eastern population.

In the course of investigation the main historical methods – comparative, synchronous, systematic – were used. Comparative analysis allows retracing the gradual implementation of legal regulation with regard to different native minorities in the different territories of the Far East. Synchronous method gives an opportunity to review interrelations and interference between the governmental actions related to education and upbringing. Based on systematic approach we can conclude that political and legal development of native minorities is a subsystem integrated into the general line of the Soviet governance, while the system is formed by the general directions of its policy.

3 Discussion

Chronologically the historiography on the subject can be divided into three main groups: prewar period (A. G. Bazanov, I. F. Fedorov); postwar period: 1950's–1960's (M.A. Sergeev, E. V. Yakovleva, V. G. Balitskiy) and 1970's–1980's (V.A. Zibarev, V. S. Lukovtsev, V. N. Uvachan, I. S. Gurvich, Z.P. Sokolova) and contemporary period (L. Y. Ivashchenko, A. E. Zavalishin, S. V. Bobyshev, V. P. Serkin, V.A. Kryazhkov, A.I. Gorelikov).

Many contemporary researchers of the life of indigenous population in Soviet times give estimation of the national policy within the framework of social modernization. The work by Bobyshev (2000) dealing with the content of governmental activity related to economical and cultural assistance provided to the small-numbered peoples in 1920s–1930s is also of great importance. Kryazhkov (2010) investigates the problems of regulatory and legal framework evolution for indigenous peoples of the Far North.

The monograph "Arctic Mirrors, Russia and the Small Peoples of the North" by Y. L. Slezkin (Ph.D., USA) dedicated to the history of interaction between the alien and indigenous population of the North is of exceptional interest. The author gives a staged analysis of the transformation of attitude of the Russians to the aborigines and vice versa. These processes provide an opportunity to trace the process of self-identification of indigenous peoples. The figures reflected in the "arctic mirrors" of Russian self-consciousness – a foreigner, an adherent of a different faith, a non-Russian, a national minority, a primitive communist, the last aborigine – are the product of the established complex interaction far beyond the clichéd framework of colonial rule and exploitation. A considerable part of the monograph is dedicated to the radical period between 1920's and 1930's, when the task of understanding and reformation of the life of small-numbered ethnic groups associated with the processes of the indigenization and the "Stalin's revolution" was especially difficult. The author does not just propose the original interpretation of these problems, but demonstrates biographical approach to ethnography through the example of politicians and public figures from M. Speransky to L. Sternberg and A. Skachko, who had advanced and put into practice the idea of diversity of human nature despite the predominant attitudes.

Eventually the conclusion can be made that the historiography of the period under investigation covers much ground. The investigation of interrelations between the Soviet government and native minorities is represented by the works of numerous scientists because, firstly, this subject is now becoming ever more relevant, and, secondly, this is a highly problematic period in Russian history which indicates the necessity of in-depth analysis of transformations in the development of the Far East native minorities.

4 Results

The October revolution of 1917, which cannot be overestimated for its historical value, opened a new stage of state policy concerning the indigenous small-numbered peoples of Russia. The process of establishing of aboriginal regulatory and administrative authorities formed the basis of this important period. This process was accompanied by the formation of special bodies of national and local significance to secure the protection of these peoples by the state (Babay and Kiselev 2005).

In 1917 the indigenous population lost its special status of non-Russian and was granted equal rights with the other peoples of Russia. This fact is of great importance.

The revolution declared new values including those related to national relations. The "Declaration of rights of the peoples of Russia" stated the right of the peoples of Russia to self-determination, equality and sovereignty, the suspension of national and religious privileges and restrictions; free development of national minorities and ethnic groups living in the territory of Russia (Lipatov and Savenkov 1957).

In general, the "Declaration of rights of the peoples of Russia" contained fundamental principles of national policy of the Soviet regime with respect to the peoples of the North. The only way to understand the decisions made by the Soviet government with regard to the small-numbered peoples is to perform the analysis of this document.

The primary target of the Soviet policy concerning the indigenous peoples was their integration into the Soviet society. The small-numbered peoples presented different stages of historical development; however all of them were pre-capitalistic. Marxist concept of the modes of production allowed for theoretical justification for the indigenous peoples to "skip the capitalist stage" and come over to socialism (Lipatov and Savenkov 1957).

The appeal of All-Russian Central Executive Committee and the Council of People's Commissars of Russian Soviet Federative Socialist Republic (RSFSR) dated August 16, 1919 "To workers, peasants, non-Russian population and working cossacks of Siberia" guaranteed the indigenous peoples the rights stated in the "Declaration of rights of the peoples of Russia", including the right to decide their own destiny independently. It also stated that the areas actually used by the non-Russians were not subject to any restrictions and cutbacks and became the sole social property of the indigenous peoples (Sibrevcom 1959).

The Constitution of the Russian Soviet Federative Socialist Republic dated 1918 acknowledged the areas with specific culture and ethnic composition (p. 11). Paragraph 22 stated that the establishment or the admission of special privileges and preferences based on race or ethnical identity, the oppression of national minorities and the restriction of their legal equality disagree with the fundamental laws of the Republic (The indigenous peoples of Russia 1995).

The Constitution of the Russian Soviet Federative Socialist Republic dated 1924 also stated the right of nations to self-determination and formalization of their existence within the Republic; inadmissibility of oppression of ethnic minorities or restriction of their equal status and exclusion of privileges for particular nations; the right to free use of national language at conventions, in court, in school, in administration and social life (p. 13) (Kryazhkov 2010).

The People's Commissariat of Nationalities of the Russian Soviet Federative Socialist Republic established in accordance with the Constitution of 1918 had exercised the administration of the Far North till 1924. However this administration was only token. Without local bodies the People's Commissariat of Nationalities had no significant effect to the development of the indigenous peoples of the North.

The People's Commissariat of Nationalities was dismissed on April 9, 1924 in accordance with the resolution of the All-Russian Central Executive Committee and the Council of People's Commissars. In 1924 it was replaced with the Committee for development of the peoples of the North under the Presidium of the All-Russian Central Executive Committee (or the Committee for Affairs of the North). In accordance with RSFSR All-Russian Central Executive Committee and the Council of People's Commissars Decree dated February 2, 1925 "On approval of the Charter of the Committee for development of the peoples of the North under the Presidium of the All-Russian Central Executive Committee", the purpose of the Committee was to assist stable development of economical, political and cultural life of the indigenous peoples (The collection of RSFSR laws 1925).

Further implementation of measures on administration of national relationships in the Northern regions can be divided into two stages. The first period had lasted from 1926 to 1929, when the Councils were created according to a tribal principle. The second period had lasted from 1932 to 1936 when nomadic and village Councils were created.

The first period of local government authorities formation in the northern national territories was based on "Temporary provision on administration of aboriginal ethnic groups and tribes in the Northern territories of RSFSR" (The collection of RSFSR laws 1926, p. 575) approved by the All-Russian Central Executive Committee and the Council of People's Commissars Decree dated October 25, 1926.

On June 14, 1927, based upon the Temporary provision, the Far Eastern Committee of the All-Union Communist Party (of Bolsheviks) and the Far Eastern Executive Committee made the following decision: the establishment of "aboriginal" administrative institutions by local administration must conform to ethnic and territorial criteria, while the tribal principle must be used only when such "aboriginal" organization has remained effective and the aboriginal population wishes to retain it. By the end of 1920-s there had been established 9 district executive committees and 127 "aboriginal" Councils in the Far Eastern region (State Archive of the Russian Federation, 281).

These administrative bodies can not be considered proper governmental authority institutions. Their activities were limited to the affairs of aborigines. The affairs of trade, economic and other organizations located in the region had remained out of their influence area (Bobyshev 2005).

In general the institution of tribal principle-based Councils can be considered as a consequence of cooperation between the peoples of the North and more developed peoples with the Soviet form of socialist statism.

In this case the external influence preconditioned the establishment of Councils as socialist political bodies, while the internal conditions resulted in their specific tribal nature.

Councils establishment initiative was not launched by the peoples of the North, therefore even the tribal nature of Councils could not make them clear to the population especially since the official form of the Russian government that had existed before the revolution was associated with oppression (Popkov 1990).

The organization of tribal Councils involved many other problems caused by the fact that the Councils were the result of cooperation between classless and socialist national communities. As a result they covered both primitive communal and (to some extent) socialist relations. Misunderstanding of the fundamentals of the socialist system was the main problem. Thus, the collective authority of the Council was seen as a personal authority of its chairman who in turn identified himself with the Council.

Lack of actual data on decomposition level of primitive communal relations of different peoples had sometimes resulted in establishment of tribal-type local bodies upon the abolishment of imperial governing bodies despite the break-up of tribal ties and the development of social differentiation. In these circumstances the introduction of tribal-type Councils often contributed to consolidation of power of the former foremen and rich community members who frequently chaired the Councils. In the Councils formed this way the shamans and the kulaks took the leading role. For the most part these were the same people that had been used by the czarist regime in the administrative system of the peoples of the North.

Participation in their work allowed the indigenous population to obtain qualities required to overcome inactivity and to make autonomous political, economical and other decisions.

Between 1920's and 1930's due to the striving of the Soviet regime to introduce radical changes to all spheres of life of the peoples of the North, the use of this special system of administration had been stopped and replaced with the common Soviet system of administration (Vakhtin 1990).

This replacement was followed by the period of "Sovietization" of life of the indigenous peoples. During this period local ethnic administration was substituted by state administration and necessary measures required to accelerate integration of these peoples into the Soviet system of social relations were initiated.

The development of the indigenous peoples under the abovementioned conditions could continue only under the fundamentals of the Soviet state with its governmental property in land, forests, waters, mineral resources, means and instruments of production, its planned economy, its centralized system of government bodies that exercises power based on communist ideology and its supranational structure – the Communist Party of the Soviet Union – acting without any legal restraints and based on the two main principles – division into classes and party membership.

The establishment of national areas in the territory populated by the small-numbered peoples of the North can be considered the first step in the specified direction. The system of administration consisting of national areas, national districts and national Councils was established in 1929–1932.

The system of tribual Councils was abolished in 1930. They were replaced with the common institutions of local Soviet government from village Councils to national areas. This implied the end of "aboriginal conservation areas" that had been defended by Russian ethnographists in the early years of the Soviet regime. The aborigines were politically and administratively connected with the territories of metropolitan country.

The tribual principle-based system of Councils could not satisfy the Soviet government. This system left intact many aspects of the previously existed tribal structure. The influence of genearchs remained very strong and dictated decisions of the Councils during the tribal meetings. The poor were apolitical – the population in general did not understand the fundamentals of the new regime. From the perspective of the Soviet government the main goal was to make Councils able to meet competition with tribal principle-based institutions as administrative and political bodies.

For this reason the administrative system of the peoples of the North was drastically reorganized in the first half of 1930's. The changes consisted in establishment of national areas and village nomadic Soviets.

National area is a type of Soviet autonomy corresponding to a national administrative-territorial unit. National areas formed at the suggestion of the Committee of the North can be considered the last attempt to reach a compromise between the necessity of preservation of native minorities of the North and the inevitably approaching industrial development of the North (Vakhtin 1990).

In 1930 the following areas were established: Chukotka, Okhotsk (Even), Koryak national areas as well as Zeya-Tura (Even) and Dgeltulak (Even) national districts.

However the totalitarian system was aiming at countrywide uniformity of economical and social life of the peoples of the Soviet state. It resulted in disestablishment of Okhotsk-Even area and some other national units as early as in the middle of 1930's. Koryak and Chukotka areas were integrated into the Kamchatka region of The Far Eastern territory while the national districts of Okhotsk-Even area became a part of Nizhneamurskiy region. Apart from the areas there was a number of independent national districts in the territory (The collection of RSFSR laws 1932).

The specific political and legal nature of a national area at the development stage consisted of the following:

- it had no constitutional legal platforms but was established by a special act of the highest agency of State power of RSFSR in accordance with the political directives; it was interpreted as a national administrative community; was a part of a region, territory;
- it defined the areas of settlement of native minorities and the boundaries of these areas; demonstrated a special status of the indigenous peoples;
- it was ethnically non-uniform by default but named after the indigenous people prevailing on the corresponding area; the government authorities of the area acted in accordance with the "Provision on congresses of Councils and Executive Committees of national areas in the northern territories of RSFSR" dated April 20, 1932 (State Archive of the Russian Federation).

The pinnacle of the Soviet policy pertaining to the administrative and territorial division of the northern regions was reached with introduction of the USSR Constitutions dated 1936 and 1937 (Kukushkin and Chistyakov 1987).

These Constitutions did not grant any special rights to national minorities and the peoples of the North. There were only three general statements relevant to the indigenous peoples: on national areas which finally obtained the status of specific administrative-territorial entities within territories and regions, the guaranteed representation in the Council of Nationalities of the Supreme Soviet of the USSR and the acceptance of the provision on these areas (p. 102 of RSFSR Constitution); on the languages of national areas permitted for court proceedings in the Russian Federation (p. 114 of RSFSR Constitution); on equality of USSR and RSFSR citizens regardless of their nationality and race in the fields of state, economic, social, political and cultural activity; on prohibition of rights restriction and provision of special privileges based on the abovementioned principles (p. 123 of USSR Constitution, p. 127 of RSFSR Constitution).

5   Conclusions

In general, the national policy between 1920s and 1930s was based on humanitarian considerations. In this context the difference between the policy of the Soviet state and the policy of tsarist Russia becomes apparent: the policy of tsarist Russia had also been based on humanitarian principles, but the Soviet policy was more effective. The attempts of the Soviet regime to improve the life conditions of native minorities, to provide them with welfare assistance and to secure their numerous privileges were quite sincere. Soviet authorities attached great political importance to these measures in the context of furtherance of the main principles of the Soviet national policy.

The priority of political and ideological considerations over the economical interests in the national policy of the Soviet government can be considered as the most important positive factor for the indigenous peoples of the Far East. From the very beginning the approach of the Soviet government to the problems of the peoples of the North was based on political perspective rather than on profitability of investments.

The Soviet state considered maintenance of traditional activities of these peoples the highest political, ideological and cultural interest. It encouraged their traditional production mainly by fixing high prices for the goods produced by these peoples. This economical measure guaranteed preservation and development of traditional production. For example, this policy was completely opposite to the measures taken towards the indigenous population of North America. Upon destruction of traditional social and economic system the Indians were provided only with material assistance that resulted in degradation of the peoples.

The Soviet state gave high priority to reorganization of traditional economy of the peoples of the North. In accordance with Marxist-Leninist theory, economical changes predetermine all other social changes. For this reason the Soviet government commenced from production force changes as a new society development prerequisite. This activity included the upgrading of the instruments of labor, mechanization and introduction of industrial society achievements to the indigenous peoples. The objective of the Soviet government was the economic modernization of the indigenous peoples. It included process improvements, labor management changes and orientation of conventional wisdom of native minorities in line with new conditions. This objective follows from the Marxist ideology where economical changes precede and determine all other changes. The Bolsheviks made an attempt to create a new culture based on the new economic system.

Being a center link of the system of administrative bodies of national areas in 1920's – the first half of 1930's, the Committees for Affairs of the North were abolished as a result of a change of the ideological task with respect to native minorities. Originally the activity of the Soviet government was focused on emphasizing of traditional peculiarities of aboriginal population. In the second half of 1930's the orientation of ideological line was changed towards the forced integration of small-numbered ethnic groups into the Soviet society resulting in introduction of modernization activities into the social and economic sphere and the sphere of national and territorial division of the national regions of the Far East. The way of life and the traditions of indigenous ethnic groups were leveled off by the shortfall policy of regional authorities approving the transformation of ethnocultural development of aboriginal population under the guise of formation of the integral Soviet society.

Acknowledgments. This research study was supported by the Ministry of Education and Science of the Russian Federation, the Agreement No. 14.Z56.16.5304-MK (The Grant of the President of the Russian Federation).

References

On approval of the Charter of the Committee for development of the peoples of the North under the Presidium of All-Russian Central Executive Committee: RSFSR All-Russian Central Executive Committee and Council of People's Commissars Decree dated February 2, 1925. The collection of RSFSR laws, No 12: paragraph 79 (1925)

On approval of the Temporary provision on administration of aboriginal ethnic groups and tribes in the Northern territories of RSFSR: All-Russian Central Executive Committee and Council of People's Commissars Decree dated October 25, 1926. The collection of RSFSR laws, vol. 73: paragraph 575 (1926)

On strategy of the national policy of the Russian Federation until 2025: Edict of the President of the Russian Federation dated December 19, 2012 No 1666. Legislation Bulletin of the Russian Federation, No 52: 7477 (2012)

State Archive of the Khabarovsk region. Archive fund 35. Series 1. File 45, p. 76

State Archive of the Russian Federation. Archive fund 3977. Series 1. File 279, p. 34

State Archive of the Russian Federation. Archive fund P-3977. Series 1. File 598, pp. 2–6

State Archive of the Russian Federation. Archive fund 3977. Series 1. File 280, p. 281

State Archive of the Russian Federation. Archive fund P-3977. Series 1. File 1043, pp. 12–13

The indigenous peoples of Russia: self-government, land and natural resources. The review of laws and other regulatory documents of the 19th–20th centuries, 46 p. Russian Historical Society of the Central office of the State Duma, Moscow (1995)

The Provision on congresses of Councils and Executive Committees of national areas in the northern territories of RSFSR dated April 20, 1932. The collection of RSFSR laws, No 39: paragraph 176 (1932)

The Siberian Revolutionary Committee (Sibrevcom), August 1919–December 1925: the collection of documents and materials. Novosibirsk book house, Novosibirsk (1959)

Akhmetova, A.V.: The problems of implementation of the international law on the rights of the native minorities of the North into the Russian legal system. Chelyabinsk State University. Reporter, vol. 279(25), pp. 21–24 (2012)

Babay, A.N., Kiselev, E.A.: Constitutional and legal status of the native minorities of the Amur River basin, 196 p. The Far Eastern Juridical Institute of MIA of RF, Khabarovsk (2005)

Bobyshev, S.V.: The Committees of Eastern Siberia and the Far East, p. 239. Far Eastern Federal University publishing office, Vladivostok (2000)

Bobyshev, S.V.: The Soviet state and the native minorities of Siberia and the Far East in 1920s–1930s. Komsomolsk-on-Amur: Federal State-financed Educational Institution of Higher Professional Learning Komsomolsk-na-Amure State Technical University (2005)

Gavrov, S.N. Modernization of Russia: the post-imperial transit, 269 p. MSUDT (2010)

Kryazhkov, V.A.: Native minorities of the North in Russian law, p. 560. Norma, Moscow (2010)

Kukushkin, Y.S., Chistyakov, O.I.: An Essay on the Soviet Constitution, 367 p. Politizdat, Moscow (1987)

Lipatov, A.A., Savenkov, N.T.: The history of the Soviet Constitution. Documented, 1917–1956, 1046 p. Gosyurizdat, Moscow (1957)

Popkov, Y.V.: The process of internationalization of the peoples of the North: theoretical and methodological analysis, 200 p. Nauka, Novosibirsk (1990)

Vakhtin, N.B.: The indigenous population of the Northern territories of the Russian Federation, 95 p. European publishing house, Saint Petersburg (1993)

Slezkin, Y.: Arctic Mirrors, Russia and the Small Peoples of the North, 512 p. New literary review, Moscow (2008)

Flier, A.I.: The Selected Works on Culture Theory, p. 289. "Soglasie" publishing house LLC; "Artem" publishing house, Moscow (2014)

Human Resources Make All the Difference

Olga Klimovets[(✉)]

Academy of Marketing and Social Information Technologies – IMSIT,
Krasnodar, Russia
new_economics@mail.ru

Abstract. People with higher education in Russia, including those who have graduated from university relatively recently, 5–10 years ago, possess tremendous advantages on the labour market compared to other employees. Unemployment rates are about 2 times less than the national average, the earnings – much higher, social status over the last 15 years has not deteriorated, up to date the vast majority of these people are working in the fields, which make them belong to three social groups: executives and specialists of higher and middle skill level.

Keywords: Employment · Unemployment · Workforce · Knowledge economy
New economy · International business

JEL Classification Codes: E24 · F22 · J24 · P51

Today, the Russians get higher education massively, but the demand for their knowledge is poor as far as economy is focused on resource exploitation. Russia exports its high quality human capital, but the one that remains within the country is not of high demand.

Unfortunately, the knowledge economy does not value this remaining human capital and doesn't provide incentives to learn "difficult" professions. This is evidenced by the results of the study. The doctor earns on average 20% more then the driver. For comparison: in the USA the difference is 261%, in German – 172% and even in a developing Brazil – 174% [1].

Over 20 years of comprehensive reforms, from 1995 to 2015, the structure of the labor market in Russia has changed insignificantly from the point of view of separation between public and private sectors, new and old companies.

The main employer is still the public sector - nowadays in the form of state companies (the share of those employed in the public administration even more than doubled), compared to small and medium enterprises, large companies such as *Yandex* and international companies, where less than a third of all employees are engaged.

Business bears the "social burden" and has to maintain employment. Unemployment in Russia is one of the lowest in the world and does not react to GDP changes. In most countries worldwide if GDP is falling, unemployment is growing. In Russia the unemployment can even decrease is similar GDP pattern. In such an environment, even if a person has the knowledge and skills required for new economy, there will be no area for application. Every modern professional can get out of demand in the future.

© Springer International Publishing AG, part of Springer Nature 2018
E. G. Popkova (Ed.): HOSMC 2017, AISC 622, pp. 315–320, 2018.
https://doi.org/10.1007/978-3-319-75383-6_40

The new economy requires new competences – not only theoretical knowledge, programming and data skills, but also creative, analytical thinking, communication skills and ability to work in conditions of uncertainty.

Business is looking for such competences. Only 17% of those employed has more than half of creative or analytical tasks at their workplace, almost half of the employed has to complete routine tasks. Almost the same ratio is in Brazil, but in developed countries the proportion of people that solve analytical problems is significantly higher: for example, in Germany – 29%, in Greet Britain – 45%.

The educational system concentrates young people on the technical and routine work by teaching them to act according to instruction, which is applicable, for example, in the civil service. However, this is consistent with the public mood of the majority of Russians who seek stability and prefer to work in the public sector or state-owned companies.

Public higher education doesn't improve the quality of human resources: the demand for graduates in small – many receive diplomas and work where it is not needed. The employers claim that many professions do not need long learning. 26% of graduates would as well learn less than five years [5]. However, the education system focuses on enrollment figures, not on the real needs of business.

1 What is Taught in Universities?

People with higher education in Russia, including those who have graduated from university relatively recently, 5–10 years ago, possess tremendous advantages on the labour market compared to other employees. Unemployment rates are about 2 times less than the national average, the earnings – much higher, social status over the last 15 years has not deteriorated, up to date the vast majority of these people are working in the fields, which make them belong to three social groups: executives and specialists of higher and middle skill level.

The share of people with higher education who work in low-skilled areas is relatively the same as in the OECD countries – 20%. The demand for people with higher education was growing faster than the supply since 2000s and still grows, but this trend can change with the growth of supply: the share of people with completed and uncompleted higher education grew from 26% in 2000 to 37% nowadays, and it could reach up to 45% in 2030.

According to a UN report, in 2016 Russia was among the countries with very high human development level, the problem is not in the education system, but in the economy. Obsolete and inefficient enterprises cannot create normal working conditions for graduates, and they will quickly leave, after which such enterprises are moving back into the labour market. But educating engineers in Russia is really a serious problem, the failure is observed at the level of advanced training and continuing professional education: there is no sufficient demand for such programs in the economy, employers do not have enough resources for retraining employees, and the employees themselves, as a rule, has lack of time and finance. Many employers have noted the need to create platforms of requalification and retraining of personnel for the new challenges, enterprises today are experiencing such conditions that they are not about innovation, they are more about survival.

2   How to Shift a Small Business from Law Quality Mass Consumption to Innovation?

In Germany and employment high-tech, well-funded, export-oriented small and medium enterprises ensure economic growth.

Ask a question to Russian friends, what's a small business in your view? The answer is: shopping pavilion, a barber shop, homemade cakes, taxi or repairs to the house, cafés and restaurants, design services, online shopping and site creation. Asking the same question to a resident of Germany, Austria or Switzerland, you will hear in response, "der Mittelstand". Mittelstand in Germany is more than 3.5 million small businesses, of which hundreds of thousands of engineering, and many of them are so-called hidden champions in their niches and industries. That's 15 million high-tech jobs in Germany alone. This is what drives the economy of the European Union together with giant corporations, conducts research and expands the scope of services. Mittelstand (a lot of small and medium businesses) makes engineering calculations and produces individual parts and components, creates the electronics for such automakers as BMW, Volkswagen or Daimler, helps to develop the electronics and electrical equipment for Siemens, creates IT solutions and projects for various spheres of activity, developing medical and biotechnology, teaches people. Of course, cafes, Breweries, hair salons and Internet-shops in Germany, too, but they do not provide economic growth. This task take on large corporations and innovative, well-funded and export-oriented of Mittelstand.

How does small and medium business operate in Germany? What helps small businesses to develop and to survive? And why the economic troubles of the EU have little impact on the German labour market?

Up to 100,000 new engineers and scientists recently graduated from universities flows in the German economy. But young people without higher education also receive vocational training and practical skills in special schools. The modern system of education provides the German economy with a reliable influx of skilled workers and engineers, which are so important for business, and small and medium enterprises benefit from high quality training of the workforce most of all.

The rights of owners of German companies are protected by Federal law. The rights of employees are protected by trade unions, which negotiate with business owners on clear rules of remuneration and the size of tariffs. These agreements take into account the interests of all parties, which allows avoiding strikes and negative social effect on small businesses. The Federal Employment Agency ensures that all employees of the enterprises were registered in the offices of the pension and social insurance, to which the employers and the employees themselves make certain proportions of contributions [9].

The state also performs continuous monitoring, regularly checking enterprise compliance with standards and regulations, but this is done so as not to harm their current activities. The company also financially interested to address the shortcomings identified by inspectors as soon as possible. The phrase "to scare business" simply cannot take place in the German vocabulary, because there is classification and order.

Despite the fact that many small and medium businesses are financially independent, there are over 200 programs to support entrepreneurs at both the regional and

Federal level in Germany. Their funding is conducted through the state Bank KfW (assets – EUR 489 billion) or its subsidiary DEG - the Corporation for investment and development.

Only in 2014, the KfW group has financed the companies with 74.1 billion euro and allocated 26.6 billion euro on programmes to protect the environment. It's more than the entire volume of the Russian National Welfare Fund, on which our largest corporations lay so much hope. And compare it with the volume of state support of small and medium business in Russia. The conclusion, unfortunately, is obvious.

The foundation of the German economic model is based on advanced technology and export orientation. Industrial competition in the world is very high, so annual R&D spending in Germany is 70 billion euros – more than in any other country.

Its economic power, Germany is also obliged to the cooperation of the University system, and industrial research laboratories. There is a network of publicly funded research institutions such as the Max Plank Scientific Research Society and Fraunhofer institutes for applied research, which, in partnership with the business provide the development of future technologies and products.

R&D is financed mainly by large corporations, and the task of Mittelstand is to implement in production the results of these scientific and technical studies in the first phase when producing a new product innovation is a risky business. This is a global trend – up to 85% of all venture projects in the world belong to a small high-tech companies. Mittelstand uses advanced technology in export oriented industries. Many family firms have sales offices, service centers and even factories abroad.

Relatively small but highly successful companies of small and medium businesses earned the title of "hidden champions". They have narrow specialization, but a high degree of innovation and quality makes their products popular worldwide. Some of the "hidden champions" was released outside of Germany and work worldwide. For example, Omicron NanoTechnology (about 100 employees) is a world leader in the field of analytical methods of surface physics, a manufacturer of tunneling microscopes and systems for creating nanostructures; wind turbines manufacturer Enercon; the medical technology company BrainLab, a manufacturer of technological equipment for processing of poultry products and livestock Rud. Baader and well-known to keen car enthuisiasts Webasto [10].

Manufacture and export of electrical equipment in Germany is growing annually by 12%, and this is higher than the rate of economic growth of China.

After the implementation in Germany of the Federal target program on development of biotechnology in the late 90s — early 2000s, the number of biotech companies has increased five times, and now the country is inferior in this area only to the United States.

Mittelstand is the largest employer. Approximately 70% of jobs, 57% of GDP, 40% of exports and 52% of value added in the economy are generated by small and medium business in Germany. Taxes in Germany are high. In average 50% of the profits of Mittelstand goes to the Federal and regional budgets. But the state creates the conditions for business development, implementing investment funding, supports education, provides incentives to start-ups and creates the infrastructure.

In Russia, the official tax burden on business is 35.6%, but in fact, according to Russian Union of Industrialists and Entrepreneurs and the World Bank, reaches 54%.

And if we add the inflation "tax" and unimaginable corruption costs, the numbers are again in favor of the German business. You say, Germany has no natural resources, so entrepreneurs are forced to base their business on knowledge. Yes, it is absolutely true. But also our small and medium business doesn't have access to natural resources, does it?

We live in the 21st century. There are many technologies that are successfully developed, or will be in demand. For example, the industry where Russian small and average business can take production and export niche - robotics. It doesn't need to have a lot of raw materials to create robots, but it requires intelligence, knowledge of theoretical mechanics, programming, and mathematics - and this we have always been strong. Robots, in turn, will increase productivity, freeing up people in heavy work and production for more intellectual activities.

Experts are stricken by the fact that the Ministry of Labour of the Russian Federation is planning to spend hundreds of billions of rubles to attract the country's millions of migrant workers in the coming years, instead of using these funds for the improvement of vocational education. Russia still has labor traditions, the education system, which is not destroyed yet, and even the system of state control, and most importantly - human potential.

References

1. Klimovets, O.V.: New marketing technologies in international business. In the collection: living economics: yesterday, today, tomorrow. In: The International Scientific and Practical Web-Congress of Economists and Jurists. ISAE "Consilium", pp. 25–30 (2017)
2. Pozdnyakova, U.A., Dubova, Y.I., Nadtochiy, I.I., Klimovets, O.V., Rogachev, A.F., Golikov, V.V.: Scientific development of socio-ethical construction of ecological marketing. Mediterranean J. Soc. Sci. 6(5S1), 278–281 (2015)
3. Klimovets, M.V.: Practice of outsourcing for strategic purposes by Russian and foreign companies. Mediterranean J. Soc. Sci. 6(36), 193–200 (2015)
4. Klimovets, O.V.: Review for monograph of D. Sc of Economics. In: Agabekyan, R.L. (ed.) Russian Labour Market: Peculiarities, Problems and Perspectives, 208 p. Publishing House-YUG, Krasnodar (2014)
5. Klimovets, O.V.: Research modern processes of Russian labour market (Review for monograph of D. Sc of Economics). In: Agabekyan, R.L. (ed.) Russian Labour Market: Peculiarities, Problems and Perspectives, 208 p. Publishing House-YUG, Krasnodar (2014)
6. Klimovets, O.V., Fundy, K.V.: Analysis of system development staff motivation hotels. Discussion 8(49), 97–102 (2014)
7. Klimovets, O.V.: Marketing of territories as a tool of formation of investment attractiveness of the region. In the collection: latest developments and the success of the development of economics and management. In: Collection of Scientific Papers on the Results of International Scientific-Practical Conference, pp. 55–58 (2017)
8. Klimovets, O.V.: Regional peculiarities of the implementation of the policy of import substitution. In the collection: modern scientific research: historical experience and innovations. In: Proceedings of the International Scientific-Practical Conference, pp. 6–12 (2017)

9. Klimovets, O.V.: Russian multinational corporations on the markets of Asia and Latin America. In the collection: cooperation between China and Russia in the framework of the initiative "One belt, one road". In: Collection of Materials of International Scientific-Practical Conference, pp. 150–156 (2017)
10. Klimovets, M.V.: The formation of international outsourcing in the new economy. In the collection: modern scientific research: historical experience and innovations. In: Proceedings of the International scientific Practical Conference, pp. 27–31 (2017)

Problems and Perspectives of Improving the Process of Innovations' Commercialization in a Modern University

Svetlana E. Sitnikova[1]([⊠]), Lyubov A. Halo[2],
and Natalia S. Polusmakova[3]

[1] Volgograd State Medical University, Volgograd, Russia
ses1113@yandex.ru
[2] MFUA, Serpukhov, Russia
halo@volgodon.ru
[3] Volgograd State University, Volgograd, Russia
polusmakova@volsu.ru

Abstract. The purpose of the work is to determine the problems and perspectives of improving the process of innovations' commercialization in universities of modern Russia. The theoretical and methodological basis of the research is the new institutional theory and the systemic approach to conduct of research within the economic theory. The work uses the method of structural and functional analysis, statistical analysis, and formalization. As a result of processing of statistical information on the innovational activities of Russian universities in 2016, we schematically presented the process of creation and commercialization of innovations in modern Russian universities in 2016. Based on the performed result, we determine the "institutional trap" of innovative activities of modern Russian universities. Its effect could be characterized as "innovations for innovations". In other words, Russian universities are separated from the national economic system and seek their own interests. The interests of scholars related to increase of authority in the scientific society and maximization of reward for innovative activities do not correlate with the national interests, oriented at import substitution of innovations and increase of innovational activity of domestic companies. The authors prove that the main reason of low effectiveness of the process of innovations' commercialization in modern Russian universities is underdevelopment of the institutional basis. Practical recommendations for solving the above institutional problems and improving the algorithm of innovations' commercialization in a modern university are offered – they will allow increasing the number of innovations that are successfully commercialized, i.e., patented and passed for implementation to the companies, and that are reflected in modernization of business processes and manufacture of innovational products.

Keywords: Improvement of the process of innovations' commercialization
Modern university · Russia · "institutional trap"

E. G. Popkova (Ed.): HOSMC 2017, AISC 622, pp. 321–327, 2018.
https://doi.org/10.1007/978-3-319-75383-6_41

1 Introduction

Importance of dynamics innovational development of modern economic systems could not be overestimated. High innovational activity of domestic companies allows supporting global competitiveness of economy and ensures its sustainability against the fluctuations of the world markets and crises. Due to these reasons, innovational development is established as the main priority of modern countries of the world.

Implementation of this priority in practice requires coordinated work of all participants of the innovational process, its systemic integrity, and high effectiveness. The main participants of this process are R&D organizations that create innovations and economic companies that implement innovations. Modern university performs an important strategic mission in the socio-economic system, being a source of innovations and forming the foundation for creation of innovational economy.

However, in the conditions of high level of universities' separation and their weak connection with commercial structures, the process of commercialization of innovations, which are created by universities, becomes complicated, which does not allow them to perform their function as suppliers of innovations in economic systems, oriented at development of innovational economy. This emphasizes topicality of studying the process of innovations' commercialization in a modern university.

The authors prove the hypothesis that universities in modern Russia face the problems that hinder successful commercialization of their innovations. The purpose of the article is to determine the problems and perspectives of improving the process of innovations' commercialization in modern Russia's universities.

2 Materials and Method

The theoretical and methodological basis of the research is the new institutional theory and the systemic approach to conduct of research within the economic theory. The work uses the method of structural and functional analysis, statistical analysis, and formalization. The initial information for the research was taken from the materials of reports of the Federal State Statistics Service for 2016 (Table 1).

Table 1. Statistics of innovational activity of Russian universities in 2016.

Indicator	Total	Universities
Number of universities	–	702
Share of universities in the number of R&D organizations, %	–	19.48
Personnel of universities dealing with scientific research, people	–	142,652
Number of leading production technologies that are potentially created by groups of scholars five people each	–	28,530
Number of created leading production technologies	1,398	272
Patented leading production technologies	34,706	6,760
Implemented patented leading production technologies	9,249	1,801
Number of imported patented leading production technologies	2,336	455

Source: compiled by the authors on the basis of Federal State Statistics Service 2016.

3   Discussion

The process of innovations' commercialization in modern universities is studied in detail in multiple works of such authors as Popkova et al. 2016a, Ragulina et al. 2015, Bogoviz et al. 2017, Orudjev et al. 2016, Bogdanova et al. 2016, Popova et al. 2016b, Kuznetsov et al. 2016, Kostikova et al. 2016, Simonova et al. 2017, Sitnikova 2016. However, despite the high level of elaboration of the set problem, there are perspectives of its further study, related to development of methodological and practical recommendations for optimizing the process of innovations' commercialization in modern universities.

4   Results

As a result of processing of statistical information on innovational activity of universities of Russia in 2016, we present the scheme of creation and commercialization of innovations in modern Russian universities in 2016 (Fig. 1).

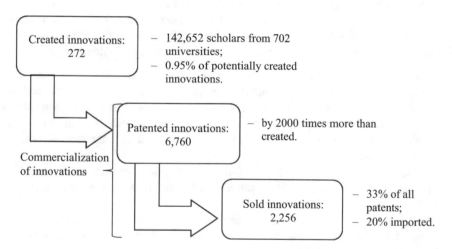

Fig. 1. The process of creation and commercialization of innovations' in modern Russian universities in 2016 Source: compiled by the authors.

Based on the performed analysis, we determined "institutional trap" of innovational activities of modern Russian universities. The essence of its effect is that modern universities have a lot of workers who participate in scientific research. The potential of creation of innovations by modern Russian universities is very high. There were 28,530 leading production technologies in 2016.

However, modern universities created 272 leading production technologies in 2016 – i.e., their innovational potential is realized by less than 1% (0.95%). This shows weak motivation of scholars who conduct scientific R&D in modern Russian universities, as even in view of the norm of potentially impossible scientific research such efficiency is very low.

It is obvious that patent activity is more important criterion of efficiency of a modern Russian scholar, as 6,760 patents were registered in 2016, which is by 2000 times more that the number of created leading production technologies. Only 33% of patented Russian innovations were used in practice (2,256), of which 20% were imported – i.e., instead of stimulating innovational development of domestic economy, the competitive advantages of other countries are increased.

The effect of "institutional trap" of innovative activities of modern Russian universities could be characterized as "innovations for innovations". In other words, Russian universities are separated from the national economic system and seek their own interests. The interests of scholars, related to increase of authority in the scientific society and maximization of reward for their innovational activity, do not correlate with the national interests that are oriented at import substitution of innovations and increase of innovational activity of domestic companies.

However, despite the fact that Russian universities do not fully perform the function of being sources of innovations for development of the national economic system, creating unpopular innovations or importing them, they continue to receive financing from the state. The main contradiction is that there is no interest and responsibility of universities for commercialization of created innovations.

The main reason of low effectiveness of the process of innovations' commercialization in modern Russian universities is underdevelopment of the institutional basis. The obstacles on the path of its formation are the following problems:

– absence of stimuli and requirements to innovations' commercialization for modern Russian scholars, as the most important indicators of efficiency of R&D activities are the number of participations in scientific events (conferences, symposiums, etc.), number of scientific publications, and number of patents, but their implementation in practice is not mandatory and is not taken into account, without any reward envisaged;
– low interest of the management of modern Russian universities in successful commercialization of created innovations, as it is not taken into account during the ranking assessment of universities;
– weak connection between universities and entrepreneurial structures and low demand for innovations from business, predetermined by import of innovations and low innovational activity.

Perspectives of improving the process of innovations' commercialization in a modern university are related to solving these problems. For this, we offer the following practical recommendations:

– introduction of the system of stimulating the scholars who work in universities for commercialization of the created innovations. One of the most significant criteria of evaluation of efficiency of R&D activities should be successful sales of innovations to companies and their implementation into economic practice. This is to be confirmed by agreement (contract) for transfer of rights for the objects of intellectual property, which describes material reward for scholar/scholars who created the innovations which is the object of the contract;

– an important element of the system of ranking evaluation of activities of modern universities should be commercialization of created innovations and efficiency of the work of universities' management should be evaluated through the prism of this indicator – for ensuring the interest in this process;
– it is necessary to correct the system of state (primarily, tax) stimulation of innovational activities of economic subjects in favor of stimulating not just implementation of innovations but innovations that are created by Russian universities – for stimulation of demand for them and ensuring import substitution of innovations.

As a result of implementation of the offered recommendations, it will be possible to overcome the determined "institutional trap" and to optimize the algorithm of innovations' commercialization in a modern Russian company, which, in our opinion, should have the following way (Fig. 2).

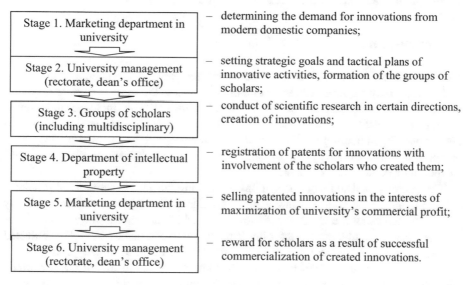

Stage 1. Marketing department in university	– determining the demand for innovations from modern domestic companies;
Stage 2. University management (rectorate, dean's office)	– setting strategic goals and tactical plans of innovative activities, formation of the groups of scholars;
Stage 3. Groups of scholars (including multidisciplinary)	– conduct of scientific research in certain directions, creation of innovations;
Stage 4. Department of intellectual property	– registration of patents for innovations with involvement of the scholars who created them;
Stage 5. Marketing department in university	– selling patented innovations in the interests of maximization of university's commercial profit;
Stage 6. University management (rectorate, dean's office)	– reward for scholars as a result of successful commercialization of created innovations.

Fig. 2. Perspective algorithm of innovations' commercialization in a modern university Source: compiled by the authors.

As is seen from Fig. 2, the offered perspective algorithm of innovations' commercialization in a modern university is conducted in six consecutive stages. At the first stage, the acting subject is the department of marketing, which conducts marketing research and determines the demand for innovations from modern domestic companies and, if possible, concludes agreements for further creation of innovations for them.

At the second stage, the university's management (rectorate and dean's offices) set strategic goals and tactical plans of innovative activities, form the groups of scholars from the university's employees for satisfying the determined demand for innovations and execution of the contracts with companies. At the third stage, the groups of scholars (including multidisciplinary, i.e., from different faculties) conduct scientific research in certain directions and create innovations.

At the fourth stage, the department of intellectual property in the university registers patents for innovations with involvement of the scholars who created them, i.e., the main subject in the process of patenting of innovations is the department of intellectual property, and the scholars provide support for it. At the fifth stage, the department of marketing sells the patented innovations in the interests of maximization of the university's commercial profit – i.e., selects the optimal terms of the contract for transfer of rights for the objects of intellectual property.

At the sixth stage, the university's management (rectorate, dean's offices) rewards the scholars as a result of successful commercialization of created innovations, thus stimulating their further interest to successful innovations' commercialization. The advantage of the offered algorithm, as compared to the applied algorithm of innovations' commercialization in universities of modern Russia, is the high level of labor division, which allows reducing the load on the scholars and ensuring their interest in successful commercialization of created innovations.

5 Conclusions

It is possible to conclude that the process of innovations' commercialization in a modern university requires further institutionalization, as the most important obstacles on the path of its optimization are institutional problems. This includes underdevelopment of the institute of stimulations for scholars for commercialization of created innovations, institutional gap with stimulations for universities for innovations' commercialization through the system of ranking score, and weak institutional connections between universities and companies.

The offered practical recommendations for solving the above institutional problems and improving the algorithm of innovations' commercialization in a modern university will allow increasing the number of innovations, which are successfully commercialized, i.e., patented and transferred for implementation to companies, as well as reflected in modernization of business processes and production of innovational goods.

It should be acknowledged that a certain limitation of the results of the performed research is usage of the experience of modern Russia in the sphere of innovations' commercialization in university, which does not allow determining the whole specter of the problems that emerge in the countries of the world. Conduct of transnational studies that cover the groups of countries according to the regional principle or to the principle of common progress in socio-economic development determines the perspectives of further scientific research.

References

Bogdanova, S.V., Kozel, I.V., Ermolina, L.V., Litvinova, T.N.: Management of small innovational enterprise under the conditions of global competition: possibilities and threats. Eur. Res. Stud. J. **19**(2 Special Issue), 268–275 (2016)

Bogoviz, A.V., Ragulina, Y.V., Kutukova, E.S.: Ways to improve the economic efficiency of investment policy and their economic justification. Int. J. Appl. Bus. Econ. Res. **15**(11), 275–285 (2017)

Kostikova, A.V., Tereliansky, P.V., Shuvaev, A.V., Parakhina, V.N., Timoshenko, P.N.: Expert fuzzy modeling of dynamic properties of complex systems. ARPN J. Eng. Appl. Sci. **11**(17), 10601–10608 (2016)

Kuznetsov, S.Y., Tereliansky, P.V., Shuvaev, A.V., Natsubize, A.S., Vasilyev, I.A.: Analysis of innovate solutions based on combinatorial approaches. ARPN J. Eng. Appl. Sci. **11**(17), 10222–10230 (2016)

Popova, L.V., Popkova, E.G., Dubova, Y.I., Natsubidze, A.S., Litvinova, T.N.: Financial mechanisms of nanotechnology development in developing countries. J. Appl. Econ. Sci. **11** (4), 584–590 (2016)

Ragulina, Y.V., Stroiteleva, E.V., Miller, A.I.: Modeling of integration processes in the business structures. Mod. Appl. Sci. **9**(3), 145–158 (2015)

Simonova, E.V., Lyapina, I.R. Kovanova, E.S., Sibirskaya, E.V.: Characteristics of interaction between small innovational and large business for the purpose of increase of their competitiveness. In: Russia and the European Union Development and Perspectives, pp. 407–415 (2017)

Popkova, E.G., Chechina, O.S., Abramov, S.A.: Problem of the human capital quality reducing in conditions of educational unification. Mediterranean J. Soc. Sci. **6**(3), 95–100 (2016a)

Federal State Statistics Service: Russian in Numbers: Short Statistical Bulletin. Federal State Statistics Service, Moscow (2016)

Sitnikova, S.E.: Peculiarities of the institutional infrastructure of innovations' commercialization of developed countries' universities. Econ. Entrep. **11**-2(76-2), 742–745 (2016)

Looking Inside Things

Olga V. Dybina[(⊠)]

Tolyatti State University, Tolyatti, Russia
dybinaov@yandex.ru

Abstract. The article studies the approaches to pre-schoolers' acquaintance with objective environment. The theoretical studies are analyzed, according to which the child's life does not "naturally" guarantee the elements of creative beginning. It is obvious that their emergence and development require specially and scientifically substantiated socio-pedagogical practices. The issues of educating and teaching are very delicate and require large care in their application. The article shows that the art of pedagogue consists in finding a limit between "can", "want", and "must" with a child. Pedagogical practice should be conscientious and structured as to the one who studies. Assimilation and accommodation are universally fit for building the system of studying. At that, the corresponding (specifically organized) consumption of items can develop creative beginning with a child. Special attention is paid to the stages of pre-schoolers' acquainting with objects with the help of which they can look inside the world and participate in its transformation, showing independence and creativity. These problems are of a many-sided character.

Keywords: Material environment · Transformation
Stages of acquaintance with the material world · Developing environment
Transformation situations

1 Introduction

"In medias res" is the Latin for "inside the things". This expression perfectly fits the topic of this research – consideration of the role of the material aspect for a human.

As a matter of fact, many of us wonder at babies – namely, at what happens with them during the first year of life. Psychologists emphasize: this is the period of active development of a baby's attitude towards material world. Here's a latest example: a ten-month baby tries to kick a ball – just like its father showed him. Positive emotions in relations with an adult will form with this baby a positive attitude towards himself and inquisitive attitude to the surrounding world. Revealing the material world, a baby reveals itself within this world as a material creature.

Our research showed that a baby easily perceives the material surrounding in its changes, movement, and development, which gives a push to appearance of forecasting view on the artificial world (things). Children's acquaintance with practical transformation of things significantly influences their creative activity – in particular, pre-schoolers' striving to create something original and new.

Capability for creative transforming activities will give a creative person the very best that is on the Earth. For example, what can a four-year child know about glass?

© Springer International Publishing AG, part of Springer Nature 2018
E. G. Popkova (Ed.): HOSMC 2017, AISC 622, pp. 328–333, 2018.
https://doi.org/10.1007/978-3-319-75383-6_42

Only that it can drink tasty juice from a glass, that flowers are put into a vase, that glass is transparent, cold, and fragile, that it produces sounds, etc. It is quite possible that later it will learn of Alexander the Great who – according to the ancient scrolls – went underwater in a specially constructed glass box (probably, that was a prototype of a diving bell).

An inquisitive and creative human, who has knowledge and strives for transformation of the surrounding world, will not heed the limits in his findings, urges, and travels. He will reach the depths and heights in cognitive and creative activities.

During analysis of own and other authors' materials, the main peculiarity wasn't omitted: for a pre-schooler, the material world is a sphere of realization of himself and his personality, and adult is the one who knows and evaluates the pre-schooler's achievements.

Material surrounding performs not only utilitarian functions – it plays an important esthetic and moral role. Moral function of the material world is vividly expressed in works of art. Let us note that such feature of character as taste is formed under the influence of material environment.

2 Methodology

Orientation at deep interest of Russian and foreign scholars in the nature of the material world allowed us to distinguish the essential and universal content of philosophical and other ideas of things. A book has to bring pedagogue to the child's being in the world of things. With crossing of such huge essential trajectories as "child-thing" and "child-adult", we can expect happy appearance of creative potential. Also, a theoretical model that we built during the research sets qualitative solution of psychological practice for formation of a creative origin with a child.

As we can see, child's life cannot guarantee the "development" of elements of creative origin. It is obvious that their emergence and development require special and scientifically substantiated socio-pedagogical practices. According to V.M. Bekhterev, the issues of education and training are very subtle and require huge care in their application.

We think that the art of pedagogue consists in finding in each specific case the measure between "can", "want", and "must" of a child. Pedagogical practice of education should be responsible and structured regarding the educated person. Assimilation and accommodations are universal for development of the training system. At that, corresponding (specifically organized) consumption (assigning) of things can actively develop a creative origin with a child.

Success depends not so much on things that a pedagogue uses as on his professional approach to them and skillful use of a key named "need-capability" (firstly, formation of a need for creativity, and then starting the development of creative capabilities).

Having the largest arsenal of practical developments, we're sure that during formation of attitude towards the material world, it is necessary to use such "subtle" means as acquaintance with creative transformations of the material world: things of the past, present, and future. Full-scale and interesting acquaintance expands children's ideas of transforming types of activities. At last, a moment comes when they cannot but shout "me too!" – that it the strength of a child's wish to do something itself and of creative light in its eyes.

Let us emphasize that the basis of creative activity is set in pre-school age. At this, it is important to know – according to K.D. Ushinsky – the activities fir for a child and enrich it with means leading to unlimited activities.

3 Main Results

Analysis of a wide circle of philosophical and psychological & pedagogical literature and own study allowed for determination of the system of scientific & pedagogical provisions that consist theoretical concept of creativity formation with pre-schoolers in the process of their acquaintance with the material world:

- about necessity for consideration of creativity through ontology of human being and anthropological orientation, which expands conceptual understanding of creativity and logic of its development and allows stating that creativity should be revealed as general capability of a child in totality of knowledge, skills, wishes, etc.;
- about moving connection between creativity and copying, which allows pedagogical practice to found not on separate conditions but their comprehensive entity and interaction between them during formation of creativity;
- about connection between creativity, material world, and acquaintance with it, which is reflected during development of model's structural component;

Based on the viewed provisions and analysis of age peculiarities of children of 3–4, 4–5, and 5–6 years old, we developed a model of formation of creativity through acquaintance with the material world.

Creation of a model of creativity formation that consists of three interconnected sub-systems (information block, action & thinking block, and transformation block) in the process of preschoolers' acquaintance with the material world seeks the main goal – rational organization of experimental work.

Based on structural components of the model and content and organization of acquaintance with the material world, the chain of interaction of adult and child was created which includes three stages – at each of which the character and content of interaction were changes.

The first stage of a new experiment included setting the task; during solving the task, the children copy their role models from adults (tranformers of the material world) and copy their means of activities and creative manifestations of item transformation.

For this purpose, the children were led (according to their age possibilities) to understanding that the thing has the past and the future, and, this, to realizing the connection "human-thing" and creative features with a human; the children were attracted to participation in transformational activities together with adults. Pre-educational activities of a child were determined by the environment, adult and child's experience, and their activity as for motives and means of action. In the process of transfer of experience of transformation, the system of targeted and organized interactions was used, which had dual purpose: teaching children the means of the thing's transformation and influencing the establishment of experience of creativity; for practical transformation, the emotional material was used which activated development of new motives (striving for transformation) in pre-schooler's activities.

The task of the second stage of experiment is formation with children of means of actions by creation of creative "field" (developing environment) for independent actions of different character, manifestation of cognitive, practical, and creative activity and establishment of interaction "child-thing" in social reality. Creative "field" is an indirect sample that stimulates creative manifestations of children. A huge role here belongs to enrichment of environment: museums of things, landscapes, collections, pieces of artificial world, laboratories, transformation centers etc. were created. Didactic games were included into creative "field".

Didactic games combine cognitive and interesting activities of a preschooler. Our games are based on experimental actions, which significantly complicates the conduct of direct educational activities – as senior preschoolers who observed the demonstration of the experiment with huge pleasure achieve independent execution of each experiment – but this does not always coincide with a pedagogue's capabilities. It is obvious that experimenting is one of the most complicated actions for a preschooler.

Didactic games begin with discussion of the topic and main tasks for different actions. Such start stimulates development of attention and concentration of children. Each member of the discussion obtains the experience of discussion conversation and a skill to protect his opinion and listen to a friend's opinion. Observation, discussion, and joint selection of the direction for future actions – all this ensures success of each didactic game. The leading role in these games belongs to the pedagogue, as these experimental actions are too complicated for beginning researchers.

In the course of experiment, creative manifestations of each child are fixed and controlled by adults. Internal capability of a child to solving creative situations, a skill to set a creative task and to solve it with available method (application of the action of research, modeling, experimental, and algorithmic character), etc. is determined.

Acquaintance with new items not only pushes the limits of the surrounding world but ensures its development – cognitive and creative. A preschooler not only fixed external features of things, understands their name and purpose but can offer another way of their application, replacement of one item with another, and even transformation of certain things.

At this stage, children get acquainted with various things. The problems of materials' features, their quality and possibility of use in the process of the game are always interesting for preschoolers. Having fixed the main examples of items' features in their mind, it is possible to pass to understanding materials' characteristics. This is the "beginning of beginnings" – getting acquainted with materials' features. For the games with materials, the number of research objects is limited: metal, wood, glass, plastic, fabric, and leather. Children constantly use things of them, without giving any thought to their peculiarities and possibilities of their use in various situations. What will be the main task for solving this problem? Surely, consideration of features of the enumerated materials according to the principle – "everything is relative".

Behavior of items of various materials in water is a classic of the genre. These experiments allow for determination and comparison of the item's mass as to the feature that is interesting for children: it is lighter or heavier than water. Experiment:

"mass verification" of objects in a transparent plastic bottle is performed: two similar balls of glass and foam are put in it, the bottle is filled with water and closed. With each turn of this "item", the balls switch places.

The children are very proud by the fact that they have a hand-made toy and can explain why the balls switch places with turning of the bottle. It is worth noting that the depth of the knowledge depends on the child's age. Becoming older, it acquires experience, preserving the wish to continue experiments. After performing the active and interesting actions with balls in the water, each young researcher has new "items": a transparent plastic bottle has two minute glass bottles with rubber corks and several small colored pebbles inside. They behave exactly like the first "items" with balls, and it is difficult to understand why the bottles switch places with turning the "item" upside down. Our young experimentalists will try to keep the answer secret, and it's easy to understand them – they're only five or six years old, and they have already made such a "wonder" themselves!

Answer: One minute bottle has water inside, the water is transparent and cannot be distinguished in water environment, and its mass – as compared to another bottle with air – is larger (air is lighter than water). It is a "focus", and children do not forget such things and dream of new experiments with various items.

Such experiments and pedagogical situations should be included into direct educational activities. At that, important role belongs to pedagogue and his attentive attitude towards a child's position. Senior preschoolers, who have certain experience of getting peculiarities of the surrounding world, sufficient volume of skills for search for necessary information, and even conduct of elementary experimental actions, often do not even want to listen to long and detailed explanations of the pedagogue.

The most impatient child tries to shift the pedagogue's attention to the experiment that it has performed or is going to perform at home. Perhaps, it is a "very important" question, the answer to which it cannot find in the books recommended by the pedagogue or even on the Internet. If this sudden question is close to the topic of the future research, it is possible to thank the "hasty" pupil and bring everyone's attention to expedience of mandatory acquaintance with the topic of the new research and preparation of question that demonstrate their interest in ways of solving the set task. The questions that do not coincide to the topic of educational activities should not always be refused. Sometimes, the answer to the question allows setting a "task" for the young researchers and directing them to the new ways of search for information.

The third stage of the forming experiment includes the tasks that require independent creative actions for transformation of things. The pedagogical process includes creative tasks, situations of transformation, didactic games, etc. Special attention is paid to transformation games and organization of transformation activities.

4 Conclusions

An obvious result is the positive dynamics of preschooler's wish to get acquainted with the material world and participate in its transformation. Forecasting and forestalling were expressed in a specific form, manifested by a word, drawing, movement, etc. The work complicated with each stage, depending on children's age, level of creative manifestations, and individual differences. By the end of experiment, each age group of the children features the change of the creativity level (it increased), as compared to the control groups' children, which shows effectiveness of the developed methodology.

"Tempora mutantur, et nos mutamur in illis" – "times change, and we change with them", said the ancient people. The contemporaries say, "Things change, and we change with them". Today this saying is topical and ambiguous and modern. New time requires things according to dialectics of a new mode.

References

Drobnitsky, O.G.: The World of Live Objects. M.: Nauka (1967). 321 p

Druzhinin, V.N.: Psychology of General Capabilities. SPb.: Piter Publ. (1999). 368 p

Druzhinin, V.N., Khazratova, N.V.: Experimental study of forming influence of environment on creativity. Psychol. J. (4), pp. 83–93 (1994)

Dybina, O.V.: Creativity is an essential feature of human being. Monograph. M.: Russian pedagogical society (2001). 96 p

Dybina, O.V.: Material world as a means of development of children's creativity. In: Theory and Practice of Public Development, vol. 2, pp. 157–159 (2014)

Dybina, O.V.: Cultural foundation of being in ontogenesis. In: Theory and Practice of Public Development, vol. 7, pp. 151–158 (2015)

Dybina, O.V.: Methodology of formation of creativity with children with the means of the material world. In: Scientific Opinion, vol.2, pp. 99–103 (2014)

Krulekht, M.V.B.: Preschooler and man-made world. SPb.: Detstvo-Press (2002). 160 p

Savenkov, A.I.: Cradle of a genius. M.: Pedagogical society of Russia (2000). 224 p

Ushinsky, K.D.: Education of a Human: Selected Works. M.: Karapuz Publ. (2000). 256 p

Venger, N.B.: Path to development of children creativity. In: Preschool Eduction, vol. 11 (1983), pp. 32–35

A Systemic Approach to Development and Implementation of Key Performance Indicators in a Non-government Healthcare Institution

S. Blinov[1]([⊠]) and V. Blinova[2]

[1] Organization of Public Health and Public Health,
Medical University "Reaviz", Samara, Russia
Sblinov@me.com
[2] Road Clinical Hospital at Samara Station of the Open Joint-Stock Company
"Russian Railways", Samara, Russia

Abstract. The use of strategic management tools and program-targeted management ensures the adoption of quality and timely management decisions in the long-term and medium-term. A systemic approach to the development and implementation of key performance indicators will ensure the unity of goals and objectives, the necessary centralization and standardization, a comprehensive analysis of the institution as a whole, of each structural unit and employee.

Keywords: Systemic approach · Balanced scorecard
Key performance indicators · PEST · SWOT · SMART · Cost reduction
Attached contingent

1 Introduction

Changes in macro-, mezo- and microeconomic situations require the use of new management tools and ways to effectively adapt the health care institution as a whole and staff to the changes that are taking place. It is necessary to ensure not only the quality of medical services – diagnostic, therapeutic, rehabilitation, preventive; Social and economic efficiency of medical institutions, their competitiveness and financial stability; but also in the long term – strengthening the health of the nation. Heads of non-governmental health care institutions (NGHCI) of JSC «Russian Railways» perform an additional corporate order – ensuring the safety of the transportation process; improving the quality of medical care for employees of Russian Railways and their family members, the company's retirees, preventing occupational diseases and injuries to employees, and ensuring the professional longevity of the company's employees.

Achieving the planned values of the indicators of medical and non-medical activities set by the founder and centralized orders; the introduction of the principles of lean production and the process approach orient the leaders of the NGHCI to the use of innovative tools for conducting medical and non-medical activities, optimizing the staffing and management structure of the institution, including modern methods of strategic and operational planning, budgeting and controlling, motivation and

© Springer International Publishing AG, part of Springer Nature 2018
E. G. Popkova (Ed.): HOSMC 2017, AISC 622, pp. 334–342, 2018.
https://doi.org/10.1007/978-3-319-75383-6_43

encouragement of employees, analysis Results of medical and preventive and administrative-economic activities.

Formation of a list of key performance indicators (KPI), built in accordance with the principles of strategic management, Management by Objectives (MBO) and SMART-technologies, will provide quality planning, comprehensive analysis, monitoring and control of the results of medical and financial and economic activities of the healthcare institutions in general, of each institution, structural units and employees.

The research was based on conceptual approaches to solving problems of effective management of such world-famous scientists and practitioners as M. Armstrong, G. Dessler, R. Kaplan, D. Norton, L. Porter, F. Taylor, A.A. Thompson, A.J. Strickland, and others.

2 Materials and Methods

The theoretical and methodological basis of the study was the conceptual provisions of the fundamental works of foreign and domestic scientists in the sphere of management and management of employees; monographs, articles in leading scientific journals; federal and regional legislative and regulatory acts governing the development of public health in Russia; local acts of JSC «Russian Railways»; publications in periodicals and materials of scientific and practical conferences.

The author's research is based on a systemic approach. To solve the problems, the methods and tools of formal, logical, process, functional research were applied; strategic, statistical and comparative analyses; expert evaluation, forecasting and regulatory planning.

The materials of the Ministry of Health of the Russian Federation, the Ministry of Health of the Samara Region, the current documentation and statistical reports of the non-governmental public health institution «Road Clinical Hospital at Samara Station of the open joint-stock company «Russian Railways»; normative and methodical materials; official information and reference data posted on Internet sites.

The empirical basis for the study was the results of observations and studies carried out personally by the authors.

3 Results

The list of the main tasks of the development of the healthcare facilities of the holding company is the provision of medical security for the transportation process, the provision of medical assistance to employees of Russian Railways, members of their families, non-working pensioners of railway transport, examination of occupational fitness, prevention of occupational diseases (poisoning) and occupational injuries of workers, On railway transport, ensuring the professional longevity of company employees are determined the Central Health Directorate (previously – the Department of Health) and are adjusted annually based on actual results achieved.

The solution of these problems requires appropriate training at the level of each institution. In the second half of the year 2015, the hospital at st. Samara has developed special organizational documents, first of all, the draft Regulation on the application of key performance indicators (KPI) activity in the Clinical Hospital at the Samara Railway Station of the open joint-stock company «Russian Railways» [1]. The situation determined the basis for the formation of a system of key performance indicators and the purpose of their implementation – the translation of strategic objectives and a long–term program of institution development into the form of specific management indicators. Indicators should be suitable for the translation and promotion of network management solutions throughout the vertical, including the labor collectives of the hospital, and to assess the current state of their achievement and make timely management decisions in the long and medium term. The actual results of each institution's activities allow us to determine the dynamics of integration into regional health care, to evaluate the activity of managers, to participate in network construction and the formation of a personnel reserve.

The draft regulations specify: the purpose of the KPI system and the requirements to it; tasks and list of basic KPI and peculiarities of their application; algorithm of approval, algorithm for monitoring and controlling their execution.

Purpose of the KPI system:

- determination of the performance of the economic entity at any level of the organizational hierarchy – the institution as a whole, the enlarged structural units (ESU)/ structural units (SU), employees (Table 1);

Table 1. Three–level KPI system

Three–level KPI system	
1 level	KPI hospital
2 level	KPI ESU
	KPI SU
3 level	KPI employees

- increasing the efficiency of the business entity (institution in general, ESU/SU, employees);
- ensuring a reliable evaluation of the result and effectiveness of the activity;
- motivation of the business entity (institutions in general, ESU/SU, employees).

The main objectives of the KPI system are:

- monitoring and monitoring the implementation of the strategy;
- assessment of achievement of strategic goals;
- creation of appropriate motivation taking into account the orientation of employees to achieve priority development goals.

The main tasks in the development and implementation of the KPI system are:

- development of the KPI list in accordance with the strategic development goals;
- determination of the methodology for calculating KPI values;
- determination of the mechanism for setting the target KPI values;

- development of procedures to ensure the implementation of the process of preparing initiatives to achieve strategic goals (the «strategic initiatives plan»);
- formation of a matrix of goals and KPI reflecting the alignment of strategic development objectives and key performance indicators across different planning horizons and organizational hierarchy levels – hospital, ESU/SU, employees;
- development of a matrix of authority and responsibility in the implementation of the development strategy and implementation of KPI at the institution level, ESU/SU and employees;
- development of procedures to monitor actual KPI values.

Simultaneously with the development of the Regulations on the application of key performance indicators (KPI), other organizational documents regulating the process of establishing and applying KPI were developed:

- draft Regulation on the commission/subcommittee on strategic planning;
- methodological recommendations for setting the target KPI values in the hospital;
- procedure for approval of the planned KPI values;
- algorithm of monitoring and control of achievement of planned KPI values;
- regulations for the formation of plans for strategic initiatives;
- matrices of authority and responsibility distribution at the hospital, ESU/SU level, officials and employees;
- approximate individual plans of employees;
- the procedure for monitoring the actual KPI values and possible deviations;
- the project of scheduling.

In December 2015, in the hospital at St. Samara a business game was conducted on the formation of a list of KPI 2016 for three levels – institution, enlarged structural units, structural units and employees. The game was attended by deputy chief physicians, heads of structural units of all levels and staff members included in the category of personnel reserve, total 38 human.

In determining significant KPI, the participants analyzed the external and internal environment (PEST/SWOT) of JSC «Russian Railways», the network of non-governmental health care institutions (NGHCI) of the holding company and hospital; peculiarities of staff motivation and problems of interaction between medical and non-medical structural divisions.

At the first stage of team work, the following were agreed:

- hierarchy of strategic objectives and KPI for the hospital at st. Samara;
- list and graduation of ESU/SU;
- strategic goals of the hospital at st. Samara (level 1 hierarchy), ESU/SU (level 2), heads of ESU/SU and employees (level 3);
- strategic maps of the development of the hospital, taking into account the principles of MBO, KPI, BSC;
- matrix of goals and KPI, taking into account the principles of MBO and the SMART methodology.

When the hierarchy and the list of indicators were agreed upon, the following were taken into account:

- strategic development goals of the holding of JSC «Russian Railways» for the period up to 2030;
- the hierarchy of the strategic objectives of the holding, the non-governmental organization of JSC «Russian Railways», the hospital at st. Samara, ESU, SU, executives;
- opportunities and prospects for the development of the hospital at st. Samara on the principles of MBO, BSC, KPI;
- specific activities, goals and objectives of the hospital, ESU/SU and staff;
- presence of forecasted strategic indicators of KPI development (list of long-term, medium-term and short-term goals);
- availability of planned indicators of KPI operational activities.

In determining specific lists of KPI of the hospital at st. Samara, ESU/SU, managers of ESU/SU and employees took into account the methodological requirements to performance indicators:

- transparency;
- measurability;
- minimum sufficiency;
- the possibility of a comprehensive description of activities;
- coherence of operational performance indicators and strategic development goals of the institution;
- consistency of KPI;
- focus on improving the indicators of medical and financial and economic activities in accordance with the development strategy of JSC «Russian Railways»;
- strengthening the competitiveness of the institution as a whole, ESU/SU, employees;
- the ability to broadcast strategic goals and indicators from the highest organizational level to the lowest.

The basic principles of the Balanced Score Card (BSC) [2, 3], the tools of the program-target management – Management by Objectives (MBO) and the SMART methodology [4] were applied to determine the KPI list of the institution as a whole, ESU/SU and employees. All indicators were determined taking into account the following qualitative characteristics:

- S – specific;
- M – measurable;
- A – attractive–achieved–agreed;
- R – resourced;
- T – time.

The developed system of the hospital at st. Samara took into account long-term (5–10 years and more), medium-term (3 to 5 years) and short-term (1–3 years) plans for medical and financial-economic activities of the hospital [5].

In accordance with the draft Regulation on the application of key performance indicators of the KPI activity in the non-governmental public health institution «Road Clinical Hospital at Samara Station of the open joint-stock company «Russian Railways» the target values of the target KPI of the institution, ESU/SU and their weight should be established by the Strategic Planning Commission taking into account the strategy and development program and approved by the chairman of the Commission or the head physician.

The commission ensures the formation of a list of KPI, the methodology for calculating them, the target KPI values, broken down by the years of operation of the approved strategy, with the initial formation of KPI, as well as in the monthly and quarterly breakdown of the target values. Approval of the list of KPI of the non–governmental public health institution «Road Clinical Hospital at Samara Station of the open joint–stock company «Russian Railways» and ESU/SU are implemented at a meeting of the Strategic Planning Commission, the results of implementation are assessed on a monthly, quarterly and year–end basis. The basis for calculating the actual values of indicators that characterize the result of financial and business activities should be the consolidated financial statements prepared in accordance with International Financial Reporting Standards (IFRS), or consolidated statements prepared in accordance with Russian Accounting Standards (RAS).

When forming the list of KPI for 2016 within the framework of a business game at working group meetings, centralized indications were given based on the results of the previous year's activities – reduction of targeted financing, increase in profits due to revenue growth and cost reduction [6], attraction additional sources of financing while maintaining strategic objectives 2020 – ensuring traffic safety, maintaining the health and professional longevity of JSC «Russian Railways» employees:

- centralization of management;
- development of planning – strategic, tactical and operational;
- application of specific performance indicators;
- optimization of the staffing level;
- development of motivational programs.

The application of specific performance indicators allowed in 2016 significantly improve the financial result of the work, showing growth relative to 2015 – 1420% (RUB 52,230,000). The growth of the number of the attached territorial population was 20.2% compared to 2015. The number of the attached adult population was 47,888 people, incl. employees of JSC «Russian Railways» – 17,848 (37.3% of the contingent), pensioners of the industry – 5,492 (11.5% of the contingent). In 2016, 29,792 medical examinations were carried out, which is 16.3% more than in 2015; 25,009 people were examined, which is 19.2% more than in 2015. The corporate income increased by 18% compared to the previous year and amounted to RUB 811,821,000.

The list of KPI for 2017 was formed on the basis of the results of the activities of 2016 [7], taking into account the directions of development of the network of NGHCI determined by the Central Health Directorate (CDH):

– harmonization of the activities of the institutions of JSC «Russian Railways» and regional healthcare;
– reduction of targeted financing;
– increase in incomes;
– optimization of the structure of health facilities;
– standardization of activities at all levels and in all directions;
– increasing the effectiveness of the activities of the NGHCI;
– introduction of a motivation system;
– development of human resources.

The decisions of the Medical Council «Integration of Healthcare Institutions of JSC Russian Railways into the National Health System of the Russian Federation» No. –2–1, Suzdal on May 24–26, 2017, marked the corporate order and determined the growth points of the healthcare institutions of the holding company.

Corporate order – medical provision for train traffic safety, social programs, medical examination, medical examinations, medical expert commissions, medical care for pensioners of JSC «Russian Railways» and high-tech medical care. Relationships at the corporate level are built on a contractual basis for the purpose of conducting medical examinations for the safety of train traffic, voluntary medical insurance for employees of the company and providing employees of JSC «Russian Railways» with vouchers to the sanatorium and health resorts of the holding company.

As points of growth of the health care institutions of JSC «Russian Railways» were indicated:

– interaction with regions;
– industrial medicine;
– telemedicine;
– unified network standards;
– medical tourism.

The basis of the regional policy of JSC «Russian Railways» in the field of health and health care is the principle of rendering medical services to the territorial population on the basis of cost recovery from the funds of compulsory and voluntary medical insurance and the provision of paid medical services. The development of relations with local authorities assumes integration with the regional health care systems of the Russian Federation and the conclusion of treaties on the joint maintenance of specialized healthcare institutions.

Centralized management of processes at all levels of «RZD MEDICINE» with the use of the KPI system will contribute to the effective medical security of the transportation process; improving the quality of medical care for employees of JSC «Russian Railways», members of their families, non-working pensioners of railway transport, attached to the population; prevention of occupational diseases and occupational injuries, ensuring the professional longevity of the company's employees.

Use of the «Regulation on the application of key performance indicators in a non-governmental public health institution «The Road Clinical Hospital at the Samara station of the open joint-stock company «Russian Railways» will provide a unified approach to employee bonuses on the basis of a holistic KPI hierarchy, principles of MBO, BSC and SMART.

4 Discussion

Noting the variety of fundamental research that allowed solving many methodological and applied problems of effective management, it should be said that specific issues related to improving the economic efficiency of the healthcare system of the holding company JSC «Russian Railways» with the use of topical tools of strategic management, program-target management, a system of balanced indicators and key performance indicators are practically not affected in domestic research.

Moreover, the issues of the expediency of applying a system of balanced indicators and key performance indicators in planning and making managerial decisions in the long and medium term continue to be debatable not only for healthcare institutions, but also for many other domestic organizations.

The current stage in the development of the system and management technologies of JSC «Russian Railways» determines the need to apply the planned values of the target performance indicators that will allow to build an integral centralized system of operational management indicators suitable for assessing the current status of their achievement, making managerial decisions, on the specified indicators.

5 Conclusions

Achievement of strategic goals and qualitative indicators defined by the Health Development Concept of the Russian Federation and the Development Strategy of JSC «Russian Railways» until 2030 is possible only if there is a unified approach to building an effective management system in each healthcare institution of the holding. Application in the daily medical and financial-economic activities of the algorithms proposed by the authors for the development and implementation of performance indicators, the matrix of distribution of powers and responsibilities and other documents could ensure achievement of the planned development targets in the near future.

6 Recommendations

The author's version of the system of the KPI activity was approved by the non-governmental public health institution «The Road Clinical Hospital at the Samara station of the open joint-stock company «Russian Railways» and can be used in other healthcare institutions of the holding company. In addition, practical recommendations can be useful for leaders of other health organizations, primarily non-governmental/private, since in the process of reforming the health care system of the Russian Federation they will contribute to raising the efficiency of activities and material incentives for employees of all structural units in providing high-quality medical care.

References

1. Revina, S.N., Kuzmina, N.M., Blinov, S.V.: Formation and implementation of a system of key performance indicators and the evaluation of the quality of medical services of the Russian Railways (by the example of the DKB at the Samara station of JSC RZD). Economic Sciences **6**, 20–25 (2016)
2. Kaplan, R., Norton, D.: Strategic Maps. Olimp–Business (2005)
3. Kaplan, R., Norton, D.: Balanced Scorecard. From Strategy to Action, 2nd edn., 320 p. ZAO Olimp–Business, Moscow (2008). Rev. and additional
4. Thompson, A.A., Strickland, A.J.: Strategic management: concepts and situations for analysis, 928 p. Williams Publishing House, Moscow (2005). Trans. With the English
5. Revina, S.N., Kuzmina, N.M., Blinov, S.V.: Development of a staff motivation system based on the application of key performance indicators (by the example of the DKB at the Samara station of JSC «RZD»). Economics **6**, 29–34 (2016)
6. Decision of the Medical Council of the Department of Health of JSC «Russian Railways», no. –2–1, 14–15 April 2016) «On the results of the activity of the Scientific Research Institute of JSC Russian Railways in 2015 and the main tasks for 2016»
7. Decision of the Medical Council "Integration of Healthcare Institutions of JSC «Russian Railways» into the National Health System of the Russian Federation», no. –2–1, Suzdal, 24–26 May 2017

New Approaches to Formation of Innovational Human Capital as an Element of Institutional Environment

Lyubov I. Vanchukhina[✉], Tatiana B. Leybert, Elvira A. Khalikova, and Artur R. Khalmetov

Ufa State Petroleum Technological University, Ufa, Russia
BUA1996@yandex.ru, ydacha6@yandex.ru,
lejjbert@mail.ru, khalmetov@gmail.com

Abstract. The article is devoted to new approaches to formation of innovational human capital as an inseparable element of institutional environment, viewed from the position of professional characteristics of human resources, capable of generating ideas and creating innovations, and the level of technical equipment of a work place, characterized as high-technology. Special attention is paid to Russian and foreign approaches to evaluation of human capital, used by economic subjects.

The authors characterize innovational human capital, systematize the criteria of assigning human capital to innovational capital, determine the main problems during creation of innovational human capital, and offer the organizational & economic mechanism for risk management during creation of innovational human capital in economic systems as an element of the institutional environment, which includes the algorithm of managerial actions aimed at reduction and neutralization of risks.

Keywords: Economic system · Institutional environment
Innovational human capital · Highly-effective work place · Risk management
Algorithm of managerial actions · HR management

1 Problem Setting and Its Connection to Important Scientific and Practical Tasks

The institutional environment of economic system is determined by the general state institutes, among which the most important is the strategy of development of industrial spheres.

In the conditions of technological and institutional transformation, one of the state's strategic courses is modeling the process of managing the innovational human capital, which is a component of innovational economy and is required by the character of the innovational process, motivated at implementing innovations into all spheres of activities. Innovational human capital is a so called basis for realization of innovational ideas and new technologies into production, sales, and management.

© Springer International Publishing AG, part of Springer Nature 2018
E. G. Popkova (Ed.): HOSMC 2017, AISC 622, pp. 343–352, 2018.
https://doi.org/10.1007/978-3-319-75383-6_44

Formation of innovational human capital, which supposes creation of highly-effective work place, capable of ensuring manufacture of competitive products with minimal expenses of time and equipped with highly-technological equipment and highly-qualified personnel, is accompanied by large risks.

Diversity of risks, appearing in the process of formation of innovational human capital, supposes the necessity for the complex approach to their evaluation and minimization of the possibility of emergence of the problems, danger, and threats, as well as the necessity for formation of organizational & economic mechanism of risk management during creation of innovational human capital.

As the analysis shows, there's a lot of discussion regarding the categorical apparatus, criteria of assigning human capital to innovational type, and risks management during creation of innovational human capital. In the conditions of integration of the national economy in the global economic system, studying new approaches to formation of innovational human capital, as an inseparable element of the institutional environment, becomes very topical.

2 Analysis of Recent Publications on the Topic

There are a lot of works devoted to the problems of evaluation of human capital as an element of the institutional environment.

The foreign approach to evaluation of human capital was founded by American economists Gary Becker and Theodore Schultz.

Schultz proved that human capital is formed only when financial capital is invested into education, which stimulates the development of intellect with a person, who is then capable of providing the growth of added capital of companies [1].

Continuing the theory of human capital, Becker makes an analogy between human capital and physical capital, which can wear out and require certain expenses. According to him, human capital should take its cost through amortization, as the company realizes a long-term and expensive project, related to formation of intellectual human capital [2].

There's also a concept of the theory of human capital at the corporate level - Human Resources Accounting, developed by Eric Flamholtz. In his studies, he used the cost method to evaluation of human capital and formulated the problem of keeping personnel at the company, which is a result of preservation and increase of human capital [3].

Development of approaches to evaluation of intellectual human capital in the conditions of innovational development of institutional экономики was studied by Russian scholars: V.P. Bagov, V.P. Barancheev, B.B. Leontyev, L.I. Lukicheva, O.V. Loseva, et al.

In the scientific works, devoted to development of the theory of human capital and its interconnection to innovational component at the meso-level, Loseva views the contents of human capital from the intellectual point of view as economic category that integrates two interconnected components - intellectual potential and results of intellectual and innovational activity [4].

She notes that a necessary condition for modernization of economic system is its capability to generate new knowledge, received by intellectual results, which is treated as human intellectual capital.

In the previous studies made by Vanchukhina et al. [5], a problem of evaluating human resources of economic systems was determined, related to increase of labor efficiency and formation of highly-effective jobs at the micro-level.

3 Goals of the Research

The purpose of the research is to develop organizational & economic mechanism of risk management during creation of innovational human capital (IHC) in economic systems as an element of the institutional environment.

The issues of categorical apparatus of "innovational human capital" are viewed, specific types of risks peculiar for formation of innovational human capital are grouped, the exiting tools of determining the characteristics of human resources are evaluated, and the algorithm of managerial personnel's actions for identification, minimization, and elimination of risks during formation of innovational human capital is determined.

4 Main Results of the Research and Their Substantiation

Innovational human capital is a type of the resource of an economic system that can manufacture highly-effective product that provides the largest growth of added value with least expenses of labor and time. Provision of the set parameters requires two components – highly-qualified human resources (intellectual human capital, capable of generating ideas and innovations) and high-tech work place, equipped with leading (latest) technologies.

After studying extensive scientific and periodic materials in the sphere of assigning human capital, the criteria according to which human capital can be considered innovational are systematized (Table 1).

During creation of innovational human capital, economic system is open to certain risks that are multiple and can differ in specific spheres of activity. They could be divided into groups, within which a more detailed consideration and determination of ways for their reduction and neutralization will be performed.

Grouping of risks related to creation of innovational human capital in the sphere of oil and gas business, is presented in Table 2.

Viewing these risk groups, which are considered to be main, it should be noted that their neutralization requires the work with personnel, which means double expenditures for creation of IHC – creating all conditions and making sure that employees have adapted and use their potential and potential of provided additional resources in the best possible way.

For substantiation of decision making in the process of creation of innovational human capital and minimization of the above risks, the authors offer the organizational & economic algorithm of risk management during creation of IHC.

Table 1. Criteria of assigning human capital to innovational

Criterion	Characteristic
Efficiency	Expressed in exceeding a certain indicator of labor efficiency, showing a special status of a work place, which allows for such efficiency, or in the volume of revenue that exceeds a certain average indicator
Wages	Expressed in high evaluation of labor cost of a specialist, who possesses the corresponding capabilities and competences for performing the work on the spot
Decent conditions	Expressed in providing a worker with proper comfort, created for the purpose of increasing his effectiveness, motivation, and maximum usage of existing resources
Provision with resources	Expressed in totality of characteristics of work place, which allow employee to achieve the results that are above the average for the sphere and are a combination of leading technologies and human management

Table 2. Grouping of risks related to creation of innovational human potential in the sphere of oil and gas business

Group title	Short characteristic
Risk of technological unpreparedness	Very topical for high-technology spheres which include the oil & gas sector. IHC requires for employees to be provided by the newest technologies, which are not easy to learn for the workers with proper skills and who are not used to technological and program innovations. Also, a lot of unique technologies are not always accessible
Risk of moral unpreparedness	IHC means increased expectations from worker. Despite the fact that professional worker has to be confident in his abilities, a lot of workers can experience stress, which leads to decrease of efficiency and failures in work
Risk of inability of performing large volumes of work	IHC often means that worker get a possibility to perform large volumes of work over lesser terms and with worse conditions, due to new technologies and methods; however, worker may see the situation only in the context of growth of the volume of his duties, which might complicate the adaptation
Risk of desynchronization of departments	It is obvious that any production is a complex system with a lot of stages. Highly-effective work place at one stage can disrupt the system's balance, which neutralized the useful effect and requires organizational improvements for normalization of the system
Risk of incapability for studying	Despite the fact that specialists in the sphere of management are easy to teach, it is necessary to take into account that transfer of work place into the form of highly-effective and high-technology may be accompanied by large changes of the work process, which will require full preparation and feedback from the worker

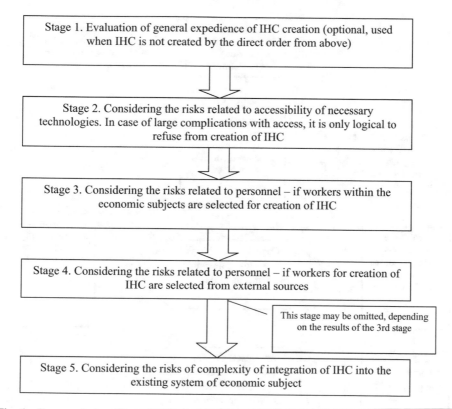

Fig. 1. Supposed algorithm of managerial actions in the sphere of risk management during creation of IHC

The offered algorithm of risk management during creation of IHC consists of two stages – evaluation of general expedience of IHC creation and consideration of risks related to accessibility of the necessary technologies (Fig. 1).

Schematically, algorithmization of the first stage of the managerial process for reducing the risks of IHC creation is shown in Fig. 2.

The peculiarity of the first stage of the algorithm of managerial process for reducing the risks related to creation of IHC consists in the fact that any risks are to be temporarily ignored as it performs the task of additional verification of expedience of IHC creation. At that, before the managerial decision on creation of IHC, verification by the persons who make the decision should be made.

The advantage of the first stage of the algorithm is that it allows evaluating the general expedience of IHC creation, in order to avoid involvement of economic subjects into ineffective and doubtful project and forms the data base for the following stages of the algorithm.

Schematically, algorithmization of the second stage of the managerial process for reducing the risks related to IHC creation is presented in Fig. 3.

At the second stage, the aspects related to accessibility for the economic systems of the technologies necessary for IHC creation and their further use are viewed.

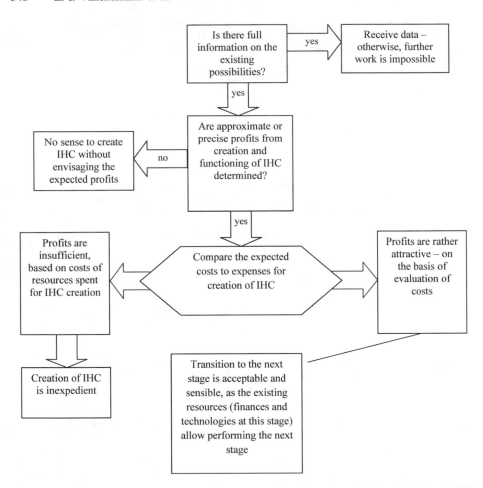

Fig. 2. Algorithm of the first stage of the managerial process for reducing the risks of IHC creation

This stage is peculiar for its importance during transition to the following components of the algorithm. That is, at the first stage of the algorithm, unfavorable ratio of expenses and supposed profit from IHC creation can be ignored by the subject of management in favor of the fact that creation of IHC will serve some other purposes, which are invisible within a purely economic comparison, but at the second stage of the algorithm, the transition is possible only under clear conditions.

This stage of the algorithm is peculiar for the fact that it directly evaluates the possibility for creation of IHC according to the danger factor and allows for clear determination of the further position on the basis of the economic system's possibilities.

The third stage of the algorithm of risk management during IHC creation is devoted to the search for human resources for work at the created highly-effective work places at the economic subjects. While the presence or absence of technologies is very

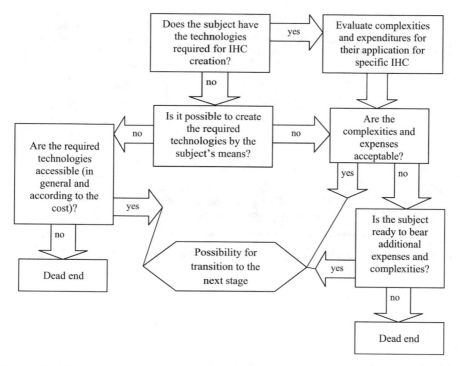

Fig. 3. Algorithm of the second stage of the managerial process for reducing the risks that accompany IHC creation

important (it is impossible to work without proper equipment), complexities with personnel are less important and could be solved in many ways, as any device or technology created by a human can be used by a human. The real issue is the one of knowledge and qualification, or the wish and possibilities to acquire the qualification. One way or another, the economic subject will be looking for workers at IHC – its own personnel.

Schematically, algorithmization of the third stage of the managerial process for reducing the risks related to creation of IHC is presented in Fig. 4.

This stage of the algorithm is peculiar for dead end options and the largest number of further possible actions. If the required personnel are already within the economic subjects, the search for personnel at the external labor market is ignored. If personnel can be additional trained, it is possible to pass to evaluation of external labor market to compare the possible expenses or, by the management's decision, to pass to the fifth stage (which is not recommended, as it is not correct). If the personnel within the economic subjects are absent and cannot be trained, a transfer to the external labor market is necessary.

This algorithm component's advantage is the fact that it is possible to distinguish maximum possible variability that is to be open in the complicated form of the stage and to provide the economic subject with a lot of ways of solving the possible problem of human resources.

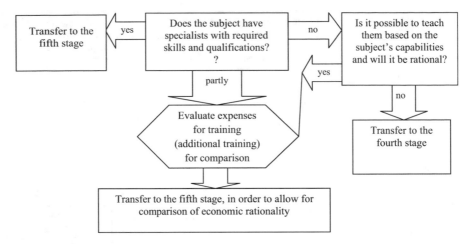

Fig. 4. Algorithm of the third stage of the managerial process for reducing the risks related to creation of IHC

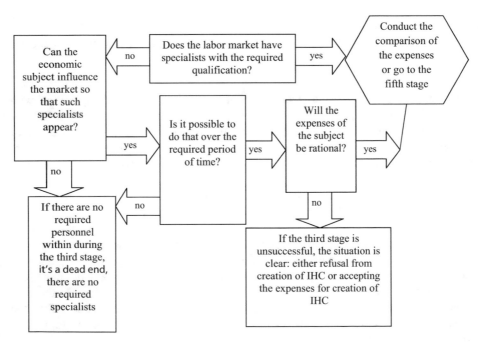

Fig. 5. Algorithm of the fourth stage of the managerial process for reducing the risks related to creation of IHC

The difficulty may consist in insufficiently correct evaluation of the skills required during the initial formation of requirements to applicants for IHC and in the relative character of the notion of rationality during deciding which of the variants is the best for a specific subject.

Schematically, algorithmization of the fourth stage of the managerial process for reducing the risks related to creation of IHC is provided in Fig. 5.

This stage is peculiar for the fact that is depends on the company's plans for its further development – i.e., the largest companies may make a decision on the more expensive variant in the form of influencing universities for creation of the required new training programs, if this is a good contribution into the future of their own HR personnel, or decide that new specialties are not required and be satisfied with the existing human resources, with only additional training.

This algorithm stage's advantage is that it clearly shows the possibilities for the economic subject within the work with external labor market during search for personnel for IHC creation.

Difficulties may arise as well – despite the clarity of each specific end, in case of several possible alternatives the decision will have a character of a complex strategic decision, not a local project one.

The fifth stage of the algorithm of risk management during creation of innovational human capital consists in integrating IHC into the existing system of the economic subject. Certain difficulties are possible during conduct of this process, systematization of which is presented in Table 3.

Table 3. Main possible problems during integration of innovational human capital into the existing system of an economic subject

Problem	Probability of appearance	Description
IHC will work quicker/more effectively, which will cause discord in the whole system	Almost 100%	The sense of creation of IHC is to increase the effectiveness. The adjacent work places are to be corrected
Special needs for equipment/resources will arise	Depends on specifics of IHC	Search for new suppliers
Working schedule will change	High	Compilation of new plans and norms

5   Conclusions

Based on the performed analysis of foreign and Russian approaches to formation of human capital, the authors substantiated the criteria of determinining innovational human capital as an inseparable element of the institutional environment. This notion is based on symbiosis of two economic categories – human capital as a resource and as a result of innovative activities. It is specified that creation of innovational human capital is related to specific forms of risks, for which there are no specialized mechanisms of reduction of risks.

The authors develop the organizational & economic mechanism of risk management during creation of innovational human capital (IHC) in economic systems, which includes the process of algorithmization of managerial decisions for reduction of risks related to creation of innovational human capital.

References

1. Schultz, T.: Investment in Human Capital: The Role of Education and of Research, New York (1971)
2. Becker, G.S.: Human Capital: A Theoretical and Empirical Analysis. Columbia University Press for NBER, New York (1964)
3. Dobrynin, A.I., Dyatlov, S.A., Tsyrenova, E.D.: Human Capital in Transitive Economy: Formation, Evaluation, and Effectiveness of Use. SPB, Nauka (1999)
4. Loseva, O.V.: Formation of the methodology of evaluating human capital in the innovational activity. Doctoral thesis, 387 p. (2013)
5. Vanchukhina, L.I., Leybert, T.B., Khalikova, E.A.: Methodological approaches to evaluation and analysis of labor efficiency in the spheres of fuel and energy complex. J. Environ. Manag. Tourism 1, 585–594 (2016)
6. OECD: OECD Investment Policy Reviews, Botswana, 26 p. (2014). http://tinyurl.com/lenua3h. Accessed 02 Apr 2016
7. Highly-efficient work place in Russian regions (analytical note). TPP Inform LLC, 28 p. (2013)

Methodology of Information & Psychological Safety of Human and Society: Epistemological Aspects

Natalia Z. Aliyeva[1]([⊠]) and Elena B. Ivushkina[2]

[1] Don State Technical University, Rostov-on-Don, Russia
alinatl@yandex.ru
[2] Institute of Service and Entrepreneurship (Branch) of Don State Technical University, Shakhty, Russia
ivushkina62@mail.ru

Abstract. The process of human interaction and information environment, the formation of personal threats and the society is determined by external social and political conditions and global informatization, and by internal conditions in the form of psychological factors and mechanisms. Such a fusion of external and internal determinants of the development of society leads to the emergence of the phenomenon of information & psychological convergence, which determines not only the information & psychological impact on a person, but its information & psychological security. The study of information & psychological convergence requires methodology and a system forming basis for the development of the concept of information & psychological security of the individual. The study helps to solve the problem, which consists in creating a methodology and tools for designing an integral concept of information & psychological convergence, which causes the convergence of elements within the security system and other regulators of relationships in society, as well as develop recommendations aimed at improving the information and psychological convergence in the implementation of the security of the individual and society.

In this article, the authors analyze the concepts of the methodology in general and specifically the methodology of information & psychological convergence. Based on the analysis of principles and methods of convergence, the authors formulate the definition of information & psychological convergence. Particular attention is paid to the influence of globalization factors on the formation and application of methods of information & psychological convergence.

Keywords: Convergence · Information & psychological safety
Methodology · Epistemological aspects

1 Introduction

Information & psychological convergence determines not only the information & psychological influence on a human but also his information & psychological safety. It is determined by not only external factors of society, politics, and globalization of the information processes, but also by internal factors of psychological influence on human

© Springer International Publishing AG, part of Springer Nature 2018
E. G. Popkova (Ed.): HOSMC 2017, AISC 622, pp. 353–358, 2018.
https://doi.org/10.1007/978-3-319-75383-6_45

and society. Such internal and external similarity and interaction of the information и psychological phenomena and their form are a complex and ambiguous process. The largest influence on the process of convergence is performed by globalization – primarily, information globalization. Globalization is appearance of commonness in the main systems of the world: political, social, and economic. Globalization is based on technological revolution that creates previously unknown communications, technique, and technologies, including information ones. Information & communication technologies, as well as appearance and popularization of the Internet, became an accelerator of the globalization processes, including in the social sphere. R. Robertson connects globalization to the transformation of the global consciousness, "globalization is related to the compression of the world and intensification of the global consciousness as a single whole, to the specific global interdependence and realization of the global whole" (Robertson 1998). W. Anderson defines globalization as a "flow of converging powers that create truly comprehensive world" (Anderson 2001).

Quick acceleration of information & communicative processes and convergence with psychological influence on a human and society's consciousness takes them to a new qualitative state: information & psychological convergence, which became the objects of our research. This situation opens not only new possibilities but also new threats. Due to using these possibilities and opposing the challenges, each national society requires the systemic strategy of provision of safety, which has to include the information & psychological safety. Keeping in mind the "global society of risk" and the large conflict potential of the modern world, it is possible to state that the modern variant of the process of globalization contradicts the safe development of the world (Beck 2001).

2   Theoretical Basis of the Research

The modern stage of society's development is characterized by the growing role of the information sphere, which performs the role of the systemic factor of the society's life. The necessity for protecting the individual and mass consciousness from the negative information & psychological influences and creation of the scientific foundations and the methodology of information & psychological safety in view of the modern realia determine the topicality and theoretical significance of this study. A necessity arises for development of the mechanisms that will not allow for information & psychological influence on the consciousness of human and society. It should be noted that the information society cannot consider only the information influence, as the information's influence leads to transformation of the psychological state of human and his ideology. Therefore, the influence should not be divided into information and psychological.

Their convergence led the authors to creation of the notion of information & psychological convergence, which is treated as a process of approximation and integration of two structures – information and psychological into a new single system. The main idea of the information & psychological convergence is emergence in the communication environment of the targeted transfer of specially selected information that influences the subject (human, group, or society) and leads him to the sense and meaning of the offered information. Secondly, this influence leads to the programmed psycho-emotional response from the object, which changes the character of the received information and creates the ideology and picture of the world that is required by the subject of the influence (Dolgov 2016).

Another sphere of information & psychological convergence is information & psychological safety, for which the basic notion is information & psychological influence applied in the form of operations, measures, struggle, and war between countries and within one country in different spheres of activity: international, economic, political, and information. "Information & psychological struggle is conducted during peace and war between states to protect one's interests, for political influence zones, sales markets, debatable territory, property, strengthening of the defense sphere, etc." (Barishpolets 2013). The task of provision of the national safety could be solved only in the complex way, by convergence of all forms and means of struggle, including the information & psychological.

In the information & psychological opposition, special role belongs to information & psychological wars, as well as to manifestations of terrorism, including information wars, information terrorism, and cyber terrorism (Taran 2009). Terrorism increases the opposition between different states, thus reducing the international safety on the whole. Information & psychological war, as an intensive and aggressive information & psychological conflict, poses a threat to the modern society. Information terrorism is a psycho-intellectual diversion, a threat to the individual and public consciousness. It is based on false information, which allows creating the contradiction between ideas, negative nature of opinions, and false understanding of events. The result of such disorientation of consciousness is the change of reality in the form of weakening of the existing society's foundations. The mechanisms of managing the information terrorism are the processes of forming the public opinion with the help of mass media (TV, Internet, and press).

In its turn, cyber terrorism is defines as attack on an information system. Thus, cyber terrorism is a part and a tool of the information terrorism on the whole. Cyber terrorism is used for destroying information infrastructure of a state, creating new regimes in a state, and creating the social environment that leads to crises and catastrophes.

It is possible to state that evolution of various forms of influence, including terrorism, goes from the information to the information & psychological convergence. The psychological component, which includes the aspects of behavior, motives, and goals of terrorists, allows determining the technologies of fighting by finding and influencing the pain points. The technologies of information & psychological manipulation could be used against terrorism and terroristic threats both in external and internal political activity (Shkhagapsoev and Tarchokov 2016).

3 Methods and Object of the Research

The methods of studying the very methodology are viewed – in particular, the methodology of information & psychological safety of human and society. Such research is of a complex character, and it is predetermined by interdisciplinary nature of approaches to study of the problem. Philosophical study of the problem of information & psychological safety of human and society as a complex and ambiguous process of realization of the information & psychological interaction is based on using the whole complex of philosophical and general scientific methods: analytical, phenomenologic, principles of objectivity, general connection, contradiction, methods of comparative analysis and synthesis, and scientific generalization. The study uses the conceptual & logical analysis of the theoretical terminology for the purpose of determining the term "information & psychological safety" for the socio-cultural context. For this, the methods of historical & genetic analysis of the scientific methodology for analyzing its re-orientation from the formal & theoretical to socially significant one are used. The primary methodological idea of the research is the dialectical and systemic approaches.

4   Results

Information & psychological convergence could be presented in the form of a complex and ambiguous process of internal and external approximation of information & psychological phenomena. The methodology of information & psychological convergence is a new and ambiguous notion – as well as the methodology of science on the whole. Despite the study of methodology as an object of scientific analysis, it is treated differently (Stefanov 1967). Due to this, in order to define the methodology of information & psychological convergence, let us view the basic notion of methodology as a whole. Analysis of methodology in scientific and reference sources defines methodology ambiguously. Firstly, methodology (Greek methodos – way, means of research, teaching, explaining) is a study of means and methods of cognition and execution of activity (9). Secondly, it is a study of methods of scientific cognition and totality of methods applied in science. Thirdly, it is a totality of means, methods, and operations of cognition and practical activity. Fourthly, it is a system of principles of methods of organization and development of theoretical and practical activity, as well as study of this system (10). Fifthly, it is an algorithm of search for the goal and a set of means, methods, and principles of achieving the goal.

The theoretical goal of methodology is creation of a model of knowledge in its ideal (in specific conditions). The practical goal is the algorithm of means and methods necessary for achieving a practical goal according to the true knowledge. As is seen, in the reference literature the notion "methodology" is viewed primarily from the point of view of study of method, together with methods used in a certain activity, which does not fully reflect the sense of the viewed notion. On the one hand, this position "impoverishes" methodology, on the other hand, it leads to identifying methodology to methods, which lead to a scientific discussion.

The notion "methodology" constantly expands. A.I. Rakitov sees it as an independent science, K.D. Petryaev sees it as an ideology. According to D.A. Kerimov, "methodology is an integral phenomenon, which unites a range of components: ideology and fundamental general theoretical concepts, philosophical laws and categories, and general and specific scientific methods" (Kerimov 2001). According to S.A. Lebedeva, it is methodological concepts and specific methods of specific theories.

Excess of the given concepts of methodology allows for the following conclusions: 1. Methodology is not one method, but a study on the methods of cognition of a scientific object. 2. Methodology is not science, but its part on the scientific tools. 3. Methodology includes not only study and methods, but also ideology, principles, notions, and categories. 4. Methodology is a system comprised of methods, principles, and categories that are connected to each other and interact with each other. 5. Methodology is based on three pats: special, interdisciplinary, and general scientific – which provide result, as a whole. 6. Methodology contains the theoretical and practical levels of cognition, which are peculiar for different elements that allow creating the abstract picture of the studied phenomenon in the first case and the real picture of the studies subjects and phenomena – in the second case. 7. The subject of cognition determines the content of methodology, which depends on its competence, the initial knowledge basis, etc. 8. Methodology is a basis of any research, as it determines the quality of result. 9. Methodology justifies its Greek definition – it is really a way, which has to be worked in order to receive the knowledge on the studied object. The first stage of the way includes gnoseological study of the phenomenon for obtaining the conceptual knowledge of the object. The next stage includes practical study for receiving the real picture of the studied phenomenon. 10. Methodology is a whole complex of components that allows receiving knowledge in theoretical and practical activity of the studied object. As a result, methodology could be defines as a system of "elements that comprise the theoretical basis and the tools of the research in a certain sphere, as well as the system of methods of practical influence on the imagined or true reality" (Tretyakova 2010).

These studies of designing the information & psychological convergence could be presented in the forms of the following provisions.

Information & psychological convergence is the process of interaction of information & psychological elements within the social systems, characterized by their approximation and a certain level of coordination of influence of these elements on social relations (Alieva 2013).

The methodology of information & psychological convergence could be defined as a study of the means of organization of various subjects' activity on approximation of information and psychological systems and their realization in the concept of the information & psychological safety.

The classification model of information & psychological methods of convergence includes two groups of methods. The first group includes the methods that are based on the general information & psychological influence on human and social relations. The second group includes the methods of information & psychological safety.

The object of information & psychological convergence is activity for approximation of various information и psychological phenomena that appear as a result of convergence and lead to emergence of social relation, that allow managing the information & psychological phenomena.

Depending on the methods, it is possible to distinguish negative information & psychological convergence, which leads to threats and negative influence on human and society, and positive information & psychological convergence, which ensures information & psychological safety.

5 Conclusions

The authors conduct the categorical analysis and specification of the categorical apparatus necessary for development of the methodology and tools of designing the concept of information & psychological convergence. The above provisions could be the basis for development of the comprehensive model of convergence of information and psychological influence on a human and society. Practical significance of the research consists in the fact that it provides a new approach to creation of the model of information & psychological system of safety.

References

Robertson, R.: Globalization, 28 p. (1998)

Anderson, W.: All Connected Now. Life in the First Global Civilization, Oxford, pp. 122–123 (2001)

Beck, W.: What is globalization? Mistakes of globalism – answers to globalization. Progress-Traditsiya, 304 p. (2001)

Dolgov, M.I.: Classification of the information & psychological influences and evaluation of the level of their threat to public consciousness. Society: politic, economics, law, no. 2 (2016). http://cyberleninka.ru/article/n/klassifikatsiya-informatsionnopsihologicheskih-vozdeystviy-i-otsenka-stepeni-ih-ugroz-dlya-obschestvennogo-soznaniya

Barishpolets, V.A.: Information technologies. Appl. Inf. Psychol. Influ. Radioelectron. Nanosyst. Inf. Tech. 5(2), 62–104 (2013)

Taran, A.-A.V.: Classification of information threats to the modern society. Bull. Russ. Univ. Peoples' Friendsh. Polit. Sci. (2), 37–40 (2009)

Shkhagapsoev, Z.L., Tarchokov, B.A.: Organization of the information & psychological opposition to terroristic activity. Hist. Socio-Educ. Thought 8(4-1), 115–117 (2016)

Stefanov, N.: Theory and method in public sciences, 271 p. (1967)

Methodology of philosophy. http://sargelas.com/metodologiya-filosofii

Methodology as a system of principles. A short dictionary of philosophic terms. http://nenuda.ru

Kerimov, D.A.: Methodology of Law. Subject, Functions, and Problems of Philosophy of Law (Monograph), 2nd edn., Avanta+, 560 p. (2001)

Tretyakova, O.D.: Epistemology of the methodology of law convergence: economic measure. Legal science and practice. Bull. Nizhny Vodgorod Acad. MIA RF 1(12), 61–65 (2010)

Alieva, N.: Convergence of science, technologies and society: research-on-research aspects. World Appl. Sci. J. 28(12), 2271–2275 (2013)

Theoretical and Methodological Aspects of Human Capital Management

Galina V. Golikova[1(✉)], Valery G. Larionov[2],
Svetlana I. Verbitskaya[3], Tatiana E. Fasenko[4],
and Dmitry V. Kokhanenko[4]

[1] Voronezh State University, Voronezh, Russia
ggalinal23@yandex.ru
[2] Bauman Moscow State Technical University, Moscow, Russia
vallarionov@yandex.ru
[3] Smolensk Branch of Plekhanov Russian University of Economics,
Smolensk, Russia
verb.svet@mail.ru
[4] Financial University under the Government of the Russian Federation,
Barnaul Branch, Barnaul, Russia
{TEfasenko,dvkohanenko}@fa.ru

Abstract. In the modern conditions, competitive advantages of economy and possibilities for its modernization are determined by the accumulated and realized human capital. Human capital can be determined as a totality of knowledge, competences, and qualities, embodied individually in each employee, which stimulate the creation of personal, social, and socio-cultural economic well-being of the society. Intellectualization of economy destroys the foundations of the anthropogenic civilization – motives and goals that formed its integrity and were the driving force of the progressive movement. The imperative of maximization of material well-being was replaced by realization of the possibility for self-affirmation through possession and use of knowledge.

Keywords: Human capital · Labor · Competition · Human · Economic sense
Capital · Resource · Investments · Education · Economic development

JEL Code: I200 · I210

1 Introduction

Starting from 1960's, a special attention in the theory of managerial thought has been paid to the issue of formation of human capital. Gary Becker was the first to move the notion "human capital" to the micro-level; in 1964, he formulated the definition "human capital – totality of knowledge, skills, and capabilities of a human" [1]. At that, the main investments in human capital are the company's expenditures for education and training of personnel. In his work "Theory of Human Capital", Becker calculates the economic effectiveness of education and substantiates the interconnection between labor efficiency and investments into education of the company's employees. Becker puts the main emphasis on the necessity for investments into the sphere of education, i.e., receipt of knowledge and professional skills of employees, the presence of which will lead to economic growth of the company.

While studying the evolution of the notion "human capital", it should be noted that a lot of scientific efforts were aimed at evaluation of economic effectiveness of investments in the organizational resource and at the possibility of calculation of an employee's cost for the company in terms of money. Thus, for example, W. Petit, A. Smith, D. Ricardo and others supposed that not workforce but labor was sold. The scholars tried to form the methodology of evaluation of employee's effectiveness in the organization. The basis of the evaluation was comprised of humans (human) capable of providing competitiveness of the organization in the external environment by continuous improvement of qualitative and quantitative characteristics of their labor.

2 Methodology

For development of the directions of human capital management, it is necessary to systematize various approaches to determining the content of this notion (Table 1).

Table 1. Approaches to determining the notion of human capital

Authors	Contents of the notion
T. Schultz, G. Becker [1]	Human capital is a valuable resource, which is more important than natural resources or accumulated wealth
J.R. Walsh	Viewing investments into education from the point of view of receiving profit, he saw higher and postgraduate education as a basis for long-term successful economic realization of the individual's capabilities. He showed that the value of education, received in college, exceeds its cost. The value of higher education is connected to such indicators as individual's satisfaction with his own life scenario, his capability to manage his time, mobility, and demand in the labor market, etc.
P. Neumann	Education is distinguished as a special element of human capital. Its main components include cultural & ethnical peculiarities; general education; professional education, key qualification qualities
S.A. Dyatlov	In the process of reproduction of human capital, the following stages are distinguished: micro-cycles, local cycles, and macro-cycles of the human capital turnover
A.I. Dobrynin [5]	Human capital is the existing stock of health, knowledge, skills, capabilities, and motivations that stimulate the growth of his labor efficiency and influence the growth of income; he states that post-industrial society is peculiar for reproduction of the production forces of human not in the commodity form but in the form of human capital

(continued)

Table 1. (*continued*)

Authors	Contents of the notion
M.M. Kritsky	Human capital was initially a comprehensive specific form of life activity, which assimilates the preceding forms and is realized as a result of historical movement of human society to its modern state
N.N. Koshel	It is necessary to have correspondence of human capital to forms and organization of human's activity, in which his activity and energy could be manifested

Generalizing various treatments of the notion "human capital", it is possible to distinguish three main components:

– health, which, being psychosomatic, physiological state of a human, turns into a certain mode of life, which supports its health;
– culture, which sets stereotype models of behavior and values of the individual, determining the character of its realization in the process of labor activity;
– education, which determines maximally achievable social status and creates objective and subjective foundations for a person's career growth, where the objective foundations are requirements for a certain type of activity and office, related to education, and subjective – the person's capability to use the acquired skills in the production activity, which increase his capability for creativity and innovative activity.

In the modern economic theory, there are various approaches to treatment of the notion "human capital" [1–3, 5, 7, 11]. Generalizing them, it is possible to offer the following definition, "Human capital is a totality of knowledge, skills, and capabilities of an employee, which includes the accumulated storage of qualitative and quantitative characteristics of his labor, necessary in the sphere of public production, which are capable of increasing the labor efficiency and ensuring the growth of economic effectiveness of the activity of the company, region, or country on the whole".

3   Results

Human capital should be viewed as the most important resource that forms investment attractiveness. This statement is confirmed by the fact that in the foreign countries investments into human capital are viewed as the most important and effective investments. The strategic characteristics of human capital are reflected in Fig. 1.

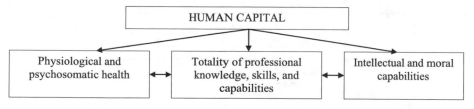

Fig. 1. Strategic characteristics of human capital

At that, the main condition that ensures their effectiveness is formation of the motivation for further education, professional development, and career growth with the employees.

According to Korchagin, these characteristics are formed with the help of investments into increase of quality of life and living standards of the population, education, health, knowledge, etc. The investors that conduct investments into development of personnel are guided by the idea that human capital is the main factor of production of goods and services. As a matter of fact, modern technical means of production, top-quality resources, etc. will remain at the level of "dead weight", until a human hand touches them – but this touch should be professional.

The main investors during formation of qualitative characteristics of human capital are states and companies that seek the goal of receiving the corresponding economic effect, in the form of growth of profit (at the micro-level) or gross domestic product (at the macro-level). At that, human can view both the object and the subject of investment activity.

Figure 2 shows the scheme of human capital formation.

Being an object of investments, human is influenced by investors, and, as a subject – participates in the process of investment activity.

Investments into human capital are aimed at increase of qualification of employees, and, therefore, are more effective at the level of region and state than other types of investments. Highly-qualified personnel can provide larger economic return from the material means of production and stimulate further scientific and technical development of society, which leads to formation of large socio-economic effect.

It should be noted that the modern conditions of development of the society led to formation of a completely new type of economy of macro- and meso-levels, which set their requirements to organization of economic activity at the micro-level – a separate economic subject. In this case, we speak of economy of intellectual labor and knowledge & innovations economy, based on implementation of achievements of the scientific and technical progress, which supposes using the labor of the highest qualification – among the managerial apparatus by the economic subject, and among the main employees that ensure the production process.

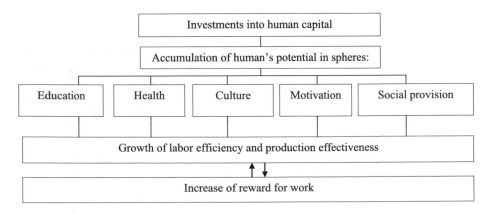

Fig. 2. Scheme of formation of human capital with investing

For the policy of management of human capital development, an important issue is evaluation of its current state and role in provision of socio-economic development. In the theory of human capital, evaluation of its state is most complex, as there is no unambiguous methodology. The economic theory has developed various approaches to evaluation of human capital. L. Sharok offered to systematize the existing approaches and divide them into the methods of evaluation of human capital and the methods of evaluation of effectiveness of its use [4, 8] (Table 2).

Additional indicators of evaluation of human capital include the following: number and share of researchers in the population, number of people with scientific degrees, share of the people involved in the non-material investment complex (R&D and

Table 2. Classification of the methods of human capital evaluation

Direction of evaluation	Indicators	Content of indicators
Evaluation of human capital	Natural quantitative and qualitative indicators of human capital	Number of able population
		Number of years of study
		Average life span
		Share of population with higher education
		Living standards, etc.
	Cost indicators of evaluation	Cost method
		Pricing method
		Index of value of human capital, etc.
	Integral evaluation of human capital	Index of development of human capital
		Index of development of cultural potential
		Index of living standards, etc.
Effectiveness of use of human capital	Effectiveness of use of human capital at the regional level	Level of population's employment
		Gross regional product
		Labor efficiency for the spheres of regional economy
	Effectiveness of use of human capital at the level of the economic subject	Labor efficiency (general, but for separate structural departments)
		Profitability of human capital
		Internal norm of profitability
		Kirkpatrick model
		Return of investments according to the Phillips model
		Index of innovativeness of human capital, etc.

education) [6, 9, 10], and the indicators of coverage of various levels of education of the corresponding age categories, number of undergraduates, postgraduates, and doctoral candidates, share of expenditures for education and science in GDP, etc. According to I. Soboleva, a drawback of these indicators of evaluation of human capital is their incapability to perceive and assess the parameters of quality, as mostly the existing system of formal education is evaluated.

Systematizing the existing approaches to evaluation of human capital and determination of its role in socio-economic development, we offer the following methodology (Table 3).

Table 3. Classification of methods for human capital evaluation

Level of human capital management	Offered methods of evaluation
Individual (1st level)	Cost approach Pricing approach Index of return of investments on the basis of the Phillips model
Micro-level (2nd level)	Model of individual cost of employee Method of calculation of direct expenditures for personnel Kirkpatrick model Comparative method (indicators of labor efficiency, level of correspondence of education to the occupied position, etc.) Index method (index of human capital, index of innovational development of personnel, etc.)
Meso-level (3rd level)	Comparative method (level of population's employment, share of population with higher education, share of innovational directions of study in the universities of a city, region, etc.) Index method (HDI, index of quality of life, etc.)

Analysis of perspectives of innovational development in the Russian Federation shows a lot of obstacles on this path. Lack of highly-qualified employees is one of the main barriers for development of science-drive production, especially in the long-term. Negative influence is performed by the general priority of the resource sector, which is not related to development of the processing industry spheres, primarily science-drive; orientation at short-term goals, large underestimation of human capital; violation of succession of scientific and technical knowledge. The unclear border between managers and workers leads to "deflating" of the traditional hierarchy and distribution of responsibility, which was previously concentrated on the upper floor of the hierarchy, at the whole organization. This, the opposition to changes is reduced: "plain" hierarchical structures stimulate the organizational study.

Destruction of the foundations of private property, which is based on formation of ownership for information and knowledge. According to the Western scholars, "… movement of information challenges the exclusive and individual ownership". Firstly, this is caused by the fact that knowledge and information are related to those who produce them, and capital cannot be taken from labor. Secondly, even if knowledge is sold, it remains with its manufacturer, so knowledge is "collective benefit".

4   Conclusions

Thus, progress of knowledge leads not only to change of balance between material and non-material factors of production but to "erosion" of private property, emergence of pluralistic relations, and relations that lead to new quality.

Formation of new priorities of personality and society, which means transition from maximization of material consumption to "quality of life": humanistic, ecological, and "non-monetary" criteria of being.

References

1. Becker, G.S.: Human behavior: economic approach. In: Kapelyushnikova, R.I. (ed.) Selected Works on the Economic Theory, pp. 360–374. SU HSE, Moscow (2003)
2. Bezrukova, T.L., Chugunova, E.V.: The structure of intellectual capital in innovational organizations. J. Forest. **4**(4), 130–134 (2011)
3. Bezrukova, T.L., Kuznetsova, T.E., Chugunova, E.V.: Formation of intellectual potential in innovational organization. In: Igolkin, S.L., Bezrukova, T.L., Akhmedova, A.E. (eds.) Perspectives of Innovational Development of the Modern Global Society: Economic & Legal and Social Aspects: Materials of the International Scientific and Practical Conference-2012, Voronezh, 24–27 April 2012, vol. 3, pp. 92–97 (2012)
4. Bezrukova, T.L., Bikulova, D.U.: A comparative analysis of methodological approaches to evaluation of population's living standards in economic studies. In: Current Directions of Scientific Research of the 21st Century: Theory and Practice: Collection of Scientific Works by the Materials of the Scientific and Practical Conference, Voronezh, no. 3, p. 1 (8-1), pp. 345–352 (2014)
5. Dobrynin, A.I., Dyatlov, S.A., Tsyrenova, E.D.: Human capital in the transition economy, pp. 296–300. SPb., Nauka (2000)
6. Mangutov, I.S.: Theoretical and methodological issues of labor potential of personnel as an object of functional management. Bull. A.S. Pushkin Leningrad State Univ. **6**(4), 89–101 (2010)
7. Nesterov, A.A., Forrester, S.V.: Problems of human capital in the modern economy: monograph, pp. 169–179. Samara State Technical University, Samara (2010)
8. Pavlova, E.A., Smirnova, L.A.: Evaluation of competitive advantages of a company on the basis of analysis of its innovational potential. Mod. Issues Sci. Educ. **1**, 584–591 (2015)
9. Rybalkina, Z.M.: Development of managerial potential as a factor of increase of company manageability. Curr. Issues Hum. Nat. Sci. **7**(1), 155–157 (2014)
10. Stepanova, T.E.: Knowledge economics: a methodological aspect: monograph, pp. 53–57. Saratov University Publications, Saratov (2004)
11. Stepchenko, N.A.: Development of human capital in the conditions of globalization of the world economy. Finan. Credit **35**, 50–52 (2005)

Service Labor and Problems of State Management of the Russian Middle Class Development

Konstantin V. Vodenko[1](✉), Sergey S. Chernykh[1],
Tatiana I. Barsukova[2], Roman K. Ovcharenko[3],
and Olga S. Ivanchenko[1]

[1] M.I. Platov South-Russian State Polytechnical University,
Novocherkassk, Russia
vodenkok@mail.ru
[2] North Caucasus Federal University, Stavropol, Russia
[3] South-Russian Institute of Management, Russian Academy of National
Economy and State Service, Moscow, Russia

Abstract. The object of the research is value orientations of representatives of middle class, viewed through the prism of influence of the service labor sub-institute, which determines the character of the basic institutional model of the Russian society. The object of the research is cultural peculiarities and value orientations of middle class in Russia and trends of its further economic development and political participation in the process of establishment of the Russian version of civil society in view of the general vector of Russia's modernization. It is concluded that sub-institute of service labor and the labor sub-culture continue to play an important role in the structure of value orientations of the Russian middle class, hindering its political activity and fixing its representatives to their job, thus slowing down the social mobility. As a result, the value orientations of Russian middle class help to form negative position as to the Western path of development, which orients its representatives to the practices of communal solidarity.

Keywords: Service labor · Middle class · Russian society
Value orientations · Institutional matrices

1 Introduction

Analysis of the Russian society allows determining the culturally determined specifics of its institutional environment, which forms peculiarities and significant differences of Russian middle class from its Western analogs. The topic of middle class is closely related to the problem of sub-institute of service labor, which forms redistributive relations in the structure of employment [2, 6]. However, the value orientations of representatives of Russian middle class – at least at the declarative and ideological levels – contradict the dominating practices of service labor, which might lead to protests and confrontations with public authorities in the sphere of the viewed group of population.

© Springer International Publishing AG, part of Springer Nature 2018
E. G. Popkova (Ed.): HOSMC 2017, AISC 622, pp. 366–371, 2018.
https://doi.org/10.1007/978-3-319-75383-6_47

The object of the research is value orientations of representatives of middle class, viewed through the prism of influence of sub-institute of service labor, which largely determines the uniqueness of the basic institutional model of the Russian society.

The object of the research is cultural peculiarities and value orientations of middle class in Russia and trends of its further economic development and political participation in the process of establishment of the Russian version of civil society in view of the general factor of the country's modernization.

The level of scientific elaboration of the topic includes analysis of scientific sources, concentrated around the value aspects of the Russian modernization and leading trends of development in Russia: institutes of civil society, middle class, and the system of socio-labor relations [10, 11].

2 Materials and Methods

The theoretical basis includes the settings of "understanding" sociology of M. Weber, structural and functional analysis of T. Parsons, and cultural sociology of J. Alexander. The article uses the theory of neo-institutionalism of D. North, theory of institutional matrices of S.G. Kirdina, and the Marxist approach on the basis of analysis of B.Y. Kagarlitsky in the context of consideration of the modernizing function of middle class [1, 3, 5, 6].

Within the Russian science on the boundary between economics, social science, social philosophy, and history – namely in the works of O.E. Bessonova, S.G. Kirdina, P.I. Smirnov, et al. – the notion "service labor" was conceptualized, which formed in the context of its contradiction to exchange (buy-sell) in market economies of the Western type. S.G. Kirdina notes that "the attribute of service labor, which differentiates it from the influence of the institute of hired labor, is striving for *full employment of population*" [7, p. 117]. This shows that the viewed sub-institute functions in economies of redistributive type, which are peculiar for the centralized mechanism of labor relations' regulation.

The authors speak of service labor as of sub-institute, not institute, which shows its subordination as to market interrelations of exchange and emphasizes that practices of service labor have informal character. That's why it is possible to speak of a special sub-culture of service labor, which forms special relations between managers and subordinates, which actually pierces labor relations that are formally determined as market relations.

3 Results

Long evolution of the Russian society in the conditions of state paternalism led to strong dependence of the whole social system on the decisions of authority's hierarchy. As a rule, the characteristics of sub-institute of service labor are peculiar for their fixation to a certain place and their mandatory character [7]; mandatory character, departmental organization, hierarchical nature, and nomenclature [2]; service labor activity – activity for "another" – society, group, clan, and person, unlike hired labor, where the main purpose consists in satisfying the needs of a performer [8].

Obviously, the influence of the above sub-institute on society is so large that it stimulates formation of communal attitudes in public conscience, which determine the priority of the whole over the part, put the system's goals above the individual's goals, and form a special subjection to public authorities, as well as servility, which is peculiar for the Russian labor mentality.

Middle class is within the field of influence of service labor's practices, though its segments have large differences – at that, its domination grows with transition from representatives of small business to employees of the budget sphere. In Russia, the state continues to perform the dominating role in the sphere of distributive relations. Due to this, service labor continues to play the key role in the system of public relations; according to P.E. Sheregi, there's a strong fact that "out of 143.3 mln of population of the Russian Federation, only 52.9% are economically active, and only 17.8% are employed in the sphere of material production. In simple words, if the latter indicator is rounded to 20%, state budget is a direct or indirect source of consumption for at least 80% of the Russian population. As a matter of fact, it is the Soviet variety of government natural distribution – however, now it has been replaced by the money equivalent, which creates a visibility of equal monetary exchange" [9, p. 35]. This creates additional structural pre-conditions for larger dependence of representatives of middle class on the state's redistribution system and stimulates the reproduction of the service labor's practices.

The value orientations of middle class in Russia constitute the national model of civil society, which strives for practices of solidarity and reproduction of social relations on the basis of civil participation. "A peculiarity of economic institutes of redistribution X-economy (which include relative ownership, redistribution, service labor, and limitation of costs, or X-effectiveness) is the different state of economic subjects as compared to the market institutional environment. They are different due to independence, as well as "involvement" into economic cultures. Such "connection" determines specifics of mass economic behavior, which supposes not so much struggle as co-existence with these structures" [6, p. 70]. That's why the sub-institutes of service labor hinder the possibilities of the national middle class to become the subject of opposition to bureaucratic authorities, which controls budget flows.

Therefore, Russian middle class is not well-prepared for protecting its own economic interests at the individual level – especially, as compared to the Western societies, where middle class has a large experience of fighting for its rights. "It is possible to see that half of the population think that strengthening of civil rights and liberties, as well as civil society, may ensure Russia's well-being. Nevertheless, a third of the population thinks that this might be stimulated by "strengthening of the authorities' vertical" [6, p. 70]. At that, a lot of representatives of middle class show negative attitude towards the above labor relations, considering them to be vivid relics of fiefdom and the Soviet totalism. At that, the latter are often assigned to supporters of the West or to liberals, though such definitions are not correct, as striving of middle class to emancipation is a universal social phenomenon. That's why protest mood among representatives of middle class as to current public authorities and their bureaucratic machine should be evaluated as to the level of acceptance – non-acceptance of the above type of labor practices.

4 Discussion

In the conditions of economic crisis, which led to reduction of revenues, including for representatives of middle class, class contradictions aggravate, which leads to increase of the opposition. Thus it is necessary to take into account the historic duality of middle class (bourgeoisie), which may be a stalwart of stability and order, under the conditions of compliance with the class compromise and growth of well-being, and, quite on the contrary, it may perform the role of the subject of bourgeois (civil) revolutions in case of large crisis in the system of distribution of resources, public authorities, and management. Geopolitical influences from outside can only increase these tendencies, especially in case of high contradictions and conflicts among the elites.

With the general decrease of everyday consumption, refusal from former practices of consumerism may lead to frustration among large groups of population – especially in middle class, which now leads to reduction of its loyalty to current public authorities. Middle class, which possesses high level of self-consciousness and capability for private initiative, preserves the protest potential, which could be channeled by anti-corruption protests. That's why middle class preserves a positive possibility to influence the process of the country's modernization by increasing the quality and strengthening the civil control.

In the countries with a high level of systemic corruption, which face the crisis of system of management in the spheres of politics and economy, middle class is a potential realistic scenario [5]. The culture of middle class produces and supports the values of the legal state and democracy. That's why the attempts of ideological shift of its active representatives to "post-materialistic" values, compiled according to the image of new morality of "patriotic Orthodoxy", cannot compensate for the existing consumer settings. It is especially difficult to conduct such a shift due to qualitative growth of information technologies, which have already led to unprecedented level of social and political transparency in Russia, providing additional resources to the Russian oppositionists.

Dependence of a large part of the Russian society on the redistribution system of national economy and domination of socio-cultural practices of service labor hinder extinguish the political activity of middle class and its capability to present revolutionary projects. At that, while not ignoring institutional determinants (peculiarities of institutional matrices) of development of the Russian society, it is necessary to note formation of innovational value orientations as to interaction between the state and the civil society. Thus, it is necessary to treat seriously the forecast of M. K. Gorshkov and N. E. Tikhonova that "development of this tendency may lead to refusal from standing the aggravation of the position in the course of another economic crisis – refusal not at the individual level, but at the scale of society, from the norm and necessity to "repay the debt to the country" by military service, etc." [4, p. 52]. Therefore, the sub-institute of service labor in the conditions of the Russian realia may contradict the value orientations of representatives of middle class who wish to show their initiative and implement their professional functions in the conditions of market relations.

5   Conclusions

The sub-institute of service labor and the labor sub-culture, which is based on it, continue to play an important role in the structure of value orientations of Russian middle class, hindering its political activity and tying to the work place, thus slowing down the social mobility. As a result, value orientations of Russian middle class help to form the negative position as to the Western path of development, which orients its representatives for the practices of communal solidarity. At that, the value orientations of middle class are peculiar for a clear contradiction between the values of modern time, which produce nonconformity, internal locus control, initiative, and communal values of solidarity, supported by the bureaucratic system of power. Domination of service labor, predetermined by the redistribution character of the Russian economy, forms a special national model of civil society, where the main value settings suppose not the struggle but co-existence with deep structures of power and management. At that, the perspectives of development of the Russian model of civil society on the basis of increase of middle class's subjectivity also preserve their topicality due to economic crisis and reduction of consumption, as well as growth of civil control and increase of the level of transparency in the country.

Acknowledgments. The article was prepared within the grant of the President of the Russian Federation for state support for young Russian scholars – doctors of science – on the topic "Cultural and ideological foundations of formation of the national model of regulation of socio-economic and scientific and innovational activities" (MD-651.2017.6).

References

1. Alexander, J.: The Meanings of Social Life: A Cultural Sociology (2013). 640 p
2. Bessonova, O.E.: Transfer Case: The Institutional Theory of Economic Development of Russia. Novosibirsk (1999). 152 p
3. Weber, M.: Selected Works: The Protestant Ethic and the Spirit of Capitalism. SPb (2014). 656 p
4. Gorshkov, M.K., Tikhonova, N.E.: Socio-cultural factors of consolidation of the Russian society (2013). 54 p
5. Kagarlitsky, B.Y.: Revolt of Middle Class. (2012). 224 p
6. Kirdina, S.G.: Civil Society: Retreat from Ideologeme. Sociological Studies. 2, 63–73 (2012)
7. Kirdina, S.G.: Institutional Matrices and Development of Russia: Introduction into the X-Y-Theory. SPb (2014). 468 p
8. Smirnov, P.I.: Stages of social development as ideal types. Bulletin of SPbSU 6, Issue 3, P. 45 (1998)
9. Sheregi, F.E.: Education as a social institute. Edges of Russian Education (2015). 644 p
10. Vodenko, K.V., Cherkesova, E.Y., Shvachkina, L.A., Fateeva, S.V., Erosheva, I.Y.: The specifics of the socio-cultural determination of the current economic activity. Int. J. Econ. Financ. Issues **6**(S1 (Theory and Practice of Organizational and Economic Problems of Territorial Development and the Effectiveness of Social and Economic Systems)), 206–210 (2016)

11. Vodenko, K.V., Shevchenko, O.M., Barsukova, T.I., Hubuluri, E.I., Mishina, N.V.: Modern institutes and regulatory forms of social and labor relations in Russian society. Int. Rev. Manag. Mark. **6**(S6 (Special Issue on "Management of Systems of Socio-Economic and Legal Relations in Modern Conditions of Development of Education and Society")), 185–190 (2016)

The Problems of Preclusive Effect of Judicial Acts of Special and Simplified (Short) Civil and Criminal Proceedings

Andrey V. Yudin$^{(\boxtimes)}$, Vyacheslav V. Ivanov, and Ilya D. Simonov

Samara University, Samara, Russia
udin77@mail.ru

Abstract. The purpose of the article is to determine the regularities of preclusive effect of judicial acts of special and simplified civil and criminal proceedings. The authors found on the functional similarity in establishment of special and simplified civil and criminal proceedings, which allowed determining general methodological approaches to solving the set task. As a result of the research, the authors come to the conclusion that differentiation of preclusive effect of judicial acts of special and simplified civil and criminal proceedings is not logical. Court proceedings in a certain procedural order, including special or simplified proceedings, does not diminish the force of adopted judicial act and all manifestations of its legal force.

Keywords: Preclusive effect · Intersectorial prejudgement
Special proceedings · Simplified procedure · Summary procedure

1 Introduction

1.1 Establishing a Context

The problems of preclusive effect of judicial acts, made in various types of court proceedings, are a subject of scientific discussion and attract researchers' attention. At that, certain aspects of manifestation of preclusive effect of judicial acts are not studied very well, which makes their research very topical. One of such important problems is analysis of the effect of prejudgement in special and simplified legal and criminal proceedings.

The institute of prejudgement has the intersectorial value and is important from intrasectorial positions, as, being a component of evidence law, sees its own dependence on the change of the procedure – primarily, on differentiation of the types of proceedings, creation and termination of procedural orders of consideration of civil and criminal cases.

Implementing a new procedural order of consideration of legal conflicts (at the level of type of proceedings, category of case, etc.), the legislator does not usually model the special rules of action of preclusive effect of judicial acts, adopted in these types of proceedings – though, there are opposite examples. Thus, on the basis of Article 90 of the Criminal Procedure Code of the Russian Federation (hereinafter – RF CPC), the circumstances, determined by the effective sentence, excluding the sentence

© Springer International Publishing AG, part of Springer Nature 2018
E. G. Popkova (Ed.): HOSMC 2017, AISC 622, pp. 372–379, 2018.
https://doi.org/10.1007/978-3-319-75383-6_48

made by the court according to Article 226.9 (proceedings on a criminal case with investigation in the simplified form), 316 (special order of court proceedings with the accused person's agreeing with accusation), or 317.7 (sentencing with defendant with a plea deal) of the RF CPC, are acknowledged by court, prosecutor, investigator, and interrogator without additional inspection. In most cases, it is supposed that the general provisions on prejudgement are enough for the purpose of law-enforcement activities.

Based on materials of two classic processes – civil and criminal – the general regularities of the effect of prejudgement in separate proceedings of civil and criminal cases should be determined. This caused the structure of the research: (1) the problems of prejudgement in special, simplified, and summary procedures of civilized process; (2) problems of prejudgement in the special order of court proceedings in criminal procedure; (3) general regularities of manifestation of prejudgement in special and simplified proceedings of a legal process.

It should be noted that the authors do not discuss the issues of the notion, content, limits, and system of special, simplified, and short proceedings in civil and criminal proceedings; The authors consider that from the functional side in civil and criminal proceedings there are orders of consideration of cases with exceptions from the general order, which is expressed in a special object of judicial activities, special order of consideration, etc. Validity of this is confirmed by studies in the sphere of legal process.

1.2 Literature Review

Prejudgment was the object of research by Soviet jurisprudents, is studied in modern Russia and foreign countries.

The history of development of the laws that regulate preclusive effect of judicial acts was viewed by Eremkina (1970), Gurvich (1976), Bezrukov (2007), Gromoshina (2010), Dudanova (2011), Dikarev (2011), Golovko (2012), Lang (2017) et al.

1.3 Establishing a Research Gap

The previous studies do not allow for complex consideration of the problems of preclusive effect of judicial acts, made in special and simplified civil and criminal proceedings.

The research uses the rather-legal aspect and is aimed at analysis of the problems of preclusive effect of judicial acts made in special and simplified civil and criminal proceedings.

1.4 Aim of the Study

The purpose of the research is to determine and open the problems of preclusive effect of judicial acts made in special and simplified civil and criminal proceedings. This is necessary for determining whether the judicial acts made in special and simplified civil and criminal proceedings have preclusive effect.

2 Methodological Framework

2.1 Research Methods

The following methods were used during the research: analysis, synthesis, comparison, and generalization; historical and legal and rather-legal – which allow for complex consideration of the problems of prejudgement in various types of court proceedings.

2.2 Research Basis

The basis of the research is scientific studies and publications of the Russian jurisprudents that study various aspects of the institute of prejudgement.

2.3 Research Stages

The problem was studied in two stages:

First stage: analysis of existing scientific literature on the topic of the research and procedural laws; determining the problem, purpose, and methods of the research.

Second stage: formulating the conclusions received in the course of analysis of scientific literature and the laws, preparing the publication.

3 Results

3.1 Problems of Prejudgement in Special, Simplified, and Summary Procedures of Civilized Process

While for civilized process the special proceedings is traditional, the history of simplified and summary procedures has a shorter history; thus, summary procedure in civil procedure appeared in 1996 (Chapter 11.1 "Court order" was introduced by the Federal Law dated 30.11.1995 No. 189-FZ "On certain changes and additions into the Civil Procedure Code of the RSFSR"), in arbitrary process - in 2016 (Chapter 29.1 "Summary procedure" was introduced by the Federal Law dated 02.03.2016 No. 47-FZ "On certain changes into the Arbitration Procedure Code of the Russian Federation"); simplified procedure in arbitrary procedure appeared in 2002 (together with adoption of the RF APC in 2002), in civil procedure – in 2016 (Chapter 21.1 "Simplified procedure" was introduced by the Federal Law dated 02.03.2016 No. 45-FZ "On certain changes in the Civil Procedural Code of the Russian Federation and the Arbitrary Procedural Code of the Russian Federation").

Procedural law does not contain any exceptions for preclusive effect of judicial acts made in the proceedings other than adversary. However, simple logic shows that judicial acts made within adversary proceedings are different from judicial acts made in "special" proceedings from the point of view of fullness of studying the evidential material, motivation of conclusion, procedure of decision making, and other components.

Traditionally, "special proceedings" are seen as special proceedings in civil and arbitrary procedures, as well as proceedings on cases that emerge from administrative and other public and legal relations in arbitrary procedure; in this case this term is used in the meaning "non-adversary" – the one not belonging to adversary from the formal point of view (as, for example, proceedings which is simplified in its nature is adversary).

It is possible to suppose that any proceedings with their specifics are based on *refusal* from the elements that are usual for adversary proceedings. Therefore, it is necessary to determine how these elements are significant for preclusive effect of judicial acts, and, therefore, how the refusal from them may influence preclusive effect of judicial acts made in "special" proceedings. If prejudgment is connected to existence of such elements, refusal from them is refusal from prejudgement.

Preclusive effect is based on inadmissibility of repeated ascertainment of the circumstances that already have been a subject of judicial ascertainment with the same parties that participate in averment. The circumstances that are ascertained with preclusive effect are reflected in a judicial act and are the result of the court's analytical activity.

Is it possible to state that in the special, summary, and simplified proceedings the legislator frees the parties from the obligation of proving, and the court – from the necessity to determine the circumstances of the case? Is it possible to state that judicial acts on such proceedings are made "blindly", not on the basis of determined circumstances of the case? This is not acceptable.

Averment take place in special proceedings – at that, the court has more active role that in adversary proceedings; in summary and simplified proceedings, the court also studies the evidences provided by the court. There are no reasons to doubt in sound quality of such evidentiary activities – in any case, until the interested party announces his right, debtor refuses (or requires replacement) of the judicial order, and defendant or another party starts stating insufficiency of simplified procedure for determining the real circumstances of the case. During determining the circumstances in all above cases, the evidential load is not reduces, and the court cannot be futile.

The court decision on the case of special proceedings is also rather meaningful: for example, during the court's satisfying the announcement on determining the fact that has a legal meaning, the resolutive part of the decision states the presence of the fact with a legal meaning and states the determined fact (Clause 2 of Article 222 of the Arbitrary Procedure Code of the Russian Federation – hereinafter RF APC).

Moreover, with cases of special proceedings it is possible to see decisions the facts for which have a quality of "increased prejudgement". They are peculiar for the fact that even the parties that do not participate in the case for which such circumstances were ascertained, have to take them into account. Thus, M.A. Budanova states that "due to special specifics of certain categories of civil cases, such facts, which are ascertained by effective decision, the court act, cannot be avoided by persons who participated in the case or by any other persons. This exception from the general rule on subjective limits of prejudgement refers to the categories of cases, as the result of consideration and solving of which the court determines a legal status of a party (e.g., for cases of debtor acknowledged to be a bankrupt, dissolution of marriage, annulment of marriage, acknowledging minor sui juris, etc.)".

The court order states the law on the basis of which the requirement is satisfied; volume of money to be levied or movable property to be demanded with specification of its cost; amount of penalty, if levying it is presupposed by the federal law or agreement, as the volume of fines if they are due; the period for which the levied debt appeared for the obligations that suppose execution in parts or in the form of periodic payments (Clauses 5–7, 10, Part 1, Article 127 of the Civil Procedure Code of the Russian Federation (hereinafter – RF CPC), Clauses 5–7,10, Part 1, Article 229.6 of the RF APC).

As is known, court decision on the case of simplified procedure according to the general rule is made by announcing the resolutive part (Part 1 of Article 232.4 of the RF CPC, Part 1 of Article 229 of the RF APC) and only the claim of the persons participating in the case leads to the court's preparing the motivated decision (Part 2 of Article 232.4 of the RF CPC, Part 2 of Article 229 of the RF APC).

With a motivated decision on the case of simplified procedure, there are not doubts in preclusive effect of the decision.

As to court order or in case of "short" decision on a case of simplified procedure, it is necessary to note the following. Prejudgement is traditionally projected at the circumstances related to the motivational part of a judicial act – for these circumstances can further form the decision on another case related to the already viewed one. At the first glance, there's no sense in acknowledging preclusive effect of the court decisions given in the resolutive part, for, firstly, they are mandatory not due to prejudgement but due to general binding nature of a judicial act; secondly, due to exclusiveness of a judicial act, the court cannot repeatedly consider the case in which it would have to reproduce the similar resolutive part – this would mean violation of ban for repeated consideration of the similar lawsuit.

However, it is possible to see that the minimum factual basis is present in this case as well – the court ascertains and provided in a judicial act the basis for emergence of a debt, period, volume of the main debt, sanctions, and legal qualification of legal relations in dispute. Due to objective reasons, prejudgment is limited by one obligation and is brought down to stating the debt of debtor before creditor – however, this comes from the object of the viewed proceedings. If we say that if a dispute arises in the future regarding the authenticity of the agreement, the party that achieved the levy will not be protected by prejudgment – we will not be logical, for levy according to the agreement in adversary proceedings does not guarantee in the future its invincibility during a claim on contestation of agreement only due to the fact that it was a basis for debt enforcement.

It is possible to offer a lot of examples of using prejudgement of judicial order or short decision on the case of simplified procedure. For example, during consideration in adversary proceedings of a case on penalty recovery due to non-fulfillment of agreement, the fact of delay could be motivated by court with a reference to a court order or decision on a case of simplified procedure, with which the main debt was levied.

3.2   Problems of Prejudgement in Special Order of Court Proceedings in Criminal Procedure

Prejudgment in criminal procedure has a range of peculiarities. As was said above, in 2015 the text of Article 90 of the RF APC was corrected and the legislator specified that additional inspection is not required for acknowledging the circumstances determined by the court decision in a criminal, civil, and administrative case, except for the sentence set during consideration of a criminal case finished by investigation in a shortened form of interrogation (Article 226.9 of the RF CPC) or in case of the accused person's agreeing with the accusation (Article 316 of the RF CPC), or with a plea deal (Article 317.7 the RF CPC). Thus, there appeared a situation when the court decisions made within various procedures (common and special) lead to different legal consequences, expressed in presence or absence of their preclusive character.

In all above cases, verdict is made without studying the evidences, for the court's conclusion on the defendant's guilt is based on his own acknowledgment. In 2002, the procedure of special order of court decision with defendant's agreeing with the accusation, which did not suppose studying the evidences, was introduced in Russia. The initial mistrust from law enforcers to the new procedure has passed, and 2/3 of criminal cases are viewed by courts within this special order.

Special order of court proceedings does not suppose the judge studying and assessing the evidences gathered for the criminal case. The court does not ascertain the factual circumstances of the case – it is a result of Part 7 of Article 316 of the RF CPC – but compares the circumstances in the form as they were ascertained by the preliminary investigation bodies with existing evidences.

In 2009, plea deal was introduced into the RF CPC, and the criminal defendants received more possibilities for building their protection strategy for the purpose of reducing the form and volume of punishment. As this procedure could be applied only in cases of group crimes, in certain cases the materials of a criminal case as to individual defendants are separated from the criminal case for independent investigation and further consideration by the court. Therefore, if sentences for such cases are made in the order of special proceedings, the circumstances ascertained by these sentences are deemed ascertained for the main case without verification.

As a result, sentence for the main case could be made on the basis of the testimony of the convicted for the separate case. This contradicts the principle of freedom of assessment of evidences, according to which no evidences can have prearranged force (Part 2 of Article 17 of the RF CPC) and part 2 of Article 77 of the RF CPC, which sets that the defendant's testimony that are not confirmed by other evidences, cannot be a basis for judgment of conviction.

Obviously, judgments on the cases that are considered in a special order are exceptions from this rule. However, these exceptions are true only for the defendant – for he is convicted on the basis of his testimony. Convicting other people on the basis of sentence based on the testimony of the person convicted for another judgment, when the defendants cannot challenge the evidence, is not acceptable.

This shows the necessity for weighted approach to provision of preclusive effect to judicial acts made within special, simplified procedures.

3.3   General Regularities of Manifestation of Prejudgement in Special Proceedings of a Legal Procedure

Beginning from 2009, prejudgment of decisions on civil and criminal cases became equal - regardless of the category of the case and procedure within which it is considered and solved, the circumstances ascertained by court were acknowledged to have preclusive effect in criminal and civil procedures. These changes were criticized by a lot of authors. Thus, L.V. Golovko wrote that "criminal procedure, in which the issue of guilty – not guilty and the issue of taking a person's freedom are solved, unlike civil and arbitrary procedure, provides a maximal number of guarantees: prosecution has to prove the defendant's guilt, and the latter has a right for protection and presumption of innocence, the complainant has his right, etc. So it was considered that ascertaining of the facts mandatory for criminal court through other procedures is ignoring these guarantees, for neither arbitrary nor civil procedure requires such guarantees, for they have different tasks. When it was said that the decision of civil court has not influence on a criminal case, it was not supposed that the judge on the criminal case should be indifferent towards it but it was supposed that it had a value of a common evidence for him. That it, you may use as the evidence the decision of civil or arbitrary court, but you cannot ascertain the facts that have the criminal and legal meaning with prejudgment through other processes and without guarantees".

Differentiation of preclusive effect of judicial acts, made in special and simplified legal and criminal proceedings, is not very consistent. Consideration of a case by the court in a certain procedural order does not diminish the force of the made judicial act as a justice act and all manifestations of its legal force. Establishing the defendant's debt for a civil case on the basis of the documents studied by court, which prove the debt, as well as establishing the defendant's guilt on the basis of his confession does not change the nature of factual circumstances ascertained by the court which should be adopted without additional verification for a different case; otherwise, levying the money and convicting a person would have been unjust, as it would have happened only because this fact was not argued by anybody.

4 Discussions

The positions that are brought down to refusing prejudgement for the final judicial acts, made in special and simplified proceedings, are motivated by a relatively easy cancelling/changeability of such acts or by unstable character of the order of proceedings within which they are made. Thus, special proceedings can take place only if there's no dispute on right; court order could be cancelled due to debtor's claim; court may pass from simplified procedure to adversary procedures in case of procedural difficulties, etc.

However, could this be a basis for refusing preclusive effect for judicial acts made in the viewed proceedings? We think it could not. If it could, the debtor would refuse from execution of court order or decision on the case of simplified procedure with a reference to incompleteness of the procedural order applied for his case; or the interested party may doubt in the facts ascertained by the court in the order of special proceedings. Obviously, such references would be deemed inconsistent. Therefore, refusal of preclusive effect of judicial acts for the viewed cases will be deemed inconsistent as well.

If due to some reasons the judicial act is canceled, the effect of prejudgement will disappear; still the similar order exists for decisions on cases of adversary proceedings.

A different case is objectively narrower sphere of effect of preclusive effect of judicial acts for the analyzed cased, which is the result of limitation of the applied orders of proceedings and the possibility of opening them only in cases envisaged by the law on the basis of certain criteria (thus, cases of special proceedings are given in the law; the list of cases of adversary proceedings is given in the codes; there are also criteria for consideration of case in simplified order), while the case of adversary proceedings are opened "universally", due to plaintiff stating the emergence of a legal dispute before court (Article 3 of the RF CPC, Article 4 of the RF APC).

5 Conclusion

During modeling of the orders of proceedings that are different from adversary the legislator follows the general legal logic and does not neglect the elements important for emergence of preclusive effect of judicial acts – proving, ascertaining of circumstances of the case, court judgments. Thus, there are not foundations for refusing preclusive effect of judicial acts made on the proceedings of civilized and criminal procedures other than adversary.

References

Bezrukov, A.M.: Preclusive Effect of Judicial Acts, 144 p. Walters Kluwer, Moscow (2007)

Budanova, M.A.: Procedural exemptions in proving in civil court proceedings: Ph.D. thesis. Saratov, 25 (2011)

Golovko, L.V.: Russian judicial procedure is archaic (2012). https://pravo.ru/review/view/73251/

Gromoshina, N.A.: Differentiation, unification, and simplification in civil court proceedings, 409 p. Prospekt, Moscow (2010)

Gurvich, M.A.: Court decision. Theoretical problems, 176 p. Legal literature, Moscow (1976)

Dikarev, I.S.: Debatable issues of prejudgement in criminal procedure. Peace justice **2**, 30 (2011)

Eremkina, A.P.: Prejudgement in Soviet civil process. Ph.D. thesis, Moscow, 16 p. (1970)

Lang, P.P.: Institute of special proceedings in a legal process. Ph.D. thesis, Kazan, 39 p. (2017)

Blended Learning in Teaching EFL to Different Age Groups

Maria V. Arkhipova[1]([⊠]), Ekaterina E. Belova[1], Yulia A. Gavrikova[1],
Natalya A. Lyulyaeva[1], and Elina D. Shapiro[2]

[1] Nizhny Novgorod State Pedagogical University named after Kozma Minin
(Minin University), Nizhny Novgorod, Russian Federation
arhipovnn@yandex.ru, natachaluna@yandex.ru,
belova_katerina@inbox.ru, y.a_gavrikova@mail.ru
[2] Nizhny Novgorod State Linguistic University named N.A. Dobrolyubov,
Nizhny Novgorod, Russian Federation
shapiro_elina@mail.ru

Abstract. The article investigates how blended learning inspires students of different ages to be more motivated in the process of acquiring receptive and productive foreign language skills. The authors analyze the neurobiological and psychological characteristics of various age groups of people learning a foreign language. The results of this analysis lead to the conclusion that the age of a student is a defining factor of the necessity of information technologies use in the learning process. The authors argue that the younger the learners are the more IT-oriented they are. It means that neglecting new technology in teaching a foreign language to children, teenagers and young adults is not an option any more. Older students used to traditional methods of teaching a foreign language appreciate the use of online resources and tools. The combination of student-centered methods and modern technology is the quintessence of blended learning which serves as an effective teaching tool for EFL students.

Keywords: Blended learning · IT-technologies · Age groups · Teaching EFL
Electronic course · Online resources · Foreign language skills

1 Introduction

Being the international language English is widely used as a means of communication in the international relationship and commonly used in all branches of knowledge. This fact made it necessary to include English into school curriculum starting from the primary level. Nowadays the study of English as a foreign language (hereinafter EFL) by young children in Russia is definitely on the rise. The demand for English among adults is huge too. Among teenagers there is a need for English when taking various examinations as it represents a contribution factor, whereas among adults to know and improve English skills constitutes an additional opportunity for promotion and career development. An effective learning environment for children and adults is sure to be different. Thus, methods, and techniques applied to teaching dissimilar learning groups

© Springer International Publishing AG, part of Springer Nature 2018
E. G. Popkova (Ed.): HOSMC 2017, AISC 622, pp. 380–386, 2018.
https://doi.org/10.1007/978-3-319-75383-6_49

are entirely different propositions and should be based on diverse approaches, biological and psychological aspects related to age impact.

As pointed by Asl and Valipour (2015) age plays a crucial role in what to teach and how to teach it, since a young learner class is different from an adult or a teenager class in terms of the learners' language learning needs, the language competences emphasized, and the cognitive skills addressed. Harmer (2007) emphasizes that learner groups are to be kept in mind in the process of teaching as well as the fact that every learner is unique.

In accordance with this fact the goal of our study is to provide theoretical bases on teaching English as a foreign language to different age groups. The following questions constitute the foundation of our research: (1) What are the differences of teaching EFL to different age groups in terms of biological and psychological aspects? (2) Can blended learning promote students' motivation towards learning a language?

2 Theoretical Fundamentals of the Research

Teaching English to young learners at elementary and secondary school is different from that to adults, the main differences of which being the following:

(1) Language perception

Children perceive messages better if they are presented in natural chunks and language-acquisition theory bears this out. The necessity to break these chunks into individual pieces is of much less relevance for children than for adult learners, because (a) children are more accepting and tend not to analyze, and (b) children's ability to mimic is extremely good (Abe 1991). On the contrary, adults can engage with metalanguage and can talk about abstract issues (Harmer 2007).

Junior students have weak capacity for memorizing the material, first of all, lexical. The more complicated the training material is, the slower the educational process becomes.

(2) Attention span

Ruff and Lawson (1990) point out that the time spent consciously on doing a task is different according to different ages. Children have a short attention span; the amount of time they can concentrate and focus on learning materials varies from 5 to 10 min after which they can easily get bored (Harmer 2007). Dukette and Cornish (2009) state the range of concentration of about 5 min for a two-year-old child, to 20 min of teenagers. Adults have a longer concentration span to continue an activity and are more self-disciplined.

(3) Styles of learning

Children understand better when they interact, touch, hear, and see. Flash cards, pictures, various objects and realia, games make lessons alive and resemble more closely the outside world. According to Harmer (2007), adults can benefit from their abstract thought, life experiences; they have expectations about the learning process and their individual patterns of learning.

(4) Attitude to mistakes

Children, as a rule, are sociable and learn a foreign language more quickly without worrying about their mistakes. In comparison with children, most adults care about being corrected and this makes them feel anxious and, therefore, leads to avoiding speaking a foreign language in public (Asl and Valipour 2015).

Some students' fear to make a mistake in case of extremely demanding teachers, who reduce the mark for any tiny slip/mistake, complicates the educational process. It is especially unacceptable with pupils, because this attitude of teachers' may lead to the students' antipathy to the subject, namely, the English language.

(5) Motivation

Adults for the most part are highly motivated. They attend classes of their own free choice usually at some personal and financial sacrifice. They lack the uninhibited enthusiasm of small children (K.S. Joan).

The latest research in Russia in the field of motivation has revealed its continuing decline. At primary school children are eager to learn a foreign language (I.N. Andreeva, I.A. Zimnyaya, and I.B. Minayeva). But later the accumulation of the basic material and overcoming various difficulties reduce their counter activity. Psychophysiological characteristics of teenagers (the fear of making a mistake, the lassitude provoked by the uniformity of lessons presentation manner) contribute significantly to motivation reduction. The research of students' motivational aspect, related with the foreign language study, also shows the process of its permanent decline (M.V. Arkhipova, I.L. Belykh, L.V. Garibova, N.N. Kasatkina, and N.V. Shutova). The scientists have stated a sharp decline in the sphere of motivation for foreign language education among university students as well. First and second year University students demonstrate a high level of anxiety when studying foreign languages. Oral speech and the likelihood of making mistakes represent the most widespread stressors (E.A. Sedova). Third year students consider the possibility of negative consequences in case of failure as the dominant factor while learning a foreign language (M.V. Arkhipova, I.A. Bakhtina, E.A. Sedova, and N.V. Shutova). After conducting the research in the sphere of students' motivation for foreign languages learning we have revealed a similar tendency both among school and university students.

The dependence of studying on motivation was validated by multiple treatises (V.G. Aseev, L.I. Bozhovich, V.K. Vilunas, V.S. Ilyin, A.N. Leyontiev, A.K. Markova, A. M.Mateshkin, Y.I. Meshkov, A.A.Rean, P.V.Simonov, V.A. Yakunin and other disquisitions). The studies revealed the capacity of motivation to represent the compensatory factor in case of lacking aptitude. Meanwhile, even the paramount abilities fail to recompense the lack or low level of motivation for studying, thus being a stub track to success in education.

Taking into consideration all the above mentioned differences, it is of vital significance to find appropriate teaching methods for children and adults with respect to their biological and psychological capabilities.

The question of methods of teaching English as a foreign language has always attracted attention of scholars. Recent studies show that language learners may benefit when the process of language learning includes *blended learning*.

According to the empirical research (S. Fridman), the traditional lecture has proved to be inefficient in today educational environment. Having studied the so-called Internet generation's attitude to such traditional educational form as lecture, S.G. Krylova (2015) discovered that the strongest stimulus to attend lectures is the necessity to follow teachers' or educational establishment rules and regulations. Academicians attend more willingly the lectures supported by electronic presentations and gladly accomplish the tasks involving interactivity.

One of the urgent tasks of foreign language tuition has become the introduction of blended learning. The term *blended learning* means the combination of online digital media use with traditional classroom methods. Only blended learning is able to satisfy the needs and requirements of today education. Teachers' irresponsiveness to the necessities of current education requirements puts at risk the efficiency and effectiveness of the teaching/learning process as a whole, creating a growing gap between a teacher and a student, between a trainee and a mentor.

3   Methodology of the Research

The material for the research has been collected for many years of teaching a foreign language. The descriptive methodology was used as the main tool. It was carried out in the forms of observation and survey.

The first stage included study and analysis of various literature on methods of teaching English as a foreign language to children and adults with respect to their biological and psychological characteristics. In the second stage the survey results were studied and compared. These two stages lead to their subsequent conclusion and definition of the prospective and potential research.

4   Analysis of Survey Results

The aim of blended learning approach consists in ascertaining a harmonious balance between the traditional in-class education, online access to the lessons and independency of the foreign language studying process. The use of online digital media is based on visualization, which helps to activate all the analyzers and connected with them mental processes of sensation, perception and analysis. As a result, a rich empirical basis arises for cumulatively analytical mental activity not only of children but adults as well. Online resources give access to authentic materials and native speakers' oral and written practices, which is crucial for communicative and other skills training. Online authentic resources appeal to both students and professors due to certain characteristics: demonstrativeness; structured and laconic character of the theoretical material or lesson content; familiar and user-friendly interface; simplicity of the usage; up-to-date authentic content allowing the necessary or required skills development; and independent or individual learning.

Some of the features should be dwelt upon at length.

The human's mind receives its nourishment primarily from visual rather than verbal sources, and visualization is rightly considered one of the underlying principles in the process of education. A narrative rendered in visual images attracts all ages, especially junior students. The blended learning method is of great help here, as it may keep you from droning the material to the class or group. Instead, they make the class more interesting and stimulate students' imagination. Complete visual images and sophisticated ideas are provided while giving the teacher time to organize their discussions. They also have controversial topics and present a myriad of viewpoints. The teacher can gradually expand not only viewing time and attention span, but also analytical abilities.

The use of IT-technologies being an integral part of the blended learning method may be extremely beneficial for computer-minded students and those, who are passive at the lesson. Their passive participation in the educational process is conditioned by their inferiority complex, their extreme shyness to ask questions in front of their groupmates in case they do not understand something.

The study revealed the following sample tips of developing an electronic course:

1. The modularity principle presupposes splitting the material into sections consisting of modules that are closed in content; what is equal to children's easiest perception of material in chunks.
2. The principle of completeness means the presence of a theoretical core, control questions, tasks and exercises for independent solution, control tests with answers, references and comment;
3. The principle of regulation, which gives the student the ability to independently control the course, thus making it possible for the learner to feel comfortable;
4. The principle of adaptability signifies that the course allows adaptation to the needs of a particular user in the process of learning activities, variation of the depth and complexity of the material studied, as well as its applied orientation which depends on the future specialty of the student.

Blended learning has certain evident advantages because (1) students are confronted with electronic devices almost from birth nowadays, (2) they have learned to read the sophisticated visual language but have not developed the cognitive ability to distinguish the real world from the virtual one, (3) students who have been exposed to electronic devices are one year ahead of their unexposed peers in vocabulary, (4) students exposed to electronic devices have a fragmented and confused idea of reality and a shorter attention span.

In order for students to develop cognitively it is necessary for them to be able to interpret things they encounter in the real world. It is also necessary for teachers not to foster the misconception that school is only a socializing experience. In order for the school to be a humanizing experience, it is necessary to return to the primary duty of teaching the younger generation via the powers of the subject how to remain human despite dehumanizing elements in their environment. To do this, teachers should explore the world that the students live in every day. It is not necessary to like *The Sponge Bob*, but it is necessary to know who he is and why he is a superhero. An approach to incorporating blended learning into a classroom does not need to do away with the traditional method, but a teacher has to invest more time and pay careful attention to how all material is to be used.

The humanizing experience can be obtained with the atmosphere of trust between the teacher and the learner. If they expect the teacher to refuse, criticize their impressions or opinions and their expectations come true, the teacher is likely to fail to ever feel their further respond to the material offered in any format. If the students feel that the teacher expects only his/her own viewpoint to be supported, the class will be counterproductive. Teachers will have to learn to take it into consideration and make the process of teaching student-oriented.

5 Conclusions

Information and computer technologies, which play a growing part in current educational process and stimulate research, represent one of the most efficient ways of educational environment organization. They allow academic process participants to reap the benefits of modern electronic educational sphere by using the latest pieces of research in diverse domains and formats, trying collaborative activities which significantly boost motivation and introduce personalized approach into learning. Information and computer technologies facilitate communication between trainees and mentors. Furthermore, the blended learning method intensifies various grammar and speaking activities forming deep and reliable skills. Introducing the latest IT forms and achievements within the blended learning method into the class boosts students' motivation and creativity.

Thus, we see the prospect of our further studies in conducting an experiment with the aim to research the impact of the blended learning method on the process of teaching EFL to different age groups. The experiment is supposed to provide evidence supporting our hypothesis: if to organize the process of teaching that will meet biological and psychological needs of different age students' groups, it is possible to increase motivation to learn a foreign language, influence students' performance so that it could enhance academic achievements.

References

Abdulova, E.V., Krylova, S.G., Minyurova, S.A., Rudenko, N.S.: Virtual educational environment – a means of intercultural interaction. In: Okushova, G.A. (ed.) Connect-Universum – 2014: Collection of the Materials of the 5th International Scientific and Practical Internet Conference, pp. 46–50. Tomsk State University, Tomsk (2015)

Arkhipova, M.V., Zhernovaya, O.R.: Experimental study of the peculiarities of training motivation of modern pupils during studying foreign languages. Hist. Socio-Educ. Thought **8** (6–1), 177–179 (2016)

Belova, E.E., Minasyan, V.A.: Regarding easy and difficult languages. In: Belova, E.E. (ed.) Theoretical and Practical Aspects of Linguistics, Literature Study, and Methodology of Foreign Languages Teaching: Collection of Articles on the Materials of the International Scientific and Practical Conference, 15 April 2015, pp. 11–18. Minin University, Nizhny Novgorod (2015)

Gavrikova, Y.A.: Meliorative communicative strategies as a mechanism of influencing the complex axiological vocabulary. Bull. N.I. Lobachevsky Nizhny Novgorod Univ. **6**, 347–352 (2010)

Lyulyaeva, N.A.: Algorithmization of foreign languages teaching in higher school. Problems of modern pedagogical education. Series: Pedagogics and psychology. Collection of scientific articles: Yalta: RIO GPA, (55). – P. 11. pp. 63–73 (2017)

Filippova, S.Y., Shapiro, E.D.: Certain results of the experiment on implementation of blended learning in N.A. Dobrolyubov NNLU. In: Bwlova, E.E. (ed.) Language and Language Education in the Modern World: Collection of Articles on the Materials of the International Scientific and Practical Conference, 13 April 2016, pp.266–268. Minin University, Nizhny Novgorod (2016)

Abe, K.: Teaching english to children in an EFL setting. J. Teach. Engl. Outs. United States **XXIX**(4), 6–7 (1991). English Teaching Forum, 11

Asl, H., Valipour, V.: Teaching english as a foreign language to persian children vs. adults. Indian J. Fundam. Appl. Life **5**(S1), 5199–5204 (2015)

Harmer, J.: The Practice of English Language Teaching, 4th edn. Pearson Longman, Essex (2007)

Hobbs, R.: Measuring the impact of media education on student skills and teacher performance. In: Forum International de Recherchers "Les Jeune et les medias demain. Problematiques et Perspectives", p. 25. UNESCO-GRREM, Paris (1997)

Hobbs, R.: Media literacy's effect on viewing motivations. Telemedium J. Media Lit. **46**(1), 15 (2000)

Joan, K.S.: Teaching English to Young Learners, 62 p. English Language Center University of Maryland, Baltimore County

Masterman, L.: A Rational for Media Education. Media Literacy in the Information Age, pp. 15–68. Transaction Publishers, New Brunswick and London (1997)

Kaznacheeva, S.N., Bondarenko, V.A.: Specifics of motivation learning a foreign language by students of non-language directions in Minin University. Bull. Minin Univ. **3**, 1 (2016)

Ruff, H.A., Lawson, K.R.: Development of sustained, focused attention in young children during free play. Dev. Psychol. **26**(1), 85–93 (1990)

Dukette, D., Cornish, D.: The Essential 20: Twenty Components of an Excellent Health Care Team. RoseDog Books, pp. 72–73 (2009)

Experience of Approbation and Introduction of the Model of Management of Students' Independent Work in the University

Olga V. Bogorodskaya$^{(\boxtimes)}$, Olga V. Golubeva, Marina L. Gruzdeva,
Alexandra A. Tolsteneva, and Zhanna V. Smirnova

Nizhny Novgorod State Pedagogical University Named After Kozma Minin
(Minin University), Nizhny Novgorod, Russian Federation
olgluzd@yandex.ru, gololga@yandex.ru,
gru1234@yandex.ru, tolstenev25@yandex.ru,
z.v.smirnova@mininuniver.ru

Abstract. In Nizhny Novgorod State Pedagogical University named after Kozma Minin in the period from 2015 to 2017, the project "Creating a system for managing the independent work of trainees" was developed and implemented as part of the program for the modernization of pedagogical education. In the article, approaches creation and results of approbation of a control system of independent work of students are considered. A model for managing the independent work of students in a university is proposed, which includes: Analysis of normative documents, target component, conditions of implementation; Technology implementation, the object management, the resulting component of the model. The main tool for managing the independent work of the students is the syllabus of discipline. The most effective forms and methods of teaching in the implementation of the model are considered, preference is given to the use of the electronic educational environment of the university and active methods of learning. A new form of organization of student's work is offered - independent self-employed work. The role of a tutor in a university is defined when accompanying the independent work of students. A significant place in the article is given to the organization of reflexive activity of students, tools for organizing reflexive activity are offered - diaries of individual and group reflection. The pedagogical efficiency of the proposed model was assessed according to the selected criteria: Motivation to realize independent work, the level of mastering disciplines, the level of formation of general educational activities among students.

Keywords: Model of management of independent work of students
Syllabus of discipline · Tutorial support · Reflexive activity
Pedagogical effectiveness

1 Introduction

The urgency of the development and approbation of the project "Creating a system for managing the independent work of students" is conditioned by modern conditions for training students in the higher education system, namely, reducing the classroom

© Springer International Publishing AG, part of Springer Nature 2018
E. G. Popkova (Ed.): HOSMC 2017, AISC 622, pp. 387–397, 2018.
https://doi.org/10.1007/978-3-319-75383-6_50

workload and increasing the amount of independent work of students. Work on the development, approbation and implementation of the model for managing the independent work of students was conducted in 2015 at Kozma Minin Nizhny Novgorod State Pedagogical University.

2 Theoretical Bases of Research

We understand the management of the independent work of students as a set of processes implemented by the University's structural divisions and scientific and pedagogical workers aimed at the development of educational information, its transferring and controlling learning by students in the conditions of auditor and extracurricular independent work that ensure effective formation of the general cultural and professional competencies of trainees. In addition, any management process assumes the existence of feedbacks in the system that provide the necessary correction of the process.

In determining the approaches to the solution of this problem, we relied on the studies of contemporary Russian [4, 6, 7] and foreign [12–14] authors whose main idea was the interpolation of modern management methods in the educational process in the system of general and vocational education.

We propose a model for managing the independent work of trainees (Fig. 1).

The problem of managing the independent work of students is super disciplinary, which makes it possible to set a *goal - the formation of strategies for self-study and self-education as the basis for future professional activity, which is reflected in the educational and professional standard of the teacher.*

The conditions for the implementation of the model were:

– Presence of normative and methodological support: Provisions on the organization of independent work of students in the university; Methodical manual for teachers [8].
– Use of the electronic educational environment providing access to educational resources.
– Development and creation of a new design of the educational environment integrating the auditorium and recreational space for independent work, the possibility of collective, group and individual activities of students, both under the guidance of the teacher and independently [9].
– Involvement of all necessary services of the University in the process of managing the independent work of students. In addition, the creation of the institute of tutoring.

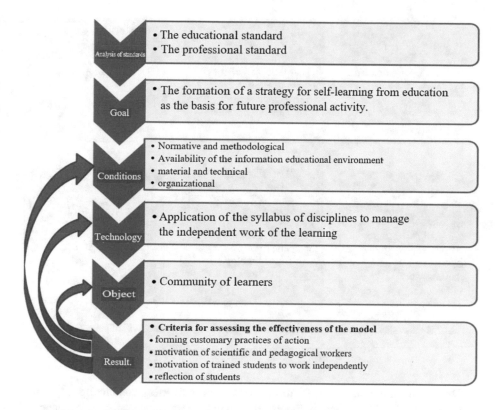

The educational standard
The professional standard

The formation of a strategy for self-learning from education as the basis for future professional activity.

Normative and methodological
Availability of the information educational environment
material and technical
organizational

Application of the syllabus of disciplines to manage the independent work of the learning

Community of learners

Criteria for assessing the effectiveness of the model
forming customary practices of action
motivation of scientific and pedagogical workers
motivation of trained students to work independently
reflection of students

Fig. 1. Structure of the management model of independent work of students

3 Methodology of the Study

During the research, we used theoretical and practical methods: Analysis of literature on the topic of research, modeling, pedagogical experiment, testing, questioning, observation, methods of mathematical statistics.

The syllabus of discipline is applied as the main technology for the implementation of the model - this is a "plan of action" for students, which can be a kind of outline-outline of the academic subject aimed at helping in the organization of educational activities [3].

Let us represent the structural elements of the syllabus (Fig. 2).

The choice of forms and methods of training in the implementation of the management system of independent work of trainees was based on:

- Analysis of the requirements of educational and professional standards to manage the independent work of students;
- Studying the forms and methods of managing independent work, proposed by teachers and implemented in the course of the project;

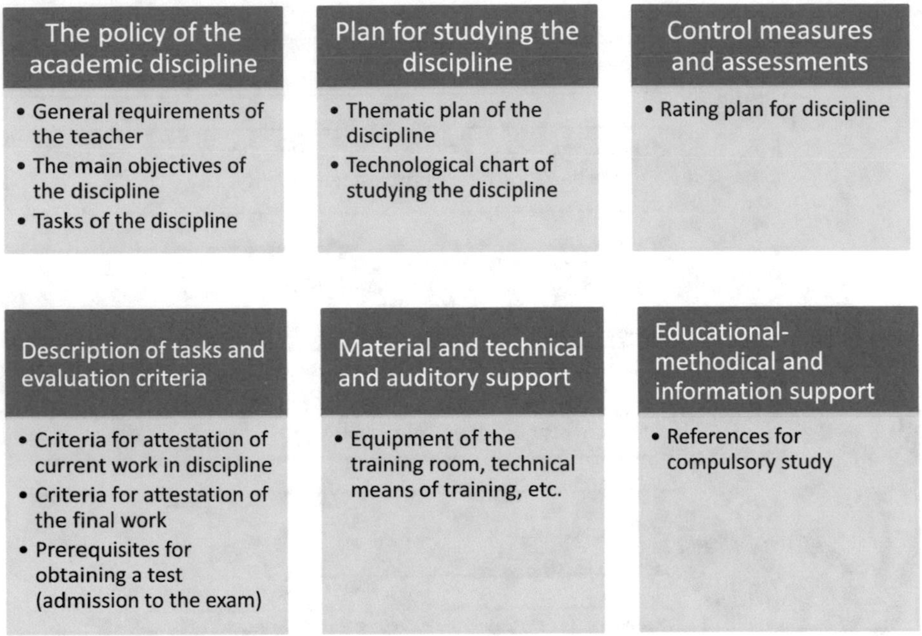

The policy of the academic discipline	Plan for studying the discipline	Control measures and assessments
• General requirements of the teacher • The main objectives of the discipline • Tasks of the discipline	• Thematic plan of the discipline • Technological chart of studying the discipline	• Rating plan for discipline

Description of tasks and evaluation criteria	Material and technical and auditory support	Educational-methodical and information support
• Criteria for attestation of current work in discipline • Criteria for attestation of the final work • Prerequisites for obtaining a test (admission to the exam)	• Equipment of the training room, technical means of training, etc.	• References for compulsory study

Fig. 2. The main sections of the syllabus

During the implementation of the project, a new form of conducting classroom activities was offered - self-study independent work - performed under the guidance of the teacher and on his instructions or under the guidance of the tutor when they were present in the classroom. The time devoted to independent work amounted to 30% of practical (laboratory, seminar) classes [11]. Among the most suitable methods for conducting self-study by teachers are:

- project training,
- Informational-interactive training,
- information and communication technologies,
- case study,
- content analysis,
- game simulation methods,
- hermeneutics of the text etc.

The creation of a system for managing the independent work of students required the creation of an institute of tutors. When organizing the system of work of the tutor, they relied on a number of principles:

1. Openness is the management of students with their own cognitive and educational activities.
2. Continuity - ensuring a timely, consistent, cyclical process in the development of cognitive and interest of students.

3. Flexibility - the orientation of students to expand social contacts, support initiatives in choosing ways of doing things.
4. Individualization - the account of personal inquiries, individual features and interests of students.

The choice of forms of tutoring was directly related to the courses taught and had academic or educational and research character. Directly the forms of independent work of the students were determined by the departments (teachers of the department) when developing the working programs of disciplines (modules) and syllabuses of disciplines (modules). Particular attention was paid to individual work with students on the formation of positive motivation.

The analysis of the results of the work carried out during the implementation of the project line "Managing student's independent work" made it possible to draw the following conclusions.

Tutorship activity was carried out on various scales. The vectors of the tutor's activity were [2]:

1. Social - the adaptation of the student in the university and the educational space of the university and the society in general, moral mentoring, the resolution of conflict situations.
2. Anthropological - mastering of techniques and technologies of development of personal qualities necessary in education, including self-education.
3. Cultural-objective - the disclosure of the potential of academic disciplines, the choice of courses, the back of independent work, supervising independent self-employed work; Work with small groups of students performing specific (group) assignments; Development of recommendations for effective learning using information and communication technologies.

In addition, it should be noted such a direction of the tutor, as the accompaniment of the teaching staff.

Within the framework of this direction, the tutor can perform the following functions:

1. Consulting work to manage the independent work of students.
2. Consultative work on the individualization of student learning.
3. Providing feedback between the learners and teachers.
4. Organization and solution of technical issues arising from the teacher when organizing independent and individual work with students.
5. Preparation for conducting full-time and virtual educational events.
6. Control, duplication and distribution of documents related to the management of independent or individual work of students.

Summing up the work done at this stage, it can be noted that the spectrum of the tutor's responsibilities is very wide. It should be remembered that the main task of the tutor would be the formation of an individual educational program for students. In addition, it should be noted that such a direction of the tutor's activity as a tutor support of the teaching staff.

One of the tasks of the project was the organization of professional reflexive activity of students. Professional self-awareness is built based on reflection of the experience gained in the process of teaching and activity, therefore the development of reflection of student's learning activities is currently a key task of building the process of training future professionals [1].

Reflexion (from the late reflexio - reversal) is the comprehension by the subject of himself, the content of his personality (values, interests, motives, emotions, and deeds), his knowledge and states [5].

The effectiveness of the process of reflexive self-management is impossible without the most important reflexive skill: Conscious goal setting. Professional self-awareness requires purposeful development. The ability to set conscious smart goals both in terms of their activities and in terms of organizing the reflection of this activity is an important task already at the stage of training. It is this ability of conscious goal setting in the reflection of activities. Ensures the basic readiness of the graduate to the profession makes him free in choosing his professional path and be effective. Undoubtedly, the development of the capacity for reflexive activity must be in a harmonious combination of actual practical experience. A.K. Markova defines the formation of the capacity for goal setting as a key element in the development of the professionalism of a specialist in subject activities.

Proceeding from this, the organization of reflexive activity of students should be based on the mechanisms of conscious goal setting, which was realized in the framework of approbation of the project "Creation of a system for managing the independent work of students". The organization of reflexive activity of students included the following goals and objectives:

The goal is the formation of the reflective thinking of the student as the basis for conscious self-regulation of the process of forming psychological readiness for professional activity.

As a tool for the reflexive activity of students, a group reflection sheet and an individual reflection journal were developed and criteria for evaluating reflexive activity were singled out:

- Awareness and ability to formulate and analyze goals (their own, disciplines, modules),
- Ability to determine and evaluate their activity and contribution to the development of discipline (module),
- The ability to highlight key events, ideas, people and assess the extent and mechanism of their impact on one's own professional development,
- The ability to analyze what experience has been obtained and to identify needs that are not yet satisfied with professional activities,
- Ability to analyze failures and their causes, determine potential ways to overcome failure.

At the preliminary stage of assessing the students themselves and assessing the teachers, the first two criteria had the greatest number of difficulties, they, in our view; determine the priority areas for further work on organizing and teaching reflexive students.

A qualitative analysis of the results of the control segment of student's reflection on the ability to understand and analyze goals has shown that students see the goals of teachers most often in the context of informing and creating the conditions for mastering the disciplines of the module, and their own in gaining knowledge and new information. Thus, one of the main problems in the reflexive activity of students, affecting the effectiveness of mastering the module and acquiring competencies, is an insufficiently conscious goal setting. In connection with this, the urgent task is to actively use work methods aimed at increasing awareness of one's own goals. Mastering the methods of correction, including from the point of view of goals, his personal professional plan, the ability to plan tasks for its implementation, the definition of his personal goals in the process of mastering each module, discipline in the professional educational program contributes to the expansion of prospects and options for professional development.

To overcome the difficulties encountered, webinars and master classes with the participation of students and teachers "The Art of Reflexive Questions in Awareness of Professional Activities" were held. Because of the content analysis, it can be concluded that in the groups of models of management of independent work of students:

- Students higher evaluate their skills of reflective activity,
- Goals are prescribed more specifically,
- The formulation of goals coincides as a whole for the group,
- They give more specific answers, noting their contribution to training,
- They give more diverse answers, listing various forms of activities that they implemented on their own initiative in order to master the disciplines.

Thus, the work on the development of reflection using special technologies (Diary of Reflection, a sheet of group reflection), based on the creation of a system for managing the independent work of students, favors greater awareness of professional activity, a clearer understanding of the goals and position in the process of mastering the materials of various disciplines.

4 Analysis of the Results of the Study

To assess the pedagogical effectiveness of the model in general, we have defined the following criteria:

- Level of formation of general educational activities;
- The level of mastering the content of educational disciplines;
- Motivation of scientific and pedagogical workers to manage self-study
- of students;
- Motivation of students to realize independent work.

We determined the following types of general educational activities formed during the implementation of the model:

- organizational - the ability to set goals and objectives for the implementation of independent work; Plan independent work and carry it out in a timely manner;

Organize the necessary conditions in the workplace; Collaborate in solving learning problems, perform tasks both individually and in a group; To exercise self-control and self-analysis of educational activities;

- information - the ability to work with sources of educational information, to search for and use reference and additional sources; Own at the required level of information and communication technologies; To conduct selection and grouping of information on given topics; Make plans of various types and create texts of various types; Perform reproductive tasks using methodological materials;
- educational and intellectual actions - the ability to independently analyze, synthesize, generalize information and other types of intellectual operations; Independently choose tools and methods for solving problems, ways of doing the work; Give a detailed answer and justify it, argue their position; To carry out independent work, carried out in the form of educational-research and project assignments, which allow obtaining new information; To present the results of independent work.

Generalized levels of the formation of general educational activities can be as follows:

0th level - the trainee does not own this action at all or did not start the action;

1 st level - inability to perform an independent action;

2nd level - the student is sufficiently free to perform the action when consulting and correcting the actions of the teacher or tutor;

3rd level - independent performance of the action, an objective assessment of their own achievements.

The experimental data are shown in Fig. 3.

The data obtained indicate a significant increase in the number of students who reached the second and third levels and the reduction of students at the zero and the first levels.

The level of mastery of the content of the academic disciplines was evaluated based on the results of the intermediate certification of students in accordance with the rating system used at the university. Figure 4 shows the average score of students on the basis of the exam and examination session.

The data obtained demonstrate a higher average score for students participating in the experiment.

Motivation of students to perform independent work and motivation of scientific and pedagogical workers to manage the independent work of trainees was evaluated by questioning. Summarizing the obtained data, it should be noted:

- Increasing the level of importance of independent work for the organization of the educational process;
- Raising the level of information and methodological support of disciplines and material and technical support of the educational process in the implementation of the model for managing the independent work of students;
- Preference for the use of group forms of organization of independent work;
- Wide use of the information educational environment.

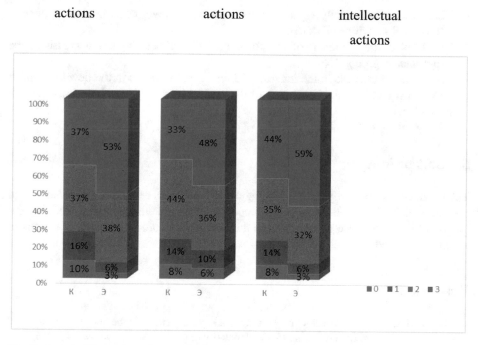

Fig. 3. Level of formation of general educational activities of students of the experimental and control groups.

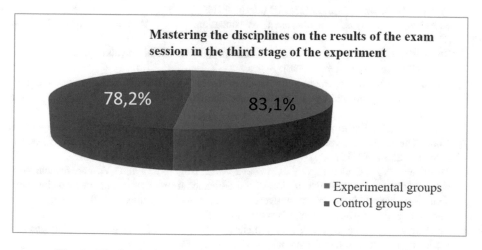

Fig. 4. The level of development of the content of educational disciplines.

In this way:

- There was a statistically significant increase in the level of formation of general educational activities of students;
- The level of development of the content of disciplines did not decrease in the experimental groups;
- There was an increase in the interest of trainees and teachers in the development of new forms and the implementation of the project to manage the independent work of students.

5 Conclusions

The obtained results testify to the pedagogical effectiveness of the proposed model and the possibility of its replication in the preparation of students, as well as the dissemination of the experience gained through the system of improving the qualifications of scientific and pedagogical workers.

References

1. Bogorodskaya, O.V.: Conscious goal setting as a key competence in the reflection of students' learning activities. Bulletin of the Minin University of Nizhny Novgorod, vol. 4 (2016)
2. Golubeva, O.V., Shlyakhov, M.Y.: The role of the tutor in the management of independent work of students. Bulletin of the University of Minin, vol. 4 (2016). http://vestnik.mininuniver.ru/
3. Gruzdeva, M.L., Tukenova, N.I.: Sillabus as a means of organizing independent work of students. Bull. Univ. Minin 1(13), 19 (2016)
4. Korotkov, E.M.: Quality management of education, p. 320. Academic project, Moscow (2007)
5. Karpenko, L.A., Petrovsky, A.V., Yaroshevsky, M.G.: Brief Psychological Dictionary. PHOENIX, Rostov-on-Don (1998)
6. Morozova, N.V.: Management of non-audit independent work of students in the conditions of the rating system. Publishing House Kazanskiy Publishing House, Kazan, pp. 324–326 (2010)
7. Solostina, T.A.: Ensuring the quality of education of students by means of independent work. Thesis for a Candidate Degree in Candidate of Pedagogical Sciences, 13.00.01, Moscow, 250 p. (2015)
8. Tolsteneva, A.A., Golubeva, O.V., Gruzdeva, M.L., Bogorodskaya, O.V., Perevoshchikova, E.N., Smirnova, Z.V.: Modernization of the educational process: management of independent work of students in the university: educational and methodical manual, 72 p. The University of Minin, Nizhny Novgorod (2016)
9. Tolsteneva A.A., Smirnova Z.V., Guryanicheva E.N.: Model of management of independent work of students: results of approbation. Bull. Univ. Minin, 4 (2016)
10. Shkunova, A.A., Pleshanov, K.A.: Experience in modeling the syllabus of independent work of students. Bull. Univ. Minin 3(16), 8 (2016)

11. Smirnova, Z.V., Gruzdeva, M.L., Chaykina, Z.V., Terekhina, O.S., Tolsteneva, A.A., Frolova, N.H.: The role of students' classroom independent work in higher educational institutions. Indian J. Sci. Technol **9**(22), 95568 (2016)

12. Kezar, A.: Understanding and Facilitating Change in Higher Education in the 21st Century. ERIC Digest. George Washington University, Washington DC (2001). http://www.ericdigests.org/2002-2/21st.htm

13. Zeitlinger, B.V.: Managing schools towards high performance: linking school management theory to the school effectiveness knowledge base. In: Visscher, A.J. (ed.) 354 p. Swetz, Lisse, The Netherlands (1999)

14. Mendenhall, R.W.: Technology: creating new models in higher education. Western Governors University, Salt Lake City, Utah. http://www.nga.org/Files/pdf/HIGHEREDTECH.pdf

Application of Multi-level Analysis in the Process of Advertising Industry Trends Study

Valery A. Borodin$^{(\boxtimes)}$, Sergey B. Prianichnikov, Anna I. Galushkina,
Elena A. Nagaeva, and Sergey V. Ustinkin

Nizhny Novgorod State Linguistic University Named After N.A. Dobrolyubov,
Nizhny Novgorod, Russian Federation
borodin@lunn.ru, pryanichnikovsb2013@yandex.ru,
galuanna@yandex.ru, lena.nagaeva@mail.ru,
sv.ustinkin@gmail.com

Abstract. The long-term steady development of enterprises under the contemporary economic conditions is to a considerable extent determined by the possibility of output not only for the domestic advertising market, but also on the world. The today universal market for advertisement rapidly is developed, moreover mainly, due to the countries of Asia and Latin America. Whereas the Russian market for advertisement lags behind the world leaders, and its portion remains small. Like many of market, the market for advertisement is globalized, and explosion the Internet advertisement makes possible to leave to this segment of market for companies practically from any country. In the article is given the analysis of the trends of development of advertising market in the world, whose understanding will allow Russian companies to develop their strategies of output to the markets for other countries, which will, in turn, be important competitive advantage and will increase the competitive ability of these companies. With the estimation of the effectiveness of the solutions about the output for one or other advertising market or another it is necessary to consider different criteria and indices of micro and macro of the media, which sometimes can contradict each other. Therefore as the basic method in the article it is proposed to use multilevel analysis.

Keywords: Advertising industry · The advertising markets
Advertising BTL and ATL markets · Marketing communications
Media agencies · The tendency of regional advertising markets

1 Introduction

The formation of the sustainable economic development of the Russian enterprises and organizations is largely determined by the ability to sell their products, where the large role is played by formed system of marketing communications and development of the advertising industry. In this connection, the analysis of the status and possible trends in the advertising industry is an important direction of promotion on the markets of domestic goods and services as well as enhancing the competitiveness of Russian enterprises.

2   Theoretical Foundations of the Research

Advertising industry is an industry that defines advertising as a product of economic activity providing public demand for advertising services. The development and transformation of promotional activities led to a special social institution, which performs an important function in ensuring the public needs for advertising services. Organization of the advertising industry is aimed at the development of specialization and the formation of a variety of advertising organizations producing complex promotional product and/or rendering only advertising services, as well as involvement in the advertising process of different economic actors.

Among foreign experts who contributed to the advertising industry, to the creation of conceptual bases for analyzing the advertising market in the world, one can mention S. Zu, S.T. Kavusgil, J. Volz, D. Griffith, G. Yalchinkaya, D. Yugenheim, L.D. Kelly, J. Hudson, S. Bradley, etc. In particular, Zu and Kavusgil developed the theory of global marketing strategy (GMS), which justifies the need for standardization of marketing strategies implemented in different countries by transnational corporations [10]. J. Volz considered the possibility of applying GMS theory to the analysis of advertising campaigns carried out by TNCs on world markets. Griffith and Yalchinkaya suggest that the firm's activities in the global advertising market should be viewed from the standpoint of resource theory, within which each resource is assessed in terms of its contribution to the firm's overall competitiveness [8]. Jugenheim et al. provides a description of the various tools and resources that can be used to analyze the advertising industry in the world [9].

The state of the advertising industry can be examined on the basis of a comparative analysis of the advertising market in developed countries, the volume and dynamics estimation of the advertising market in Russia as a whole as well as analysis of the advertising industry Russian metropolises. Nowadays, well-known analytical and consulting companies, media agencies analyze the global advertising industry [1, 2].

Analytical materials provided by Russ Outdoor, Zenith Optimedia (ZO), Carat Russ Media, Ad Age, GroupM, PricewaterhouseCoopers, TNS Media Intelligence Initiative, ACAR (Association of Communication agencies of Russia), JeSPAR analyst, data of annual financial reports of leading world OOH-operators, official statistics Rosstat are considered to be of interest.

It is important to note that every company has its own approach to prepare analytical materials, for example, the company Russ Outdoor estimating media market and outdoor advertising industry Russia additionally considers the costs to print because the poster in outdoor advertising is a means of distributing content. However, data of ESPAR-Analytic Company and Russ Outdoor Company are of interest for calculations of the outdoor advertising industry volume.

3 Methodology of Research

Modern methodological tools research trends in the advertising industry is character-ized by the use of wide range of different economic, sociological and even psycho-logical methods. However, while the external environment is being transformed, various factors influence the development of the advertising industry. Thus, it is important to examine the level of demand, to study the market capacity as well as to determine the prospects for its development. Therefore, a multileveled analysis of all aspects of the functioning of this market should be applied.

In the last decade, there has been a trend of constant growth of total global advertising spending. Situation on the global advertising market remained relatively stable: global spending in traditional media: television, radio, press, outdoor adver-tising, cinema, the Internet grew. This trend should persist for the near term, however, the development of the advertising market will largely depend on the overall economic situation in the world.

4 Analysis of Research Results

Starting from 2009, there has been a steady growth of total global advertising expenditure, and for the period 2014–2016 the following eight countries contributed to the growth of the advertising industry: the USA - $20,945 million, China - $12,990 million, the United Kingdom - $5,080 million, Japan - $3,138 million, India - $2,706 million, Indonesia - $2,475 million, Brazil - $2,470 million, and Argentina - $2,077 million. Additionally, according to analysts, the average annual growth of global advertising expenditure for the same period amounted to 0.6% in Eastern and Central Europe, 2.5% in Japan, 3.2% in Central Europe, 3.7% in North America, 6.9% in Latin America and 9.1% in Asian developing countries.

The developing countries in Asia and Latin America showed the greatest growth and they rightly expect to reduce the economic gap with developed countries through increased promotional activities.

In 2014–2016, the global investment growth in advertising amounted to the fol-lowing: 4.7% in 2014, 4.2% in 2015, 5.0% in 2016 and 4.5% in 2017 (forecast). It is characteristic that GDP grew proportionally: 5.0% in 2014, 4.7% in 2015, 5.9% in 2016, 6.1% in 2017 (forecast). Thus, there is a direct correlation of GDP growth and increased investment in advertising.

The analysis of the proportion of expenditure on the world market media showed that advertising on TV is still leading (with declining shares). Comparative data are as follows:

1. TV 39.1% (2014) and 36.8% (2016);
2. Internet 5.1% (2014) and 12.9% (2016);
3. Newspaper 14.9% (2014) and 11.7% (2016);
4. Magazines 7.4% (2014) and 5.9% (2016);
5. Cinema 0.5% (2014) and 0.5% (2016);

6. Radio 6.8% (2014) and 6.2% (2016);
7. Outdoor advertising 6.8% (2014) and 6.2% (2016).

The trend is the growth of Internet advertising and the decline in the share of other advertising media. The main reason is the large number of audience and the relative cheapness of Internet advertising. Thus, during the last decade, the situation in the global advertising market remained positive: the costs of the most common media (TV, radio, press, outdoor advertising, cinema, Internet) grow.

However, the development of the advertising market significantly depends on the global economic situation. So Zenith Optimedia forecasts that in the coming years global advertising costs are expected to increase and in general, this forecast is close to estimates of other leading media agencies.

However, a number of experts reduced the forecast of the world advertising market development in the coming years and this is due, above all, to unstable Euro zone economies and the existence of economic sanctions. Therefore, in a number of European countries, there is a tendency to reduce advertising costs, while the situation in the euro area had an impact on the growth forecast in the entire European region. It is reduced in Western Europe and Central and Eastern Europe. The same tendency is observed in the Asia-Pacific region and Latin America.

It is safe to say that the main world advertising market growth should be provided by emerging markets (all markets that are on the level of advertising expenditures below North America, Western Europe and Japan). According to experts, their contribution to the growth of the global market will make up more than 60%, thus, BRIC countries (China, Brazil, Russia, India) will make a significant global growth of the advertising market.

Today, the list of largest national advertising markets is headed by the United States (about 30%), Japan (10–11%), China (6–7%) Germany (5–6%), United Kingdom (over 4%), Brazil (about 3.5%).

The highest dynamics of growth of expenditure on advertising, according to Zenith Optimedia is observed in Latin America, the smallest one - in North America. In the Middle East and Africa, a significant drop in the advertising market is observed, which is associated with unstable political and economic situation in a number of countries in the region, and Russia's share of the global advertising market is about 2.0%. However, there will be very characteristic changes in the largest advertising powers list in the coming years. China is closer to the advertising budget of Japan, Brazil will join the five largest advertising powers in the world, and Russia must approach Canada.

A large part of the costs among all media traditionally falls on TV (about 40%) and the press (about 30%). According to Zenith Optimedia, the proportion of the world's outdoor advertising segment will remain stable in coming years and will be at the level 6–7% with the trend of gradual reduction. The main small reallocation of budgets will occur among segments and Internet use.

The state of the advertising industry in Russia can be summarized as follows [3–7]. The amount of the advertising market in Russia in the past 10 years has a tendency of growth; however, the share of the advertising market in Russia is still relatively small. According to experts, in recent years 2013 was the best year for the advertising market in Russia in terms of volumetric indicators (dollars). In 2014, there was a decline in total advertising market (up to RUB 340 billion). The part of the advertising market in media was estimated at 8.5 billion US dollars, that is 17.5% less than 2013 estimates. This decrease was due to the average annual ruble exchange rate vis-à-vis the United States dollar, which decreased in 20% in 2014 compared to 2013.

Nowadays, the television advertising remains the largest segment of the advertising market in media. ACAR experts evaluated that advertising on television amounted to RUB 160 billion in 2014 that is by 2% higher than in 2013. This in view of the fact that "media inflation" (the rise of instrumentation cost allocation) according to experts ranged from 7 to 12 per cent in 2014. There has been a downward trend in the proportion of television advertising by an average of 5–9%, it is important to note that the share of the Russian television advertising accounts for 2% of the world total television advertising. However, the comparative analysis should take into account that Russia's population is about 2% of the world population and GDP in the last decade estimated at 2.5–3.5% of global GDP.

As in the global advertising market, Internet takes the second place in Russia as for advertising. For example, in 2014, ACAR experts appreciated that the volume of Internet advertising reached RUB 84–85 billion. The placement of the banner ("display") Internet advertising fell by 5% to RUB 19 billion, and contextual advertising grew by 27% to RUB 65–66 billion. Thus, the share of the Russian Internet advertising in the global advertising market is 1.8%–2% of global spending on Internet advertising.

As far as outdoor advertising, its share in the Russian market does not change substantially for a long time and is estimated by experts at RUB 40–41 billion. Stagnation in the market of outdoor advertising is due to its excessively high share in the total volume of media in the middle of the last decade. So in 2005–2008 outdoor advertising share exceed 16% of the total volume of advertising in the media, however, this situation is explained by the relatively liberal approach of regulators to the placement of outdoor advertising. After 2009 the requirements of urban authorities (especially in large cities: Moscow, St. Petersburg, Ekaterinburg, Novosibirsk, Kazan, etc.) to this kind of advertising has noticeably increased. But in 2014, the year the share of outdoor advertising has fallen to only 12% of total advertising in media, almost twice the world average.

In the past 5–6 years in Russia there was a decrease in the volume of advertising in newspapers and magazines, while in value terms, the press has lost about 11%. For example, compared to 2008 (RUB 57 billion) press advertising market fell by more than 40% in 2014, and its share of total advertising in the media dropped more than twice from 21.5% in 2008 to 9.7% in 2014. For example, in 2004, the share of advertising in the press was more than 30% of the total volume of Russian advertising in the media.

Russian press advertising is only 0.7%–0.8% of costs (income) of advertising in the press in the world advertising market. Consumer interest to newspapers and magazines falls down a clear trend. It is partially the fault of magazines and newspapers publishers, who inflate costs, make a bad choice of advertising location.

Analytical materials published by the international company Zenith Optimedia, show that the proportion of press advertisement in worldwide advertising costs decreased in the last decade from 42.7% to 22.5%, though these 22.5% are more than twice Russian 9.7%. For comparison, the proportion of the press advertising expenditure (income) in the United States is 16% of national advertising costs.

The average annual amount of advertising on the radio in the past three years is estimated at RUB 17–18 billion, while the share of radio ads over the last 10 years is held stable at around 5% of the total advertising market. The world market of radio advertising has a similar situation.

Currently, the global advertising market estimated at more than 500 billion dollars. Therefore, we cannot allow the decline in the share of the Russian media of advertising product up to 1% of the world as according to experts it could lead to substantial losses in Russian advertising market and move Russia to the periphery of the world advertising industry. It is appropriate to point out that even in good years (2012–2013) the Russian market of advertising failed to enter the top ten most powerful advertising markets in the world.

Especially it is necessary to assess the condition of the Russian BTL market, because this type of advertising product allows you to interest unobtrusively the audience while the purchaser shall decide alone on buying, and nobody and nothing "pressure" him or her.

BTL includes direct marketing; events and contests; POS materials; stimulation of selling from dealers and buyers. BTL appeared in the middle of the 20th century and today in Russia this way of advertising has about 24% of the advertising budget, while in Western countries its share amounts to more than 50%. We can point out the main advantages BTL advertising: it is not obtrusive; less expensive, television with ATL oversaturated with advertising, to circumvent prohibitions on certain kinds of advertising (alcohol, tobacco).

Currently, major advertising agencies, marketing communications agencies become media agency and work closely on BTL advertising. While advertising agencies open host departments accommodation, media planning, design studios. According to ACAR, 20% of the largest media agencies have 60% of advertising in the media; the 20 largest BTL-agencies have less than 20% of the total BTL market.

There is a tendency of BTL market growth in Russia. In 2013, this market amounted to RUB 91 billion, or $2.85 billion at an average 2013 rate. In 2014, the BTL market grew approximately 10% up to RUB 100 billion. Compared to the market of advertising in media, the higher BTL growth is explained by the Sochi Olympics. In the first quarter of 2014, the BTL budget significantly increased due to the Olympics and participation of dozens of major Russian companies-producers of consumer goods and services as sponsors. For example, the BTL market growth is estimated at 20–25% compared to the previous year. However, the absence of similar events in 2015 changed the stats. The BTL market decline in 2015 compared to the previous year is due to several factors:

1. lack of comparable events as the Sochi Olympic Games;
2. reduced rates for accommodation in the media made media advertising relatively more affordable for large and medium-sized advertisers and they're cutting budgets in general, redistribute them in favor of the media, especially the Internet;

3. restrictions on "marketing services" provided by suppliers of retailers (introduced in 2013 the new edition of the law "On trade") have led to a reduction in supply of these services and have a negative impact on the dynamics of the BTL market.

After BTL were included in the advertising market, marketing services include:

(a) marketing research services;
(b) marketing consulting, consulting on marketing strategies, product and assortment policy, branding;
(c) public relations-business PR.

5 Conclusions

Thus, it may be noted that the modern market of the advertising industry is characterized by major trends:

- globalization, i.e. it serves an integral part of the global economy;
- integration, i.e. combining its economic actors as well as deepen their interaction and linkages between them;
- concentration, i.e. strengthening the processes of mergers and acquisitions market players;
- networking, i.e. the development of a global network of advertising agencies;
- diversification of the activities of the actors of the market, i.e. the output of their work beyond the core business.

Along with the Russian market, the advertising industry is characterized by a significant time lag from world leaders. The development of the Russian market of advertising industry is influenced by government regulation, in particular by the negative impact of multiple restrictions at both Federal and local levels.

References

1. Review of entertainment industry and the media. Encyclopedia of marketing. http://www.marketing.spb/travel/entertal. Accessed 07 June 2017
2. Review of entertainment industry and the media: the prediction for 2016–2020 years. PWC. http://www.pwc.ru/assets/(e)-media-ou. Accessed 07 June 2017
3. Vartanov, S.A.: The dynamics of media industry of Russia in 2000–2014: General trends and correlation with macroeconomic elements. Electron. Sci. J. "Mediaskop". http://www.medioscop.ru/183/. Accessed 10 June 2017
4. Advertising in printed media: status and prospects.http://www.adindex.ru/analiticx. Accessed 10 June 2017
5. Rating federal providers of UN-and Indoor-advertising-2017-Adindex. http://www.adindex.ru/analiticx. Accessed 12 June 2017
6. Russian advertising Yearbook 2015-the Association of communication agencies of Russia. http://www.akarussia.ru/download/rre15. Accessed 10 June 2017
7. Review of the mobile advertising market in Runet 2015-IAB Rossia. http://www.labrus.ru/files/study. Accessed 14 June 2017

8. Griffit, D., Yalcinkaya, G.: Resource-advantage theory: a foundation for new insights into global advertising research. Int. J. Advert. **29**(1), 15–36 (2010)
9. Jugenheimer, D.W., Kelley, L.D., Hudson, J., Bradley, S.: Advertising and Public Relations Research. Routledge, New York (2014)
10. Zou, S., Cavusgil, S.T.: The GMS: a broad conceptualization of global marketing strategy and its effect on firm performance. J. Mark. **66**(4), 40–56 (2002)
11. Zou, S., Volz, Y.Z.: An integrated theory of global advertising. An application of the GMS theory. Int. J. Advert. **29**(1), 56–84 (2010)

Preparation of Bachelors of Professional Training Using MOODLE

Marina N. Bulaeva[✉], Olga I. Vaganova, Margarita I. Koldina,
Anna V. Lapshova, and Anna V. Khizhnyi

Nizhny Novgorod State Pedagogical University Named After Kozma Minin
(Minin University), Nizhny Novgorod, Russian Federation
bulaevamarina@mail.ru, vaganova_o@rambler.ru,
ritius@mail.ru, any19.10@mail.ru, xannann@yandex.ru

Abstract. The object of the research in this article is Moodle – a system for distant learning. At the legislative level, the need to use this type of education in the educational activities of vocational schools is fixed. The electronic medium is used for blended learning, and is a tool with full set of resources for online courses. The electronic environment of Moodle is characterized by modularity, special flexibility in managing the learning process, easy publishing of training materials and their support in the format of international standards, management of user groups, the use of Web 2.0 services and the ability to integrate with other web applications. The article presents a study on the possibilities and the analysis of the experience of the use of distance learning for students in an electronic environment Moodle applied in the Pedagogical University. The experience of the implementation of e-learning courses on the example of "General and vocational pedagogy". The experience of creating electronic training courses for their implementation proves the promise of this direction. Its introduction of the educational process of professional educational institutions will improve the quality of training specialists. The dynamism of the platform management and the modular structure of the training makes it possible to create the organization of the educational process, taking into account the individual requirements of the students to contribute to improving the learning outcomes of students.

Keywords: E-learning · Bachelor of professional training
Information technology · E-learning · Mixed education · Distance learning
Educational technologies

1 Introduction

To improve the quality of the educational process in vocational schools e-learning is being promoted. Requirements for the use of various educational technologies, including e-learning, are enshrined in the Federal Law the Russian Federation dated by December 29, 2012 No. 2733-FZ "On Education in the Russian Federation." At the legislative level the need for such kind of educational facilities is fixed in vocational schools.

© Springer International Publishing AG, part of Springer Nature 2018
E. G. Popkova (Ed.): HOSMC 2017, AISC 622, pp. 406–411, 2018.
https://doi.org/10.1007/978-3-319-75383-6_52

Electronic training is an important component of the educational process in vocational schools, provides ample opportunities, access to educational resources and management to a new level; thus significantly increases the opportunities of the education system [6].

The urgency of e-Learning technologies usage is defined by the following factors: the introduction of new federal education standards that focus on the implementation of competence-based approach; increase of independent work of students; implementation of the principle of "learning throughout life"; Freedom in the choice of the place of study by the entrant; IT implementation in education; the rapid development of information and communication technologies; promoting new opportunities pits in the educational process.

2 Theoretical Bases of Research

E-learning research takes both domestic researchers (it is worth noting the work M.Y.U. Bukharkin, M. Moses, E.S. Polat [12] Robert I. [11]), and foreign (M. Barber [13], M. Rosenberg, E. Masie, T. Anderson, E. Hanushek [14]). The study of the practice of using Moodle environment in the educational process is presented in D.S. Kostylev and [5] E.K. Samerkhanova, W. Rice, H. Foster, J. Cole, R. Jirmann. The implementation of the requirements of the competency approach to the preparation of teachers is, presented in the works O.V. Akulovoh, V.A. Adolf, V.A. Bodrov. I.S. Batrakovoh, G.A. Bordovsk, E.V. Baranovoh A.K. Markovoh, N.F. Radionovoh N.N. Surtaevohoh A.P. Tryapitsyn, Z.I. Kolychev and N.V. Chekalev and etc.

3 Research Methodology

Methodological basis of research supports the competence and modular approach to studies. The approach focuses attention on educational results, requires the ability to solve professional problematic and non-standard situations [15]. The world educational practice separated educational practices and highlights the concept of "competence" as the concept of competence that allows combining intellectual skills and single formation. Secondly, it reflects the idea of designing the content of education, based on learning outcomes. Thirdly, the key competences integrate closely related skills and knowledge [1]. The electronic environment Moodle allows realizing in practice the principles of content-competence and practice-oriented preparing when creating electronic educational complex discipline [9].

Moodle is characterized by modularity, increased flexibility in the management of the educational process, easiness of training materials and publications of their support in international standards format, management of user groups, the use of Web 2.0 services, and the ability to integrate with other web - applications. All of these factors determine its effectiveness [8].

It should be noted that the main objective of the project on creation of Moodle is to provide effective tools for managing learning process. In this case, Moodle has the ability to scale, that is, may increase the number of students to a few hundred or a thousand, and can be used in elementary school or for individual self-study. Most often Moodle is used as a platform for online courses – to provide blended learning [2].

4   Analysis of the Research Results

Now let us consider the aspects of the vocational training (on branches) on the subject "General and vocational pedagogy". The course is available for registered users, such as university students, or other users who passed a special registration.

E-learning course "General and vocational pedagogy" is designed to provide vocational training to bachelors. The educational material is presented in to be studied at an individual pace, a sufficient number of internal and external links have been created that allow you to create efficient and quick access to necessary information. The content of the course implements the requirements of a rating system and activity-oriented, personal-oriented approach to bachelor's education. The course is located in the official website Minin University in the section "Distance Courses" http://moodle.mininuniver.ru/course/view.php?id=898.

This course is a combination of a clear logic of the discipline, a balanced theoretical material and practical-oriented jobs that allow you to identify the level of formation of professional knowledge and students' skills. The course contains necessary examples of material to support theoretical material.

The distance learning course "General and professional pedagogy" was developed on the basis of the principle of interaction at a remote distance between a student and a tutor (teacher). The educational material is presented in the course fully and clearly presented on the course for the convenience of self-study at an individual pace, a sufficient number of internal and external links have been created that allow you to create efficient and quick access to the necessary information.

Electronic content of "General and vocational pedagogy" is built on a modular basis, each module is a complete system, complete fragment having its didactic tasks and direction in the formation of students' professional competencies and their applications. Practical activities are included [3]. A student is provided with a set of electronic theoretical training and reference material such as instructions and etc.

The structure of the developed course of "General and vocational pedagogy" includes 5 modules

Introductory module (news forum; abstract; Instructions for students to study the discipline; Educational and methodological support of the discipline; Glossary);

Module 1. Professional pedagogy as a branch of pedagogical knowledge;
Module 2. Theoretical bases of vocational training of workers;
Module 3. Characteristics of a holistic education process in a vocational school;
Module 4. Subjects of educational process; Materials for certification.

Each module is filled with information resources and interactive elements - theoretical materials, lectures, presentations to lectures, practical tasks and tasks, tests, links to Internet resources.

Reflection planning on the learning procedure of the discipline includes the training sessions of all types and control measures [15].

The course provides guidance on the types and forms of activity that are in the e-course - for practical work, for self-study. They include instruction organization of learning and assessment criteria. Recommendations in the key are: advice on planning and organization of independent work of the student: types of work and the description of the workflow, the implementation rules, criteria and evaluation indicators; instructions for the implementation of practical and coursework; advice on working with scientific or special text; tips for working with literature and etc.

The course has training materials that are provided in various file formats supported by Moodle such as text and web-pages and links to files.

Element "lecture" allows you to divide the lecture into pieces, add to the text of the lecture quizzes, links to external sources and illustration [6]. The lecture material is presented in the "Glossary" on each item. The important thing is respect of the copyright, so the lectures are links to copyrighted material borrowed in the form of pictures, charts and tables [9].

In the description of task type, name and purpose, a task or a group of level assignments, order fulfillment and recommendations for implementation of these tasks are provided. Algorithms and examples of assignments or solutions of typical tasks, indicators and criteria for evaluating all work according to the student's rating plan are given.

With the help of interactive elements "chat" and "forum" individual consultations are held, organized by the judgment of the course work, each participant may speak about any topic. The possibilities of interactive elements (different types of jobs, glossaries, forums, tests) emphasize students' individual fragments of the studied material, ability to check their level of knowledge, the organization of interaction of participants with each other and with the teacher [1].

Educational outcomes of students in e-learning system are considered as an integrated single procedure implemented via plurality of means and methods of assessment. Using the evaluation procedures determined in formation of educational results, general cultural and professional competences are enrolled. Estimates are presented as assignments and tests containing questions for the input control, helping to master course content better; job training and tests containing questions for self-control, finding the correct answers with explanations, tips, technique; quizzes and tests are designed to replicate, consolidate and control the protection of the reports on the practical and independent work, to prepare for other accreditation procedures, essays and case assignment.

In tests organization control is provided with the help of questions of various types, for example, "in Selecting the missing words", "random question for compliance", "image choice", "multiple choice", "n and the correspondence" "in false answers," "true/false", "in calculations", "diagram relations", "short answer", "numeric answer" [7].

Using an essay evaluation means evaluating higher levels of productive development of educational results, the ability to analyze selected for reflection subject. Solution case assignments allow you to simulate a professional situation, to form students' professional position, to choose their own way of solving professional problems. The development of directional control system takes into account the time frame, the logics of the construction of the educational material, the level of training, the complexity of the previous assignments.

Distance learning course "General and vocational pedagogy" has been tested in the learning process in Minin University to teach bachelors 44.03.04 Vocational training (on branches). The course was taught as a mixed training, involving the combination of classroom training with the elements of e-learning.

5 Conclusions

Experience in creating e-learning courses proves the efficiency of this trend. Its introduction into educational process in vocational schools will improve the quality of training. Moodle is a powerful tool for establishing effective rates with the possibility of the adaptation to any student, whether full-time, part-time or distance learners. Moodle provides interaction and implementation of ongoing communication between students and teachers. Moodle is a flexible and effective tool in the educational process. The dynamism of the platform management and the modular structure of the training make it possible to create the organization of the educational process, taking into account the individual requirements of the students, to contribute to improving the learning outcomes of students.

References

1. Alexandrova, N.M., Markova, S.M.: Problems of development of vocational teacher education. Vestnik Mininskogo Universiteta 1(9), 11 (2015). (in Russian). http://www.mininuniver.ru/mediafiles/u/files/Nauch_deyat/Vestnik/2015-04-16/Aleksandrova.pdf
2. Gorlova, V.G., Markova, S.M.: Model professional pedagogical education teacher of vocational training. Vestnik Mininskogo Universiteta 1(5), 21–24 (2014). (in Russian). http://www.mininuniver.ru/mediafiles/u/files/Nauch_deyat/Vestnik/2014-07%201/markova_Gorlova.pdf
3. Kostylev, D.S., Salyaeva, E.Y., Vaganova, O.I., Kutepov, L.I.: The implementation of the requirements of the federal state educational standards to the functioning of the electronic information and educational institute environment. Azimuth Res.: Pedagogy Psychol. 2(15), 80–82 (2016). (in Russian). T. 5
4. Krivonogova, A.S., Tsyplakova, S.A.: Technology project activities teacher of vocational training. Vestnik Mininskogo Universiteta 1(9) (2015). (In Russian). http://www.mininuniver.ru/mediafiles/u/files/Nauch_deyat/Vestnik/2015-04-16/Krivonogova.pdf
5. Kutepova, L.I., Mukhina, M.V., Smirnova, Z.V.: Realization scientific and practical conference about results of students practice is a effective way for increasing role and meaning of practic for students which studied in service specialty. Vestnik Mininskogo Universiteta 2(6), 16 (2014)
6. Kutepova, L.I., Nikishina, O.A., Aleshugina, E.A., Loshkareva, D.A., Kostylev, D.S.: Organization of independent work of students in the conditions of the information-educational environment of high school. Azimuth Res.: Pedagogy Psychol. 3(16), 68–71 (2016) (in Russian). T. 5

7. Lapshova, A.V.: Criteria and indicators of teacher professionalism in additional education system. Vestnik Mininskogo Universiteta **4**(8) (2014). (in Russian). http://www.mininuniver. ru/mediafiles/u/files/Nauch_deyat/Vestnik/2014-12-4/Lapshova_A.V..pdf
8. Lapshova, A.V.: Professional training of bachelors in terms of modernization of the university. Vestnik Mininskogo Universiteta (2) (2013). (in Russian). http://www. mininuniver.ru/mediafiles/u/files/Nauch_deyat/Vestnik/2013-06%202/lapshova.pdf
9. Markova, S.M.: Technologization of the pedagogical process of professional education. World Sci. Discoveries **3**(51), 296–301 (2014). (in Russian)
10. Markova, S.M., Tsyplakova, S.A.: Designing of pedagogical process on the basis of the process. Vestnik Mininskogo Universiteta **3**(7) (2014). (in Russian). http://www. mininuniver.ru/mediafiles/u/files/Nauch_deyat/Vestnik/2014-09%203/Markova_S.M._ Cyplakova_S.A..pdf
11. Polunin, V.Yu.: Designing technology for professional training of workers and specialists. World Sci. Discoveries **3**(51), 373–378 (2014)
12. Robert, I.V.: Didactics education informatization period. Teach. Educ. Russia **8**, 110–119 (2014). (in Russian)
13. Smirnova, Zh.V., Muhina, M.V.: Modernizacija processa podgotovki studentov vuza s primeneniem modul'nogo obuchenija [The modernization process of preparation of students of high school with the use of modular training]. In: Mezhdunarodnyj zhurnal prikladnyh i fundamental'nyh issledovanij, no. 4–4, pp. 827–829 (2016). (in Russian)
14. Smirnova, Zh.V., Parshina, A.: Rol' informacionnyh tehnologij v modernizacii obrazo-vatel'nogo processa vuza [The role of information technology in the process of modernization of the educational institution]. In: Integracija informacionnyh tehnologij v sistemu professional'nogo obuchenija sbornik statej po materialam regional'noj nauchno-prakticheskoj konferencii [The integration of information technology in training a collection of articles on materials of regional scientific-practical conference], pp. 45–47. NGPU im. K. Minina, Nizhny Novgorod (2016). (in Russian)
15. Polat, E.S., Buharkina, M.Y., Mosaic, M.V.: Theory and practice of distance education. ucheb.pos. for students. vyssh.ped.uchebn.zavedeny. Academy, Moscow (2004). (in Russian)
16. Tsyplakova, S.A.: Theoretical bases of design training of students in high school. Vestnik Mininskogo Universiteta. **1**(5) (2014). http://www.mininuniver.ru/mediafiles/u/files/Nauch_ deyat/Vestnik/2014-07%201/Cyplakova.pdf
17. Barber, M., Donneliy, K., Rizvi, S.: An Avalanche is Coming. Higher Education and The Revolution Ahead. Institute for Public Policy Research, London (2013)
18. Hanushek, E.A.: The economic value of higher teacher quality. Working Paper No. 56, National Center for Analysis of Longitudinal Data in Education Research (2010). http://www. urban.org/UploadedPDF/1001507-Higher-Teacher-Quality.pdf. Accessed 03 Apr 2017
19. Markova, S.M., Sedhyh, E.P., Tsyplakova, S.A.: Upcoming trends of educational systems development in present-day conditions. Life Sci. J. **11**(11s), 489–493 (2014)
20. Smirnova, Z., Vaganova, O., Shevchenko, S., Khizhnaya, A., Ogorodova, M., Gladkova, M.: Estimation of educational results of the bachelor's programme students. IEJME. Math. Educ. **11**(10), 3469–3475 (2016)

Social Component of Modern Human

The Value-Based and Cultural Matrix as a Component of the National Model of Social Development

Konstantin V. Vodenko[1]([⊠]), Valentina I. Rodionova[2],
Lyudmila A. Shvachkina[2], and Marina M. Shubina[2]

[1] M.I. Platov South Russian Polytechnic University, Novocherkassk, Russia
vodenkok@mail.ru
[2] Don State Technical University, Rostov-on-Don, Russia

Abstract. The article is devoted to introduction of the notion "value-based and cultural matrix" into the discourse of socio-humanitarian sciences. The methodology that allows constructing this theoretical concept is substantiated. The main theoretical and methodological principles are the axiological approach – as a universal methodological principle, it allows analyzing various social phenomena; institutional theory (D. North); theory of institutional matrices (S.G. Kirdina). Value-based and cultural matrix is treated as a totality of national, religious, cultural, educational, and family traditions, the established legal custom, and the generally recognized moral values, as a stereotype of life of a specific society in a separate period of time and in the territorial frameworks of a specific region, sanctioned by the national law.

Keywords: Culture · Values · Religion · State · Law · Civilization
Institutes

1 Introduction

Studying the specifics of the models of social development is very topical, as these specifics predetermine the processes of socio-economic and scientific and technological development. During studying the national cultures, it is necessary to pay attention to their components that are ideological attitudes which determine the specifics of the processes of development, society, economy, and science. In this article, we try to substantiate the notion "value-based and cultural matrix", which connects axiology and social ethics and cultural, legal, educational, and family traditions and law. Such task on creation of this theoretical institute is caused by the fact that, based on the obvious processes, generally known historical facts, and stereotype notions, which are recognized in humanitarian sciences of various profiles, it is necessary to produce a new notion that generalizes certain social phenomena in their totality. This theoretical construct could be used as a methodological approach in studies of various socio-humanitarian sciences.

© Springer International Publishing AG, part of Springer Nature 2018
E. G. Popkova (Ed.): HOSMC 2017, AISC 622, pp. 415–421, 2018.
https://doi.org/10.1007/978-3-319-75383-6_53

2 Materials and Methods

For studying value-based and cultural matrix as a theoretical construct and social reality, it is possible to use a range of theoretical and methodological approaches. Thus, the axiological approach, as a universal methodological principle, allows analyzing various social phenomena [1, 2]. A lot of social phenomena's ontological source is moral values, which are components of a certain ideology. For example, development of certain economic events in the world depends on the hierarchy of values in the ideology of the people who make decision on application of certain financial levers for achieving their goals. The purposes of application of the economic levers directly depend on the hierarchy of values of the people who form them. It is true for any other social phenomena that take place in various spheres – political, cultural, military, family, and others. However, for using the axiological approach during an attempt to analyze the challenges of the modern world, it is necessary to create additional theoretical constructs that have to explicate the axiological component in social phenomena, stimulate the ordered course of the research, and systematize and differentiate large ideological material.

Studying value-based and cultural matrices is impossible without using the institutional theory. Thus, according to its founder D. North, institutions "are the humanly devised constraints that structure political, economic, and social interaction" [3, p. 18], or, "rules of the game" in the society that created the stimulating structure of society. D North distinguishes three main components in institutions: informal limitations (traditions, customs, social norms); formal rules (constitutions, laws, judicial precedents, administrative acts); mechanisms of constraint that ensure observation of rules (courts, police, etc.). Another author, J. Hodgson, defines institutions as systems of stable and generally acknowledged social rules, which structure the social interactions [4, pp. 1–25]. In particular, he criticized D. North for not including informal rules into institutions. Thus, institutions are the regulating basis which determines the order of interaction between the subjects of economic relations.

Institutions are usually divided into formal (e.g., the Constitution of the USA) and informal (e.g., the Soviet "telephone law"). Informal institutions are generally accepted rules and ethic codes of behavior of people. These norms, "laws", or habits are the result of joint existence of people. Due to them, people learn what others want from them and understand each other very well. These codes of behavior are formed by culture. Formal institutions are the rules that are created and supported by authorized people (state officials).

The methodology of this research is based on the theory of institutional matrices of C.G. Kirdina. Institutional matrix is defined by D. north as a complex of interconnected formal and informal rules and limitations that determined the content and dynamics of economic political institutions of each specific society [3, pp. 129, 147–148]. In his institutional and evolutional theory, he determined the historical dependence of the present on the past in the form of the selected trajectory of institutional development, and social institutions – as a set of formal rules, informal limitations, and mechanisms of their compulsive implementation. At that, formal rules could be changed quickly, and informal rules are very slow to change.

Institutional matrices are a sustainable, historically formed system of basic institutes which regulate the economic, political, and ideological sub-systems of the society [5, p. 17]. Institutional matrices, denoted as X- and Y-matrices, correspond to the Eastern and Western types of civilizations. Each society was peculiar for one institutional matrix during its historical development. For example, X-matrix of the Eastern type for Russia (same as for the Asian counties and Latin America), while for most of European countries and the USA it is Y-matrix. Matrix cannot be replaced by another – even in the age of globalization, informatization, and global transformations.

However, interchange is possible between institutional matrices; due to this, at a certain historical stage, one institutional matrix is enriched with the elements of another institutional matrix (in the form of new traditions, norms, rules, models of behavior, and values), which are assimilated by the basic matrix with time.

3   Results

In the widest sense of the words, "value-based and cultural matrix" is the generally accepted tradition of existence of a specific society in specific period of time, which is established or not forbidden in the state's law. In a more specified form, the notion "value-based and cultural matrix" supposes totality of the national, religious, cultural, educational, and family traditions, established legal custom, and generally acknowledged moral values, as a stereotype of life of specific society in a certain period of time within the territorial limits of a specific region, sanctioned by the state law.

The term "stereotype" in the notion "value-based and cultural matrix" is used in a wide sense. It means not only the final scheme of perception of a certain event or phenomenon but also certain sustainable traditions, inclinations, and ideologies that are expressed in consciousness and actions of certain people and groups. As long as this public stereotype is expressed in the actions of social units, it could be sanctioned or not sanctioned by the national law.

It is necessary to pay attention to the fact that the type of value-based and cultural matrix directly depends on the territorial limits of existence of a specific society. The territorial interconnection of "place of residence, work, religious activities, political organization, and other various factors" is emphasized by Parsons [6]. He thought that territorial factor should be taken into account during supporting the normative order and control over behavior.

It is possible to distinguish the value-based and moral component, cultural and social component, and legal component of value-based and cultural matrix of society.

Value-based and moral component of value-based and cultural matrix is the axiological basis of the generally accepted tradition of existence of a specific society in a specific period of time in a specific region. It could be distinguished and studied separately, but in practice it pierces all other components of value-based and cultural matrix, being the leading criterion of acceptance, morality of a certain tradition, culture element, or education, and the regulator of the society's legal custom.

The cultural and social component of value-based and cultural matrix is the generally   accepted   traditions   of   behavior,   ethics,   morality,   family   and   business interrelations, education, and social and national & cultural peculiarities. The cultural and social component also includes the leisure traditions of the society.

The components of value-based and cultural matrix could be transformed in various periods of time. Thus, in the same region the type of value-based and cultural matrix could be different in different periods of time. Also, it is necessary to take into account that adjacent value-based and cultural matrices could be very different as to their type. The family, educational, financial, and other legal relations in the society, being sanctioned by the national law, are a legal component of value-based and cultural matrix of the region.

Let us analyze the interconnection between the components of value-based and cultural matrix. It is obvious that while changing, the value-based and moral component, as the axiological basis of the whole value-based and cultural matrix, leads to changes of other components; as a result, the type of value-based and cultural matrix is changed. What can be a source of changes of the value-based and moral component of the matrix? Obviously, the cardinal changes of the value-based and moral component are possible only as a result of targeted influence on the society. An example is distribution of Islam in Dagestan, related to Arabic conquests, and distribution of Christianity in the Rus, which was caused by the actions of Prince Vladimir. Also, an example of the influence on society and quick changes of the value-based and moral component of value-based and cultural matrix of the region is implementation of state-backed atheism in Soviet Russia. The issue of the process of natural transformation of the value-based and moral component of value-based and cultural matrix of the region should be studied further.

The cultural and social component depends on the value-based and cultural component, which is shown in the example of religious culture. For example, a large role in development of the religious Orthodox values in the everyday life of the Russian people belonged to the Church calendar, which connected together the dogmatic ideas of Eastern Christianity to the cultural and ritual aspects of people's life. This is proved by existence of so called "people's calendar", where observations over weather, family traditions and customs, and periods of field works were connected to the dates of memory of Christian saints and religious holidays. Organization of the connection of the feasting and cultural people's traditions of heathen Slavic tribes to the Church Christian calendar became an effective means of quick turning of people to the Church in the times of Prince Vladimir. Efficiency of the liturgical calendar as a guide of Christian values into people's life was used (with a negative sign) in Russia in the Soviet period. Together with wide implementation of the atheistic ideology into society through mass media and pedagogical programs, the ideologists of the USSR used temporal connection of new "communistic" holidays to the most reverent Christian feasts, which allowed secularizing the social life.

Therefore, it could be concluded that the influence on the social and feasting traditions of society is very important in formation of the type of value-based and cultural matrix of the region.

The legal component of value-based and cultural matrix is based on axiological content of the value-based and moral component and on the generally recognized traditions of cultural and social component of value-based and cultural matrix. Its social sustainability and public authority are ensured by sanctionining in written national law.

The people's legal traditions, forbidden in the national law, lose their power and disappear from the society's life.

This could be seen by the example of actions of the Russian Orthodox princes [7, pp. 108–113]. Not trying to replace the Slavic custom of blood revenge by the Christian ideals of mercy, they prohibited the revenge kills, setting the tradition of weregilds. Thus, the blood tradition was given the vector of development from the lynch law to the state measures of punishment, determined by a court, not by personal wish of the victim's relatives. This step led to the desired consequences and was established in the law which prohibited blood feud. At that, a subtle educational step by adapting the traditions to new rules led not to external obedience of the society to the established law but changed the sign of the civil legal consciousness from minus to plus: from heathen values of lynch law and revenge to the Christian values of court decisions, legality, and justice for the victim, balanced by mercy to the criminal and acknowledging the guilt of the one who performed a crime, leaving his relative and descendants aside – which was reflected in later legislative acts [8, p. 4]. The same historical example is seen in Islamification of Dagestan, when the usual legal norms were replaced by the norms of the Sharia law – though preserving authority in certain regions, they were based on the general Islamic principles [9, pp. 4–5].

Therefore, based on the above, it is possible to conclude that it is possible to change the national law and therefore change the legal component of value-based and cultural matrix, thus influencing the type of value-based and cultural matrix.

4 Discussion

It is necessary to study the issues of the types of value-based and cultural matrices and the issues of their similarities and differences. The issues of typologization and classification of value-based and cultural matrices requires additional studying – however we can offer certain conclusions. As in the theoretical construct "value-based and cultural matrix" the leading component is the value-based and ethic component, which is reflected in the cultural and social component and is fixed in the legal component, the main typologization should be conducted by the type of the value-based and ethic component. The additional attribute is the leading peculiarity of the cultural and social component.

Similarity and difference of value-based and cultural matrices could be determined by the value-based and ethic component. Thus, for example, traditional religions offer similar moral value and ethic norms for their followers, which will be reflected in the legal component in the form of similar legal norms. These matrices will be different by the peculiarities of the national culture, which are vivid but not substantial in state management. For example, value-based and cultural matrix of the Russian Orthodox state (18th – 19th centuries) and value-based and cultural matrix of Islamic Dagestan of the same period are similar in the axiological content of their components and their reflection in law, as compared to value-based and cultural matrix of a modern Islamic state and value-based and cultural matrix of a modern liberal and democratic state. The main difference of the matrices is not so much in culture (as it is possible to find the followers of any national cultures in a liberal and democratic state) as in the axiological content of the value-based and ethic component of matrices and its reflection in the legal component in the form of legislative prohibition or permission of certain actions – e.g., divorces, abortions, euthanasia, etc. [10, 11].

Then it is necessary to consider similarities and differences of the notion "value-based and cultural matrix" as compared to the opinions popular in modern knowledge. In particular, let us consider its correlation to the ideas of the Russian sociologist N.Y. Danilevsky that are provided in his book "Russia and Europe". N.Y. Danilevsky classified the cultural and historical types of society as to the totality of interconnected features of a large social organism. As the main characteristic of the cultural and historical type, N.Y. Danilevsky used national culture, and the cultural and historical type was named by the national and territorial feature. Besides, the list of features included the peculiarities of the social, religious, scientific, industrial, political, and artistic development of peoples [12].

Similar to the cultural and historical type, the notion "value-based and cultural matrix" is a comprehensive system determined by interconnected features of the social organism. However, the leading feature here is the value-based and moral component, the axiological content of which is the foundation of two other components of the matrix – cultural & social and legal.

Thus, N.Y. Danilevsky assigned religion to the cultural activities of peoples, but in the notion "value-based and cultural matrix of region" religion is one of the main sources of axiological content of the value-based and moral component. Apart from religion, in the conditions of modern realia, the source of axiological content of the value-based and moral component could be various non-religious philosophical ideas, shared by large social masses. An example could be the value of freedom of religion, shares by all peoples of Europe and North America [12].

5 Conclusions

The cultural and historical type, together with the national and cultural characteristic, is closely connected to the place of residence of peoples. Value-based and cultural matrix is also territorially connected to the areas of various peoples, as, despite active migration and geopolitical processes, the modern existence of societies is still peculiar for connection of a certain nation to a certain territory. However, unlike the cultural and historical type, value-based and cultural matrix could be described and studied in any region – from a small territorial social unit which occupies a small region (e.g., local national diaspora) to continental sectors of the Earth. The narrower the region of study of value-based and cultural matrix, the more authentic sociological data are obtained; the wider the region of study, the more average data are obtained. However, for the tasks of state management it is necessary to study large regions, as they allow determining the existence of similar value-based and cultural matrices on the basis of their similar axiological content.

Acknowledgments. The article was prepared within the grant of the President of the Russian Federation for state support for young Russian scholars – doctors of sciences – on the topic

"Cultural and ideological foundations of formation of the national model of regulation of socio-economic and scientific and innovational activities" (MD-651.2017.6).

References

1. Dokuchaev, I.I.: Value and existence. Foundations of historical axiology of culture. M.: Science, 595 p. Axiology (2009)
2. Vodenko, K.V., Polozhenkova, E.U., Matyash, T.P., Burmenskaya, D.N., Shvachkina, L.A.: Socio-cultural context and theological sources of the modern European sciences formation: theoretic-methodological ideas and approaches. Mediter. J. Soc. Sci. (Italy) **6**(no. 5, Supp. 3), 99–107 (2015). MCSER (Mediterranean Center of Social and Educational Research), Rome
3. North, D.: Institutions, institutional changes, and functioning of economy. M. (1997)
4. Hodgson, G.M.: What arc institutions? J. Econ. Issues **40**(no. 1) (2006)
5. Kirdina, S.G.: Institutional matrices: macro-sociological explanatory hypothesis
6. Parsons, T.: The notion of society: components and their interrelations. Thesis, issue 2, pp. 99–122 (1993)
7. Vorotilin, E.A., Leyst, O.E.: The history of political and legal sciences. In: Leyst, O.E. (ed.) M.: Zertsalo, 568 p. (2006)
8. Zyubanov, Y.A.: Christian foundations of the Criminal Code of the RF: a comparative analysis of the norms of the Criminal Code of the RF and Scripture. M.: Prospekt, 416 p. (2007)
9. Shikhsaidov, A.R.: Distribution of Islam in Dagestan/Islam and Islamic culture in Dagestan. In: Gadzhiev, M.S., Kaymarazov, G.S., Shikhsaidov, A.R. (eds.) M.: Oriental Literature of the RAS, 198 p. (2001)
10. Vodenko, K.V.: The resource of interethnic agreement among the Russian youth. Humanitarian of the Russia's South, no. 3, pp. 203–214 (2017)
11. Lubsky, A., Volkov, Y., Denisova, G., Voytenko, V., Vodenko, K.: Civic education and citizenship in modern Russian society. Indian J. Sci. Technol. **9**(36) (2016)
12. Danilevsky, N.Y.: Russia and Europe. M.: Kniga, 576 p. (1991)

Comparison of Approaches to Development of Industrial Production in the Context of the Development of a Complex Product

Ekaterina P. Garina[1]([⊠]), Alexander P. Garin[1], Viktor P. Kuznetsov[1],
Elena G. Popkova[2], and Yaroslav S. Potashnik[3]

[1] Nizhny Novgorod State Pedagogical University named after Kozma Minin
(Minin University), Nizhny Novgorod, Russian Federation
e.p.garina@mail.ru, rp_nn@mail.ru,
kuzneczov-vp@mail.ru

[2] Volgograd State Technical University, Volgograd, Russian Federation
210471@mail.ru

[3] Nizhny Novgorod State Technical University,
Nizhny Novgorod, Russian Federation
yaroslav.sandy@mail.ru

Abstract. The article analyzes the theoretical and methodological basis for the design and development of a complex industrial product in the machine-building industry, when the task of product development is solved through the dominance of the concept of product development management over technological and production solutions. The authors considered the classical project management methodologies in the context of product development as the conditions for making managerial decisions to ensure the sustainable development of industrial enterprises. Several directions in the field of industrial product development have been studied, among which at the level of industrial production the most widespread are: (1) organizational aspect (organization of the production process), (2) engineering design of systems and (3) operational management. A set of solutions for product design/development, differing in the level of abstraction, critical success factors, and used variables has been formed. It is determined that the development of production in the context of product development is built either through: drawing up of road maps - allocation of certain functional tasks in the process of product creation with the subsequent selection of tools for their implementation and organization of production, or through a cross-functional solution of the problem within the framework of project management. That the typologization of solutions for the development of production through the creation of a product that allows carrying out the prolongation of the results does not exist. Each project is developed for certain production, which predetermines the need for further research on the issue.

Keywords: Development of production · Product of production
Production system · Engineering design · Operational management

© Springer International Publishing AG, part of Springer Nature 2018
E. G. Popkova (Ed.): HOSMC 2017, AISC 622, pp. 422–431, 2018.
https://doi.org/10.1007/978-3-319-75383-6_54

1 Introduction

In the modern economy, Russian industrial enterprises are increasingly confronted with increasing competitive pressure, determined by the globalization of value increment processes; shortening the life cycles of the product; an endless growth of technical, market and organizational interdependencies. The complexity is further increased in industrial production, where the pace of technical development, competition and increased demands from consumers determines the need for new technical solutions of the product. As a consequence, the tasks of the manufacturer are shifted towards reducing the volumes of aggregate output against the backdrop of a significant increase in the functionality of the product; increase the flexibility of production and reduce the time for product development. In such complex and dynamic conditions, each new production process is already considered as a "unique" and dynamic action, and the task of product development is increasingly being solved through the dominance of the concept of product development management over technological and production solutions. The process of creating a product in this perspective solves the overall strategic task of the enterprise, and the production system ensures effective interaction of information flows, materials, personnel, capital within the set business objectives; market uncertainty is reduced through careful planning of projects for the development and development of industrial products.

Thus, the task of forming a theoretical and methodological basis for the development and development of a complex industrial product in the machine-building industry is an urgent task requiring the study of appropriate approaches that determine the new hierarchy, logic, principles and criteria for making managerial decisions aimed at ensuring the sustainable development of industrial enterprises.

2 Theoretical Basis of the Study

Product development is a complex set of activities, the primary stage of projects that serve PD processes (Table 1).

"Reference points" of projects: product design, development costs, terms, quality of the product received, etc. depend on the number of individual tasks and the complexity of their implementation and are determined by the complexity of the project, the presence/absence of technical innovations; coordination links, the scope of the project. The final values of the indicators depend on the project environment or the set goals for the project implementation. Large differences in projects for the creation/development of products according to their targets, sizes, customers, suppliers, partners and the required characteristics of the product complicate the task of systematizing projects. Product development covers activities that translate knowledge of market needs and technological capabilities into information for the manufacturer. Information is transformed into product concept, models, technical characteristics, layouts, prototypes, engineering drawings, design processes, tools, equipment and software.

In the world practice, there are several areas in the field of design/development of industrial products, among which at the level of industrial production, the most widespread: (1) organizational aspect (organization of production process), (2) engineering

Table 1. Overview of classical methodologies in the context of product development [6]

Project integration management	PD-process is divided into several projects implemented simultaneously. Integration of projects is carried out in the field of managing the "resource pool" (MSP). The effectiveness of PD projects varies from the minimum during the stage of concept formation to the model that grows in subsequent stages
Scope management	Most projects are aimed at specific "windows" of opportunities due to the high uncertainty of the project, which is difficult to achieve in the production of a new product. In an attempt to solve the problem, the methods PERT, CPM; management methods in the methodology of the "quality gate": Gantt diagram, etc. The approach is based on the realization of the idea of parallel product design. The task is to form and distribute the scope of the project work to small subtasks (decomposition of works). To obtain the result, it is necessary that the subtasks are relatively independent of each other, which in practice is rarely achieved. In addition, the volume of forthcoming work is difficult to foresee in the context of the new project
Cost management	It involves estimating the cost of the project based on an assessment of the base time costs, and then the cost control activities that occur during the project implementation stage. Studies show that the focus on target costs may not be appropriate in terms of product differentiation by technology, time to market, or satisfied needs for the consumer. That is, the manufacturer's narrow focus on the cost of creating and selling a new product can distract designers from the essence of the product itself. In conditions of high uncertainty, the amount of costs generated is ambiguous
Quality management	A set of quality criteria and its measurement processes is formed. Originally implemented in the field of engineering product design (TQM). With regard to the PD process, the product is considered in a system of functional, aesthetic and technological indicators. The following methods are used: analytical hierarchy «leanprojectmanagement»; just-in-time and others. Dedicated tasks, more often in production planning and supply chain, are implemented in parallel to reduce the overall duration of the project. Management style NPD-projects identified as an element of success

design of systems and (3) operational management (operation management). A variety of approaches predetermined the multiplicity of decisions of product design/development, often differing in terms of abstraction, critical success factors, used variables (Table 2).

From an organizational point of view, with the development of industrial production in the context of product design/development, attention is mainly focused on the determinants of the "success project" in the organization of production and technological processes. And given that the architecture of a modern product is usually the result of the development of several firms that form a modular supply system to the OEM, from the organizational point of view, the coordinating tasks of product

Table 2. Comparison of approaches to the development of industrial production in the context of product development [2, 4]

	Process/production organization	Engineering design	Operational management
Product view	The product is the result of a production process built by creating a federation of systems	The product is a complex system of interacting components	The product is a sequence of stages of development of production (technological) processes
Typical performance indicators	"Success" of the project	Form and function of the product. Technical capabilities/performance. Innovativeness of the product. Sometimes direct costs	Overall efficiency of the project. Cost of production. Execution time. Coefficient of capacity utilization
The dominant paradigm	There is no dominant paradigm	Geometric characteristics of the model. Product Specifications	Parametric model of the execution process. Technology system
Variables	Stages of product development. Product structure	Size, shape, configuration, function of the product	Technological process, production schedule. Differentiation of the production process
Critical success factors	Organizational alignment. Team characteristics	Creative concept. Product configuration. Optimizing performance	Logistics, design, production
Reference points of the project	formation of terms/sequence of activities; identification of project participants: Manufacturers/suppliers of components, assemblers of components; the formation of a communication mechanism between team members; chaining of supplies (configuration, location of individual points); technological process of product assembly; development of assembly technologies	The concept of the main product; target attributes of the product (speed, price, reliability, power, etc.); product architecture; product variants; unified components; shape, configuration and industrial design of the product; individual components, platforms, modules; values of key parameters of the project for product development; the configuration of components and the determination of priorities for their selection; detailed design of components, platforms, modules; material support and the process of selecting components	Prototyping the product; prototype technology; testing the market, launching the product on the market; forming a production plan, launching, implementing

development are highlighted as systems of integration interaction between participants, through the idea of federation of systems [4]. The structure, being a form of the system, is determined by its content, i.e. processes that occur both in the production system and in the supply chain.

More often than not, product creation is a design solution within the engineering design of systems - industrial product design: when assembling the product (DFA); when designing product components (DFM), modules, platforms [6]. The study of the issue within the framework of this approach is at a more detailed level - on individual engineering solutions: design and modeling of the product, building the production system. At the same time, the indicators of the effectiveness of the project are: architecture and functionality of the product, technical capabilities of the manufacturer/labor productivity, product innovation and the costs determined by it, the cost of the product.

Operational management focuses on the alignment, development of production (technological) processes, business processes of the enterprise, on the parametric model of execution processes, the production schedule; on the management of operations, control of production and the sequence of the project.

3   Analysis of the Results of the Study

Let us dwell in more detail on the highlighted solutions.

Engineering design

The problems of the architecture of engineering systems are dealt with by Alexander, Simon; the interconnection of product architecture is considered by Ulrich; Organic design issues are handled by Sanchez, Mahoney; the development of individual industries was considered in the works of Baldwin, Clark et al. The five main decisions in the framework of the product development concept are outlined: target values of product attributes (for example, speed, price, reliability, power); product concept; product variants; product architecture; general physical form and industrial design of the product. In addition, the concept considers advanced elements of the product creation system, such as the life cycle of services and after-sales deliveries. And the attributes of the product take into account both customer needs, and technical characteristics (engineering characteristics) of the product, design and modeling. The concept of development includes the implementation of attributes in a certain technological approach, as a result of which the concept of the main product is determined. If the technological approach is changed, then the concept of a new generation is worked out.

The concept of the product:

– is static and is developed "to details", after that the design of a product is realized. In modern conditions, the development of products is carried out using several concepts, and the design is already being developed in the process of their implementation. Completion of specifications can be carried out later, already in dynamic environments. In this case, manufacturers prefer the standardization in the design and production of the product. A set of components partly determines the architecture of the product;

- includes both decisions on technical characteristics, and on the basic physical parameters of the product configuration, the appearance of the product within the framework of industrial design.

Decisions within the framework of "product development" can be as follows:

- product planning strategy: the formation of a product portfolio to achieve certain performance indicators, the effectiveness of the manufacturer; the formation of a map of components that determine the functionality of products, based on the company's core competencies;
- product development: the definition of the technologies used in the production of the product; choice of enterprise management model; formation of project staffing; determination of project performance/effectiveness indicators; definition of the investment component of the project; identification/selection of processes planned for use (for example, Stage-Gate);
- supply chain management and logistics. The "supply chain and logistics" category covers both incoming and outgoing streams of materials, intellectual property, and firm services. Supply chain design solutions for product creation include vendor selection, product and production system design questions, component set definition, supply chain configuration, process and business process mix, process equipment selection, and so on.

With respect to the aggregate of components, most new engineering solutions involve the use of a list of standard (already existing) components. If the product contains components developed for it, then within the supply chain decisions are made about who will develop these components, produce and test them. Researchers of management operations also pay great attention to the development of physical supply chains; industrial design, that is, differentiation in the performance of orders. Optimization of the configuration of the dedicated supply chain systems of a certain community of "platform" products uses and extends the concept of the "Generic Specification" (GBOM) of the product family [6].

The supply chain configuration covers not only the alternative choice of suppliers and delivery modes for the end products of this specification, but also aspects of the manufacturing process such as the way of processing materials/components/semi- finished materials, production of materials, production time or time to market, under the project of the release of new products. These solutions may differ depending on the characteristics of the final products in the course of their changes. In addition, the availability of alternative production processes in turn affects the final solutions of the product and (or) the product family.

Until the early 1990s design solutions are usually implemented in the geometry of the models of nodes and components, in the list of materials, in the production documentation, in the values of the design parameters and the desired characteristics of the product.

Specialization in the field of design (it is the intensity of the deployment of specialists at the design stage) can be carried out in two ways:

1. Percentage of project resources allocated to a specialist;
2. The number of specialists developed products. Analysis of projects in engineering shows a high correlation between directions - 0.87.

Work on product design is closely motivated by the needs of production. Thus, the development of product attributes within marketing leads to a large amount of work on the analysis of parametric characteristics, design problems, and work on optimizing the design.

Process/Production Organization

Design solutions related to the "customization" of the product, its production and launch include solutions for the target market segment, product range, priority setting, resource allocation and technology selection. When developing products, as a rule, product characteristics, terms and cost of development are determined and a prototype of the product is created. Product performance may contain several dimensions, such as quality, innovation, manufacturability, etc. Describes the time that passes between the beginning and the end of the project, and measures to save resources. These factors have a significant impact on the likelihood of economic success. A roadmap (usually a diagram illustrating the timing of planned projects) prescribes specific solutions, including: target market of the company, product portfolio, specific product development projects, assets common to all products; technologies planned for use; Efforts that need to be made to coordinate the above solutions in the company's corporate finance, marketing and operational strategies. Approval of the plan is a product, that is, product planning is a set of solutions that guarantee the company growth in a strategic aspect.

Projects of product portfolio choice are considered in the works of Aiello et al. Determining what opportunities exist for product realization, Christensen and Bower are exploring potential pitfalls in existing markets. Their research shows that successful firms often fail to recognize technological or other changes in the market, because there is a time lag of product planning. The working version of solutions for the product line is offered in the work of Green and Krieger, which includes making decisions on the line of product models that have maximum objective functions, such as social load, maximum profit and others. Several procedures were developed to solve this combinatorial problem. A number of researchers approached the question from the perspective of the cost structure.

When the product is launched, the firm also decides the timing and sequence of product introduction. Moorthy and Png argue that it is in the interest of the firm to introduce end products with medium characteristics, coupled with high-tech products. Padmanabhan, Bhattacharya suggest that it is more appropriate to introduce products with average characteristics before high-tech products (for example, the introduction of network devices and exogenous technological improvements). Adler and others are studying solutions for developing product lines in parallel and sharing resources in various projects. They identified the "traps" of effects arising from the development of several product projects in parallel; "Traps of maximum utilization" of production capacity; the effects of variability in the development of projects. According to

Nobeoka et al. sharing of resources can also lead to more effective use of resources, reducing labor intensity, and improving the education system in different projects. Substantial exchange of assets in the production of a number of products leads to the development of a product platform. Most of the work on the platforms, however, focuses only on the benefits of developing "product platforms". In the framework of this topic Robertson and Ulrich indicate that the client loses the perception of differentiation due to the existence of product platforms. In turn, Krishnan and Gupta discuss the issue of low final cost of products through the creation of product platforms. Where the key component of product planning is the decision on which technologies to include in the new product. It is proved that advanced technologies increase the risk level of the process of developing new products. Also an "approach" is widespread, in which products are collected using proven technologies. This approach makes the development process more manageable, but the conditions of competitiveness may require the development of technologies and products simultaneously.

Closely related to the planning of development work decisions on the types of communication. Cross-functional links (for example, between marketing and engineering) are widely considered in the literature [5]. Prospects are also developments to strengthen cross-functional relationships in order to obtain the greatest added value of the product.

Operational management
After the design and development of the product, solutions are found for the production of the product that meet the main criterion for evaluating the process - the required performance in the given production constraints. Despite efforts to design and ensure optimum production and installation characteristics, no production system is able to account for all possible variations in input materials; possible production processes; actual skills of workers; environmental factors (for example, temperature, humidity) and other. There are several strategies for ensuring the required product performance:

- *Process management.* The process can be either on the control (the effects the producer receives from monitoring the identified causes of the discrepancy) or out of control. Effects from unrecorded causes/factors can be eliminated through proper process design (for example, using Taguchi methods), implemented offline, or as a result of constant monitoring on the production line.
- *Maintenance and repair.* In addition to the statistical control of processes, some tests and test methods are aimed at ensuring compliance of products with: *verification and testing of* raw materials, parts and components after receipt from suppliers, at any point in the production of the final product; *audit,* i.e. periodic inspection of the production process or quality/product results; *checking of specialized products* - the product undergoes a series of tests (specified in the contract) until the moment of product transfer to the client.

Various tricks in the process of creating and developing a product predetermine the allocation of a subset of solutions. In the 1950s, the XX century. Clustering of these solutions for product development was built in accordance with the functional logic - through the allocation of certain functional tasks in the process of creating a product with the subsequent selection of tools for their implementation and organization of production [1]. In the late 1990s - early 2000s when developing a product, they begin to focus not on the expression of traditional functional links, but on the coordination of operations/process stages-an approach is developed for "cross-functional study of the problem". For example, researchers' attempts to formulate a "product line of product models through industrial design" are complemented by studying the conditions and limitations of traditional marketing. And the inconsistency of decisions between product differentiation, its design and operational complexity can be "removed" through the transformation of the product architecture.

There are other possible criteria for clustering the solution, for example, statistical analysis and optimization. The ability to combine approaches is seen through the development of design methods for products developed by marketers and incorporating a set of technological constraints. In fact, this is the creation of industrial designs, a certain form and style. This design can be one of the most important factors in explaining consumer preferences in some commodity markets, including cars, household appliances and furniture. At the same time, the absence of an industrial design does not allow us to reflect the inherent difficulties in modeling the relevant factors. However, product planning solutions, development of product metrics are atypical in production practice. For example, there are several research results on the use of a product platform with market advantages of a wide variety of products. Nevertheless, the integration of the market request, the new product project and technological considerations for the solution of the product lines are increasingly developed, beginning with the level of sharing resources.

4   Conclusions

1. There is a wide variety of approaches to the development of production through the creation of a product. The differences between them are predetermined by the chosen focus, the technologies used, the production capacity and the existing hierarchy of systems and subsystems, enterprise processes. In the world practice at the level of industrial production, the most widespread: (1) organizational aspect (organization of production process), (2) engineering design of systems and (3) operational management (operation management).
2. The diversity of approaches also predetermines the multiplicity of product design/development decisions, often differing in terms of abstraction, critical success factors, and variables used.
3. Clustering solutions is built either through: (a) drawing up of road maps - allocation of certain functional tasks in the process of product creation with the subsequent selection of tools for their implementation and organization of production, or (b) through a cross-functional solution of the problem within the framework of project management.
4. Typologization of solutions for the development of production through the creation of a product that allows carrying out the prolongation of the results does not exist. Each project is developed for certain production, which predetermines the need for further research on the issue.

References

1. Krishnan, V., Ulrich, K.T.: Product development decisions: a review of the literature. Manag. Sci. **47**(1), 1–21 (2001). Design and Development http://www.jstor.org/stable/2661556. Accessed 16 Mar 2013
2. Lee, H.L.: Effective inventory and service management through product and process redesign. Oper. Res. **44**(1), 151–159 (1996)
3. Mizikovsky, I.E., Bazhenov, A.A., Garin, A.P., Kuznetsova, S.N., Artemeva, M.V.: Basic accounting and planning aspects of the calculation of intra-factory turnover of returnable waste. Int. J. Econ. Persp. **10**(4), 340–345 (2016)
4. Kuznetsov, V.P., Romanovskaya, E.V., Vazyansky, A.M., Klychova, G.S.: Internal enterprise development strategy. Mediterr. J. Soc. Sci. **6**(1, S3), 444–447 (2015)
5. Romanovskaya, E.V., Garin, A.P., Dalidovich, K.N.: Features of the process of making managerial decisions. Bull. Univ. Minin. **3**(11), 7 (2015)
6. Chase, R.B., Equiline, N.J., Jacob, R.F.: Industrial and Operational Management, 8th edn. Translated from English. Publishing House "Villamé", Moscow (2004). 704 p.

Actual Issues of Improving the Methodological Approaches to Estimation of Population's Living Standards: A Regional Aspect

Pavel N. Zakharov(✉), Karina V. Nazvanova,
and Artur A. Posazhennikov

Vladimir State University, Vladimir, Russian Federation
pav_zah@mail.ru, kalateya_flower@mail.ru,
zzarturzz@yandex.ru

Abstract. The urgency of studying and evaluating such a social and economic category as "living standards" is due to the need to comply with the standards established within the framework of the Concept of the Social and Economic Development of the Russian Federation until 2020, as well as in the Main Directions of the Government of the Russian Federation until 2018. Within the framework of the set tasks, it is necessary to ensure a stable and dynamic improvement of the living standards, as well as the solution of demographic, social and environmental problems. Different regions of such a large-scale state differ in the profile of the population, the degree of accessibility of certain resources, and, therefore, the living standards. The regional differentiation of the Russian Federation is so great that the definition of the population's living standards throughout its territory does not lend itself to comparable quantities, which justifies the difficulty of identifying the most significant criteria that determine this indicator. Thus, when assessing the Russia's population's living standards, it is necessary to use the toolkit that would emphasize such a pronounced differentiation of incomes in the country. The article analyzes the category "population's living standards", examines the most significant, from the author's point of view, modern techniques and identifies their positive and negative sides for the purpose of the further development of an effective system for assessing the living standards in the Russian Federation.

Keywords: Assessment of living standards · Methodology · Population
Indicators of living standards · Region

1 Introduction

The world community recognized that the living standards is one of the main indicators characterizing the economic development of the region. Thus, "the living standards" is a highly effective method for assessing the socio-economic well-being of the population of the region, the people's life potential and the ability to meet their needs" [1]. Increasing interest in the study of this indicator gives rise to more and more controversy about what exactly includes a category such as "living standards" and what methods should be used to evaluate and measure it.

© Springer International Publishing AG, part of Springer Nature 2018
E. G. Popkova (Ed.): HOSMC 2017, AISC 622, pp. 432–439, 2018.
https://doi.org/10.1007/978-3-319-75383-6_55

Taking into account the high differentiation of the regions of the Russian Federation in terms of territory, climatic characteristics, population density and density, the level of economic development, etc. The authors of the article reviewed existing methods for assessing the living standards in a regional context with the aim of identifying promising areas for improving the methodological bases for researching this problem.

2 Theoretical Basis of the Research

Most researchers, when characterizing the phenomenon of "living standards," pay great attention to the economic side, the material security of life of the population [7, 8, 10]. Thus, at the beginning of the twentieth century, at the initial stage of the development of this concept, the German theoretical scientist I. Steffen pointed out that the "living standards" must be considered as a conscious formation of the economy in the interests of the majority. There is also an opposite point of view, according to which the living standards is the most integrated social indicator. For example, George Galbraith defined the living standards as "the possibility of consuming goods and services" [9], thereby bringing this judgment beyond the limits of quantitative, economic indicators. A.I. Subbeto unites the above approaches and considers the living standards as a system of qualities of spiritual, material and socio-cultural, ecological and demographic components of life [6].

As part of the socio-economic approach, there is the problem of association with the standard of living. The opinions of specialists are divided, some believe that the concept of the standard of living is a component of the "living standards" (Nobel laureates, Stiglitz and A. Sen), others - that the living standards and the standard of living in practice should not be divided. The most widely spread view was that the broad concept of the living standards is historically a branch of the standard of living.

The similarity of the concepts "standard of living" and "living standards" is based on the overall economic component of the categories. However, if we take as a basis a comprehensive assessment of categories, it turns out that the living standards is only a social component of the standard of living. Thus, the standard of living is an economic category of the living standards and in practice, these indicators are closely related.

3 Research Methodology

The methodological basis for the study was general scientific and special methods, namely the system approach, the method of analysis and synthesis, the method of comparison and analogy, methods of econometric.

4 Analysis of Research Results

The definition of an effective methodology for assessing the living standards for the Russian Federation is even more controversial in connection with the distinction of income among the population that is noticeable and distinguished among many states.

According to Rosstat in 2015, 10% of the wealthiest Russians account for about 29% of total cash income and 10% of the wealthiest - about 2% [2]. Thus, when assessing the Russia's population's living standards, it is necessary to use the toolkit that would emphasize such a pronounced differentiation of incomes in the country.

The list of indicators for determining the level of population's living standards is quite extensive, which leads to the need to understand the "living standards" - as a systemic concept that determines the life activity of people in a certain territory, as well as the set of conditions in which it occurs.

To date, there are a number of different methods for assessing the population's living standards. Together, these methods should be divided into 2 categories: *Subjective and objective*. Nevertheless, today the definition of a high level of living standards includes a number of indicators that are universal for all methods [3].

Objective methods of assessing the population's living standards are based on the analysis of statistical data, and therefore, for the most part, are simple in systematization and calculations. However, they use rather averaged and generalized data, which does not always reflect the population's living standards of the region as a whole, especially in the Russian Federation, where there is a fairly strong differentiation of the population in terms of income, access to resources, health, working conditions and other criteria.

Subjective methods for assessing the living standards include various ways to find information on the degree of satisfaction of needs by different segments of the population. Such methods are based on sociological surveys and expert assessments, and reflect objective indicators in the subjective views of people. Such a fact is the merit of the presented system of assessing the population's living standards, since such data are more difficult to politicize in the interests of the authorities, and reveal the weaknesses of specific regions being analyzed for the life of the population and the study is directed. The disadvantage of subjective methods can serve as an incorrectly compiled evaluation program. Mainly, the reliability of the results with a subjective assessment of the population's living standards is affected by the degree of ownership of the data by the researchers themselves, and, however close to the results obtained by such residents in the regions themselves, the data obtained during the research using objective methods are considered more reliable.

It should also be noted that in the process of increasing interest in the study and assessment of the population's living standards of different states and the constituent entities included in their composition, there have also emerged cumulative techniques that combine the signs of objective and subjective assessments.

The objective component of assessing the living standards is a combination of statistical indicators, as well as the variation calculation based on them have been developed by this area of activity indicators. The subjective evaluation is represented by the results of sociological surveys and the results of the questioning of the population. Subjective indicators focus on the satisfaction of the population with their own lives, their subjective feelings, but studies show that the relationship between living conditions and life satisfaction is small, which significantly complicates the measuring and comparative apparatus of the methodology of researching the population's living standards.

Such *combined methods* allow not only to obtain an assessment of the living standards based on indicators applied to different population groups and taking into account the territorial feature, but also to analyze differences in such indicators and averaged values for a certain region [5].

The methodologies presented in Table 1 include a number of particular indicators, each of which is an analysis of the criteria of the methodologies and an assessment of their practical applicability. It should be noted one of the main drawbacks - an indefinite set of criteria, depending on the subjective representation of conducting research. A positive characteristic of all-Russian methods is their extensive range of study of the sides of the life of the inhabitants of the region, which is not taken into account in the case of the HDI methodology.

Table 1. Review of existing methods for assessing the population's living standards

No. in order	Name	Territorial application	Main content
1	Calculation of the HDI (human development index) by the UN methodology	International	A standard tool for assessing and comparing the living standards of different countries in terms of living standards, literacy and longevity
2	Gini Coefficient	International	Statistical indicator of the degree of stratification of the society of a given country or region in relation to any studied trait
3	Integral Welfare Ratio	RF	Calculation of the welfare of the population of the region and the economic potential of its inhabitants
4	Integral assessment of the living standards	RF	A six-block system for assessing the population's living standards of a subject of the federation
5	Regional methodology for assessing the living standards (designed to assess the living standards in Yaroslavl Oblast)	Subjects of the Russian Federation	Quantitative assessment of quality and living standards (from 0 to 1), based on the living standards index ( ) and the index of satisfaction with the living standards ( ), their subsequent matrix comparison

Speaking about the methodology for assessing the living standards in Yaroslavl Oblast, we note the existence of both a mathematical principle for studying the living standards and an extensive sample of respondents who can demonstrate the real moods of society for research. Techniques that combine quantitative and qualitative indicators differ with the instruments used; one cannot unequivocally state the effectiveness and fairness of their application for all regions. It should be noted that combined methods are recognized as the most capacious, which is explained by the combination of diverse methods of assessing the living standards included in them.

The Human Development Index (HDI) is a method of objective assessment of the population's living standards, an integral indicator capable of determining the human development in the area of the study. Beginning in 1990, this index finds its application in the United Nations "Human Development Reports" (UN). In these reports, we are talking about how to relate the living conditions of different countries, how to promote human development so that it enhances the level and living standards. When calculating the HDI, the following indicators are taken into account (average among three):

- The average duration of life expectancy at birth (ADLEB) measured by the statistical indicator of the life expectancy of an individual at birth (the minimum value in this area is considered 25 years, the maximum - 85);
- The level of adult literacy (2/3) and the cumulative share of students (1/3 of the indicator), including not only the received and documented education, but also the intellectual abilities of the individual, grounded in various psychological and sociological methods;
- The standard of living, estimated through GDP per capita and purchasing power (PPP), then the standard of living of citizens, objectively measured by the level of GDP per capita, which in turn determines the purchasing power of a particular citizen.

Formulas for calculations are shown in Table 2.

Table 2. Calculation of the HDI according to the data of the Vladimir region for 2015

1	Lifespan index	
2	Education index	2/3ALI + 1/3GEI
	The adult literacy index (ALI)	
	Index of student population aggregate (GEI)	
3	GDP Index	where = 5990 + 2 (GRP per capita - 5990)½

Where LE is the life expectancy at birth; ALR - level of literacy of the adult population, %; GGER - the cumulative share of students; - GRP per capita at purchasing power parity, in dollars. USA [1]. - the average constant for the world community.

The calculation of the HDI of the Vladimir region in 2015 was carried out as follows:

1. I (PW) = (70−25)/(85−25) = 0.75
2. I (sample) = 2/3 * 0.521 + 1/3 * 0.09 = 0.38
   ALI = 52.1%/100 = 0.521
   GEI = 9.13%/100 = 0.09
3. = 5990 + 2(11666−5990)½ = 6140,7

GDP Index = (6140.7−100)/(6311−100) = 0.97
HDI = (0.75 + 0.38 + 0.97)/3 = 0.7

The Gini coefficient is a statistical indicator, that component of the living standards assessment, which allows determining the degree of economic differentiation of a particular region. The advantages of this methodology can also be recognized in a region that is diverse in characterization and size, supplementing data on GDP and per capita income, the ability to compare the trait in different population groups and the visibility of the dynamics of the unevenness of the trait, as well as anonymity.

Integral welfare factor is a calculation of the welfare of the population of the region and the economic potential of its inhabitants: [1]

1 - GRP (gross regional product per capita, rubles/person); 2 - PP (purchasing power, the ratio of per capita income with the subsistence minimum per capita,%); 3 - PR (poverty rate - the proportion of the population with incomes below the subsistence level,%); 4 - CF (coefficient of funds - income ratio 10% of the most and 10% of the most well-off population, times); 5 - PL (poverty level - the share of food expenditures in the total amount of consumer spending,%); 6 - MPS (the marginal propensity to save - the ratio of the growth of savings of the population on ruble accounts to the total increase in monetary incomes, times). These indicators are calculated in relation to federal data (region/RF or RF/region).

To illustrate the population's living standards in the Vladimir region, we present an analysis of the changes in the proposed indicators for several years (and 2015) according to data provided by the Federal State Statistics Service (Table 3).

Table 3. Calculation of integral coefficients of the population's living standards of the Vladimir region

Year	Indicators of living standards				
	GRP (Region/RF)	PP (RF/region)	PR (RF/region)	CF (RF/region)	PL (RF/region)
2015	**0.58**	**1.3**	**0.82**	**1.4**	**0.81**

Integral methodology can be considered the most general and complex, which underlines its name. As already noted above, the integral methodology for assessing the population's living standards is based on 6 main blocks of indicators:

- 1 block (the level of economic development of the region) includes the total expenditures of the consolidated budgets - CB, the costs of the CB for social policy and investments in fixed assets;
- 2 block - indicators characterizing the material well-being and the degree of consumption of goods and services by the population;
- 3 block - socio-demographic indicators;
- 4 block - indicators of the labor market;
- 5 block - indicators of social tension (indicators characterizing the incidence of alcoholism, drug addiction, as well as crime);
- 6 block - indicators characterizing the ecological situation.

It is assumed that the indicators comprising the presented blocks cover all indicators of the population's living standards. In addition, their statistical accounting is quite simple. Based on the results of calculations, the region is given a rank according to the living standards index of the population.

Nevertheless, the dimension of the final integral index of the development of the population's living standards does not allow bringing it to comparable with other considered methods in the article, which implies that we cannot compare the result and evaluate it in comparison with other methods.

The calculation of particular indexes for the methodology proposed in Yaroslavl Oblast is divided into the calculation of particular objective and subjective indexes. Objective indexes are subdivided into sub-indexes such as the population quality index, the index of livelihood opportunities, the index of the level of economic welfare, the index of the level of development of the social sphere, the index of the level of safety of life. Subjective indexes are based on the results of a sociological survey, which involves respondent's assessment of the same five indicators of satisfaction with the living standards.

The wide application of the methodology is limited by the fact that the calculation algorithm is presented in the program documents of the administration (government) of Yaroslavl Oblast, and also involves the application of the expert assessment method for the sample of respondents.

Analysis of methods for assessing the living standards allows us to say that the indicators characterizing the population's living standards of the region vary unevenly, their dynamics are barely noticeable, which ambiguously characterizes the policy of the region. The leadership of the Vladimir region managed to maintain the value of the subsistence minimum and income of the population, which is demonstrated by the size of the PP (this indicator is one of the leading indicators for the regions of the Central Federal District), which also explains the increase in the poverty coefficient (PC) - its increase is mainly due to an increase in the subsistence minimum by Compared with other areas. You can also note the increase in the savings capacity of citizens (the MPS indicator), which positively notes the superiority of income levels over expenditure levels. However, the integral indicators of the living standards in the region do not show a noticeable positive dynamics, which indicates an inefficient or stagnant nature of the policy pursued by regional authorities in the area of implementing reforms to improve the population's living standards.

5 Conclusions

The most effective system for assessing the population's living standards should be addressed by minimizing negative factors and integrating quantitative and qualitative indicators. It should be considered that the differentiation of the subjects of the Russian Federation is so great that it is not possible to determine the average indicator of the population's living standards throughout the territory, which justifies the difficulty of identifying the most significant criteria that determine this indicator, since the determining criterion for a better life support for some may be insignificant for others. This is the complexity and the need to develop a new effective method of determining the population's living standards of the Russian Federation.

References

1. Kulikov, N.I., Vdovina, E.S.: State Educational Institution of Higher Professional Education "Tambov State University". Tambov, C.: Assessment of the population's living standards of the region. Issues of modern science and practice **30**(7–9). University named after V.I. Vernadsky (2010)
2. http://www.gks.ru/ - the official portal of the Federal Service of State Statistics of the Russian Federation
3. Zubets, A.N.: Origins and history of economic growth. A.N. The tooth, M.: Economics (2014). 463 p.
4. Correction of the methodology for a comprehensive assessment of the development of regional economic systems in the conditions of forming an innovation-type economy. Zakharov, P.N., Nazvanov, K.V.: Regional economy: theory and practice, vol. 33, pp. 35–47 (2014)
5. Rossoshansky, A.I.: Assessment of the population's living standards: review of methodological approaches. Young Sci. **11**, 440–445 (2013)
6. Subbeto, A.I.: Living standards management and human survival. Stand. Qual. **1**, 63–65 (1994)
7. Balakrishnan, N.: Parametric and Semiparametric Models with Applications to Reliability, Survival Analysis, and Living Standards. Springer Science + Business Media, LLC (Birkhäuser Basel Publ.) (2004)
8. Bock, H.H., Hayashi, E.C., Yajima, K., Ohsumi, N., Tanaka, Y., Baba, Y. (eds.) Data Science, Classification, and Related Methods: Proceedings of the Fifth Conference of the International Federation of Classification Societies (IFCS 1996), Kobe, Japan, 27–30 March 1996. Springer (1998)
9. Galbraith, J.: The Affluent Society. Houghton Mifflin, Boston (1958)
10. Kyritsis, Z., Papadopoulou, A.: The living standards via semi Markov reward modeling. Methodol. Comput. Appl. Probab. (2016)

Interconnection Between the Categories of Region's Self-development and Population's Living Standards

Zhanna A. Zakharova[1]([✉]) and Valery V. Bogatyrev[2]

[1] Vladimir Branch of the Russian Academy of National Economy
and Public Service under the President of the Russian Federation,
Vladimir, Russian Federation
zjane77@mail.ru
[2] Vladimir State University, Vladimir, Russian Federation
mp_ved.vlsu@mail.ru

Abstract. The article investigates the relationship between the categories of the region's self-development and the quality of life of the population. The purpose of writing this article was to test the hypothesis of a stable relationship between the categories of the region's self-development and the quality of life of its population. The authors based on the review of existing developments proposed a methodological approach to quantifying the parameters of the region's self-development and the quality of life of the population. The method of quantitative assessment proposed by the authors envisages the use of official statistics and suggests the definition of two integrated integral indicators reflecting the level of self-development of the region and the quality of life of the population. Approbation of the methodology was carried out on the example of five regions of the Central Federal District of the Russian Federation (Vladimir, Ivanovo, Kaluga, Kostroma and Yaroslavl Oblasts), typical of central Russia and comparable in key parameters of social and economic development. The results of the study are of interest to public authorities at the regional level, the teaching staff of higher education institutions, post-graduate students and students studying in economic areas of training. The authors also determined the directions for further exploratory research in terms of clarifying the list of direct and inverse indicators that reflect the parameters of the region's self-development and the quality of life of the population.

Keywords: Economy · Region · Self-development · Quality of life
Population · Integral indicator

1 Introduction

From the point of view of the performance of any social and economic system (municipal formation, region, country), the quality of life of the population as a whole and of a single individual is of undoubted importance. In this regard, it is necessary to identify those factors that have a decisive influence on improving the quality of life of the population of the territory. The degree to which public needs can be satisfied in

© Springer International Publishing AG, part of Springer Nature 2018
E. G. Popkova (Ed.): HOSMC 2017, AISC 622, pp. 440–447, 2018.
https://doi.org/10.1007/978-3-319-75383-6_56

general cannot be considered separately from the satisfaction of the needs of a single individual in basic and derivative needs (higher-level needs).

The purpose of writing this article was to test the hypothesis of a stable relationship between the categories of the region's self-development and the quality of life of its population.

2 Theoretical Basis of the Research

In the modern specialized scientific literature, several approaches to the classification of regions are singled out according to the criteria of self-development [1, 5, 7, 8, 11–15].

From the point of view of one of the approaches, where the criteria mainly consider the indicators of budget provision and the dynamics of the change in the gross regional product (hereinafter - GRP), those regions in which the growth of the regional product is provided with only own or reduced use are the self-developing ones external sources.

According to another approach, researchers consider the self-development of the economy as its development based on economic (commercial) calculation, the postulates of which are self-sufficiency, self-financing and profitability. Therefore, one of the main indicators of self-development is the share (in percent) of profitably operating enterprises and economic organizations of their total number. At the same time, it is noted that the self-development of the economy and the region is possible only when the share of profitable organizations and enterprises will be approximately 80–90%. The second indicator suggests the share of small business in the gross regional product, which, for comparison, in the industrialized countries reaches 40–60% [3].

However, although the authors of these approaches also speak about the need to improve the quality of life of the population of the region in conditions of self-development, they do not offer any indicators in its evaluation. We, like many other researchers, adhere to this approach, that the self-developing regions are those that, while growing GRP, are at the same time socially oriented. For example, Tatarkin A.I., Tatarkin D.A. and Sidorova E.N., correlate the process of evolution of underdeveloped territories from the state of subsidization to financial self-sufficiency and self-development with a change in the level and quality of life of the population in the social security system, defining the main indicator of the latter as the social sphere in GRP. Naturally, according to the authors, there is no direct relationship between the GRP per capita of the region and the indicators characterizing the share of the social sphere in GRP [10] or indicators reflecting the quality of life of the population of the territory. Moreover, the inverse relationship is more common - the higher the GRP level, the lower the share of the social sphere. As the authors explain, this may well be connected with the policy of inter-budget equalization, since in the subsidized regions, through the redistribution of financial resources, the level of financing of social services rises to a certain average Russian level in absolute terms, while the comparison base - GRP per capita - remains at the same low level. This situation can be explained by the presence of a disproportionately high share of the social sphere in relation to the level of development of the regional economy [10].

From our point of view, self-development of the region is an opportunity to use the resource component in terms of providing the region with the necessary amount of resources (both existed and delivered), provided that the potential of the region's self-development (internal and external sources) is available in terms of opportunities that can be realized in the future and existence of motives for subjects of economic activity of the region to sustainable social and economic development in the long term [6].

At the same time, not improving the qualitative characteristics of the well-being of the population with the growth of GRP regions having the potential for self-development or seeking to develop themselves can enhance the potential for self-development of other regions that are to some extent already self-developing. Thus, the movement of skilled labor resources from the region with a low level of wages, or unfavorable living conditions (insufficient and poor development of the social infrastructure) into the region capable of satisfying the demands (both in terms of wages and living conditions) of the fluid part of enterprising people [2] can strengthen the potential for self-development of a self-sufficient region.

When we correlate the region's self-development and achievement of the quality of life indicators of the population of the region, we do not focus on the share of the social sphere in the GRP of the region, and first, we single out the block of self-development indicators of the region, taking into account both internal sources of self-development and interaction with the external environment, which reflects the level of the well-being of the population, the degree of satisfaction of the needs of the population of the region, as well as the ecological effect.

Therefore, within the framework of our research the self-development block includes such criteria as:

- the share of self-developing industries, types of economic activity in the GRP structure of the region[1], in%; The value of this indicator is formed as a share of economic activities "Wholesale and retail trade; repair of motor vehicles, motorcycles, household products and personal items" and "Real estate transactions, rental and provision of services" in the GDP structure of the region.
- the degree of openness of the regional economy, which can be measured by such indicator as the ratio of the volume of foreign trade turnover of the region to GRP, in%. The region is not a closed socioeconomic system. Increasing the degree of openness of the regional economy allows more actively attracting external resources in the form of investments, technologies, information, etc. Other things being equal, the openness of the region's economy increases its competitiveness and efficiency;
- indicators of self-organization of economic activity of the population of the region: The number of economically active population; the number of individual entrepreneurs and small business organizations per 10,000 people in the territory;
- the proportion of loss-making enterprises in the total number of registered enterprises in the region, as an indicator reflecting the efficiency of enterprises in the region (less than 20%).

---

[1] We refer industries and types of economic activity to self-developing ones, which, in a market economy, can independently by accumulating domestic resources and internal motives (high profitability, the possibility of increasing sales volumes and increasing market share, etc.) respond to changes in demand (or changes in the external environment), ensuring its stable state in the medium and long term.

To indicators reflecting the quality of life of the population of the territory, we include:

- unemployment rate;
- per capita income of the population;
- share of the population with incomes below the subsistence level;
- retail turnover per capita;
- volume of paid services per capita;
- security with housing;
- The volume of emissions of pollutants into the atmosphere (thousand tons per million RUB. GRP);
- morbidity per 1000 people of the population;
- share in the costs of citizens of the costs of payment for services (communal, health, etc.)

3 Research Methodology

The methodological basis of the research is general scientific research methods (analysis and synthesis, deduction and induction, normative and positive methods, etc.), quantitative econometric methods (summary and grouping of statistical data, regression analysis, and statistical indexes).

4 Analysis of Research Results

In the course of the research, a methodology was developed to assess the impact of the region's self-development on the quality of life of the population. The proposed methodology provides for the formation of a list of two groups of indicators. The first group of indicators reflects the level of self-development of the region's economy. The second group of indicators reflects the quality of life of the population of the region. Taking into account the fact that the indicators are multidimensional and multi-directional, the methodology assumes conducting normalization in order to bring them to a comparable form. The implementation of the above procedures allows us to calculate the integral indicators of the self-development of regions and the quality of life of the population. Comparison of integrated indicators allows quantifying the tightness of communication and the nature of the impact of the region's self-development on the quality of life of the population.

As an empirical basis for the study, the official statistics of 2015 were used for five regions of the Central Federal District of the Russian Federation (Vladimir, Ivanovo, Kaluga, Kostroma and Yaroslavl Oblasts), which are typical for the general population and are comparable in the main socioeconomic indicators.

At the first stage of the study, we summarized the indicators reflecting the parameters of sustainable development of the above regions of Russia, presented in Table 1.

Table 1. Indicators of regional self-development[a]

No.	Indicator/Region	Vladimir Oblast	Kaluga Oblast	Ivanovo Oblast	Kostroma Oblast	Yaroslavl Oblast
1	GRP total, RUB billion	327.89	324.94	151.05	146.31	388.14
2	Specific weight of self-developing types of economic activity in the GRP structure of the region, %	25.1	25.9	27.7	21.9	27.4
3	Number of PI per 10000 population	131	135	198	102	172
4	The share of unprofitable enterprises (in % of the total number of enterprises in the region)	32.4	30.8	40.3	35.6	36.6
5	Degree of openness of the region's economy (foreign trade turnover of the region to GRP), in %	34	71.05	24.69	17.13	21.98

[a]Constructed by: [9]

Taking into account the multidimensional of the indicators presented in Table 1, the authors carried out a normalization in accordance with the approach proposed by Zakharov and Nazvanova [4]. The results of normalization are given in Table 2.

Table 2. Normalization of the indicators of the region's self-development[a]

No.	Normalized indicator/Region	Vladimir Oblast	Kaluga Oblast	Ivanovo Oblast	Kostroma Oblast	Yaroslavl Oblast
1	GRP	0.8448	0.8372	0.3892	0.3770	1.0000
2	Specific weight of self-developing types of economic activity in the GRP structure of the region, %	0.9061	0.9350	1.0000	0.7906	0.9892
3	Number of PI per 10000 population	0.6616	0.6818	1.0000	0.5151	0.8687
4	Specific weight of unprofitable enterprises	0.9506	1.0000	0.7643	0.8652	0.8415
5	Degree of openness of the region's economy	0.4786	1.0000	0.3500	0.2400	0.3100
	Integral indicator of the region for self-development	0.7456	0.8820	0.6351	0.5024	0.7412

[a]Constructed by: [9].

The results of normalization made it possible to determine the integral index of the region's self-development, namely: Vladimir Oblast is 0.7456, Kaluga Oblast is 0.8820, Ivanovo Oblast is 0.6351, Kostroma Oblast is 0.5024, and Yaroslavl Oblast is 0.7412.

The next stage of the study was to determine the quality of life indicators in the regions included in the sample (Table 3).

Table 3. Indicators of the quality of life in the regions

No.	Indicator/Region	Vladimir Oblast	Kaluga Oblast	Ivanovo Oblast	Kostroma Oblast	Yaroslavl Oblast
1	Unemployment rate, %	5.6	4.3	5.6	5.3	5.3
2	Average per capita monetary income of the population, RUB/month	23732	27550	22560	22466	27369
3	The share of the population with incomes below the subsistence minimum,% of the total population of the subject of the Russian Federation	14.1	10.9	15.8	14	10.5
4	Retail turnover per capita, thousand RUB	139.8	167.22	143.8	131.3	160.8
5	The volume of paid services per capita, RUB/person	47744	43042.00	37212.00	36823.00	37997.00
6	The volume of emissions of pollutants into the atmosphere (thousand tons per billion RUB. GRP)	0.09	0.08	0.21	0.30	0.23
7	Morbidity per 1000 people of the population;	937.7	712.40	877.50	768.90	857.50
8	Share in the costs of citizens of the costs of payment for services (communal, health, etc.), %	28.5	24.50	23.40	22.20	27.90
9	Housing security (area of living accommodation per 1 inhabitant), sq. meters.	27.2	28.30	25.20	26.60	26.00
10	Specific weight of household expenditures on housing and communal services, %	10.5	8.60	10.30	7.70	10.40

The results of normalization of the indicators reflected above are given in Table 4.

Table 4. The results of the normalization of the quality of life indicators of the population

No. in order	Indicator/Oblast	Vladimir Oblast	Kaluga Oblast	Ivanovo Oblast	Kostroma Oblast	Yaroslavl Oblast
1	Unemployment rate	0.7679	1.0000	0.7679	0.8113	0.8113
2	Income of the population, per capita	0.8614	1.0000	0.8189	0.8155	0.9934
3	Share of the population with incomes below the subsistence level	0.7447	0.9633	0.6646	0.7500	1.0000
4	Retail turnover per capita	0.8694	1.0399	0.8943	0.8165	1.0000
5	Volume of paid services per capita;	1.0000	0.9015	0.7794	0.7713	0.7958
6	The volume of emissions of pollutants into the atmosphere	0.8889	1.0000	0.3810	0.2667	0.3478
7	Morbidity per 1000 people of the population	0.7597	1.0000	0.8119	0.9265	0.8308
8	The share in the expenses of citizens of the costs of payment for services	0.7789	0.9061	0.9487	1.0000	0.7957
9	Security with housing;	0.9611	1.0000	0.8905	0.9399	0.9187
10	Specific weight of household expenditures on housing and communal services, %	0.7333	0.8953	0.7476	1.0000	0.7404
Integral indicator of the region on the quality of life of the population		0.8319	0.9694	0.7508	0.7693	0.7946

As a result of the computational procedures, the following integral estimates of the quality of life of the population in the regions of the Russian Federation were obtained: Vladimir Oblast is 0.7456, Kaluga Oblast is 0.8820, Ivanovo Oblast is 0.6351, Kostroma Oblast is 0.5024, and Yaroslavl Oblast is 0.7412.

5   Conclusions

The study confirmed the hypothesis of authors about the existence of a causal relationship between the categories of the region's self-development and the quality of life of the population. At the same time, the potential for self-development is not always fully realized in order to improve the quality of life of the population (Ivanovo Oblast, having a higher potential for self-development, is inferior to the Kostroma Oblast in terms of an integral characteristic of the quality of life of the population). The authors also draw attention to the absence of significant discrepancies between the integral estimates for each group of indicators of individual regions.

The directions of the future exploratory researches of the authors are refinement and addition of lists of used indicators, as well as further approbation of the methodology for other regions of the Central Federal District of Russia. And the authors also plan to understand under what conditions the self-development of the region is ensured and the achievement of indicators of the quality of life of the population, and with which on the contrary reduction. The definition of these conditions will make it possible to develop recommendations for public authorities on the achievement of indicators of the quality of life of the population of the region in conditions of using the opportunities for self-development of these regions.

References

1. Abramova, E.A.: Problems of the formation and functioning of the mechanism of self-development of regional socioeconomic systems. Modern Sci. Intensive Technol. 1(41), 17–24 (2015). Regional appendix
2. Babayev, B.D.: Donor or "vampire"?!: Monograph, 271 p. Ivanovo, Moscow (2009)
3. Buvaltseva, V.I.: Essence and content of the process of self-development. In: All-Russian Scientific Conference "Science and Education". Collection of Scientific Papers, pp. 25–29. KemSTU, Belovo (2002)
4. Zakharov, P.N., Nazvanova, K.V.: Correction of the methodology for a comprehensive assessment of the development of regional economic systems in the conditions of forming an innovation-type economy. Reg. Econ. Theory Pract. 33(360), 35–47 (2014)
5. Zakharova, Z.A.: Approaches to sources of regional self-development. Multilevel Public Reproduction: Questions of Theory and Practice, vol. 12(28) (2017)
6. Zakharova, Z.A.: Optimization of the conditions for attracting private investment in the infrastructure of the regions. Reg. Econ. Theory Pract. 22(349), 49–56 (2014)
7. Zakharchuk, E.A., Pasynkov, A.F., Nekrasov, A.A.: Classification of regions of the Russian Federation according to the criteria of self-development. Econ. Reg. 3, 54–62 (2011). Molchanova M.U.
8. Molchanova, M.Y.: Criteria for the self-development of regions as the basis for the formation of intergovernmental fiscal relations. Fundam. Res. 6(1), 131–135 (2013). https://www.fundamental-research.ru/en/article/view?id=31429. Accessed 12 Apr 2017
9. Regions of Russia. Socioeconomic indicators. Digest of articles/Rosstat, 1326 p. (2016). ISBN 978-5-89476-428-3
10. Tatarkin, A.I., Tatarkin, D.A., Sidorova, E.N.: Transformation of the subsidized regions of Russia into self-developing. Federalism 4, 39–52 (2012)
11. Bielikova, N.V., Ivanova, O.Y.: Formation of regions financial self-sufficiency in the context of economic reforms implementation. Actual Probl. Econ. 177(3), 219–228 (2016)
12. Karpenko, T.V., Zaloznaya, D.V., Volodina, T.V., Belousova, L.F., Breusova, E.A.: Prospective mechanisms of peripheral areas investment and innovation potential formation. Asian Soc. Sci. 11(20), 112–118 (2015)
13. Akberdina, V.V., Grebyonkin, A.V., Bukhvalov, N.Y.: Simulation of innovative resonance in the industrial regions. Econ. Reg. 4, 289–308 (2015)
14. Animitsa, E.G.: Outlines of the theory of urban agglomerations self-development. Econ. Reg. 1, 231–235 (2012)
15. Rumyantsev, A.A.: Science and innovation space of a macro region: Prospects of innovative territorial development. Stud. Russian Econ. Dev. 26(4), 379–387 (2015)

Research Trends of HR Management in Tourism

Svetlana N. Kaznacheeva[✉], Antonina L. Lazutina,
Tatyana V. Perova, Jeanne V. Smirnova, and Elena A. Chelnokova

Nizhny Novgorod State Pedagogical University Named After Kozma Minin
(Minin University), Nizhny Novgorod, Russian Federation
cnkaznacheeva@gmail.com, lal74@bk.ru,
perova_tatyana83@mail.ru,
z.v.smirnova@mininuniver.ru, chelnelena@gmail.com

Abstract. The authors turn to the consideration of personnel management in tourism as the most important sphere of activity of the modern tourist organization. The authors point out the fact that at present the tourism industry is one of the fastest growing spheres. Personnel management reflects the importance of the sphere of people management in organizations, as it affects one of the main aspects of the organization's activities, personnel management. The authors present the interpretation of the concept "HR management", define the goals and tasks of personnel management. The peculiarities of tourist activity are considered. The article presents data of average salary proposals for the tourism industry in 2016 and dynamics of average salary offers for the cities of the Russian Federation for 2016. The authors outline the main tendencies of personnel management in the sphere of tourism: the staff turnover in Russia in 2016 reached peak values; in the future among candidates for employment in the tourist company will be more and more representatives of the Millennium generation; many tourist organizations are losing capital because of the low emotional intelligence of line managers; tourism enterprises are looking for employees in social networks; a good social package is a serious advantage of a tourist organization; tourist organizations are focused on the internal staff development, rather than looking for ready-made professionals; charity is an occasion for pride of employees for the company; The loyalty of Russian employees towards the employer is growing more actively than the Western colleagues; expected the growth of competition in the regional labor market of tourist services.

Keywords: Management · Personnel management · Tourism
Tendencies of personnel management in the sphere of tourism

1 Introduction

In any sphere of human activity, one of the important conditions for the success of an organization is the effective use of human resources. To properly manage personnel and create an environment for employees in which they could realize their potential is the task of personnel management.

© Springer International Publishing AG, part of Springer Nature 2018
E. G. Popkova (Ed.): HOSMC 2017, AISC 622, pp. 448–455, 2018.
https://doi.org/10.1007/978-3-319-75383-6_57

2   Theoretical Basis

The notion of «cadre management» came into scientific circulation at the end of the 20th century. The concepts of «personnel management» and «HR management» are very close to each other, but «HR management» reflects the importance of the sphere of people management in organizations deeply. Leadership in the formulation of the problem of personnel management belongs to E. Mayo. He considered personnel management as an installation for professional competence. D. McGregor believed that human resources management is aimed at ensuring social control of the organization's daily work. K. Levin, F. Herzberg dealt with the possibilities of personnel management in the formation of initiative and self-realization of personnel. In general, the Western theory of personnel management implies individualization of organizational standards to strengthen corporate loyalty. Personnel management is understood as «a combination of organizational, socio-psychological and psychological means (forms and methods) allowing to solve various «human» problems and tasks» [5]; «a direction of management studying the development of the company's personnel potential for the purpose of hiring and training effectively employees in the company» (Bazarov and Eremina) [1]; purposeful activity of the management structure of the organization, managers and specialists of the personnel management system, including the development of the concept and strategy of personnel policy, principles and methods of personnel management [6].

Under the personnel management we understand the system of planning, organization, motivation and control of personnel necessary for the formation and achievement of the objectives of the company.

The objectives of personnel management are:

- «the ability to work with people, to select and evaluate them properly, to seek their interest in improving their qualification level» [8];
- «satisfaction of the organization's need for qualified personnel and to use them effectively with consideration of opportunities for self-realization of each employee within the framework of this organization» [12, p. 4].

The tasks of HR management change depending on the stage of development of the organization. At any stage of the organization's development, tasks such as labor market analysis, employee adaptation, motivation, corporate culture, career management and so on are solved. During the stabilization period, the main tasks of personnel management: analysis of activities, detection of sources of losses and construction of personnel work; regular evaluation procedures.

3   Research Methodology

Personnel management is necessary in any organization, especially in the organizations of the tourist industry, as this industry develops quickly today. One of the main objectives of the development of tourism business in Russia is the need to increase the efficiency of using the human resources of tourist enterprises.

Features of tourism activities which determine the nature of work in this area [11]:

- «a relatively large proportion of living labor, which makes it difficult rationing;
- a high degree of influence on the process of production and sales of tourist products subjective factors both from employees of the tourist enterprise, the firm, and the client;
- the complexity of the production of a tourist product, which is the result of well-coordinated work of independent collectives, whose work is subordinated to the same goal - the satisfaction of the client's needs;
- the availability of productive and unproductive labor;
- features of labor in tourism assume other approaches to personnel policy, organization of recruitment, selection, reception of personnel, career guidance and adaptation, training and management of a business career, professional and professional promotion, etc.»

As analysts note, the demand for work in the sphere of tourism is decreasing, and although the number of vacancies in most regions of the country until recently grew, there were fewer people wishing to work in the industry. For example, in St. Petersburg in 2016 the number of vacancies in tourism increased by 13% compared to the same period last year. Positive dynamics of the growth of vacancies was also observed in the Moscow and Sverdlovsk regions, in the Republic of Bashkortostan, Krasnodar region and Moscow. The maximum share of vacancies in the tourism of the country falls on Moscow (39%) and Petersburg (13%), however, comparing with 2010, the number of vacancies in them also decreased. Employers usually do not focus on specialized education in the field of tourism.

To the education in the tourist sphere, the applicants are more likely to choose themselves, which purposefully choose tourism as a profile of future activity. In connection with the absolutely undeveloped sphere of tourism in the Soviet time, specialists born no earlier than 1980 have a profile higher education. The received education in the sphere of tourism for the employer is only a sign that the person intends to grow and develop exactly in the tourist business.

According to HeadHunter, tourism still remains a female profession: 82% of applicants are women.

According to analysts, candidates for work in this sphere are well-educated, experienced and active, as almost half of them are between the ages of 26 and 35, 38% have more than 6 years of experience, two-thirds have higher education, and a third speaks English fluently [9].

Knowledge of one or more foreign languages is more important for many companies, especially if the company actively develops such areas as Spain, Italy, England, Germany, Latin America, USA, Canada, when it is important to know the national language. When working on other «non-math areas», such as Asia, Cuba, Africa, China, discussion are conducted in English. But experience is also important here. It usually takes a conversational level, as well as keeping correspondence in a foreign language [10].

The most popular requests for personnel in 2016 were: a sales manager, a tourism manager, a hotel administrator, a chef, a waiter (Table 1).

Table 1. Average salary offers in December 2016 (RUB) [4]

City	Tourism manager	Sales manager (hotel services)	Hotel administrator	Cook	Chef
Moscow	48 000	52 000	30 000	35 000	70 000
St. Petersburg	37 920	41 080	23 700	27 650	55 300
Novosibirsk	31 200	33 800	19 500	22 750	45 500
Yekaterinburg	32 160	34 840	20 100	23 450	46 900
Nizhny Novgorod	28 800	31 200	18 000	21 000	42 000
Kazan	27 360	29 640	17 100	19 950	39 900
Samara	29 760	32 240	18 600	21 700	43 400
Chelyabinsk	29 760	32 240	18 600	21 700	43 400
Rostov-on-Don	30 720	33 280	19 200	22 400	44 800
Ufa	29 760	32 240	18 600	21 700	43 400
Krasnoyarsk	30 720	33 280	19 200	22 400	44 800

In general, on the labor market the salary offers of employers are at the same level as a year ago. Negative dynamics indicates an increase in the vacancy rate to less qualified personnel (waiters, maids). Positive dynamics indicate a change in demand in favor of vacancies in managerial positions (Table 2).

Table 2. Dynamics of proposals in 2016 [4]

City	Average salary offers in the sphere of tourism and hotel business, December 2015, RUB	Average salary offers in the sphere of tourism and hotel business, December 2016, RUB	Growth/fall, %
Moscow	32 100	34 400	7.2
St. Petersburg	26 900	28 900	7.4
Novosibirsk	29 700	28 500	−4.0
Yekaterinburg	23 800	25 300	6.3
Nizhny Novgorod	22 100	24 300	10.0
Kazan	20 800	21 500	3.4
Samara	28 700	27 300	−4.9
Chelyabinsk	27 600	26 300	−4.7
Rostov-on-Don	27 400	27 200	7.2
Ufa	23 500	24 800	7.4
Krasnoyarsk	28 200	27 300	−3.2

Today tourism managers in the personnel market are quite a lot. But for the tourist industry, the sale of tours is a key position in the range of services, so experienced tourism managers are always needed. For this position, for a specialist the most important thing is to have a highly specialized experience in sending tourists to one country or one direction. Also, the feature of this position can be called the fact that the employer will soon recruit a young specialist with work experience of about 1–2 years than a competent manager with extensive experience after 35–40 years. It is believed that having experience in tourism for 5 years or more, you can take leadership positions or set up a business.

According to analysts from Job.ru and Hr Data Center, a lot of recent events in the political field and causing huge damage to the tourism industry will bring many difficulties to market participants next year. Those who can successfully reorient to new directions, actively master the domestic market, it may be possible to avoid staff reductions within companies. The labor market in the hotel segment seems to be more stable in this respect, as the demand for domestic tourism will increase, in this case, it is possible to expect an increase in the number of vacancies for new employees [4].

So, the management of the tourist organization needs to not only implement innovations, but also adapt to modern staff, for the effective operation of the enterprise, because the newly admitted employees are representatives of the Millennium generation, which cannot be maintained with the usual methods.

4 Analysis of Research Results

The main tendencies of personnel management in the sphere of tourism:

Trend 1. The turnover of personnel in Russia reached a peak in 2016. According to forecasts of Hay Group experts, a global consulting company in the field of management, staff turnover reached its maximum in 2016. It was about 28%. Such trends are typical not only for the Russian labor market, by 2018 the turnover of personnel on a global scale will grow to 23–24%. In general, about 190 million people will be replaced by employers. Therefore, already now it is very important to start working with the staff. Moreover, the level of staff turnover is 54% lower in organizations that monitor the level of staff involvement and create favorable working conditions [3]. From this it follows that new personnel need to be interested in work.

Trend 2. Among candidates in the tourist company will be more and more representatives of the Millennium generation. The Millennium generation is people involved in digital technology. These are impulsive and ambitious people. They are not confused by the frequent change of work. Self-realization is important for this generation. They are in constant search of themselves and are not afraid to take on the difficult work. Such young people are ready to spend personal time for an interesting project. Representatives of the Millennium generation live today and do not plan for a long time [7]. Under the influence of the Millennium generation, such major Internet resources as YouTube and Facebook, have began to develop. Therefore, if you do not put new interesting tasks before such an employee, then the employee can easily leave, despite the high salary.

Trend 3. Many tourist organizations are losing capital because of the low emotional intelligence of line managers. About half of all workers are dissatisfied with their salary. Many employees complain about the lack of connection between the size of payments and the work done. In most cases, workers do not have the opportunity to develop within the organization. Therefore, the main reason for the employee's replacement of the place of work is the head. To retain valuable personnel, the HR manager should be assisted, established with the employees of communication, encouraged and directed them.

Trend 4. Tourism enterprises are looking for employees in social networks. According to the statistics of Kelly Services (Fig. 1), the number of people who are looking for work through social networks has increased by 5% [2]. From this it follows that the placement of vacancies in such Internet resources is quite effective. Also, a personal page gives an accurate psychological portrait of a person and shows the true working motives and interests.

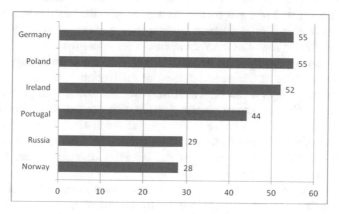

Fig. 1. Percentage of people who are looking for work through social networks [10]. Source: Kelly Services data

Trend 5. A good social package is a serious advantage of a tourist organization. In the conditions of the current crisis, for young specialists it is very important to get some benefits in addition to basic earnings. Experts believe that now it will be increasingly difficult to attract employees to those companies that can not offer such conditions. For young workers, the most attractive organization is an organization where they can get an opportunity to improve their skills.

Trend 6. Tourist organizations are focused on the internal staff development, instead of looking for ready specialists. This trend can be traced in 48% of Russian companies and 73% of foreign firms. The main aim is to increase the efficiency of employees. It should be noted, the costs of finding and recruiting staff are declining year by year, spending on internal training has been reduced in about 15% of organizations. Nevertheless, business is focused on the internal development of personnel. Most firms have programs to identify and develop the abilities of employees. Many successful top-managers of the world devote time to personal participation in the development of talents of their subordinates.

Trend 7. Charity is an occasion for the pride of employees for their company. HeadHunter – the Russian company of Internet recruitment, job search. According to its research, every third Russian organization is engaged in charity. Most often this is support for socially disadvantaged groups of the population. Business should not be configured only for profit, it should benefit society. Most employees are proud that the organization in which they work is engaged in charity.

Trend 8. Loyalty of Russian employees towards the employer grows more actively, than at the western colleagues. According to the results of the study of the international recruitment company Kelly Services, recently in Russia the loyalty of subordinates towards the employer has increased substantially. At the moment, 37% of Russian specialists stated full trust to their employer, while in Europe the percentage of loyalty is 26% [9].

Trend 9. Competition is expected to grow in the regional tourism labor market. Most companies want to outsource individual functions to regions. This is connected with the development of regional centers, as well as with the fact that organizations are looking for replacement of employees from distant corners, since such people have greater purposefulness and efficiency compared with urban residents.

So, to solve the problem with personnel, first of all, it is necessary to conduct a comprehensive analysis of the current system of motivation. It needs to be adapted to young employees, taking into account their interests and values. Also, it is necessary to change the hierarchical system of communication with personnel on a linear-personal. Only the right communication with a young specialist will allow him or her to develop, show initiative and be responsible for the result of his or her actions. It is worth noticed that the managers of the company copy the behavior of their bosses, so it is very important for them to pay attention to communication with staff. It is also important to use social networks to find staff. They are convenient for placing detailed information about the company, as well as for finding really worthy personnel.

References

1. Bazarov, T.Y., Eremina, B.L.: Personnel Management: Textbook, 2nd edn., Pererab. And additional. Unity, Moscow (2002)
2. Research Kelly Global Workforce Index (KGWI): more and more Russians are looking for work through social networks (Electronic resource). http://www.kellyservices.ru. Accessed 10 Apr 2017
3. Researching the level of staff effectiveness (Electronic resource). www.haygroup.com. Accessed 10 Apr 2017
4. Personnel management (Electronic resource). http://www.managementnews.ru/termin/40/. Accessed 12 Apr 2017
5. Human resources management (Electronic resource). https://psyera.ru/kadrovyy-menedzhment-1911.htm. Accessed 13 Apr 2017
6. Personnel management as the most important factor of the company's survival in the conditions of the formation of market relations in the Russian Federation (Electronic resource). http://www.learnmanage.ru/lmans-712-1.html. Accessed 13 Apr 2017
7. Kaznacheeva, S.N., Chelnokova, E.A.: The brand as a tool for creating competitive advantages of the company. Mod. Sci.-Intensiv. Technol. Regional Annex 2(42), 16–21 (2015)

8. Mnushko, Z.N., Pestun, I.V.: Personnel management: principles, tasks, directions, efficiency (Electronic resource). http://www.provisor.com.ua/archive/2004/N10/art_27.php?part_code=73&art_code=4176. Accessed 13 Apr 2017
9. Staff 2015: Do not you have to cut down? (Electronic resource). http://www.kellyservices.ru. Accessed 13 Apr 2017
10. Trends in personnel management: 9 trends in headhunting and personnel management (Electronic resource). http://www.gd.ru. Accessed 13 Apr 2017
11. Nature of labor in the field of tourism (Electronic resource). http://www.kukiani.ru/index.php?page=content&subpage=s&r=7&p=16&s=54. Accessed 30 Mar 2016
12. Shchekin, G.V.: The Fundamentals of Personnel Management: A Textbook, 5th edn., 280 p. The stereotype. MAUP, Kyiv (2004)
13. Sakharchuk, E.S.: Analysis of practice centered aspects of educational programs in the sphere of tourism and hospitality. World Appl. Sci. J. 27(Education, Law, Economics, Language and Communication), 305–308 (2013)

The Model of Effective Work
of a Transport Node

Anatoly G. Kitov[✉], Artem A. Sirotkin, Vladimir N. Nosakov,
Anatoly A. Permovsky, and Alexander I. Fedoseev

Nizhny Novgorod State Pedagogical University Named After Kozma Minin
(Minin University), Nizhny Novgorod, Russian Federation
anatolykitov@yandex.ru, arsirotkin@rambler.ru,
nosakovvladimir@yandex.ru, ttpis@yandex.ru,
fai_44@list.ru

Abstract. In the modern conditions, great attention is paid to finding opportunities to increase the efficiency of use of vehicles and handling equipment in transport nodes, as well as increasing the standards of cargo operations. In many ports that function as transport nodes, the increase in vehicle handling standards is achieved mainly through the opening of internal reserves, Whereas the cardinal solution of this problem is connected with the perspective of development of the berthing front. From how the maintenance of vehicles in transport nodes is coordinated, how the reloading technique is used depends on the duration of unproductive downtimes of vehicles and handling equipment. The most effective way of transshipment of goods in transport nodes is the transshipment of goods under the direct option: A vehicle is a vehicle without temporary storage. The interaction of various modes of transport is considered when implementing the operational planning of the transport node. At the same time, the moments of arrival of the adjacent modes of transport in the transport node and the work of the handling equipment in the port are coordinated. To ensure the efficiency of cargo transshipment in transport nodes, it is proposed to use a probabilistic model that takes into account the interaction of various types of transport and reloading equipment in the transport node.

Keywords: Transport node · Cargo handling · Direct option
Probabilistic model · Efficiency

1   Introduction

The transport node is the point of intersection and branching of the communication routes of several modes of transport. Therefore, the transport nodes include whole sets of devices of these types of transport: railway, river, sea and automobile nodes. Transport nodes are connected with transport systems and are distinguished by a rather high level of infrastructure complexity [9]. In the transport system, the nodes play a regulating function.

One of the variants of the transport node is the seaport, in which work is carried out, for example, with container loads [8, 11, 12].

© Springer International Publishing AG, part of Springer Nature 2018
E. G. Popkova (Ed.): HOSMC 2017, AISC 622, pp. 456–461, 2018.
https://doi.org/10.1007/978-3-319-75383-6_58

The most important problem of transport nodes is the optimization of the inter-action of traffic flows in order to minimize the mutual expectation of vehicles.

A variety of industry structures deal with the issues related to transportation and other forwarding services provided at transport nodes, for example, in the OJSC "Research Institute for Road Transport" this is the cargo transportation administration [5, 7].

At the same time, in [7, 11] it is noted that "forwarding services by various types of freight forwarding companies affect the distribution of functions included in the standard freight forwarding service between the links of the companies, transport nodes and customers that interact during the delivery process".

The main purpose of this study is to analyze the quality of operational planning for the interaction of traffic flows in a transport node with the identification of factors that affect unproductive idle vehicles. The researchers set and tried to solve the following problems:

- analyze the factors affecting the quality of vehicle interaction in transport nodes;
- to develop a mathematical model of cargo interaction of conveyances in a transport node;
- calculate optimal characteristics of the berthing front and handling equipment;
- to test the mathematical model on a real transport node.

2 Theoretical Basis of the Research

The transport node is characterized by the following features [4, 10, 12]:

- The organization of uninterrupted operation of all types of transport, which consists in the timely and full satisfaction of cargo owners needs in transportation;
- Complexity of the processes implemented in the node (interaction of various modes of transport, sorting, loading and unloading of a large volume of goods of different nomenclature, integrated passenger services [3, 6, 11]);
- The possibility of dividing a node into a large number of interconnected and interacting subsystems and elements whose functioning is subject to a common goal;
- Hierarchical structure of links of individual subsystems of the node and performance criteria;
- The availability of a control system that ensures the intensive use of technical devices and throughput, as well as transportation and transshipment of cargo at minimal cost;
- Resistance to the effects of fluctuations in traffic and other parameters.

Therefore, any failure in the operation of one such node can lead to problems in the entire transport system [11, 12].

The efficiency of the transport node is associated with the synchronization of the receipt of vehicles for cargo handling and the performance of the handling equipment of the unit [2, 13].

3   Research Methodology

The studies were carried out based on a system analysis of the existing technology of interaction of traffic flows in rivers and seas of Russian ports. The results of the analysis made it possible to substantiate a mathematical probabilistic model of the functioning of the port as a transport node. In the future, the probabilistic model was used to develop optimal parameters for the entire berthing complex of the transport node.

If we take a look at the transport node at which transshipment of goods from road transport to river vessels and vice versa is performed, and the efficiency of the system "car - river port vessel" will be estimated by the minimum of the total reduced costs for the vehicles participating in this process and the transport node [2, 14]:

$$C = C_{vehicles} + C_{tr.node}, \ \text{rub/tons}. \tag{1}$$

Where

$C_{vehicles}$ - specific reduced costs for vehicles, rub/ton;
$C_{tr.node}$ - reduced costs for the transport node, rubles/ton.

$$C_p = C_c + C_m, \ \text{rubles/tons} \tag{2}$$

Where

C_c - reduced costs for the coastal economy of the transport node, rubles/ton;
C_m - reduced costs for reloading mechanization of the transport node, rubles/ton.

Based on the studies carried out in [1], it is possible to represent Eq. (1) as follows

$$C = \frac{C_{pr} * (1 + {}'Y_{grr})}{B_{grr}Q_{year}} + \frac{n_{be} * (1,15 * L_{pc} * C_{pr} + n_m * C_m)}{\phantom{B_{grr}Q_{year}}}, \ \text{rub/ton}. \tag{3}$$

Where

C_{pr} - the resulted expenses on the maintenance of vehicles on parking, rub/day;
${}'Y_{grr}$ - relative waiting time of the transshipment vehicle;
B_{grr} - the rate of reloading, tons/day;
N_{be} - number of berths, units;
1,15   - factor, taking into account the distance between the vessels at the berth;
L_{pc} - is the length of a river ship, m;
C_{pr} - annual reduced costs for the maintenance of one running meter of the quay, rubles/m;
N_m - number of operating reloading mechanisms on one berth, units;
Q_{year} - annual volume of cargo processing, tons;

$$'Y_{grr} = \frac{k\upsilon * Qcday}{Qday + Bgrr} \; ; \tag{4}$$

$$K\upsilon = \frac{1 + \upsilon2}{2} \; ; \tag{5}$$

$$B_{grr} = n_{pr} * n_m * q_m, \; t/day; \tag{6}$$

$$T = \frac{Qday}{Bgrr} \; , \tag{7}$$

$$Q_{year} = Q_{day} * t_{year}, \; tons \tag{8}$$

Where

Q_{day} - daily volume of cargo processing, tons/day;
υ^2 - coefficient of variation in the duration of cargo handling;
Q_m - design capacity of one reloading mechanism, t/day;
T_{year} - the operational period of river transport during the year, days;
τ - is the utilization factor of the berth capacity

We transform the formula (3) with allowance for (4)–(8)

$$C = \frac{3\text{ст}*\tau}{Q\text{сут}} + \frac{k\upsilon*3\text{ст}}{Q\text{сут}} * \frac{\tau2}{\tau+1} + \frac{1,15*L\text{тс}*3\text{пр}+n\text{м}*3\text{м}}{t\text{год}*n\text{м}*q\text{м}} \; , \text{rub / ton} \tag{9}$$

We differentiate expression (9) and, equating first derivative to zero, we obtain the following expression

$$0.5 * \tau^4 + \tau^3 + \frac{(1 - A)}{3 + \upsilon2} \tau^2 - \frac{2 * A}{3 + \upsilon2} \tau - \frac{A}{3 + \upsilon2} = 0 \tag{10}$$

Where

$$A = \frac{Qday*(1, 15 * Lpc * Cpr + nM*Cm)}{Cpr * tyear*nm*qm} \tag{11}$$

Solve Eq. (10) by the Newton method [2].
Finding the real roots of Eq. (10), you can determine the optimal rate of cargo works, as well as the required number of berths

$$N_{pr} = \frac{Bgrr}{nm*qm} \; , \text{units.} \tag{12}$$

4   Analysis of Research Results

Approbation of the proposed methodology was carried out on the example of processing grain cargo in the Volgograd transport node. As a result of the implementation of mathematical modeling of the flow of traffic during September 2015, the optimal rate of reloading works was determined. It should increase by 14% (from 6,380 tons per day to 8,765 tons per day). In addition, the required number of berths is recommended to increase from two to three. However, due to the tightness of the water area in the Volgograd transport node it is not possible to extend the quay wall, it is necessary to install an additional handling device with a capacity of 3,176 tons per day to ensure the optimum rate of reloading operations. Because of these measures, the reduced costs for the fleet being processed and berths can be reduced by 13.7%, which is about 2.3 million rubles a year of profit.

5   Conclusions

In this way, the proposed method allows to determine the optimal parameters of the interaction of transport streams in the transport node, including the capacity of the reloading equipment and the required number of berths, which will generally improve the efficiency of the transport hub.

References

1. Verzhbitsky, V.M.: Numerical methods. Linear Algebra and Nonlinear Equations, 386 p. Vysshaya shkola, Moscow (2000)
2. Voronin, V.F., Kogan, V.E.: Conditions for the effectiveness of the system pushed structure – port. In the Collection: Optimal Planning of the Cargo Fleet: Works of GIIVT, the city of Gorky, vol. 173, pp. 72–80 (1980)
3. Devyatov, D.M., Permovsky, A.A.: Decrease in the quality of passenger traffic - the problem of modern urban transport. In: The collection Industrial Development of Russia: Problems, Perspectives Proceedings of the 12th International Scientific and Practical Conference of Teachers, Scientists, Specialists, Postgraduates, Students, 3 volumes, pp. 24–30 (2014)
4. Kitov, A.G., Fedoseev, A.I.: Evaluation of reliability of cargo transshipment by direct option in the transport node. In the Collection: Social and Technical Services: Problems and Ways of Development a Collection of Articles on the Materials of the III All-Russian Scientific and Practical Conference. Nizhny Novgorod State Pedagogical University named after Kozma Minin, pp. 205–207 (2017)
5. Kitov, A.G., Vakhidov, U.S., Shapkin, V.A., Shapkina, U.V.: Application of an innovative system of numerical simulation (fem) for the study of vibroacoustic characteristics (nvh) of automobile parts. Vestnik of the University of Minin, vol. 1, no. 1, p. 19 (2013)
6. Permovsky, A.A., Repina, R.V.: Determination of the quality criteria for passenger road transport. In the Collection: Social and Technical Services: Problems and Ways of Development a Collection of Articles on the Materials of the II All-Russian Scientific and Practical Conference. Nizhny Novgorod State Pedagogical University. K. Minina, pp. 216–219 (2015)

7. Sirotkin, A.A., Kitov, A.G.: Forwarding services on land transport: Current state and development prospects: Textbook, Moscow (2016)
8. Sirotkin, A.A., Kitov, A.G.: Forwarding service of container transportations (course design): Teaching-methodical manual, Nizhny Novgorod (2014)
9. Sirotkin, A.A., Repina, R.V.: Optimization of freight traffic in the context of a comprehensive analysis of the transport system. In the Collection: Industrial Development of Russia: Problems, Perspectives Collected Articles on the Materials of the XIII International Scientific and Practical Conference of Teachers, Scientists, Specialists, Postgraduates, Students. Department of Economics of the NGPU Named K. Minina, pp. 206–211 (2015)
10. Fedoseev, A.I., Kitov, A.G.: Development of a work plan for drivers. In the collection: Social and Technical Services: Problems and Ways of Development a Collection of Articles on the Materials of the II All-Russian Scientific and Practical Conference. Nizhny Novgorod State Pedagogical University. K. Minina, pp. 232–236 (2015)
11. Cherchenko, D.O.: Current state and directions of development of multimodal transport in the Rostov transport hub. In: International Scientific and Practical Conference World science, E. 1, vol. 12, no. 16, pp. 53–55 (2016)
12. Konings, R., Van Der Horst, M., Hutson, N., Kruse, J.: Comparative strategies for developing hinterland transport by container barge analysis for Rotterdam and U.S. ports. Transportation Research Record, no. 2166, pp. 82–89 (2010)
13. Brockel, H.C.: Today's port – a trade and transport hub. Ann. Am. Acad. Polit. Soc. Sci. 345(1), 95–102 (1963)
14. Fisenko, A.I., Kuleshova, E.A.: The role of the Russian Far East in the system of East-West international transport corridors. In: Problems and Trends in the Economy and Management in the Modern World Proceedings of the International Conference, pp. 605–609 (2012)

Analysis of Deceptive Communication Speech Acts in Linguistic Examination

Svetlana V. Kozmenkova[(⊠)], Timur B. Radbyl, Viktor I. Tsyganov, and Vasily A. Yumatov

N.I. Lobachevsky Nizhny Novgorod State University,
Nizhny Novgorod, Russian Federation
buhuchet@iee.unn.ru, sudexpert2011@mail.ru,
ziganovw@yandex.ru, yumatovva@yandex.ru

Abstract. Purpose of the research. The purpose of the study is to identify the differential signs of verbal acts of unfair information that are significant for conducting a linguistic expert study in disputed verbal and written texts.

Theoretical bases, methods and materials of research. The method of logical analysis of felicity conditions in the framework of the scientific instruments of the postclassical theory of speech acts was used in the work. The article summarizes the experience of 13 linguistic expertise's in law enforcement practice under clause 152 of the Civil Code of the Russian Federation "On protection of honor, dignity and business reputation" and 7 researches of the expert on the analysis of texts of mass-media in out-of-court sphere (for revealing signs of "black" PR in the election campaign) for the period from 2007 to 2016.

Findings. Important theoretical issues of delineation of speech acts of sincere error and unfair informing are considered. A method is proposed for detecting speech acts of unfair communication based on the detection of diagnostic signs: *substantial* (logical contradictions, excessive detailing of the message, omissions of important details, reporting "extra" information, references to rumors, unverified sources, etc.); *proper language* (inadequate tone expression, excessive language means uncertainty impersonal and passive constructions, WE-exposure, speech strategy *de re*, and expressive syntax pr.).

Practical implications. The results of the research can be used as methodological recommendations for the identification and qualification of various kinds of implicit information in the production of linguistic examinations and in the production of examinations in cases of humiliation of honor and dignity, extremism, unfair and unreliable advertising,

Scientific value. The original method of revealing the content and actual language features of the speech act of dishonest information as opposed to the verbal deed of good faith was tested.

Keywords: Linguistic expertise · Speech act theory · Felicity conditions
Deceptive communication · Honest mistake · Deceptive information speech act

© Springer International Publishing AG, part of Springer Nature 2018
E. G. Popkova (Ed.): HOSMC 2017, AISC 622, pp. 462–468, 2018.
https://doi.org/10.1007/978-3-319-75383-6_59

1 Introduction

Recently, the application of modern linguistic theory, the need for which in the modern information age has become increasingly important due to the expansion of spheres and areas where it became possible, has acquired particular urgency. One of such areas is linguistic expertise (Galyashina 2003).

Linguistic forensic examination is defined as a "procedurally regulated expert linguistic study of verbal and (or) written text, resulting in a written opinion on issues whose resolution requires the use of special knowledge" (Rosinskaya (*ed.*) 2001, p. 124). One of the most important requirements in the theory and practice of legal proceedings for a linguistic expert conclusion is the scientific validity that "presupposes the scientific, logical and methodological literacy of the conducted study and the presentation of results" (Belchikov *et al.* 2010, p. 34). Scientific validity is largely determined by the adequacy of the methods and methods used, based on certain theoretical positions of linguistic science. Often it is the right choice of linguistic theory as the methodological basis of research that gives the desired result.

Thus, among the most popular categories of requests for a linguistic expertise are the tasks of determining and qualifying the content and verbal features of deceptive communication for the analysis of textual materials. In our opinion, the methodological tools of the theory of speech acts can significantly help in solving this problem. A.N. Baranov indicates the importance of using the methodological tools of the theory of speech acts in linguistic expertise, in particular: "**Linguistic pragmatics is** no less important for the examination of the text (here *and further it is highlighted by the author - A.B.*). In particular, **the speech act theory**, which determines the communicative orientation of the utterance (illocutionary force) and the essence of such important speech categories as affirmation, evaluation and appeal" (Baranov 2011, pp. 18–19).

2 Theoretical Basis of the Research

The founder of the theory of speech acts (speech act theory) is the English philosopher-analyst and logic J.L. Austin, who presented the basic ideas of the theory of speech acts in the course of lectures at Harvard in 1955, which were then published in the book "How to do things with words?" In 1962 (Austin 1962). Subsequently, these ideas were developed by the American philosopher and logician J.R. Searle in the work "What is a speech act?" (Searle 1971, pp. 39–53).

The speech act is a purposeful speech action performed in accordance with the principles and rules of speech behavior adopted in this society. The structure of the speech act in main features reproduces the model of the ordinary, non-verbal action: there is intention, purpose and effect (result). A person is responsible for both verbal and non-verbal actions if they violate accepted norms of behavior.

In the modern theory of speech acts, much attention is paid to speech acts that violate the fundamental principle of <u>speech communication</u> (the Cooperative Principle) and postulates of communication (conversational maxima) put forward by H.P. Grice in the work "Logic and Conversation" in 1975 (Grice 1975, pp. 41–58), and also the

Principle of Politeness (Politeness Principle), formulated in the well-known book G.N. Leech "Principles of Pragmatics" (Leech 1983).

3 Materials of Research

The article summarizes the experience of 13 linguistic expertise's in law enforcement practice under clause 152 of the Civil Code of the Russian Federation "On protection of honor, dignity and business reputation" and 7 researches of the expert on the analysis of texts of mass-media in out-of-court sphere (for revealing signs of "black" PR in the election campaign) for the period from 2007 to 2016.

4 Methods of Research

Therefore, in the framework of the theory of speech acts, in addition to cooperative speech acts, which can be called honest communication speech acts, we can speak of deceptive communication speech acts. The latter include such non-cooperative speech acts as lies, slander, defamation, compromising, incomplete information, biased coverage of facts in the right speaker, concealment of important factual circumstances, the dissemination of unconfirmed rumors and unverified opinions, distortion of facts by incorrect or inconsistent presentation etc.

It is clear that many of these speech acts can not be determined in a purely linguistic way. For example, a verbal act of lies or slander can be qualified as such only when it is found that the information contained in the text material does not correspond to the actual side of the matter, which is beyond the competence of the linguist.

However, the modern theory of speech acts still has sufficient tools for identifying meaningful (conceptual and logical), communicative-pragmatic and verbal signs of unfair information in the analyzed text. This applies primarily to the mechanisms for identifying the communicative intention of the author (which in the requests for expertise is often referred to as the "communicative orientation" of the statement or text), features of compositional construction of the text and its actual side, as well as the presence of purely linguistic ("verbal") signals of dishonest information.

It is methodologically important that, according to external linguistic features, the speech acts of unfair informing do not differ from other types of incomplete or distorted information, in particular, from speech acts of the so-called "conscientious delusion" caused simply by the lack of information from the speaker, incomplete knowledge of the facts or other reasons, but in no way the absence of his goodwill or the establishment of intentional non-cooperative speech communication.

It is necessary to distinguish between these two fundamentally different types of speech acts - honest error and unscrupulous communication. In this paper, the distinction between the two types of speech acts is based on the method of analyzing the so-called "success conditions" within the framework of the methodological apparatus of the theory of speech acts.

So, in addition to the presence of a certain communicative intention, which in the theory of speech acts is commonly called "illocutionary force" or "illocutionary function" of utterance, speech acts are characterized from the point of view of the so-called "felicity conditions" put forward in the already cited work of J.R. Searle (Searle 1971).

The conditions for the success of a speech act are conditions that are necessary to recognize a speech act as appropriate, and failure to comply with one or more success conditions, on the contrary, leads to communication failures.

J.R. Searle identifies four main groups of conditions for success:

(1) Preparatory preconditions are conditions that reflect objective (situational) and subjective (psychological) prerequisites, compatible with the nomination of this illocutionary goal, i.e. the circumstances of the speech act, in the absence of which it will suffer a communicative failure;

(2) Essential conditions are conditions that directly correspond to the communicative intention of the speaker (expression in a specific linguistic form of assertion, question, motivation, commitment, etc.), which the speaker seeks to convey to the listener's consciousness by means of his utterance;

(3) The condition of sincerity is a condition that reflects the internal (psychological) state that can be attributed to the speaker, based on the assumption of the sincerity and seriousness of the speech act (for the purposes of our study, it is important that, unlike the preliminary conditions, violation of the terms of sincerity by the speaker usually happens unnoticed for the addressee and therefore does not directly entail a communicative failure, although the counterfeit, falsity of this speech act can be exposed in the long run);

(4) Propositional content conditions are conditions that impose restrictions on the choice of language facilities for the performance of a speech act (for example, the act of communication can not include the verb in the imperative mood, and the act of the request is a verb in indicative mood in the past tense, etc.).

5   Analysis of the Results of the Study

From the point of view of the theory of speech acts, delusion should be regarded as a special kind of speech acts of representatives, or constatives messages. All speech acts of this type unite the communicative intention of the speaker to inform the addressee of any information so that the addressee takes note of it and draws conclusions from it himself.

At the heart of **delusion** as a speech act of an <u>inauthentic message</u> lie the following conditions for the success of a speech act:

(1) Preparatory conditions: (a) communication intent of the message - the speaker informs the recipient information about a certain fact, event, person, without having knowledge about its conformity to reality; (b) the speaker is convinced of the truth of this information. refers to reliable, in his opinion, or simply authoritative, respectable sources, or trusts data from his own experience, or relies on false analogies, etc.;

(2) Essential condition: the speaker confirms the truth of the information being communicated to the addressee;

(3) The condition of sincerity: The speaker observes the condition of sincerity; sincerely considers his information to be a reliable and worthy attention of the addressee;

(4) The condition of propositional content: the speaker uses the standard language form to report facts (indicative, nouns in specific reference use, etc.); from a linguistic point of view, a delusion has the form of a proposition - an affirmation or opinion, and does not practically differ from linguistic features.

Because of the speech act of sincerity conditions, and error nontermological and tautological referred to in some sources "**honest mistake**" (a *bona fide ignorance*) (while any delusion, by definition, good faith, because unscrupulous errors do not happen, it is difficult to assume a sincere intention of the speaker to be mistaken, deceived, not to receive reliable information about the facts, etc.).

From the speech acts **of inaccurate messages** should distinguish a similar shape, but differ on communicative intentions of the speech act **of unfair posts** (*deceptive information*) - false (knowingly misleading the recipient, unlike the self-delusion of the author), slander, flattery, hypocrisy, manipulation speech acts of other types.

From the speech act of an unreliable message ("bona fide error"), the speech act of dishonest communication differs in two parameters:

(1) In the field of preparatory conditions - the condition varies (a): there is a communicative intention of deliberately misleading the addressee or concealing from him the true state of affairs;

(2) In the field of the condition of sincerity, a **breach** of the terms of sincerity for the speakers, which is precisely characterized by a contradiction between the declared speaking purpose-falsified communicative intention and the actual communicative intention that the speaker hides.

Note that hiding the true communicative goal by the speaker is the main feature of the manipulative speech strategy. In the Russian linguistic tradition, in such cases it is customary to speak of "language demagogy" (Nikolaeva 1988, pp. 154–165).

However, it is not difficult to see that in terms of content and speech (neutral) style of presentation, the speech act of bona fide error is not too different from the usual statement or opinion.

It follows that in order to correctly classify a speech act as a "bona fide error" or an unfair manipulation (lie, slander, etc.), it is necessary to know the extralinguistic conditions of communication. For example, whether the speaker is aware that the information he provides is untrue, is whether the true goal of the speaker is to provide the addressee with some information for evaluating it objectively, whether the speaker is sincere in his intention to inform the audience of certain facts, etc. It is clear that these requirements go beyond the scope of the possibilities of linguistic analysis.

However, in linguistic pragmatics and the theory of speech acts, nevertheless certain principles of differentiation of sincere and insincere speech strategies have been worked out. This primarily relates to the ways of presenting information in the message, its quality and quantity. In particular, such speech acts, as a rule, contain various meaningful and verbal means of evasion from the truth, i.e. evasion of the speaker from responsibility for the utterance, which, following J. Lakoff, should be treated as "fence" words (*hedge (s)/hedging*) - words and expressions whose function is to create some kind of semantic blurriness, fuzziness - like *in general or approximately*, etc. (Lakoff 1972, p. 195).

On the basis of these principles, we developed a **methodology for detecting** violations of the condition of sincerity in the analyzed text at the content and actual language levels, which consists in the detection of the following *diagnostic signs*:

- **Significant features:** (1) logical contradictions; (2) excessive detailing of the message; (3) on the contrary, excessive "conciseness", omitting important details; (4) violation of the postulate of "quantity" (H.P. Grice), i.e. the message of "superfluous", irrelevant information; (5) excessive "pedaling" of unreliable sources of information (references to rumors, unverified sources, etc.);
- **Actually language features:** (1) atypical for speaking vocabulary; (2) a special, inadequate tone of utterance; (3) the abundance of linguistic means of uncertainty (vague pronouns and adverbs, hypothetical particles, subjunctive moods of verbs, etc.); (4) the presence of means of expression or evaluation, inappropriate for the speech act of the message; (5) means of evasion of the speaker's responsibility for utterance (impersonal and passive constructions, WE-discourse, etc.); (6) use of the speech strategy *de re* instead of *de dicto*; (7) means of expressive syntax instead of neutral style of presentation (parcellation, rhetorical questions and exclamations, question-answer form of presentation, ironic quotes, defaults, etc.).

6 Conclusions

Thus, both in the field of author's intentionality and in the choice of speech strategies that correspond to it, one can still see some significant content and speech differences between "conscientious delusion" and dishonest information.

So, a genuinely erring speaker/author does not refer to rumors in support of his opinion, since he is sincerely convinced of his truth. He does not need the means of expression and evaluation, does not give a lot of details unnecessary for presenting his position and irrelevant, in his opinion, facts, does not put "hedges" that insure against possible charges of dishonest reporting, does not use means of evasion from responsibility for own utterance, i.e. from the I-presentation (I-discourse), from the first person, etc.

In general, the conducted study showed that it is possible to identify verbal acts of unscrupulous communication by linguistic methods. The scientific tools of the theory of speech acts, in particular, the scientific concept of "conditions of success," make it possible to adequately determine non-cooperative communicative intentions and corresponding content and speech attributes in the text if it is possible to take into account the totality of extra linguistic circumstances of the generation of the analyzed text.

7   Practical Implications

The results of the research can be used as methodological recommendations for the identification and qualification of various kinds of implicit information in the production of linguistic examinations and in the production of examinations in cases of humiliation of honor and dignity, extremism, unfair and unreliable advertising.

References

Baranov, A.N.: Linguistic Examination of the Text: Theoretical Grounds and Practice: Study Allowance, 3rd edn., p. 592. Science, Flint (2011). (In Rus.)

Belchikov, Y.A., Gorbanevsky, M.V., Zharkov, I.V.: Methodical recommendations on the issues of linguistic examination of disputed media texts: Collection of materials. IPK "Informkniga", p. 208 (2010). (In Rus.)

Galyashina, E.I.: Basics of judicial speech: Monograph, p. 236. STENSY (2003). (In Rus.)

Nikolaeva, T.M.: Linguistic Demagogy. In: Pragmatics and Intensionality Problems, pp: 154–165. Science (1988). (In Rus.)

Rosinskaya, E.R. (ed.): Forensic Examinations in Civil Proceedings: Organization and Practice, p. 535. Yuright Publishing House; ID Yuright (2001). (In Rus.)

Austin, J.L.: How to do Things with Words: The William James Lectures delivered at Harvard University in 1955, p. 179. Clarendon Press, Oxford (1962). Urmson, J.O. and Sbisà, M. (ed.)

Grice, H.P.: Logic and conversation. In: Cole, P., Morgan, J. (eds.) Syntax and Semantics. IH. Speech Acts, pp. 41–58. Academic Press, New York (1975)

Lakoff, G.: Hedges: A study in meaning criteria and the logic of fuzzy concepts. In: Papers from the Eighth Regional Meeting of the Chicago Linguistic Society, pp. 183–228 (1972)

Leech, G.N.: Principles of Pragmatics, p. 250. Longman, London (1983)

Searle, J.R.: What is a speech act? In: Searle, J.R. (ed.) The Philosophy of Language (Oxford Readings in Philosophy), pp. 39–53. Oxford University Press, Oxford (1971)

Problems of Formation of Perspective Growth Points of High-Tech Productions

Tatyana S. Kolmykova[1](✉) iD, Ekaterina A. Merzlyakova[1],
Vladimir V. Bredikhin[1], Tatyana O. Tolstykh[2],
and Oksana P. Ovchinnikova[3]

[1] Southwest State University, Kursk, Russia
t_kolmykova@mail.ru, ek_mer@mail.ru, bvv00l@mail.ru
[2] Industrial Management, National University of Science and Technology
(MISIS), Moscow, Russia
tt400@mail.ru
[3] Peoples Friendship University of Russia, Moscow, Russia
oovchinnikova@yandex.ru

Abstract. The current socio-economic situation in Russia is characterized by the impact of complex foreign policy conditions, the use of international sanctions, and the intensification of global competition. Internal features of Russia's economic development are associated with the exhaustion of the export-raw material model of development, the impact of a new technological and technological structure, which puts the industrial sector in the task for transition to an innovative development paradigm. On background, high-tech science-intensive enterprises should become the locomotive of the growth of the Russian economy. In this article, the role of high-technology sectors in initiating the processes of innovative development of the economy based on the structural-dynamic analysis of macroeconomic parameters is substantiated, and the main criteria for classifying industries as high-tech sectors are outlined. Economic and statistical analysis of the activity of enterprises of the high-tech sector of the Russian economy was carried out, taking into account international comparisons. The estimation of the basic tendencies of formation of the world's added value of hi-tech manufactures is given. The main tendencies in formation of perspective points of growth of high-tech sector are revealed. The key factors for successful implementation of the tasks of innovative development of enterprises in the high-tech sector of the economy are identified. In addition, the significance of this study is to develop theoretical and methodological support and justification of scientific and practical measures aimed at the formation of a high-tech sector of the innovation economy.

Keywords: Innovation · Innovation management · Innovative development
High-tech production · High technology

1 Introduction

The Russian economy overcomes the recession and stagnation, which replaced the active development.

© Springer International Publishing AG, part of Springer Nature 2018
E. G. Popkova (Ed.): HOSMC 2017, AISC 622, pp. 469–475, 2018.
https://doi.org/10.1007/978-3-319-75383-6_60

First, the situation of a sharp decline in prices for natural resources once again demonstrates the country's export-raw material dependence. The Russian economy has been developing over the past decades largely thanks to the raw material sector, in particular the oil and gas sector. This sector plays the main role in the formation of the country's balance of payments and provides the main share of export earnings. According to the Federal State Statistics Service, revenues from the export of minerals account for more than 60%. At the same time, about 40% of all export earnings come from crude oil and natural gas. One of the negative consequences of this economic structure is the obstacle to creating incentives for the development of high-tech industries.

Secondly, in the world political space, tendencies towards multi-polarity are increasingly being traced. Strengthening the struggle of states for spheres of influence provokes the growth of regional instability and the exacerbation of world security problems. The negative geopolitical situation and the sanctions regime make their adjustments to the functioning of the Russian economy. There are significant restrictions in relations with traditional Western partners. This leads to a complication and a reduction in the interaction on the transfer of technology. A viable alternative and a promising solution to this kind of problems is the activation of the domestic research and development sector, which is the driver of innovative development.

The world community enters the period of formation of a new paradigm of scientific and technological development, which is connected with the transition to the sixth technological order and the beginning of new revolutionary industrial changes. Lagging in the pace and scale of Russia's transition to a new paradigm will naturally lead to a loss of competitiveness, both in the world and in the domestic market [7]. In this context, the role of the high-tech sector of the economy is obvious. Since it is where the main production forces of the sixth technological order are located. In this way, the problem of development and formation of perspective points of growth of high-tech industries is actual.

The purpose of the study is to identify the main trends in the formation of promising points of growth of high-tech industries on the basis of analysis and evaluation of high-tech industries taking into account international comparisons.

2   Theoretical Basis of the Research

The importance of the innovative component of the growth of the Russian economy is underlined by the strategic documents adopted within the framework of the goal setting at the federal level. The strategy of social and economic development of the Russian Federation, the National Security Strategy of the Russian Federation, the Strategy for Scientific and Technological Development of the Russian Federation, and a number of sectoral strategies contain certain goals and priorities for innovative development. Thus, in terms of ensuring national security in science, technology and education, the need to integrate science, education and the high-tech industry is indicated.

According to Rosstat's normative documents, the criterion for classifying high-tech industries is a high level of technological development. It is determined by the ratio of R&D costs to gross value added. However, there is no single approach that makes it possible to classify branches of the economy in terms of their level of manufactura-bility, in world or domestic practice. The OECD, the US National Science Foundation and the United Nations (within the Standard International Trade Classification) clas-sification are universally recognized [1].

The current grouping of industries based on technological development, Rosstat develops science intensity because of the recommendations of Eurostat, and OECD based on NACE Rev.1.1, taking into account the national characteristics of the development of industries. The list of economic activities included in the group of high-tech industries is normatively fixed and includes:

– Production of pharmaceutical products;
– Production of office equipment and computers;
– Production of electronic components, equipment for radio, television and communications;
– Manufacture of medical products; measuring instruments, monitoring, control and testing; optical devices, photo and cinematographic equipment; watches;
– Production of aircraft, including space vehicles.

The concept of high technology is associated with the category of science intensity, which is generally understood as the degree of communication with scientific research's and developments. Legislative, however, the criterion for classifying the sector as a science-intensive one is the proportion of people with a high level of vocational education in the number of employees. It should be noted that high-tech innovations are the "highest" form of innovation activity. They contribute to a sig-nificant social and economic impact, as well as the development of new industries.

3 Research Methodology

Today, the global share of medium- and high-tech activities in the added value of manufacturing industries is about 50%. At the same time, in some countries higher values are also fixed: Singapore - 81%, Switzerland - 65%, Germany - 60%, the USA - 51%. The world volume of added value created by industrial enterprises of high-tech industries is also increasing - 1.8 trillion dollars US in 2014 [10].

This trend is not typical for all countries. Unconditional leader in the production of high-tech products are the United States. However, if in 1999 the share of the United States accounted for 37.1% of the global added value of high-tech industries, in 2014 - less than 29%. The main reason for this dynamic is the rapid innovation growth of China's economy. Its share in the global production of high-tech products increased from 3.4% in 1999 to 27.3% in 2014. Thus, it can be said that at present the USA and China occupy leading positions in the production of high technologies.

It should be noted that China's experience in building and developing an innovative economy is unique. It is recognized as one of the most successful in the world. In the shortest time, the country has evolved from an economic system with a low level of technological development to one of the world's innovative centers open to interna-tional trade and technology exchange.

On this picture, Russia's positions are still insignificant. Despite the positive dynamics, the country's share in the global production of high technologies is just over 1%. The same is true for high-tech knowledge-intensive innovations. There are a number of factors hindering the creation of high-tech innovations:

- Insufficient funding of research activities both from the state and from the private sector;
- Lack of effective incentives for R&D;
- Time factor (the process of developing new technologies can take several years).

Specialists note that the development of innovative activities of high-tech enterprises directly depends on the volume and sources of funding [6]. Analysis shows that gross domestic expenditure on R&D in Russia has a steady upward trend. Their share in GDP increased from 0.85% in 1995 to 1.13% in 2015. In the United States, the same indicator in 2015 was 2.75%, in China - 2.05%. In Japan it is -3.59% [8].

Significant financial resources were directed at supporting the research sector and reviving the research infrastructure, attracting talents and re-conquering leadership positions. A significant part of gross domestic expenditure on R&D (about 20%) is directed to supporting applied research, which serves as a basis for high-tech knowledge-intensive innovations.

4 Analysis of Research Results

As for the structure of high-tech production in Russia, here the undisputed leaders are aircraft manufacturing enterprises, including spacecraft. They account for almost 50% of all high-tech production in the country (Fig. 1, [8]).

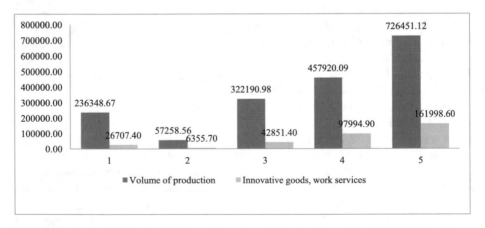

Fig. 1. Production in Russia of high-tech products, including innovative, 2015, RUB million.

(1 - production of pharmaceutical products, 2 - production of office equipment and computers, 3 - production of electronic components, equipment for radio, television

and communications, 4 - production of medical products, measuring, control and testing instruments, optical instruments, and cinema equipment, watches, 5 - production of aircraft, including space vehicles)

This state of affairs is natural. The cosmos has long been recognized as a strategic priority in the struggle for world economic leadership. Prospects for using space and space technologies in many ways are able to influence the trajectories of the development of the world economy and the economy of Russia. Aeronautical and space-rocket technologies have a strong influence on the formation of the world economy by importing space technologies into the civilian sphere of activity, a sharp increase in the volume of production, a change in the structure of the world market towards innovative high-tech industries. This predetermines the economic effect of the development of aerospace industries.

Russia historically has the highest potential in such industries as space and nuclear technologies. At the same time, space technologies are one of those areas of high technology, where Russia still maintains a leading position on world markets. Enterprises that create aviation and space-rocket technologies have a powerful scientific and technical base, unique experimental equipment, traditionally high organization and responsibility, technological discipline, well-established programs for the development and production of complex types of science-intensive technology. According to the researchers, they have highly skilled cadres of workers, technicians, engineers, production organizers, economists and scientists [2].

According to the Thomson Reuters report on innovation, five of the ten most innovative aerospace companies in Europe are in Russia for the period (Table 1).

Table 1. Top 10 European innovation leaders in the aerospace industry [9]

Company	A country	Number of inventions
Airbus	France	225
RSC Energia	Russia	113
ISS M.F. Reshetnev	Russia	80
Thales	France	73
Center Nat Etud Spatiales	France	42
Institute of Engineering Science. A.A. Blagonravov RAS	Russia	61
Khrunichev State Research and Production Space Center	Russia	41
Deut Zent Luft Production Center	Germany	32
Snecma	France	24
Moscow Experimental Design Bureau "Mars"	Russia	19

The domestic aerospace industry is characterized by a high level of centralization and state control. On the one hand, this approach allows achieving breakthrough scientific and technological results, on the other hand it is not effective enough that in the process of introducing into the production and launching of the technologies and solutions being developed.

Space technologies and infrastructure are usually associated with the launch of carrier rockets, the management and maintenance of satellites, manned spaceflight missions and other important tasks. However, the traditional landscape of the aerospace industry is changing rapidly, and now the industry is on the threshold of a new era that can lead to a significant expansion in aerospace applications and the emergence of new markets [4, 5].

Today, the state, the educational sector and a number of private innovative companies are experimenting with new approaches to the creation and deployment of groupings of spacecraft and the provision of new services. The changes concern the characteristics of the satellites that are put into orbit. In 2015, slightly less than half of the nearly three hundred launched satellites weighed 10 kg and less. In 2014, a Russian carrier rockets was launched, which deployed a mix of 33 small satellites and CubeSat in low-Earth orbit [4].

New technologies allow creating and managing satellite groupings with the possibility of full coverage by monitoring the Earth's surface. Solutions for remote sensing can be used to study climate and environmental monitoring, in agriculture and forest industry, for monitoring road traffic and identifying emergencies.

Along with space technologies, biotechnologies and pharmaceuticals are highly concentrated in industries with a high degree of concentration. In these industries, several large international companies dominate, such as Pfizer, Glaxo Smith Kline, Abbott Laboratories and Eli Lilly. However, it should be noted that the competitive environment in the field of biotechnology is also changing. Biotechnological start-ups attract more and more investments. According to a study by Hiroko Tabuchi, published in The New York Times, in the first half of 2014, international investment in biotech start-ups grew by 26% to approximately $3 billion., which could exceed the 2008 record of $5 billion dollars US [6]. New areas are emerging, such as personal genomics and genetic engineering. More and more attention is paid to biotechnology in agriculture.

5   Conclusions

Thus, today, Russia is actively investing in high-tech and knowledge-intensive innovations and is still competitive in leading areas, however, to expand the sphere of leadership and maintain current positions, it is necessary to make a big breakthrough. Key factors for the success of this task are attracting talents and increasing involvement in international research activities.

On the way to achieve and maintain the leading position in the world market of high-tech and science-intensive innovations, it seems expedient to develop higher education institutions and introduce new models for organization of research activities, expand cooperation between business and scientific institutions, create global innovation centers, and integrate Russia into a global innovation network.

To achieve the stated goals, the maximum concentration of efforts is needed in those industries in which the country already has a substantial reserve (aviation and space-rocket technologies), as well as in the sectors of high-tech in the world markets (biotechnology and pharmaceutics).

References

1. Balatsky, E.V., Ekimova, N.A.: The doctrine of high-tech jobs in the Russian economy, 124 p. Edithus, Moscow (2013)
2. Belousov, A.I., Maslova, A.G.: Features of aerospace engineering and modern aerospace engineering education. Bull. Samara State Aerosp. Univ. **360**(5), 333–338 (2012)
3. Erygina, L.V., Serdyuk, R.S.: The state of the Russian rocket and space industry and its development trends. Vestn. Siberian State Aerosp. Univ. **53**(1), 207–211 (2014)
4. Isupov, A.M.: Perfection of state regulation of aircraft-building clusters of the Russian Federation: strategic aspect. Vesnik of the Samara State Univ. **105**(4), 27–33 (2013)
5. Kolmykova, T., Telizenko, A., Lukianykhin, V.: Problems of modernization and development for industrial complex. Probl. Perspect. Manage. **11**(4), 27–33 (2013)
6. Kolmykova, T., Kazarenkova, N.: International criteria for the country's banking system efficiency assessment. Econ. Ann. XXI **157** (3–4(1)), 97–99 (2016)
7. Ljevčenko, A.S., Rudičev, A.A., Kuznecova, I.A., Nikitina, J.A.: Competitive strategy as instrument of increase of business activity of the industrial enterprise. J. Appl. Eng. Sci. **13** (1), 19–24 (2015)
8. Voynilov, U.L., Gorodnikova, N.V., Gohberg, L.M., et al.: 2017 Indicators of science: Statistical collection. National Research University, Higher School of Economics, 304 p. HSE, Moscow (2017)
9. State of Innovation Report: Thomson Reuters (2016). http://stateofinnovation.com/2016-state-of-innovation-report
10. United Nations Industrial Development Organization: Industrial Development Report 2016. The Role of Technology and Innovation in Inclusive and Sustainable Industrial Development. Vienna (2015). http://www.unido.org/

Tax Planning as a Basis of the System of Corporate Tax Management

Lyudmila S. Kirina[✉] and Natalia A. Nazarova

Financial University under the Government of the Russian Federation,
Moscow, Russia
kirina304@yandex.ru

Abstract. Topicality of the studied problem is predetermined by the necessity for improving the management of tax expenses of an economic subject, caused by the peculiarities of development of the taxation process in Russia and, as a consequence, by the taxpayers' wish to optimize their tax liabilities. Thus, a need for theoretical and practical knowledge on formation of tax expenses and management of their volume on the basis of improving the tax strategy of an economic subject arises, which predetermines the emergence of independent sphere of financial management – tax management of organization (corporate tax management), which task is to optimize the management of tax liabilities at the micro-level. The purpose of the article is to determine and open the general tendencies and peculiarities of functioning and development of tax management at the corporate level. The leading approaches to studying this problem are the historical & legal and comparative & legal approaches, which allow for complex consideration of the process of functioning of tax management as an element of financial management and the object of economic & legal regulation. Results: the article views the causes of emergence of tax management: opens the general tendencies and peculiarities of its functioning; determines the most significant directions; reflects the stages and methods of tax planning; offers the algorithm of calculation of planned liabilities for the year and the variants of determining the effectiveness of tax management. The materials of the article are of practical value for specialists in the sphere of taxation and tax law; persons who conduct scientific research in the sphere of economic & legal regulation of tax policy of an economic subject; persons who conduct entrepreneurial and legislative activities.

Keywords: Taxes · Tax management · Financial management
Tax planning · Tax expenses · Tax liabilities · Tax budget · Tax load

1 Introduction

1.1 Establishing a Context

With the growth of qualitative and quantitative parameters of business, conducted by an economic subject, the need for the structured form of management of financial flows and tax expenses for improving the financial state and the corresponding processing of the increasing flow of information grows. Corporate tax management contributes into profitability of business and optimization of tax liabilities. Due to this, the search for

© Springer International Publishing AG, part of Springer Nature 2018
E. G. Popkova (Ed.): HOSMC 2017, AISC 622, pp. 476–484, 2018.
https://doi.org/10.1007/978-3-319-75383-6_61

the methods of optimization of tax payments is one of the most important directions of tax planning performed in the interests of business, which shows the topicality of the viewed problem.

1.2   Literature Review

The problems of financing and development of tax management are studied in the works of economists and lawyers in Russia and abroad.

The history of development of tax management and tax planning, as its main function, was studied by Bablenkova et al. (2009); Barulin et al. (2008); Vylkova (2017); Evstigneev (2004); Kucherov (2016); Melnik (2000), et al.

The establishment of tax management as a form of practical activity was studied in the works of Tikhonov and Lipnik (2004); Pimenov and Rodionov (2017), et al.

The history of development of financial management was studied by such foreign authors as Braley and Mayers (2008); Van Horn and Vahovich (2006); Mintzberg (2009), et al.

1.3   Establishing a Research Gap

The previous studies do not allow for complex consideration of the process of formation and functioning of tax management as an element of intra-corporate relations and the object of economic & legal and tax regulation in different periods of establishment of the Russian tax system.

This work views the economic & legal aspect and is aimed at analysis of preconditions for emergence of tax management and perspectives of its development in Russia.

1.4   Aim of the Study

The purpose of the study is to determine and open the general tendencies and peculiarities of formation and functioning of tax management in Russia. This is necessary for answering the question on the possibility of development of new economic & legal relations at the macro- and micro-levels in the sphere of taxation that would take into account the specifics of the Russian tax laws and the possibilities of its application for optimization of tax liabilities of economic subjects.

2   Methodological Framework

2.1   Research Methods

The following methods were used in the process of the research: general scientific – analysis, synthesis, comparison, and generalization; specific scientific: historical & legal and comparative & legal, which allow for complex consideration of the process of formation and functioning of tax management as an element of financial management and the object of economic & legal regulation in Russia. The historical & legal method was used for studying the stages of establishment and development of companies' tax management. The comparative economic & legal analysis was used for determining the peculiarities of formation and regulation of this phenomenon in the Russian tax law at different stages of development of the Russian tax system.

2.2   Research Basis

The research basis includes the scientific studies and publications of the Russian and foreign economists and lawyers who study various aspects of tax management and tax planning in organization.

2.3   Research Stages

The problem was studied in two stages:

First stage: analysis of the existing scientific literature on the topic of the research, as well as the laws in the sphere of economic & legal regulation of tax management; determination of the problem, goal, and methods of the research.

Second stage: formulation of profits received in the course of analysis of scientific literature and law, preparation of publication.

3   Results

3.1   Emergence of Tax Planning as a Form of Practical Activity

Tax planning - from the position of taxpayer - as one of the elements of corporate tax management, is an inseparable part of its financial and economic activity. This process is very important, as optimization of taxation allows for simultaneous allocation of financial assets for their capitalization. Due to this, there is a task of wide and legal reduction of tax load for each taxpayer.

Based on this, it is possible to determine the main directions of tax planning: studying tax law (existing and for perspective); evaluation of tax consequences for the performed actions; development of the methods of reducing the tax expenses of economic subject; evaluation of the possible tax risks of the planned events; calculation of tax load before and after implementing the means of tax optimization.

The performed actions include: selection of organizational & legal form of business, conclusion of deals and agreements, selection of the sources of tax payment (by means of own or borrowed assets), determination of the terms of tax payments, etc.

Unlike other stages of tax management, tax planning emerges before the registration of economic subject and the beginning of conduct of the production activities, as it includes the selection of the organizational & legal form of a legal entity. Due to this, it is possible to argue with certain authors who state that tax planning is peculiar only for reduction of tax payments, as the taxpayers hadn't had any organizational & legal form and production activities, and then they appeared. Thus, the tax payments didn't reduce but increased.

However, despite the wide distribution of the practice of legal optimization of tax payments, the quality of tax planning is improving very slowly. The reason for this lies not only in the instability of tax law: the very sphere of legal provision of tax processes in unstable. Tax law in a very important sphere of science and practice, which should become a systemic element of professional knowledge for the persons who deal with the problems of taxation.

It should be noted that tax planning, which is a part of the system of corporate tax management, includes determination of the most proper legal means of reducing the tax liabilities and compilation of the optimal tax budget. Due to this, taxes and payments into the non-budget funds are separated into a block as an independent part of the financial budget.

The positive aspect of this approach is that the attention is paid not only to reduction of tax expenses but to realization of the financial strategy of economic subject.

The organization may have its own succession of calculation of the tax budget. One of the possible variants is given in Table 1.

Table 1. Stages of calculation of budget taxes

Stages	Content of the stage
1.	Collection of information for calculating tax load of organization (factual and planned) and the indicators of tax load of the main tax objects
2.	Determining the sensitivity of taxes (coefficient of elasticity)
3.	Optimization of tax payments (normatives of amortization payments, subsidies, taxes, and contractual policy)
4.	Compilation of tax calendar
5.	Compilation of the operational budget for taxes

It should be noted that the Table reflects not the process of tax budgeting but presents the stages of corporate tax planning in the system of tax management during management of tax payments, and budgeting is presented at the last – fifth – stage of the budget compilation. At that, the first stage includes the systematization of information on taxes and tax payments for the previous year, evaluation of tax payments in the current years in the comparable (as to the studied year) terms of taxation. These data are used for calculation of the tax load on organization – ratio of the accrued and paid taxes during the year to profit of the organization. At the second stage, the coefficient of taxes' elasticity is determined – it shows the change of the value of a certain tax during the change of the tax base or tax rate. At the third stage, the tax payments are optimized. At that, apart from the possibility of applying the normatives and subsidies, it is necessary to pay attention to optimization of corporate property tax and usage of the legally set delays of tax payments into the budget. At the fourth stage of tax planning, the tax calendar is compiled in which it is specified what taxes and when should be paid by the organization. This will allow avoiding the delays and, therefore, penalties. The last stage – compilation of the tax budget. The budget on taxes is effective not as such but within the existing system of financial budgeting. As one of the operational budgets, it should be coordinates to the budget of money assets' movement. Connection of the tax budget to the budget of financial assets budget is aimed at the balance of financial income and payments into the budget so that the level of sufficient pure money flow is achieved before payment of the taxes.

The comparative analysis of the offered stages of corporate tax management shows that certain authors unify the 1st, 2nd, and 3rd stages, i.e., collection and analysis of information, indicators of tax load in different financial & economic situations, comparison and ranking of the received data are unified in the collection of information for calculating the tax load on the organization (factual and planned) and the indicators of taxation of the main objects.

The next stage is calculation of elasticity, which coincides with the authors; the stage of optimization coincides as well. The following stages can be switched – tax calendar and tax budget.

Thus, it is possible to conclude that the stages of tax planning in the system of tax management coincide (Fig. 1).

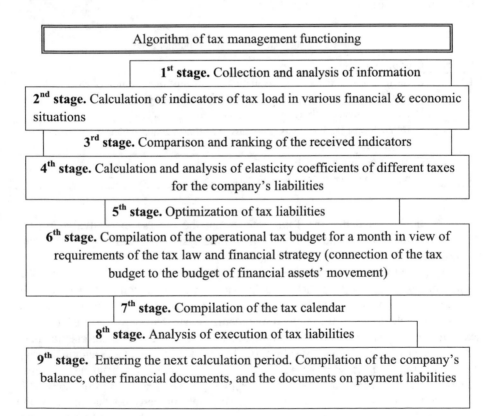

Fig. 1. Contents of the stages of corporate tax management

However, the existing methods of tax planning in the modern understanding do not ensure the possibility of operative tax management; due to this, it is inexpedient to use the methods of managerial accounting with distinguishing the tax component. For that, all taxes and other mandatory payments to the budget should be viewed as business's expenses without connection to the financial accounting classification of expenses. Taxes in the pure form are not considered in the managerial accounting and are not taken into account during calculation of expenses. It could be explained by the difficulties of the methodological order that emerge during the attempt to combine the approaches to analysis of taxes and the ones that are traditionally used during analysis of production expenses. Due to this, the changes in the methodology of operational analysis (as a component of managerial accounting) of tax expenses for managing the financial resources are expedient.

The offered changes of the methodology will allow controlling the taxes, i.e., controlling the volume of payments into the budget by influencing the assortment of the issued products and the character of the use of the consumed resources.

For that, the organization's management's attention should be paid to control over correctness, completeness, and timeliness of payment of taxes for the purpose of avoiding penalties, fines, and other sanctions, and then – to the possibility of the legal reduction of tax payments (tax planning). Regardless of the internal structure and the system of corporate management, tax information is provided in the data of tax, financial, and managerial accounting.

It may seem that tax information is provided in full in the tax accounting, but this form of accounting is aimed for the taxation purposes only. Regardless of the fact, whether it is distinguished by the organization into a separate type of accounting (we think it means that the organization has a separate structural department and/or staff of the organization includes one or several employees who deals only with the tax issues) or not, the organization has to pay the taxes and perform a range of other functions – e.g., register with the tax authorities, file tax calculations, declarations, and other documents in due time and due form, and have the tax registers with the set form. Besides, the organization may develop and use its own registers for the tax purposes. However, we think that this is a weak spot of tax accounting during the attempt to use these data for the purpose of tax expenses management.

Firstly, similarly to financial accounting, the data of tax accounting are oriented at the external user. Moreover, while in case with financial accounting there are several groups of financial accounting's users, in tax accounting there's only one user – tax bodies (and, probably, the organization for the internal purposes). That's why the information in tax accounting satisfies not the organization's but the tax bodies' needs.

Secondly, the data of tax accounting are an object of commercial and tax secret. In tax accounting, there's no possibility to compare the indicators of various organizations. The possibility of various comparisons is very important for management; tax accounting does not provide such a possibility. Tax accounting provides such a possibility to the employees of tax bodies but not to the wide audience or the organizations.

Thirdly, preparation and filing of tax accounting are performed after some time, and the data of the accounting may grow old by the moment of managerial decisions. The corporate structure should have the operative tax accounting. Taking into account high sensitivity of tax information to the precision of the data and practical impossibility of comparisons to other economic subjects, operative tax accounting loses its value for the purposes of managerial decisions.

Thus, the data formed in the tax accounting pose a certain value during managerial decisions but do not allow solving the tasks of tax management. In particular, tax accounting does not provide information on what products and when should be produced for the tax expenses to be minimal.

At that, the circle of issues to be solved by managerial accounting is not limited as it includes all issues that are not properly reflected in other forms of accounting.

We share the point of view of managerial accounting peculiar for the foreign countries. Firstly, because in case of such approach managerial accounting is not something immense but has its specific object and methods of research and as such interacts with other forms of accounting for solving the set tasks. Secondly, because managerial accounting studies the costs and other issues (primarily, the ratio expenses/profit), not limiting by the set frameworks, as is true in case of financial, tax, and statistical form of accounting, but with such level of depth and coverage that is necessary for specific organization, which is very important for supporting our knowledge at the modern level.

Tax planning, as was noted, is a part of financial planning and its task includes legal reduction of the absolute value of tax and other payments into the budget. In its turn, financial planning is an inseparable part of organization's financial management, which includes tax management. That's why the issues of provision of future payments by financial resources and reflection of the change of the organization's position under the influence of concluded contracts, which are not yet completed, should be assigned to the joint sphere of financial and managerial accounting.

In case if there are direct and reverse ties between the elements of tax management, this stimulates the timely and efficient formation of the tax process, and regulation is deemed organic and effective.

4   Discussion

It should be noted that the modern literature has certain differences in using the terms "tax planning" and "tax management". Some authors identify them – which is not correct. The English word "management" is a synonym for the Russian word "управление" and supposes the conduct of analysis, planning, control, and regulation. Thus, the notion "tax management" includes the notion "tax planning". Therefore, it is possible to agree with Galimzyanov, who divides tax management into organization of tax accounting, control over the correctness of calculation of taxes, minimization of taxes within the law, and structural transformations. A similar point of view is supported by practical specialists by large consulting companies who note the following main elements of the effective system of tax management:

– comprehensive tax calendar;
– strategy of optimization of tax liabilities and the corresponding plans of their realization;
– wise compliance with obligations;
– good state of financial accounting.

As is seen, optimization of tax liabilities is subjected by the above authors to the more general task – tax management. This, tax management, tax control, and tax administering could be seen as synonyms – at that, they are correlated to tax planning as the general with the specific.

5 Conclusions

Based on the performed research of the process of formation and functioning of tax management as an element of financial management and the object of economic & legal and tax regulation, it is possible to make the following conclusion. Formation of the company's tax management has been conducted gradually, and its emergence has specific economic and legal preconditions, namely:

- Russia's transition to market relations, which predetermined creation of a new tax system that played an important role in provision of the revenues part of the consolidated budget of Russia;
- a large number of economic subjects are interested in increase of net profit by reducing tax expenses and tax load, without violation of current tax laws;
- development of new forms of activity in the sphere of management of organization's finances, aimed at reduction of tax liabilities of the economic subject;
- the necessity for acquisition and application by the company's specialists of the means of optimization of tax liabilities that allow improving its financial state.

References

Braley, R., Mayers, S.: The Principles of Corporate Finance, 2nd edn. Olimp Business Publication (2008)

Van Horne, J.K., Vahovich, J.M.: Foundations of Financial Management. Williams, Wantage (2006)

Mintzberg, H.: Management: the Nature and Structure of Organizations as Seen by Guru. EKSMO, Moscow (2009)

Garner, E., Owen, R., Conway, R. Attracting capital (1995)

Bondarchuk, N.V.: Financial Management of Organization (Methods of Harmonization of Taxes, Money Flows, and Prices). Ekonomika Publication, Moscow (2008)

Vylkova, E.S.: Encyclopedia of Managing the Taxation of Economic Subjects. St. Petersburg State University of Economics Publication, St. Petersburg (2017)

Dadashev, A.Z., Kirina, L.S.: Tax planning in organization (2004)

Evstigneev, E.N.: Foundations of tax planning: corporate tax management of the directions of taxes minimization. The imitational model of tax planning. Tax risks, Piter, Moscow (2004)

Melnik, D.Y.: Tax management. In: Finance and Statistics (2000)

Tikhonov, D., Lipnik, L.: Tax planning and minimization of tax risks. Alpina Business Books, Moscow (2004)

Barulin, S.V., Ermakova, E.A., Stepanenko, V.V.: Tax Management: Study Guide. OMEGA-L, Moscow (2008)

Kirina, L.S., Gorokhova, N.A.: Tax management in organizations: study guide for masters: for students of higher educational institutions, specialty: economics ; Financial University with the Government of the RF. Yurait, Moscow (2014)

Pimenov, N.A., Rodionov, D.G.: Tax management in organizations: study guide for bachelors: for students of higher educational institutions, specialty: economics. Financial University with the Government of the RF, Yurait, Moscow (2017)

Kucherov, I.I.: Tax law. The General Part. Study Guide for Academic Bachelors. Yurait, Moscow (2016)

Bablenkova, I.I., et al.: Forecasting and Planning in Taxation: Study Guide. Ekonomika, Moscow (2009)

Problems and Prospects for Implementing Inter-dimensional and Inter-industry Projects in Digital Economy

Tatyana O. Tolstykh[1(✉)], Natalia N. Kretova[2],
Anna A. Trushevskaya[3], Elena S. Dedova[4], and Marina S. Lutsenko[2]

[1] Industrial Management National University of Science
and Technology (MISIS), Moscow, Russia
tt400@mail.ru
[2] Voronezh State Technical University, Voronezh, Russia
nnkretova2016@yandex.ru, luchiksan@rambler.ru
[3] St. Petersburg State University of Aerospace Instrumentation,
St. Petersburg, Russia
trushevskaya.anna@mail.ru
[4] Plekhanov Russian University of Economics, Voronezh, Russia
dedova.es@gmail.com

Abstract. The purpose of this study was to review the current issues of project management and their impact on the development of regions and industries in the new communication in environment, formed by the features of the digital economy. Concepts related to management of inter-branch, interregional projects are specified, their stages of management are considered. Paying attention to methodological problems related to the study of factors affecting the success of inter-branch and interregional projects in the context of an updated postindustrial society, an integral part of which is the digital economy. Express diagnosis of the status of the most important of them for projects implemented in Russia. The theoretical and practical basis was the regulatory provisions of the Russian Federation, research papers of Russian and foreign scientists related to problems of project management, as well as developed foreign standards. Methodical basis of the study were chosen system analysis, logical analysis methods and statistical methods. The developments outlined in this article represent an original contribution to improving the efficiency of development of territories and industries based on benchmarking of current global trends and Russian experience in managing globalization projects in the digital economy.

Keywords: Digital technologies · Inter-dimensional and inter-sectorial projects
Stakeholders · Expert analysis · Success factors

1 Introduction

According to the Strategy of Scientific and Technological Development of the Russian Federation, the key factors determining the competitiveness of national economies are the high rate of mastering a new knowledge and creating innovative products [8].

© Springer International Publishing AG, part of Springer Nature 2018
E. G. Popkova (Ed.): HOSMC 2017, AISC 622, pp. 485–493, 2018.
https://doi.org/10.1007/978-3-319-75383-6_62

The most significant factors of the scientific and technological development of the Russian Federation in the context of the rapid change in the macro and mezzo environment under the Strategy are: reduction of the innovation cycle, erosion of territorial and sectorial boundaries; the emergence of fundamentally of new tools related to IT technology in the conduct of research and development; the importance of intellectual innovative human potential, the growing role of international standards. In accordance with the tasks set in the Strategy, the role of project management in the implementation of the scientific and technological development of the Russian Federation is becoming the key one.

Russian branches of territorial development are attractive for foreign investors, which was repeatedly confirmed at the Eurasian forums. Investors are ready to come with their finances, specialists, advanced technologies and localize them in this or that production in Russia. However, these tasks are encumbered by insufficient knowledge and thoroughness of this issue on the part of the Russian Federation, which naturally reduces the effectiveness of efforts [1, 2, 9].

Thus, the main goal of this research is to focus on methodological study of the priority aspects of managing inter-sectorial, interregional projects that potentially or really reduce the pace of their implementation in Russia.

2 Theoretical Basis of the Research

Let us turn to the concept of inter-industry, interregional project and project management. They are quite diverse, which is dictated by various standards existing today, for example, PMBOK, GOST R 54869-2011, DIN 69901, PRINCE2 [6, 7]

In our opinion, the main characteristics of the project management of the future are the speed of taking managerial decisions and erasing the sectorial and territorial boundaries. Therefore, we combined the class of inter-dimensional and inter-branch projects under the name *Spatial reactive projects (SRP)* [5]. *SRP* projects are a set of interrelated activities aimed at creating one or several innovative products or services in time and resource constraints based on several industries and territories, leading to their multiplicative development effect.

Note that the implementation of SRP occurs in the period of intensive development of the digital economy, including Russia. The digital economy is a new paradigm of accelerated economic development, which assumes the comprehensive integration of digital information technologies and real economic processes at the level of states, markets and companies. In Russia, the share of the digital economy in GDP is 2.8%, or $ 75 billion. Given the scale of SRP, the digital economy can have a significant impact on the implementation process. Analyzed the practice of implementation of inter-branch, interregional projects, we will outline the following stages:

1. Analysis of problems and motivations of the industry in general through expert analysis, compilation of various kinds of forecasts using the opportunities of online and innovative digital technologies.
2. Inventory resources of territories, industries and enterprises based on their potentials.

3. Formation of criteria, goals, motivation for enterprises, industries and territories.
4. Analysis of existing business processes and information flows between enterprises and territories in the digital economy.
5. Building relationships between enterprises and creating a single information space structure. Formation of "rules of the game" and organization of information flows in the digital economy.
6. Risk assessment of SRP in a digital economy.
7. Estimating the cost of SRP in a digital economy.
8. Financing.
9. Implementation.

Therefore, the speed of changing the macro environment does not allow us to focus on the long-term development of new technologies and the phased implementation of such projects. Reactivity should characterize the project management of the future and be the basis of the methodology for implementing inter-dimensional and inter-industry projects [4]. This is possible if you know in advance the problem areas (the most significant implementation factors) of such projects and take proactive actions for effective management, taking into account the features and capabilities of the digital economy.

3 Research Methodology

In the framework of this study, an assessment of the factors influencing the success of SRP implementation in the digital economy with the use of appropriate economic and mathematical methods, the possibilities of online and digital technologies. Express diagnostics of the current state of the most important of them was carried out, which gives an idea of the potential success of SRP in Russia.

The information was collected based on an expert survey conducted using a special assessment sheet, which lists the parameters that affect the level of success of the project. It was suggested to evaluate the significance of each parameter on a five-point scale, with "5" meaning that the parameter is very important, "4" has an important meaning, "3" has a rather important meaning, "2" does not have a great meaning, "1" does not really matter.

The tasks of organizing the survey included the development of questionnaires and selection of experts. The study was completed in a questionnaire, which was used in the survey of experts. The expert group was formed from leading specialists with extensive work experience. When conducting the survey, each expert was sent a questionnaire and complete list of factors that took into account, including the movement of information in the digital economy, such as the consistency of stakeholder interests, the completeness and reliability of information on the resource potential of the territories. The task of the experts was to fill out questionnaires according to the set of rules. Communication was carried out online, which allowed attracting experts from different regions to the survey.

The results of the survey were processed using economic-mathematical methods of rank correlation. Preliminary survey data for each of the experts participating in the evaluation were reduced to one table-matrix of ranks.

To assess the degree of consistency of expert's opinions, the concordance coefficient was calculated by formula (1) with intermediate calculation of the indicators according to formulas (2), (3) and (4)

$$W = \frac{S(d^2)}{\left(\left(\frac{1}{12}\right)n^2(m^3 - m) - n\sum_{j=1}^{n} T_j\right)}, \tag{1}$$

Where

$S(d)$ is the sum of the squares of the differences in rank;
T_j is an indicator characterizing equal ranks;
m   total number of factors being evaluated;
n   is the total number of experts involved in assessing the significance of the factors

$$S(d^2) = \sum_{i=1}^{m} d_i^2 = \sum_{i=1}^{m} \left[\sum_{j}^{n} (x_{ij}) - X\right]^2, \tag{2}$$

Where d is the deviation of the sum of ranks from the j-th parameter from the average value of the sum of ranks.

$$X = \frac{1}{m}\sum_{i=1}^{m}\sum_{j=1}^{n} X_{ij} \text{ Or } X = \frac{n(m+1)}{2}, \tag{3}$$

$$T_j = \frac{1}{12}\sum_{t_j} (t_j^3 - t_j), \tag{4}$$

Where:
X_{ij} the rank of the i-th parameter and the j-th expert;
T_j is the number of identical ranks for the j-th expert.

When conducting an expert survey, the expert's opinions were completely independent, which contributed to an increase in the objectivity and reliability of the results of the examination.

As a result, we obtained a matrix of the values of the parameters of factors that affect the success of the project. They are presented in Table 1.

Table 1. Matrix of the values of the parameters of factors affecting the success of SRP in the digital economy

Factors	Expert assessments						
	1	2	3	4	5	6	7
1. Security assurance of return of funds to investors	5	5	5	5	5	5	5
2. State support	5	5	5	5	5	5	5
3. Availability of competent project personnel	5	4	4	5	5	5	5
4. A stable and transparent legal environment	4	4	4	5	5	5	5
5. Political stability	5	5	5	5	5	4	4
6. Consensus of stakeholder interests	5	5	5	5	5	5	4
7. Completeness and reliability of information on the resource potential of the territories	5	5	4	5	5	5	4
8. The presence of competitors	3	3	3	3	3	3	3
9. Variability of exchange rates	4	5	4	5	4	5	5
10. Favorable demographic environment	4	4	4	4	4	4	4
11. Favorable economic environment	4	4	5	4	4	4	4
12 Climate	3	3	4	4	4	4	4
13. Social and cultural environment	3	2	3	3	3	3	3

(A source: was compiled by the authors on the basis of the data of the expert survey conducted)

The results of the survey were processed using economic-mathematical methods of rank correlation. Preliminary survey data for each of the experts participating in the evaluation were summarized in one table-matrix of ranks, which is shown in Table 2.

Table 2. The matrix of the ranks of factors affecting the success of SRP in the digital economy

Factors	X	Sigma	V	Absolute values	Relative values
1. Security assurance of return of funds to investors	5	0	0	72.5	0.114
2. State support	5	0	0	72.5	0.114
3. Availability of competent project personnel	4.714	0.488	0.104	62	0.099
4. A stable and transparent legal environment	4.571	0.535	0.117	57	0.089
5. Polytic stability	4.714	0.488	0.104	61.5	0.097
6. Consensus of stakeholder interests	4.857	0.378	0.078	67	0.105
7. Completeness and reliability of information on the resource potential of the territories	4.714	0.488	0.104	61.5	0.099
8. The presence of competitors	3.000	0.000	0.000	12	0.019
9. Variability of exchange rates	4.571	0.535	0.117	56.5	0.089
10. A Favorable Demographic Environment	4.000	0.000	0.000	35	0.055
11. Favorable economic environment	4.143	0.378	0.091	40.5	0.064
12. The climate	3.714	0.488	0.131	28.5	0.045
13. Social and cultural environment	2.857	0.378	0.132	10.5	0.016
Coefficient of concordance	0.752093				
Pearson's criterion	63.1758				
Number of degrees of freedom	12				
Coefficient of concordance	0,72212				

(A source: was compiled by the authors on the basis of the data in Table 1)

The values of the concordance coefficient and Pearson's criterion, given in Table 2, indicate that the results of the expert survey processing can be considered reliable.

Following the logic of the analysis, we will also perform the Spearman rank correlation, which will allow us to study the congruence between each of the experts. [10] The results presented in Table 3 indicate that the consistency of opinions of the experts of the conducted survey is not accidental.

Table 3. Rank correlation of the Spearman factors affecting the success of SRP in the digital economy

	Expert 2	Expert 3	Expert 4	Expert 5	Expert 6	Expert 7
Expert 1	0.858	0.744	0.878	0.916	0.768	0.595
Expert 2		0.752	0.881	0.773	0.773	0.603
Expert 3			0.668	0.702	0.539	0.501
Expert 4				0.922	0.922	0.810
Expert 5					0.834	0.693
Expert 6						0.862

(A source: was compiled by the authors on the basis of the data in Table 2)

4   Analysis of Research Results

According to the calculation in Table 3, we can say with confidence that all experts are similar to each other in opinion; there are no directly opposite opinions

Therefore, based on the analysis, we can conclude that the most significant factors for reducing the risks of an inter-territorial, inter-sectorial project are:

- governmental support;
- security assurance of return of funds to investors
- consistency of stakeholder interests.

We will perform an express diagnostics of the state of these factors.

Guarantees for the safety of return of funds to investors are provided by the legislation of the Russian Federation and international treaties in force on the territory of Russia (in particular, the Law on Foreign Investments). State support can be implemented both at the financial and organizational levels. To date, among the ways of financing SRP can be identified the following means:

- Means of the National Welfare Fund,
- Budgetary allocations, loans,
- Own net worth

In Russia, there is an example of this kind of financing. It is carried out, for example, within the framework of the implementation of the Interdisciplinary project "Eastern polygon" in Siberia and the Far East, which affects the development of the Baikal-Amur Mainline and the Trans-Siberian Railway and creating a global transport corridor. In the development passport of the Baikal-Amur Mainline and the

Trans-Siberian Railway with total financing of 560 billion rubles. It is shown that of them 110 billion rubles. - Federal budget funds; 150 - the National Welfare Fund; 302 billion rubles. - The investment program of Russian Railways. Today, this is the largest infrastructure project.

As for the organizational component, here it is worth mentioning the development of project offices in Russia designed to simplify the work of stakeholders of projects, including interregional and inter-branches. This work has recently been quite active, for example, in the Belgorod, Leningrad region and other areas.

Continuation of the organizational factor is the coordination of interests of stake-holders. The openness of the space formed by the digital economy, potentially provides unlimited attraction of resources in the form of investor interest, on the other hand is a potential source of problems, as the number of stakeholders is widening, which leads to certain difficulties in reconciling different and sometimes even divergent interests [11].

The general classification of SRP stakeholders is shown in Fig. 1.

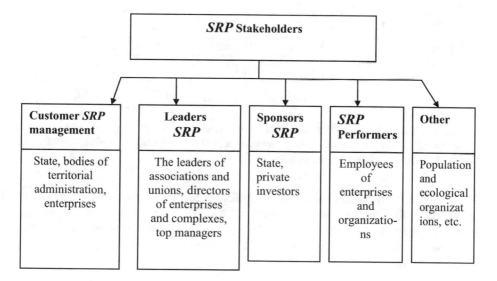

Fig. 1. SRP Stakeholders. A source: compiled by the authors

The SRP consists of projects of several hierarchical levels, which complicates the relationship between its participants. The enlarged SRP stakeholder scheme for the example of the development of high-speed water passenger and freight-and-passenger transportations in the Volga-Caspian and Azov-Black Sea basins is examined in Fig. 2. From it is clear that the potential risk of consistency of stakeholder interests in the project under consideration is certainly an environmental component, since the implementation of a project of this scale affects the ecosystem of the entire region.

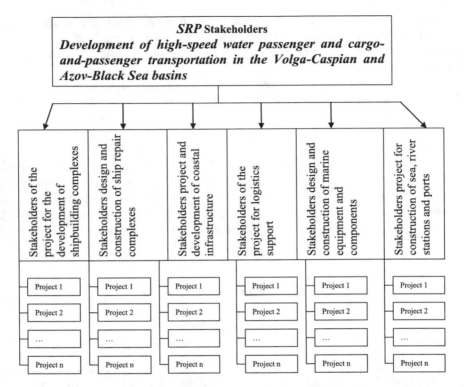

Fig. 2. Enlarged scheme of SRP stakeholders on the example of development of high-speed water passenger and cargo-and-freight transportation in the Volga-Caspian and Azov-Black sea basins. A source: compiled by the authors

It should be noted that the issue of building communication in the process of coordinating the interests of stakeholders of SRP, despite its importance, remains poorly understood. The authors believe that the expectations of stakeholders, linked to the strategic objectives of the project, should be placed at the center of the management of these communications [3].

5   Conclusions

Accelerated development of Russian territories is possible only with a comprehensive approach that bringing together the interests of businesses of different levels and territorial entities. The analysis shows that the implementation of projects of various levels of globalization without sufficient methodological elaboration of this issue reduces the effectiveness of efforts.

The importance of the presented study is to systematize data in the field of inter-branch and inter-dimensional projects on the basis of analysis of various scientific and practical studies.

In the course of the work, the concept of inter-branch and inter-dimensional projects was clarified, their participants were specified and the stages of SRP implementation in the conditions of the digital economy were formulated.

The success of the SRP implementation depends on various factors, which can be both potential opportunities and project risks. Based on the exclusively applied aspect of this study, an analysis was made on the success factors of SRP in Russia. As a research method, expert analysis was selected with the calculation of the verification coefficients of concordance, Pearson's criterion and Spearman's rank correlation.

Based on the analysis, significant factors of the success of inter-territorial and inter-sectorial projects were analyzed. Reviving that along with traditional business security factors, factors related to the characteristics of the digital economy have a potential serious impact on the success of SRP.

In general, the issue in question requires further close interaction between authorities and business, and intensification and institutionalization of regional cooperation is a necessary condition for ensuring the sustainable development of the Russian economy.

References

1. Anokhina, L.V.: Project management in the modernization of the regional industrial complex. Econ. Entrepreneurship **1–2**(66–2), 1070–1073 (2016)
2. Komarova, A.V.: Project management: characteristics and trends of development. Problems Econ. Manag. Oil Gas Complex **4**, 44–51 (2011)
3. Kretova, N.N.: Use of integrated communications in modern enterprises. Econominfo **10**, 72–75 (2008)
4. Cooke-Davies, T.J., Arzymanow, A.: The maturity of project management in different industries – an investigation into variations between project management models. Int. J. Proj. Manag. **21**(6), 471–478 (2003)
5. Posledov, S.V.: Approaches to the management of interregional and intersectorial projects. In: Posledov, S., Tolstikh, T.O. (eds.) Series: Economics and Management. Vestnik of Voronezh State University, vol. 1, pp. 24–29 (2016)
6. Razy, M.L. et al.: Program and project management: 17-module program for managers "Management of the development of the organization." Module 8, 320 p (2000)
7. A Guide to the Project Management Body of Knowledge (PMBOK Guide) [E- resource]: a giude (2008). http://www.novsu.ru/file/1213136
8. The strategy of scientific and technological development of the Russian Federation, approved by the Decree of the President of the Russian Federation of 01 Dec 2016
9. Tatarkin, A.I.: Program-project management as a condition for innovative development of socio-economic systems. Socio-econ. Sci. **2**(23), 25–42 (2014). Vestnik of the Perm National Research Polytechnic University
10. Ward, S., Chapman, C.: Transforming project risk management into project uncertainty management. Int. J. Project Manag. **21**(2), 97–105 (2003)
11. Eskerod, P., Vaagaasar, A.L.: Stakeholder management strategies and practices during a project course. Project Manag. J. **45**(5), 71–85 (2014)

Approaches to Developing a New Product
in the Car Building Industry

Viktor P. Kuznetsov[✉], Elena V. Romanovskaya,
Anastasia O. Egorova, Natalia S. Andryashina, and Elena P. Kozlova

Nizhny Novgorod State Pedagogical University of K. Minin,
Nizhny Novgorod, Russian Federation
keo.vgipu@mail.ru, alenarom@list.ru, nesti88@mail.ru,
natali_andr@bk.ru, elka-a89@mail.ru

Abstract. The work analyzes the engineering and marketing approach to the development of a new product in the automotive industry. The authors consider the engineering and marketing approaches and the essence of these approaches is defined. The possibility of using the modular method at the stage of developing new models was studied. The idea of this method is to develop a new unit or module when using the engineering and commercial potential of component manufacturers. The authors made an attempt to systematize a complex of marketing, organizational, economic and engineering processes through the industrial system of an industrial enterprise. The authors set a goal - to identify the features of creating a new product in an industrial enterprise using a marketing approach and a modular production method. The authors studied the possibility of applying a modular production method at the product development stage, the essence of which is to develop a new module using the development potential and commercial-engineering solutions of the parts manufacturers. The authors consider the transition to a modular method of design and production of goods; consolidation of changes in the development and release of a new product in the company when using a corporate production system based on the example of PJSC GAZ. The scientific significance of the study lies in the proposed method for creating a new product in the automotive industry. The main provisions and conclusions of the article can be used in scientific activity when considering questions about the essence of creating a new product in an industrial enterprise.

Keywords: New product · Approach · Production · Automotive
System · Strategy

1 Introduction

The control system of the process of planning and product research in the US automotive industry began to be studied in the late 1950s. A huge role in the study of the difficulties in coordinating the design and development of products of the automotive industry was made by foreign scientists D. Laiker, K. Prahalad, F. Kotler, L. Gelloway et al. In Russia, significant scientific and practical experience on this problem has been

© Springer International Publishing AG, part of Springer Nature 2018
E. G. Popkova (Ed.): HOSMC 2017, AISC 622, pp. 494–501, 2018.
https://doi.org/10.1007/978-3-319-75383-6_63

accumulated in G.Y. Goldstein, O.G. Turovets, O.V. Aristova, N.I. Novitsky, A.I. Prigogine, N.I. Lapina, Y.A. Ushanov and other specialists.

Involving future buyers in the development of new products is one of the important moments of stimulating innovation. This approach was most widely used in aviation, machine-building, automotive industries, as well as in instrument making in virtually all countries [3]. Consumers - customers of the latest products - had an important impact on the research and production programs of equipment suppliers [5]. Despite this, the problem of managing the stages of designing and developing new products implies an additional theoretical basis.

2   Theoretical Basis of the Study

Particular difficulties in solving this problem are due to the current situation in the economy, a decrease in the interest of the majority of commodity producers in the development of products. To analyze the state of theoretical elaboration of the system for creating a new product, we will conduct a study of the methodological basis that characterizes the role of marketing in the innovation activity of an industrial enterprise. There are two main approaches to the role of marketing in the development of a new product [1]:

1. Engineering: involves the implementation of innovative developments, then the production of the product is carried out and only in conclusion a complex of marketing events
2. Marketing: is the preliminary study of the market segment, the formation of the concept of a new product and the further request for innovative developments.

The engineering approach is necessary at the initial stage of the organization of production. This stage is carried out in accordance with the technical requirements, that is, it is determined normatively. This approach is relevant, first, in the military-industrial complex to preserve the strategic superiority of military developments.

For goods of mass demand, the marketing approach is more applicable. With it, the product development process is carried out in accordance with the developed business plan, while the technical requirements can only play the nature of the constraints.

3   Methodology of the Study

Theoretical and methodological basis of the research are fundamental provisions and scientific works of foreign and domestic scientists in the field of the theory of product creation and organization of production through corporate production systems [6].

In Russian and foreign practice, a sufficiently large number of methods for creating and launching a product on the market have been developed [7]. In our opinion, the general methodology for creating a new product in industry can be presented as follows:

1. Development of a marketing strategy for the introduction of a new product.

This step involves studying the market situation with the definition of the most promising sectors of the market. All the necessary information can be obtained as a result of a complex of marketing research, such as focus groups, panel market research and questionnaires. This complex allows you to find "problem zones" of the marketing strategy and determine the most promising niches in the market for the customer's product. The logical conclusion of this step is the process of developing one or several promotional strategies for promoting the product to the market.

2. The choice of the optimal concept of a new product.

At this stage, there is a synthesis of ideas about the concept of a new product with the help of such research tools as brainstorming with experts, creative group discussions and interviews with consumers. Selection and implementation of the concepts of a new product are carried out both by the firm's developer and by the consumers of the product. All collected information is analyzed in accordance with the SWOT-analysis scheme for studying the situation on the market while promoting the product.

3. Developing a formula for a new product.

This step requires the synthesis of qualitative and quantitative research results. Each result solves certain research tasks. Quantitative research is used to refute or confirm hypotheses that arise in the process of qualitative research, or hypotheses about the product, regardless of the results of qualitative research. Most often, quantitative research is underestimated by most industrial companies, which quite often justify their hypotheses with the results obtained on the results of focus groups. However, the quantitative studies make it possible to select the most optimal of several product formulas.

4. Creating a brand idea for a finished product.

At this stage, the concept of a new product is defined, a formula for a new product is developed, and the development of reinforcing elements, the so-called marketing mix, is necessary. In order to obtain reliable information from clients, we can, in our opinion, use focus groups and interviews with consumers. Because of these studies, primary reactions are received, decisions are taken "what to fix," and with a minimum number of optimal solutions, quantitative testing is carried out.

5. System testing of the brand.

Final testing before entering a new product on the market helps the client company to make a final decision about the optimality of the concept of implementing a new product and bringing it to the market.

However, it should be noted that the refusal to enter a new product will not be a loss of money and time. The cost of advertising such a product is many times higher than the cost of the entire research cycle. It should also be noted that most Russian companies practically do not use the modular design method, which significantly increases the time and financial costs of creating a new product. Therefore, in our opinion, it is advisable to use the modular design method at the concept stage of a new product.

The idea of this method is to develop a new unit or module when using the engineering and commercial potential of manufacturers of components for cars. Despite the fact that the idea of developing car parts has the form of creating a whole cycle, but the complex of development tools used in this case does not differ from the technical solutions that were traditionally used earlier.

4 Analysis of the Results of the Study

Proceeding from the above, it can be concluded that this method is aimed at transferring works on the regulation of compatibility of automotive components among themselves, which have traditionally been carried out inside the automotive organization, their partners and suppliers. To implement this mechanism, it is necessary to select a parent company from the manufacturers of automotive components of one module, which will assume the role of coordinator of actions of all participants. In this case, from the introduction of modular production, the manufacturers of the final product, in this case the car, receive the following results:

- A reduction in the number of component parts leads to an increase in production efficiency;
- Shifting part of the work on the development and assembly of car kits, external manufacturers will allow grouping forces on the process of developing the engine or body structure, this in turn will lead to a reduction in development time;
- Reduction in the volume of work on inventory management in warehouses, delivery and transportation of parts will reduce management costs.

With the introduction of modular production, the automotive company builds relationships with automotive component manufacturers both with business partners, while taking into account their technical level of development and the ability to develop proposals, rather than their belonging to their own group. Therefore, in our opinion, the process of searching for manufacturers of automotive components from the external environment will be faster and more profitable. Such a measure will enable smaller firms among the suppliers to participate in the tender and win it. The order of the automaker stimulates the innovative processes at the enterprise, and gives an opportunity to expand the business and even enter new markets. In addition, modular production assumes that the company, which won the tender for the production of the component of the car, will further actively participate in the design of the new model of the car, taking into account its technical capabilities. The company will get additional opportunities - signing contracts for long-term supply of components, and as a result, an increase in the amount of profit. The pursuit of modular production will give automotive component manufacturers additional business expansion opportunities, but a number of challenges need to be addressed. First of all, it should be noted that there is a possibility that there is no clear agreement on the division of responsibility between car manufacturers and car part manufacturer companies for the defect of any part of the module. If the manufacturer of components for a new car agrees to give a guarantee on the quality of the module, it will mean that it must take all the costs associated with training workers and development work for those elements of the car module for which their own technologies are not developed fully. At the same time, in the context of strengthening the requirements of the automotive company to reduce production costs, the module vendor is trying to multiply the added value of its own products, most demonstrating its capabilities in developing new types of goods. Modular production increases the load of module suppliers, but automobile enterprises implement the principle of adaptability to this method of production organization as one of the criteria for finding component suppliers.

We will carry out a study of the organization of production at PJSC GAZ in terms of the modular production method and the corporate production system. There is a toolkit that allows you to implement practical solutions to achieve the greatest efficiency of the production system [8]. The general mechanism is shown in the Fig. 1.

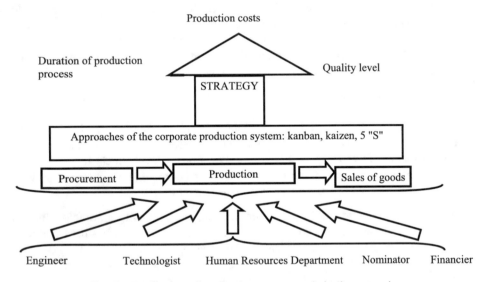

Fig. 1. Application of production system tools in the enterprise

The main task in the organization of production with maximum efficiency is to eliminate all discrepancies in the vehicle with technical regulations that will reduce customer satisfaction with the purchase. To achieve this goal, the corporate production system of PJSC GAZ has introduced a classification of production costs that do not add value to the product from the point of view of the final consumer of cars. This type of cost in the terminology of the production system of PJSC GAZ is called "losses".

To exclude them, work is carried out in the following areas: modernization of packaging; improvement of working carts, shelving; improving the layout of the operator's workplace and the layout of materials. Modernization of the tool and its location, as well as technological equipment significantly reduces the time required to find the right tool. A special place should be taken to optimize the sequence of operations on the production line, as well as their integration and redistribution, depending on the workload of the worker on the assembly line.

It is important to remember that in production, there is a flow of cost and a flow of value and at the end of the production process, the value stream must exceed the cost stream. This difference, in fact, represents the profits of the corporation. In order for the product value to be high for the customer, in our opinion, it is necessary to constantly improve the quality of the car.

One of the important characteristics that affects the achievement of an ideal quality is the consistency of assembly operations on the conveyor. As a rule, achieving the standardization of the operation increases the quality of cars and components. The constancy of production will depend on stability in all its components: production technology, production equipment and materials. In turn, the level of the employee's qualification acquires special importance, which depends on the ability to perform work on the standards. The permanence of the production equipment will depend on the absence of breakdowns, defects, and on proper maintenance. The consistency of the quality of raw materials and materials determines the absence of defects, the fixed quantity and the absence of a deficit. The technological state of the production process depends on the standard algorithm of work execution and management decisions.

At this stage of economic development, companies that are planning to increase their sales require a rapid response to market demands. For this, in our view, it is necessary to correct the revolutionary and evolutionary changes in the company. It should be noted that the principle of kaizen in the production system of PJSC GAZ provides for permanent improvements, but is not a method of evolution in the company [2]. If the production equipment is morally worn out, then the firm will not be able to reach the world level of their product. Accordingly, the client will not acquire it. In the production system of the company "Toyota", it is specified that as a result of constant improvements it is possible to achieve a technological jump-kayrio. It involves the modernization of both the main and auxiliary production processes using innovative production technology. Leading global corporations distinguish between these kinds of changes in the functioning of the company and the organizational environment, clearly allocating the different types of resources that are necessary for the realization of each.

Let us analyze in more detail the instrument of kaizen offers. Kaizen aims to improve all aspects of the company's activities from processes and production relationships related to the receipt of materials and components, to the processes of their processing and ways of interacting with distribution systems and end-customers.

The fundamental principle of kaizen is the philosophy of systematic improvement of all processes. With this approach, each process is evaluated by the following key indicators [4]:

- The required time;
- Resources used;
- Product quality.

Involvement of employees in the process of systematic improvements helps to solve problems existing in the company and the effective use of labor resources. For the employees' interest, it is necessary to create a corporate culture and organizational environment in which both management and operators act as a single team. Thus, the Kaizen principle can consolidate changes in the company's production system when developing and releasing a new product. The production system of PJSC GAZ includes the use of the kanban system.

At the initial stage of the introduction of the system of equalization of production, it is necessary to organize the work of the conveyor from the development of areas for warehouses of expeditions, where the selection of details takes place. The kanban system made it possible to arrange expeditions near the production line. Changed accordingly and the placement of components that are located near the sites where they are installed on the car.

Thus, the storekeeper spends less time on movement, which reduces the duration of the production cycle. The saved minute at each stage of production allows reducing the duration of the production cycle from 10% to 20%. This system works based on the principle: the required number of parts at the right time in the right place. The number of parts delivered to the workplace is determined by the two-hour interval of the employee's work. With a standard supply system, the parts were brought into the work area by a container, which required additional operations for the accounting and sorting of the car components. With the use of the kanban system, all the components are on the expedition, where their assortment and quantity are clearly known. On the conveyor №3 of PJSC "GAZ" each storekeeper serves six operating zones of the assembly line. Bypassing the working area, the storekeeper looks, where what parts are missing, what needs to be replenished, and brings the necessary assortment and the number of components for the car from the expedition warehouse. A special plate is affixed to the container, which indicates which parts, their number and on which operating zone they need to be delivered.

As a result of changes in the corporate production system with the aim of consolidating the work on the creation of a new product, we proposed to use the modular method of production, both at the stage of developing a new product, which involves the use of the potential of manufacturers of automotive components, and at the production stage using the kaizen and kayrio system kanban, which allow to reduce production costs and unite all participants in the production system with a common corporate philosophy.

5 Conclusions

In conclusion, it should be noted that the authors identified the features of creating a new product in an industrial enterprise, in particular, a general methodology for creating a new product in the industry, which is based on the marketing approach to the creation of a new product. To implement this technique, it is necessary to move to a modular method of designing and manufacturing a car using engineering and marketing potential, both the corporation itself and the automotive component manufacturers. To combine changes in the development and release of a new product, the production system needs to be transformed to the client needs, which can increase the economic efficiency of production with minimal organizational costs, which is especially important in the current conditions of a slowdown in economic growth.

References

1. Andryashina, N.S.: Modern approaches to the creation of a new product in machine building. Bull. Univ. Minin **1**(5), 1 (2014)
2. Egorova, A.O., Kuznetsov, V.P., Andryashina, N.S.: Methodology of formation and realization of a competitive strategy of machine building enterprises. Europ. Res. Stud. J. **19**(2, Special Issue), 125–134 (2016)
3. Garina, E., Kuznetsova, S., Semakhin, E., Semenov, S., Sevryukova, A.: Development of national production through integration of machine building enterprises into industrial park structures. Europ. Res. Stud. **XVII**(Special Issue), 267–282 (2015)
4. Klychova, G.S., Kuznetsov, V.P., Trifonov, Y.V., Yashin, S.N., Koshelev, E.V.: Upgrading corporate equipment as an asian real option. Int. Bus. Manag. **10**(21), 5130–5137 (2016)
5. Shushkin, M.A.: The development of enterprises of the automotive industry on the basis of industrial partnership strategies: Doctoral thesis. Penza State University, Penza, 45 p (2013)
6. Festel, G., Würmseher, M.: Challenges and strategies for chemical/industrial parks in Europe. J. Bus. Chem. (2013). http://www.businesschemistry.org/article/?article=173. Accessed 30 Jun 2017
7. Fujita, M., Thisse, J.F.: Economics of Agglomeration Cities, Industrial Location, and Regional Growth. Cambridge University Press, Cambridge (2002). pp. 104–106
8. Vumek, D., Jhons, D.: Lean Production: How to Get Rid of Loss and Achieve the Company Flourishing; Translated from English- 4th edn. Alpina Business Books, Moscow (2013). p. 472

Advantages of Residents of Industrial Parks (by the Example of AVTOVAZ)

Svetlana N. Kuznetsova[✉], Elena V. Romanovskaya,
Marina V. Artemyeva, Natalia S. Andryashina,
and Anastasia O. Egorova

Nizhny Novgorod State Pedagogical University Named After Kozma Minin
(Minin University), Nizhny Novgorod, Russian Federation
dens@52.ru, alenarom@list.ru, vershinina82@mail.ru,
natali_andr@bk.ru, nesti88@mail.ru

Abstract. The article offers new advantages for residents of industrial parks, and argues for the need to revise approaches to the development of the institute of industrial parks, which will minimize the risks of integrated development of regions and the country in general.

In the article, the authors turn to the question of the real economic efficiency of industrial parks. The authors of the article point to the significance of increasing regional economic efficiency, which will allow the regions of the Russian Federation to take a new approach to the issues of expediency of creating industrial parks. The authors identify the most promising advantages for investors: Reduction of time from investment, reduction of administrative risks, reduction of production costs, and profit from tax benefits. The authors point out that the main indicators of the creation of industrial parks and the result for the regional economy are the following: Growth of investment attractiveness, volume of investments, growth of tax revenues, creation of high-productive jobs, and improvement of innovative development. Speaking about the importance of industrial parks for the region, the authors suggest using the experience of AVTOVAZ. The procedure for using the infrastructure and technological capabilities of AVTOVAZ is considered. This approach allows organizing and carrying out work on preparation of production of new products at AVTOVAZ, as well as creating new jobs.

Keywords: Industrial park · Resident · Concentration · Cooperation
Cluster · Economy · Investor · Advantages · Efficiency · Infrastructure

1 Introduction

Studies show that the modern economic environment is characterized by the transformation of economic relations arising in domestic industrial corporations:

- Expansion of economic independence of corporations;
- Centralization on a voluntary basis of strategic management functions in order to obtain maximum profit;
- Contractual relations between structural units at all levels.

© Springer International Publishing AG, part of Springer Nature 2018
E. G. Popkova (Ed.): HOSMC 2017, AISC 622, pp. 502–509, 2018.
https://doi.org/10.1007/978-3-319-75383-6_64

At the same time, the mechanisms of functioning of the organization of economic entities that have developed over the past two decades do not fully correspond to the situation in the Russian industry [1].

2 Theoretical Basis of the Research

In the opinion of the authors, the practical aspect of the formation of industrial parks and the characteristics of the modern Russian economy has been poorly studied. It should be emphasized that foreign authors have extensively studied the practical and theoretical parts of the formation of industrial parks. However, if you introduce the experience of foreign partners to the Russian market, there is a need for additional study of many issues, one of which is the search for effective coordination of industrial parks in Russia.

Interest in structuring industrial parks in Russia was formed later than in some foreign countries. However, Russian scientists have done a great job in the development in this direction. However, at the same time, widely worked theoretical aspects are not provided with practical recommendations for determining the most effective industrial parks [2].

All of the above reasons determined the relevance of the article.

This problem was investigated by such domestic scientists as V.I. Nekrasov, O.I. Botkin, V.N. Eremin, A.N. Pytkina, A.I. Tatarkina, K.M. Pirogov and others. In their works, they emphasize the low growth of territorial production complexes.

3 Methodology of the Study

Formation of industrial parks will contribute to the improvement of the production sector. Such parks are one of the types of contractual inter-firm production network of small and medium-sized enterprises located in a specially created and under the management of an industrial zone with a unified engineering infrastructure and technologically connected with a large enterprise engaged in the development and production of finished products [3].

Organization of work of enterprises in the format of industrial parks has many advantages for the regional economy in general, and for residents in particular. In addition, industrial parks are economically attractive for investors.

If we consider a regional economy, then the formation of industrial parks will have the following advantages:

- Growth of investment attractiveness and volume of investments;
- Increase of funds from payment of taxes to the budget;
- Increase in the number of high-tech work areas;
- Introduction of innovative technologies in the work of enterprises in the region [4].

Investment attractiveness is explained by the presence of the following significant features:

– Reduction of the period, the starting point of which is the time of investment by the depositor, and the final one - the output of the product;
– Reduction of administrative risks;
– Reduction of costs of manufactured products;
– Receiving additional profits through tax incentives

For residents, industrial parks are of particular interest, since they make it possible to create a cluster through concentration and cooperation, complementary productions. The creation of a cluster, in turn, reduces the costs of engineering services and transportation, and, consequently, reduces the cost of production.

For residents, the main advantage of industrial parks is the concentration in one place of communal, transport and technological infrastructure, buildings, structures, structures (Table 1).

Table 1. Advantages of placing on the territory of industrial parks

Benefits	Indicators
Interest in filling customers with their infrastructural sites, using an administrative resource to more effectively enter the market of the region	- State financing of 50% of the spent funds for the purchase of equipment - Reimbursement of up to 25% of expenses for the purchase of real estate in an industrial park (up to 5 million rubles) - Reimbursement of up to 25% of the costs of renting an industrial park property (up to 1 million rubles) - State financing of up to 50% of the funds expended for equipment purchase under leasing agreements - State financing of spent funds to pay % on loans - State financing of energy saving, energy audit and connection to power supply networks - State financial support and provision of benefits to participants in industrial clusters
Industrial sites have the necessary infrastructure as a result of which, they are the most suitable for the development of production and production services	- Zero rate of land lease for the period of construction - Re-registration of land in ownership - Privileges for connection to power supply networks (0 for electricity networks, 0 for heating networks, 0 for gas lines)

(*continued*)

Table 1. (*continued*)

Benefits	Indicators
In industrial parks services are offered - assistance in the sale of products, there is a raw material and production cooperation, public facilities, transportation and warehouse logistics services	- Privileges for local taxes (lowering coefficient to the rate of rent for land or land tax) - Privileges for regional taxes (reduction of the property tax rate from 2.2% to 0.1%) - Privileges on federal taxes (privileges in the part of the profit tax credited to the regional budget - lower by 4.5% of the item) - Privileges on customs duties and payments
Customers of the industrial park have state financial support, as well as the provision of benefits from the state	- Profit tax 0-5% (without TOP 20%) - Land tax 0% (without TOPs on average 1.5%) - Corporate property tax 0% (without TOP 2.2%) - Insurance contributions to state non-budgetary funds 7.6% (without TOP 30%)

To date, there is an active development of industrial parks. The specialists make the following forecasts for 2020:

- The share of occupancy of objects by companies-tenants of the industrial park - at least 70% of the total area;
- The amount of tax deductions to the federal budget - not less than the total aggregate amount of funds requested in the form of subsidies;
- The number of high-performance jobs: For industrial parks - no less than 1500; for techno parks - not less than 500;
- Index of budgetary profitability: For industrial parks - not less than 2; For techno parks - not less than 1.5;
- An indicator of the effectiveness of the productive activity of people - not less than 1.05.

The development of the Institute of Industrial Parks allows managing the risks of complex development of regions and the country as a whole in the following areas [5]:

- Reduction of innovative risks: It is ensured by the use of modern production facilities in industrial parks, which presuppose introduction of advanced technologies;
- Reduction of commercial and technological risks: It is ensured by the involvement of companies that are international leaders in the design, construction and management of industrial parks;
- Reduction of social and personnel risks: Is ensured through the development of professional knowledge, skills and skills of employees through the organization of internships, post-graduate training of personnel, publication of methodological, information and educational materials and programs;

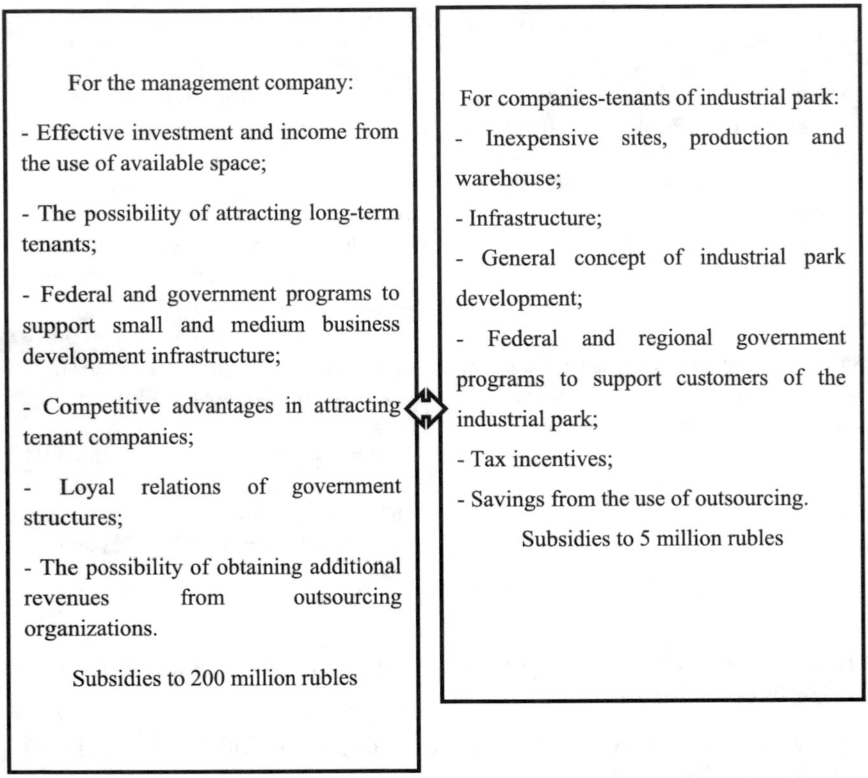

For the management company:

- Effective investment and income from the use of available space;

- The possibility of attracting long-term tenants;

- Federal and government programs to support small and medium business development infrastructure;

- Competitive advantages in attracting tenant companies;

- Loyal relations of government structures;

- The possibility of obtaining additional revenues from outsourcing organizations.

Subsidies to 200 million rubles

For companies-tenants of industrial park:

- Inexpensive sites, production and warehouse;

- Infrastructure;

- General concept of industrial park development;

- Federal and regional government programs to support customers of the industrial park;

- Tax incentives;

- Savings from the use of outsourcing.

Subsidies to 5 million rubles

Fig. 1. Advantages of industrial park [14]

– Distribution of risks between the private partner-investor and the state: Is provided through the work of special mechanisms of public-private partnership when investing in industrial parks. (Figure 1).

Modern international policy creates such conditions for investment, in which the risk of investment is very high. Therefore, to make Russia attractive for investment, it is necessary to have very high returns on capital. If we consider this problem from a practical point of view, then it can be noted that the Russian Federation has regions that are open to attracting capital to industries such as manufacturing, agriculture and many others, but a unified development program and a unified approach that will be accessible for understanding to foreign investors [6, 7].

This approach has become a standard technology for creating industrial parks, but it is necessary to realize that we need those investments that bring high technologies to Russia (Fig. 2). It is the investment in domestic production that should become the main one, since the opening of foreign productions will not be effective for the development of Russia and its regions, since there is a risk that foreign goods can become dominant over domestic goods [8, 9].

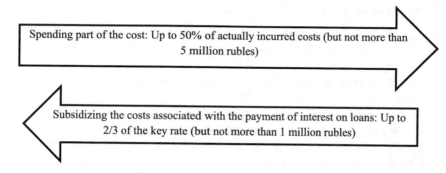

Spending part of the cost: Up to 50% of actually incurred costs (but not more than 5 million rubles)

Subsidizing the costs associated with the payment of interest on loans: Up to 2/3 of the key rate (but not more than 1 million rubles)

Fig. 2. Opportunities of industrial parks and residents [14]

It is necessary not only to create a mechanism for attracting investments, but also an effective regulatory mechanism that ensures maximum positive return on investment for the regions and minimizes the corresponding risks by ensuring the protection of public interests.

4   Analysis of the Results of the Study

As various studies show, the real economic efficiency of industrial parks exists, but the low regional effectiveness of industrial parks creation remains a constraining factor. An increase in regional production efficiency will allow the regions of the Russian Federation to assess the need for the development of existing and construction of new industrial parks on their territories [10, 11].

As the data show for 2015–2016, there are more than 1.9–2.0 million of empty industrial sites with utilities. To eliminate this problem, it is necessary to ensure the attraction of funds from the federal budget for the development of infrastructure. At the same time, in the forecast period, it is necessary to continue work on maintaining the infrastructure of industrial parks through investment programs [12, 13].

Actual is the problem that constrains the spatial development of the Russian Federation; these are unoccupied production capacities (Fig. 1). To solve this problem, it is necessary to develop a network of industrial parks, actively fill in with resident companies.

For example, the AVTOVAZ industrial park provides for the customers renting premises for any type of activity on a long-term basis, using the infrastructure and technological capabilities of AVTOVAZ.

According to AVTOVAZ statistics, the Russian automotive market will remain at the level of 2016, but a slight change in demand (within 5%) is possible. During 2016, the number of copies sold was reduced to 1,425,791 in Russia, which is 11% less than the result of 2015. Also in 2017 it is planned to increase the percentage of exports of goods by 50%.

The main tasks of the AVTOVAZ industrial park are [14]:

- Formation of working areas for suppliers in the non-operating facilities of AVTOVAZ, including for additional placement in one particular place of various components of the production of cars;
- Formation of working zones for other companies in the non-operating facilities of AVTOVAZ;
- The formation of working areas for AVTOVAZ and its subsidiaries associated with new products and services in the automotive and other industries;
- Preparation, coordination and implementation of various works on the production of new products at AVTOVAZ;
- Performance of works previously performed for AVTOVAZ by third-party contractors not based in the local region, under contractual agreements.

The effect of the creation and development of an industrial park for tenant companies is expected in the following areas:

- Implementation of investment and innovation policies in the regions;
- Formation of organizational and legislative bases for maintenance of small and medium business;
- Strategic cooperation with domestic partners and with partners from abroad for small and medium-sized businesses to introduce new programs and projects in the region;
- Cooperation with investors and investment funds for the implementation of projects and programs in the region based on long-term prospects;
- The formation of a suitable place for the production of competitive, high-tech products and services;
- Modernization of foreign economic relations and increase in exports;
- Attraction of foreign capital and technologies, development of industry in the regions;
- Guaranteeing the economic security of the region;
- Maintenance of business in the regions at the expense of budgetary and extra-budgetary funds with guarantee of protection of the state interests in the implemented investment and innovative projects;
- Provision of sites for the implementation of various business ideas;
- Increase the level of employment and life of the population of the region;
- Increase in budget revenues;
- Increasing the impact of local authorities on regional companies.

References

1. Belenov, O.N., Smolyaninova, T.Y., Shurchkova, Y.V.: Industrial parks: essence and basic characteristics. Reg. Econ. Manag. Electron. Sci. J. 1(33), 66–76 (2013)
2. Berkovich, M.I., Antipina, N.I.: Features and classification of industrial parks: regional aspect. Bull. Kostroma State Tech. Univ. 1(3), 25–28 (2013). (Economics)

3. Kuznetsova, S.N.: Development of the organizational and economic mechanism for the formation of industrial parks: Ph.D. thesis. The Ivanovo State University, Ivanovo, p. 24 (2012)
4. Lenchuk, E.B.: Formation of the institutional environment for industrial development in the context of import substitution problems. Bull. Inst. Econ. Russ. Acad. Sci. **6**, 7–21 (2014)
5. Maltseva, A.A., Chevychelov, V.N.: World trends in the development of technopark structures: selective analysis. Probl. Anal. State-Manag. Des. Cent. Probl. Anal. Public-Manag. Des. **5**(2), 29–42 (2012). Moscow
6. Sysoev, E.V.: Actual aspects of improving the state industrial policy. Transp. Bus. Russ. **5**, 193–194 (2013)
7. Shpak, N.A.: Prospects for the development of Russia's innovative infrastructure. Mod. Probl. Sci. Educ. **5**, 421 (2014)
8. Fei, Yu., Han, F.: Zhaojie Cui evolution of industrial symbiosis in an eco-industrial park in China. J. Clean. Prod. **87**, 339–347 (2015)
9. Fujita, M., Thisse, J.F.: Economics of Aglomeration Cities, Industrial Location, and Regional Growth, pp. 104–106. Cambridge University Press, Cambridge (2002)
10. Shi, H., Chertow, M., Song, Y.: Developing country experience with eco-industrial parks: a case study of the Tianjin Economic-technological development area in China. J. Clean. Prod. **18**, 191–199 (2010)
11. Henderson, J.V.: Urban Development. Theory, Fact and Illusion. Oxford University Press, Oxford (1988). pp. 325–350
12. Industrial Estate: Principles and practices: Vienna Australia. UNIDO, p. 46 (1997)
13. Jacobs, J.: The Economy of Cities, p. 268. Random House, New York (1969)

Features of the Population's Savings Transformation into Investments at the Present Stage

Olga G. Lebedinskaya$^{(\boxtimes)}$, Alexander G. Timofeev,
Elvira A. Yarnykh, Nina A. Eldyaeva, and Sergey V. Golodov

G.V. Plekhanov Russian University of Economics, Moscow, Russian Federation
{Lebedinskaya.OG, Timofeev.AG, Yarnykh.EA, Eldyaeva.NA,
Golodov.SV}@rea.ru

Abstract. In conditions of limited access to external borrowed sources, the savings of the population that can be transformed into investments become one of the main internal reserves for stimulating economic growth and modernizing the Russian economy. The article presents an analysis of current trends in investment behavior of the Russian population, its motivations and reasons restraining the most complete involvement of a part of household income that is not used for current consumption into economic turnover. The authors outlined the problems of attracting investments in the country's economy through the redistribution of effective consumer demand from the consumer market to the sphere of turnover of the company's cash resources.

It is noted that one of the significant factors in the transformation of the population's savings into investment is the lack of a "saving" culture of the population. High-income differentiation and a low saving rate and their target character, and the predominance of "short" investor sentiment.

The hypothesis advanced by the authors of the article about the existence of regularities in the savings behavior of the population in an unstable economic situation made it possible to draw conclusions about the preferred forms of savings on the part of the population, and also on the quality of which institutions need to be developed to stimulate the investment activity of the population.

Keywords: Savings · Investment resource
Transformation of savings of the population
Investment potential of the government · Investment savings · Savings motives
Investment activity

1 Introduction

The new economic system of Russia has led to global changes in the mechanisms of management and the process of social reproduction. First, this refers to investment activity, since a fundamentally new mechanism for distributing gross domestic product has been created, which is used mainly for investment, rather than for final consumption. The emergence of independent economic entities and change in property

E. G. Popkova (Ed.): HOSMC 2017, AISC 622, pp. 510–518, 2018.
https://doi.org/10.1007/978-3-319-75383-6_65

relations caused the decentralization of the investment process. Now, the role of the population is changing, which through its savings can become one of the main participants in investing.

New industrial relations led to the emergence of other social relations, under which every citizen is responsible for his future. If before there was a government system of social protection of the population, it is now necessary for everyone to make a decision about the distribution of disposable income for consumption and further saving, providing education, medical care and old age. Therefore, the implementation of ways to use the savings of the population is one of the most important problems at the macro and micro levels.

2 Theoretical Basis of the Research

Issues of theoretical and practical research on the transformation of the savings of the population into investments at different times were touched upon by Russian and foreign scientists, such as Sibirskaya et al. (2015), Yarnykh et al. (2015), Aliev and Taisumova (2011), Zaretskaya and Kondratieva (2011), Darda and Sadovnikova (2013), Campbell (2006), Nagy and Obenberger (1994), Shanthikumar and Malmendier (2003).

The problems of searching for the possibility of using the savings of the population as a source of investment are devoted to the work of V.A. Yadov (Yadov, I.A.) Prigozhin (Prigozhin), A. Tversky, D. Kaneman. The authors noted that most of the research in the modern economy is because people are rational agents trying to maximize welfare with minimal risk; however, in some situations the operating is irrational, as evidenced by the low percentage of "lucky" investors. The financial behavior of people (and its variety - the savings) depends on the individual perception of risk, while remaining social. People act in accordance with their values and patterns of behavior in a particular environment

In the model of rational expectation (Grossman and Stiglitz, 1984), investors are focused either on the acquisition of high-value assets (see potentially yielding significant potential income) or on passive investment (often in the form of a deposit). In Kyle's model Kyle (1985); the importance is given to "noise", under the influence of which an informed insider decides to invest.

Barber et al. (2009) notes the growth of systemic rather than individual errors in making investment decisions, implying that the work and news feeds have a significant impact on the decision to invest. Olsen (Olsen) distinguishes two cornerstones of the financial behavior of the population - the problems of cognitive psychology and the limits of arbitrage (limits of market inefficiency). In this regard, the work outlines the specificity of savings behavior in the context of the economic crisis, which has been one of the central socio-economic events in recent years.

3 Research Methodology

At present, the savings rate of the population can be determined based on the following methodological approaches, which are not sufficiently coordinated among themselves and give different results:

- Balance of money incomes and expenses of the population of the Russian Federation;
- Use of disposable income account for the household sector in the System of National Accounts (SNA);
- Data on a sample survey of households (mainly CIRCOM, VCIOM);
- The data of the Central Bank of the Russian Federation regarding the calculation of the savings rate, etc.

The analysis of individual forms of accumulation of population savings, the characteristics of certain aspects of which are presented in Table 1, made it possible to identify two of their basic forms - organized and unorganized.

Table 1. Institutions and technologies of accumulation of money savings of the population for investment purposes

Financial and investment institutions	Technologies of accumulation of savings
Banks and banking institutions	- Opening of term deposits - Issue of deposit and savings certificates - Trust management of household savings - Placement of coupon bonds, - Placement of preferred shares
Mutual (mutual) investment funds	Acquisition of investment units under a contract with a management company
Investment companies	Attracting the population's funds through issuance and placement of shares
United banking management funds	Attracting the population's funds by providing an equity certificate in exchange for the funds transferred
Regional and municipal investment-loan systems	Issue and placement of bonds of targeted regional and municipal loans
Pension funds	Use of the accumulative part of deductions for portfolio investments of collective operators of the stock market
Insurance companies	Placement of the accumulated portion of insurance premiums in portfolio of investment instruments
Stock exchanges	Stock market tools

Traditionally, organized people understand those savings that are accumulated in the deposit accounts of the population, they are invested in securities, real estate, precious metals and stones, antiques, transferred to trust management, to the united funds of bank management, as well as directed to the funded part of the pension. In other words, these monetary resources have already been accumulated by the subjects of the financial market and can be easily transformed into investments. Unorganized (unsystematic) savings are savings that are in the hands of the population, which in their essence are a potential investment resource, but practically not realized.

The mechanism of transformation of savings into investment can be considered as the sequence (set) of states of the relationship of savings, investment, and income, correlated in a special way based on the adopted model of behavior of economic agents, and formalized by a certain institutional complex. The effectiveness of this mechanism determines the balance of savings and investment.

The equation of such a balance has the simplest form: 1a + 2a = 16, where 1a is the aggregate of the subjective motives of saving (from precaution, the desire to create an interest income in the future, to ensure the welfare of descendants, to provide the necessary amount for speculation, etc.) - subjective savings, 2a - the totality of objective motives Change in income, interest, prices, tax policy, etc.), 16 -system of investment factors.

This equation shows that in the "low" phases of the economic cycle (recession and depression), the predominant part of savings for investment is formed by a system of subjective motives, in "high" phases (recovery and boom) - objective motives. Statistics show that often enough, when making a decision about investing, the population uses "mental labels" that significantly worsen the results of an investment decision. Excessive self-confidence of an individual investor (the belief that he will be right when choosing an investment object) leads to an extremely low differentiation of options for saving behavior. As a result, an increase in the share of private investors "escaping from risk" during the reduction of deposit rates will not only not cause a loss of savings for investment (reduce the volume of deposits attracted), but, conversely, increase their aggregate volume.

An important role in the transformation of savings into investment is performed by the so-called institutional system of transformation (ICT), the elements of which are the tax and banking systems, the stock market. Despite the fact that she herself is not capable of producing the impulses of transformation processes, her action leads to the necessary combination of savings-investment motives. The analysis shows the presence of several types of such models: With the dominant role of the stock market in the ICT structure, with the dominant role of the banking sector (in the whole sector and separately, with the dominant role of state banks and, accordingly, private sector).

4 Analysis of Research Results

The dynamics of changes in the population's savings behavior is shown in Figs. 1 and 2.

The one shown in Fig. 1 dynamics and the coefficient of structural shifts calculated by the authors. Gatev (k = 0.11) makes it possible to state that in the period under study (1990–2015) there have been significant changes in the structure of the population's savings behavior: The share of financial assets in the structure of investments has significantly increased (19.5%), and the acquisition of real estate is losing attractiveness. For example, savings in deposits and securities account for about 1.5%, expenses for the purchase of real estate - 1.6–2%, purchase of foreign currency - 2.1–2.5% in the costs of the population.

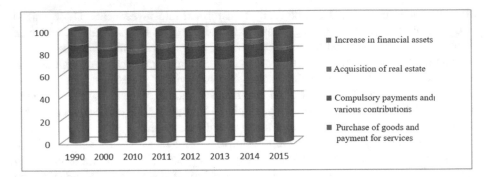

Fig. 1. Dynamics of changes in the structure of household savings

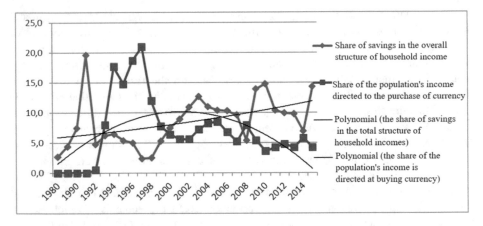

Fig. 2. Dynamics of monetary expenditure on accumulation.

The results of the analysis of the dynamics of expenditures have shown that there is an obvious correlation between savings and investments in the pre-crisis (100–2007), the crisis (2008–2009) and the post-crisis period (after 2009). The fact that, starting from 1996, the gap between the index of the savings potential (%), calculated as the difference between the per capita income and the two-fold size of the subsistence minimum and the dynamics of real money incomes of the population (%) has practically decreased, only increases the need to understand the motives for saving and investing the population.

The analysis allowed revealing the existence of a multitude of differently directed motives, indicating that the savings of the population are inherently diverse and the degree of their "suitability" for investment is different. The investment behavior of the population, regardless of belonging to a particular class, may depend on the type of risk. Therefore, we identified three groups: Savings riskofily, savings riskophobes and neutral savers whose motives for saving are presented in Table 2.

Table 2. Types of savings behavior of the population and motives

Riskofily	Neutral type	Riskophobes
Purchase of goods and services, the payment of which requires the accumulation of money (real estate, summer vacation, education)	"Amortization" motive, the reason of which is the need to replace obsolete durable goods (car, motorcycle, etc.)	Investment (speculative) motive, i.e. purchase of securities with the purpose of their further resale
The creation of a "safety cushion" in the event of a sudden drop in revenues	A motive of comfort, suggesting the desire to have with him considerable money to meet his needs at any time	
Creation of "reserve funds of financing" of the expected events (wedding)	Savings "out of habit"	
Pension motive		

From the point of view of the effectiveness of the economy, we are interested in investment-oriented savings, the initial goal of which is to obtain additional income. The analysis showed that whatever efforts the state made to stimulate investment, the population will always wipe between "extreme" efforts - maximum or minimum, refusing neutral behavior, which in turn causes a certain set of motives, any stimulation of which will not give a meaningful effect in terms of the transformation of savings into investment. This statement confirms the fact that cash in the structure of savings priorities of the population remains in the first place (Table 3).

Table 3. Savings priorities of the population

Forms of savings of the population	Share in the structure, %
1. Cash	36.5
2. Cash currency	12.7
3. Deposits in Sberbank in ruble equivalent	5.2
4. Deposits in Sberbank in currency equivalent	0.5
5. Securities	11.0
6. Deposits in commercial banks in ruble equivalent	3.4
7. Deposits in commercial banks in currency equivalent	0.7
8. Other forms of savings	30

Nevertheless, the analysis of the dynamics of the share of savings of the population shows that the pre-savings level of the savings rate has practically recovered in the country (the 2012 level), but the recovery is accompanied by a significant transformation of the structure of organized savings (Fig. 3).

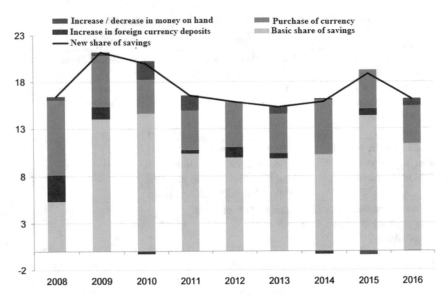

Fig. 3. Income-weighted share of savings as a change in money on the hands, foreign currency deposits and the purchase of currency in the income of the population, % (quoted from the Bulletin of the Department of Research and Forecasting of the Central Bank of Russia)

However, analysis of the general dynamics of household savings (organized and unorganized) indicates a slightly negative dynamics.

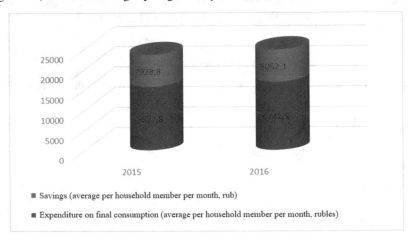

The share of savings in the structure of income according to the Federal State Statistics Service decreased in 2016 compared to 2015 from 33.66% to 32.47% in Russia in general, and in the rural area, the savings rate fell to 29.9%. Locomotives of dynamics are households consisting of one adult without children, in which, on average, per month, per household member, the savings rate is 24,875 rubles or 31.1%. The share of those who believe that "the funds are enough to buy everything that they consider necessary" by the beginning of the second quarter of 2017 decreased to 2.6% (in the 1st quarter of 2016 it was 3.2%). This trend indicates an increase in unorganized household savings, accompanied by a negative overall dynamics.

One cannot but agree with the opinion that the savings potential of the population is characterized by four parameters that also include an assessment of investment potential:

- Labor potential;
- Institutional capacity;
- Financial potential;
- Consumer potential.

This is the situation in general in the Russian Federation. In some years, the excess of savings over investments was up to 19% of consolidated GRP. The population's funds play a big role in the formation of investment resources.

5 Conclusions

The analysis made it possible to identify the key features of the mechanism of the transformation of savings into investments in Russia:

- The average savings rate in early 2017 recovered to the level of 2012–2013 and amounted to 24.3%; However, it underwent significant changes in the ratio of forms of savings: The share of unorganized savings with the overall structure was only 3%:
- The underdevelopment of socially significant ways of transforming the savings of the population into investment, which is confirmed by the fact that the population is not in a hurry to transfer their pension savings to private capital, despite the higher profitability they offer.
- The inability of the banking system to increase the volume of lending to investment projects, as evidenced by their still high growth rates;
- low attractiveness of the domestic stock market for national "blue chips", which place on it no more than 15% of issued investment securities;
- A narrow range of financial instruments available to the public.

The authors singled out three blocks of causes that cause "failures" in transforming the population's savings into investments.

The first block is "savings". The authors noted that savings, firstly, are affected by current and accrued income, the level of interest rates; on the value of which, in turn, the value of borrowed capital and, consequently, the total level of income. The size of investment resources is directly dependent on the level of income of the population and accumulated savings, which depend on this income. If we consider the indicators of the structure of the use of income by object of expenditure, which can be directed at investing the economy, then their share is insignificant.

The second block is "investment": Disintegration of the national economy, the lack of a system of equilibrium prices, inefficient taxation system.

The third block - institutional - excessive monopolization and outright speculation of the banking system and the stock market is a low level of development of Internet technologies, which largely determines the degree of development of transformational processes.

In the circumstances, the important role in the formation of an effective mechanism for transforming the savings of the population in Russia is, first, the activities of the government aimed at:

- Ensuring consistency in the implementation of pension reform;
- Creation of favorable conditions for the development of institutional investors; Improvement of the tax mechanism;
- Development and adoption of laws that promote the functional and institutional development of the banking system; Development of the stock market.

Acknowledgments. Article is prepared with financial support of G.V. Plekhanov Russian University of Economics.

References

Barber, B.M., Odean, T.: All that glitters: the effect of attention on the buying behavior of individual and institutional investors. Rev. Financ. Stud. **21**, 785–818 (2008)

Barber, B.M., Lee, Y., Liu, Y., Odean, T.: Just how much do individual investors lose by trading? Rev. Financ. Stud. **22**, 609–632 (2009)

Campbell, J.Y.: Household finance. J. Financ. **61**, 1553–1604 (2006)

Eldyaeva, N., Sibirskaya, E., Yarnykh, E., Dubrova, T., Oveshnikova, L.: Strategy of systemic development of entrepreneurial infrastructure of regional economy. Eur. Res. Stud. **XVIII**, 235–258 (2015)

Kyle, A.S.: Continuous auctions and insider trading. Econometrica **53**, 1315–1336 (1985)

Nagy, R.A., Obenberger, R.W.: Factors influencing investor behavior. Financ. Anal. J. **50**, 63–68 (1994)

Shanthikumar, D., Malmendier, U.: Are small investors naive? Stanford University Working Paper (2003)

Sibirskaya, E., Khokhlova, O., Eldyaeva, N., Lebedinskaya, O.: Statistical evaluation of middle class in Russia. Mediterr. J. Soc. Sci. **6**(36), 125–134 (2015)

Sibirskaya, E.V., Khokhlova, O.A., Oveshnikova, L.V., Stroeva, O.A.: An analysis of investment-innovation activity in Russia. Life Sci. J. **11**(7s), 155–158 (2014)

Yarnykh, E.A., Oleynikov, B.I., Tsylina, I.T.: Statistical methods for the analysis of demographic processes within the region. In: 8th International Scientific Conference Science and Society, pp. 117–123, 24–29 November 2015

Aliev, B.H., Taisumova, H.V.: Formation of the mechanism of transformation of savings of the population into investments for the development of the economy on an innovative basis. Financ. Credit **30**(462), 12 (2011)

Bulletin of the Department of Research and Forecasting of the Central Bank of Russia, no. 2(14) (2017)

Darda, E.S., Sadovnikova, N.A.: Methodological aspects of the formation of intermediate consumption in the financial sector: domestic and international experience. Fundam. Res. (10-3), 610–614 (2013)

Zaretskaya, V.G., Kondrat'eva, Z.A.: Savings as a potential source of investment in the region. Financ. Credit **2**(434), 65–66 (2011)

Grossman, S., Stiglitz, J.: On the impossibility of informationally efficient markets. Am. Econ. Rev. **70**, 393–408 (1984)

Human as an Economic Subject

Economic and Statistical Analysis in Evaluating the Perspectives of Structural Changes of Regions' Economy

Yuri I. Treshchevsky[1]([✉]) [iD], Valeri P. Voronin[2],
Maria B. Tabachnikova[1], and Galina N. Franovskaya[1]

[1] Voronezh State University, Voronezh, Russia
{utreshevski, fgnvrn}@yandex.ru, gmasha3@gmail.com
[2] Voronezh State University of Engineering Technologies, Voronezh, Russia
voronin070441@mail.ru

Abstract. The purpose of the article is to determine the perspective directions of structural changes in Russia's regions on the basis of economic and statistical analysis.

Economic and statistical methods of research include a wide range of specific actions, operations, and algorithms. The authors offer and approbate – by the example of specific data of the municipal entities of Volgograd Oblast – the method of cluster analysis. The authors used this method for analysis of various processes – innovational, investment, and structural changes at the level of Russia's regions.

Using the methods of cluster analysis allowed determining the underrun of all municipal entities of Voronezh Oblast for specific directions of the economic and social development. The obtained data allowed offering for each group of municipal entities the perspective directions of structural transformations in the economic and social spheres.

Keywords: Economic and statistical analysis · Cluster analysis
Structural changes

GEL Classification Codes: R 15 – general spatial economics
Econometric and input-output models · Other models

1 Introduction

In the modern conditions, intensive changes in the socio-economic structure of the Russia's regions take place. These changes are very contradictory. On the one hand, new technologies expand the possibilities of territorial diversification of modern productions and create new, highly-efficient jobs. On the other hand, the reducing demand for traditional professions directs the financial, material, labor, and entrepreneurial resources into large cities. Therefore, the state of the social sphere differentiates as well – the number of free spots in educational and medical establishments in the centers of concentration of people decreases as well. Expenditures for social establishments in the municipal entities with reducing population become too high. The large number of

© Springer International Publishing AG, part of Springer Nature 2018
E. G. Popkova (Ed.): HOSMC 2017, AISC 622, pp. 521–529, 2018.
https://doi.org/10.1007/978-3-319-75383-6_66

municipal entities in Russia's regions sets the task of determining their groups in which the similar socio-economic processes that require the unified methods of management take place.

2 Research Methods

Cluster analysis was used as a method of studying the structure. The theoretical foundations of cluster analysis were set by Oldenderfer and Blashfield (1989), Hartigan and Wong (1979). Further on, theoretical substantiation of application of this method in the regional studies was developed by Golichenko and Shchepina (2008, 2009). The method was also approbated in the study of investment processes (Kruglyakova and Treshchevsky 2012) and innovational development (Risin and Treshchevsky 2011).

As to the municipal entities, the method of Myasnikova was used (Myasnikova 2015). The method of cluster analysis was used in this article for evaluation of the state of socio-economic development of municipal entities and the level of functional and spatial balance of the region. During selection of the clustering options, division into 4, 5, and 6 clusters was used with inclusion of Voronezh and Novovoronezh into municipal entities. From the point of view of statistical significance of the indicators that characterize the clusters, the best division is division into 5 clusters with preservation of all municipal entities of Voronezh Oblast.

The calculations were performed in 2012 and 2015, which allows determining the dynamics of changes during transition from the period of "calm" macro-economic situation to its aggravation at present.

As a result, the massive of 12 indicators formed; they show the differences between the clusters (Table 1).

Table 1. Indicators of socio-economic development of municipal entities of Voronezh Oblast.

var 1	Number of subjects of small and medium entrepreneurship per 10,000 people of the population
var 2	Share of average number of employees (without part-timers) of small and medium m companies in the average number of employees (without part-timersй) of all companies and organizations, %
var 3	Share of profitable agricultural companies in their total number, %
var 4	Monthly average nominal accrued wages of employees of small and medium companies and non-profit organizations of city district (municipal region), RUB
var 5	Share of children aged 1–6 who receive a pre-school educational service and (or) service for their support in municipal pre-school educational establishments, in the total number of children aged 1–6 years, %
var 6	Share of municipal general education institutions that correspond to certain requirements of education, in the total number of municipal general education institutions, %
var 7	Total area of residential premises per one person, total sq.m.
var 8	Annual average number of constant population, thousand people
var 9	Total coefficient of natural increase, per mille
var 10	Total birth rate, per mille
var 11	Number of newcomers (total migration), people
var 12	Number of departed (total migration), people

A part of the indicators is poorly differentiated. Though this does not prove either high or low level of development of poorly differentiated sub-systems of the region, they should not be viewed as top-priority directions of structural changes.

3 Structural Differences Between the Municipal Entities of Voronezh Oblast, 2012

The statistical characteristics of clusters that were formed during division of 34 municipal entities into 5 clusters are presented in Tables 2 and 3. Most of the values of indicators are significant at the 5% level, Var 2 and Var 7 – at the 10% level.

For determining peculiarities of the clusters, they are ranked according to the sum of average normed values. Accordingly, they were marked "A", "B", "C", D", and "E". The state of the analyzed characteristics (parameters) of clusters is presented in Table 2.

Table 2. Average values of clusters' parameters (2012)

Indicators	Cluster A	Cluster B	Cluster C	Cluster D	Cluster E
Var 1	1.000000	0.334845	0.453457	0.254106	0.183068
Var 2	0.448543	0.050904	0.348008	0.533682	0.317924
Var 3	0.000000	0.000000	0.885790	0.965675	0.912671
Var 4	0.563847	1.000000	0.208637	0.099282	0.108420
Var 5	0.759908	1.000000	0.437486	0.307444	0.275564
Var 6	0.652582	0.834832	0.478949	0.814009	0.461772
Var 7	0.217391	0.214674	0.317571	0.586889	0.373447
Var 8	1.000000	0.016402	0.041626	0.014619	0.019589
Var 9	0.978571	0.928571	0.770714	0.355357	0.500000
Var 10	0.608696	0.565217	0.793478	0.358696	0.447205
Var 11	1.000000	0.018312	0.021921	0.013299	0.010992
Var 12	1.000000	0.038288	0.029185	0.020387	0.019307
Sum	8.229537	5.002045	4.786822	4.323445	3.629961

Analysis of the data presented in Table 2 allows for the following conclusions.

The strongest indicators of cluster "A" (the value of normed indicator – 1.0) are the following: number of the subjects of small and medium entrepreneurship per 10,000 people; annual average number of population; number of newcomers (migration); number of departed (migration).

The indicator's value "total coefifcient of natural increase" is rather high.

As to all other indicators, cluster "A" is ranked 2nd and lower.

Absence of the agricultural sector in the municipal entity that belongs to cluster "A" led to zero value of the indicator "share of profitable agricultural companies".

The cluster is ranked 2nd as to the following indicators: share of average list number of employees of small and medium companies in the average number of employees;

monthly average nominal accrued wages of employees of small and medium companies and non-profit organizations; share of children aged 1–6 who receive a pre-school educational service and (or) service for their support in municipal pre-school educational establishments, in the total number of children aged 1–6; total coefficient of birth rate. As to this indicator, cluster "A" is behind cluster "B".

The cluster is ranked 3rd and lower as to the following indicators: share of municipal educational establishments that correspond to certain requirements of education, in the total number of municipal entities; total area of residential premises per one resident.

The highest values of indicators of cluster "B" (1.0): monthly average nominal accrued wages of large and medium companies and non-profit organizations; share of children aged 1–6 who receive a pre-school educational service and (or) service for their support in municipal pre-school educational establishments, in the total number of children aged 1–6.

High value (1st among the clusters) – share ofя municipal general education institutions that correspond to certain requirements of education, in the total number of municipal general education institutions (0.83).

High value (2nd among the clusters and a significant gap from all others) – total coefifcient of natural increase (0.92).

Low values: share of profitable agricultural companies (absence of agricultural companies in the cluster); number of subjects of small and medium entrepreneurship; share of average list number of employees of small and medium companies; total area of residential premises per one resident; number of population; number of newcomers and departed.

The average indicator's value – total coefficient of birth rate.

As to the parameters of cluster "B", it is possible to conclude the following.

No parameter achieved the maximum possible value.

The relatively strong points of the cluster are the following: total coefifcient of natural increase (0.77, 3rd position among the clusters after "A" and "B", excessive increase of the values of the indicators of clusters "D" and "E"); the number of subjects of small and medium entrepreneurship per 10,000 people (0.45, 2nd position after cluster "A", exceeding the values of the indicator of other clusters); annual average number of population (2nd position among the clusters, after cluster "A"); the number of newcomers (2nd position – after cluster "A").

Low values of the indicator – share of municipal general education institutions that correspond to certain requirements of education (0.47, 4th position among the clusters with certain advance of cluster "E" and substantial underrun from other clusters).

The average values of the indicators for all other parameters:

- share of average list number of employees of small and medium companies in the total number of employees;
- share of profitable agricultural companies (internal position – 0.88), but among the clusters this position is low (3rd position among the clusters, in view of zero values of the indicator with clusters "A" and "B", this position is low);

- monthly average nominal wages (0.21 – 3rd position among the clusters, advance as compared to the clusters "D" and "E");
- share of children aged 1–6 who receive an educational service and (or) service for their support in municipal pre-school educational establishments (3rd position among the clusters, advance as compared to the clusters "D" and "E");
- area of residential premises per capita (3rd position among the clusters, large underrun from cluster "D");
- number of departed (0.03, 3rd position among the clusters, large advance of the indicator as compared to clusters "D" and "E").

Cluster "D" was characterized in the following way (as of 2012). Strengths:

- share of profitable agricultural companies (indicator's value does not reach 1.0, but it is still high – 0.96, first place among the clusters);
- share of average list number of employees of small and medium companies (indicator's value – 0.53, but it exceeds the values of the indicator of other clusters);
- share of municipal general education institutions that correspond to the modern requirements (0.81, small advance as compared to cluster "A", "C", and "E", small underrun from cluster "B");
- area of residential premises per capita (0.58, larger value, as compared to other clusters).

Weaknesses:

- monthly average nominal wages (0.099), the lowest value among the clusters);
- share of children aged 1–6 who receive a pre-school educational service or service for support in the municipal pre-school educational establishments (0.30, fourth position, slight advance as compared to cluster "E");
- annual average number of constant population (0.014, the lowest position among the clusters);
- total coefifcient of natural increase (0.35, the lowest position among the clusters);
- total coefficient of birth rate (0.35, the lowest indicator among the clusters).

The values of the indicators that characterize migration processes are medium.

Cluster's "E" strength is total area of residential premises per capita (2nd position among the clusters).

The weakest points of the cluster (underrun from all other clusters): number of subjects of small and medium entrepreneurship; share of children aged 1–6 who receive a pre-school educational service and (or) service for their support in municipal pre-school educational establishments; share of municipal general education institutions that correspond to modern requirements; indicators that characterize migration processes.

4 Structural Differences Between the Municipal Entities of Voronezh Oblast, 2015

The structure of clusters changed a little during 2012–2015. On the whole, the core of the clusters (totality of municipal entities that do not change the position in clusters) are rather wide. As of 2015, the clusters included the following municipal entities (Table 3).

Table 3. The structure of clusters created by the municipal entities of Voronezh Oblast

Cluster	Structure of cluster
A	Voronezh, Novovoronezh
B	Nizhnedevitsky and Ramonsky municipal regions
C	Borisoglebsky city district, Bogucharsky, Kalacheevsky, Kantemirovsky, Liskinsky, Novousmansky, Pavlovsky, Rossoshansky, and Semiluksky municipal districts
D	Anninsky, Bobrovsky, Buturlinovsky, Verkhneamonsky, Novokhopersky, Petropavlovsky, Repyevsky, Khokholsky, and Ertilsky municipal districts
E	Verkhnekhavsky, Vorobyevsky, Gribanovsky, Kamensky, Kashirsky, Olkhovatsky, Ostrogozhsky, Paninsky, Povorinsky, Podgorensky, Talovsky, and Ternovsky municipal districts

As of 2015, the state of parameters of the clusters somewhat changed, but the general relative characteristics preserved. The exception is cluster "A" which included two cities – Voronezh and Novovoronezh – in 2015. The average values of the indicators that characterize the clusters in 2015 are shown in Table 4.

Table 4. Average values of the parameters of clusters (2015)

Indicators	Cluster A	Cluster B	Cluster C	Cluster D	Cluster E
Var1	0.618897	0.345043	0.382066	0.226298	0.149299
Var2	0.301363	0.408632	0.474480	0.751052	0.354976
Var3	0.000000	0.866650	0.948300	0.980878	0.943708
Var4	0.764590	0.287307	0.177598	0.073166	0.119487
Var5	0.860772	0.332137	0.477712	0.429613	0.249306
Var6	0.892543	0.645332	0.511320	0.651465	0.296682
Var7	0.243119	0.974261	0.183033	0.510420	0.402226
Var8	0.507805	0.009877	0.050022	0.015564	0.012512
Var9	0.986111	0.230556	0.609877	0.314198	0.420833
Var10	0.955357	0.357143	0.367064	0.216270	0.296131
Var11	0.506146	0.018358	0.048486	0.020885	0.014033
Var12	0.508167	0.011483	0.052823	0.022214	0.019525
Sum	7.144870	4.486778	4.282781	4.212022	3.278719

Comparisons of the values of clusters' parameters in 2012 and 2015 show that the general "ratio of strength" did not change. At the same time, there were changes in the structure of clusters and values of certain parameters.

The strongest indicators of cluster "A" (values of normed indicators – 1.0) are absent. A significant role in this transformation belonged to unification of the above municipal entities into one cluster.

The cluster's strengths in 2015 were as follows: number of subjects of small and medium entrepreneurship per 10,000 people; monthly average nominal accrued wages; share of children aged 1–6 who receive a pre-school educational service and (or) service for their support in municipal pre-school educational establishments; share of municipal educational establishments that correspond to the modern requirements; annual average number of population; total coefifcient of natural increase; total coefficient of birth rate; number of newcomers (migration); number of departed (migration).

A formal weakness is share of profitable agricultural companies – 0.000.

As is seen from the data presented in Table 4, the *strong indicators of cluster "B"* are as follows: average monthly wages (lower than in cluster "A" but higher than in other clusters); total area of residential premises per capita (the highest indicator among the clusters, very good statistical characteristics that allow viewing cluster as comprehensive system).

Low values of the indicator include share of profitable agricultural companies (the lowest value among all the clusters, except for "A" and "B").

As is seen from the data presented in Table 4, *the strengths of cluster "C" in 2015* were as follows: total coefifcient of natural increase (2nd position after cluster "A" and higher than in other clusters); total coefficient of birth rate (2nd position after cluster "A" and higher than in other clusters); indicators of migration (large number of newcomers and departed).

Weaknesses of the cluster: total area of residential premises per capita (however, statistical characteristics is unsatisfactory – with the average value 0.183033, standard deviation is 0.120735); share of municipal general education institutions that correspond to modern requirements (last but one position among the clusters, acceptable state of standard deviation – 0.229960 with the average value 0.511320).

As is seen from the data presented in Table 4, the *strengths of cluster "D"* are as follows: share of average list number of employees of small and medium companies; share of profitable agricultural organizations.

Weaknesses: average monthly wages of employees of large and medium companies and non-profit organizations (the lowest indicator among the clusters); total coefficient of birth rate (the lowest value among the clusters); indicators that reflect the migration processes (low – but higher than in cluster "E").

As is seen from the data presented in Table 4, *cluster "E" didn't have any strengths in 2015.*

The weakest positions (last position among the clusters): number of subjects of small and medium entrepreneurship per 10,000 people; share of children aged 1–6 who receive a pre-school educational service and (or) service for their support in municipal pre-school educational establishments; share of municipal general education institutions that correspond to the modern requirements; number of newcomers (migration).

5 Conclusions and Offers

The above material allows for the following conclusions.

Achievement of spatial balance of socio-economic development requires setting specific goals aimed at elimination of underrun of municipal entities from the leaders. At that, it is necessary to found not on the peculiarities of development of each separate municipal entity but on the state of their homogeneous groups. It is especially true in cases when their number is large (Voronezh Oblast has 32 municipal regions and 2 city districts).

Analysis showed that there are five groups of municipal entities (virtual clusters) on the territory of Voronezh Oblast; these entities are peculiar for a certain underrun in one or another sphere of activities. At that, here we speak not of production specialization but of provision with jobs, conditions for entrepreneurial activities, and social services.

Thus, for the most developed cluster "A" the most perspective directions of structural changes are as follows: increase of provision of the population with residential premises, and, accordingly, expansion of residential construction; expansion of the material basis of pre-school education establishments. Additional direction of structural changes in one of the municipal entities of the cluster (Novovoronezh) – development of small business.

For cluster "B", which includes two municipal regions, the most important direction of structural changes is their financial aspect – increase of effectiveness of agricultural companies' activities.

For cluster "C", increase of the share of municipal general education institutions that correspond to the modern requirements is topical.

For cluster "D", it is important to raise wages; expand employment by means of creating jobs in various sectors of economy; reduce the outflow of young and middle-aged population.

For cluster "E" the following aspects are important: development of small business; expansion of the network of municipal pre-school and general education institutions that correspond to the modern requirements.

References

Oldenderfer, M.S., Blashfield, P.K. Cluster analysis/factor, discriminant, and cluster analysis. In: Enyukov, I.S. (ed.) Finances and Statistics, 215 p. (1989)

Hartigan, I.A., Wong, M.A.: Algoritm AS 136: a K-means clustering algorithm. J. Royal Stat. Soc. Ser. C (Appl. Stat.) **28**(1), 100–108 (1979)

Golichenko, O.G., Shchepina, I.N.: Analysis of efficiency of innovative activities of Russia's regions. Econ. Sci. Modern Russ. **1**(44), 77–95 (2009)

Golichenko, O.G., Shchepina, I.N.: The system of characteristics for complex analysis of innovative activities at the regional level. Econ. Sci. Modern Russ. **1**(13), 89–91 (2008)

Kruglyakova, V.M., Treshchevsky, Y.I.: Basic strategies of development of investment activities in regions of the CFD. Modern Econ. Probl. Solut. **3**(27), 27–38 (2012)

Risin, I.E., Treshchevsky, D.Y.: Typologization of innovational development of Russia's regions on the basis of stage-by-stage clustering. Bull. South-West. State Univ. Ser. Econ. Soc. Sci. Manag. **1**, 20–27 (2011)

Myasnikova, T.A.: Strategizing of Socio-Economic Development of Municipal Entities in Russia's Regions: Theory, Methodology, and Methodological Provision, p. 271. Scientific book, Voronezh (2015)

Factors and Conditions of Functioning and Development of Regional Socio-Economic Systems

Olga A. Shaporova(✉), Ekaterina I. Mosina, Irina V. Kuznetsova, Elena E. Semenova, and Natalya A. Baturina

Orel State University of Economics and Trade, Orel, Russia
super-ya-57@mail.ru, ecaterinamosina@yandex.ru,
kuzma_79@mail.ru, osuet@mail.ru, 1278orel@mail.ru

Abstract. Economic space of the Russian Federation is undergoing certain significant changes. Firstly, due to appearance of the system theory of economic development, the subjects of the economic environment were divided within socio-economic systems. Secondly, the key directions of the post-industrial paradigm of development became obsolete due to appearance of new factors and conditions of interaction between economic relations at the national and global levels. These preconditions influence the regional socio-economic systems that occupy a rather important position in the development of the national economy of the Russian Federation. According to this, we study the factors and conditions of functioning and development of regional socio-economic systems. The purpose of the article is to consider the conditions and factors of development of regional socio-economic systems through modeling these processes within the current and required state of functioning. It is necessary to solve the following tasks for implementation of the set goal: consider the theoretical and methodological aspects of study of the regional socio-economic systems; determine the key approaches to wide study of factors and conditions within socio-economic systems; offer the models of regional socio-economic systems, formed on the basis of current factors of functioning and required conditions of development. The methodology of the research includes the methods of modeling, theoretical comparison, analysis, synthesis, and comparative evaluation. Within this research, the key moment is creating the model of development of regional socio-economic systems. This study poses substantial interest from the position of determining the theoretical regularities of the categorical machine and modeling of the current situation for comparison with the required conditions of development of regional socio-economic systems.

Keywords: Regional socio-economic systems · Factors · Conditions
Models · Progress · Stability · Balance · Region · Approaches
Concepts

© Springer International Publishing AG, part of Springer Nature 2018
E. G. Popkova (Ed.): HOSMC 2017, AISC 622, pp. 530–541, 2018.
https://doi.org/10.1007/978-3-319-75383-6_67

1 Introduction

Directions and priorities of development of economy in the model of overcoming growth are losing their significance. Lack of favorable situation of the resources market, manifestation of the stagnation processes of the national economy, and lack of the complex of internal and external stimuli of quick growth lead to changes of the conditions of functioning and development of socio-economic systems. These changes influence the subsidized territories that are formed of the regional socio-economic systems. Regional socio-economic systems are a totality of separately formed economic processes that exist within a specific region. The conditions and factors of functioning of each regional socio-economic system influence the development of a specific subject of the Russian Federation. This thesis is confirmed by the following postulates. Firstly, only through regional socio-economic systems is it possible to transform the region for the conditions of change of the imperatives of the global economic space. Secondly, accumulation in regional socio-economic systems, limited by the investment and innovational resources, allows for the "breakthrough leap" in the technological development of the subjects of the Russian Federation. Thirdly, the production and economic interconnections, formed within regional socio-economic systems, increase the region's competitiveness at the national level. Fourthly, improvement of the financial and economic state of each element of the socio-economic system stimulates the regional development on the whole. Besides, it is important that regional socio-economic systems apply and approbate market approaches to regulation of relations and thus require administrative management for the formation of factors and conditions of effective functioning. This requirement confirms the topicality of the selected topic of the research.

The purpose of the article is to consider the conditions and factors of development of regional socio-economic systems through modeling of these processes within the current and required state of functioning. Implementation of the set goal requires solving the following tasks:

- considering the theoretical and methodological aspects of studying regional socio-economic systems;
- determining the key approaches to joint studying the factors and conditions within socio-economic systems;
- offering the models of regional socio-economic systems formed on the basis of current factors of functioning and required conditions of development.

The research tools consist in using the method of modeling, theoretical comparison, analysis, synthesis, and comparative assessment.

For obtaining the comprehensive idea of regional socio-economic systems as a subject of the research, let us view theoretical and methodological aspects of the categorial tools, concepts of formation, and typologization of forms.

2   The Theoretical and Methodological Aspects of Studying Regional Socio-Economic Systems

Studying the establishment and development of regional socio-economic systems was formed on the basis of distinguishing the categorical tools of the categorical definition and concepts in the sphere of development of this system. At that, a lot of authors tried to substantiate the conditions of appearance and the factors that influence the development of regional socio-economic systems.

The origins of studying the issue of development of regional socio-economic systems in the modern conditions of functioning were reflected in the works of Holling (2001), Reggiani et al. (2002), Zhang et al. (2000), Komarevtseva (2016), and Sherstnev (2008). The above authors enriched the concepts and models of formation of the adaptive conditions for effective development of regional socio-economic systems. At that, these studies were based on start of application of managerial methods in the process of regulation of the necessary conditions for functioning of various regional systems.

Further on, the approaches to considering regional socio-economic systems shifted into the aspect of studying the factors that influence the establishment and functioning of regional socio-economic systems, which were viewed in the articles by Li et al. (2000), Stroeva et al. (2015), Burkaltseva (2017), Gorokhov and Ivanov (2013), Sibirskaya and Stroeva (2010), and Shatalov (2017). The importance of notes in the sphere of distinguishing the definitions "region" and "regional socio-economic system" stimulated the emergence of a new direction of the economic thought – the system paradigm of within the regional economic space. This paradigm allowed distinguishing the peculiar differences between the notions "region" and "regional socio-economic system", dividing their objects for full-scale studies.

First of all, these definitions differ in the sphere of spatiality of research. A regional socio-economic system reflects the spatial environment that is based on various economic and legal processes and is regulated by economic relations within this definition. The region is a territorial organizations established by means of the normative and legal regulation of the relation between the territorial subjects that interact within this system. At that, the region is formed under the influence of homogeneous natural conditions that reflect the essence of development of this definition. A regional socio-economic system includes a totality of differentiated sub-systems that form the homogeneous functional environment of development. The existing resources are redistributed within this system. The natural conditions of the region are reflected in exploitation of natural resources redistributed among the systems of higher hierarchy. This thesis allows for the conclusion that region is a dependent system with "primary" interaction between hierarchical subjects. Simply speaking, the region distributes resources in favor of higher territorial subjects (Federation). A regional socio-economic system is often an independent element with "secondary" interaction. According to this, distribution of resources is conducted only on the basis of mutual flows of accumulation and restoration of substructural systems. However, it should be noted that subordination within regional socio-economic systems also exists.

The peculiar categorical features of "regional socio-economic system" were the basis of the concept of formation and origin of this definition. At present, there are two key economic schools the concepts of which view the formation of regional socio-economic system according to the following positions:

1. The economic school of market concept of formation of regional socio-economic systems.
2. The economic school of competitive concept of formation of regional socio-economic systems.

The main aspects of the market concept of formation of regional socio-economic system are viewed in the works Cabezas et al. (2003), Elgazzar (2003), Gurnovich and Ostapenko (2015), and Filyushin (2011).

According to this concept, the market forms the regional socio-economic system. At the initial stage the subjects (external and internal suppliers of goods and services, consumers, transport and logistics companies) create a comprehensive market structure on a certain territory (territorial structure). Over a certain time, this structure begins to integrate with larger territorial structures and is divided according to the principle of general market. The formed markets require regulation of social and economic relations. According to this requirement, a regional socio-economic system is created.

The economic school of the competitive concept of formation of regional socio-economic system, which is realized under the influence of the ideas of Chub (2014), Li et al. (2003), Arashukov (2012), Konovalova (2015), Ozdoeva et al. (2012). The position of these authors is brought down to the fact that the regional socio-economic system could be created only in the conditions of competition.

Under the influence of tough competition between financial organizations, entrepreneurs, and human resources, the system of adaptive conditions is transformed. Each of the presented subjects reconsiders their own activities in the conditions of change of the technological paradigm and restructures production for provision of leadership positions within the competitive struggle. Based on the generated model of development, the subjects enter the new stage of competitive struggle in various differentiated systems. Under the influence of the formed economic relations and social interactions, a regional socio-economic system is organized.

The above schools are a regional socio-economic system – the model formed under the influence of external processes of constant struggle. We think that regional socio-economic system consists of the totality of generic sub-systems that interact on the basis of typologization (Fig. 1).

The regional socio-economic systems presented in Fig. 1 include seven main types:

- economic – functions through interrelations between economic subjects;
- territorial – divides the integrated regional socio-economic system into smaller territorial entities;
- subjective – reflects sub-systems within the interaction between the population that lives on the territory of this system;
- market – combines the sub-systems of implementation of market relations;
- sectorial – includes a lot of differentiated spheres within a certain sector;

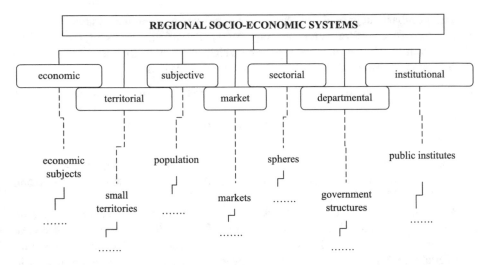

Fig. 1. Certain types of regional socio-economic systems (the authors' typologization)

- departmental – forms regulative relations between public authorities within the regional socio-economic system;
- institutional – requires the presence of public interaction for important socio-economic issues.

Let us note that the hierarchy that is presented in this typologization of the main means of formation and subjection within regional socio-economic systems.

The above concepts and types of typologization of regional socio-economic systems allow forming a narrow idea of solution to the problem of effectiveness within the factors and conditions of functioning and development of this system. Let us determine the key approaches to joint study of the factors and conditions within socio-economic systems.

3   The Key Approaches to Joint Study of the Factors and Conditions Within Socio-Economic Systems

In order to consider the approaches to joint study of factors and conditions within socio-economic systems, let us determine the main regularities of these categories. Factor is a moving force by which a socio-economic system is created or developed. The conditions are viewed as a totality of processes and relations necessary for creation and change of internal and external structures of a socio-economic system. Thus, there's a duality of the primary and secondary nature of these categories. Firstly, the conditions allow forming and changing not only the socio-economic systems but also the factors. Secondly, the factors stimulate the transformation of conditions according to adaptation to the new realia of existence of the socio-economic system. This article does not set the task of determining the primary and secondary influence of economic categories (factors and conditions) on development of socio-economic systems.

It is more important to consider the interconnection between the factors and conditions for creation of the effective model of development of regional socio-economic systems.

The main approaches to joint study of the factors and conditions within socio-economic systems are based on establishment of interconnection between these categories within the global economic systems: industrial and post-industrial economy.

Approach #1. The system of top-priority factors in the conditions of industrial economy. Within this approach, the regional socio-economic system is viewed as a structural element of the industrial economy. According to the hierarchy, this system is the lowest one. The regional socio-economic system is peculiar for such factors as natural resources, production capacities, human resources, research potential, and attractive elements. Balancing these factors in the industrial economy allows forming attractive conditions for functioning of the regional socio-economic system.

At that, this approach viewed only the conditions of existence of regional socio-economic systems – development, which is a drawback of this approach.

Approach #2. Factors of the life cycle in the conditions of post-industrial economy. In this approach, regional socio-economic system is viewed as the resource basis of post-industrial economy. According to this, factors (resources) on the basis of the life cycle form the conditions of development of regional socio-economic system. Presence of the production resources leads to interaction between the institutes within the production cycle. These factors allow forming the conditions of development for the regional socio-economic system. After that, the conditions within the regional socio-economic system change in favor of accumulation of capital. At that, the main factors are services and capital. This peculiarity is seen vividly in the capitalistic model, within which the regional socio-economic systems function on the basis of the key factor – capital. After that, the factor "capital" loses its economic features, being replaced by the factor "exclusive post-industrial product".

On the whole, the approach that considers regional socio-economic systems through the factors of the life cycle in the conditions of post-industrial economy is the most complex one for understanding the perception, as it is used mainly in the Western capitalist systems. The simplest and the most popular approach for the Russian model of management is consideration of a regional socio-economic system through the factors of internal and external environment.

Approach #3. The factors of internal and external environment in the conditions of regional socio-economic system's formation. In this approach, the conditions and factors are equal. Thus, for example, the institutional factor forms institutional conditions of development of the regional socio-economic system. At that, it should be noted that the conditions of functioning of the regional socio-economic system are limited by the internal, external, and mutually dependent environments. Despite this limitation, the formation of regional socio-economic system takes place both in the process of creation of general conditions of functioning and through the specific environment.

The above approaches allowed for the conclusion that factors and conditions perform a very important role in the process of formation and development of the regional socio-economic systems. However, inconsistency of the current and required conditions, as well as lack of substantial factors, does not allow transforming the created regional socio-economic systems. On the basis of this thesis, we deem it necessary to offer the proprietary models of regional socio-economic systems, formed on the basis of the current factors of functioning and required conditions of development.

4 The Models of Regional Socio-Economic Systems, Formed on the Basis of Current Factors of Functioning and Required Conditions of Development

The authors' idea of development of regional socio-economic systems is brought down to creation of the models that reflect the current and required development of these systems on the basis of created conditions and influencing factors. It should be noted that these models are average prototypes of the most regional socio-economic systems. These models of development are applicable for "progressing" and "unstable" regional socio-economic systems. This division is caused by the fact that in the conditions of adaptation to new realia of economic development a lot of regional socio-economic systems do not stand a certain level of competition and show the results of stagnation activities. These regional socio-economic systems are defined by the economic category as "unstable". On the contrary, other systems pass through the quick level of adaptation to changes and begin to transform according to the future realia of development. These regional socio-economic systems are defined by the economic category as "progressing". Let us consider the proprietary model of development of "progressing" regional socio-economic systems.

The model of development of "progressing" regional socio-economic systems presented in Fig. 2 (through formation of factors and conditions) functions in two conventional dimensions:

- bonuses, in which the socio-economic system that is higher according to the hierarchy stimulates the regional socio-economic system for observing the rules of market behavior;
- limitations, in which the socio-economic system that is higher according to the hierarchy limits the regional socio-economic system for the purpose of preservation of economic influence.

The purpose of the model of development of "progressing" regional socio-economic systems in the current state is preservation of these conditions. Simply speaking, the conditions of subjection of the regional socio-economic system to a stronger system. For this, the regional oscillation of management is weakened, only the current risks are regulated, federal institutes of development of influence the regional socio-economic systems, high revenues within "progressing" regional socio-economic systems, as compared to other systems, are preserved, and the economy of the regional socio-economic systems as a supplement of federal socio-economic systems is developed. We think that these factors take the "progressing" regional socio-economic systems close to self-absorption. This thesis is confirmed by the fact that when achieving a certain level of development and giving the necessary resources to other system, the subject (regional socio-economic systems) cannot constantly restore the required sources and means. After consuming the stock and the impossibility to manufacture new resources, the

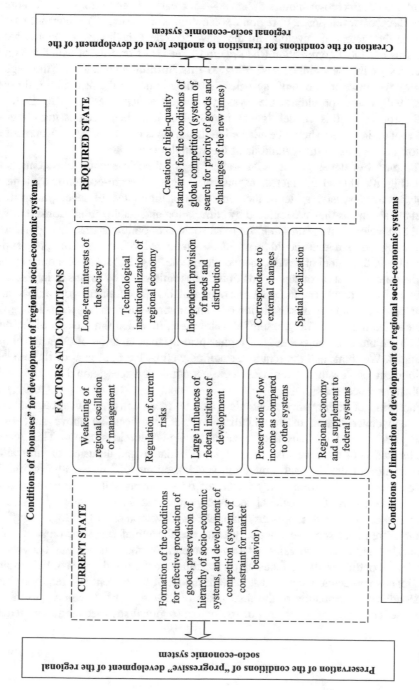

Fig. 2. Model of development of "progressing" regional socio-economic systems (through formation of factors and conditions)

process of self-absorption starts. In order to prevent such situation, we offer – as a main goal of regional socio-economic systems – to create the conditions of transition to another level of development of regional socio-economic system. This purpose requires changing the factors of development that have to reflect qualitative standards for the conditions of global competition (the system of search of priority of benefits with the challenges of the new time). According to this, institutionalization of the regional economy within socio-economic systems should be aimed at the technological aspect, which will allow providing the system's needs independently. The factor of self-absorption in this model is eliminates. Besides, the "progressing" regional socio-economic system should be adapted to the changes of the global environment and transform according to the conditions of the spatial localization.

The model of development of "unstable" regional socio-economic systems is different (Fig. 3). Due to the fact that "unstable" regional socio-economic systems are prototypes of depressive systems, the conditions of limitations of development in this model are absent, as they are replaced by stimulation and stabilization. Considering the model of development of "unstable" regional socio-economic systems, its importance in determining the final goal should be noted. As compared to the model of "progressing" development, the model of development of "unstable" regional socio-economic systems is in the state of creation of connection of the economic space by means of intensity and density of interaction between the factors of stimulation. The ultimate goal of this model of deepening the priority of stimulation of a larger socio-economic system. This goal is similar to the model of "progressing" development, in which the process of "stimulation" on the basis of subjection is of higher priority than creation of other developing conditions. We think that the required state, as well as the final goal of the model of development of "unstable" regional socio-economic systems should be directed at stabilization processes.

Formation of high-quality economic space by means of diversification, differentiation, and concentration of the stabilization factors reflect the above state of the development model of "unstable" regional socio-economic systems. This requires implementation of the concepts of the regional initiative and diffusion of innovations, diversification of the regional production specializations, and observation of the balance of priorities. The formed key factors of development will create modern conditions for adaptation of the regional socio-economic systems to changes.

Thus, the presented proprietary models of development of "progressing" and "unstable" regional socio-economic systems are the initial stage for creation of the large-scale model of well-balanced development. At that, interaction between the factors and conditions allows focusing on the fundamental peculiarities of development of regional socio-economic systems. Determination of the current and required state is a precondition of formation of the final goals of development. According to this, the models reflect certain key features of the object (regional socio-economic system).

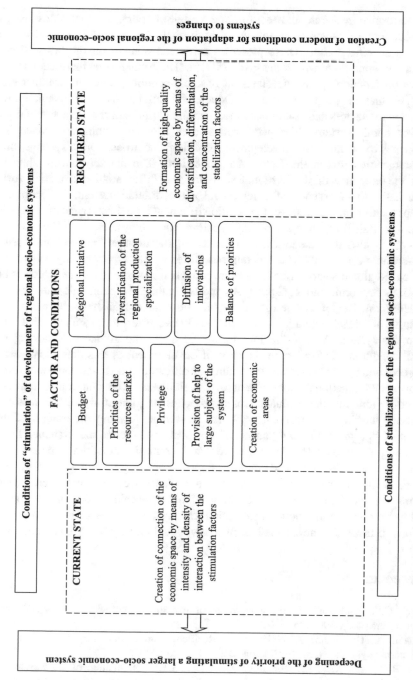

Fig. 3. The model of development of "unstable" regional socio-economic systems (through formation of factors and conditions)

5 Conclusions

The performed research allowed forming certain important conclusions. Regional socio-economic systems are a totality of high-quality economic processes that exist within a specific region. These processes form and change on the basis of certain factors and conditions within the given system. The influence of these economic categories on functioning and development of socio-economic systems is rather significant. This thesis is supported by a lot of studies by Russian and foreign authors, which determine the factors and conditions as fundamental origins of development of regional socio-economic systems. The performed studies allowed forming the concepts and approaches to creation and development of regional socio-economic systems through the interaction between the factors and conditions. The market concept shows that under the influence of the territorial structure and differentiation a market forms – which leads to emergence of a regional socio-economic system. The competitive concept focuses on the process of generation of a new model by means of competition, which then transforms into the regional socio-economic system.

There is also the categorical significance of the notions "region" and "regional socio-economic system". These definitions differ in the aspect of the sphere of the research. Regional socio-economic system reflects the economic space which is based on various economic and legal processes and is regulated by economic relations within this definition. Region is a territorial organization that establishes – by means of the normative and legal regulation – the relations between territorial subjects that interact within this system. At that, "region" is formed under the influence of homogeneous natural conditions that reflect the essence of development of this definition. Regional socio-economic system includes the totality of differentiated sub-systems that form the homogeneous functional environment of development. The determined regularities allowed conducting the evaluation of approaches to joint study of the factors and conditions within socio-economic systems, as well as forming the models of development of "progressing" and "unstable" regional socio-economic systems. The presented proprietary models reflect the current and required state of development of these systems on the basis of created conditions and influencing factors. On the whole, the models of development of "progressing" and "unstable" regional socio-economic systems are the initial stage for creation of a large-scale model of well-balanced development that will allow transforming the system of regional socio-economic relations in favor of a new paradigm of innovational and technological development.

References

Holling, C.S.: Understanding the complexity of economic, ecological, and social systems. Ecosystems **4**(5), 390–405 (2001)

Reggiani, A., De Graaff, T., Nijkamp, P.: Resilience: an evolutionary approach to spatial economic systems. Netw. Spat. Econ. **2**(2), 211–229 (2002)

Zhang, R.Q., Xie, X.G., Liu, S.X., Lee, C.S., Lee, S.T.: Computation of large systems with economic basis set: systems involving weak sodium-organic interaction. Chem. Phys. Lett. **330**(3–4), 484–490 (2000)

Komarevtseva, O.O.: Development of the model of effective management of changes on the basis of the DEA methodology in the economic systems of municipal entities of Orel Oblast. Econ. J. **3**(43), 41–58 (2016)

Sherstneva, N.L.: System analysis of the strategy of development of a municipal economic system. Bull. State Univ. Manag. **1**(22), 170–173 (2008)

Li, W., Szidarovszky, F., Kuang, Y.: Notes on the stability of dynamic economic systems. Appl. Math. Comput. **108**(2–3), 85–89 (2000)

Stroeva, O., Lyapina, I.R., Konobeeva, E.E., Konobeeva, O.E.: Effectiveness of management of innovative activities in regional socio-economic systems. Eur. Res. Stud. **XVIII** (Special issue), 59–72 (2015)

Bukkaltseva, D.D.: Points of economic and innovational growth: the model of organization of region's effective functioning. MID (Mod. Innov. Dev.) **8**(1(29)), 8–30 (2017)

Gorokhov, A.V., Ivanov, K.I.: The system approach to the study of socio-economic systems. Bull. Volga State Technol. Univ. Ser. Econo. Manag. **2**(18), 24–34 (2013)

Sibirskaya, E.V., Stroeva, O.A.: The methodology of evaluation of the process of investing of innovational activities of regional economic systems. Financ. credit. **15**(399), 16–23 (2010)

Shatalov, M.A.: Formation of the tools of state regulation of sustainable socio-economic development at the regional level. Регион: системы, экономика, управление **1**(36), 66–69 (2017)

Cabezas, H., Pawlowski, C.W., Mayer, A.L., Hoagland, N.T.: Sustainability: ecological, social, economic, technological, and systems perspectives. Clean Technol. Environ. Policy **5**(3–4), 167–180 (2003)

Elgazzar, A.S.: Applications of small-world networks to some socio-economic systems. Phys. A: Stat. Mech. Appl. **324**(1–2), 402–407 (2003)

Gurnovich, T.G., Ostapenko, E.A.: Management of sustainable development of regional socio-economic systems: foreign experience and Russian practice. Econ. Entrep. **12-3**(65-3), 432–436 (2015)

Filyushin, A.V.: System of indicators and criteria of effectiveness of interaction between the economic systems in the region. Probl. Modern Econ. **3**, 257–259 (2011)

Chub, A.: The integral indicator of the sustainable development of the regional socio-economic systems: the structure, the methodology of the formation, the direction of the application. Life Sci. J. **11**(8), 177–183 (2014)

Li, W., Rychlik, M., Szidarovszky, F., Chiarella, C.: On the attractivity of a class of homogeneous dynamic economic systems. Nonlinear Anal. **52**(6), 1617–1636 (2003)

Arashukov, A.S.: The system aspects of evaluation of socio-economic systems' development. Polythemat. Netw. Online Sci. J. Kuban State Agrar. Univ. **84**, 692–702 (2012)

Konovalova, K.Y.: Development of the methodology of indicator assessment of a regional socio-economic system's sustainability. Bull. Expert Counc. **3**(3), 15–21 (2015)

Ozdoeva, D.M., Durdyeva, D.A., Zhemukhov, I.R., Israilov, S.K.: The system approach to studying the functioning of regional economic system. Terra Econ. **10**(1–3), 155–159 (2012)

Threats to Food Security of the Russia's Population in the Conditions of Transition to Digital Economy

Gilyan V. Fedotova[1]([✉]), Natalia N. Kulikova[2], Artur K. Kurbanov[3], and Anastasia A. Gontar[1]

[1] Volgograd State Technical University, Volgograd, Russia
g_evgeeva@mail.ru, 261984@mail.ru
[2] Volgograd State University, Volgograd, Russia
kulikovanata72@yandex.ru
[3] General A.V. Khrulev Military Academy of Material and Technical Provision, St. Petersburg, Russia
kurbanov-83@ya.ru

Abstract. Topic/subject. Topicality of the research is predetermined by the necessity for studying the issues of provision of Russia's food security by such main criteria as food accessibility and product quality of consumption by Russians. The object of the research is the level of food security in view of modern global threats during the transition to digital economy and its influence on the population's living standards.

Goal/tasks. The purpose of the research is to theoretically and practically substantiate the necessity for search and development of additional mechanisms of provision of the high-quality level of food security for Russia's population. According to this goal, the authors formulate and solve the following tasks: short analysis of criteria of food independence, evaluation of the results of the work of import substitution programs, analysis of the consumer sector, evaluation of dynamics of real income of Russians, conclusion on the growth of the number of low-income citizens, substantiation of the necessity for targeted food help.

Methodology. The authors use the method of statistical analysis, financial analysis, horizontal and vertical analysis of data for analyzing the threshold criteria of food security.

Results. Increase of the level of food security of Russia needs provision of access to high-quality food for the country's population, which is impossible without the targeted government food help.

Conclusions/significance. The significance of this work consists in the emphasis on internal threats to the state's food security which grow despite the increase of food independence on external import.

The problem of famine has always been one of the global problems of humanity; it influenced the quality of social environment and life of the whole society. The problem of famine was first determined by Thomas Malthus in 1798; he wrote of existence of a gap between the growth of population of the Earth and growth of production of food [2]. This gap is a reason for a lot of social conflicts. In the time of high technologies, conquest of nano- and bio-technologies, the problem of food security still exists but in

© Springer International Publishing AG, part of Springer Nature 2018
E. G. Popkova (Ed.): HOSMC 2017, AISC 622, pp. 542–548, 2018.
https://doi.org/10.1007/978-3-319-75383-6_68

another, wider, aspect. Therefore, we shall analyze the existing threats to food security of the population.

The demographic situation in the country directly depends on qualitative provision and economic accessibility of foods products for the population, volumes of manufacture of these products, and quality of citizens' consumption. As of now, the problem of qualitative increase of food security of Russia is rather urgent – it is caused by the necessity for providing the acceptable level and quality of consumption of main food products by various social groups of population [1]. According to the norms of consumption, for the purpose of food independence from external import, the share of domestic products should constitute at least 80% in the structure of consumption, but this indicator is determined by the Doctrine of Food Security of Russia (hereinafter - Doctrine), signed by the Decree of the President of the RF dated January 30, 2010, No. 120, which specifies the indicators of the state of food security (Fig. 1).

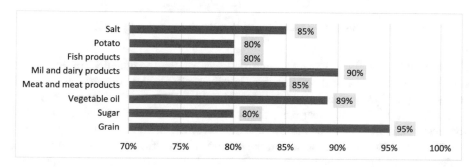

Fig. 1. Low threshold values of the production volumes for the main food products according to the Doctrine.

According to this document, provision of food security requires the states volumes of production for the eight main food groups. The problem of provision of the physical volume of domestic food production has become urgent in August 2014, after special measures implemented by Russia in response to the anti-Russian economic sanctions from the USA and the EU [10]. The Decree of the President of the RF dated August 6, 2014, No. 560 "On application of special economic measures for the purpose of provision of security of the RF", food embargo for a range of imported food products from the EU and the USA was implemented. The embargo reached its goal, as import of products from these countries reduced by three times – from $60 billion to $20 billion in 2016, according to the Ministry of Agriculture of the RF. For compensation of the reducing import, the Road map of import substitution in the agriculture was development, the aggregated result of which are presented in the graph (Fig. 2).

Figure 2 shows the growth rates of food production as a result of financing of development of agricultural production in 2015–2016, according to which it is seen that the dynamics of growth of agricultural products in Russia in the post-crisis period remained positive due to subsidies from the budget.

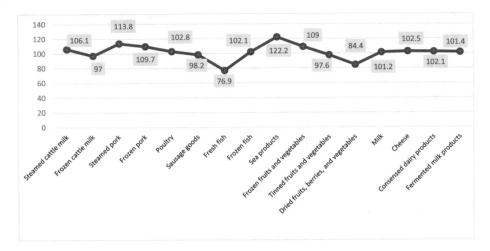

Fig. 2. Results of import substitution in 2016 as compared to 2015, % [12].

Thus, it is possible to see that the volumes of food production in Russia are sufficient for provision of the population, but the main problem consists in economic accessibility of food products, which is another side of the state's food security.

According to the statistical data, the economic accessibility of high-quality food products is unsatisfactory for the Russians, as over the period of 2014-2016 the real income of the population reduced, and the sanctions led to growth of prices in the retail market. The food market became one of the growth factors for inflation, together with oil prices and depreciation of the ruble [3]. According to the experts of the Center of agri-food policy of the Institute of applied economic research of the Russian Presidential Academy of National Economy and Public Administration under the President of the Russian Federation, the share of embargo in the inflation shock constitutes around 20% [11]. Dynamics of the index of consumer prices for 2012–2016 in Russia proves high level of inflation for the category "Food products" after the introduction of trade embargo in 2014 (Fig. 3).

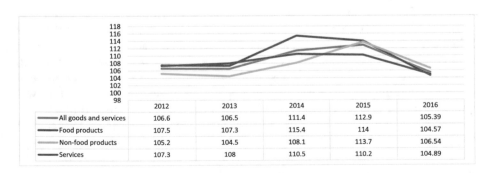

Fig. 3. Dynamics of the index of consumer prices for 2012–2016, % [12].

The presented negative dynamics of the price indices (Fig. 3) is aggravated by the continuous rates of reduction of real revenues of the population for the viewed period and reduction of their purchasing power, as well as growth of the number of citizens with money income below the subsistence level and deficit of money income (Fig. 4).

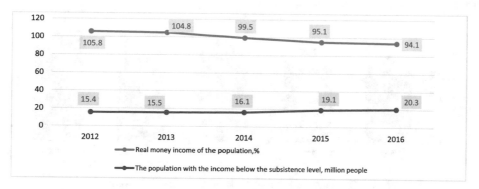

Fig. 4. Rates of real income and purchasing power of the population for 2012–2016 [12].

The negative dynamics of real income of the population that started in 2014 continues in 2016, which proves the fact of threat to food security of the people of Russia, due to growth of prices and reduction of real purchasing power of the population. The studies of the fund "Public opinion" showed that more than 63% of the Russian started saving during selection of food products, 40% started purchasing food products of cheaper brands, and 30% reduced the volumes of purchases or even refused from certain products. Such negative dynamics of the quality of consumption does not stimulate the increase of food security of Russia, as against the background of growth of food production the citizens reduce the quality of food consumption due to its high cost and reduction of real purchasing power of their income [4, 5].

Thus, there is the following picture of food security for the main criteria of evaluation [6]:

(1) growth of food production due to subsidizing of the programs of import substitution in the agricultural complex spheres and increase of export of food products of the Russian origin, which increases food independence of the state;
(2) reduction of real purchasing power of the population, increase of the number of people with income below the subsistence level;
(3) food embargo from Russia led to artificial deficit of food products, which stimulated the growth of prices in the sector of consumer products;
(4) due to growth of the index of consumer prices, a lot of Russians reduce their expenditures for food, refuse from certain food products, and purchase cheaper brands of food products, i.e., reduce the quality of consumption.

The above arguments show that the level of food security remains unsatisfactory for such criteria as quality of consumption and economic accessibility of food products for Russians.

As was mentioned above, the number of Russians with the income below the subsistence level increase, while the subsistence level has been growing over the whole analyzed period (Fig. 5).

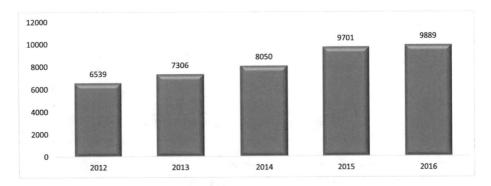

Fig. 5. Dynamics of subsistence level in 2012–2016, RUB [12].

The presented dynamics of the subsistence level reflects the tendencies of the general growth of prices for the consumer sector products. Absence of growth of real income of the population, as well as insufficient social guarantees, set by the laws of the RF, do not stimulate the increase of the index of consumer confidence of Russians (Table 1).

Table 1. Dynamics of the index of consumer confidence of the Russia's population, % [12].

Year	Total	According to sex		According to age		
		Males	Females	Below 30	Aged 30–49	50+
2012	−8	−7	−9	−2	−7	−13
2013	−11	−11	−11	−5	−10	−15
2014	−18	−17	−18	−14	−17	−20
2015	−26	−26	−27	−21	−27	−28
2016	−18	−17	−19	−12	−17	−21

Analysis of dynamics of consumer confidence of the Russians shows that the citizens' confidence in the future reduces due to lack of stable income and high inflation expectations. The largest reduction of the index was observed in 2015, and the people of the pension age constitute the social group that is not condifent in the future – as they do not have other income apart from the state pension, which is very small.

The tendency in reduction of the quality of food consumption of the Russians requires from the state structures the thorough reconsideration of the social support programs for the poor groups of the population, increase of the volumes of targeted financial help, and increase of social guarantees in such spheres as education, medicine, and accommodation. We think that the existing expenditures for food, which constitute the large share (up to 38%) in the structure of consumer expenditures of the population, should be compensated by the state for the categories of the people with the income below the subsistence level (Fig. 4). The number of such Russians reached 20.3 million, which constitutes 14% of the total number of the citizens. This part of the citizens, who live with the income below the subsistence level and constitute 14%, is the most unprotected category in terms of food; it is necessary to develop the mechanisms of targeted food help for them. The mechanisms of additional food help should increase the calories consumption and supplement the poor rations with high-quality domestic product that are useful for health [7, 9].

Such mechanisms are discussed at the legislative level. For example, there is the initiative of the Ministry of Industry and Trade of the RF, which developed the plan of a food program for the poor people in February 2016. The plan consists of two stages:

- 1st stage – introduction of food stamps (16 million) for subsidizing the purchase of certain food products for RUB 1,400 per month. The expenditures will constitute RUB 140 billion;
- 2nd stage – organization of social catering in special cafes and canteens in 2018–2020. The expenditures will constitute RUB 214.2 billion.

These measures were offered for confirmation by the Government of the RF, but still there's no result. The Ministry of Industry and Trade of the RF is working on the project of implementation and is conducting the preliminary negotiations with the National System of Payment Cards JCS for food subsidies being provided through the Mir bank cards. Thus, the national payment system will control the system of food help to the poor citizens and farmers for purchase of seeds and saplings [13]. Still, this project has not yet started.

It should be noted that such practice of food support for the poor is practiced by a lot of countries. In particular, the Supplemental Nutrition Assistance Program (SNAP) has been practiced in the USA since 1977 – it is aimed at subsidizing the purchases of food products for the poor, includes the programs of school lunches and breakfasts, and the programs of additional catering for children, pregnant women, and nursing mothers [8]. Such help stimulates the national economy, creates additional jobs, and supports local farmers.

This practice should be used in Russia, as the state is aimed at development and support for manufacture of all necessary products within the country, increase of the living standards, and preservation of the national food independence from external import.

References

1. Vertakova, Y., Plotnikov, V., Fedotova, G.B.: The system of indicators for indicative management of a region and its clusters. Procedia Econ. Finan. **39**, 184–191 (2016)
2. Maltusc, T.: Studying the law of human population. Kyiv: Foundations, 535 p. (1998)
3. Fedotova, G.V., Sibagatulina, L.M.: Cooperation and integration in the agro-industrial complex. Finan. Analytics: Probl. Solutions **29**, 14–22 (2016)

4. Fedotova, G.V., Sibagatulina, L.M.: Dependence of the agro-industrial complex on the economic development of the country. Bull. South-West State Univ. Ser. Econ. Sociol. Manag. **4**, 50–54 (2015)
5. Fedotova, G.V., Dubinina, E.D.: Analysis of the financial state of the agricultural economy of the Russian Federation. Finan. Anal. Probl. Solutions **19**, 17–27 (2016)
6. Federal Law dated December 1, 2014 No. 384-FZ "On the federal budget for 2015 and the planned period of 2016 and 2017". https://www.consultant.ru/document/cons_doc_LAW_171692/. Accessed 10 Dec 2016
7. Overview of the Center for macro-economic research of Sberbank of Russia. Effectiveness of state expenditures in Russia, 13 January 2011. http://www.sberbank.ru/common/img/uploaded/files/pdf/press_center/Review_20110113.pdf. Accessed 30 Jan 2017
8. Supplemental Nutrition Assistance Program (SNAP). Food and Nutrition service. https://www.fns.usda.gov/snap/supplemental-nutrition-assistance-program-snap. Accessed 30 Jan 2017
9. Sazonov, S.P., Fedotova, G.V., Sibagatulina, L.M.: Analysis of state programs of import substitution. Finan. Anal. Probl. Solutions **9**, 12–21 (2016)
10. Jha, M., Medetsky, A.: Russia Dominates Wheat Market as Soggy Fields Push Out France. https://www.bloomberg.com/news/articles/2016-07-19/russia-dominates-wheat-market-as-soggy-fields-push-out-france. Accessed 30 Jan 2017
11. Who was punished by the anti-sanctions? http://www.gazeta.ru/business/food/2015/08/03/7667813. Accessed 30 Jan 2017
12. The data of the official web-site of the Federal State Statistics Service of the RF. http://www.gks.ru/. Accessed 30 Jan 2017
13. The data of the official web-site of the Ministry of Industry and Trade of the RF. http://minpromtorg.gov.ru/press-centre/news/?tag_18%5B%5D=16. Accessed 30 Jan 2017

Economic Relations and Economic Systems

Svetlana N. Revina[✉], Anna V. Sidorova, Aleksei L. Zakharov,
and Grigory F. Tselniker

Samara State University of Economics, Samara, Russia
29.revina@mail.ru, an.sido@bk.ru, b0707@mail.ru,
grigorij-celniker@ya.ru

Abstract. Topicality. As is known, the economic theory is based on the fact that any economy is based on production and consumption of good and benefits that are necessary for people's lives. At that, economy is based on people's needs and economic interests, which make them act in one way or another. As a result, there appear relations of production (creation), exchange, consumption, and acquisition of life goods, which are called economic relations. It is important to understand that implementation of the needs and interests, which are systemically interconnected, predetermines the existence of not just a totality of public relations but an economic system.

Purposes and tasks of the research. The purpose of the research is to study economic relations and economic systems. The authors set the following tasks: studying the notion "economic activities" and its correlation to the notion "entrepreneurial activities"; studying the structure of an economic system; determining the position on the debatable issues that emerge during conduct of the research.

Conclusions. The Constitution of the RF formulates the principles of market economy (competitiveness, equality of ownership forms, integrity of economic space, freedom of entrepreneurial activities, etc.), so the Russian economy is a market economy, not mixed economy. Market economy is the most flexible and effective.

Keywords: State and economy · Classification of economic systems
Management methods · Market and command economies
Economic connections

1 Introduction

Structure of economic system is the subject of economic research; in the theory of law this issue is not viewed or is reproduced in the system of random notions and categories. During description of the structure of economic system it is noted that its main elements are the following: (1) socio-economic relations that are based on existing ownership forms for economic resources and results of economic activities, which exist in each economic system; (2) organizational forms of economic activities; (3) economic mechanism, i.e., the means of regulation of economic activities at the macro-economic level; (4) the system of stimuli and motivations that guides the participants of the economic life; specific economic connections between the companies and organizations (Economy 1999).

© Springer International Publishing AG, part of Springer Nature 2018
E. G. Popkova (Ed.): HOSMC 2017, AISC 622, pp. 549–554, 2018.
https://doi.org/10.1007/978-3-319-75383-6_69

2 Methodology

During the research, the authors used the following methods: logical methods, analysis and synthesis, systemic methods, modeling, etc. All these methods were used in a complex.

3 Results

Modern science divides economic systems into market, command, and mixed. In our opinion, there are no purely market economies, as the state participates in all economic processes. This influence became widespread in the end of the 19th century, and development of the scientific and technical revolution and the production and social infrastructure it only grew.

4 Discussion

The width of coverage of the system of economic connections cannot be overestimated. They cover all the stages of reproduction: production, distribution, exchange, and consumption of goods (works, services). For characteristics of the system of economic connections, the purpose of activities in the system of reproduction or direction of efforts of the participants of the corresponding process are not important. Of course, the direction of economic activities at receipt of profit cannot be considered its attribute. The theory includes the idea that the category "economic activities" is wider than such notions as economic, entrepreneurial, or commercial activities (Ruchkina 2003).

In its turn, it would be a mistake to compare economic activities to entrepreneurial activities. The Laptev's idea that "economic activities is a wider notion than entrepreneurial activities. Economic activities - are the activities on manufacture of products, production of work, and provision of services. There exist also such forms of economic activities that are not aimed at receipt of profit – the main criterion of determination of entrepreneurial activities, i.e., the notion of economic activities is a generic one as to the notion 'entrepreneurial activities'" (Laptev 1997) is correct.

In view of all terminological peculiarities of definition of economic connections, it is necessary to remember the special role of entrepreneurial activities in the whole system of economic relations. The target orientation of entrepreneurial activities at receipt of profit makes it possible for the market to appear and function.

If the approach to determination of the structure of economic system in the science has formed, the problem of their classification still remains urgent - despite the fact that the classification factor is the attribute of approach and methods of solving the main economic problems (Nikolaev 2004).

Division of economic systems into market, command, and mixed became traditional for the modern science. Such classification is provided by the American economists Paul A. Samuelson and William D. Nordhaus – they distinguish the market, command, and mixed economies. At that it is emphasized that classification is mainly conventional, and the distinguished types of economic systems do not exist in the pure form in any country of the world. All countries have mixed economies – economies with the elements of market and the command form of management (Samuelson et al. 2000). This classification is supported by the Russian science (Brodsky 2002).

There are also other approaches to classification of economic systems. Thus, Brodsky divides social systems into three groups: totalitarian systems; liberal systems; social systems with mixed (Brodsky 2002).

The command economy is peculiar for public ownership for all material resources; collective economic decisions via centralized economic planning; monopolization and bureaucratization of economy; all large decisions that concern the volume of the used resources, the structure and distribution of products, and organization of production are made by the central planning body (Voytov 1999; Fomin 1999).

These features are implemented in the whole system of economic connections. Thus, companies are the property of the state and conduct production on the basis of government directives – in other words, production plans are set by the planning body for each company. The ratio in the national product of the means of production and means of consumption is set in the centralized way, and distribution of consumer goods among the population is performed in the same way. The production means are distributed among the spheres on the basis of long-term priorities which are set by the central planning body. The central planning body is responsible for the decisions – which goods are to be produces, how they are to be produces, for whom, etc. – i.e., it is responsible for the classic questions (Nosova 1998). All its decisions are of the directive character. Foundation only on centralized planning is a peculiar feature of the command economy. The administrative and command economy is peculiar also for lack of independence of economic subjects, which led to production of the unpopular goods, freezing of capital investments and lack of desire to implement new technologies, management of economic processes with the command and administrative methods without consideration of objective economic laws, and the privileged position of the government elite as a peculiarity of distribution of materials goods and services.

Despite the attractiveness of arguments of A. Smith, there not purely market economies. The state actively participates in economic processes. The object of application of state regulation economy is so called "magical quadrangle": inflation – unemployment – growth rates – balance of the payment balance (Fomin 1999). The measures performed by the state for regulation are diverse: tax policy, direct regulation of prices, creation of the legal basis of economy, subsidizing the unprofitable spheres, science, defense, and protectionist policy (creation of conditions for competitiveness of the national capital), social provision of the lowest groups of the social pyramid, and ecological policy (Kirichenko 1999).

Moreover, the late 19th century have a start to expansion of the government interference into economy. After that, with development of the scientific and technical revolution and production & social infrastructure, this process accelerated. As a result, new economic mechanisms, organizational forms, and economic ties between the subjects formed in the second half of the 20th century, i.e., a new system was created: the modern market economy (modern capitalism). This became possible because market system is the most flexible and can take revolutionary transformations and changes.

However, stating the fact of the government's participation in regulation of economic relations, it is necessary to note that each country has either market or command economy; the term "mixed economy" is ambiguous, for it is impossible to combine totally different principles and methods of management in one country, though it is impossible to deny the presence of the state sector and the role of the state in regulation of economic processes. The Constitution of the RF formulates the principles of market economy (competitiveness, equality of ownership forms, integrity of economic space, freedom of entrepreneurial activities, etc.), so in terms of laws the Russian economy is a market economy, not mixed economy.

The market system of economy is the most flexible of all existing systems. Most developed countries have the system of the modern capitalism.

The main advantages of market include the following: self-regulation of the structure of production, its adaptation to the structure of public consumption; effective distribution of resources between the spheres and regions; socially necessary conditions of production; stimulation of manufacturer for increase of efficiency of labor and quality of products, implementation of the achievement of the scientific and technical revolution into production; flexible reaction to demand; effective sanitary function; provision of the freedom of economic activities; precondition for development of democracy in the country, and provision of personal freedom.

Despite the fact that as compared to all other economies, the market economy turned to be the most flexible, it has a range of drawbacks, which, in their turn, play a significant negative role in the economic life of society.

The drawbacks of the market system of economy include the following: tendency for random setting of balance in economy goes through constant violations of this balance, which leads to public losses of labor, unemployment, etc.; market is a tough system in the social aspect; market does not ensure solving the ecological problems, does not stimulate the development of fundamental science and culture; a lack of manageability appears, and it is difficult to direct the development of economy at achievement of the national goals (strengthening of the geo-political position of the country in the world, provision of the scientific and technical, socio-economic, cultural, spiritual, and moral progress of the society, and decent life of citizens) (Klotsvog 2006).

Formation of the market in Russia was complicated by specific circumstances: lack of the agreement on the issues of the market mechanism formation, strong opposition to this process; fight between branches of public authorities, weakness of the federative structure, growth of centrifugal tendencies; long domination of the market system; high level of militarization and monopolization of economy; aged capital of existing companies; lack of temporary production infrastructure; social tension in the Russian society; apathy and indifference of the people to the things that surround them.

Certain economists characterize the market reforms of 1990's as a shock therapy (Gubanov 2006; Lvov 2007). Others evaluate the reforms positively. Thus, Mau thinks that overcoming the communism is ensured with minimal "blood" and losses (Mau 2002). The "shock" therapy, which started in January 1992, was an attempt for a quick transition to market economy. The corresponding program, compiled with support from the Western economists, the World Bank, and the IMF, began to be realized after the dissolution of the Soviet Union. A lot of Western scholars, while comparing the reforms during transition to market economy in the former socialist countries of the

Eastern Europe, see the largest failure of this model in Russia (Kornai 2000). The most difficult consequences of the liberal reforms were observed in the social sphere. The scholars think that the initial market macro-economy in Russia was created as a system of capitalization of not the main capital (socialist main production funds) but state revenues. Liberalization of prices was performed without the corresponding macro-economic preparation – in particular, it was not connected to formation of the system of money and financial turnover. Mass stick and money markets did not become widespread, and the unprecedented depreciation of ruble took place (Evstigneeva and Evstigneev 2006). Gubanov calls these reforms liberal and capitalistic. According to him, there could be no compromise with comprador capital, for the cost of such compromise is complete destruction of the country (Gubanov 2006a). Hence his offer of a transition to state capitalism is logical (Gubanov 2006). Bogomolov thinks that the "oligarch capitalism" was formed in Russia in 1990's, which is a deviation from the world practice of economic development.

It is necessary to dwell on the role of state in the modern economy. In the Western literature, there is a popular approach to determining the role of the state on the basis of the theory of "market defects" or "market gaps". According to this theory, the state should perform the functions that the economic systems is to able to perform – i.e., eliminate the drawbacks of market economy. Thus, an American study guide by McConnell and Brue determines the most important tasks of the state as: (1) provision of the legal basis and public atmosphere that stimulate effective functioning of the market system; (2) protection of competition; (3) redistribution of revenues and wealth; (4) correction of distribution of resources for the purpose of changing the structure of the national product; (5) stabilization of economy, i.e., control over the level of employment and inflation causes by the fluctuations of the situation, and stimulation of economic growth (McConnell and Brue 1992).

Yasin distinguishes the following functions of the state in the sphere of economy: (1) formation of the laws, provision of law and order, solving the disputes on the basis of the laws (court), constraint for compliance with the laws and court decisions (the state's work as a "night watch"); (2) provision of macro-economic stability – prevention of inflation, sustainability of the national currency; (3) stimulation of economy's development and effective changes in its structure; (4) prevention or elimination of the so called market "gaps"; (5) protection of socially vulnerable groups of the population, provision of social guarantees, etc. (Yasin 2006).

Mau states that the role of state in regulation of economic life grew by strengthening the institutes of state power and increase of state's direct interference into economy.

5   Conclusions

From the position of the modern state of economy and society on the whole and in view of the Soviet experience, social rights and guarantees of the citizens, as well as existing state paternalism, the model of socially-oriented market economy fits Russia the most.

According to Zhilinsky, economic system is socially oriented – or "economy for human" – when production relations and economic mechanisms ensure subjection of production to increase of the living standards and development of personality (Zhilinsky 1998).

References

Brodsky, M.N.: State and legal regulation of economic activities in the modern Russia (theoretical model and practical implementation). Doctoral thesis. – SPb. 23 (2002)

Voytov, A.G.: Economy: General theoretical course (fundamental theory of economy), Study guide (1999)

Gubanov, S.: Politics of a new stage: goal and means. Economist. **11**, 5 (2006a)

Gubanov, S.: Russia's way in the basic coordinates of the age. Economist **7**, 5–6 (2006b)

Evstigneeva, L., Evstigneev, R.: Transformation risks of the Russian economy. Issues Econ. **11**, 11 (2006)

Zhilinsky, S.E.: The legal basis of entrepreneurial activities (entrepreneurial law), 14 (1998)

Kirichenko, V.: Increase of state regulation: deepening or termination of reformation transformations? Russ. J. Econ. **2**, 3–13 (1999)

Klotsvog, F.: The key approaches to regulating the economy. Economist **11**, 17 (2006)

Kornai, J.: The road to a free economy: ten years later (reconsidering the experience). Issues Econ. **12**, 46 (2000)

Laptev, V.V.: Entrepreneurial Law: Notion and Subjects, vol. 21 (1997)

Lvov, D.: Regarding the strategy of Russia's development. Economist **2**, 4 (2007)

McConnell, C.R., Brue, S.L.: Economics: 2 V 1, 94 (1992)

Mau, V.: Russian reforms: modern challenges. J. Shareholders **8**, 2 (2002)

Nikolaev, M.V.: The theoretical and methodological issues of formation of effective economic systems. Kazan (2004)

Nosova, S.S.: Economic theory in questions and answers: study guide. Rostov-on-Don (1998)

Ruchkina, G.F.: The notion and content of economic activities and its ratio to economic and entrepreneurial activities. Yurist 2 (2003)

Samuelsson, P.A., Nordhaus V.D.: Economics, pp. 50–51 (2000)

Fomin, G.N.: Foundations of the Economic Theory, vol. 97 (1999)

Bulatov, A.S. (ed.): Economics, vol. 15 (1997)

Kamaev, V.D. (ed.): Economic Theory, pp. 12–15 (2001)

Yasin, E.: State and economy at the stage of modernization. Issues Econ. **4**, 5 (2006)

Improvement of the Economic Mechanism of State Support for Innovational Development of the Russian Agro-Industrial Complex in the Conditions of Import Substitution

Aleksei V. Bogoviz[1]([✉]) [iD], Alexander N. Alekseev[2],
and Denis A. Chepik[3] [iD]

[1] Federal State Budgetary Scientific Institution
"Federal Research Center of Agrarian Economy and Social Development
of Rural Areas—All Russian Research Institute of Agricultural Economics",
Moscow, Russia
aleksei.bogoviz@gmail.com
[2] G.V. Plekhanov Russian University of Economics, Moscow, Russia
alexeev_alexan@mail.ru
[3] Russian Research Institute of Agricultural Economics, Moscow, Russia
denis_chepik@mail.ru

Abstract. The purpose of the article is to develop the offered for improvement of the mechanism of state support for development of the agro-industrial complex in the conditions of import substitution and provision of food security of Russia. For evaluation of effectiveness of realization of the mechanism of subsidizing within the policy of import substitution in the agro-industrial complex, a proprietary formula was developed, which is presented in the article. The authors substantiate the thesis that the policy of import substitution is necessary, as refusal from its realization with return to competition with foreign suppliers of agricultural products will complicate the problem of food security. However, the mechanism of implementation of the policy of subsidizing is not effective, so achievement of the trajectory of innovational development of the Russia's agro-industrial complex requires another, more complex, mechanism. Thus the authors suggest the executive bodies of public authorities, which form the Russian agrarian policy, pay attention to the system of measures that includes various means of innovational development of companies of the agro-industrial sphere, namely: development of territorial and sectorial clusters, technological parks, innovations-oriented economic areas, and public-private partnership.

Keywords: Innovations in agro-industrial complex · Import substitution
Food security · State support · Companies of agro-industrial complex
Russian Federation

1 Introduction

Topicality of solving the problem of food security is explained by the fact that during the economic crisis the situation in the global economic system is subject to constant changes, which makes the previously set international economic ties insufficiently

© Springer International Publishing AG, part of Springer Nature 2018
E. G. Popkova (Ed.): HOSMC 2017, AISC 622, pp. 555–561, 2018.
https://doi.org/10.1007/978-3-319-75383-6_70

reliable. High risk of emergence of national food crises is a serious reason for increase of the volumes of domestic manufacture of agricultural products and compulsory temporary refusal from implementation from the principle of international division of labor in the agro-industrial complex (AIC).

For the purpose of national food security, the Government of the Russian Federation implements the policy of import substitution in the agro-industrial complex of the country. This policy led to successful ousting of foreign food suppliers in the national market, but it is not yet possible to speak of successful solution of the problem of modernization and technical & technological transformation of the Russia's AIC. This is a reason for concerns regarding short-term effect from the Russian policy of import substitution and expectations of reduction of competitiveness of the Russian AIC companies in the mid-term, which may lead either to deficit of products of the AIC or to return to dependence on its import.

This argument is used as a proof in favor of weakening or cancelling of the policy of import substitution in the AIC and return to competition, including with foreign suppliers. Our hypothesis within this article consists in the fact that the policy of import substitution is correct, and refusal from its implementation with return to foreign competition will complicate solution of the problem of provision of food security. At the same time, in order to achieve innovational development of the Russian AIC, it is necessary to improve the economic mechanism of state support, including subsidizing of innovations-active companies in the sphere of agro-industrial production.

2   Materials and Method

The current mechanism of subsidizing of technical and technological modernization and innovational development within the policy of import substitution in the Russia's AIC is based on the Order of the Ministry of Agriculture of Russia dated June 25, 2017, No. 342 "Regarding the concept of development of agrarian science and scientific provision of the agro-industrial complex of Russia until 2025" and the Decree of the Government of the Russian Federation dated December 27, 2012, No. 1432 "Regarding establishment of the Rules of provision of subsidies to manufacturers of agricultural equipment) (Ministry of Agriculture of the Russian Federation 2017).

In order to evaluate effectiveness of implementation of this mechanism, we offer the following formula:

$$Eis = \Delta Siac/\Delta Smid, \text{ where} \tag{1}$$

Eis – effectiveness of implementation of mechanism of subsidizing within the policy of import substitution in AIC;
$\Delta Siac$ – annual growth of the share of innovations-active companies of the AIC;
$\Delta Smid$ – annual growth of the volume of subsidies into technical and technological modernization and innovational development within the policy of import substitution in the AIC.

As is seen from the formula, effectiveness of implementation of the mechanism of subsidizing within the policy of import substitution in the AIC is evaluated by finding the ratio of the results of its application, expressed in annual growth of the share of innovations-active companies of AIC, to costs of its achievement, expressed in annual growth of the volume of subsidies for technical and technological modernization and innovational development.

The indicators are in relative expression (annual growth) for provision of their compatibility, as the share of innovations-active companies in absolute expression is measures in per cent, and the volume of subsidies – in rubles. It should be emphasized that efficiency of import substitution in the AIC is assessed through the prism of innovational activity of companies, as the share of import of the AIC products in Russia is not large, which is a certain signal for transfer from ousting the foreign rivals to the stage of holding the market positions by domestic companies of the AIC through their innovational development.

The values of the indicator of effectiveness are treated with a traditional method. If it is below 1, effectiveness is low, and the larger its values as to 1 the larger the effectiveness. The information support for this research includes the materials of the official statistics of the Federal State Statistics Service of the Russian Federation (Federal State Statistics Service) and the Ministry of Agriculture of Russia for 2012–2016, presented in Table 1.

Table 1. Indicators of effectiveness of implementing the mechanism of subsidizing within the policy of import substitution in the Russia's AIC

Indicators	2012	2013	2014	2015	2016
Share of innovations-active companies, %	9.6	9.3	9	10.3	10.2
Volume of subsidies aimed at compensation of direct costs of creation and modernization of the objects of the agro-industrial complex and purchase of technology and equipment, RUB billion	10.1	10.5	10.8	11.1	11.4

Source: compiled by the authors on the basis of:
1. (Federal State Statistics Service 2016);
2. (Ministry of Agriculture of the Russian Federation 2017).

3 Discussion

There is no single approach to defining the notion and sense of import substitution in the modern scientific circles. Certain researchers consider this notion in a wide sense, seeing import substitution as ousting foreign rivals from the market. Within this approach, import substitution is equaled to protectionism and it is considered that long application of this tool damages the functioning of the market, slowing down its development. This approach is described in the works (Bogoviz and Mezhov 2015; Popkova et al. 2016; Sadovnikova et al. 2013; Popova et al. 2015).

We favor the second approach – according to which import substitution is viewed in a wide sense and includes the measures for development of domestic entrepreneurship, which allow it ousting foreign rivals and keeping the conquered market positions in the long-term.

This approach supposes that import substitution includes not only protectionist measures but also market means of implementation of the practice of import substitution, the most important of which is stimulation of innovative activity of companies of the managed market. This allows implementing the practice of import substitution without violation of the market mechanism. It is shown in the works (Bogoviz et al. 2017; Sandu et al. 2017; Przhedetskaya and Akopova 2015).

4   Results

The results of assessing the effectiveness of implementation of mechanism of subsidizing within the policy of import substitution in the Russia's AIC for 2013–2016 are shown in Table 2.

Table 2. Results of evaluation of effectiveness of implementation of the subsidizing mechanism within the policy of import substitution in the Russia's AIC in 2013–2016

Indicator/symbol		2013	2014	2015	2016
Growth of the share of innovations-active companies of the AIC	ΔSiac	0.97	0.97	1.14	0.99
Growth of the volume of subsidies for innovational development of the AIC	ΔSmid	1.04	1.03	1.03	1.03
Effectiveness of implementation of mechanism of subsidizing in the AIC	Eis	0.93	0.94	1.11	0.96

Source: compiled by the authors.

As is seen from Table 2, effectiveness of implementation of the mechanism of subsidizing within the policy of import substitution in the Russia's AIC was low in 2013 and 2014. In 2015, it exceeded 1, constituting 1.11, and in 2016 it dropped below 1. This shows inexpedience of application of the mechanism of subsidizing and confirms the offered hypothesis on the necessity for search for new mechanisms. We think that it should be a complex mechanism that includes various means of innovational development of the agro-industrial complex of the country:

– formation of territorial and sectorial clusters which will allow ensuring effective interaction and cooperation between the companies of the AIC with preservation of their competition;
– development of technological parks that ensure promotion of innovational culture among the companies of the AIC and provision of their interaction with the leading R&D centers and universities;

– increase of the number of innovations-oriented special economic areas that create favorable conditions for acceleration of innovational development of the AIC by means of necessary infrastructure;
– wider application of the principles and mechanisms of public-private partnership for the purpose of provision of access to the companies of the AIC to state production capacities for their effective development.

This mechanism should be based on internal competition among the Russian companies of the AIC, ensured by the effective anti-monopoly policy. The principle of its work consists in state's creating the possibilities for high innovational activity by companies of the AIC allows for their innovational development, and competition stimulates practical implementation of this possibility (Fig. 1).

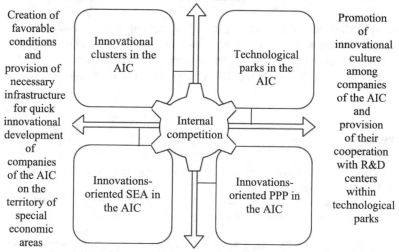

Fig. 1. Innovations-oriented mechanism of import substitution in the AIC in the interests of provision of food security of Russia Source: compiled by the authors.

As is seen from Fig. 1, as a result of implementation of the developed mechanism, the level of innovative activity of the AIC companies and their competitiveness are achieved in the long-term. This will allow preserving the leading positions in the Russian market and starting forcing the world food markets, developing export of the AIC products.

Eventually, the policy of import substitution will be replaced by the policy of development of export-oriented entrepreneurship in the AIC, which will ensure food security of Russia and turn the agro-industrial complex into a new vector of growth and development of national economy.

5 Conclusion

Thus, the offered hypothesis shows that successful implementation of the policy of import substitution in the AIC and provision of food security of Russia require improvement of economic mechanism of state support. It should be directed at financing of technical and technological modernization of the AIC companies and formation of a favorable environment in which they will be able to develop their innovational potential, as well as supporting the market stimuli that push the companies to this. Diversity of accessible methods of support for innovative activity of companies of the AIC is an important condition for increase of their effectiveness and competitiveness.

During further scientific research, it will be expedient to concentrate on development of practice-oriented offers and recommendations for most effective application of the methods of implementation of the offered innovations-oriented mechanism of import substitution in the AIC for solving the problem of provision of food security of Russia.

References

Bogoviz, A., Mezhov, S.: Models and tools for research of innovation processes. Mod. Appl. Sci. **9**(3), 159–172 (2015)

Popkova, E.G., Shakhovskaya, L.S., Abramov, S.A., et al.: Ecological clusters as a tool of improving the environmental safety in developing countries. Environ. Dev. Sustain. **18**(4), 1049–1057 (2016)

Popova, L., Popova, S.A., Dugina, T.A., Korobeynikov, D.A., Korobeynikova, O.M.: Cluster policy in agrarian sphere in implementation of concept of economic growth. Eur. Res. Stud. J. **18**, 27–36 (2015)

Przhedetskaya, N., Akopova, E.: Institutional designing of continuous education in Russia under the conditions of neo-economy and globalization. Reg. Sect. Econ. Stud. **15**(2), 115–122 (2015)

Sadovnikova, N., Parygin, D., Gnedkova, E., Kravets, A., Kizim, A., Ukustov, S.: Scenario forecasting of sustainable urban development based on cognitive model. In: Proceedings of the IADIS International Conference ICT, Society and Human Beings 2013, Proceedings of the IADIS International Conference e-Commerce 2013, pp. 115–119 (2013)

Bogoviz, A.V., Ragulina, Y.V., Shkodinsky, S.V., Babeshin, M.A.: Factors of provision of food security. Agric. Econ. Russ. **2**(1), 2–8 (2017)

Ministry of Agriculture of the Russian Federation: Information guide on measures and directions of state support of the agro-industrial complex of the Russian Federation (2017). http://www.gp.specagro.ru. Accessed 21 July 2017

Federal State Statistics Service: Industrial production in Russia 2016: statistical collection. Federal State Statistics Service, Moscow (2016)

Sandu, I.S., Bogoviz, A.V., Ryzhenkova, N.E., Ragulina, Y.V.: Formation of innovational infrastructure in the agrarian sector. AIC Econ. Manag. **1**(1), 35–41 (2017)

Leading Tools of State Regulation of Regional Economy

Aleksei V. Bogoviz[1]([✉]) [iD], Vladimir S. Osipov[2,3] [iD],
and Tamara G. Stroiteleva[4] [iD]

[1] Federal State Budgetary Scientific Institution
"Federal Research Center of Agrarian Economy and Social Development
of Rural Areas—All Russian Research Institute of Agricultural Economics",
Moscow, Russia
aleksei.bogoviz@gmail.com
[2] Lomonosov Moscow State University, Moscow, Russia
vs.ossipov@gmail.com
[3] Russian State Agrarian University - Moscow Timiryazev Agricultural
Academy, Moscow, Russia
[4] Institute of Supplementary Vocational Education of Altai State University,
Barnaul, Russia
stroiteleva_tg@mail.ru

Abstract. The purpose of the article is to substantiate the necessity for application of the leading tools of state regulation of regional economy in modern Russia and to develop practical recommendations for achieving high effectiveness of this regulation. The evidential base of the research consists of the authors' development of the official statistical information of the Federal State Statistics Service of the RF on dynamics of regional economy. The authors use the method of structural and trend analysis. As a result of the research, the authors conclude that traditional tools of state regulation of regional economy – such as tax, infrastructural, and social policy – do not allow achieving high effectiveness of management. They should be replaced by the leading tools, the choice of which should be directly determined by the top-priority goals of development of region's economy. These goals are creating innovations-oriented economy in the region, supporting high competitiveness of regional economy, overcoming ecological problems, and increasing the population's living standards. They should be achieved with the help of such tools as innovational & investment policy, prioritization of entrepreneurship, territorial marketing, management of sustainability of region's economy's development, and policy of social responsibility.

Keywords: State regulation · Regional economy
Competition between regions

1 Introduction

Modern region's economy functions in a completely new environment, as compared to the previous years, let alone the previous century. Stable but often unprofitable state companies have been replaced by highly-effective private business. Marketing relations

© Springer International Publishing AG, part of Springer Nature 2018
E. G. Popkova (Ed.): HOSMC 2017, AISC 622, pp. 562–567, 2018.
https://doi.org/10.1007/978-3-319-75383-6_71

have become so deeply embedded into the structure of economic systems that even the spheres of economy that present socially significant goods and services are passed into private management and private property.

Simultaneously, the process of globalization and integration of the regional economy takes place. Similarly to this process at the level of countries, regions enter the fight for new possibilities and resources that allow realizing the potential of their development. In these conditions, it is impossible to focus on the internal processes, as even regional residents, investors, and companies can move along the international migration, investment, and entrepreneurial flows. That's why a modern region should fight for external sources of growth and development, as well as preservation of their own assets and residents.

These tendencies changed the face of the modern region and transformed the sense of the processes that take place at the level of regional economy. This requires reconsideration of approaches to state management of regional economic systems. It is a serious problem in modern Russia, as regional authorities have rebuilt the programs of regional development of and have set new goals and tasks, with new priorities of management of this process. However, this did not ensure economic growth or achievement of sustainability of development of Russia's regional economy.

Our hypothesis consists in the fact that consideration of the concept of managing the regional economy is not enough for successful and full adaptation of the region to new economic conditions – this requires application of completely new managerial tools. In this paper, our aim is to substantiate the necessity for application of leading tools of state regulation of regional economy in modern Russia and to develop practical recommendations for achieving high effectiveness of such regulation.

2 Materials and Method

The evidential base of the research consists of the authors' development of the official statistical information of the Federal State Statistics Service on dynamics of development of regional economy (Table 1). The object of the research is Volgograd Oblast, as it could be characterized as an average statistical region of Russia, being neither a leading nor an underdeveloped region in terms of economy.

Table 1 shows not only the absolute values of the indicators but also relative values. Thus, we calculated the shares of various categories of revenues and expenditures in the general structure with application of the structural analysis method, as well as dynamics of the indicators' change, namely the rate of growth, with the help of trend analysis.

Table 1. Selected economic statistics of Volgograd Oblast of Russia for 2006–2016

Indicators	2006	Share, %	2011	Share, %	Growth, %	2016	Share, %	Growth, %	
								To 2011	To 2006
Total income	29,562.7	100.0	75,590.8	100.0	155.7	100,282.9	100.0	32.7	239.2
Corporate tax	5,857.6	19.8	17,555.9	23.2	199.7	19,488.0	19.4	11.0	232.7
Income tax	7,572.3	25.6	18,455.8	24.4	143.7	27,219.1	27.1	47.5	259.5
Property tax	3,623.2	12.3	7,906.6	10.5	118.2	12,330.5	12.3	56.0	240.3
Uncompensated receipts	5,651.7	19.1	16,090.5	21.3	184.7	22,628.5	22.6	40.6	300.4
Total income	30,044.8	100.0	78,240.5	100.0	160.4	107,519.0	100.0	37.4	257.9
General national issues	3,104.8	10.3	7,310.0	9.3	135.4	9,129.1	8.5	24.9	194.0
National economy	4,766.6	15.9	9,654.1	12.3	102.5	17,900.4	16.6	85.4	275.5
Housing and utilities infrastructure	3,785.0	12.6	7,065.7	9.0	86.7	5,974.7	5.6	−15.4	57.9
Socio-cultural events	16,817.9	56.0	47,281.3	60.4	181.1	69,982.8	65.1	48.0	316.1
Budget balance	−482.1	–	−2,649.7	–	449.6	−7,236.1	–	173.1	1,401.0

Source: compiled by the authors on the basis of the materials: (Federal State Statistics Service, 2016).

3   Discussion

Various theoretical & methodological and applied issues of state regulation of regional economy are viewed in multiple works of modern authors, which include (Franco and Ali 2017), (Belov and Kravets 2013), (Bogoviz et al. 2016), (Lui 2017), (Popkova 2013), (Ragulina et al. 2015), (Wolfe and Bramwell 2016), (Przhedetskaya 2014), (Osipov et al. 2016a), (Osipov and Skryl 2016b).

The performed literature overview on the topic of the research showed that the main tools state regulation of regional economy, according to the authors, are tax policy, aimed at maximization of regional budget's revenues, and infrastructural and social policy, oriented at development of the region with minimal budget expenditures.

Despite a high level of scientific elaboration, effectiveness of these tools of state regulation of regional economy in the modern economic conditions is not sufficiently studied, which requires further research in this sphere of scientific knowledge.

4   Results

The results of the performed statistical analysis of data from Table 1 allowed for determination of the main tendencies of development of regional economy of Volgograd Oblast in 2006–2016:

- growth of dependence on federal support: share of uncompensated receipts in the general structure of the regional budget's revenues constituted 22.6% in 2016, as compared to 21.3% in 2011 (growth in absolute expression constituted 40.6%) and 19.1% in 2006 (growth in absolute expression constituted 300.4%), which shows the increase of the region's dependence on external sources of financing and its status of subsidized region;

- termination of infrastructural project: share of expenditures for housing and utilities infrastructure in the general structure of regional budget's expenditures constituted 5.6% in 2016, as compared to 9% in 2011 (reduction in absolute value constituted 15.4%) and 12.6% in 2006, which shows transition of infrastructural projects to the rank of less important ones and general reduction of region's expenditures for their realization;

- expansion of social programs: the share of expenditures for socio-cultural events in the general structure of region's expenditures constituted 65.1% in 2016, as compared to 60.4% in 2011 (growth in absolute value constituted 81.1%) and 56% in 2006 (growth in absolute value constituted 316.1%), which shows growth of priority of social programs in the region;

- increase of the budget deficit: deficit of the regional budget grew by 1.7 times in 2016, as compared to 2011, and by 14 times, as compared to 2006, which characterizes the region as lossmaking, not capable for self-provision.

This proves the offered hypothesis that the traditional tools of state regulation of regional economy – such as the tax, infrastructural, and social policy – do not allow achieving high effectiveness of management. They should be replaced by the modern leading tools. We think that the choice of such tools should be directly determined by the top-priority goals of region's economy's development.

One of the most important goals is creation of innovations-oriented economy in the region. The tools for achieving it are innovations & investment policy and prioritization of entrepreneurship. They should be oriented at formation and support for favorable investment climate in the region (low corporate tax, modern infrastructure, etc.) and stimulation of innovational activity of entrepreneurship (tax subsidies, subsidized crediting, etc.).

Another goal is support for high competitiveness of the regional economy. For its realization, we recommend to use such tool as territorial marketing. This tool should be aimed at emphasizing and developing the unique features of the region. For example, in case of Volgograd Oblast, it is rich history, geographical location on the Volga, etc.

Another goal is related to overcoming the ecological problems that grow with increase of industrial growth of economy. A top-priority tool for its realization is management of sustainability of region's economy's development. Its application supposes establishment of high ecological standards and strict control over their observation in the interests of preservation of the environment.

The goals include also the increase of the population's living standards. For that, we recommend to use such perspective tool of state regulation of regional economy as the policy of social responsibility. Instead of traditional social policy, which supposes state financing of social programs in the region, we recommend to use the policy of social responsibility which allows attracting business for solving the region's social problems. For that, regional authorities should stimulate entrepreneurship for social responsibility through its propaganda, creation of rakings of companies as to the level of social responsibility, and provision of various tax and non-tax privileges.

Realization of the offered recommendations and application of these leading tools of state regulation of regional economy will allow ensuring a high rate of stable and sustainable economic growth of regional economy, as well as reducing the load on the regional budget and solving the problem of its deficit, which reflects their high effectiveness. The model of state regulation of regional economy with application of the offered leading tools is presented in Fig. 1.

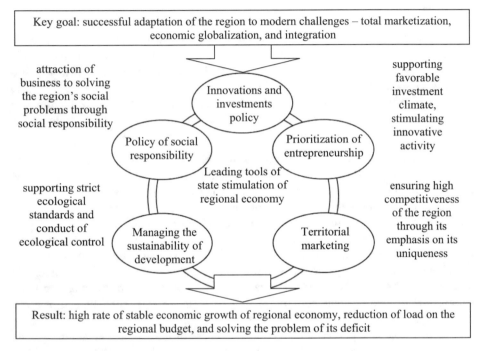

Fig. 1. Model of state regulation of regional economy with application of the offered leading tools Source: compiled by the authors.

As is seen from Fig. 1, the leading tools of state regulation of regional economy ensure successful adaptation of the region to modern challenges – total marketization, economic globalization, and integration.

5   Conclusions

It is possible to conclude that state regulation of regional economy in the modern market conditions should include the elements of marketing and be maximally flexible. New tendencies in development of regional economy form challenges and threats and provide new possibilities for its growth. That's why management of a region should not fight the influence of external factors but use them in the region's interests.

The offered leading tools of state regulation of regional economy are developed in view of these new challenges and allow taking the load for development of regional economy from the state to entrepreneurship and ensuring its complete and full character. However, in view of dynamics of change of the external environment, these tools require periodic reconsideration for supporting their effectiveness and topicality.

References

Franco, I.B., Ali, S.: Decentralization, corporate community development and resource governance: a comparative analysis of two mining regions in Colombia. Extr. Ind. Soc. **4**(1), 111–119 (2017)

Belov, A.G., Kravets, A.G.: Business performance management in small and medium businesses and functional automation. World Appl. Sci. J. **24**(24), 7–11 (2013)

Bogoviz, A.V., Ragulina, Y.V., Kutukova, E.S.: Economic zones as a factor of increased economic competitiveness of the region. Int. J. Econ. Financial Issues, **6**(8 Special Issue), 1–6 (2016)

Lui, L.T.P.: Institutions do matter: exploring the problem of governance in the Hong Kong Special Administrative Region from the perspective of executive-legislative relations. Asian Educ. Dev. Stud. **6**(1), 72–82 (2017)

Popkova, E.G.: Marketing strategy to overcome the "underdevelopment whirlpool" of the Volgograd region. In: Conference of the Eurasia-Business-and-Economics-Society (EBES), Russian Academy of Sciences, Institute of Economics, Ural Branch, Ekaterinburg, 12–14 September 2013, pp. 52–61 (2013)

Ragulina, Y.V., Stroiteleva, E.V., Miller, A.I.: Modeling of integration processes in the business structures. Mod. Appl. Sci. **9**(3), 145–158 (2015)

Wolfe, D., Bramwell, A.: Innovation, creativity and governance: social dynamics of economic performance in city-regions. Innov.: Manag. Policy Pract. **18**(4), 449–461 (2016)

Przhedetskaya, N.V.: Design of marketing management of innovational model of education in the conditions of development economy. Bull. Tula State Univ. **4**(1) (2014)

Federal State Statistics Service. Regions of Russia. Socio-economic indicators. 2016: statistical collection. Rosstat, Moscow (2016)

Osipov, V.S., Skryl, T.V., Evseev, V.O.: An analysis of economic issues of territories of priority development. Res. J. Appl. Sci. **11**(9), 833–842 (2016a)

Osipov, V.S., Skryl, T.V.: The strategic directions of the modern Russian economic development. Int. Bus. Manag. **10**(6), 710–717 (2016b)

Modeling the Management of Innovational Processes in Regional Economy

Yulia V. Ragulina[1]([✉]), Aleksei V. Bogoviz[1],
and Alexander N. Alekseev[2]

[1] Federal State Budgetary Scientific Institution
"Federal Research Center of Agrarian Economy and Social Development
of Rural Areas—All Russian Research Institute of Agricultural Economics",
Moscow, Russia
julra@list.ru, aleksei.bogoviz@gmail.com
[2] G.V. Plekhanov Russian University of Economics, Moscow, Russia
alexeev_alexan@mail.ru

Abstract. The purpose of the article is optimization modeling of management of innovational processes in regional economy of modern Russia. The authors use the method of problem and systemic analysis, plan-fact analysis, and method of modeling of socio-economic processes and systems, as well as the optimization method. The authors study the course of execution of the Strategy of innovational development of the Russian Federation until 2020, determine the existing model of managing innovational processes in regional economy of modern Russia, and substantiate perspectives and develop recommendations for optimization of this model. Based on studying the practice of managing innovational processes in regional economy of modern Russia, the authors determine the key problems of the applied model. It is determines that effectiveness of managing the innovational processes at the level of regional economy is at a low level due to founding on financial support for innovative activity, implementation of which is not possible due to deficit of the regional budget. The authors offer recommendations for optimization of this model, which are aimed at development of non-financial measures for stimulation of innovative activity in the region, and develop the optimization model of managing innovational processes in regional economy of modern Russia.

Keywords: State management · Innovational processes · Regional economy

1 Introduction

The modern global economy has entered the age of innovations, which are the key landmark in development of economic systems, being a source of their sustainable economic growth, factor of social well-being, and condition of achieving high international status. Despite acknowledgment of the necessity for creation of innovational economy by all participants of international economic relations, certain countries have succeeded in this process, while others have faced the obstacles, which became a reason for their underrun and their receiving the status of outsiders of the global economic system.

© Springer International Publishing AG, part of Springer Nature 2018
E. G. Popkova (Ed.): HOSMC 2017, AISC 622, pp. 568–573, 2018.
https://doi.org/10.1007/978-3-319-75383-6_72

Russia has also entered the path of creation of innovational economy. The proclaimed course at modernization started this process, but there are not significant results in this direction – though, more than a half of the assigned term (2011–2020) has already passed. This shows the existence of systemic problems in implementation of the Strategy of innovational development of the RF until 2020.

The working hypothesis of the research consists in the fact that one of the most important reasons for low innovative activity of economic subjects and slow rate of modernization of modern Russia's economic system is ineffectiveness of managing the innovational processes at the level of regional economy. The purpose of the article is optimization modeling of managing the innovational processes in regional economy of modern Russia. The following tasks are solved for this:

- studying the course of execution of the Strategy of innovational development of the RF until 2020;
- determining the existing model of managing the innovational processes in regional economy of modern Russia;
- substantiating the perspectives and developing the recommendations for optimization of this model.

2 Materials and Method

The information and analytical basis of the research includes the materials of the official statistics of the Federal State Statistics Service and the Strategy of innovational development of the RF until 2020. The following data are used for studying the course of execution of this strategy (Table 1).

Table 1. Initial data for the research

Indicators of the course of implementation of the Strategy of innovational development of the RF until 2020	2012		2014		2016	
	Plan	Fact	Plan	Fact	Plan	Fact
Growth of the number of small innovational companies, %	5	3.98	8	4.90	12	6.49
Share of innovational products in the total volume of sales, %	20	18.92	32	28.74	45	34.61
Volume of sold innovational products, RUB billion	25	19.53	46	33.30	50	32.90
Number of annually created elements of the infrastructure of the national innovational system	30	19.35	60	28.74	90	23.04
Volume of investments into the companies of the high-tech spheres, RUB billion	5	3.85	10	7.14	15	9.65

Source: compiled by the authors on the basis of Government of the Russian Federation (2011), Federal State Statistics Service (2016).

The authors use the method of problem and systemic analysis, method of plan-fact analysis, and method of modeling of socio-economic processes and systems, as well as the optimization method.

3 Discussion

The theoretical basis of the research consists of the materials of fundamental and applied studies of modern authors on the issues of managing the innovational processes in regional economy, among which are Pradel-Miquel (2015), Wolfe and Gertler (2016), Wolfe and Bramwell (2016), Harris et al. (2013), Ibrahim et al. (2014), Shala et al. (2015), and Veselovsky et al. (2015).

4 Results

The results of the performed analysis, based on the plan-fact analysis, are shown in Table 2.

Table 2. Results of plan-fact analysis

Indicators of the course of implementation of the Strategy of innovational development of the RF until 2020	Underrun of factual values of the indicators from the targeted ones, %		
	2012	2014	2016
Growth of the number of small innovational companies, %	20.5	38.7	45.9
Share of innovational products in the total volume of sales, %	5.4	10.2	23.1
Volume of sold innovational products, RUB billion	21.9	27.6	34.2
Number of annually created elements of the infrastructure of the national innovational system	35.5	52.1	74.4
Volume of investments into the companies of the high-tech spheres, RUB billion	23.1	28.6	35.7

Source: compiled by the authors.

As is seen from Table 2, in the first year (2012) of implementation of the Strategy of innovational development of the RF until 2020, all factual values of all indicators were behind the planned ones. This underrun accumulated and reached 74.4% for the indicator of innovational infrastructure development by 2016.

Based on studying the practice of managing innovational processes in the regional economy of modern Russia, we determined the following problems of the applied model:

– deficit of budget resources of the region, which limits the possibilities in the sphere of supporting innovational activity of economic subjects;

- absence of the specialized regional support (tax, credit, investment, or other) for innovational entrepreneurship;
- absence of measures for stimulating the demand for innovational products of regional companies, which is a reason for low demand for it;
- absence of measures for creation of profitable conditions for venture investments into regional companies of high-tech sectors and development of the national innovational system.

Thus, this model supposes low involvement of regional authorities into implementation of the national strategy of creation of innovational economy – it is limited by propaganda of innovational activity without creating the favorable conditions. At the federal measures are not enough, the process of implementation of this strategy is very slow, and it is no wonder that factual results are behind the targeted (planned) ones.

The perspectives of increase of effectiveness of managing the innovational processes in the modern Russia's regional economy are related to expansion of activity in the institutional sphere. In the conditions of deficit of the regional budget, the main efforts in the sphere of such management should be brought down to attraction of private investments into the innovational sphere of the region. For that, we offer the following practical recommendations.

Firstly, it is necessary to implement measures for creation of profitable conditions for venture investing into the regional companies of high-tech sectors and into development of the infrastructure of the national innovational system. Such measure could be placing information on venture investors and their investments into the region's innovational sphere on the official web-site of the regional administration. This will stimulate formation and strengthening of business reputation (brand) of such private and corporate investors, thus ensuring commercial attractiveness of venture investments for them.

Secondly, it is necessary to implement measures for stimulation of demand for innovational products of regional companies, due to which it is not popular. For that, we recommend to start the system of certification of innovational products of regional companies and formation of regional brands of innovational products under the aegis of regional authorities. The state-managed brand will increase attractiveness of innovational products of regional companies and will raise its sales.

Thirdly, it is necessary to implement measures for non-financial support for innovational activity of the region's companies through creation of favorable conditions. Separate companies, especially small business, do not possess sufficient resources for manifestation of high innovational activity. That's why we recommend stimulating implementation of cluster initiatives in entrepreneurship through consultation support and simplification of registration procedures. Creation of regional sectorial innovational clusters will allow unifying the resources of region's companies, thus enabling them to implement innovations, and participation of universities in clusters will ensure quick commercialization and diffusion of innovations.

The recommended measures will allow stimulating development of innovational processes in the region with minimal expenditures of the regional budget, which will ensure their high effectiveness. Based on the above, we developed the optimization model of managing the innovational processes in regional economy in modern Russia, which is shown in Fig. 1.

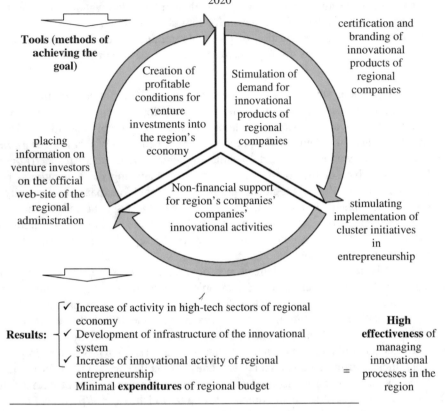

Goal: optimization model of managing innovational processes in the regional economy for stimulating the implementation of the Strategy of innovational development of the RF until 2020

Tools (methods of achieving the goal)

placing information on venture investors on the official web-site of the regional administration

Creation of profitable conditions for venture investments into the region's economy

Stimulation of demand for innovational products of regional companies

certification and branding of innovational products of regional companies

Non-financial support for region's companies' companies' innovational activities

stimulating implementation of cluster initiatives in entrepreneurship

Results:
- ✓ Increase of activity in high-tech sectors of regional economy
- ✓ Development of infrastructure of the innovational system
- ✓ Increase of innovational activity of regional entrepreneurship
- Minimal **expenditures** of regional budget

= **High effectiveness** of managing innovational processes in the region

Fig. 1. Optimization model of managing the innovational processes in regional economy of modern Russia. Source: compiled by the authors

As is seen from Fig. 1, in the offered optimization model the non-financial measures of supporting innovational activity in regional economy are preferred, which leads to minimum expenditures of the regional budget. Together with substantial results in the form of increase of activity in high-tech sectors of regional economy, development of infrastructure of the innovational system, and increase of innovational activity of regional entrepreneurship, high effectiveness of managing the innovational processes in region is achieved.

5   Conclusions

Thus, it is possible to conclude that within the current model of managing the innovational processes at the level of regional economy its effectiveness is at a low level due to founding on the measures of financial support of innovational activity, implementation of which is not fully possible due to deficit of the regional budget. That's why the offered recommendations for optimization of this model are aimed at development of non-financial measure of stimulation of innovational activity in the region.

Theoretical significance of the results of the performed research consists in development of conceptual provisions of the theory of state management of innovational processes at the level of regional economy. Practical significance of the offered recommendations consists in stimulating the achievement of the goals of the Strategy of innovational development of the RF until 2020 and creation of innovational economy in the national economic system.

References

Government of the Russian Federation: Innovational Russia – 2020 (Strategy of innovational development of the Russian Federation until 2020) (2011). http://cluster.hse.ru/cluster-policy/docs/Инновационная%20Россия%202020%20-%20Коллектив%20авторов.pdf. Accessed 22 June 2017

Federal State Statistics Service: Russian in Numbers: Short Statistical Colleciton. Federal State Statistics Service, Moscow (2016)

Veselovsky, M.Y., Pogodina, T.V., Idilov, I.I., Askhabov, R.Y., Abdulkadyrova, M.A.: Development of financial and economic instruments for the formation and management of innovation clusters in the region. Mediterr. J. Soc. Sci. **6**(3), 116–123 (2015)

Shala, M., Hajrizi, E., Hoxha, V., Stapleton, L.: Cost-oriented agile innovation for mechatronics management in less developed regions. IFAC-PapersOnLine **48**(24), 150–152 (2015)

Ibrahim, S., Sukeri, S.N., Rashid, I.M.: Factors influencing the diffusion and implementation of management accounting innovations (MAIS), Malaysian manufacturing industries in Nothern Region. Adv. Environ Biol. **8**(9), 504–512 (2014). Special issue 4

Harris, R., McAdam, R., McCausl, I., Reid, R.: Knowledge management as a source of innovation and competitive advantage for SMEs in peripheral regions. Int. J. Entrepreneurship Innov. **14**(1), 49–61 (2013)

Wolfe, D., Bramwell, A.: Innovation, creativity and governance: social dynamics of economic performance in city-regions. Innov.: Manag. Policy Pract. **18**(4), 449–461 (2016)

Wolfe, D.A., Gertler, M.S.: Growing Urban Economies: Innovation, Creativity, and Governance in Canadian City-Regions, pp. 1–420 (2016)

Pradel-Miquel, M.: Making polycentrism: governance innovation in small and medium-sized cities in the West Midlands and Barcelona metropolitan regions. Environ. Plan. C: Gov. Policy **33**(6), 1753–1768 (2015)

New Challenges for Regional Economy at the Modern Stage

Aleksei V. Bogoviz[1](✉) (iD), Elena I. Semenova[1] (iD),
and Alexander N. Alekseev[2]

[1] Federal State Budgetary Scientific Institution "Federal Research Center
of Agrarian Economy and Social Development of Rural Areas—All Russian
Research Institute of Agricultural Economics", Moscow, Russia
aleksei.bogoviz@gmail.com
[2] Plekhanov Russian University of Economics, Moscow, Russia
alexeev_alexan@mail.ru

Abstract. The purpose of the research is to study the influence of new challenges on the Russia's regional economy at the modern stage and to develop the framework strategy of its reaction to these challenges for simultaneous protection of regional interests and stimulating the implementation of the national course of socio-economic development. The authors study the connection between national and regional indicators of economic development. These indicators form three categories that reflect the most topical landmarks of development of modern economic systems: economic development, sustainability, and innovations. For studying this connection, the authors use the methods of regression and correlation analysis. The results of the performed analysis contradicted the existing theoretical ideas on a priori positive influence of positive global tendencies on the regional economy and showed that these tendencies influence the national and regional levels in a different way. That's why the strategy of regional development should be prepared not by the model of the national strategy but in view of the region's peculiarities. For that, the authors offer the corresponding recommendations that allow regional economy to adapt to new modern challenges.

Keywords: Regional economy · Modern challenges · Sustainable growth
Innovational development

1 Introduction

For most economic systems, the 21st century became the age of changes: from the change of the economic mode to change of the course of socio-economic development. The model of free market economy, which was considered to be a model, showed its flaws. That's why the countries with transitional economy – such as modern Russia – faced a new crisis, which was a sign of transition to the new model of economic growth and development.

E. G. Popkova (Ed.): HOSMC 2017, AISC 622, pp. 574–580, 2018.
https://doi.org/10.1007/978-3-319-75383-6_73

This formed unprecedented conditions, in which economic systems found themselves. The influence of new conditions on the national economy has been thoroughly studies and is being studied now. Specifics of functioning and development of regional economy under the influence of these conditions are not paid enough attention, though the regional level is the level of implementation of economic reforms and achievement of the initial results, of which the national socio-economic progress consists.

The working hypothesis of this research consists in the fact that the influence of the global tendencies at the regional level could be manifested in a way that is different from the national economy, which requires correction of the course of regional development, which should be connected to but not identical to the national course. In modern Russia, regional strategies of socio-economic development are equal to the national strategy.

We think that instead of harmonization of all levels of the national economic system, this might be a reason for its long imbalance and crisis. The purpose of the research is to study the influence of new challenges on the Russia's regional economy at the modern stage and to develop the framework strategy of its reaction to these challenges for simultaneous protection of regional interests and stimulation of implementation of the national course at the socio-economic development.

2 Materials and Method

For verification of the offered hypothesis, this work studies the connection between the national and regional indicators of economic development. These indicators form three categories that reflect the most topical landmarks of development of modern economic systems: economic development, sustainability, and innovations. The initial statistical data, used during conduct of this research, are systematized and presented in Table 1.

Within the existing scientific concept, it is supposed that globalization is a universal tool, which is very effective for economic development. That is, growth of the index of globalization of national economy should potentially lead to growth of GRP, real financial income, and employment and business activity. Growth of ecological effectiveness at the national level theoretically supposes increase of sustainability of regional economy – that is, reduction of polluting emissions into atmosphere.

The index of innovativeness of national economy should lead to increase of innovational activity of companies at the level of regional economy. That is, there should be a strong direct connection between x_1 and y_1, y_2, y_3 and y_4, x_2 and y_5, and between x_3 and y_6. For studying this connection, this work uses the method of regression and correlations analysis.

For measuring the direction of connection, coefficient b is used in the model of paired linear regression of the type $y = a + bx$, where positive sign of the coefficient shows direct connection, and negative sign shows the reverse connection of the indicators. Correlation coefficient r^2 is used for measuring strength of connection. Connection is strong if this coefficient is over 0.9.

Table 1. Dynamics of national and regional indicators of economic development in 2006–2016

Indicators and their symbol		Values of indicators for years						
		2006	2011	2012	2013	2014	2015	2016
x_1	Index of globalization of economy, points	51.6	51.7	52.8	50.1	52.0	53.7	52.1
x_2	Index of ecological effectiveness, points	77.5	78.1	79.3	80.1	81.3	82.4	83.5
x_3	Index of innovativeness of economy, points	34.3	35.1	36.4	37.2	39.1	39.3	38.5
y_1	Economic growth of regions (average growth of GRP), %	1.0	2.1	1.2	1.1	1.1	1.1	1.1
y_2	Growth of real financial income of population, %	111.7	105.4	101.2	105.8	104.8	99.5	95.9
y_3	Level of population's employment, %	61.3	62.7	63.9	64.9	64.8	65.3	65.3
y_4	Growth of business activity (number of employees), %	1.0	1.0	1.0	1.0	1.0	1.0	1.0
y_5	Pollutant emissions into atmosphere, thousand ton	20.4	19.1	19.2	19.6	18.4	17.5	17.3
y_6	Innovational activity of companies (share of innovations-oriented companies), %	9.9	9.5	10.4	10.3	10.1	9.9	9.3

Source: compiled by the authors on the basis of: (Yale Center for Environmental Law and Policy 2017), (INSEAD 2017), (KOF Swiss Economic Institute 2017), (Federal State Statistics Service 2016).

3 Discussion

Specifics of development of regional economy at the modern stage and the challenges it faces, as well as possibilities and problems of its development are viewed in multiple works of various, among which are (Belov and Kravets 2013), (Bogoviz et al. 2016), (Popkova 2013), (Ragulina et al. 2015), (Przhedetskaya 2014), (Khairullov et al. 2016), (Kiseleva et al. 2016), (Tkachenko et al. 2016), (Afonasova 2015), (Corpakis 2012), and (Gibbs and O'Neill 2017).

4 Results

Based on the data of Table 1, we performed regression and correlation analysis, the results of which are shown in Table 2.

As is seen from Table 2, growth of the influence of globalization processes on the modern Russia's national economy at the level of regions leads to the effect opposite to economic development. Thus, gross regional product decreases, as well as real financial income of population, and growth of employment and business activity is very low. At that, the values of the correlation coefficient for all these models are very low – that is, connection between these indicators is low.

Table 2. Results of regression and correlation analysis

Dependent variables	Their dependence on independent variables					
	x_1		x_2		x_3	
	b	r^2	b	r^2	b	r^2
y_1	−0.03	0.008	–	–	–	–
y_2	−2.32	0.25	–	–	–	–
y_3	0.26	0.04	–	–	–	–
y_4	0.001	0.02	–	–	–	–
y_5	–	–	−0.46	0.84	–	–
y_6	–	–	–	–	0.002	0.99

Source: compiled by the authors.

The value of correlation coefficient for the indicators of sustainability is below 0.9, but still it is rather high – 0.84. However, growth of ecological effectiveness at the national level by 1 point leads to reduction of sustainability of regional economy by 0.46% - that is, it performs reverse influence, which contradicts the basic theoretical idea. The level of correlation of the index of innovativeness of national economy and innovational activity of companies at the level of regional economy is very high (0.99), but the influence is minimal.

Thus, the results of the performed analysis contradicted the existing theoretical ideas on a priori positive influence of positive global tendencies on the regional economy. In other words, these tendencies influence the national and the regional economy differently. That's why the strategy of regional development should be developed not according to the scale of the national strategy but in view of the region's peculiarities. For that, we offer the following recommendations.

Firstly, as globalization may negatively influence development of the region, it needs to be limited. It should be noted that in the conditions of the national course at freetrading such limitations cannot have a vivid barrier character (for example, be implemented in the form of custom barriers). Therefore, these should be modern market measures.

For supporting regional entrepreneurship, regional authorities should stimulate strengthening of the corresponding brand. This supposes formation and development of the brand by the regional authorities with further distribution on the products of regional companies. Consumers' high loyalty to the brand of regional products will allow reducing their interest to imported products, thus limiting the influence of globalization on the regional economy.

Secondly, as the national measures for supporting the environment are not sufficient for achievement of high ecological effectiveness of entrepreneurship at the level of regional economy, it is necessary to take additional measures at the regional level of an economic system. For that, it is recommended to adopt regional ecological standards that take into account peculiarities of the environment and entrepreneurship of the specific region.

The additional measure in this direction is implementation of mandatory and voluntary ecological standards, observation of which will be a condition for distribution of a regional brand for the company's products. This will allow implementing the offered recommendations and increasing the achieved effect.

Thirdly, in view of the fact that development of the innovational economy at the national level does not necessarily suppose high innovational activity of entrepreneurship at the region's level, there's necessity for additional measures for stimulating such activity. For this, we recommend – according to the above recommendations – to introduce the requirements for evaluation of innovations' implementation for assigning the regional brand to company's products.

This will allow replacing administrative measures (tax subsidies, etc.) with market measures (branding) for stimulating innovational activity of regional entrepreneurship, thus increasing favorable influence of competition on regional economy and saving the assets of the regional budget. These recommendations are presented in the form of framework strategy of regional economy's reaction to new modern challenges (Fig. 1).

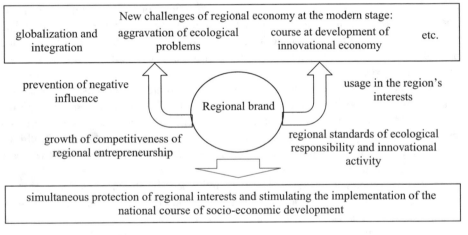

Fig. 1. Framework strategy of regional economy's reaction to new challenges of modern times. Source: compiled by the authors.

As is seen from Fig. 1, the central tool of implementation of the offered strategy is formation and strengthening of the regional brand. The developed strategy allows regional economy to adapt to the viewed challenges and to any challenges of modern times, which ensures its universal application. Due to application of market measures instead of administrative ones, this strategy ensures simultaneous protection of regional interests and stimulation of implementation of the national course of socio-economic development.

5 Conclusions

It should be concluded that the modern stage of development of the global economic system is characterized by a high level of uncertainty due to acknowledged instability of the model of free market economy and lack of new, improved model which is to replace it. That's why regional economy faces new challenges. The developed and presented framework strategy reflects the logic of regional economy's reaction to new modern challenges.

At that, the most important condition of its successful application in practice and achievement of its high effectiveness is constant monitoring of internal and external environment, in which regional economy functions, and analysis of factors of its development. As this research showed, only flexible and adaptable market approach to managing the regional economy will allow it to overcome contradictions with national economy and maximizing total economic effect at all levels of the economic system.

References

Afonasova, M.A.: The concept of regional economy restructuring in the context of a problem concerning overcoming of the social and economic development inertia. Soc. Sci. (Pakistan) 10(9), 2206–2212 (2015)

Belov, A.G., Kravets, A.G.: Business performance management in small and medium businesses and functional automation. World Appl. Sci. J. 24(24), 7–11 (2013)

Bogoviz, A.V., Ragulina, Y.V., Kutukova, E.S.: Economic zones as a factor of increased economic competitiveness of the region. Int. J. Econ. Finan. Issues 6(8), 1–6 (2016). Special Issue

Corpakis, D.: A European regional path to the knowledge economy: challenges and opportunities. In: Intellectual Capital for Communities: Nations, Regions, and Cities, pp. 213–225 (2012)

Gibbs, D., O'Neill, K.: Future green economies and regional development: a research agenda. Reg. Stud. 51(1), 161–173 (2017)

INSEAD: Global Innovation Index (2017). http://www.globalinnovationindex.org. Accessed 10 July 2017

Khairullov, D.S., Galeeva, G.M., Absalyamova, S.G., Butov, G.N.: Challenges of sustainable development of regional economy in the conditions of WTO market. In: Social Sciences and Interdisciplinary Behavior - Proceedings of the 4th International Congress on Interdisciplinary Behavior and Social Science, ICIBSOS 2015, pp. 261–264 (2016)

Kiseleva, N.N., Tikhomirov, A.A., Lyapuntsova, E.V., Sklyarenko, S.A., Gukasova, N.R.: Problems of economic security of the industrial sector of the regional economy at the present stage of development. Res. J. Pharm. Biol. Chem. Sci. 7(1), 2214–2221 (2016)

KOF Swiss Economic Institute: Index of Globalization (2017). http://globalization.kof.ethz.ch/query/. Accessed 10 July 2017

Popkova, E.G.: Marketing strategy то overcome the "underdevelopment whirlpool" of the Volgograd region. In: Conference of the Eurasia-Business- and-Economics-Society (EBES). Russian Academy of Sciences, Institute of Economics, Ural Branch, Ekaterinburg, Russia publ., 12–14 September 2013, pp. 52–61 (2013)

Ragulina, Y.V., Stroiteleva, E.V., Miller, A.I.: Modeling of integration processes in the business structures. Mod. Appl. Sci. 9(3), 145–158 (2015)

Tkachenko, E., Rogova, E., Bodrunov, S.: Regional models of the management of knowledge economy development: the problem of measurement and assessment. In: Proceedings of the European Conference on Knowledge Management, ECKM, January 2016, pp. 881–889 (2016)

Yale Center for Environmental Law and Policy: The Environmental Performance Index (2017). http://epi.yale.edu/. Accessed 10 July 2017

Przhedetskaya, N.V.: Designing marketing management of the innovational model of education in the conditions of development economy. Bull. Tula State Univ. 4(1) (2014)

Federal State Statistics Service: Regions of Russia. Socio-economic indicators, 2016: statistical bulletin. Federal State Statistics Service, Moscow (2016)

Economic Analysis of Labor Resources Usage in Regional Markets

Tatiana N. Gogoleva[1(✉)], Pavel A. Kanapukhin[1],
Margarita V. Melnik[2], Irina Y. Lyashenko[1],
and Valeriya N. Yaryshina[1]

[1] Voronezh State University, Voronezh, Russia
tgogoleva2003@mail.ru, pavkan72@yandex.ru,
55irina@mail.ru, yaryshina@econ.vsu.ru
[2] Financial University under the Government of the Russian Federation,
Moscow, Russia
aik@fa.ru

Abstract. The purpose of the article is to substantiate the necessity for considering the territorial peculiarities of using labor resources employed at harmful and hazardous productions, determined on the basis of the economic and statistical analysis, during formation of the optimization model and the procedure of decision making for selecting the programs of the planned control of the employment sphere for labor safety quality.

There are various methodologies of labor safety, oriented at characterizing this process within a specific company and without a direct connection to management of labor safety within the region. T.V. Azarnova developed the approach to evaluation of effectiveness of regional management of labor safety quality and formed the optimization stochastic model of the system of planned control of the employment sphere in the labor safety sphere which is oriented at the differentiated management of the control functions as to the groups of companies that are distinguished depending on the complexity of the structure and level of implementation of the labor safety processes. The authors offer to use – apart from studying the situation at specific companies and groups of companies – the economic and statistical analysis of regional peculiarities of the influence of the production environment and the labor process factors for statistically determined groups of companies on the labor conditions of employees.

Results: using the economic and statistical analysis of regional peculiarities of the labor conditions of employees allowed determining the specifics of the influence of various factors at the level of federal districts of Russia and among the regions of the Central Federal District. The obtained results became the basis of the offers that are to be taken into account during development of the effective system of regional management of labor safety quality.

Keywords: Labor market · Labor conditions
Statistical analysis of labor at the regional market
Regional system of management of labor safety quality

GEL Classification Codes: C19 Econometric and Statistical Methods
and Methodology: General. Others · J08 Labor Economics Policies

© Springer International Publishing AG, part of Springer Nature 2018
E. G. Popkova (Ed.): HOSMC 2017, AISC 622, pp. 581–590, 2018.
https://doi.org/10.1007/978-3-319-75383-6_74

1 Introduction

In the modern conditions, labor quality and labor safety are the factor of effectiveness of labor resources usage – and the quantity of resources reduces in Russia. Quality of labor safety and related processes are regulated by multiple laws and bylaws. At the same time, the preventing activities are still not developed, and violations in the sphere of labor safety tend to repeat and perform huge material and moral damage to the society and the state. Thus, optimization of labor safety quality by means of optimization of state control in the sphere of labor law and labor safety becomes very topical.

The system of management and control over economic subjects depends on connection between the goals of the managed object and the possibilities of the subsystems. Effectiveness of management is influenced by the measures for optimization of the whole system of management [2]. One of the directions of optimization is formation of the programs of planned control in the sphere of employment for quality of labor safety, during development of which it is necessary to take into account the peculiarities of companies and external socio-economic conditions and institutes that determine the environment of their functioning and form the regional peculiarities of the state of employees' labor conditions. Thus, there's a necessity to consider the internal characteristics of organization of production processes at companies and the external environment of their development.

2 Research Methods

At the legislative level, the issues of labor safety in the RF are regulated by the Labor Code and the Law "On special evaluation of labor conditions", which determine the main norms of the state of labor conditions [9]. At present, there are various analytical approaches to evaluation of the state of labor safety, which authors - O.U. Drozhchanaya, E.A. Krasnoshchekaya, M.P. Gandzyuk, R.P. Kerb, and E.V. Spatar [3–6, 8] – focus on determining the system of indicators that would help to conduct the evaluation, emphasizing on the situation in the analyzed sphere at the company level. Thus, the main and sole object of evaluation is company as an economic subject. At the same time, the problem of labor quality and labor safety should not be fixed to the company level, as it is more complex and includes diverse external conditions in which the company's activities are conducted. These conditions, created by the external environment, should be taken into account at the public level of considering and solving the problems of labor quality and labor safety – at the regional level. Such problems are not viewed by the above authors.

The works of Azarnova [1, 2] set the question of managing the quality of labor safety at the regional level and its assessment – for which the optimization stochastic model of the system of planned control of the employment sphere and the algorithm of finding the optimal intensity of control were formed – which ensure the differentiated management of the control functions as to the companies. This model is oriented at development of recommendations for the regional programs of control over labor safety quality as to the groups of companies with a certain complexity of the structure and the level of implementation of the labor safety processes. Thus, the emphasis was shifted to the level of homogeneous groups of companies of one region which are controlled by the regional authorities. However, determination and consideration of the regional peculiarities of the external environment, which form the factors that influence the state of labor conditions, are not envisaged. This aspect of the problem should be studied in detail on the basis of analysis of the official statistics data.

For evaluation of the state of labor conditions at companies depending on their territorial location, it is necessary to consider several groups of factors: natural and geographic, technological (factors of the production environment factors of the labor process), and socio-economic factors.

Natural and geographic factors determine the peculiarities of functioning of separate spheres that are located on various territories. These peculiarities are reflected in the corresponding territorial codes (СЗВ-СТАЖ, СЗВ-КОРР, СЗВ-ИСХ). At the same time, there are no methods of accounting of territorial peculiarities of various productions within the same group of spheres (e.g., processing productions are represented by various spheres in Belgorod and Voronezh Oblasts, so the state of labor conditions will be determined by various factors).

Technological factors that determine the state of labor conditions are the object of attention of annual statistical observations. Each company fixes its indicators by determining the number of employees who are under the influence of the production environment factors (chemical, biological, influence of sprays, noise, vibration, neon and ionizing rays, micro-climate, light environment, etc.) and the factors of the labor process (work load and work tension). Labor conditions, according to the tasks of statistical observation within the official state statistics in the RF, are studied at large and medium companies of the following types of economic activities: agriculture, hunting and forestry (as well as provision of services in these spheres); mineral production; processing productions; production and distribution of electric energy, natural gas, and water; manufacture of food products; textile and sewing production; construction; transport and communications. As these indicators are unified for all companies of Russia, and the official statistics provides their generalized values in the territorial aspect, they are used in this research for comparative analysis of external environment that determines the state of labor conditions.

Socio-economic factors that determine the territorial peculiarities of the state of labor conditions include the relations that emerge between employees, employers, and the society on the whole as to formal and informal execution of the normative requirements. In the official statistics this group is not determined, so this research studies its influence only in general features.

In the course of the analysis a task was set of determining the types of activities and risk factors for which the studied regions have the most vulnerable positions from the point of view of the state of labor conditions. This analysis was performed on the basis of the official statistics that characterizes the situation in the federal districts of the RF and Oblasts of the Central Federal District (CFD) [7].

2.1   Peculiarities of the State of Labor Conditions of the Employees in View of the Federal Districts of Russia (as of Year-End 2016)

Comparative analysis of the general situation that characterizes the state of labor conditions in various territorial entities of the RF was conducted at the level of federal districts. The task of this stage was to determine the peculiarities in the labor conditions in various regions on the whole and as to the above sectorial groups.

The general situation that characterizes the state of labor conditions in the federal districts of the RF regardless of the sectorial groups and the factors of hazardous influence is shown in Fig. 1. It should be emphasized that the regions are different as to the indicators of average Russian value (38.5% of the employees work in abusive and (or) hazardous labor conditions.

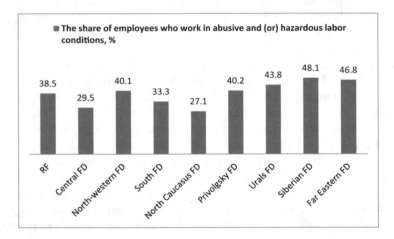

Fig. 1. The share of employees who work in abusive and (or) hazardous labor conditions, conduct activities in agriculture, hunting, forestry, extraction of minerals, processing productions, production and distribution of electric energy, natural gas and water, construction, and transport and communications in federal districts of the RF (as of year-end 2016)

Figure 1 brings us to the conclusion of a large scatter of the indicator that characterizes the share of the employed in the abusive and (or) hazardous labor conditions, as compared to the average Russian level (the lowest indicators – 27.1%; 29.5% of the employed against the highest indicators 46.8%; 48.1% in the corresponding federal districts). The share of regions in which the studied indicator is below the average indicator (the Central and North Caucasus Federal Districts) account for 27.0% of the average number of employees of the RF. In its turn, in the Urals, Far Eastern, and Siberian Federal Districts, where the indicator is higher than the average Russian indicator, 31.1% of all employees of the RF are employed.

Further analysis of territorial peculiarities of employment in the abusive and (or) hazardous labor conditions supposes their determination at the sectorial level. The existing statistics allows doing this for the sectorial groups (Table 1).

Table 1. Share of the employed in abusive and (or) hazardous labor conditions who conduct activities for separate sectorial groups in the federal districts of the RF (2016), %

Sphere	RF	Central FD	Northwestern ФО	South FD	North Caucasus FD	Privolgsky FD	Ural FD	Siberian FD	Far Eastern FD
Agriculture, hunting	30.8	24.3	35.5	30.5	26.3	31.5	43.4	35.2	31.3
Forestry	31.2	18.5	43	11	5.5	17.6	15.8	41.2	35.8
Processing productions	55.6	67.6	59.7	52.4	52.7	57.4	42.9	69.9	58.8
Processing productions	42.2	31	43.8	38.8	30.5	45.9	52.5	51.3	48.5
Food productions	31.9	28.9	32.7	31.7	29.4	35.9	29.2	33.1	42
Textile and sewing productions	29	28.3	31.5	37.8	12.3	27.5	10.6	29.3	16.9
Construction	37.9	29.6	37.9	32.4	28.9	42.7	39.7	45.3	49.3
Transport and communications	31.1	29.2	33.3	27.3	24.9	28.6	30.8	38.2	38

It seems that the causes of the differences for the sectorial groups between the federal districts are related to the peculiarities of the used technologies (technological factors) and geographic conditions (natural and climatic factors).

Comparative analysis of the role of various factors of the production environment and the labor process (technological factors) showed that the influence of noise, air ultrasound, and infrasound, and work load are most popular in Russia.

The "anti-leaders" as to the share of the employed influenced by noise, air ultrasound, and infrasound (the most popular unfavorable factor of the production process) are the Siberian (23.3% of the employed), Far Eastern (22.5%), and Ural (22.3%) federal districts. The smallest share of the employed is subject to the influence of this factor in the North Caucasus Federal District (10.1%). Analyzing the share of the workers employed in the conditions of large work load – which is the second most widespread unfavorable factor – it is possible to note that the largest share of the employed in the conditions of this factor is observed in the Siberian Federal District (24.1%), the smallest share – in the Central Federal District (12%).

The peculiarities of the role of separate factors that determine abusive (hazardous) conditions of labor in the sectorial view are as follows:

- of all the employed in agriculture, the largest share of the workers in the unfavorable conditions is observed in the Ural Federal District (40.8%), the smallest share – in the Central (24%) and the North Caucasus (25.7%) Federal Districts. The most widespread unfavorable factors are noise, air ultrasound, infrasound, and chemical factor, which requires special attention within the measures on management of labor safety.
- mineral production is peculiar for the large share of the employed in the abusive and (or) hazardous labor conditions, which constitutes 55.6% in Russia. The average Russian indicator is exceeded in the Siberian (69.9%) and the Central (67.6%) Federal Districts. It is rather high in the North Western (59.7%), Far Eastern (58.8%), and Privolgsky Federal Districts (57.4%). This situation could be

explained by the sectorial specifics and geographic location of the companies of this group of spheres.

- in the sector of production and distribution of electric energy, natural gas, and water, the largest share of the workers in the abusive and hazardous conditions is observed in the Siberian (54.5%) Federal District, the smallest – in the North Caucasus Federal District (28.4%).
- in processing industries, the maximum share of the workers in the hazardous conditions is observed in the Ural Federal District (52.5%). This district is also the "anti-leader" for such significant factors as influence of noise, ultrasound, and infrasound (29.9%), as well as work load (24.3%). The smallest share of workers in hazardous conditions – as to the level of noise, ultrasound, and infrasound – is observed in the North Caucasus Federal District (30.5% and 13.1%, accordingly). The minimum share of the employed in the analyzed sector who work in the hard labor conditions is observed in the Central Federal District (11.3%).
- in the construction sector, 49.33% of the workers of the Far Eastern Federal District work under the influence of hazardous conditions. This is the maximum value of this indicator for the whole country (with the average Russian value 37.94%). The best situation is observed in the North Caucasus (28.92%) and the Central Federal Districts (29.62%). The most significant unfavorable factor for whole Russia and for all federal districts in this sectorial group is work load of the labor process: 20% of all workers experience the influence of this factor. The second most widespread factor is noise, ultrasound, and infrasound.
- in the transport sector, 39.29% of the employed work under the influence of abusive and (or) hazardous conditions. The largest share is observed in the Siberian and Far Eastern Federal Districts (47.99% and 45.7%, accordingly). The smallest share is observed in the Ural (36%) and South (36.6%) Federal Districts. The leaders are such unfavorable factors as work load (the largest share of the employees subject to this factor is observed in the Siberian Federal District – 22.71%, the smallest share – in the Central Federal District – 13.5%); noise, ultrasound, and infrasound (the largest share of the employees subject to this factor is observed in the Far Eastern Federal District – 21.8%, the smallest share – in the Central Federal District – 10.95%).
- the communication sector is peculiar for a small share of the employees working in hazardous conditions. It varies from 2.65% in the South Federal District to 7.98% in the Urals Federal District. The main reason for that is the technological specifics of the sectorial group.

The received results allow for the following conclusion: at the level of federal districts there are substantial differences of the influence of factors that determine the state of labor conditions of employees; at that, the main factors belong to the natural & climatic and technological. The smallest sectorial differences between federal districts are observed if the sphere in the country is peculiar for a relative homogeneity of technological processes (the example with the technological group "communications").

2.2   The State of Labor Conditions of Employees in the Regions of the Central Federal District (as of Year-End 2016)

In order to identify the peculiarities of the influence of abusive and (or) hazardous factors that are peculiar for specific regions, as compared to the situation in the federal districts, it is necessary to conduct analysis of the results of statistical observations at the level of a separate federal district and spheres. This reduces the influence of the natural & climatic factor on the regional peculiarities of the studied influences and increases the role of the technological and socio-economic characteristics of specific regions. For this purpose, the statistical data that characterize the influence of abusive and (or) hazardous factors on employees in the Central Federal District and its oblasts were studied (Fig. 2).

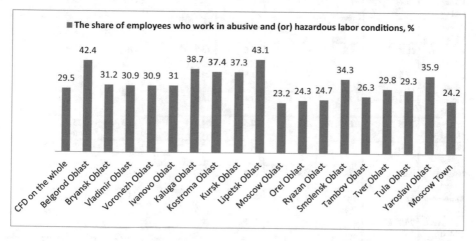

Fig. 2. The share of employees who work in abusive and (or) hazardous labor conditions, conduct activities in agriculture, hunting, forestry, extraction of minerals, processing productions, production and distribution of electric energy, natural gas and water, construction, and transport and communications for the regions of the Central Federal District of the RF (as of year-end 2016).

The least "hazardous" – from the point of view of labor conditions (Fig. 2) are Moscow, Orel, and Ryazan Oblasts, and Moscow, and the most "hazardous" are Belgorod and Lipetsk Oblasts. This is due to the structural peculiarities of these regions' economies and their geographical peculiarities (technological and natural & climatic factors in our classification).

Table 2 shows the general results of comparison of the oblasts of the CFD (excluding Moscow) as to the indicator "share of workers in the abusive and (or) hazardous labor conditions" in the sectorial view. It should be noted that all spheres are rather close as to the studied indicator in sectorial groups "processing productions" and "transport and communications". Large differences and the presence of indicators that deviate from the average values are peculiar for the following sectorial groups:

Table 2. Share of workers in abusive and (or) hazardous labor conditions in the oblasts of CFD for large groups of spheres, 2016.

Oblasts of the CFD	Agriculture, hunting	Mineral extraction	Processing industries	Food production	Textile and sewing industries	Construction	Transport and communications
CFD on the whole	24	67.6	31	28.9	28.3	29.6	28.2
Belgorod Oblast	24.1	74.6	49.9	46.1	2.6	46	25.1
Bryansk Oblast	17.9	94.7	41	35.2	19.4	31.6	27.3
Vladimir Oblast	35	58.8	31.9	29.5	34.6	41.9	29.3
Voronezh Oblast	27.6	59.3	32.7	29.6	20.9	38	27
Ivanovo Oblast	29.2	57	39.4	24.1	42.6	11.5	15.8
Kaluga Oblast	35.4	66.1	44.1	29.4	20.3	31.8	24.6
Kostroma Oblast	26.4	54.8	43.5	34.6	45.9	17.3	31.9
Kursk Oblast	23.6	82.9	32.8	32.9	24.4	45.9	27.4
Lipetsk Oblast	31.1	65.3	52	32.1	15.9	49.1	27
Moscow Oblast	20.6	21.8	23.8	22.5	24.2	18.3	25.2
Orel Oblast	15.9	–	31.5	24.3	7.2	35.3	20.6
Ryazan Oblast	4.2	45.4	25.7	15.8	7.2	30.3	27.4
Smolensk Oblast	18	8.8	35.3	44.6	22.2	35	33.1
Tambov Oblast	25.7	–	30.1	27.8	26.1	38	25.9
Tver Oblast	18.7	–	35	30.4	36.1	21.5	24.5
Tula Oblast	21.4	48.7	31.6	23.9	6.3	36.8	25.2
Yaroslavl Oblast	36.9	42.7	37.4	33.7	38.2	46.4	30.3

- agriculture and forestry, hunting: the share of workers who work in abusive and (or) hazardous conditions is much lower in Ryazan, Orel, and Bryansk Oblasts. Such main factor of the production environment as "noise" is not observed here, and the influence of the "work load" factor of the labor process is much lower. If the levels of development of the agricultural production technologies in the oblasts of the CFD are more or less equal, the average shares of the workers at hazardous productions are the results of influence of socio-economic factors that are not taken into account by statistics;
- processing productions: the oblasts of the CFD have different types of processing productions or lack companies belonging to this group of spheres (Orel, Tver, and Tambov);

- food productions: in this group, high indicators are peculiar for Belgorod and Smolensk Oblasts – at that, the structure of food industry of these regions is traditional for the CFD, as well as the technologies used; at the same time, the factors "noise" and "work load" in these oblasts are higher than the average indicator for the CFD; the reason of deviations lies in the socio-economic factors that are not considered by statistics;
- textile and sewing production: the lowest indicators are peculiar for Belgorod, Ryazan, Orel, and Tula Oblasts, which is peculiar for the factors of the production environment and labor processes;
- construction: in the range of oblasts (Belgorod, Kursk, Lipetsk, and Yaroslavl) there is substantial increase of indicator as compared to the average indicators for the CFD; it seems that these oblasts use obsolete technologies in this group of spheres.

3   Conclusions and Recommendations

There's a certain differentiation as to the quantity and quality of employment in abusive and hazardous labor conditions between various regions of Russia, which is caused by objective reasons (natural & climatic peculiarities and peculiarities of the sectorial structure of the studied groups of sectors as to specific districts). Processing industry is the most vulnerable one from the point of view of the state of labor conditions – on average, 55.6% of the workers are in abusive and (or) hazardous conditions. At the same time, as for homogeneous – from the point of view of the natural & climatic and technological factors that influence the labor conditions – regions, there are also substantial differences, which are explained by the influence of socio-economic factors that are to be taken into account during formation of the regional programs of control over labor safety, which will raise the effectiveness of the system of regional management labor safety.

Management of control functions should be differentiated as to companies not only depending on the complexity of their structure and level of implementation of the labor safety processes but also in view of regional and sectorial peculiarities of the production processes organization.

As the performed analysis showed, out of three groups of factors distinguished in this study (natural & climatic, technological, and socio-economic) that influence the state of labor conditions, the least studied and statistically uncontrolled are external socio-economic conditions and institutes that determine the environment of functioning of companies as the main economic subjects and form regional peculiarities of the state of labor conditions of employees, at which the regional programs of labor quality control are oriented. Expert technologies should be used for evaluating the influence of this group of factors.

Acknowledgments. The study was conducted with the financial support of the Russian Foundation for Basic Research (project No. 16-06-00535-a).

References

1. Azarnova, T.V.: The method of processing of the expert and statistical information during determining the optimal intensity of planned control over organizations in the sphere of labor safety. In: The Theory of Active Systems: Works of the International Scientific and Practical Conference, pp. 254–261. IPU RAS (2007)
2. Azarnova, T.V.: Optimization of management by the labor market and population's employment in the region on the basis of the models of the functional and marketing effectiveness and quality: doctoral thesis, Voronezh, 13 May 2010. 395 p.
3. Gandziuk, M.P., Kupchik, M.P., et al.: Foundations of Labor Safety, pp. 68–70. Osnova, Kyiv (2000). 416 p.
4. Drozhchanaya, O.U.: Improvement of management over labor safety at a company. Bull. Kurgan SAA **4**, 29–31 (2013)
5. Kerb, L.P.: Foundations of labor safety. KNUE, Kyiv, pp. 18–20 (2003). 215 p.
6. Krasnoshchekova, E.A.: Methodologies of evaluating the socio-economic state of labor safety at Russian companies. Bull. Saratov State Tech. Univ. **2**(1–55), 279–283 (2011)
7. The state of labor conditions of employees that conduct activities in agriculture, hunting, forestry, mineral production, processing production, production and distribution of electric energy, natural gas and water, construction, and transport and communications in the RF in 2016. Federal State Statistics Service (Rosstat), Chief transregional center (CTC), vol. II, pp. 1–3 (2016)
8. Spatar, E.V.: Evaluation of labor safety with various methods. In: Equipment, Technologies, Engineering, vol. 1, pp. 5–9 (2016)
9. Federal law "On special evaluation of labor conditions" dated December 28, 2013, No. 426-FZ. http://www.consultant.ru/document/cons_doc_LAW_156555/

Strategy of Risk Management in the Process of Formation of Innovations-Oriented Regional Economy

Larisa Kargina[1]([✉]), Sofia L. Lebedeva[2], and Olga S. Semkina[3]

[1] Moscow State University of Railways Engineering, Moscow, Russia
Larisa-kargina@yandex.ru
[2] Institute of Economics and Finance,
Moscow State University of Railways Engineering, Moscow, Russia
Sl-lebedeva@mail.ru
[3] Financial University under the Government of the Russian Federation,
Moscow, Russia
Semkina@yandex.ru

Abstract. The purpose of the article is to develop the strategy of risk management in the process of formation of innovations-oriented regional economy. Volgograd Oblast has been selected as the objects of the research. For assessing the influence of risk component on the process of formation of innovations-oriented economy in Volgograd Oblast, the authors study intermediary results of implementation of the Strategy of socio-economic development of Volgograd Oblast until 2025. They are studied with the help of the methodology of time series analysis. The main conclusion of the research is that in addition to high risk level, which accompanies innovational activity, additional risks emerge in the process of formation of innovations-oriented regional economy. This leads to high risk component, which is a serious obstacle in this process. The authors perform assessment of the risk component on the process of formation of innovations-oriented regional economy by the example of Volgograd Oblast, determine the key risks that emerge in the process of formation of innovations-oriented regional economy, and develop the strategy of risk management in the process of formation of innovations-oriented regional economy.

Keywords: Risk management · Innovations-oriented economy
Risks of innovations · Regional economy

1 Introduction

Under the pressure of global competition, the tendency for formation of innovations-oriented economy was distributed to the regional level of economic systems. Russia's regions adopted the strategy of innovational socio-economic development, but the first results were not satisfactory. Instead of quick growth and innovational development of regional economy, the first half of the time row, given to implementation of these strategies, was marked by ineffective spending of allocated assets of the federal and regional budgets.

© Springer International Publishing AG, part of Springer Nature 2018
E. G. Popkova (Ed.): HOSMC 2017, AISC 622, pp. 591–596, 2018.
https://doi.org/10.1007/978-3-319-75383-6_75

In this research, the authors offer the hypothesis that a serious restraining factor on the path of implementation of strategies of modern Russia's regional economy's innovational development is high risk component. Our goal in the context of this article consists in development of the strategy of risk management in the process of formation of innovations-oriented regional economy.

The object of the research is Volgograd Oblast, as it presents the average statistical Russian region, being neither the donor not the recipient. The set goal should be achieved in the context of successive solution of the following main practical tasks:

- assessment of influence of the risk component on the process of formation of innovations-oriented regional economy by the example of Volgograd Oblast;
- determination of the key risks that emerge in the process of formation of innovations-oriented regional economy;
- development of the strategy of risk management in the process of formation of innovations-oriented regional economy.

2 Materials and Method

For assessment of influence of the risk component on the process of formation of innovations-oriented economy in Volgograd Oblast, the authors study intermediary results of implementation of the Strategy of socio-economic development of Volgograd Oblast until 2025 (Table 1).

Table 1. Dynamics of results of implementation of the Strategy of socio-economic development of Volgograd Oblast until 2025

| | Values of indicators for years, ratio of the current year to 2008, times | | | | | | | |
| | 2010 | | 2012 | | 2014 | | 2016 | |
	Goal	Result	Goal	Result	Goal	Result	Goal	Result
Volume of investments into fixed capital in compatible prices	1.5	1.0	2.0	0.9	2.7	0.7	3.5	0.6
Volume of expenditures for R&D at the end of the period, as compared to GRP	1.3	1.1	1.7	1.0	2.2	0.9	3.0	0.7
Number of innovational companies	1.6	1.1	2.1	1.5	2.8	1.9	3.7	2.1
Share of innovational companies in the total structure of business	1.2	1.1	1.4	1.2	1.9	1.4	2.3	1.9
Volume of innovational products	1.8	1.5	2.6	1.8	3.2	2.1	4.4	2.6
Share of innovational products in the total structure of production of goods	1.3	1.2	1.5	1.4	1.8	1.6	2.2	2.0

Source: compiled by the authors on the basis of Volgogradskaya Pravda (2008), Federal State Statistics Service (2016).

They are studied with the help of the methodology of time series analysis. Table 1 contains the results of trend analysis of dynamics of these indicators, on the basis of which plan-fact analysis is performed. During the research, the authors use such scientific methods as systemic and logical analysis, synthesis, deduction, induction, and graphic representation of data.

3 Discussion

The theory and practice of risk management of regional economy development are discussed in the works (Popkova 2013; Belov and Kravets 2013; Przhedetskaya 2014; Ragulina et al. 2015). Specifics of formation of innovations-oriented regional economy are studied in publication (Pradel-Miquel 2015, Wolfe and Gertler 2016; Wolfe and Bramwell 2016; Harris et al. 2013; Ibrahim et al. 2014; Shala et al. 2015; Veselovsky et al. 2015).

4 Results

The performed plan-fact analysis of dynamics of the results of implementation of the Strategy of socio-economic development of Volgograd Oblast until 2025 showed that the achieved results are below the planned results – at that, the difference between the targeted and received results grows with time.

The largest underrun is observed for the volume of investments into fixed capital (83% in 2016, as compared to 33% in 2010) and for the volume of expenditures for R&D as to GRP (77% in 2016, as compared to 16% in 2010). This is explained by increase of deficit of financial resources with companies and regional authorities, which was a reason for reduction of investment activity.

The largest underrun is observed for such indicators as the share of innovational products in the general structure of commodity production (9% in 2016, as compared to 7% in 2010) and the share of innovational companies in the total structure of business (17% in 2016, as compared to 8% in 2010). However, in view of absolute changes of these indicators (without connection to the general structure), the underrun of which constitutes 43% and 41% in 2016, accordingly, this shows reduction of general business activity in the region, not the increase of innovational activity of entrepreneurship.

Thus, inaccessibility of the set targeted values of the indicators of efficiency of the process of formation of innovations-oriented regional economy in Volgograd Oblast shows the influence of the risk component on this process. Its influence is seen at the corporate (micro) and the regional (meso) levels.

On the one hand, at the corporate level high risk component of innovational activity restrains innovational activity of entrepreneurial structures. It may be characterized in the following way. The state offers stimuli for manifestation of innovational activity ("tax holidays", subsidies, support for sales of innovational products, etc.).

However, these stimuli cannot be received in practice due to complex procedures of receipt of the announced benefits (large lines, large set of documents, constantly changing requirements, etc.) – institutional risk – or unexpected change of state innovational policy, due to increase of deficit of the regional budget – which leads to cancelling the state measures for stimulation of innovational activity – political risk.

On the other hand, at the regional level the risk component of state innovational policy has not been taken into account at the stage of its development, so there are unexpected risks at the stage of its implementation – which, without proper management, hinder the implementation of this strategy.

Thus, the measures for implementation of the regional innovational policy could be ineffective due to unfair attitude of business to innovative activities, related to copying of existing innovations from rivals or non-implementation of innovations with false accounting on implementation of innovations for receipt of state support – risk of false innovations – and/or due to low innovative activity of business due to lack of possibilities, when state support is not sufficient, and unattractiveness of such support – risk of low demand for stimuli.

It should be noted that this work does not view generally known risks that accompany innovative activity, as our task consists in determination of additional specific risks, peculiar for the process of formation of innovations-oriented regional economy. In this context, the risks are connected to non-execution of obligations by participants of the relations within the process of formation of innovations-oriented regional economy (entrepreneurship and state). We offer the following recommendations for management of the determined risks:

- developing the system of electronic documents for access to subsidies for innovational activity of business and advertising this system for increase of its reputation in the region's business circles;
- providing stability of reigon's innovational policy, for dissolving the doubts of entrepreneurship in regional authorities' plans and volume of provided stimuli for innovational activity of business;
- toughening control over innovational activity of business through establishment of clear criteria of its assessment and creation of a special control comission;
- conducting sociological surveys for determining the demand for state stimuli for innovational activity of entrepreneurship and collection of feedback (tips and wishes) for correcting the policy of such stimulation.

According to the above, we developed the following strategy of risk management in the process of formation of innovations-oriented regional economy (Fig. 1).

As is seen from Fig. 1, business identifies risks, evaluates them, and provides the gathered information to the state (regional authorities). In its turn, it identifies the risks and evaluates their importance through stress testing. If the current level of one, several, or all risks exceeds the set acceptable limits (appetite for risk), which is a function of the state, the corresponding measures in the sphere of managing these risks are implemented. As a result, risks are kept at the acceptable level, for them not to hinder formation of innovations-oriented economy in the region.

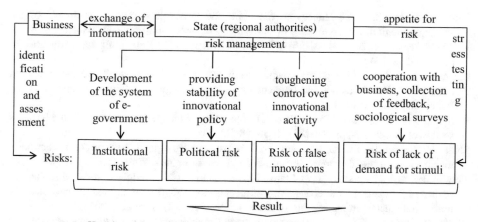

Fig. 1. Strategy of risk management in the process of formation of innovations-oriented regional economy. Source: compiled by the authors.

5 Conclusion

Thus, the main conclusion of the research is that in addition to a high level of risks that accompany innovational acitvity, additional risks appear in the process of formation of innovations-oriented regional economy. This leads to a high risk component, which is a serious obstacle in this process. For its normalization, a strategy of risk management in the process of formation of innovations-oriented regional economy is offered.

Despite the connection of the performed calculations for verifying the offered hypothesis in the economic practice of Volgograd Oblast, the authors' conclusions and recommendations are oriented at the Russia's regional economy on the whole, and the developed strategy of risk management in the process of formation of innovations-oriented regional economy is accessible for implementation in any region of modern Russia.

References

Belov, A.G., Kravets, A.G.: Business performance management in small and medium businesses and functional automation. World Appl. Sci. J. **24**(24), 7–11 (2013)

Harris, R., McAdam, R., McCausl, I., Reid, R.: Knowledge management as a source of innovation and competitive advantage for SMEs in peripheral regions. Int. J. Entrepreneurship Innov. **14**(1), 49–61 (2013)

Ibrahim, S., Sukeri, S.N., abd. Rashid, I.M. (2014). Factors influencing the diffusion & implementation of management accounting innovations (MAIS), Malaysian manufacturing industries in Northern Region. Adv. Environ. Biol. **8**(9, Special Issue 4), 504–512 (2014)

Popkova, E.G.: Marketing strategy to overcome the "underdevelopment whirlpool" of the Volgograd region. In: Conference of the Eurasia-Business-and-Economics-Society (EBES), 12–14 September 2013. Russian Acad. Sci., Inst. Econ., Ural Branch, Ekaterinburg, Russia, pp. 52–61 (2013)

Pradel-Miquel, M.: Making polycentrism: governance innovation in small and medium-sized cities in the West Midlands and Barcelona metropolitan regions. Environ. Planning C: Gov. Policy 33(6), 1753–1768 (2015)

Ragulina, Y.V., Stroiteleva, E.V., Miller, A.I.: Modeling of integration processes in the business structures. Modern Appl. Sci. 9(3), 145–158 (2015)

Shala, M., Hajrizi, E., Hoxha, V., Stapleton, L.: Cost-oriented agile innovation for mechatronics management in less developed regions. IFAC-Papers Online 48(24), 150–152 (2015)

Veselovsky, M.Y., Pogodina, T.V., Idilov, I.I., Askhabov, R.Y., Abdulkadyrova, M.A.: Development of financial and economic instruments for the formation and management of innovation clusters in the region. Mediterr. J. Soc. Sci. 6(3), 116–123 (2015)

Wolfe, D., Bramwell, A.: Innovation, creativity and governance: social dynamics of economic performance in city-regions. Innov. Manage. Policy Pract. 18(4), 449–461 (2016)

Wolfe, D.A., Gertler, M.S.: Growing urban economies: Innovation, creativity, and governance in Canadian city-regions. In: Growing Urban Economies: Innovation, Creativity, and Governance in Canadian City-Regions, pp. 1–420 (2016)

Volgogradskaya Pravda: The Law of Volgograd Oblast, 21 November 2008, No. 1778-od "Regarding the strategy of socio-economic development of Volgograd Oblast until 2025" (2008). http://docs.cntd.ru/document/819076044. Accessed 23 June 2017

Przhedetskaya, N.V.: Designing marketing management of the innovational model of education in the conditions of development economy. Bull. Tula State Univ. 4(1) (2014)

Federal State Statistics Service; Regions of Russia. Socio-economic indicators, 2016: Statistical Bulletin. Federal State Statistics Service, Moscow (2016)

Systemic Contradictions in Development of Modern Russia's Industry in the Conditions of Establishment of Knowledge Economy

Aleksei V. Bogoviz[1](✉) ⒾⒹ, Yulia V. Ragulina[1] ⒾⒹ,
and Natalia V. Sirotkina[2]

[1] Federal State Budgetary Scientific Institution
"Federal Research Center of Agrarian Economy and Social Development
of Rural Areas—All Russian Research Institute of Agricultural Economics",
Moscow, Russia
aleksei.bogoviz@gmail.com, julra@list.ru
[2] Voronezh State University, Voronezh, Russia
docsnat@yandex.ru

Abstract. The purpose of the work is to determine and analyze systemic contradictions in development of modern Russia's industry in the conditions of establishment of knowledge economy and to determine the perspectives of overcoming them. In order to achieve it, the authors use the method of comparative analysis, which allows comparing the values of indicators of industry's development in Russia and in the countries with developed knowledge economy and evaluating the level of their deviation. Based on study and content analysis of authoritative scientific literature on the topic of the research, the key characteristics of industry that determine the basis of knowledge economy are distinguished and the evaluation of their correspondence to modern Russia's industry is performed. As a result, the authors conclude that Russia's industry is developing in the direction that is opposite to the set target course on creation of knowledge economy. The authors offer recommendations and perspectives for overcoming the determined contradictions in development of modern Russia's industry in the conditions of knowledge economy development.

Keywords: Industry · Knowledge economy · High-tech spheres of economy
Modern Russia

1 Introduction

Modern Russia presents itself as one of the leading countries of the world, striving to and capable of influencing the global economic processes and be a leader and model for other, less developed countries. For strengthening of this status, the Russian government takes active measures for provision of global competitiveness of the national economic system, the most important conditions for which is creation of knowledge economy.

© Springer International Publishing AG, part of Springer Nature 2018
E. G. Popkova (Ed.): HOSMC 2017, AISC 622, pp. 597–602, 2018.
https://doi.org/10.1007/978-3-319-75383-6_76

However, these measures are not effective, which is a reason for knowledge economy being at the initial stage of formation in Russia. This leads to the problem of deprivation, related to the fact that Russia has a large potential of development (large territory and resources), which is a reason for its claims for the global authority and which determines close attention from other countries, while this potential has not been realized over the recent decades.

Difference between expectations and reality, deepened by the attempts of the Russian government to present facts in a more profitable light, leads to mistrust to Russia and instead of leadership in the global economic system leads to its becoming an outsider. One of the essential differences of knowledge economy from post-industrial economy is refusal from full specialization in the service sphere and founding on the real sector, which allows supporting sustainability of an economic system. This determines central position of industry in knowledge economy.

That's why industry was selected as the object of the research. Our hypothesis is that systemic contradictions in development of industry hinder the establishment of knowledge economy in modern Russia. The purpose of the work is to determine and analyze systemic contradictions in development of industry of modern Russia in the conditions of establishment of knowledge economy and determination of perspectives for overcoming them.

2 Materials and Method

Based on studying and content analysis of authoritative scientific literature on the topic of the research, we determined two key characteristics of industry, which determine the basis of knowledge economy: large share of high-tech spheres in economy and high level of employment in these spheres, as well as high innovative activity of industrial enterprises. They are described in Anokhina et al. (2016), Bogoviz and Mezhov (2015), Dudukalov et al. (2016), Duman and Kurekovà (2016).

These characteristics were selected as criteria of evaluation of correspondence of industry of modern Russia to the standards of the countries with developed knowledge economy. Such standards are determined on the basis of studying the experience creation of knowledge economy in the countries of the world, reflected in materials of the research by Korobkin et al. (2015), Malyshkov and Ragulina (2014), Popkova (2017), Przhedetskaya and Akopova (2015). The information and analytical basis of the research is given in Table 1.

In order to verify the offered hypothesis and to determine the systemic contradictions in development of modern Russia's industry in the conditions of establishment of knowledge economy, the authors use the method of comparative analysis, which allows comparing the indicators of development of industry in Russia and the countries with developed knowledge economy and evaluating the value of their deviation.

Table 1. Indicators of development of industry in Russia in 2016 and standard values for the countries with developed knowledge economy

Indicators of development of industry	Values of indicators	
	In Russia	In countries with developed knowledge economy
Turnover of high-tech industrial enterprises, RUB billion	3,189.4	–
Share of high-tech spheres in the structure of industry, %	5.90	≥ 70
Share of high-tech spheres in economy, %	2.25	≥ 30
Share of industrial enterprises that conduct technological innovations, %	6.76	≥ 50
Share of industry in the structure of GDP, %	38.1	–
Share of involved in industry in the structure of employment, %	26.15	–
Level of employment among technical specialists, %	52.6	≤ 5

3 Discussion

Fundamental and applied issues of creation and development of knowledge economy in various countries of the world are studied in the works of such authors as Przhedetskaya and Panasenkova (2014), Shakirtkhanov (2017), Trindade et al. (2016), Veselovsky et al. (2017). The experience of modern Russia in creation of knowledge economy is studied in the works Khan (2017), Fujii and Managi (2016), Kinahan (2016).

Despite the detailed study of the concept of knowledge economy, industry – its important structural element and indicator of development – is not paid sufficient attention, which determines the necessity for further study of this topic in this direction.

4 Results

Based on analysis of data from Table 1, we determined three important systemic contradictions in development of industry of modern Russia in the conditions of knowledge economy. The firs one is related to weak development of high-tech spheres of economy. These include space, pharmaceutical, and telecommunication sphere, as well as production of computer equipment and high-precision and optical equipment.

Instead of domination in the structure of industry, high-tech spheres in Russia constitute 5.90%. Their share in the general structure of Russia's economy is very small and constituted 2.25% as of 2016. In the countries with developed knowledge economy, the share of high-tech spheres in economy exceeds 30%, and in industry – 70%. Turnover of high-tech companies of Russia constituted RUB 3,189.4 billion (3.71% of GDP) in 2016.

Another contradiction is low innovational activity of industrial enterprises of Russia. The share of Russian industrial enterprises that implement technological innovations constituted 6.76% in 2016. For comparison, in the countries with developed knowledge economy this indicator exceeds 50%. This shows strong underrun of Russia from the countries with developed knowledge economy as to this indicator.

The third contradiction consists in a low level of employment in the Russian industry. Despite the quick development of the service sphere, the structure of modern Russia's economy is peculiar for high share of industry (38.1%). Universities of Russia provide technical specialists for industrial companies each year. This is caused by technical specialization of a lot of Russian universities, which emerged in the period of industrial economy in the 20th century, and large state order for preparation of technical specialists due to prioritization of industry.

However, due to low innovational activity, Russian industrial companies lose competitiveness. As of now, a lot of them exist only due to state order and support from territorial authorities, balancing on the edge of bankruptcy. A lot of Russian industrial companies close each year. While in 2010 there were 898,400 industrial enterprises in Russia, their number in 2016 reduced by 40%, constituting 545,400.

This causes the problem of employment of technical specialists, the offer of which has been exceeding the demand. This leads to a high level of unemployment of among technical specialists, the level of which constituted 52.6% in 2016, while in the countries with developed knowledge economy the value of this indicator does not exceed 5%. The share of the employed in industry constitutes 26.15% in the structure of total employment in Russia.

In order to determine the means of overcoming the systemic contradictions in development of modern Russia's industry in the conditions of establishment of knowledge economy, it is necessary to analyze the reasons for their emergence. The main reason for emergence of the first contradiction is high market barriers for entering the high-tech spheres of economy. As private business cannot enter this market, the state has to show the initiative.

Development of high-tech spheres in Russia's economy should be performed on the basis of public-private partnership. Within this mechanism, the state initiates creation of high-tech companies, and private business provides investments for their development and conducts effective management of companies. This is done under thorough control of the regulating bodies of public authorities.

The key reason for the second and closely related third contradiction consists in deficit of financial resources for modernization of Russia's industrial enterprises. Overcoming it is seen in the context of development of the leasing mechanism, which allows updating equipment and technologies of industrial enterprises with a minimum rusk for the leasing organizations.

The sense of the process of overcoming the systemic contradictions in development of modern Russia's industry in the conditions of establishment of knowledge economy is reflected by Fig. 1.

As is seen from Fig. 1, as a result of implementation of the offered recommendations, the targeted (corresponding to the countries with developed knowledge economy) and factual indicators of economic system's development will be balanced in modern Russia. Elimination of systemic contradictions in development of industry will

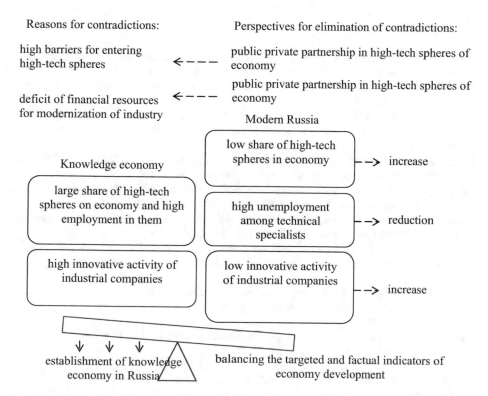

Fig. 1. The sense of the process of overcoming systemic contradictions in development of modern Russia's industry in the conditions of establishment of knowledge economy. Source: compiled by the authors.

set foundations bringing indicators of development of other spheres of national economy in correspondence with the targeted values of other spheres, thus stimulating the establishment of knowledge economy in modern Russia.

5 Conclusions

Thus, Russia's industry is developing in the direction opposite to the set course for development of knowledge economy. The results of the research showed three systemic contradictions in development of industry of modern Russia in the conditions of establishment of knowledge economy. They include low share of high-tech spheres in the structure of industry, innovational passivity of industrial enterprises, and high level of unemployment in industry.

The offered recommendations showed perspectives of overcoming these contradictions. At the same time, practical realization of the offered perspectives could be complicated due to institutional problems. That's why studying the process of development of industry of modern Russia in the conditions of establishment of knowledge economy with the help of the tools of the institutional theory is a perspective direction of further scientific research.

References

Anokhina, M., Zinchuk, G.M., Petrovskaya, S.A.: The development of the concept of economic growth of the agro-industrial complex. J. Internet Bank. Commer. **21**(3), 224 (2016)

Bogoviz, A., Mezhov, S.: Models and tools for research of innovation processes. Mod. Appl. Sci. **9**(3), 159–172 (2015)

Duman, A., Kurekovà, L.: The role of state in development of socio-economic models in Hungary and Slovakia: The case of industrial policy. Changing Models of Capitalism in Europe and the U.S., pp. 99–120 (2016)

Dudukalov, E.V., Rodinorova, N.D., Sivakova, Y.E., et al.: Global innovational networks: sense and role on development of global economy. Contemp. Econ. **10**(4), 299–310 (2016)

Fujii, H., Managi, S.: Economic development and multiple air pollutant emissions from the industrial sector. Environ. Sci. Pollut. Res. **23**(3), 2802–2812 (2016)

Khan, B.Z.: Human capital, knowledge and economic development: evidence from the British Industrial Revolution, 1750–1930. Cliometrica, 1–29 (2017)

Kinahan, K.L.: Design-based economic development: understanding the role of cultural institutions and collections of industrial and product design. Econ. Dev. Q. **30**(4), 329–341 (2016)

Korobkin, D., Fomenkov, S., Kravets, A., Kolesnikov, S., Dykov, M.: Three-steps methodology for patents prior-art retrieval and structured physical knowledge extracting. CCIS, vol. 535, pp. 124–136 (2015)

Malyshkov, V.I., Ragulina, Y.V.: The entrepreneurial climate in Russia: the present and the future. Life Sci. J. **11**(6), 118–121 (2014)

Popkova, E.G.: Guest editorial. Int. J. Educ. Manage. **31**(1), 2 (2017)

Przhedetskaya, N., Akopova, E.: Institutional designing of continuous education in Russia under the conditions of neo-economy and globalization. Reg. Sectoral Econ. Stud. **15**(2), 115–122 (2015)

Przhedetskaya, N.V., Panasenkova, T.V.: Business education: concept and evolution of development in knowledge economy. World Appl. Sci. J. **2**(1) (2014)

Shakirtkhanov, B.R.: Innovative development of industrial enterprises of Kazakhstan in the conditions of economic growth downturn. Int. J. Econ. Res. **14**(7), 121–133 (2017)

Trindade, J.R., Cooney, P., de Oliveira, W.P.: Industrial Trajectory and economic development: dilemma of the re-primarization of the Brazilian economy. Rev. Radical Political Econ. **48**(2), 269–286 (2016)

Veselovsky, M.Y., Khoroshavina, N.S., Bank, O.A., Suglobov, A.E., Khmelev, S.A.: Characteristics of the innovation development of Russia's industrial enterprises under conditions of economic sanctions. J. Appl. Econ. Sci. **12**(2), 321–331 (2017)

The Concept of Provision of Regional Economy's Global Competitiveness

Viktor P. Khorev[(⊠)], Alexandr I. Tarasov, and Sergey A. Golubtsov

Military University of the Ministry of Defense of the RF,
Moscow, Russian Federation
horev52@mail.ru, tarasov.a.i@mail.ru,
golubcov2008@mail.ru

Abstract. The purpose of the article is to verify the offered hypothesis and to develop the concept of provision of regional economy's global competitiveness by the example of modern Russia. For that, the authors determine competitiveness of Russian regions and conduct statistical analysis, aimed at determination of differences, their depth and dynamics of change. The proprietary methodology of transferring the national values of the competitiveness index into global values is used. As a result of the research, the authors come to the conclusion that modern Russian regions, despite the national integrity, are characterized with a high level of differentiation. On the one hand, their peculiarities emphasize their uniqueness, and, on the other hand, they cause different levels of attractiveness from the point of view of territorial marketing – for residents, investors, employees, entrepreneurs, tourists, etc. This is a reason for quick development of certain regions due to attraction of larger volume of resources of all types, while other regions are behind them as to the level of socio-economic development. This increases disproportions in development of the Russian regional economy, which is a serious problem, influencing the rate of the national economic growth and social development. In order to solve this problem, the authors develop the concept of provision of global competitiveness of the regional economy.

Keywords: Global competitiveness · Regional economy
Territorial marketing · Modern Russia

1 Introduction

In the modern global economy, tough competitive struggle take place at all levels of economic systems, including the regional level. Despite the turnover of integration and disintegration processes in the global economic space, it is already formed and is an integrated system, the structural elements of which grow or diminish.

Large diversity of economic systems, caused by their specific peculiarities due to natural and socio-economic factors, determines not only the expedience but the necessity for using the advantages of access to the global flows of resources, which are viewed in a wide sense, including material, financial, technological, human, etc.

The working hypothesis of this research is formulated in the following way. The regions of modern Russia, despite the national integrity, are characterized with a high

E. G. Popkova (Ed.): HOSMC 2017, AISC 622, pp. 603–608, 2018.
https://doi.org/10.1007/978-3-319-75383-6_77

level of differentiation. Their peculiarities, on the one hand, emphasize their uniqueness, and, on the other hand, cause the different levels of attractiveness from the point of view of place marketing – for residents, investors, employees, entrepreneurs, tourists, etc.

This is a reason for quick development of certain regions due to attraction of larger volume of resources of all types, while other regions are behind them as to the level of socio-economic development. This increases disproportions in development of the Russian regional economy, which is a serious problem, influencing the rate of national economic growth and national development. The purpose of the article is to verify the offered hypothesis and to develop the concept of provision of regional economy's global competitiveness by the example of modern Russia.

2 Materials and Method

In order to verify the offered hypothesis, the authors determine competitiveness of Russia's regions and conduct statistical analysis aimed at determining the differences, their depth and dynamics of change. For that, such indicators as dispersion, standard deviation, and variation coefficient are calculated in Microsoft Excel.

The information and analytical basis of the research consists of the materials of the ranking of competitiveness of Russia's regions for 2013 (the first year of the ranking) and 2016 (the most topical information) according to AV Group, as well as the materials of the ranking of global competitiveness for 2013 and 2016 according to the World Economic Forum.

As the index of competitiveness of Russia's regions is national, in order to determine their global competitiveness, the authors use the proprietary methodology of transferring the national values of the competitiveness index onto the global values. For that, the percentage ratio (PR) of all indicators of competitiveness in the national scale is determined (their share as to the leading region in the national ranking of competitiveness is calculated).

Then, the value of the competitiveness index in points is calculated – as the share of the national competitiveness index (SH). After that, the level of global competitiveness (GC) of the region is calculated, as the share of the highest possible value of the global competitiveness index.

Three regions from the conventional categories were selected as the objects of the research:

- regions that are leaders in the national ranking: Krasnodar Krai, Moscow Oblast and the Republic of Tatarstan;
- regions with medium place in the national ranking: the Republic of Komi, Vladimir Oblast, and the Chuvash Republic;
- regions that are outsiders in the national ranking of competitiveness: Kostroma Oblast, the Republic of Adygea, and Kurgan Oblast.

At that, statistical analysis is conducted for the whole mass of data, which includes 83 regions of Russia. The initial data for calculations are given in Table 1.

Table 1. Dynamics of the values of index of competitiveness of Russia's regions (in the national scale) and the index of global competitiveness of Russia in 2013–2016

Regions of Russia	2016		2013	
	Competitiveness index	Place in the regions' ranking	Competitiveness index	Place in the regions' ranking
Krasnodar Krai	4.11	2	2.73	6
Moscow Oblast	4.01	3	2.98	2
Republic of Tatarstan	3.87	4	2.62	9
Republic of Komi	1.40	42	2.01	40
Vladimir Oblast	1.36	43	2.02	39
Chuvash Republic	1.31	44	1.85	51
Kostroma Oblast	0.48	72	1.55	68
Republic of Adygea	0.27	74	1.27	76
Kurgan Oblast	0.06	78	1.43	73
Russia	4.50	43	4.25	64

Source: compiled by the authors on the basis of: (AV Group 2017; World Economic Forum 2017).

3   Discussion

Various aspects of provision of regional economy's global competitiveness of the theoretical, methodological, and applied nature are reflected in multiple studies by such representatives of foreign and Russian science as (Popkova 2013; Dudukalov et al. 2016; Belov and Kravets 2013; Przhedetskaya 2014; Ragulina et al. 2015; Wolfe and Bramwell 2016; Veselovsky et al. 2015).

4   Results

The national values of the indices of competitiveness of Russia's regions, shown in Table 1, are transferred into the global scale in Table 2.

As is seen from Table 2, such short period as three years showed substantial increase of differentiation of the level of global competitiveness of Russia's regions. Thus, for example, global competitiveness of Krasnodar Krai, which is one of the leaders in the national ranking of Russia, grew from 57.55% to 61.65%, that is its absolute growth constituted 4.1%, and relative growth constituted 7.1%.

Table 2. Transfer of national values of the competitiveness index into the global scale

Regions of Russia	2016			2013		
	PR	SH	GC	PR	SH	GC
Krasnodar Krai	0.822	3.69	61.65	0.81	3.45	57.55
Moscow Oblast	0.802	3.61	60.15	0.89	3.77	62.82
Republic of Tatarstan	0.774	3.48	58.05	0.78	3.31	55.23
Republic of Komi	0.28	1.26	21.00	0.59	2.54	42.37
Vladimir Oblast	0.272	1.22	20.40	0.60	2.55	42.58
Chuvash Republic	0.262	1.18	19.65	0.55	2.34	39.00
Kostroma Oblast	0.096	0.43	7.20	0.46	1.96	32.68
Republic of Adygea	0.054	0.24	4.05	0.38	1.61	26.77
Kurgan Oblast	0.012	0.05	3.75	0.42	1.80	30.14

Source: compiled by the authors

At that, global competitiveness of Kurgan Oblast, which is one of the outsiders in the Russian national ranking, dropped from 30.14% to 3.75%, that is its absolute growth constituted − 26.39%, and relative growth constituted − 88%. The estimate statistical indicators, obtained in the course of analysis of the selection for all Russian regions, are shown in Table 3.

Table 3. Estimate statistical indicators, obtained in the course of analysis of selection for all Russian regions

Estimate statistical indicators	2016	2013
direct average	1.67	1.62
dispersion	1.16	1.12
standard deviation	1.07	1.03
coefficient of variation	0.65	0.64

Source: compiled by the authors.

As is seen from Table 3, 2016 marked increase of direct average of the values of global competitiveness indices of Russia's regions (1.67) as compared to 2013 (1.62). Simultaneously, dispersion grew from 1.12 to 1.16, standard deviation grew from 1.03 to 1.07, and variation coefficient grew from 0.64% to 0.65%. This confirms the working hypothesis of this research and shows the necessity for development of the concept of provision of regional economy's global competitiveness.

The purpose of this concept, which consists in provision of global competitiveness of regional economy, in the conditions of global competition should be realized with the help of marketing tools – which acquire the form of place marketing at the regional level. The key direction of achievement of the set goal is provision of region's uniqueness with emphasis on its absolute and relative competitive advantages. This is necessary for distinguishing it against the background of other regions.

A supporting direction, which ensures reliable basis for the offered concept, is supporting stability, predictability, and manageability of the processes that take place in the region's economic system. This will allow minimizing the risk component in the regional economy, thus increasing its attractiveness for the targeted audience of place marketing: residents, investors, employees, entrepreneurs, tourists, etc.

At the top of implementation of the offered concept we see supporting the sustainability of development of regional economy, which supposes achievement of a high rate of economic growth, social well-being, development and justice, as well as protection of environment and low ecological cost of development of entrepreneurship in the region. The developed concept is shown in Fig. 1.

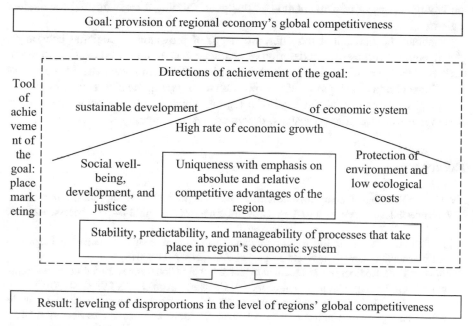

Fig. 1. The concept of provision of regional economy's global competitiveness Source: compiled by the authors.

As is seen from Fig. 1, the main directions of achievement of the goal, which is set within the offered concept, form a building with strong foundations, attractive façade, and reliable roof. That is, in terms of philosophy, it is possible to compared provision of region's global competitiveness with construction of a building. In both cases, there's a need for responsible and well-thought approach, oriented at long-term result. Realization of the offered concept leads to leveling of disproportions in the level of regions' global competitiveness.

5 Conclusions

It should be noted that the developed concept of provision of regional economy's global competitiveness was created for modern Russia, but it is universal and could be applied in any other country. However, the differences will be manifested in the methods of implementation of these directions of achieving the goal of the concept.

Another advantage of the offered concept is that despite the short character of formulations and seeming simplicity, it is rather detailed, allowing building an idea on the sense and logic of provision of regional economy's global competitiveness.

As compared to the existing analogs, the offered concept is not limited by principles of provision of regional economy's global competitiveness but provides the tools and the key directions of implementation of this goal, forming a clear manual for modern regions.

It should be noted that the offered concept of provision of regional economy's global competitiveness is of descriptive character and required adaptation to peculiarities of each specific region in which it is to be realized. In particular, it is necessary to pay attention to development of the approach to managing the elements of marketing mix of place marketing at the level of a region, which is a top-priority direction for development of scientific provisions, described in this article.

References

AV Group: Ranking of competitiveness of Russia's regions (2017). http://forumstrategov.ru/UserFiles/Files/LC-AV_%20RCI-16_Growth%20Poles%20161022-04.pdf. Accessed 28 July 2017

Belov, A.G., Kravets, A.G.: Business performance management in small and medium businesses and functional automation. World Appl. Sci. J. **24**(24), 7–11 (2013)

Dudukalov, E.V., Rodinorova, N.D., Sivakova, Y.E., et al.: Global innovational networks: sense and role on development of global economy. Contemp. Econ. **10**(4), 299–310 (2016)

Popkova, E.G.: Marketing strategy то overcome the "underdevelopment whirlpool" of the Volgograd region. In: Conference of the Eurasia-Business- and-Economics-Society (EBES), 12–14 September 2013, pp. 52–61. Russian Acad. Sci., Inst. Econ., Ural Branch, Ekaterinburg, Russia (2013)

Ragulina, Y.V., Stroiteleva, E.V., Miller, A.I.: Modeling of integration processes in the business structures. Modern Appl. Sci. **9**(3), 145–158 (2015)

Veselovsky, M.Y., Pogodina, T.V., Idilov, I.I., Askhabov, R.Y., Abdulkadyrova, M.A.: Development of financial and economic instruments for the formation and management of innovation clusters in the region. Mediterr. J. Soc. Sci. **6**(3), 116–123 (2015)

Wolfe, D., Bramwell, A.: Innovation, creativity and governance: Social dynamics of economic performance in city-regions. Innov. Manage. Policy Pract. **18**(4), 449–461 (2016)

World Economic Forum. The Global Competitiveness Report 2016–2017 (2017). https://www.weforum.org/reports/the-global-competitiveness-report-2016-2017-1. Accessed 28 July 2017

Przhedetskaya, N.V.: Design of marketing management of the innovational model of education in the conditions of development economy. Bull. Tula State Univ. **4**(1) (2014)

Digitization and Internetization of the Russian Economy: Achievements and Failures

Aleksei V. Bogoviz[1]([⊠]) [ID], Svetlana V. Lobova[2] [ID],
Alexander N. Alekseev[3], Inga A. Koryagina[4],
and Tatiana V. Aleksashina[5]

[1] Federal State Budgetary Scientific Institution
"Federal Research Center of Agrarian Economy and Social Development
of Rural Areas – All Russian Research Institute of Agricultural Economics",
Moscow, Russia
aleksei.bogoviz@gmail.com
[2] Altai State University, Barnaul, Russia
barnaulhome@mail.ru
[3] Plekhanov Russian University of Economics Russia, Moscow, Russia
[4] Plekhanov Russian University of Economics, Moscow, Russia
2001inga@mail.ru
[5] Russian University of Transport, Moscow, Russia
altavip@yandex.ru

Abstract. The purpose of the article is to study the processes of digitization and Internetization of the Russian economy through the prism of their influence on a modern human. In order to evaluate the current state and successes of modern Russia in the sphere of digitization and Internetization of economy, the methods of horizontal and trend analysis are used, with the help of which the authors study the dynamics and assess the progress of the Russian economic system in implementation of these processes. In order to study the social consequences that accompany the processes of digitization and Internetization of the Russian economy, the authors use the method of correlation analysis for analyzing the connection (correlation) between the indicators of socio-economic development of the economic system and the indicators of digitization and Internetization of economy. As a result of the research, it is concluded that sustainable growth of quantitative indicators of digitization and Internetization of the Russian economy shows the achievement in this sphere. At the same time, the authors found the failures in management of these processes, the most serious of which is orientation at improvement of macro-economic indicators. According to the existing model of management, the processes of digitization and Internetization of the Russian economy are aimed at increase of effectiveness of state management, optimization of business processes, and increase of economic growth rate. The authors developed a new model of digitization and Internetization of the Russian economy, which allows ensuring growth of the living standards of the Russian population, this stimulating the improvement of the indicators of socio-economic development.

Keywords: Digitization · Internetization · Russian economy

E. G. Popkova (Ed.): HOSMC 2017, AISC 622, pp. 609–616, 2018.
https://doi.org/10.1007/978-3-319-75383-6_78

1 Introduction

Under the influence of the global tendency of distribution of the latest information and communication technologies, which appeared as a result of the recent scientific and technical revolution – digital means of storing, transfer, and processing of information, as well as the Internet – the Russian economy has been peculiar for active digitization and Internetization. An essential feature that differentiates Russia from other developed countries is that these tendencies take place by the government initiative and are of the revolutionary character – they are not evolutional transformations of self-developing economic systems.

Taking place in the specific environment, predetermined by unreadiness and opposition from society and business, as well as lack of financial resources with all economic subjects, including households, entrepreneurial structures, and governments of all levels of the economic system, the processes of digitization and Internetization in Russia are peculiar for intensity and efficiency that are different from developed countries. At the same time, their importance and significance for modernization of the Russian economic system and support for its sustainable positions among the developed counties is very high.

This explains high topicality of studying the essence and peculiarities of the processes of digitization and Internetization of the modern Russia's economy. The authors offer and substantiate the scientific hypothesis that in contrast with other developed countries, in modern Russia these processes are slower and are accompanied by additional complications; they are also related to various negative manifestations that are not so vividly seen during the generalized macro-economic analysis, so they remain hidden and unproved. The purpose of the article is to study the processes of digitization and Internetization of the Russian economy through the prism of their influence on a modern human.

2 Materials and Method

In order to evaluate the current state and successes of modern Russia in the sphere of digitization and Internetization of economy, the methods of horizontal and trend analysis are used, with the help of which the authors study the dynamics and assess the progress of the Russian economic system in implementation of these processes. For studying social consequences that accompany the processes of digitization and Internetization of the Russian economy, the authors use the method of correlation analysis.

With the help of this method, the authors analyze the connection (correlation) between the indicators of socio-economic development of the economic system (y) and the indicators of digitization and Internetization of economy (x). E-government index, indicators of development of E-commerce – volume of electronic trade, volume of electronic payments, volume of the digital content market in the Internet, and share of E-commerce in the structure of GDP are used as the indicators of digitization of economy, and number of Internet users and share of Internet users in the structure of population are used as the indicators of Internetization.

Dynamics of initial indicators for conduct of the research is given in Table 1. The data for 2008-2017 are given with a four-year interval – for the sake of the scientific paper format, though the full data are used during the analysis.

Table 1. Dynamics of indicators of socio-economic development economic of the system and the indicators of digitization and Internetization of the modern Russia's economy in 2008–2017

Indicators		Values of indicators for years			
		2008	2011	2014	2017
Indicators of digitization and Internetization economy					
x_1	E-government index, points	0.6512	0.6714	0.7296	0.7814
–	Volume of electronic trade, RUB billion	274.1	256.3	247.8	284.9
–	Volume of electronic payments, RUB billion	278.9	278.5	264.1	268.7
–	Volume of the digital content market in the Internet, RUB billion	4.44	3.89	3.83	5.07
x_2	Share of E-commerce in the structure of GDP, %	2	4	6	8
–	Number of Internet users, million people	23.7	48.6	75.3	109.5
x_3	Share of Internet users in the structure of population, %	16.5	34.1	58.9	76.4
Indicators of socio-economic development of economic system					
y_1	GDP per capita in current prices, USD	416,155.00	416,853.00	439,508.11	432,103.81
y_2	GDP per capita in constant prices, USD	12,468.4	14,187.2	14,388	10,885.48
y_3	Index of economy competitiveness, points	3.8	4	4.2	4.5
y_4	Index of knowledge economy, points	5.64	5.78	5.82	6.01
y_5	Index of happiness, points	5.738	5.364	5.186	5.963

Source: (Miniwatts Marketing Group 2017; Data Insight 2017; The Division for Public Administration and Development Management, The United Nations 2017; The World Bank 2017; The Earth Institute (UN) 2017; World Economic Forum 2017; International Monetary Fund 2017).

3   Discussion

The issues of conceptual substantiation of the essence and meaning, as well as methodological measuring of costs and consequences of the processes of digitization and Internetization of economy are studied in multiple scientific works: (Jepsen and Drahokoupil 2017; Sood and Baruah 2017; Etemad et al. 2010; Plaksin et al. 2017), etc.

The applied issues of digitization and Internetization of the modern Russia's economy are studied in multiple works of such modern authors as (Popkova et al. 2016a; Ragulina et al. 2015; Bogoviz et al. 2017; Bogdanova et al. 2016; Popova et al. 2016b).

However, despite a lot of publications on this topic, the processes of digitization and Internetization of the Russian economy are not studied sufficiently, as in the existing scientific literature they are studied from the macro-economic point of view, while their social manifestations remain without attention and required detailed analysis.

4 Results

The results of the performed analysis are given in Table 2 and 3.

As is seen from Table 2, all indicators of digitization and Internetization of the modern Russia's economy has been showing positive dynamics over the recent years. In particular, growth of the value of E-government index in 2017 constitutes 7%, as compared to 2014, and 20% - as compared to 2008. Growth of the volume of electronic trade in 2017 constitutes 15%, as compared to 2014, and 4% - as compared to 2008. Growth of the volume of electronic payments in 2017 constitutes 2%, as compared to 2014; its reduction in 2014 constituted 4% - as compared to 2008.

Table 2. Results of the horizontal and trend analysis

Indicators	Horizontal			Trend	
	2011/2008	2014/2011	2017/2014	2014/2008	2017/2008
E-government index, points	1.03	1.09	1.07	1.12	1.20
Volume of electronic trade, RUB billion	0.93	0.97	1.15	0.90	1.04
Volume of electronic payments, RUB billion	1.00	0.95	1.02	0.95	0.96
Volume of the digital content market in the Internet, RUB billion	0.88	0.99	1.32	0.87	1.15
Share of E-commerce in the structure of GDP, %	2.00	1.50	1.33	3.00	4.00
Number of Internet users, million people	2.05	1.55	1.45	3.18	4.62
Share of Internet users in the structure of population, %	2.07	1.73	1.30	3.57	4.63
GDP per capita in current prices, USD	1.00	1.05	0.98	1.06	1.04
GDP per capita in constant prices, USD	1.14	1.01	0.76	1.15	0.87
Index of economy competitiveness, points	1.05	1.05	1.07	1.11	1.18
Index of knowledge economy, points	1.02	1.01	1.03	1.03	1.07
Index of happiness, points	0.93	0.97	1.15	0.90	1.04

Source: compiled by the authors.

Table 3. Results of the correlation analysis

y/x	x_1	x_2	x_3
y_1	0.799938188	0.788013	0.827381
y_2	−0.476924501	−0.35757	−0.34353
y_3	0.988378585	0.994377	0.988723
y_4	0.950473194	0.972796	0.955299
y_5	0.297985209	0.182122	0.160966

Source: compiled by the authors.

Growth of the volume of digital content market in 2017 constituted 32%, as compared to 2014, and 15% - as compared to 2008. Growth of the share of E-commerce in the structure of GDP constituted 33% in 2017 – as compared to 2014, and 300% - as compared to 2008. Growth of the number of Internet users in 2017 constituted 45%, as compared to 2014, and 362% - as compared to 2008. Growth of the share of Internet users in the structure of population in 2017 constituted 30%, as compared to 2014, and 363% - as compared to 2008.

This shows the substantial success of modern Russia in digitization and Internetization of economy. While in 2008 Russia was an outsider among developed countries, its indicators corresponding to the level of developing countries, now (as of 2017) it successfully implements the processes of digitization and Internetization of economy. There's also growth of the values of indicators of socio-economic development of economic system, which allows supposing the positive dynamics between the selected dependent (y) and independent (x) variables.

As is seen from Table 3, correlation of the processes of digitization and Internetization of the Russian economy and the values of the country's competitiveness index, as well as knowledge economy index, is very high. That is, these processes perform positive influence on economy of the Russia's economic system on the whole.

Detailed study of dependence of the selected indicators show that correlation of the processes of digitization and Internetization of the Russian economy and GDP per capita in current prices is rather high, while their correlation with GDP per capita in constant prices is negative. That is, growth of the values of E-government index, share of E-commerce in the structure of GDP and share of Internet users in the structure of population negatively influences the living standards of the population, leading to reduction of real disposable income.

Connection between the indicators of digitization and Internetization of the Russian economy and the value of happiness index is very low. This shows that these processes do not influence the Russia's population living standards. Based on the results of the performed analysis, it is possible to conclude that successful processes of digitization and Internetization – despite multiple and serious obstacles (social opposition, deficit of financial resources, etc.) – are achievements of the modern Russian economy.

At that, the most important mistake of managing these processes is orientation at improvement of the macro-economic indicators, due to which the consequences of these processes are left without attention. If these processes are viewed through the prism of their influence on a modern human in Russia, it is possible to see that they are related to large complications in transformations related to the transition to the digital form of storing, transfer, and processing of information, additional expenditures due to the necessity for purchase of digital devices and Internet connection – they do not simplify, accelerate, or reduce the cost of everyday actions (for consumer) and business processes (for employee and entrepreneur).

In order to eliminate the contradictions between the national economic and individual social and business interests in the process of digitization and Internetization of the Russian economy, we offer a new model of implementation of these processes which supposes their orientation at a modern human (Fig. 1).

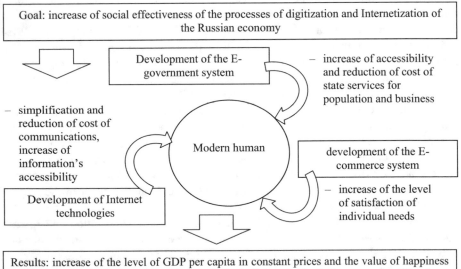

Fig. 1. The model of digitization and Internetization of the Russian economy, oriented at a modern human Source: compiled by the authors.

As is seen from Fig. 1, the offered model's goal is increase of social effectiveness of the processes of digitization and Internetization of the Russian economy, i.e., improvement of social consequences of these processes. For this purpose, it is recommended to conduct these processes in connection to a modern human.

In the process of development of Internet technologies, it is necessary to orient at simplification and reduction of the cost of communication and increase of information accessibility. During development of the E-government system, it is expedient to strive for increase of accessibility and reduction of the cost of state services for population and business. In the process of development of the E-commerce system, it is recommended to orient at the increase of satisfaction of individual needs.

As a result of practical implementation of the offered model of digitization and Internetization of the Russian economy, oriented at a modern human, it is possible to increase the level of GDP per capita in constant prices and the value of happiness index, accompanied by the growth of the population's living standards, improvement of business climate, and increase of macro-economic effect.

5 Conclusions

Thus, sustainable growth of quantitative indicators of digitization and Internetization of the Russian economy shows the achievement in this sphere. At the same time, the mistakes in management of these processes were determined in the course of the research – the most serious of which is orientation at improvement of macro-economic indicators.

According to the existing model of management, the processes of digitization and Internetization of the Russian economy are aimed at increase of effectiveness of state management, optimization of business processes, and increase of the rate of economic growth. The recommended model of digitization and Internetization of the Russian economy, oriented at a modern human, allows ensuring growth of the Russian population's living standards and stimulates the improvement of the indicators of Russia's socio-economic development.

References

Bogdanova, S.V., Kozel, I.V., Ermolina, L.V., Litvinova, T.N.: Management of small innovational enterprise under the conditions of global competition: possibilities and threats. Eur.Res. Stud. J. **19**(2 Special Issue), 268–275 (2016)

Bogoviz, A.V., Ragulina, Y.V., Kutukova, E.S.: Ways to improve the economic efficiency of investment policy and their economic justification. Int. J. Appl. Bus. Econ. Res. **15**(11), 275–285 (2017)

Data Insight. Internet trade in Russia 2017: numbers and facts (2017). http://datainsight.ru/ecommerce2017. Accessed 6 Oct 2017

Etemad, H., Wilkinson, I., Dana, L.P.: Internetization as the necessary condition for internationalization in the newly emerging economy. J. Int. Entrepreneurship **8**(4), 319–342 (2010)

International Monetary Fund. Report for specific countries and subjects. http://www.imf.org. Accessed 6 Oct 2017

Jepsen, M., Drahokoupil, J.: The digital economy and its implications for labour. 2. The consequences of digitalisation for the labour market. Transfer **23**(3), 249–252 (2017)

Miniwatts Marketing Group. Interned World Stats (2017). http://www.internetworldstats.com/stats4.htm. Accessed 6 Oct 2017

Plaksin, S., Abdrakhmanova, G., Kovaleva, G.: Approaches to defining and measuring Russia's internet economy. Foresight STI Gov. **11**(1), 55–65 (2017)

Popova, L.V., Popkova, E.G., Dubova, Y.I., Natsubidze, A.S., Litvinova, T.N.: Financial mechanisms of nanotechnology development in developing countries. J. Appl. Econ. Sci. **11**(4), 584–590 (2016)

Ragulina, Y.V., Stroiteleva, E.V., Miller, A.I.: Modeling of integration processes in the business structures. Mod. Appl. Sci. **9**(3), 145–158 (2015)

Sood, A., Baruah, A.: The new moral economy demonetisation, digitalisation and India's core economic problems. Econ. Political Wkly. **52**(1), 31–36 (2017)

The Division for Public Administration and Development Management, The United Nations. The UN Global E-Government Development Index (2017). http://www.unpan.org/. Accessed 6 Oct 2017

The World Bank. The Knowledge Economy Index (2017). http://www.worldbank.org/kam. Accessed 6 Oct 2017

The Earth Institute (UN). World Happiness Report 2017 (2017). http://worldhappiness.report/. Accessed 6 Oct 2017

World Economic Forum. The Global Competitiveness Report 2016–2017 (2017). http://www3. weforum.org/docs/GCR2016-2017/05FullReport/TheGlobalCompetitivenessReport2016-2017_FINAL.pdf. Accessed 6 Oct 2017

Popkova, E.G., Chechina, O.S., Abramov, S.A.: Problem of the human capital quality reducing in conditions of educational unification. Mediterr. J. Soc. Sci. **6**(3), 95–100 (2016a)

Economic Stimuli for Creation of Highly-Efficient Jobs on the Basis of the New Internet Technologies

Aleksei V. Bogoviz[1]([⊠]) [iD], Svetlana V. Lobova[2] [iD],
Alexander N. Alekseev[3], Galina G. Vukovich[4],
and Anna Y. Grönlund[5]

[1] Federal State Budgetary Scientific Institution "Federal Research Center
of Agrarian Economy and Social Development of Rural Areas – All Russian
Research Institute of Agricultural Economics", Moscow, Russia
aleksei.bogoviz@gmail.com
[2] Altai State University, Barnaul, Russia
barnaulhome@mail.ru
[3] Plekhanov Russian University of Economics, Moscow, Russia
alexeev_alexan@mail.ru
[4] Kuban State University, Krasnodar, Russia
kaf224@yandex.ru
[5] National State University of Physical Education Sport and Health Named
After P.F. Lesgaft, St. Petersburg, Russia
gronlund@mail.ru

Abstract. The purpose of the article is to determine the perspective economic stimuli for creation of highly-efficient jobs on the basis of new Internet technologies in modern Russia and to develop the corresponding practical recommendations. The authors use the method of factor analysis for determining the level and direction of the influence of various factors (indicators of entrepreneurial activities) on the value of the entrepreneurship index according to the U.S. News in 2010–2017. As a result of analysis, the authors come to the conclusion that the main reason for implementing new Internet technologies into activities of the modern Russian companies is insufficiency of market stimuli and existing possibilities of the modern Russian companies in the sphere of implementation of new Internet technologies. The most important factors that perform negative influence on entrepreneurship in modern Russia and are its problem areas include low accessibility of capital, low quality of infrastructure, lack of transparency of business practice (large share of shadow economy), insufficient development of the market environment (low level of competition, high entering barriers), and insufficiently strong and effective institutional provision of entrepreneurship. For solving the determined problems, the authors develop the mechanism of economic stimulation of creation of highly-efficient jobs on the basis of new Internet technologies in modern Russia. It allows supplementing the market methods with the methods of regulatory economic stimulation of creation of highly-efficient jobs on the basis of new Internet technologies in modern Russia. This will ensure the interest of modern Russian companies to creation of highly-efficient jobs on the basis of new Internet technologies and will provide such an opportunity, thus guaranteeing sustainable positive effect.

© Springer International Publishing AG, part of Springer Nature 2018
E. G. Popkova (Ed.): HOSMC 2017, AISC 622, pp. 617–623, 2018.
https://doi.org/10.1007/978-3-319-75383-6_79

Keywords: Economic stimuli · Creation of highly-efficient jobs
New internet technologies · Modern Russia

1   Introduction

New Internet technologies open wide possibilities for increase of effectiveness of modern economic systems of all levels, from separate companies to the global economy on the whole. This becomes possible due to growth of efficiency – replacement of manual labor in mechanical (routine, recurrent) operations by machines within the process of automatization allows accelerating these operations, reducing the probability of mistakes that take place under the influence of "human factor", reducing the cost of their completion, and increasing the intellectual component of labor activities of a modern specialist.

Therefore, new Internet technologies are a precondition for creation of highly-efficient jobs. Their advantage – as compared to usual jobs – apart from growth of efficiency, is larger possibilities for opening the innovational potential of employees due to provision of more time for manifestation of innovational initiatives. As a result, satisfaction with labor from the employees, profitability of business for entrepreneurs, and economic growth and social development (growth of the population's living standards due to increase of the volume of accessible benefits) for the national and the global economic system grows.

This explains high topicality of studying the perspectives of creation of highly-efficient jobs on the basis of the new Internet technologies. However, despite these advantages, new Internet technologies are implemented into the modern Russian companies very slowly, which leads to a scientific and practical problem – a large share of potential of the efficiency growth remains unrealized, slowing down the national level and rate of socio-economic growth and development.

Our hypothesis within this article consists in the fact that the main reason for emergence of this problem is insufficiency of market stimuli and existing possibilities for modern Russian companies in the sphere of implementation of new Internet technologies. This leads to the necessity for additional economic stimuli, provided by the state, which are seen in a wider sense – not only as requirements but also as support for implementation of new Internet technologies by domestic companies. The purpose of the article is to determine the perspective economic stimuli of creation of highly-efficient jobs on the basis of the new Internet technologies in modern Russia and to develop the corresponding practical recommendations.

2   Materials and Method

For verification of the offered hypothesis, the authors use the method of factor analysis. The authors determine the level and direction of influence of various factors (indicators of entrepreneurial activities) in the value of entrepreneurship index as to the U.S. News in 2010–2017.

At that, creation of highly-efficient jobs on the basis of new Internet technologies is a vector of development of entrepreneurship, and it is influenced by the same factors in the same volume (level and direction), as in case with entrepreneurship as such. Dynamics of the values of indicators for factor analysis is given in Table 1.

Table 1. Dynamics of the values of indicators and entrepreneurship index in Russia in 2010-2017 according to the U.S. News

No.	Indicator	Symbol	2010	2017
1	Accessibility of capital	AC	1.85	1.70
2	Quality of infrastructure	QI	4.45	4.40
3	Transparency of business practice (corruption)	TB	1.20	0.60
4	Level of education	LE	4.60	4.60
5	Level of human resources' qualification	LQ	5.30	5.30
6	Enterprising (development of entrepreneurial capabilities)	EC	3.70	3.70
7	Openness of economy	OE	6.20	6.90
8	Accessibility and popularity of innovations	AI	3.00	3.00
9	Market environment (level of competition, barriers for entering)	ME	8.45	8.10
10	Technological provision	TP	7.30	7.30
11	Institutional provision of entrepreneurship	IP	1.92	1.80
Index of entrepreneurship		Entr.	4.36	4.30

Source: USNews (2017).

3 Discussion

Importance and priority of growth of efficiency for provision of quick and sustainable socio-economic development of economic systems and supporting high competitiveness of entrepreneurial structures are emphasized in multiple works of various authors, among which are Popkova et al. (2016a), Ragulina et al. (2015), Bogoviz et al. (2017), Bogdanova et al. (2016), Popova et al. (2016b), Kuznetsov et al. (2016), Kostikova et al. (2016), Simonova et al. (2017).

Thus, in the modern studies and publications only separate aspects of the set problem are studied, which leads to the possibility and necessity for scientific research of all other aspects – in particular, the issues of economic stimulation of creation of highly-efficient jobs on the basis of the new Internet technologies.

4 Results

The results of the factor analysis are given in Table 2.

Based on the data of Table 2, let us evaluate the change of the value of entrepreneurship index under the isolated influence of each factor (indicator) in 2017, as compared to 2010:

Table 2. Calculation of the values of entrepreneurship index in Russia with changes values of each factor as of 2017, as compared to 2010

Indicators	2010	ΔAC	ΔQI	ΔTB	ΔLE	ΔLQ	ΔEC	ΔOE	ΔAI	ΔME	ΔTP	ΔIP	2017
AC	1.85	1.70	1.85	1.85	1.85	1.85	1.85	1.85	1.85	1.85	1.85	1.85	1.70
QI	4.45	4.45	4.40	4.45	4.45	4.45	4.45	4.45	4.45	4.45	4.45	4.45	4.40
TB	1.20	1.20	1.20	0.60	1.20	1.20	1.20	1.20	1.20	1.20	1.20	1.20	0.60
LE	4.60	4.60	4.60	4.60	4.60	4.60	4.60	4.60	4.60	4.60	4.60	4.60	4.60
LQ	5.30	5.30	5.30	5.30	5.30	5.30	5.30	5.30	5.30	5.30	5.30	5.30	5.30
EC	3.70	3.70	3.70	3.70	3.70	3.70	3.70	3.70	3.70	3.70	3.70	3.70	3.70
OE	6.20	6.20	6.20	6.20	6.20	6.20	6.20	6.90	6.20	6.20	6.20	6.20	6.90
AI	3.00	3.00	3.00	3.00	3.00	3.00	3.00	3.00	3.00	3.00	3.00	3.00	3.00
ME	8.45	8.45	8.45	8.45	8.45	8.45	8.45	8.45	8.45	8.10	8.45	8.45	8.10
TP	7.30	7.30	7.30	7.30	7.30	7.30	7.30	7.30	7.30	7.30	7.30	7.30	7.30
IP	1.92	1.92	1.92	1.92	1.92	1.92	1.92	1.92	1.92	1.92	1.92	1.80	1.80
Entr.	4.36	4.35	4.36	4.31	4.36	4.36	4.36	4.42	4.36	4.33	4.36	4.35	4.31

Source: compiled by the authors.

- ΔEntr (AC) = 4.3473−4.36009 = −0.01364. That is, under the influence of the factor "accessibility of capital", the value of entrepreneurship index in Russia decreased by 0.31%;
- ΔEntr (QI) = 4.3564−4.36009 = −0.00455. That is, under the influence of the factor "quality of infrastructure", the value of entrepreneurship index in Russia decreased by 0.10%;
- ΔEntr (TB) = 4.3064−4.36009 = −0.05455. That is, under the influence of the factor "transparency of business practice", the value of entrepreneurship index in Russia decreased by 1.25%;
- ΔEntr (LE) = 4.3609−4.36009 = 0. That is, under the influence of the factor "level of education", the value of entrepreneurship index in Russia has not changed;
- ΔEntr (LQ) = 4.3609−4.36009 = 0. That is, under the influence of the factor "level of qualification of human resources", the value of entrepreneurship index in Russia has not changed;
- ΔEntr (EC) = 4,3609−4,36009 = 0. That is, under the influence of the factor "enterprising", value of entrepreneurship index in Russia has not changed;
- ΔEntr (OE) = 4.4245−4.36009 = 0.06364. That is, under the influence of the factor "openness of economy", the value of entrepreneurship index in Russia increased by 1.46%;
- ΔEntr (AI) = 4.3609−4.36009 = 0. That is, under the influence of the factor "accessibility and popularity of innovations", the value of entrepreneurship index in Russia has not changed;
- ΔEntr (ME) = 4.3291−4.36009 = −0.03182. That is, under the influence of the factor "market environment", the value of entrepreneurship index in Russia decreased by 0.73%;
- ΔEntr (TP) = 4.3609−4.36009 = 0. That is, under the influence of the factor "technological provision", the value of entrepreneurship index in Russia has not changed;

- ΔEntr (IP) = 4.3500−4.36009 = −0.01091. That is, under the influence of the factor "institutional provision of entrepreneurship", the value of entrepreneurship index in Russia reduced by 0.25%.

Let us verify the correctness of the performed calculations. The total change (sum of all Δ) constituted: −0.01364 + −0.00455 + −0.05455 + 0 + 0 + 0 + 0.06364 + 0 + −0.03182 + 0 + −0.01091 = −0.052. The difference between the value of entrepreneurship index in 2017 and 2010 constituted: 4.3091−4.3609 = −0.052. The received values coincided, so the calculations are correct. The total growth of the value of entrepreneurship index constituted −1.19%. According to the value of entrepreneurship index, Russia was ranked 24th in 2017, having gone down by 3 positions as compared to 2010.

The performed analysis showed that the most important factors that negatively influence entrepreneurship in modern Russia are low accessibility of capital, low quality of infrastructure, lack of transparency of business practice (high share of shadow economy), lack of formation of the market environment (low level of competition, high entering barriers), and insufficiently strong and effective institutional provision of entrepreneurship.

We recommend conducting economic stimulation of creation of highly-efficient jobs on the basis of the new Internet technologies in modern Russia within three consecutive phases. The first phase supposes formation of favorable conditions for creation of highly-efficient jobs on the basis of the new Internet technologies, for which it is recommended to:

- support the implementation of new Internet technologies by domestic companies through their involvement into special economic areas, technological parks, and clusters, for which the favorable conditions are created (infrastructure, business climate, market environment, institutes);
- provision of preferences (positive stimulation) for the companies that create highly-efficient jobs on the basis of new Internet technologies in modern Russia.

The second phase supposes the necessity and requirements for creation of highly-efficient jobs on the basis of new Internet technologies, for which it is offered to:

- determined the necessity for creation of highly-efficient jobs on the basis of the new Internet technologies in modern Russia in the national strategy of long-term socio-economic development and modernization of the economic system;
- attract the interest of companies to Internet technologies by their propaganda through social advertising;
- set the national standards of efficiency for the companies that set the necessity for creation of highly-efficient jobs on the basis of the new Internet technologies in modern Russia.

The third phase supposes monitoring and control over the efficiency of creation of highly-efficient jobs on the basis of the new Internet technologies, which requires:

- determination of the evaluation criteria of efficiency of creation of highly-efficient jobs on the basis of the new Internet technologies;

– systemic evaluation and analysis of Russian companies' efficiency;
– correction of the applied economic stimuli if necessary.

The offered mechanism of economic stimulation of creation of highly-efficient jobs on the basis of the new Internet technologies in modern Russia is presented in Fig. 1.

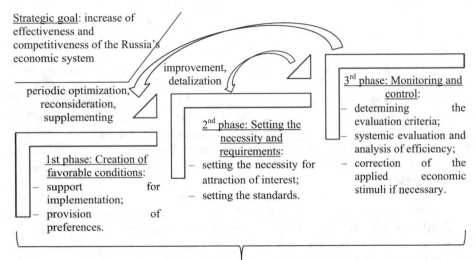

Strategic goal: increase of effectiveness and competitiveness of the Russia's economic system

improvement, detalization

periodic optimization, reconsideration, supplementing

3rd phase: Monitoring and control:
– determining the evaluation criteria;
– systemic evaluation and analysis of efficiency;
– correction of the applied economic stimuli if necessary.

2nd phase: Setting the necessity and requirements:
– setting the necessity for attraction of interest;
– setting the standards.

1st phase: Creation of favorable conditions:
– support for implementation;
– provision of preferences.

Long-term effect: creation of highly-efficient jobs on the basis of new Internet technologies by modern Russian companies, accompanied by mass automatization of business processes, growth of profitability of business, growth of innovative activity of companies, economic growth, and growth of the population's living standards

Fig. 1. The mechanism of economic stimulation of creation of highly-efficient jobs on the basis of the new Internet technologies in modern Russia Source: compiled by the authors.

As is seen from Fig. 1, this mechanism seeks the strategic goal, related to increase of effectiveness and competitiveness of the Russia's economic system. As a result of its practical implementation, a significant positive long-term effect is achieved – creation of highly-efficient jobs on the basis of new Internet technologies by modern Russian companies, accompanied by mass automatization of business processes, growth of profitability of business, growth of companies' innovative activity, economic growth, and growth of the population's living standards.

5   Conclusions

It is possible to conclude that the offered hypothesis was proved – in modern Russia, the effect of market stimuli is not sufficient for mass creation of highly-efficient jobs on the basis of the new Internet technologies; also, in case of their intensive influence on the Russian companies, a lot of them cannot implement the corresponding initiatives.

The offered mechanism allows supplementing the market methods with the methods of regulatory economic stimulation of creation of highly-efficient jobs on the basis of the new Internet technologies in modern Russia. This will ensure not only the interest from modern Russian companies к creation of highly-efficient jobs on the basis of new Internet technologies but will also provide them with such opportunity, thus guaranteeing sustainable positive effect.

References

Bogdanova, S.V., Kozel, I.V., Ermolina, L.V., Litvinova, T.N.: Management of small innovational enterprise under the conditions of global competition: possibilities and threats. Eur. Res. Stud. J. **19**(2 Special Issue), 268–275 (2016)

Bogoviz, A.V., Ragulina, Y.V., Kutukova, E.S.: Ways to improve the economic efficiency of investment policy and their economic justification. Int. J. Appl. Bus. Econ. Res. **15**(11), 275–285 (2017)

Kostikova, A.V., Tereliansky, P.V., Shuvaev, A.V., Parakhina, V.N., Timoshenko, P.N.: Expert fuzzy modeling of dynamic properties of complex systems. ARPN J. Eng. Appl. Sci. **11**(17), 10601–10608 (2016)

Kuznetsov, S.Y., Tereliansky, P.V., Shuvaev, A.V., Natsubize, A.S., Vasilyev, I.A.: Analysis of innovate solutions based on combinatorial approaches. ARPN J. Eng. Appl. Sci. **11**(17), 10222–10230 (2016)

Popova, L.V., Popkova, E.G., Dubova, Y.I., Natsubidze, A.S., Litvinova, T.N.: Financial mechanisms of nanotechnology development in developing countries. J. Appl. Econ. Sci. **11** (4), 584–590 (2016)

Ragulina, Y.V., Stroiteleva, E.V., Miller, A.I.: Modeling of integration processes in the business structures. Mod. Appl. Sci. **9**(3), 145–158 (2015)

Simonova, E.V., Lyapina, I.R. Kovanova, E.S., Sibirskaya, E.V.: Characteristics of interaction between small innovational and large business for the purpose of increase of their competitiveness. Russia Eur. Union Dev. Perspect., 407–415 (2017)

U.S. News. Entrepreneurship Rankings (2017). https://www.usnews.com/news/best-countries/russia. Accessed 23 Oct 2017

Popkova, E.G., Chechina, O.S., Abramov, S.A.: Problem of the human capital quality reducing in conditions of educational unification. Mediterr. J. Soc. Sci. **6**(3), 95–100 (2016a)

Economic Stimuli for Creation
of Highly-Efficient Jobs for a Modern Human

Aleksei V. Bogoviz[1](✉) , Yulia V. Ragulina[1] ,
Alexander N. Alekseev[2], Mikhail N. Lavrov[3],
and Elena V. Kletskova[4]

[1] Federal State Budgetary Scientific Institution "Federal Research Center
of Agrarian Economy and Social Development of Rural Areas – All Russian
Research Institute of Agricultural Economics", Moscow, Russia
aleksei.bogoviz@gmail.com, julra@list.ru
[2] Plekhanov Russian University of Economics, Moscow, Russia
alexeev_alexan@mail.ru
[3] Moscow Region State University, Moscow, Russia
mnlavrov@mgou.ru
[4] Altai State University, Barnaul, Russia
stroiteleva_ev@mail.ru

Abstract. The purpose of the article is to determine the perspective economic stimuli for creation of highly-efficient jobs for a modern human by the example of modern Russia. For studying the influence of creation of highly-efficient jobs on the socio-economic system of modern Russia, the authors use the methods of regression and correlation analysis. In the course of the research, the authors prove that creation of highly-efficient jobs in economy is accompanied by positive social changes, in particular – increase of the happiness index of the economic system and growth of innovative activities of entrepreneurship, measured as the share of innovational organizations and the share of innovational goods. Social opposition, related to social apprehension of unemployment rate growth, has not been confirmed – creation of highly-efficient jobs leads to reduction of unemployment rate. The authors offer recommendations and present a logical scheme of economic stimulation of creation of highly-efficient jobs for a modern human.

Keywords: Economic stimuli · Efficiency · Creation of highly-efficient jobs
"Knowledge economy" · Modern human

1 Introduction

The beginning of the 21st century was marked with transition of the global economic system to a new path of strategic development – "knowledge economy". This led to selection of new vectors of their development, the list of which is dominated by knowledge in the widest sense of this scientific sense, including not only formalized results of intellectual activities (useful models, patents, etc.) but also non-formalized and intermediary results of these activities that cannot be separated from their carrier-human. That's why the target object for management in the interests of development,

© Springer International Publishing AG, part of Springer Nature 2018
E. G. Popkova (Ed.): HOSMC 2017, AISC 622, pp. 624–630, 2018.
https://doi.org/10.1007/978-3-319-75383-6_80

increase, and maximum opening of the existing potential is human, who forms human capital of a modern organization.

The tool of this management is creation of highly-efficient jobs for a modern human. Despite the logic and integrity of the concept of "knowledge economy", the attempts of its implementation in practice face a lot of problems. In particular, realizing the necessity predetermined by the influence of the market, acknowledging the advantages, and having a possibility for creation of highly-efficient jobs for their employees, most employers do not start the corresponding initiatives.

While at the micro-level at the scale of a separate company this leads to insignificant negative consequences in the form of lost profit (unrealized potential of using human capital and, accordingly, lost profit), the negative consequences are more vivid and deeper at the macro-level – they are related to practical impossibility to implement the set course of strategic development of the country in the aspect of creation of "knowledge economy" and reduction of global competitiveness of economic system. This is a serious scientific and practical problem of modern times.

In this article we offer the hypothesis that the root of this problem is insufficiency of natural (market) stimuli for creation of highly-efficient jobs by modern companies. When they are supplemented by efficient economic (non-market – that is, state) stimuli, economic systems will receive highly-efficient jobs, which will stimulate the creation of "knowledge economy". The purpose of the article is to verify the offered hypothesis and to determine the perspective economic stimuli of creation of highly-efficient jobs for a modern human by the example of modern Russia.

2 Materials and Method

For studying the influence of creation of highly-efficient jobs on the socio-economic system of modern Russia, the authors use the method of regression and correlation analysis. Using these methods, the authors study the coefficients b of the models of paired linear regression of the type $y = a + b * x$, which show the character of change of y with increase of x by 1, as well as coefficients of determination (r^2) which reflect the character of connection of the studied indicators. Verification of the offered hypothesis is conducted through successive verification of the following scientific hypotheses:

– Hypothesis H_1: Creation of highly-efficient jobs in economy is accompanied by positive social changes, in particular – increase of the happiness index of an economic system;
– Hypothesis H_2: A restraining factor on the path of creation of highly-efficient jobs by modern Russian companies is social opposition, caused by public apprehension of the unemployment rate growth;
– Hypothesis H_3: Creation of highly-efficient jobs in economy leads to growth of activity of entrepreneurship, measures as the share of innovational organizations and the share of innovational goods.

The information and analytical basis for the research is the official statistical information of 2003–2016, provided in open access by the Federal State Statistics Service of the Russian Federation, Columbia University Earth Institute, at the High School of Economics of the RF (Table 1).

Table 1. Dynamics of change of the total efficiency of the Russian economy (x), index of happiness (y_1), share of unemployed in the structure of economically active population (y_2), share of innovational organizations (y_3), and share of innovational goods (y_4) in Russia in 2003–2017.

Indicators	2003	2004	2005	2006	2007	2008	2009	2010	2011	2012	2013	2014	2015	2016
Efficiency, %	107	106.5	105.5	107.5	107.5	104.8	95.9	103.2	103.8	103.3	102.2	100.7	97.8	99.8
Index of happiness	6.25	6.26	6.20	6.32	6.32	6.16	5.64	6.15	6.19	6.16	6.1	6.01	5.84	5.96
Share of unemployed, %	7.9	8.2	7.8	7.1	7.1	6	6.3	8.4	7.3	6.5	5.5	5.5	5.2	5.6
Share of innovational organizations, %	10.6	9.8	10.3	10.5	9.3	9.4	9.4	9.6	9.4	9.3	9.6	9.9	9.7	9.7
Share of innovational goods, %	4.4	4.3	4.7	5.4	5	5.5	5.5	5.1	4.6	4.9	6.1	7.8	8.9	8.2

Source: compiled by the authors on the basis of (Federal State Statistics Service of the Russian Federation 2017a), (Federal State Statistics Service of the Russian Federation 2017b), (Helliwell et al. 2017); (High School of Economics 2016).

3 Discussion

Emergence of the concept "knowledge economy" led to active scientific discussion. Some experts state that it is a completely new way that changes the essence of the approach to managing the development of economic systems, as compared to the post-industrial economy, which is opposed by "knowledge economy". Among the modern authors who develop this scientific direction is (Fathollahi et al. 2017; Antony et al. 2017).

Other scholars think that it is rather transition to a new level of socio-economic development and new quality of growth of economic systems; they determine "knowledge economy" as continuation of evolutional branch of post-industrial economy. Representatives of this scientific direction include such Russian scholars as (Popkova et al. 2016; Ragulina et al. 2015; Bogoviz et al. 2017; Bogdanova et al. 2016; Popova et al. 2016; Kuznetsov et al. 2016; Kostikova et al. 2016; Simonova et al. 2017).

The theoretical and methodological issues of measuring the efficiency and studying this scientific category, as well as the applied issues of creation of highly-efficient jobs are studied in the works (Ayadi et al. 2013; Li et al. 2017).

Various aspects of economic stimulation of development of entrepreneurial activities with the quantitative (growth of the number of companies, growth of the volumes of production, number of jobs) and the qualitative (increase of innovative activities that indirectly influences the problem of creation of highly-efficient jobs) methods are viewed in the works (Leiser 2017; Houndonougbo and Mohsin 2016).

The performed overview of literature on the selected topic showed that existing scientific studies are rather narrow. They are oriented primarily at the economic component of creation of highly-efficient jobs, related to growth of profit and competitiveness of companies, development of national economy, etc. At that, the social component of this process, related to growth of satisfaction of the employees with labor and the population's living standards, are not taken into account.

This is a reason for existence of the gaps in the system of scientific knowledge in the sphere of human capital management, as the reasons of connection between creation of highly-efficient jobs and development of "knowledge economy" are not determined. Without a clear understanding of causal connections, the scholars cannot develop the system of principles and tools of managing this process, and without the targeted management the economic systems usually develop in the direction of emergence and deepening of economic crises. That's why filling this scientific gap poses a scientific interest and has high scientific significance.

4   Results

As a result of the regression and correlation analysis, we received the following data (Table 2).

Table 2. Selected results of the performed regression and correlation analysis

Estimate data	$y_{1(x)}$	$y_{2(x)}$	$y_{3(x)}$	$y_{4(x)}$
b	0.81	−13.47	4.99	34.13
r^2	0.94	0.42	0.19	0.45

Source: calculated by the authors.

As is seen from Table 2, growth of efficiency of the Russian economy by 1% leads to growth of the happiness index by 0.81 points (indicators correlation - 94%), reduction of unemployment rate by 13.47% (indicators correlation - 42%), growth of the share of innovational organizations by 4.99% (indicators correlation - 19%), and growth of the share of innovational goods by 34.13% (indicators correlation - 45%).

Therefore, hypotheses H_1 and H_3 are confirmed, and hypothesis H_2 is not. As the positive social consequences of creation of highly-efficient jobs are substantial and include growth of social well-being and employees' satisfaction with labor, and are not accompanied by negative manifestations, leading to reduction of unemployment rate, this tool of creation of "knowledge economy" should be paid close attention from the state and requires active measures aimed at its economic stimulation. For this, we offer the following measures:

– Economic stimulation should be aimed not at provision of profits to an entrepreneur but at creation of additional possibilities for creation of highly-efficient jobs – that is, be of strictly targeted character. If this is tax stimulation, the corresponding tax privileges should cover only expenditures of the company or modernization of technologies and equipment in case of credit stimulation, but subsidized credit resources should be provided only for modernization of technologies, equipment, etc.;

– It is necessary to ensure clarity, plainness, and transparency of the goals of economic stimulation, as well as measurability and controllability of the results of their achievement. In order to avoid the formal approach from the companies that consists in correction of the corporate accounting in the necessary direction for receiving stimuli from the state, it is necessary to conduct complex evaluation of efficiency – i.e., take into account not only the number of highly-efficient jobs at a company and the level and growth of efficiency but also innovative activity as one of the most important aspects of efficiency;

– Economic stimuli for creation of highly-efficient jobs should be diverse, and the system of these stimuli should include tax, credit, customs, marketing, and other stimuli. Due to this, each company will gain access to the necessary privileges, provided by the state, and the mechanism of stimulation will cover the large part of economy.

The offered logical scheme of economic stimulation of creation of highly-efficient jobs for a modern human is presented in Fig. 1.

As is seen from Fig. 1, the central link of the system of economic stimulation of creation of highly-efficient jobs is modern human. The state conducts targeted stimulation according to the offered recommendations, and entrepreneurial structures create highly-efficient jobs and develop the potential of a modern human.

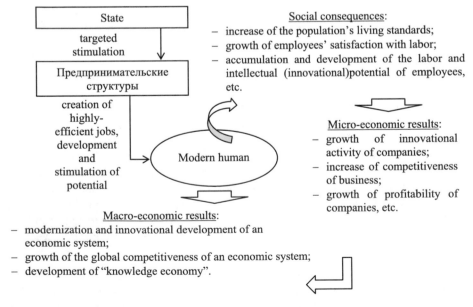

Fig. 1. Logical scheme of economic stimulation of creation of highly-efficient jobs for a modern human. Source: compiled by the authors.

Thus, the positive social consequences arise: increase of the population's living standards, growth of employees' satisfaction with labor, and accumulation and development of labor and intellectual (innovational) potential of employees. This leads to the micro-economic results: growth of innovative activity of companies, increase of competitiveness of business, growth of profitability of the companies, etc., and the macro-economic results: modernization and innovational development of an economic system, growth of the global competitiveness of an economic system, and creation of "knowledge economy".

5 Conclusions

Thus, economic stimulation of creation of highly-efficient jobs is an important step on the path of development of "knowledge economy". An essential peculiarity of this stimulation should be orientation at a modern human. The mechanism of stimulation is activated due to the emerging social effect, which leads to the further micro- and macro-level economic effects.

The performed research allowed specifying the logical connections between the elements in the system of economic stimulation of creation of highly-efficient jobs for a modern human. It is limited by a strict number of factors, which does not allow describing the social effect that emerges as a result of creation of highly-efficient jobs for a modern human.

Deep micro-economic studies on the basis of certain companies, aimed at the detailed study of causal connections between the creation of highly-efficient jobs for a modern human and emergence of the social effect and the accompanying economic effects, is a perspective direction for further scientific research in continuation of this article.

References

Bogdanova, S.V., Kozel, I.V., Ermolina, L.V., Litvinova, T.N.: Management of small innovational enterprise under the conditions of global competition: possibilities and threats. Eur. Res. Stud. J. **19**(2 Special Issue), 268–275 (2016)

Bogoviz, A.V., Ragulina, Y.V., Kutukova, E.S.: Ways to improve the economic efficiency of investment policy and their economic justification. Int. J. Appl. Bus. Econ. Res. **15**(11), 275–285 (2017)

Kostikova, A.V., Tereliansky, P.V., Shuvaev, A.V., Parakhina, V.N., Timoshenko, P.N.: Expert fuzzy modeling of dynamic properties of complex systems. ARPN J. Eng. Appl. Sci. **11**(17), 10601–10608 (2016)

Kuznetsov, S.Y., Tereliansky, P.V., Shuvaev, A.V., Natsubize, A.S., Vasilyev, I.A.: Analysis of innovate solutions based on combinatorial approaches. ARPN J. Eng. Appl. Sci. **11**(17), 10222–10230 (2016)

Popova, L.V., Popkova, E.G., Dubova, Y.I., Natsubidze, A.S., Litvinova, T.N.: Financial mechanisms of nanotechnology development in developing countries. J. Appl. Econ. Sci. **11**(4), 584–590 (2016)

Ragulina, Y.V., Stroiteleva, E.V., Miller, A.I.: Modeling of integration processes in the business structures. Mod. Appl. Sci. **9**(3), 145–158 (2015)

Simonova, E.V., Lyapina, I.R., Kovanova, E.S., Sibirskaya, E.V.: Characteristics of interaction between small innovational and large business for the purpose of increase of their competitiveness. Russia the European Union Development and Perspectives, pp. 407–415 (2017)

Popkova, E.G., Chechina, O.S., Abramov, S.A.: Problem of the human capital quality reducing in conditions of educational unification. Mediterr. J. Soc. Sci. **6**(3), 95–100 (2016)

Federal State Statistics Service of the Russian Federation (2017a). Index of labor efficiency. www.gks.ru/free_doc/new_site/vvp/vvp-god/pr-tru.xlsx. Accessed 24 Oct 2017

Helliwell, J., Layard, R., Sachs, J.: World happiness report 2017. Columbia University Earth Institute (2017). https://s3.amazonaws.com/sdsn-whr2017/HR17_3-20-17.pdf. Accessed 24 Oct 2017

Federal State Statistics Service of the Russian Federation. Population's survey on the issues of employment (2017b). http://www.gks.ru/wps/wcm/connect/rosstat_main/rosstat/ru/statistics/publications/catalog/doc_1140097038766. Accessed 24 Oct 2017

High School of Economics. Indicators of innovative activity (2016). https://www.hse.ru/data/2016/03/21/1128209282/Индикаторы%20innovative%20деятельности%202016.pdf. Accessed 24 Oct 2017

Fathollahi, Momeni, F., Elahi, N., Najafi, S.M.S.: Appropriate theoretical framework for understanding and analyzing economic issues in knowledge-based economy. J. Knowl. Econ. **8**(3), 957–976 (2017)

Antony, J., Klarl, T., Lehmann, E.E.: Productive and harmful entrepreneurship in a knowledge economy. Small Bus. Econ. **49**(1), 189–202 (2017)

Ayadi, R., Boussemart, J.-P., Leleu, H., Saidane, D.: Mergers and Acquisitions in European banking higher productivity or better synergy among business lines? J. Prod. Anal. **39**(2), 165–175 (2013)

Li, L., Liu, X., Yuan, D., Yu, M.: Does outward FDI generate higher productivity for emerging economy MNEs? – Micro-level evidence from Chinese manufacturing firms. Int. Bus. Rev. **26**(5), 839–854 (2017)

Leiser, S.: The diffusion of state tax incentives for business. Public Finance Rev. **45**(3), 334–363 (2017)

Houndonougbo, A.N., Mohsin, M.: Macroeconomic effects of cost equivalent business fiscal incentives. Econ. Model. **56**, 59–65 (2016)

Transformation of Consumers' Behavior in the Conditions of Digital Economy by the Example of Services in Cancer Treatment

Yuri V. Przhedetsky[1]([✉]), Natalia V. Przhedetskaya[2],
Tatiana V. Panasenkova[2], Viktoria V. Pozdnyakova[1],
and Olga V. Khokhlova[1]

[1] Rostov Research Oncological Institute of the Ministry of Healthcare of the RF,
Rostov-on-Don, Russia
yurypr@gmail.com, ysol@yandex.ru
[2] Rostov State University of Economics, Rostov-on-Don, Russia

Abstract. The purpose of the article is to study the essence of the process of transforming the consumers' behavior in the conditions of digital economy by the example of services in the sphere of cancer treatment and to determine the perspective of further development of these services. For the purpose of high detalization of the research and high precision and authenticity of its results, the focus in made on one subject – modern Russia. Studying the information processes in consumers' behavior in the conditions of digital economy by the example of services in the sphere of cancer treatment is performed with the help of a quantitative scientific method – correlation analysis, and a qualitative method – logical analysis of causal connections. The information and analytical basis of the research is the materials of the official statistics of the Federal State Statistics Service and the International Telecommunication Union for 1992–2016. The authors show that in the conditions of digital economy development new possibilities appear in the sphere of cancer treatment, which leads to changes in consumers' behavior. As a result of the performed research, it is possible to conclude that modern Russia is a place of active transformation processes in consumers' behavior in the conditions of digital economy by the example of services in cancer treatment. Some of them are already reflected in economic practice, and others reflect its future outlines and perspective directions of state policy in the sphere of regulation of this sphere of economy. The authors compile a system of transformation processes in consumers' behavior in the conditions of digital economy by the example of services in the sphere of cancer treatment.

Keywords: Transformation of consumers' behavior · Digital economy
Services in cancer treatment sphere

© Springer International Publishing AG, part of Springer Nature 2018
E. G. Popkova (Ed.): HOSMC 2017, AISC 622, pp. 631–637, 2018.
https://doi.org/10.1007/978-3-319-75383-6_81

1 Introduction

One of the most important tendencies of the world economy in the 21st century, which determined the uniqueness of its modern stages of development, is digitization. Digital technologies became widely accessible and covered almost all spheres of economy, with emergence of new spheres that specialize in creation of digital technologies and provision digital services, and the spheres in which digital technologies determine the form of economy – electronic entrepreneurship. Digital technologies changed the economy to such extent that a new type of economic system appeared – digital economy, in which the above processes of digitization of economic activities are vividly expressed and widespread.

Emergence of new possibilities in the sphere of production and distribution of benefits leads to the waves of innovative activity of entrepreneurial structures which are under a lot of pressure. In their turn, the consumers set larger demand for innovations, and, thus, innovations become the standards of doing business, observation of which is mandatory for achieving commercial success. That is, digital technologies are established as one of the main components of creation of benefits in economy.

These processes are peculiar not only for traditionally flexible and dynamically developing spheres of economic activities but also for the spheres that are not subject to changes and provide public benefits and are controlled by the state – e.g., healthcare. Changes in these spheres pose the largest interest due to their high social significance and unprecedented nature, which emphasizes their topicality as the object of the research – however, they are not studied sufficiently by the modern economic science.

The working hypothesis of the work is the idea that in the conditions of development of digital economy new possibilities open in the sphere of cancer treatment, which leads to changes in consumers' behavior. The purpose of this article is to verify the offered hypothesis and to study the essence of the process of consumers' behavior transformation in the conditions of digital economy by the example of services in the sphere of cancer treatment, as well as perspectives of further development of these services.

2 Materials and Method

In order to achieve high level of detalization of the research, as well as high precision and authenticity of its results, the focus is made on one object – modern Russia. Studying the transformation processes in consumers' behavior in the conditions of digital economy by the example of services in the sphere of cancer treatment is conducted with the help of a quantitative scientific method – correlation analysis, and a qualitative method – logical analysis of causal connections.

The information and analytical basis of the research is the materials of the Federal State Statistics Service and the International Telecommunication Union for 1992–2016. The data are systematized and presented in Table 1. The sign "-" denotes absence of data for the indicator in a certain period of time.

Table 1. Dynamics of changes of the indicators of cancer treatment and development of digital economy in Russia in 1992–2016

Indicators	Changes of the values of the indicators in time									
	1992	1995	2000	2005	2010	2012	2013	2014	2015	2016
Population's cancer diseases rate, thousand people	882	974	1,226	1,357	1,540	1,656	1,629	1,693	1,750	1,810
Population's cancer diseases rate, % of diseases	0.97	0.97	1.15	1.28	1.38	1.46	1.42	1.47	1.55	1.62
Number of private healthcare organizations, thousand	-	-	-	5.4	8.3	15.9	26.4	34.2	41.3	55.1
Share of private organizations in healthcare, %	-	-	-	3.12	10.20	20.10	31.40	42.56	58.90	64.07
Turnover of healthcare organizations, RUB billion	-	-	46.7	126.4	311.5	495.8	685.3	1508	1920	2310
Index of development of digital economy	-	-	-	-	5.83	5.91	6.12	6.24	6.35	6.91

Source: (Federal State Statistics Service 2016; International Telecommunication Union 2017).

3 Discussion

Various aspects of formation and development of digital economy are viewed in multiple fundamental and applied studies of modern scholars, among which the most important are Popkova et al. (2016a), Ragulina et al. (2015), Bogoviz et al. (2017), Bogdanova et al. (2016), Popova et al. (2016b), Kuznetsov et al. (2016) Kostikova et al. (2016) and Simonova et al. (2017). At that, the transformation processes in consumers' behavior, caused by the influence of digital economy, including in the sphere of cancer treatment, are not studied sufficiently and require further elaboration.

4 Results

Based on the results of correlation analysis, we determined the following formalized (widespread in practice) transformation processes in consumers' behavior in the conditions of digital economy by the example of services in the sphere of cancer treatment в modern Russia, which could be characterized as realized possibilities in the sphere of improvement of these services.

Firstly, growth of demand for services in the sphere of cancer treatment. This is shown by large annual (8% per year on average) and total (205% in 2016, as compared to 1992) growth of the number of those who sought services in the sphere of diagnostics and treatment of cancer. Coefficient of correlation of the level of population's cancer diseases rate in Russia and the index of development of digital economy constitutes 91%.

Increase of consumers' information on oncological diseases and services in the sphere of diagnostics and treatment lead to growth of the number of applications for these services. This influences the population's healthcare in a positive way, for early diagnostics of oncology diseases means larger success in their treatment. At that, it should be noted that the effect could be direct and reverse.

Thus, digital devices (e.g., cell phone, microwave ovens, etc.) are sources of radiation, which – in certain conditions (e.g., high frequency and dose) – may lead to emergence and development of oncology diseases. This phenomenon is confirmed by growth of the share of oncology diseases in the general structure of the Russian population's disease rate. Their share constituted 1.62% in 2016.

Secondly, growth of demand for services of private organizations in the sphere of cancer treatment. This phenomenon is confirmed by the growth of the number of private healthcare organizations by 49% per year on average – in 2016 it grew by 10 times, as compared to 2011, constituting 55,100. This also proves the growth of the share of private organizations in healthcare grew by 77% per year on average – in 2016 it grew by 20 times, as compared to 2011, constituting 64.07%; turnover of healthcare organizations grew by 68% per year on average – in 2016 it grew by 18 times, as compared to 2011, constituting RUB 2,310 billion.

It should be noted that due to the lack of the necessary institutional provision in Russia, there are no narrowly specialized private organizations in the sphere of cancer treatment (cancer detection centers). However, almost in all multi-profile private medical organizations there are services in the sphere of diagnostics and cancer treatment. Obviously, growth of demand for private services is caused by increase of mistrust of consumers to state service in the sphere of cancer treatment due to their low accessibility and quality.

Based on the results of logical analysis, we determined the following non-formalized (not popular in practice) transformation processes in consumers' behavior in the conditions of digital economy by the example of services in the sphere of cancer treatment в modern Russia, which could be characterized as unrealized possibilities in the sphere of improvement of these services.

Firstly, it is reconsideration of the essence of quality of services in the sphere of cancer treatment. While initially the consumers were interested in efficiency of treatment, as of now they pay attention to service (convenience and comfort). In the conditions of digital economy, wide possibilities open in the sphere of improving the services in the sphere of cancer treatment. They include the following:

– electronic appointment to a doctor, which is available in private and state healthcare organizations in Russia;
– electronic payment for medical services;
– electronic receipt of the results of analysis, etc.

Secondly, rationalization of decisions making in the sphere of cancer treatment. In the conditions of digital economy, most of information on treatment of oncology diseases is accessible on the Internet. It includes self-marketing of oncological detection centers on their official web-sites, patients' reviews, etc. Multi-functional search systems allow selecting and sorting the necessary information. Based on it, consumers make weighted decisions.

According to this, the patients' mobility grows. Patients are no longer tied to the nearest oncological detection center. They are ready to go abroad in order to receive the required services in the sphere of cancer treatment. In the conditions of digital economy, consumers have access to the medical information, based on which they can realize that the disease could be treated – despite the opposite statements in the nearest oncological detection center. That's why consumers are ready to cover large distances in order to get treatment.

We also determined new possibilities in the sphere of cancer treatment which appeared due to establishment of digital economy but have not yet been realized in practice and form further perspectives for development of these services.

Firstly, storing information on patients on digital carriers. This could be realized with the help of the system of electronic medical cards for patients. This innovation will allow increasing mobility of patients, as in order to get treatment in several healthcare organizations simultaneously they won't have to collect information as it is stored on their personal medical card.

Secondly, receipt of electronic prescriptions. People with oncological diseases often cannot even leave their homes, so they are not able to get a prescription from the doctor. At that, they always need prescriptions, as they medicines they need are not sold without prescriptions. The possibility to receive the necessary prescriptions without an actual trip to a doctor would increase the quality of services in the sphere of cancer treatment.

Thirdly, usage of digital technologies for conduct of case conferences regarding patients with oncological diseases. Unification of knowledge and experience of several doctors allows selecting the radiation dose and medicine for each patient without a lot of time and expenditures.

Based on the above, we compiled a system of transformation processes in consumers' behavior in the conditions of digital economy by the example of services in the sphere of cancer treatment (Fig. 1).

As is seen from Fig. 1, in the conditions of digital economy there open new possibilities in the sphere of cancer treatment. Transformation processes in consumers' behavior in the conditions of digital economy by the example of services in the sphere of cancer treatment stimulate growth of the quality of these services. That's why stimulation of further growth of power of consumers in the sphere of provision of services in the sphere of cancer treatment is an important direction of development of the healthcare system in modern Russia.

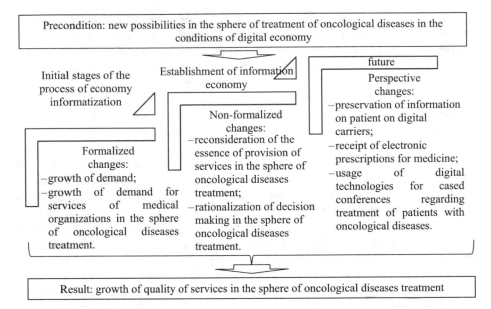

Fig. 1. System of transformation processes in consumers' behavior in the conditions of digital economy by the example of services in the sphere of cancer treatment Source: compiled by the authors.

5 Conclusion

As a result of the performed research, it is possible to conclude that modern Russia is a place of active transformation processes in consumers' behavior in the conditions of digital economy by the example of services in the sphere of cancer treatment. Some of them are reflected in economic practice, and others reflect its future outlines and perspective directions of the state policy in the sphere of regulation of this sphere of economy.

The influence of digital economy on treatment of oncology diseases is contradictory. On the one hand, new possibilities for cancer treatment open. On the other hand, distribution of digital devices leads to growth of oncology diseases. The search for the means for solving this contradiction is a perspective direction for further scientific studies.

References

Bogdanova, S.V., Kozel, I.V., Ermolina, L.V., Litvinova, T.N.: Management of small innovational enterprise under the conditions of global competition: possibilities and threats. Eur. Res. Stud. J. **19**(2 Special Issue), 268–275 (2016)

Bogoviz, A.V., Ragulina, Y.V., Kutukova, E.S.: Ways to improve the economic efficiency of investment policy and their economic justification. Int. J. Appl. Bus. Econ. Res. **15**(11), 275–285 (2017)

International Telecommunication Union: ICT Development Index 2016 (2017). http://www.itu. int/. Accessed 10 Nov 2017

Kostikova, A.V., Tereliansky, P.V., Shuvaev, A.V., Parakhina, V.N., Timoshenko, P.N.: Expert fuzzy modeling of dynamic properties of complex systems. ARPN J. Eng. Appl. Sci. **11**(17), 10601–10608 (2016)

Kuznetsov, S.Y., Tereliansky, P.V., Shuvaev, A.V., Natsubize, A.S., Vasilyev, I.A.: Analysis of innovate solutions based on combinatorial approaches. ARPN J. Eng. Appl. Sci. **11**(17), 10222–10230 (2016)

Popova, L.V., Popkova, E.G., Dubova, Y.I., Natsubidze, A.S., Litvinova, T.N.: Financial mechanisms of nanotechnology development in developing countries. J. Appl. Econ. Sci. **11**(4), pp. 584–590 (2016b)

Ragulina, Y.V., Stroiteleva, E.V., Miller, A.I.: Modeling of integration processes in the business structures. Mod. Appl. Sci. **9**(3), 145–158 (2015)

Simonova, E.V., Lyapina, I.R., Kovanova, E.S., Sibirskaya, E.V: Characteristics of interaction between small innovational and large business for the purpose of increase of their competitiveness. In: Russia and the European Union Development and Perspectives, pp. 407–415 (2017)

Popkova, E.G., Chechina, O.S., Abramov, S.A.: Problem of the human capital quality reducing in conditions of educational unification. Mediterr. J. Soc. Sci. **6**(3), pp. 95–100 (2016a)

Federal State Statistics Service: Russia in numbers: short statistical bulletin. Federal State Statistics Service, Moscow (2016)

Regress Economy vs Progress Economy: "Alternatives of Senses"

Marina L. Alpidovskaya(✉), Alla G. Gryaznova,
and Dmitry P. Sokolov

Financial University, Moscow, Russia
morskaya67@bk.ru

Abstract. The transformation of the productive forces on the basis of rapid development of information technology, accompanied by the convergence of information and nano-, bio- and cognitive technologies, enhances the importance of the workforce and of the results of intellectual work in the production of modern goods and services. In its turn, the "network effect" of modern economy determines a deep-rooted conflict between information economy and market mechanisms, as well as the distribution and redistribution, in the realities of which this effect is witnessed, basing on the supremacy of private ownership on the means of production. As a result, judging from the experience of market methods of distribution of the produced value, the jobs are released, inequality grows and the consumer demand for the goods is reduced. Finally, the existing distribution system of the national income hinders the development of productive forces, which is a precondition for its radical transformation in accordance with the dialectical logic of development of social and economic relations.

Keywords: Global economy · Systemic crisis · Transformation of capitalism
Socialization of property · NBIC - technologies

JEL Code: F6 · O3 · P1

1 Introduction

A modern American economist Ilya Stavinsky outlines in his book "Capitalism today and capitalism tomorrow", "The source of all wealth, besides the wealth of nature, is human work" (Stavinsky 1997). Such a claim is not new since W. Petty, but it was in the first half of the twenty-first century when the emphasis on the role of human labour and human as an owner of this factor became of particular importance. The basis for rethinking the role of human and his work in contemporary conditions is a "perpetuum mobile" of social reproduction, the dialectic of productive forces and production relations. People, means of production and dynamically developing technologies serve as the representatives of the first. They define, in turn, radical changes of the second, implemented in a specific mode of distribution of the results of the production between various elements of society.

The rapid development of information technologies, including the new coil of the convergence of information and nano-, bio- and cognitive technologies, enhances the

© Springer International Publishing AG, part of Springer Nature 2018
E. G. Popkova (Ed.): HOSMC 2017, AISC 622, pp. 638–646, 2018.
https://doi.org/10.1007/978-3-319-75383-6_82

role of the labour force and the results of intellectual work in the production of modern goods and services. A significant amount of material capital for the implementation of the process of production is no longer required on many expanding markets, which entails the lowering of entry barriers to those markets for new entrepreneurs. The development of markets of outsourcing and freelancing is a conjoint process. This kind of work allows the entrepreneur to fulfil its true role of resource combiner with greater ease and lower costs.

Modern economy increasingly becomes an economy of networks that permeate all spheres of human life (Alpidovskaya 2015). A side effect of the expansion of networks is a so-called "network effect". This idea was suggested by Theodore Vail, the head of the Bell Telephone company, one hundred years ago, and implied the following: the more people join the network, the more useful it becomes for each of them. According to British Economist P. Mason, the existence of this effect, determines the deep conflict between information economy and market mechanisms (Mason 2016), distribution and redistribution in which are based on the primacy of private ownership of the means of production.

Development of the productive forces on the basis of information technology leads to the expansion of the nomenclature and volumes of production (works, services) with reduction of labour expense, required for it. Consequently, on the assumption of market methods of distribution of produced value (on the contribution of production factors), there is a release of workplaces, rise of inequality and reduction in the consumer demand for manufactured goods. Compensation of a drop down of a solvent demand, still implemented at the expense of the financial mechanisms of transferring of current expenses to future periods, however, has noticeably declined since the financial crisis of 2008. Thus, the current distribution system hinders the development of the productive forces, which is a precondition for its radical transformation in accordance with the dialectical logic of development of social and economic relations.

Three aspects of the development of public relations in the light of the evolution of the information economy are consistently presented in this article. They are the following:

- the negative traits of modern private property, impeding the development of scientific and technical progress and society;
- preconditions for development of collective and public forms of economy and identification of trends of expansion of the latter in the modern world;
- strategic risks and opportunities for the development of collective and public forms of economy and their mutual influence on society and the economy.

2 Advantages and Disadvantages of Modern Private Ownership in the Light of the Development of the Information Economy

Classics of economic science (in particular, D. Locke, K. Marx, F. Engels) regarded private ownership as a form of appropriation of the results of work. Historical and legal tradition, including economic aspect valuable for us, sees in private ownership a bunch of legal authorities for a specific resource, concentrated in the hands of a single person. Institutionalists define private property as the mode of the use of a limited resource, for which a certain individual has exclusive rights.

Summing up, as an object of research by the category of private ownership we will understand an ownership of single individuals, aimed mainly at getting and multiplication of their income, i.e. public relations used for personal enrichment.

Moreover, the process of appropriation as a kind of social relations can have either labor or non-labour character. In the first case, the appropriation is carried out as a result of labour of private owner of means of production, as well as workforce-in the absence of gratuitous appropriation of someone else's labor. In the modern world this is represented by the work of many small businesses, farmers, artisans, as well as enterprises of cooperative forms of ownership (industrial and agricultural cooperatives, Communities of financial assistance, kibbutzs, communes, artels, etc.). However, the non-labour private ownership, aimed at personal enrichment of the owners (often in defiance of both the national interest and the interests of the company itself) is more important for the modern economy. In this article private ownership refers to its variant in which there is a contradiction between capital and labour.

Among the advantages of the system, under which personal benefit of the owners is made a cornerstone, one can name a certain emancipation of private initiative and the rapid development of science and technology. Despite the fact that the satisfaction of material needs of society cannot be attributed to the merits of private property due to the fact that the homogeneity of the products was the feature of the type of economy, but not of the form of appropriation. However, the actual essence of the capital as the self-expanding value, having predestined the deepening division of labour, led to the development of, among the others, the information economy, in which the tension between the public character of production and the private appropriation of the results of work is sharp.

It should be outlined that the negative effects of the domination of private ownership in its current form of the appropriation of the results of work manifest on the following three key levels:

(1) sharp increase in inequality as a result of the falling of global profit margins, the cause of which is the approach of extensively-oriented capitalist system to the limits of its development;
(2) failure of the capitalist system to deal with global threats, which it caused by itself - environmental, social, and political ones;
(3) inefficiency of private ownership in the information economy due to the mechanisms for containing scientific and technological progress and maintaining the trends of monopolization.

Let us sequentially go through the given points.

2.1   The Escalation of Inequality at the Beginning of the 21st Century

Inequality was always inherent in the capitalist system because only of the existence of dichotomy "private owner-employee". By the beginning of the 21st century other fundamental features of capitalism have started to lead to the escalation of inequality, in view of the fact that the capitalist system cannot exist in a static condition because of the need of the permanent self-expansion of the capital.

The expansion of the capitalist system logically can be implemented in two directions and their combination. The first direction is territorial expansion, through the increment of peripheral zones with lower remuneration of labour, or with a low-cost mining operations, or with extensive agricultural lands, or with accumulated national wealth, suitable for implementation at relatively low prices. For example, Russian Federation after the dissolution of the Soviet Union had all of these features.

The second direction concerns extending into the depth: the deepening of division of labour and the involvement of the spheres formerly acting according to non-market laws into exchange relations (the latter could include the commercialization of education, health care, relations of motherhood and childhood, etc.).

By the beginning of the 21st century capitalist system has approached the limits of its expansion capacity. Territorially almost all the States of the world have become involved into a single geo-economic system. The last major "breath" of the global economy happened at the end of the twentieth century with the collapse of the Soviet economic zone, and only today the excess profit, created as a result of colonial expansion to the former members of this zone, begins to dry out.

As far as deepening and widening of the division of labour is concerned, it requires two conditions: market crowded by people with higher incomes and new technologies, the output of which exceeds the expenses on its implementation. Additional sources of consumer demand turned out to be dried out: promising markets such as Chinese, are characterized by low incomes of citizens while the prospects of enhancing of consumer lending have been extremely deplorable since the crisis 2008. As for technology, the output of the IT sector was lower than the investors expected, and the profit margins in the leading industrialized countries maintained the downward trend.

Major advances in the field of information technology have led to the investment boom, based on expected income, which accelerated the productivity and promised even greater profits. However, as R. Brenner, the American economist and historian, highlights, the driving forces of the "new economy" have only aggravated the problem of chronic excessive production capacities and the reduction in the global profit margins (Brenner 2014).

The situation in the spheres of Nano-, bio-, informational and cognitive technologies appears to be quite similar. Revival of investors and large-scale investments in new sectors of the economy are also based on high expected profits, rather than on actual profits. In the conditions of hypertrophy of the financial sector the consequences of the mismatch between expectations and reality can result in even more destructive crises than those in 2000's.

Changes in distribution relations act as a compensation for the reduction of global profit margin. So, at the beginning of the 21st century there has been a tendency to polarize different social groups on the basis of income and assets: the rich (the representatives of the owners of the means of production) become richer and the poor become poorer, while the proportion of middle class reduces in the countries of both the centre and the periphery of the capitalist system (Fig. 1).

This trend can be also observed in the United States up to the year 2014 (Middle Class Shrinks).

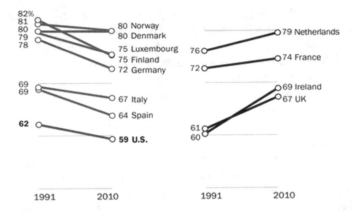

Fig. 1. Proportion of middle class in some advanced industrialized countries over the period 1991–2010 (According to (Middle Class Shrinks))

The exhaustion of the possibilities for capitalist expansion predetermines the transition from the extensive nature of the world economy to intensive one: the transition to internal operation and dismantling the foundations of "social State" in developed countries, as well as to removal of barriers to capital to maximize profits. According to A.I. Fursov, the Russian historian and social philosopher, in the long term perspective, the implementation of these processes, can lead to increased exploitation of the periphery and the masses of the population of the countries of the centre, to the turn of the achievements of information economy against society by tightening controls and information-psychological influence (*Fursov*). Thus today the existing distribution relations in a capitalist economy begin to impede social development.

2.2   Problems of Property and Global Threats

The escalation of inequality when targeting capital at a "blind" maximization of profits leads to a reproduction of a wide range of global threats, implemented at the level of social interaction, interaction between human and nature, as well as development of the personality and the body of human. Global and regional armed conflicts and economic wars, acting as a means of preserving the economic power of countries of the center and as the avoidance of the development of alternative forms of social relations, can also be regarded as the problems of public interrelations posed by modern economy. International terrorism serves as a component of this process and, at the same time, the reaction to it. The escalation of tensions entails the worsening of living standards in the "target" countries and migration influxes to the countries of the center, which cause deep cultural and institutional deformations in the donor countries.

Socio-economic instability in the late 20th and early 21st centuries largely conceals a sharp deterioration of the ecological situation on the planet. Despite the rapid development of "green" technologies of production, capital in pursuit of minimizing cost moves most dirty production from developed countries to the periphery ones, which, of course, improves the ecological situation in some regions, but in generally leads to increased pressure on the environment. According to "Global Environment Outlook", a report of the UN Environment Program (UNEP), environment is deteriorating much faster than it was previously thought. Earth's biosphere resources are decreasing every year due to increasing resource consumption, soil degradation, desertification, water pollution and contamination of the planet by the waste of "consumption economy".

In addition to the environmental damage caused by economic activities, in the twenty-first century the problem of natural climate changes arises. A modern society tends to respond to it by providing security for the elite groups, rather than the implementation of urgently necessary coherent policy to reduce damage from these processes for the whole of mankind.

The current socio-economic system is incapable of resolving the problems of human development: famine, epidemics of HIV and hepatitis, cancer. Drug and illegal weapons trafficking, human and human organs trafficking are quite high-margin activities and therefore are as attractive for capital, as disruptive to society. Problems of personality development in modern conditions are standing apart. In today's economy person is alienated from society, from other person, from labor, from politics, social organizations, culture, environment, morality and, eventually, from itself. This is manifested in the depersonalization of individuals. All the above-mentioned also violates the public interest in light of the prospects for social progress, provided by an information economy.

2.3   Transformation of Property Relations in the Process of the Development of the Information Economy

Private ownership has proved a form providing the most rapid economic and scientific development of mankind in the capitalist system of relations. Limited resources and limited product when implementing production acted as a stimulus to achieve individual and group leadership in a competitive fight, determined individualism and finding of new and emerging markets, forms of organization of the company and human needs-together with good, satisfying them. All this is true for goods and services whose utility is declining during the process of consumption -from a red apple to haircut in a hairdressing saloon.

The situation is different when information is concerned.

Object of intellectual property differs from material object by the fact that it can be used by an unlimited range of consumers. Its utility does not decrease, and in some cases even grows when the increasing "pool" of scientists takes master any scientific idea or method of research, or when the achievements in cultural environment are concerned. Until recently, the approach to intellectual property objects was similar to the ownership of a material object and operated through the system of third party access restrictions by law. This kind of approach was dominant in the economy. However, today free property objects: software, free programs and utilities, knowledge in the public domain, educational projects, music, literature etc. - are becoming more widespread. Also the form of remuneration of the use of the results of intellectual work is changing, today there are donations, mechanisms of crowd funding, advertisement placement. Products of intellectual activity are increasingly socialized, also by illegal methods that maximizes their social utility.

Organization of social relations in the implementation of the production process also changes. Familiar private ownership of the means of production, when a company is headed by one or several persons having a big part of ownership, changes under the influence of socialization of ownership. On the one hand, we are talking about the spraying of property in large companies between many shareholders with an absence of majority shareholder. Such enterprises are often actually owned by a narrow circle of people. On the other hand, the basic means of production in the information economy is a personal computer and, therefore, the contradiction between capital and labour is removed, and the production of knowledge in the modern economy ceases to be bound to one enterprise as a unifying institutional structure, it has a nature of the project activity with flexible team of creative people (in particular, such crowdsourcing projects as InnoCentive, Wikipedia, NASA Clickworkers).

Reduction of the global profit margin influences enterprises of the classic private ownership. Today knowledge economy confronts the economy of consumption. The antithesis of flexibility of creative groups is the production of an increasing range of products, the quality of which is reduced in favour of reducing the number of years of useful life. Automobile construction and electronics, where the products are often designed for use only during the warranty period and subsequent replacement, are a very bright example. This is necessary to preserve demand in the conditions of oversupply.

The realities of the development of the VI-th technological mode in the advanced industrialized countries illustrate the fact that the State, being the hub of national innovation systems, not large corporations become the locomotive of new technologies development (with the exception of the United States as a cluster of transnational corporations). In the modern world a significant part of the postindustrial economy-education, fundamental science, health care and culture - are under the control of the State. The private sector is also increasingly influenced by the State through the control of social, ecological and other spheres. Finally, in the conditions of the increasing volume of information economy, private property loses its ability to provide scientific, technical and socio-economic progress.

3 An Alternative to the Modern Private Property: Content and Trends

Modern transformation of property relations are not limited by the changes in the information economy. Trend towards socialization of ownership is observed also in the productions of real sector and in the financial sphere. The perceptions of the effectiveness of the collective and public form of ownership in scientific community has also changed. In particular, in 2009 year Elinor Ostrom was awarded by the Nobel Prize in economics. She provided proof of the high efficiency of the collective management of public ownership of natural resources (Ostrom 2013).

In the conditions of the declining global profit margin the maintenance of the orientation of the socio-economic system towards maximization of individual benefits can lead to aggravation of the problem of growing inequality and to the augmentation of the instability on the enterprises amid global financial and economic crises. Enterprises of collective forms of ownership ensure a balanced distribution of wealth in society and are more stable in times of crisis in comparison with enterprises of private ownership. In cooperative enterprises the orientation towards the maximization of profit is giving way to the objectives of the collective decision of any urgent issues and/or joint household tasks. Thus the economic activity of cooperatives is aimed at satisfaction of needs which corresponds much more to generally accepted objectives of market economy.

According to the results of the 2016, there are 1,420 cooperatives around the world with a turnover of over $100 million (The World Co-operative Monitor) in recent years, their number is increasing. Cooperatives in the spheres of agriculture and food industry, insurance, retail and finance have the largest share among modern cooperatives. In European countries, cooperatives account for about 60% of the agricultural market. The total number of members of cooperatives in the world is around 1 billion people (The World Co-operative Monitor).

Despite the progressive character of cooperatives, their development in the modern world is hampered by a lack of adequate legal frameworks, taking into consideration specifics of cooperative ownership. The case of the Russian Federation, where an enterprise of collective form of ownership is represented by closed private stock companies (people's enterprise), is illustrative. This form of organization, according to the Russian legislation, may be established only by restructuring of existing stock companies provided that holders of 50% of the voting shares have voted for it. As a result of privatization processes the ownership of the vast majority of Russian companies turned out to be concentrated in the hands of one or several holders of large blocks of shares of an organization, who is interested in retaining control over business. However, in Russia such enterprises got widespread, and like everywhere in the world, they are characterized by high stability in the time of crisis and high efficiency of activity with a rich social infrastructure created for enterprise members.

4 Conclusions

So, contemporary capitalism is no longer capable of strategic expansion due to objective laws of its development. The world community has to choose between two vectors of strategic development: either a shift from extensive type of development to the intensive one followed by the intensification of exploitation and confrontation and increasing inequality or the path of solidarity through the gradual socialization of ownership. The first path would mean the end of socio-economic development on global scale and degenerative changes in all spheres of public life, while maintaining a high level of consumption among the elite groups and layers of society, close to them. The second path is the changing of the distribution mechanisms, which will allow the society to move forward, the maximization of public benefit, not personal profit being a measure of success. The society of regress or the society of progress - that is the choice of today's world.

References

Alpidovskaya, M.L.: On the issue of inhomogeneity in socio-economic development of the countries in the era of globalization of the world economy. Natl. Interests Priorities Secur. **19** (160) (2012)

Alpidovskaya, M.L.: On the issue of goal-setting human activity and substantial character of its work in the economy of the future. Economic and legal aspects of the implementation of the strategy of modernization of Russia: a search for the model of effective socio-economic development. In: Kleiner, G.B., Sobolev, E.V., Sorokozherdev, V.V., Hashevaya, Z.M. (eds.) A Collection of Articles of the International Scientifically-Practical Conference, Krasnodar, JUIM, pp. 10–14 (2015)

Brenner, R.: The Economics of Global Turbulence: The Advanced Capitalist Economies from Long Boom to Long Downturn, 1945–2005. National Research University Higher School of Economics. House of the HSE, Moscow (2005). Russ. Ed.: A. Gusev, R. Haitkulova; Sc. Ed. I. Chubarov

Girenok, F.: What is a man? Philos. Econ. **5**(107), 165–173 (2016)

Alpidovskaya, M.L., et al.: Macroeconomics: coursebook Rostov-on-Don, Phoenix (2017). Pub. Ed. M.L. Alpidovskaya, N.V. Tskhadadze

Mason, P.: Post-Capitalism: A Guide to Our Future. Ad Marginem Press, Moscow (2016)

Sokolov, D.P., Alpidovskaya, M.L.: The genesis of property relations in Russia: a historical retrospective. J. Volgograd State Tech. Univ. **16**(11(114)), 18–26 (2013)

Sokolov, D.P., Alpidovskaya, M.L.: Content and trend of transformation of property relations in modern Russia. Natl. Interests Priorities Secur. **2**, 20–32 (2014)

Ostrom, E.: Governing the Commons: The Evolution of Institutions for Collective Action (2013). ed. T. Montyan

Stavinsky, I.: Capitalism, Today and Tomorrow. URSS, Moscow (1997)

Fursov, A.I.: "Crisis-matryoshka": dismantling of capitalism and the end of the age of the pyramids. http://www.intelros.ru/pdf/ps/02/21.pdf

Middle Class Shrinks in 9 of 10 US Cities as Incomes Fall//PEW Research Centre. https://apnews.com/d23b048519aa47caba95fd51390eff60/pew-study-sees-shrinking-middle-class-major-us-cities

The World Co-operative Monitor: Exploring the Co-operative Economy, REPORT 2016 (2016). http://www.ccw.coop/resources/world-cooperative-monitor.html

Financial and Organizational Mechanisms of Managing Innovational Development of Region's Economy

Elena I. Minakova(✉), Anna V. Krylova, Gulnara R. Armanshina,
Natalya A. Dumnova, and Svetlana A. Ilminskaya

Orel State University of Economics and Trade, Orel, Russia
osuet@mail.ru, my-orel-57@mail.ru, 1278orel@mail.ru,
super-ya-57@mail.ru, orel-osu@mail.ru

Abstract. Effective regional policy is the key factor that ensures stable conditions of development of the subjects of the RF. Recently, the development of regions' economies have been peculiar for the necessity for organizing the innovative activities as a competitive environment of the 21st century. According to this precondition, scholars and practitioners develop the mechanisms of managing the innovational development of economy, which are necessary for building close interrelations between all subjects of innovational regional environment. The purpose of scientific research is formation of the model of financial and organizational mechanisms of managing the innovational development of region's economy. Implementation of the set goal requires solving the following tasks: considering the evolutional approaches to the mechanisms of development of innovational economy; determining the peculiarities of the theoretical model of region's economy innovational development management through an objective mechanism of interaction; forming the models of organizational and financial mechanisms of managing the innovational development of region's economy. The methodological tools of the research include the methods of theoretical and evolutional analysis, modeling and graphic presentation, and determination of substantial attributes. The scientific novelty of the research consists in application of the evolutional and typological approach to formation of the proprietary mechanisms of management of region's innovational economy. Theoretical and practical significance of the research is manifested in consideration of these issues through the prism of the necessity for transforming the existing mechanisms of innovational economy of the region.

Keywords: Financial mechanisms · Organizational mechanisms
Innovations · Regional economy · Programs · Correction · External changes
Innovational development · Innovational companies · Evolution of development

1 Introduction

Economic development of regions depends on competitiveness of the territory in the conditions of global changes of the scientific and technological paradigm. Technologies become the key factor of production, manifesting their value as a product created within innovative activities. Formation of innovational environment of development of regional economy is an inevitable fact. This statement is proved by the following theses. Firstly, change of the national economic paradigm in favor of technological initiative requires from regional economies the adaptation to new conditions of existence. Secondly, formation of the information environment for provision of region's competitiveness sets requirements for creation of innovational infrastructure. Thirdly, development of the key sectors of regional economy with entering new supra-national markets is related to creation of popular innovational products. Fourthly, modernization of regional economy is impossible without creation of fundamental innovational platforms of the 21st century. Fifthly, formation of innovational economy allows the regions to take leading positions in struggle for limited resources. These theses confirm topicality and significance of the research topics.

The purpose of the article is to form the model of financial and organizational mechanisms of managing the region's economy innovational development. For this, the following tasks should be solved:

- considering the evolutional approaches to the mechanisms of development of innovational economy;
- determining the peculiarities of the theoretical model of management of region's economy innovational development through the objective mechanism of interaction;
- forming the model of organizational and financial mechanism of management of region's economy innovational development.

Implementation of the set goals requires methodological tools, which are the following:

- method of theoretical analysis – used for generalization of the key elements of the objective mechanism of interaction, which is opened in the theoretical model of management of region's economy innovational development;
- method of evolutional analysis – used for determining the key peculiarities of the mechanisms of development of region's economy on the basis of genesis and periodization of the analyzed object;
- graphic method – used for building the models of organizational and financial mechanisms of development of innovational economy of region;
- method of distinguishing significant attributes – used for determining certain methods of study within the mechanism of innovational development of region's economy;
- method of modeling – used for formation of the models of organizational and financial mechanism of development of innovational economy of region.

Study of the mechanisms of development of innovational economy of regions is related to the consideration of the issue of evolutional functioning of this topic in various periods of strategic management of national economy. According to this, let us distinguish the key periods of evolution of the mechanisms of development of innovational economy.

2 Evolution of the Mechanisms of Innovational Economy Development

Evolution of the mechanisms of development of innovational economy is connected to the change of the methodological attitude towards planning and projecting of strategic management in the national socio-economic system. The tendency of growth/reduction of socio-economic indicators transformed the national socio-economic systems according to the innovational way of development. Starting 1970, innovativeness of economy has been manifested in authomatization of the key productions of national economy, emergence of electronic technologies for provision of highly-efficient sectors of economy, and formation of the new technological paradigm within the innovational economy of knowledge. Evolutions of the mechanisms of development of innovational economy is presented in detail in Fig. 1.

For simpler consideration of the evolution of the mechanisms of innovational economy, let us apply the stage method. The stage method (Lazonick 2003) is the consecutive evaluation of the research object, performed on the basis of distinguishing the period of evolutional genesis and the main characteristics related to the key peculiarities of its development. A lot of authors do not differentiate the stage method and periodization. Periodization is peculiar for considering the essence of the single object at various historical stages of development. The stage method supposes evaluation of peculiarities of multi-form objects of the research that change according to the situation of external and internal environments. At that, the process of periodization has a large scale (centuries, millennia), unlike the stage method (month, year, decade, quarter of century).

Thus, let us consider the evolution of the mechanisms development of innovational economy.

Stage 1 – 1970's. The agent mechanism (Komarevtseva 2016) formed in the conditions of priority of technological development of capitalistic economy. Innovations in this period are viewed as a component of the automatized process of

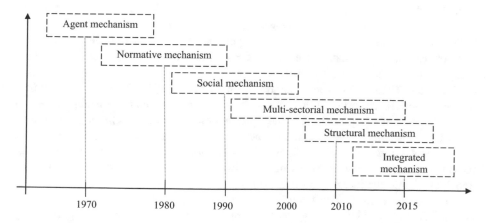

Fig. 1. Evolution of the mechanisms of innovational economy development

international corporations in the car building industry. The essence of the agent mechanism consisted in presence of institutional subjects – agents that provided the economies of developed and developing countries with additional quantity of limited resources. The factor of production was the investments of international corporations and alliances, called the global capital.

Stage 2 – 1980's. The normative mechanism (Belskaya 2009). In the process of change of market modes in favor of "open technologies", the mechanism of provision and development of innovational economy also changed. Innovations are competitive goods that ensure growth of marginal income from sales and re-selling in the international markets. The quantity of goods began to dominate over the quality. According to this, new normative approaches appeared which later turned into mechanisms. For the purpose of stabilization of economy and prevention of overproduction, the normative mechanisms that limited – on the basis of artificially created barriers – entering the formed markets were created. The key production factors were the resources. The more resources in the market, the larger the limitation within the normative mechanism. This mechanism was changed due to weakening of normatives caused by the dominating position of Chinese goods in the internal markets of the key developed countries.

Stage 3 – 1990's. The social mechanism (Kharlamov 2011). The social mechanism was based on the necessity of growth of the population's living standards together with development of national economies. Financial and stock market operations began to be viewed as innovations. The priority in the sphere of production factors shifted in favor of capital. Development of innovational economy took place on the basis of manufacture of low-quality products and formation of services market as a main tool of increase of well-being of companies and the state. Despite the fact that the social mechanism considered population as a "donor" of financial resources, limitation of the market of low-quality products was replaced by the conditions of growing demand and insufficient offer of low-quality products.

Stage 4 – 2000's. The multi-sectorial mechanism (Sawhney and Prandelli 2000). The multi-sectorial mechanism shows the necessity for differentiated development of the key spheres of innovational economy. The innovations are modernized capacities which form the exclusive technological product. The ley factor of production is human capital, aimed at creation of strategic vision of economic development. The attempts using the multi-sectorial mechanisms were successful in developed economies of early 2000's (Germany, France, and the USA). Experience of the Russian Federation showed that application of this mechanism required modernization of production funds which was partially conducted in late 2010's.

Stage 5 – 2010–2014. The structural mechanism (Borisova 2016) of development of innovational economy builds the system of innovational measures aimed at achievement of the key goal. The main production factors are technological innovations. On the whole, the structural approach is viewed as a strategy of development of innovational economy with distinguishing the goals, tasks, directions, resources, measures, processes, and volumes of financing. At the end of 2014, this mechanism transformed into the integrated (program) one.

Stage 6 since 2015. The integrated (program) mechanism (Stroeva and Lyapina 2016). The integrated (program) mechanism is based on development of economy through the means of implementation of innovational programs and projects. The production factor is intellectual initiative, which transforms into the project idea in the process of innovational development. As a matter of fact, this mechanism integrated the economic development in the aspect of innovativeness. It is possible to note that innovational development is the key goal of priority of the national and regional economy. The main elements of this mechanism are object, subject, form, method-ological tools, and the key programs of development.

As of now, priority of integrated (program) mechanism over other forms is obvious. Development of economy in view of this mechanism is built on the idea of innovative activities as comprehensive and interacting elements of integrated environment. Let us try to apply the integrated (program) mechanism to the theoretical model of manage-ment of region's economy innovational development.

3 Theoretical Model of Management of Region's Economy Innovational Development Through the Objective Mechanism of Interaction

The integrated mechanism is given as objective interaction of the key elements of program management of innovational economy. Thus, we deem it possible to deter-mine the name of this mechanism as "objective mechanism of interaction".

"Objective mechanism of interaction" is found in foreign (Cong et al. 2017) and Russian (Gerasimov 2011) literature. Most often, this mechanism is seen as integrated structure of the studied object with divided and newly grouped elements of interacting environment (Duscov 2014). Distinguishing the peculiarities of this mechanism is brought down to the presence of:

- option 1 – the system complex of adjacent elements of external environment (Qi et al. 2016);
- option 2 – well-balanced parametric elements that form qualitative preconditions of functioning of the mechanism (Sturm et al. 2006);
- option 3 – program components (object, subject, methods, forms, and key tools) (Lyapina 2011; Stroeva 2011).

We think that the most adapted idea of "objective mechanism of interaction" is given in option 3, which is related to the presence of the key elements necessary for innovational development of region's economy. Let us view them in detail.

The objects are economic categories, for development of which the model of region's innovational economy is aimed. The top-priority objects within innovational development are innovational and economic complexes that include: innovational companies (Orlova 2009; Stroeva et al. 2015), innovational complexes (Gamble 2000), innovational clusters (Dubrovsky and Shcherbakov 2012), and technological clusters (Zaytsev 2010). Presence and development of these objects reflect innovativeness of region's economy. Thus, the presented elements focus on development of objects of innovative activities. Development of region's innovational economy is secondary.

Subjects are considered as elements that form innovational development of region's economy. Based on the research on formation of the key subjects of economic development (Nikolaev 2007), let us distinguish their main types: federal and regional authorities, population, and investors. These subjects are controlling elements (public authorities bodies), financial and executive (investors), and accumulating (population). The key interconnection of these elements is tied to the block "object".

Form is seen as external cover of relations between objects and subjects of the theoretical model of development of region's innovational economy. Form is an inseparable part of projecting the external interaction, which is manifested as:

- managerial (Filatov 2010) – typical expression of controlled and subjected functions of interaction of the subjects of the model of development of region's innovational economy;
- financial (Gulzhan et al. 2015) – method of external expression of financial accumulation and distribution of assets within the interrelations of the model of development of region's innovational economy;
- organizational (Semenov and Kozin 2010) – the form of economic expression, conventionally established form of interaction between subjects and objects of the model of development of region's innovational economy.

The final interconnection of the form as an element of the model of innovational development of region's economy is tied to the block "object".

Methodological tools form innovational activity within the model of innovational development of region's economy (Conceicao et al. 1997; Stroeva et al. 2015). These tools allow improving the processes of innovational development, forming and stimulating innovative activities of region's economy, and financing of innovational and technological projects. The last item of innovational activity, despite the tie to the block "object", is closely interconnected to the programs of innovational development of region's economy. The simplest typologization of programs includes the strategy of socio-economic development and federal and regional programs of the territory's innovational development (Dahlstrom et al. 2003).

Despite the viewed key elements of the theoretical model of region's economy innovational development, it is important to distinguish the methods of study of these elements (Fig. 2).

Research methods include the method of considering and evaluating the elements of interaction of the model of innovational development of region's economy. Simply speaking, the methods allow forming the top-priority directions of development of innovative activities and analyzing the influence of factors of internal and external environments. According to the above elements, the following research methods were distinguished:

1. Objects and subjects of the mechanisms of development of region's innovational economy (Zdrazil et al. 2016) – foresight (long-term forecasting of development of the object and subject within the regional economy), factor analysis (distinguishing the key priorities of development of external and internal environments of region's innovational economy), road mapping (evaluation of expert information in the sphere of innovational activity of regional economy).
2. Forms of the mechanisms of development of region's innovational economy (Whitley 2002) – benchmarking (analysis of the model indicators of innovational

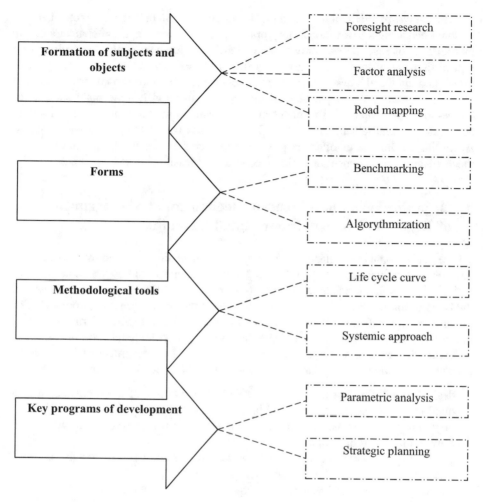

Fig. 2. Certain methods of research within the mechanisms of innovational development of region's economy

development of region's economy), algorythmization (formation of routes and modern forms of mechanisms of region's innovational economy).

3. Methodological tools of the mechanisms of development of region's innovational economy (Veshkina 2011) – life cycle curve (stage-by-stage consideration of created innovational product for the purpose of reflecting its effectiveness on region's economy), systemic approach (presenting the mechanisms of development of region's innovational economy as the systems of interacting elements).

4. The key programs of development of region's innovational economy (Booker 2000) – parametric analysis (the state of key parameters of development of region's innovational economy), strategic planning (scaling the model of development of region's innovational economy in the long-term).

The above methods allow building the theoretical model of the mechanisms of regional economy's innovational development. At that, the empirical character of the mechanisms of regional economy's innovational development could be confirmed by typical division of these tools into categories. The key mechanisms are the organizational and financial ones, which allow integrating the innovational processes into region's economy. The organizational mechanism distinguishes the main subjects of innovational development. The financial mechanism forms the blocks of distribution, accumulation, and redistribution of the key resources within the innovative activities. According to this, let us offer the proprietary models of organizational and financial mechanism of management of region's economy innovational development.

4 Organizational and Financial Mechanism of Management of Region's Economy Innovational Development

The organizational mechanism of management of region's economy innovational development is presented as a totality of the key organizational components of innovative activities that function in the conditions of external challenges and problems on the basis of interconnected work of the subjects of the innovational environment. The organizational mechanism of managing the innovational development forms according to the external challenges and problems that appear in the regional economy.

According to this, strategic landmarks and goals of development of innovational economy are formed, which have to correspond to the following parameters:

- designing and forming the system of effective interaction of the innovational production and economic systems within the region's economy;
- organizing and creating innovational standards for management of region's economy innovational development;
- conducting planning and forecasting of innovative activities within the region's economy for correction of management mechanisms;
- ensuring innovational development of qualitative information flow;
- developing and rationalizing innovative activities within the region's economy.

These parameters should be implemented by the key subjects of innovative activities, which include innovational companies, innovational complexes, innovational clusters, and technological clusters. Limitation of these subjects leads to creation of differentiated products of innovative activities. This circumstance is related to the deep nature of innovational products, observation of competitiveness and demand/offer, and prevention of overproduction of products and technologies. According to this, innovational companies create innovations for production systems and innovational complexes – intellectual technologies and innovational clusters – educational and institutional innovations, technological clusters – technological innovations. Managing this typologization of these productions and final products is related to the moderator's vision, which is regional economy. Manifestation of moderator's activities is conducted with regional management of economy, with the minimum number of additional subjects of innovative activities – regional public authorities.

The key managerial tool of the moderator within the organizational mechanism is help in development and implementation of innovations into the regional environment. For realization of this help, strategic innovational plans are created, and effectiveness of innovational development of region's economy are assessed. The methodological tools of assessment of effectiveness of innovational development include three indicators:

– innovational activity of regional economy;
– share of innovational regional structures;
– expenditures for technological innovations within the regional environment.

These indicators are reflected in the organizational mechanism for conducting its possible correction according to the emerging changes of external environment.

The financial mechanism of management of region's innovational economy seems to be different to the authors. According to the above studies of the evolutional development of mechanisms, it was determined that the integrated mechanism, which is formed on the basis of program provision of innovative activities, dominates. Regional economy within the financial activities functions through created technological and intellectual innovations, which gives the largest synergetic effect. Subjects in these models are not seen, and programs are the key indicators.

The financial program includes the directions related to application of the key tools of financing of innovational development. This program includes direct financing (algorithm of actions), direct crediting (parameters of the innovational company that receives a credit with 1% interest for development of innovative activities), subsidies (forecasted sums of allocations of assets for next years), tax subsidies (related to full refusal from payment of taxes during the planned period), pricing and tariff regulation (necessity for planning additional assets in the budget articles that are related to co-financing of innovative activities).

The managerial program of innovational development prescribes the key measures within the organization, management, control, and monitoring of the innovative activities. The managerial program includes audit and control (over development of innovational companies that receive state financial support), execution of regional projects (region is the customer, regional innovational company is the contractor), co-financing of innovative activities (region co-finances the innovational company on the conditions of borrowing experience for state unitary companies), creation of an effective system of foreign economic activities (selling innovational products in the global market), and land regulation (help in provision of land plots for implementation of innovational agro-industrial projects).

A commercial program of innovational development is the indicator of interconnected relations between the region and private investors. According to this program, the patent innovative activities (control over registration and sale of intellectual product) are regulated, investment projects (participation of innovational companies in private investment projects) are implemented, and private capital (establishment of priority of development of sectors and spheres of regional economy) is divided between the sectors.

Based on the above programs, the assets are accumulated, distributed, redistributed, and reserved within the continuous financial flow. Based on this flow, region's economy's innovative activities are corrected through adaptation to the external conditions and accounting of changes of the normatives.

5 Conclusions

The above research on the topic of financial and organizational mechanisms of management of region's economy innovational development allowed for several conclusions. Regional economy is currently in the state of transformation according to domination of the paradigm of innovational development. Creation of effective innovative activities that allow implementing strategically important directions of economy sets the tasks before the regions that are related to the search for new mechanisms of development. Assessment of evolution of the mechanisms of development of innovational economy on the basis of application of the stage-by-stage approach allowed determining the priority of the integrated mechanism. The integrated mechanism is viewed as a system of projects and programs within the innovational economy. The main elements of this mechanism are object, subject, form, methodological tools, and key programs of development. According to these elements, the authors came to the conclusion on this mechanism's similarity to the "objective mechanism of interaction", which is often viewed as the studies of foreign authors. The main elements of this mechanism are variative sets of key components:

– option 1 – the system complex of similar elements of external environment;
– option 2 – well-balanced parametric elements that form qualitative preconditions for the mechanism's functioning;
– option 3 – program components (object, subject, methods, forms, and key tools).

Based on the third option, the authors built the models of the organizational and financial mechanism of managing the innovational economy of region. These models allow integrating the innovational processes into regional economy. The organizational mechanism distinguishes the main subjects of innovational development. The financial mechanism forms the blocks of distribution, accumulation, and redistribution of the key resources within innovative activities. The built mechanism of managing the innovational economy of region allow the following: firstly, designing and forming the system of effective interaction of innovational production and economic systems within region's economy; organizing and creating innovational standards for management of region's economy innovational development; conducting planning and forecasting of innovative activities within region's economy for correction of the mechanisms of management.

References

Lazonick, W.: The theory of the market economy and the social foundations of innovative enterprise. Econ. Ind. Democr. **24**(1), 9–44 (2003)
Komarevtseva, O.O.: Economic rationality areas: a mandatory requirement in the conditions of negative changes. In: Living Economics: Yesterday, Today, Tomorrow The International Scientific and Practical Web-Congress of Economists and Jurists. ISAE Consilium, pp. 47–52 (2016)
Belskaya, G.S.: Innovational capabilities of development of Russia's economy. Bull. Tomsk State Univ. Econ. **2**, 38–46 (2009)

Kharlamov, V.I.: Innovational path of development of region's economy. Financ. Econ. **1**(2), 58–64 (2011)

Sawhney, M., Prandelli, E.: Communities of creation: managing distributed innovation in turbulent markets. California Manag. Rev. **42**(4), 24–54 (2000)

Borisova, E.V.: Stages of establishment and development of the concepts of economy's innovational development. Bull. South Urals State Univ. Ser. Econ. Manag. **10**(1), 76–82 (2016)

Stroeva, O.A., Lyapina, I.R.: Innovational potential as efficiency of managing socio-economic systems. Compet. Glob. World Econ. Sci. Technol. **9–2**(24), 172–177 (2016)

Cong, H., Zou, D., Wu, F.: Influence mechanism of multi-network embeddedness to enterprises innovation performance based on knowledge management perspective. Clust. Comput. **20**(1), 93–108 (2017)

Gerasimov, A.V.: Innovational economy – the basis of overcoming the asymmetry of Russian regions' development. Issues Econ. Legal Pract. **2**, 366–369 (2011)

Duscov, R.: Establishing a national system of innovation as a growth condition of country's competitiveness in the global economy. Economie si Sociologie: Revista Teoretico-Stiintifica, **4**, 72–84 (2014)

Qi, X., Li, X., Yan, S., Shuai, S.: Research on the innovation mechanism and process of China's automotive industry. In: PICMET 2016 – Portland International Conference on Management of Engineering and Technology: Technology Management for Social Innovation, Proceedings Technology Management For Social Innovation, pp. 1202–1210 (2016)

Sturm, F., Wohlfart, L., Wolf, P., Slagter, R., Emshanova, T.: Setting up communities of practice for innovative Russian SMES. In: IEEE International Technology Management Conference – ICE, p. 7477077 (2006)

Lyapina, I.R.: Problems of innovational development of the Russia's national economy. Sci. Notes OrelSIET **1**, 127–130 (2011)

Stroeva, O.A.: The scenario approach to managing the innovational development of regional economic systems. Reg. Syst. Econ. Manag. **3**(14), 104–110 (2011)

Orlova, N.V.: Innovational economy and innovational path of development. Bull. Saratov State Tech. Univ. **4–1**(42), 205–210 (2009)

Stroeva, O., Lyapina, I., Konobeeva, E., Konobeeva, O.: Effectiveness of management of innovative activities in regional socio-economic systems. Eur. Res. Stud. **XVIII**(Special Issue), 59–72 (2015)

Gamble, J.E.: Management commitment to innovation and ESOP stock concentration. J. Bus. Ventur. **15**(5–6), 433–447 (2000)

Dubrovsky, A.V., Shcherbakov, A.P.: Realia and perspectives of innovational development of Russia's economy. Bull. South-West State Univ. **3–2**(42), 8–13 (2012)

Zaytsev, A.S.: Assessment of topicality of development of the Russian economy in the innovational direction. Intellect, Innovations, Investments **4**, 10–14 (2010)

Nikolaev, A.B.: Certain problems of innovational development of the Russian economy. Philos. Econ. **1**(49), 200–211 (2007)

Filatov, Y.N.: Possibilities and problems of innovational development of the Russian economy. Bull. Volga State Univ. Serv. Ser. Econ. **9**, 27–30 (2010)

Gulzhan, T., Yerzhan, O., Elvira, A.: Government support of innovative and industrial development of the economy. Mediterr. J. Soc. Sci. **6**(3), 445–452 (2015)

Semenov, E.Y., Kozin, S.A.: Technological modes in economy and innovational potential of Russia's development. Bull. Irkutsk State Techn. Univ. **5**(45), 327–331 (2010)

Conceicao, P., Gibson, D., Heitor, M.V., Shariq, S.: Towards a research agenda for knowledge policies and management. J. Knowl. Manag. **1**(2), 129–141 (1997)

Stroeva, O., Shinkareva, L., Lyapina, I., Petruchina, E.: Optimization of approaches to the management of investment projects in regions of Russia. Mediterr. J. Soc. Sci. **6**(3), 87–94 (2015)

Dahlstrom, K., Skea, J., Stahel, W.R.: Innovation, insurability and sustainable development: sharing risk management between insurers and the state. Geneva Pap. Risk Insur. **28**(3), 394–412 (2003)

Zdrazil, P., Kraftova, I., Mateja, Z.: Reflection of industrial structure in innovative capability. Eng. Econ. **27**(3), 304–315 (2016)

Whitley, R.: Developing innovative competences: the role of institutional frameworks. Ind. Corp. Change **11**(3), 497 (2002)

Veshkina, E.Y.: Tendencies and priorities of development of region's innovational economy. Civ. Serv. **6**, 97–99 (2011)

Booker, E.: Success in online commerce requires more than low price. Advert. Age's Bus. Mark. **85**(11), 10 (2000)

Innovational Tools for Provision of Food Security Through State Support for the Agro-Industrial Complex in the Conditions of Digital Economy

Aleksei V. Bogoviz[1]($\boxtimes$) (iD), Pavel M. Taranov[2],
and Alexander V. Shuvaev[3]

[1] Federal State Budgetary Scientific Institution "Federal Research Center of Agrarian Economy and Social Development of Rural Areas – All Russian Research Institute of Agricultural Economics", Moscow, Russia
aleksei.bogoviz@gmail.com
[2] Don State Technical University, Rostov-on-Don, Russia
pm.taranov@gmail.com
[3] Stavropol State Agricultural University, Stavropol, Russia
a-v-s-s@rambler.ru

Abstract. The purpose of the article is to develop the innovational tools of provision of food security through state support for the agro-industrial complex (AIC) in the conditions of digital economy in modern Russia. For verification of the offered hypothesis and determination of connection between the level of national food security of modern Russia, state support for the AIC, and the use of the possibilities of digital economy, the article uses the method of correlation analysis. Also, the authors use the method of trend and horizontal statistical analysis, with the help of which the authors use the dynamics of change of the index of national food security in Russia in 2012–2017. The authors substantiate that state support for the AIC in modern Russia is not related to usage of existing possibilities of digital economy and does not fully ensure the national food security. The authors offer to supplement the traditional tools of financial support for national manufacturers of the AIC products with state subsidies, subsidized credits, etc., as well as innovational tools which is based on usage of the capabilities of digital economy. These tools allow achieving competitive advantages of domestic manufacturers of the AIC as to the quality and price. As compared to the traditional tools of state support, which belongs to the sphere of protectionism, the offered innovational tools do not limit the foreign competition and does not provide obvious preferences for domestic entrepreneurship. Instead, innovational tools increase competitiveness of domestic manufacturers of the AIC products, i.e., uses the market mechanism, leaving the choice to customers, hindering access to the market for dishonest entrepreneurs.

Keywords: Tools for provision of food security
State support for the agro-industrial complex · Digital economy
Modern Russia

© Springer International Publishing AG, part of Springer Nature 2018
E. G. Popkova (Ed.): HOSMC 2017, AISC 622, pp. 659–665, 2018.
https://doi.org/10.1007/978-3-319-75383-6_84

1 Introduction

The recent global economic crisis had a strong negative influence on the modern global economic system. The crisis influenced the spheres that produce the items of luxury and everyday usage goods, among which the central role belongs to the agro-industrial complex (AIC). The crisis led to destruction of the transnational economic relations. Due to this, the countries that export the products of the AIC faced the problem of selling these products, and the countries that import the products of the AIC faced the problem of their deficit.

In addition to this, domestic companies of the AIC reduced their business activity, and their number reduced due to decrease of the volume of paying capacity and problems with resources supply, which usually were based on international production chains in the interests of optimization of business processes by means of international division of labor. As a result, the problem of national food security became urgent in a lot of countries – especially those that import products of the AIC. Modern Russia belongs to the countries that do not specialize in manufacture of products of the AIC and import a large part of consumed products of this economic complex.

The working hypothesis of the research is based on the hypothesis that state support for the AIC in modern Russia is not connected to usage of the existing possibilities of digital economy and does not ensure the national food security of the country. The authors seek the goal of development of innovational tools of provision of food security through state support for the AIC in the conditions of digital economy in modern Russia.

2 Materials and Method

For verification of the offered hypothesis and determination of connection between the level of national food security of modern Russia, state support for the AIC, and usage of capabilities of digital economy, the authors use the method of correlation analysis.

The authors use the correlation analysis to calculate coefficients of correlation that reflect dependence of the index of national food security according to the Economist Intelligence Unit on the volume of state support for the AIC in Russia, set according to the State program of development of agriculture and regulation of the markets of agricultural products, resources, and food for 2013–2020, established by the Decree of the Government of the Russian Federation dated July 14, 2012 No. 717 and on the index of development of the information and communication technologies according to the International Telecommunication Union. Statistical information for 2012–2017 is presented in Table 1.

Also, the authors use the method of trend and horizontal statistical analysis for studying the dynamics of change of the index of national food security in Russia 2012–2017.

Table 1. Dynamics of the values of the food security index, index of development of the information and communication technologies, and the volume of state support for the AIC in Russia in 2012–2017

Indicator	Values of the indicators for the years					
	2012	2013	2014	2015	2016	2017
Index of national food security, points	68.3	60.9	62.7	63.8	63.8	66.2
Volume of state support for the AIC, RUB billion	150.22	189.23	221.26	240.07	254.15	271.12
Index of development of the information and communication technologies, points	6.74	6.77	6.81	6.85	6.91	6.92

Source: compiled by the authors on the basis of: (Economist Intelligence Unit 2017; Government of the Russian Federation 2017; International Telecommunication Union 2017).

3   Discussion

The issues of provision of national food security are studied in the works (Karandish and Hoekstra 2017; Danylenko et al. 2017; Karanina et al. 2017). The essence and specifics of the processes of state support for the AIC are viewed in the publications (Sandu et al. 2017; Mikhaylova et al. 2017). The possibilities of digital economy are analyzed in the studies (Popkova et al. 2016; Ragulina et al. 2015; Bogoviz et al. 2017; Bogdanova et al. 2016; Popova et al. 2016; Kuznetsov et al. 2016; Kostikova et al. 2016; Simonova et al. 2017).

As a result of the performed overview of scientific publications on the selected topic, it is possible to conclude that its components – the concept of national food security, state support for the AIC, and digital economy – are studied in detail. However, there is no systemic comprehensiveness of this topic's elaboration, which requires further complex studies that allow unifying the accumulated knowledge and directing them at solving the problem of provision of food security through state support for the agro-industrial complex in the conditions of digital economy.

4   Results

The results of the performed analysis are given in Table 2.

The data of Table 2 show that the values of the national food security index of Russia are not related to the volume of state support for the AIC (correlation coefficient – 0.03%) and the index of development of the information and communication technologies (correlation coefficient – 0.07%).

Moreover, despite the annual increase of the volume of state support for the AIC (its average growth constitutes 12% annually), national food security of Russia remains at the low level. Thus, it reduced by 11% in 2013, as compared to 2012, and by 3% in 2017, as compared to 2012.

Table 2. Results of analysis of the values of the national food security index

Indicator (type of analysis)	Detalization of the indicator	Value
Growth rate (horizontal analysis)	2013/2012	0.89
	2014/2013	1.023
	2015/2014	1.02
	2016/2015	1.00
	2017/2016	1.04
Growth rate (trend analysis)	2017/2012	0.97
Correlation coefficient (correlation analysis)	With volume of state support for the AIC	0.03%
	With index of development of the information and communication technologies	0.07%

Source: calculated by the authors.

This work offers the following innovational tools of provision of food security through state support for the agro-industrial complex in the conditions of digital economy. The 1st tool: optimization of transport logistics. Digital technologies may help to determine the most popular directions of movement of domestic products of the AIC. For example, the products of the AIC are usually manufactures in peripheral regions of the country, where the cost of all types of resources is lower, and then is transported into central regions, where the level of prices is higher and it is possible to sell these products with larger profit.

Determining the most popular directions of movement of domestic products of the AIC allows concentrating the state's efforts on their development. Modernization of transport logistics allows accelerating and cheapening the process of transportation of products of the AIC, thus providing its domestic manufacturers with sustainable pricing competitive advantage as compared to foreign rivals which will use other directions with less developed transport logistics.

2nd tool: introduction of the national system of bar codes. We offer to introduce the single national bar code of the products of the AIC, which allows tracking the movement on the territory of Russia and the sales online. This measure could be a means of fighting counterfeit products, thus ensuring security of its consumption by Russia's population. For the imported products of the AIC, digital bar codes will be assigned on the paid basis. Revenues from this will be used for support for domestic companies of the AIC, to the products of which the digital bar codes will be assigned for free.

3rd tool: certification of products' quality. The existing modern digital technologies allow creating special devices for instant quality control over products of the AIC. We offer to make this control mandatory on the whole territory of Russia. It will include analysis of the structure of products of the AIC for harmful additives, pesticides, conserving agents, GMO, etc. The analysis may also suppose correspondence of products of the AIC to the existing GOSTs. Similarly, for imported products of the AIC this certification should be paid, and for domestic manufacturers – free or cheap. This will allow guaranteeing high quality of products of the AIC for Russian consumers.

The mechanism of provision of food security on the basis of the offered innovational tools of state support for the AIC in the conditions of digital economy is presented graphically in Fig. 1.

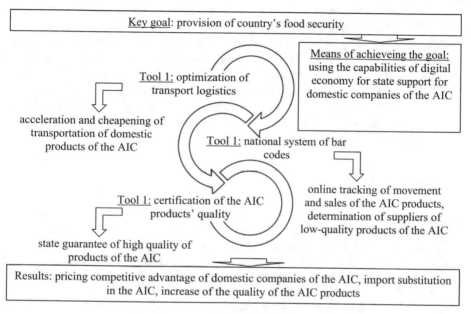

Fig. 1. The mechanism of provision of food security on the basis of innovational tools of state support for the AIC in the conditions of digital economy Source: compiled by the authors.

As is seen in Fig. 1, using the offered tools allows providing the pricing competitive advantage to domestic companies of the AIC, thus stimulating import substitution in the AIC and guaranteeing the increase of the AIC products' quality. As a result, provision of national food security in the long-term is achieved.

5 Conclusions

It should be noted that national food security is a complex notion. In practice, its provision includes not only support for domestic manufacturers of products of the AIC in the interests of ousting foreign rivals, import substitution, and reduction of import of products of the AIC. An important component in the process of provision of national food security is state control and guarantee of quality of products of the AIC.

That's why it is offered to supplement the traditional tools of financial support for domestic manufacturers of products of the AIC (state subsidies, subsidized credits, etc.) by innovational tools based on using the possibilities of digital economy. These tools allow achieving competitive advantages of domestic manufacturers of the AIC in quality and price.

As compared to the traditional tools of state support, which belongs to the sphere of protectionism, the offered innovational tools do not limit foreign competition and do not provide obvious preferences for domestic entrepreneurship. Instead, innovational tools raise the competitiveness of domestic manufacturers of products of the AIC, i.e., use the market mechanism, leaving the choice to the consumers but blocking dishonest entrepreneurs for entering the market.

A certain limitation of the results of this research is emphasis on the economy of modern Russia, as other countries of the world may have other problems in the sphere of provision of national food security and use other tools state support for the AIC. That's why determining the country analogies and verification of applicability of the offered innovational tools of provision of food security through state support for the AIC in the conditions of digital economy in various countries of the world is a perspective direction of further scientific research.

References

Bogdanova, S.V., Kozel, I.V., Ermolina, L.V., Litvinova, T.N.: Management of small innovational enterprise under the conditions of global competition: possibilities and threats. Eur. Res. Stud. J. **19**(2 Special Issue), 268–275 (2016)

Bogoviz, A.V., Ragulina, Y.V., Kutukova, E.S.: Ways to improve the economic efficiency of investment policy and their economic justification. Int. J. Appl. Bus. Econ. Res. **15**(11), 275–285 (2017)

Danylenko, A., Satyr, L., Shust, O.: Price parity in the agricultural sector as a guarantee of the national food security. Econ. Ann. XXI **164**(3-4), 61–64 (2017)

Economist Intelligence Unit: The Global Food Security Index 2012–2017 (2017). http://foodsecurityindex.eiu.com/Country/Details#Russia. Accessed 15 Nov 2017

International Telecommunication Union: ICT Development Index 2012–2017 (2017). https://www.itu.int/en/ITU-D/Statistics/Pages/default.aspx. Accessed 15 Nov 2017

Karandish, F., Hoekstra, A.Y.: Informing national food and water security policy through water footprint assessment: the case of Iran. Water (Switzerland) **9**(11), 831 (2017)

Karanina, E., Sapozhnikova, E., Loginov, D., Holkin, A., Sergievskaya, E., Zurakhovskii, A.: National aspects of food security of Russia. In: MATEC Web of Conferences, 106, 08079 (2017)

Kostikova, A.V., Tereliansky, P.V., Shuvaev, A.V., Parakhina, V.N., Timoshenko, P.N.: Expert fuzzy modeling of dynamic properties of complex systems. ARPN J. Eng. Appl. Sci. **11**(17), 10601–10608 (2016)

Kuznetsov, S.Y., Tereliansky, P.V., Shuvaev, A.V., Natsubize, A.S., Vasilyev, I.A.: Analysis of innovate solutions based on combinatorial approaches. ARPN J. Eng. Appl. Sci. **11**(17), 10222–10230 (2016)

Mikhaylova, N.A., Babich, T.V., Smirnova, O.S.: Improvement of the state support mechanism for regional agro-industrial complex under the conditions of international sanctions and Russia's membership in the WTO. Contributions to Economics, pp. 151–157 (2017). ISBN 978-3-319-60695-8

Popova, L.V., Popkova, E.G., Dubova, Y.I., Natsubidze, A.S., Litvinova, T.N.: Financial mechanisms of nanotechnology development in developing countries. J. Appl. Econ. Sci. **11** (4), 584–590 (2016)

Ragulina, Y.V., Stroiteleva, E.V., Miller, A.I.: Modeling of integration processes in the business structures. Mod. Appl. Sci. **9**(3), 145–158 (2015)

Simonova, E.V., Lyapina, I.R., Kovanova, E.S., Sibirskaya, E.V.: Characteristics of interaction between small innovational and large business for the purpose of increase of their competitiveness. In: Russia and the European Union Development and Perspectives, pp. 407–415 (2017)

Popkova, E.G., Chechina, O.S., Abramov, S.A.: Problem of the human capital quality reducing in conditions of educational unification. Mediter. J. Soc. Sci. **6**(3), 95–100 (2016)

Government of the Russian Federation: State program for development of agriculture and regulation of the markets of agricultural products, resources, and food for 2013–2020, established by the Decree of the Government of the Russian Federation dated July 14, 2012, No. 717 (2017). http://www.agro-ferma.ru/dayatelnost/rekonstruktsiya-sooruzheniy/stati/programma-razvitiya-apk-na-2013-2020-gody/. Accessed 15 Nov 2017

Sandu, I.S., Bogoviz, A.V., Ryzhenkova, N.E., Ragulina, Y.V.: Formation of innovational infrastructure in the agrarian sector. AIC Econ. Manag. **1**(1), 35–41 (2017)

Economic Stimuli for Creating Highly-Efficient Jobs: A Modern Human's View

Yulia A. Agunovich[1](✉), Yulia V. Ragulina[2],
Alexander N. Alekseev[3], Elena V. Kletskova[4],
and Pavel T. Avkopashvili[4]

[1] Federal State Budget Educational Institution of Higher Education "Kamchatka State Technical University", Petropavlovsk-Kamchatsky, Russia
agunovichО@mail.ru
[2] Federal State Budgetary Scientific Institution "Federal Research Center of Agrarian Economy and Social Development of Rural Areas—All Russian Research Institute of Agricultural Economics", Moscow, Russia
julra@list.ru
[3] Plekhanov Russian University of Economics, Moscow, Russia
[4] Altai State University, Barnaul, Russia
stroiteleva_tg@mail.ru, regcenoe@bk.ru

Abstract. The purpose of the work is to develop the concept of economic stimulation of creating highly-efficient jobs from the view of a modern human by the example of modern Russia. The authors determine the level of modern human's interest in creation of highly-efficient jobs through the prism of three various roles in socio-economic system of Russia. The information and analytical basis of the work consists of the materials of the sociological survey performed by the All-Russian Public Opinion Research Center in 2016, the materials of the report by the United Nations Development Programme on Human Development Index and the INSEAD, WIPO, Cornell University on innovational development of economy, as well as the materials of the International Monetary Fund and the Federal State Statistics Service for 2016. The data are processed with the help of the systemic, logical, dynamics (trend), and comparative analysis, as well as general scientific methods of the research, such as synthesis, deduction, induction, formalization, etc. The authors prove the thesis that the process of economic stimulation of creating the highly-efficient jobs in the economic system, studied by the modern science from the positions of entrepreneurial structures, should be viewed through the prism of human – both within and outside of these structures. The authors determine the key barriers on the path of successful economic stimulation of creation of highly-efficient jobs in modern Russia and offer practical recommendations for overcoming them, on the basis of which the concept of stimulation of creating the highly-efficient jobs from the point of view of modern human is created.

Keywords: Economic stimulation · Creation of highly-efficient jobs
Modern human

© Springer International Publishing AG, part of Springer Nature 2018
E. G. Popkova (Ed.): HOSMC 2017, AISC 622, pp. 666–672, 2018.
https://doi.org/10.1007/978-3-319-75383-6_85

1   Introduction

Under the influence of the processes of economic globalization and active regional integration, which stimulate the enlargement of economic subjects in the world economy, the basic unit of economic systems' analysis was reconsidered. At the level of the global economy, regional integration unions of countries appeared, and at the level of national economy – its regions. Also, the emphasis shifted from households (demand) to entrepreneurial structures (offer) as sources of business and innovational activity in economy.

This phenomenon influenced economics in a contradictory way. On the one hand, it supported macro-economic analysis, ensuring quick collection of information and its simplified processing, thus guaranteeing quick decision making in the sphere of state economic policy. However, on the other hand, this reduced precision and authenticity of this analysis due to its separation from the actual basic economic subject – modern human – and distorted the information used during state decisions making, thus increasing the probability of mistakes and the risk of unfavorable socio-economic consequences, including the crises of economic systems.

Consideration of a modern human as a part of entrepreneurial structure (from the aspect of demand) is narrow, because it reflects only one social role – as a worker and human resource that is used for production of goods in economy. In reality, modern human has a lot of roles in the socio-economic system, being the source of innovational potential, bearer of certain values, and consumer of goods. In view of these roles, modern human is a complex and multi-level socio-economic category, the authentic modeling of which is the guarantee of successful macro-economic analysis.

Based on the above, the authors offer the hypothesis that the process of economic stimulation of creating the highly-efficient jobs in the economic system, studied by the modern science from the positions of entrepreneurial structures, should be viewed through the prism of a human, both within these structures and outside of them. The purpose of the work is to develop the concept of economic stimulation of creating the highly-efficient jobs from the view of modern human by the example of modern России.

2   Materials and Method

For verification of the offered hypothesis, the authors determine the level of interest of a modern human in creation of highly-efficient jobs through the prism of three various roles in the socio-economic system of Russia. The first role, analyzed in this research, consists in patriotism of a modern human. Its essence is that, while being a part of the national economic system, modern human is interested in successful implementation of the designated strategic course of its development. Evaluation of the population's perception of implementation of national projects is performed on the basis of the sociological survey, conducted by the All-Russian Public Opinion Research Center in 2016.

The second role that is performed by modern human in the socio-economic system is that he is a source of innovational potential. For assessing the level of opening this potential in Russia, the materials of the report of the United Nations Development Programme on human development index and the report of INSEAD, WIPO, Cornell University on innovational development of economy are used. The data of these reports are compared for determining the differences in the potential of human development and the level of its opening (implementation) in Russia.

The third role of a modern human, studied in this article, consists in consumption of goods in economy. Sufficiency of own production of goods in Russia is evaluated through comparing the volume of gross domestic product (GDP) and GDP per capita according to the International Monetary Fund to the volume of import according to the Federal State Statistics Service. The statistical data that are the basis of the research are given in Table 1.

Table 1. Statistical data for the research

Indicators	2000	2005	2011	2012	2013	2014	2015	2016
Import, $ billion	33.9	98.7	228.9	305.8	317.3	315.3	286.7	182.4
GDP, $ billion	279.03	820.57	2170.10	2230.60	2063.70	1365.90	1280.70	1560.70
GDP per capita, $	1,905.94	5,713.36	15,145.38	15,558.80	14,388.00	9,521.08	8,928.69	10,885.48
Level of population's approval of national projects' implementation, %	–	–	–	–	–	–	–	95.64
Human development index, points	–	–	–	–	–	–	–	0.798
Innovational index of economy, points	–	–	–	–	–	–	–	38.76

Source: compiled by the authors on the basis of: (All-Russian Public Opinion Research Center 2017; Federal State Statistics Service 2016; United Nations Development Programme 2017; International Monetary Fund 2017; INSEAD, WIPO, Cornell University 2017).

The above data are processed with the help of systemic, logical, dynamics (trend), and comparative analysis, as well as synthesis, deduction, induction, formalization, etc.

3 Discussion

The economic aspect of stimulation of creating the highly-efficient jobs in the interests of accelerating the rate of economic growth, increase of competitiveness, and achievement of import substitution and other positive effects is studied in detail in the works (Popkova et al. 2016; Ragulina et al. 2015; Bogoviz et al. 2017; Bogdanova et al. 2016; Popova et al. 2016; Kuznetsov et al. 2016; Kostikova et al. 2016; Simonova et al. 2017). At that, the social aspect of this process is poorly studied by the modern economic science and requires further elaboration.

4   Results

The performed complex analysis of the data from Table 1 showed that, firstly, the level of population's approval of national projects' implementation is very high (95.64%). That is, the level of patriotic feelings of the Russia's population is rather high. That's why growth of efficiency of economy, which stimulates quick implementation of the strategic course of the national development of Russia, is one of the needs of a human in modern Russia.

Secondly, analysis also showed that import in Russia constituted 11.68% of the country's GDP in 2016. The annual average growth rate of import in Russia constitutes 27.59%, which exceed the annual average growth of GDP, which equals 12.75%. In addition to that, growth of GDP per capita constituted 71.13% in 2016, as compared to 2000, having exceeded the growth of GDP, which constituted 59.32%.

This allows stating that demand for goods in the Russian economy exceeds the existing internal offer. Therefore, Russian consumers are interested in growth of efficiency of domestic economy for increase of the volume of domestic production of goods and full satisfaction of existing internal demand.

Thirdly, the results of the analysis showed that human development index in Russia constituted 0.798 in 2016, while the maximum level of this index in the country ranking is 0.944. That is, the Russian index value equals 84.53% of the world leader. The innovational index of economy in Russia was 38.76 in 2016, while in the country ranking the maximum value of this index was 67.69. That is, the Russian index equals 57.26% of the world leader.

This shows incompleteness of implementation of the existing potential of human development in modern Russia. That's why human in modern Russia in interested in creation of highly-efficient jobs for opening the personal and innovational human potential.

Summarizing the above, it is possible to conclude that modern human is interested in creation of highly-efficient jobs through the prism of all social roles that he performed in the socio-economic system. This emphasizes the importance of economic stimulation of creating the highly-efficient jobs not only from the economic (growth of competitiveness, economic growth, etc.) but from the social (satisfying human needs) points of view.

We have determined the following barriers on the path of creation of highly-efficient jobs in modern Russia:

- high commercial attractiveness of non-productive entrepreneurship: a popular entrepreneurial scheme is formation of long chains of added value, the significant share of which supposes reselling the goods without changing them. As a result, consumers have to purchase the goods for high prices with low efficiency of entrepreneurship;
- smaller value of human resources, as compared to other types of production resources: insufficient level of knowledge on conduct of entrepreneurial activities in Russia leads to underestimation of human resources, which are valued in the world as sources of innovational development entrepreneurship;

–   inaccessibility of resources (primarily, investment and technological) for creation of highly-efficient jobs: low investment attractiveness, lack of own financial resources, and weak connection to R&D institutes predetermine the existence of this barrier.

For overcoming these barriers, the authors offer the following practical recommendations:

–   legislative establishment of limits of production chains which limit the possibilities of development of non-productive entrepreneurship;
–   conduct of a series of training and developing courses for entrepreneurs, aimed at demonstration of the value of human resources;
–   economic stimulation of highly-efficient entrepreneurship with the help of tax and credit state measures.

The concept of economic stimulation of creating the highly-efficient jobs, which reflects the influence of the offered recommendations on modern human, is shown in Fig. 1.

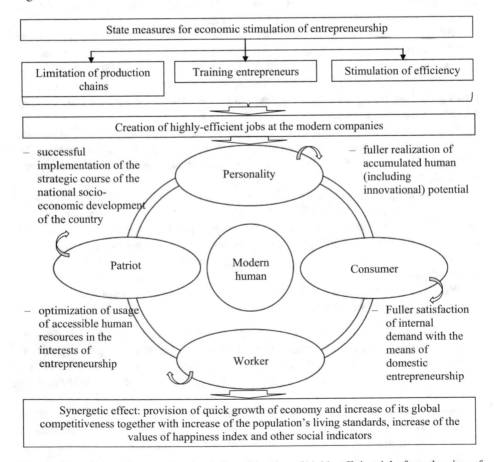

Fig. 1. The concept of economic stimulation of creation of highly-efficient jobs from the view of a modern human. Source: compiled by the authors.

As is seen from Fig. 1, implementation of the offered recommendations for economic stimulation of creating the highly-efficient jobs positively influences all levels of a modern human. As a personality, modern human takes advantages from fuller implementation of accumulated human (including innovational) potential, as a patriot – from successful implementation of the strategic course of the national socio-economic development of the country, as a human resource – from optimization of usage of accessible human resources in the interests of entrepreneurship, and as a consumer – from fuller satisfaction of internal demand by means of domestic entrepreneurship.

As a result, the synergetic effect is achieved, related to the fact that provision of quick growth of economy and increase of its global competitiveness takes place in combination with increase of the population's living standards, growth of the values of happiness index and other social indicators. That is, profit is gained at the macro-economic scale and at the scale of each separate human.

5 Conclusions

Thus it is proved that sustainable development of modern socio-economic systems requires the return to the origins of the classic economic thought and consideration of human as a basic item of analysis of these systems. The provided concept of economic stimulation of creating the highly-efficient jobs from the point of view of a modern human considers human as a central element of the system of economic stimulation of creating the highly-efficient jobs. This allows achieving economic and social positive results of this stimulation.

References

Bogdanova, S.V., Kozel, I.V., Ermolina, L.V., Litvinova, T.N.: Management of small innovational enterprise under the conditions of global competition: possibilities and threats. Eur. Res. Stud. J. **19**(2 Special Issue), 268–275 (2016)

Bogoviz, A.V., Ragulina, Y.V., Kutukova, E.S.: Ways to improve the economic efficiency of investment policy and their economic justification. Int. J. Appl. Bus. Econ. Res. **15**(11), 275–285 (2017)

INSEAD, WIPO, Cornell University: The Global Innovation Index 2017 (2017). https://www.globalinnovationindex.org. Accessed 20 Nov 2017

International Monetary Fund: Report for Selected Countries and Subjects: Russia (2017). http://www.imf.org/external/pubs/ft/weo/2017/01/weodata/weorept. Accessed 20 Nov 2017

Kostikova, A.V., Tereliansky, P.V., Shuvaev, A.V., Parakhina, V.N., Timoshenko, P.N.: Expert fuzzy modeling of dynamic properties of complex systems. ARPN J. Eng. Appl. Sci. **11**(17), 10601–10608 (2016)

Kuznetsov, S.Y., Tereliansky, P.V., Shuvaev, A.V., Natsubize, A.S., Vasilyev, I.A.: Analysis of innovate solutions based on combinatorial approaches. ARPN J. Eng. Appl. Sci. **11**(17), 10222–10230 (2016)

Popova, L.V., Popkova, E.G., Dubova, Y.I., Natsubidze, A.S., Litvinova, T.N.: Financial mechanisms of nanotechnology development in developing countries. J. Appl. Econ. Sci. **11** (4), 584–590 (2016)

Ragulina, Y.V., Stroiteleva, E.V., Miller, A.I.: Modeling of integration processes in the business structures. Mod. Appl. Sci. **9**(3), 145–158 (2015)

Simonova, E.V., Lyapina, I.R., Kovanova, E.S., Sibirskaya, E.V.: Characteristics of interaction between small innovational and large business for the purpose of increase of their competitiveness. Russia and the European Union Development and Perspectives, pp. 407–415 (2017)

United Nations Development Programme: Human Development Index 2016 (2017). http://hdr.undp.org/. Accessed 20 Nov 2017

All-Russian Public Opinion Research Center: Population's perception of realization of the national projects (2017). https://wciom.ru/research/research/socialno_ehkonomicheskie_issledovaniya/. Accessed 20 Nov 2017

Popkova, E.G., Chechina, O.S., Abramov, S.A.: Problem of the human capital quality reducing in conditions of educational unification. Mediter. J. Soc. Sci. **6**(3), 95–100 (2016)

Federal State Statistics Service: Russia in numbers: short statistical bulletin. Federal State Statistics Service, Moscow (2016)

Transformation of the Role of Human in the Economic System in the Conditions of Knowledge Economy Creation

Aleksei V. Bogoviz[1](✉) (iD), Yulia V. Ragulina[1] (iD),
Alexander N. Alekseev[2], Evgeni S. Anichkin[3],
and Viktor I. Dobrosotsky[4]

[1] Federal State Budgetary Scientific Institution "Federal Research Center
of Agrarian Economy and Social Development of Rural Areas—All Russian
Research Institute of Agricultural Economics", Moscow, Russia
aleksei.bogoviz@gmail.com, julra@list.ru
[2] Plekhanov Russian University of Economics Russia, Moscow, Russia
[3] Federal State-Funded Educational Institution of Higher Education
"Altai State University", Barnaul, Russia
rrd231@rambler.ru
[4] Moscow State Institute of Foreign Relations (MGIMO University)
of the Russian Ministry of Foreign Affairs, Moscow, Russia
dobrosotskiy@mgimo.ru

Abstract. The purpose of the work is to study the essence, consequences, and perspectives of managing the process of transformation of the role of human in the economic system in the conditions of knowledge economy creation by the example of modern Russia. For determining the dependence of traditional competitiveness of economy and successfulness of knowledge economy creation (as a new type of economy's competitiveness) on various types of resources – human, technological, material, and investment – this work uses the method of correlation analysis. The authors study the data of the United Nations Development Programme, the International Telecommunication Union, the World Bank, and the World Economic Forum for 2010–2016. Due to the usage of this method, the authors prove that human resources perform more important role in the process of knowledge economy creation than in the process of traditional provision of global competitiveness of economy, which requires reconsideration of the approach to managing them. The authors come to the conclusion that in the conditions of knowledge economy creation the process of transformation of the role of human is started. In this process, human transforms from the usual production resource (human resource), consumer of traditional goods, and entrepreneur, who uses the existing possibilities, into the leading production resource – the source of creation and implementation of innovations, as well as the consumer of innovational goods and innovations-active entrepreneur. Despite the fact that this transformation process takes place in the natural way under the influence of market forces, it could be distorted (market gaps could be formed); to prevent this, it is recommended that state manage this process. The authors present a perspective approach to managing the process of transformation of the role of human in the economic system in the conditions of knowledge economy creation and offer practical recommendations for its application in modern economic systems.

© Springer International Publishing AG, part of Springer Nature 2018
E. G. Popkova (Ed.): HOSMC 2017, AISC 622, pp. 673–680, 2018.
https://doi.org/10.1007/978-3-319-75383-6_86

Keywords: Transformation of human's role · Economic system
Competitiveness · Knowledge economy

1 Introduction

In the 21ˢᵗ century, the traditional models of competitiveness of economic systems was replaced by knowledge economy. Due to this, for the purpose of supporting high global competitiveness of economy, it is important not only to manufacture popular products and offer them in the world markets for profitable prices, manifesting high marketing activity; new characteristics of manufacture of these products – knowledge intensity and innovativeness – became very important.

Change of the success factors in the world markets led to the necessity for adapting business processes of entrepreneurial structures and the regulation mechanisms of national economic systems to a new status quo. At that, there is a scientific problem – despite the corresponding changes of formalized processes, a lot of companies and countries were not able to achieve the expected results (increasing and supporting high global competitiveness).

At the macro-level, this problem is expressed in the fact that despite starting the processes of economy's modernization, related to mass implementation of leading production technologies and equipment, as well as growth of state expenditures for stimulation of innovational activity of business, progress in development of knowledge economy remains at the same level as before implementation of these measures. At the micro-level, the essence of this problem is that acquisition of business innovations does not ensure strengthening of market positions in the expected volume.

This scientific research is aimed at studying and solving this problem. The scientific hypothesis of this research is that human resources have a more important role in the process of creation of knowledge economy that in the process of traditional provision of global competitiveness of economy, which requires reconsideration of the approach to their management. The purpose of this work is to study the essence, consequences, and perspectives of managing the process of transformation of human's role in the economic system in the conditions of knowledge economy creation by the example of modern Russia.

2 Materials and Method

For determining the dependence of traditional competitiveness of economy and successfulness of creation of knowledge economy (as a new type of economy's competitiveness economy) on various types of resources – human, technological, material, and investment – this work uses the method of correlation analysis. The authors study the data of the United Nations Development Programme, the International Telecommunication Union, the World Bank, and the World Economic Forum for 2010–2016 (Table 1).

The essence of the process of transformation of human's role in the economic system in the conditions of knowledge economy creation in contrast to the traditional

Table 1. Dynamics of values of the indicators of traditional and new competitiveness and their factors in Russia in 2010–2016

Indicators	Values of the indicators for the periods						
	2010	2011	2012	2013	2014	2015	2016
Index of human potential development, points	0.79	0.79	0.80	0.80	0.81	0.80	0.80
Index of development of information and communication technologies, points	5.61	6.00	6.48	6.70	6.91	5.52	5.87
Total natural resources rents (% of GDP)	14.15	13.90	16.15	15.56	13.73	13.47	10.31
Volume of direct foreign investments, $ billion	43.17	55.08	50.59	69.22	22.03	6.85	32.98
Index of knowledge economy, points	6.06	6.12	6.17	6.20	6.22	6.21	6.16
Index of global competitiveness economy, points	4.10	4.20	4.20	4.20	4.30	4.40	4.50

Source: (United Nations Development Programme 2017; International Telecommunication Union 2017; World Bank 2017a, b, c; World Economic Forum 2017).

type of economy's competitiveness economy is studied in this work with the help of the comparative, systemic, problem, and structural & functional analysis, synthesis, induction, deduction, and graphic presentation of information.

3 Discussion

The role of human resources in provision of global competitiveness of entrepreneurial structures and economic is studied and described in the works (Popkova et al. 2016; Ragulina et al. 2015; Bogoviz et al. 2017; Bogdanova et al. 2016), etc. The concept of knowledge economy and practical experience of its formation and development are viewed in the publications (Popova et al. 2016; Kuznetsov et al. 2016; Kostikova et al. 2016; Simonova et al. 2017), etc. At that, the gap in modern scientific knowledge is the lack of fundamental and applied research devoted to analysis of the process of transformation of human's role in the economic system in the conditions of knowledge economy creation.

4 Results

The results of the performed correlation analysis are given in Table 2.

As is seen from Table 2, the traditional index of global competitiveness of economy depends on accessibility of natural resources (correlation - 77%) and the volume

Table 2. Results of the correlation analysis

Independent variables	Dependent variables, correlation	
	Index of global competitiveness economy	Index of knowledge economy
Index of human potential development	0.53	0.99
Index of development of information and communication technologies	0.18	0.52
Accessibility of natural resources	0.77	0.04
Volume of direct foreign investments	0.62	0.32

Source: compiled by the authors.

of direct foreign investments (correlation - 62%). This index does not depend on technological (correlation - 18%) and human (correlation - 53%) resources.

The index of knowledge economy (new type of competitiveness of economic systems) does not depend on natural (correlation - 4%) and investment (correlation - 32%) resources. It is predetermined by technological resources (correlation - 52%) and depends on human resources (correlation - 99%). This starts the process of transformation of human's role in the economic system in the conditions of knowledge economy creation.

Thus, in the conditions of the traditional type of economy's competitiveness, human has a role of an ordinary production resource – a source of mechanic labor aimed at quantitative growth of efficiency (growth of the production volume) for achieving the "scale effect". Also, human is a source of entrepreneurial capability, determining the existing possibilities in the world markets and using them for organizing companies.

That is, in the aspect of studying human as an entrepreneur, emphasis is made on analysis of the current demand for goods in economy. Competitiveness is achieved with the help of creation of new (unique) combination of existing resources. Human is also the consumer of goods in economy. He sets demand for goods, thus stimulating business activity in certain (most popular) directions and being a vector of economic growth.

In the conditions of knowledge economy creation, human also has the role of innovator – the source of new and bearer of existing open and internal (corporate) knowledge and skills (competences). Activities of human resources in the conditions of knowledge economy include not only mechanic but also the intellectual, qualitative, component, within which it is aimed at creation and implementation into business processes of innovational technologies.

In the conditions of knowledge economy, human – as entrepreneur – acquires the role of source of innovations in economy and creator of innovational goods, creating new possibilities in the world markets. That is, in the aspect of studying human as entrepreneur the emphasis is made on analysis of potential (future) demand for goods in economy.

Competitiveness is achieved with the help of combining new (unique) resources. Being a consumer, human sets demand for innovations. Human is also the consumer of goods in economy. He sets demand for goods, thus stimulating business activity in certain (most popular) directions and being a vector of innovational development of economy.

The essence of the described process of transformation of human's role in the economic system in the conditions of knowledge economy creation is reflected in Fig. 1.

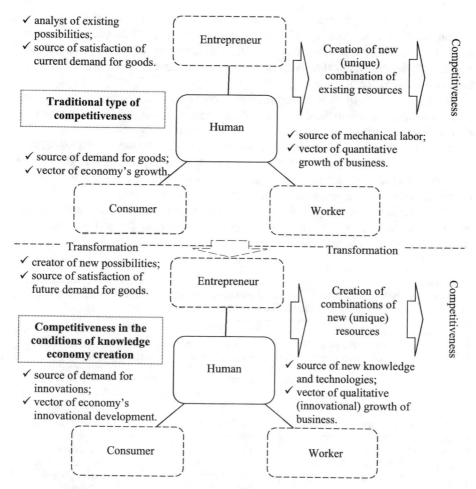

Fig. 1. Transformaiton of human's role in the economic system in the conditions of knowledge economy creation

For the purpose of successful management of the process of transformation of human's role in the economic system for supporting its high global competitiveness in the conditions of knowledge economy creation, we offer the complex of the following practical recommendations. The influence on the human, who is a consumer, should be performed through the increase of innovational literacy of population (explaining the essence of innovational goods, demonstration of their differences from traditional goods), and stimulation of demand for innovational goods (their propaganda and explaining all advantages).

For influencing human as a worker it is offered to use such tools as stimulating studying the innovations (through modernization of the educational system and co-financing of additional training) and stimulating optimization of employment – for ensuring targeted application of innovational capabilities of each employee in the economy.

During influencing human as an entrepreneur it is recommended to use such tools as stimulating the acquisition and implementation of innovations (through strengthening the connection between the scientific and business sector of economy) and stimulating the innovational activity (with the help of tax and other accessible mechanisms of stimulation).

A perspective approach to managing the process of transformation of human's role in the economic system in the conditions of knowledge economy creation, which takes into account the offered recommendations, is presented in Fig. 2.

As is seen from Fig. 2, the offered approach is based on the managerial mechanism, the subject of which are public authorities bodies that are responsible for development and implementation of state economic policy, and the object – human with all distinguished roles that he performs in the modern economic system.

This approach seeks the goal of preventing the distortion of market signals (market gaps) in the process of the economic system's transition to the path of knowledge economy creation. The method of achieving the goal is managing the process of

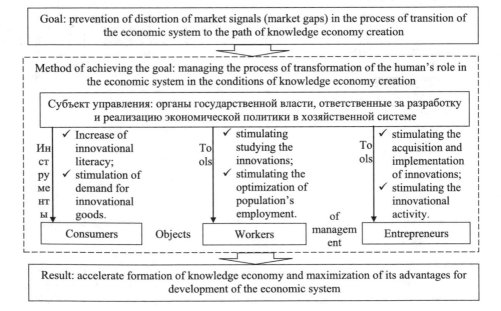

Fig. 2. A perspective approach to managing the process of transformation of human's role in the economic system in the conditions of knowledge economy creation Source: compiled by the authors.

transformation of human's role in the economic system in the conditions of knowledge economy creation on the basis of the offered practical recommendations. As a result, quick formation of knowledge economy and maximization of its advantages for development of economic system are achieved.

5 Conclusions

Summarizing the above results, it is possible to state that the offered hypothesis was confirmed – the authors substantiate that in the conditions of knowledge economy creation the process of human's role transformation is started in the economic system. In this process, human transforms from the usual production resource (human resource), consumer of traditional goods, and entrepreneur, who uses the existing possibilities, into the leading production resource – the source of creation and implementation of innovations, as well as the consumer of innovational goods and innovations-active entrepreneur.

Despite the fact that the above transformation process take place naturally under the influence of market forces, it could be distorted (market gaps could be formed); state management of this process is recommended for preventing it. The developed perspective approach to managing the process of transformation of human's role in the economic system in the conditions of knowledge economy creation allows making this process quick and smooth for all its participants and maximizing its advantages for economy.

References

Bogdanova, S.V., Kozel, I.V., Ermolina, L.V., Litvinova, T.N.: Management of small innovational enterprise under the conditions of global competition: possibilities and threats. Eur. Res. Stud. J. **19**(2 Special Issue), 268–275 (2016)

Bogoviz, A.V., Ragulina, Y.V., Kutukova, E.S.: Ways to improve the economic efficiency of investment policy and their economic justification. Int. J. Appl. Bus. Econ. Res. **15**(11), 275–285 (2017)

International Telecommunication Union: The ICT Development Index (2017). http://www.itu.int/. Accessed 19 Nov 2017

Kostikova, A.V., Tereliansky, P.V., Shuvaev, A.V., Parakhina, V.N., Timoshenko, P.N.: Expert fuzzy modeling of dynamic properties of complex systems. ARPN J. Eng. Appl. Sci. **11**(17), 10601–10608 (2016)

Kuznetsov, S.Y., Tereliansky, P.V., Shuvaev, A.V., Natsubize, A.S., Vasilyev, I.A.: Analysis of innovate solutions based on combinatorial approaches. ARPN J. Eng. Appl. Sci. **11**(17), 10222–10230 (2016)

Popova, L.V., Popkova, E.G., Dubova, Y.I., Natsubidze, A.S., Litvinova, T.N.: Financial mechanisms of nanotechnology development in developing countries. J. Appl. Econ. Sci. **11**(4), 584–590 (2016)

Ragulina, Y.V., Stroiteleva, E.V., Miller, A.I.: Modeling of integration processes in the business structures. Mod. Appl. Sci. **9**(3), 145–158 (2015)

Simonova, E.V., Lyapina, I.R. Kovanova, E.S., Sibirskaya, E.V.: Characteristics of interaction between small innovational and large business for the purpose of increase of their competitiveness. In: Russia and the European Union Development and Perspectives, pp. 407–415 (2017)

United Nations Development Programme: Human Development Index (2017). http://hdr.undp.org/en/data#. Accessed 19 Nov 2017

World Bank: Total natural resources rents (% of GDP) (2017a). https://data.worldbank.org/indicator/NY.GDP.TOTL.RT.ZS. Accessed 19 Nov 2017

World Bank: Foreign direct investment, net inflows (BoP, current US$) (2017b). https://data.worldbank.org/indicator/BX.KLT.DINV.CD.WD. Accessed 19 Nov 2017

World Bank: The Knowledge Economy Index (2017c). http://www.worldbank.org/kam. Accessed 19 Nov 2017

World Economic Forum: The Global Competitiveness Report (2017). http://reports.weforum.org/global-competitiveness-index-2017–2018/countryeconomy-profiles/#economy=RUS. Accessed 19 Nov 2017

Popkova, E.G., Chechina, O.S., Abramov, S.A.: Problem of the human capital quality reducing in conditions of educational unification. Mediter. J. Soc. Sci. 6(3), 95–100 (2016)

New Forms of State Support
for the Agro-Industrial Complex
in the Conditions of Digital Economy as a Basis
of Food Security Provision

Larisa V. Popova[1]([⊠]), Tatiana A. Dugina[1], Natalia N. Skiter[2],
Natalia S. Panova[1], and Aijan G. Dosova[1]

[1] Volgograd State Agricultural University, Volgograd, Russia
lvpopova@bk.ru, deisi79@mail.ru,
aijanraskalieva@mail.ru
[2] Volgograd State Technical University, Volgograd, Russia
ckumep@mail.ru

Abstract. The purpose of the article is to determine new forms of state support for the agro-industrial complex (AIC) in the conditions of digital economy, which are the basis of provision of national food security by the example of modern Russia. It is achieved with the help of the horizontal analysis of time rows, which supposes comparison of statistical data of the adjacent time periods; the regression analysis, which allows determining the character of dependence of variables; the method of correlation analysis, which allows determining the connection between the studied statistical indicators for the studied period. The analysis is performed on the basis of the existing statistical information on dynamics of the values of the index of national food security of Russia and expenditures of the federal budget for state support for the AIC in Russia. The authors prove that the applied forms of state support for the AIC are peculiar for low effectiveness, as they suppose high load on the federal budget and do not stimulate the provision of Russia's national food security. In the conditions of digital economy, new forms of state support for the AIC become accessible – they allow achieving larger effectiveness. The authors offer recommendations for application of these forms and provide the mechanism of action of the offered forms of state support for the AIC in the conditions of digital economy as a basis of food security provision.

Keywords: State support · Agro-industrial complex · Digital economy
National food security

1 Introduction

The recent global financial crisis has a range of peculiarities that determine the necessity to pay close attention to its analysis from the scientific society, international organizations, and state regulators. The most important ones are global coverage of the crisis, i.e., its influencing most of the economic systems of the world, the system character, which is caused by distribution the crisis for all structural elements (sectors)

© Springer International Publishing AG, part of Springer Nature 2018
E. G. Popkova (Ed.): HOSMC 2017, AISC 622, pp. 681–687, 2018.
https://doi.org/10.1007/978-3-319-75383-6_87

of the economic systems, and long duration – certain economic systems are still in the conditions of crisis, while others are still overcoming its deep consequences.

These peculiarities do not allow considering the global crisis as a positive phenomenon in the world economy, which gives a signal for reconsidering the strategic course of its development and points out the necessity for treating it as a negative phenomenon that shows ineffectiveness of the model of free market economy. Due to this, the scientific and practical problem of determining the optimal level of state interference with the economic processes for leveling the market gaps and preventing future crises becomes very topical.

While there are scientific and practical discussions regarding the necessity for state regulation of a lot of spheres of national economy, the necessity for state support for the AIC is acknowledged by most of scientific and political experts. At that, the crisis shows low effectiveness of existing and applied forms of state support for the AIC, caused by their inability to fully ensure the national food security of the countries that use them.

In this article, attention is paid to the experience of modern Russia – it is selected for the research due to its unique position in the global economic system. Thus, Russia occupies an intermediary position between developed and developing countries, possesses the highly-developed service sphere and industry, is geographically located on the border of the European and Asian regions, etc. This allows using the Russia's experience for studying and managing a lot of other economic systems, which determines large perspectives of further practical application of the results, received in the course of this research.

The basis of this research is the hypothesis that the currently applied forms of state support for the AIC are peculiar for low effectiveness, as they suppose high load on the country's federal budget and to not fully stimulate the provision of Russia's national food security. In the conditions of digital economy, new forms of state support for the AIC become accessible – they allow achieving higher effectiveness. We verify this hypothesis and seek the goal of determining new forms of state support for the AIC in the conditions of digital economy that could be the basis of provision of the national food security by the example of modern Russia.

2   Materials and Method

The offered hypothesis is verified with the help of the method of horizontal analysis of time rows, which suppose comparison of statistical data of the adjacent time periods, and the method of regression analysis, which allows determining the character of variables' dependence, as well as the method of correlation analysis, which allows determining the level of connection between the studied statistical indicators for the given time period.

Analysis is performed on the basis of the existing statistical information on dynamics of the values of the index of Russia's national food security and expenditures of the federal budget for state support for the AIC in Russia. The State program of development of agriculture and regulation of markets of agricultural products, resources, and food for 2013–2020, established by the Decree of the Government of the Russian Federation dated July 14, 2012, No. 717, supposes the following forms of this support:

- development of the infrastructural provision of business: creation of favorable conditions for preserving and restoring soil fertility, developing amelioration of agricultural lands, etc.;
- co-financing the technical and technological modernization: using the tools of tax crediting, subsidizing, etc. for providing the AIC companies with financial possibilities in the sphere of innovational development;
- information support for development of companies: optimization of information flows for development of marketing relations in the AIC (Table 1).

Table 1. Statistical information for conduct of the analysis

Indicator		Values of indicators for the years					
		2012	2013	2014	2015	2016	2017
Index of national food security, points (y)		68.3	60.9	62.7	63.8	63.8	66.2
Growth of the values of the index of national food security, %		–	−10.83	2.96	1.75	0.00	3.76
Expenditures of the federal budget for the forms of state support for the AIC, RUB billion	Development of infrastructural provision of business (x_1)	30.04	34.06	44.25	45.61	45.75	43.38
	Co-financing of technical and technological modernization (x_2)	105.15	136.25	161.52	170.45	188.07	203.34
	Information support for development of companies (x_3)	15.02	18.92	15.49	24.01	20.33	24.40
Growth of the total volume of expenditures of the federal budget for the state support of the AIC, %		–	25.97	16.93	8.50	5.86	6.68

Source: Compiled by the authors on the basis of: (Economist Intelligence Unit 2017), (Government of the RF 2017).

3 Discussion

The existing forms of state support for the AIC from the positions of fundamental economic science are described in detail in the works (Ragulina et al. 2015), (Bogoviz et al. 2017), (Orudjev et al. 2016), (Bogdanova et al. 2016), and (Popova et al. 2016b). Practical experience and the problems of provision of the national food security with application of the existing forms of state support for the AIC are reflected in the publications (Popkova et al. 2016a), (Kuznetsov et al. 2016), (Kostikova et al. 2016), (Simonova et al. 2017).

4   Results

As a result of the regression analysis, we obtained the following models of paired linear regression:

- $y = 0.68 \times -0.11$. This model means that growth of expenditures for state support for the AIC in Russia in the form of development of infrastructural provision of business in 2012–2017 leads to increase of the values of the index of national food security by 0.68 points. Correlation coefficient constitutes 8.47%, which shows statistical insignificance of the received regression model;
- $y = 0.66 \times -0.01$. This model means that growth of expenditures for state support for the AIC in Russia in the form of co-financing of technical and technological modernization in 2012–2017 leads to increase of the value of the index of national food security by 0.66 points. Correlation coefficient constitutes 2.79%, which shows statistical insignificance of the received regression model;
- $y = 0.64 \times -0.03$. This model means that growth of expenditures for state support for the AIC in Russia in the form of the information support for development of companies in 2012–2017 leads to increase of the value of the index of national food security by 0.64 points. Correlation coefficient constitutes 0.29%, which shows statistical insignificance of the received regression model;

The performed horizontal analysis of these statistical indicators showed that growth rate of the total volume of expenditures of the federal budget for the state support of the AIC, which constituted 6.68% in 2016 and 12.79% per year on average, exceeds the growth rate of the values of the index of national food security, which constituted 3.76% in 2016 and –0.47% per year on average.

Thus, it is possible to conclude that the existing forms of state support for the AIC, which are currently applied in Russia, are peculiar for low effectiveness, as expenditures for their implementation exceed the result and do not ensure the high level and high rate of the national food security growth. For solving this problem, the authors offer the following new forms of state support for the AIC in the conditions of digital economy как основа food security provision:

- online determination of urgent needs of the companies of the AIC and their satisfaction: the modern digital technologies help to create a specialized state electronic portal which will establish direct connection between the companies of the AIC and the bodies of public authorities. The systemic collection of feedback will allow for timely determination and solving of the existing problems of the AIC;
- co-financing of digitization of business processes of the AIC companies: modernization of technologies and equipment of companies of the AIC should lead not to slight improvement of business processes but to their restoration to the most leading level. That's why we recommend to prefer digital technologies during co-financing of the AIC modernization;
- electronic independent ranking evaluation of the activities of the AIC companies: digital technologies allow creating the official portal in which the interested persons will be able to leave feedback on all companies of the AIC. This will allow

strengthening the reputation (and competitiveness) of the most conscientious companies of the AIC, thus attracting interest to them from the consumers from the whole world.

The mechanism of action of the offered new forms of state support for the AIC in the conditions of digital economy as a basis of food security provision is shown in Fig. 1.

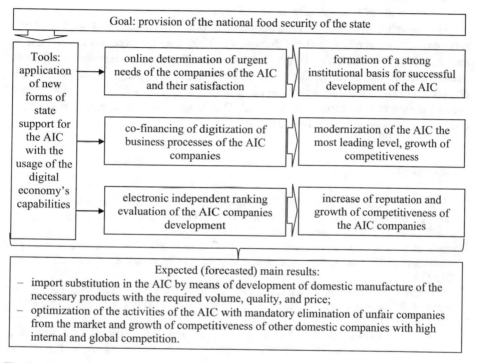

Fig. 1. The mechanism of action of the offered new forms of state support for the AIC in the conditions of digital economy as a basis of food security provision Source: compiled by the authors.

As is seen from Fig. 1, the expected (forecasted) main results of practical implementation of the offered new forms of state support for the AIC in the conditions of digital economy as a basis of food security provision include import substitution in the AIC by means of development of domestic manufacture of all necessary products with the required volume, quality, and prices.

Also, optimization of activities of the AIC is achieved by mandatory elimination of unfair companies from the market and growth of competitiveness of other domestic companies with higher internal and global competition. This allows maximizing the effect of competitive forces and thus reducing the volume of state expenditures for support for the AIC, turning in from subsidized into the leading and strategic spheres of the national economy.

5 Conclusions

It should be concluded that existing forms of state support for the AIC could be conventionally unified into one large category – forms of financial support, which could have a direct (co-financing of certain measures) or indirect (development of infrastructure, information support) character. The main drawback of these forms is short-term effect, which leads to dependence of the AIC on constant state support, and low effectiveness, caused by high expenditures of the federal budget and small feedback in the form of increase of national food security.

The offered new forms of state support for the AIC in the conditions of digital economy allow overcoming these drawbacks. They could be conventionally unified into the category of marketing forms. They allow starting (or maximizing the effect) the market mechanism (competition) and increasing the competitiveness of domestic companies of the AIC, thus activating its further independent with further reduction or even canceling of state support for the AIC. That's why these forms can be the basis of provision of the national food security.

References

Bogdanova, S.V., Kozel, I.V., Ermolina, L.V., Litvinova, T.N.: Management of small innovational enterprise under the conditions of global competition: possibilities and threats. Eur. Res. Stud. J. **19**(2 Special Issue), 268–275 (2016)

Bogoviz, A.V., Ragulina, Y.V., Kutukova, E.S.: Ways to improve the economic efficiency of investment policy and their economic justification. Int. J. Appl. Bus. Econ. Res. **15**(11), 275–285 (2017)

Kostikova, A.V., Tereliansky, P.V., Shuvaev, A.V., Parakhina, V.N., Timoshenko, P.N.: Expert fuzzy modeling of dynamic properties of complex systems. ARPN J. Eng. Appl. Sci. **11**(17), 10601–10608 (2016)

Kuznetsov, S.Y., Tereliansky, P.V., Shuvaev, A.V., Natsubize, A.S., Vasilyev, I.A.: Analysis of innovate solutions based on combinatorial approaches. ARPN J. Eng. Appl. Sci. **11**(17), 10222–10230 (2016)

Popova, L.V., Popkova, E.G., Dubova, Y.I., Natsubidze, A.S., Litvinova, T.N.: Financial mechanisms of nanotechnology development in developing countries. J. Appl. Econ. Sci. **11**(4), 584–590 (2016)

Ragulina, Y.V., Stroiteleva, E.V., Miller, A.I.: Modeling of integration processes in the business structures. Mod. Appl. Sci. **9**(3), 145–158 (2015)

Simonova, E.V., Lyapina, I.R., Kovanova, E.S., Sibirskaya, E.V.: Characteristics of interaction between small innovational and large business for the purpose of increase of their competitiveness. Russia and the European Union Development and Perspectives, pp. 407–415 (2017)

Popkova, E.G., Chechina, O.S., Abramov, S.A.: Problem of the human capital quality reducing in conditions of educational unification. Mediterr. J. Soc. Sci., **6**(3), 95–100 (2016a)

Economist Intelligence Unit: The Global Food Security Index 2012–2017 (2017). http://foodsecurityindex.eiu.com/Country/Details#Russia. Accessed 15 Nov 2017

Government of the RF: State program of development of agriculture and regulation of markets of agricultural products, resources, and food for 2013–2020, established by the Decree of the Government of the Russian Federation dated July 14, 2012, No. 717 (2017). http://www.agro-ferma.ru/dayatelnost/rekonstruktsiya-sooruzheniy/stati/programma-razvitiya-apk-na-2013-2020-gody/. Accessed 15 Nov 2017

Experience of Nizhny Novgorod State University for Conducting a Scientific Research Seminar for the Department 38.04.01 "Economics"

Julia A. Makusheva[✉], Lyudmila V. Strelkova, Oleg V. Trofimov, Olga T. Cherney, and Elena I. Yakovleva

National Research Nizhny Novgorod State University
Named After N.I. Lobachevsky, Nizhny Novgorod, Russian Federation
sjm2@yandex.ru, strelkova412@mail.ru, ovt@iee.unn.ru,
o.t.chernej@pochta.vgipu.ru,
elena.yakovleva1@gmail.com

Abstract. The article is devoted to the process of organizing research activities of undergraduates. The relevance of this topic is due to the emerging problems of students at the stage of forming their scientific results in the form of final qualification works. The problems that arise are related to the weak development of the research component in the process of mastering. The article provides an analysis of the strengths and weaknesses of the research component and the specific features of the process of organizing the research activities of students under the master's program of the specialty 38.04.01 "Economics" at the Nizhny Novgorod State University N.I. Lobachevsky. Based on the analysis, organizational solutions aimed at eliminating identified problems and improving the quality of training for masters of the specialty 38.04.01 "Economics" are proposed. Particular attention should be paid to pedagogical decisions, in particular, the introduction of innovative teaching technologies, such as interactive, modular, etc., as well as the use of active participation of students in various forms of scientific research (round tables, conferences, grants, competitions). Recommendations data on the organization of research activities of graduate students 38.04.01 "Economics" are actively introduced into the educational process, while there is a qualitative increase in the level of preparedness of graduation qualifications of undergraduates.

Keywords: Research activities · Research seminar · Competences
Research component

1 Introduction

At present, the issues of the organization of scientific research activity of undergraduates (hereinafter referred to as NID) are becoming more urgent.

The modern practice of organizing the training process of the master's program of the specialty 38.04.01 "Economics" is built on the general principles inherent in all areas of training and reflecting the competence approach. The main purpose at the same

© Springer International Publishing AG, part of Springer Nature 2018
E. G. Popkova (Ed.): HOSMC 2017, AISC 622, pp. 688–694, 2018.
https://doi.org/10.1007/978-3-319-75383-6_88

time is the expansion and consolidation of theoretical and practical knowledge obtained by undergraduates in the learning process, the acquisition and improvement of practical skills in the chosen master's program, preparation for future professional activity.

In the actualized standards, the methodology of the organization of education does not change, but some guidelines have been seriously adjusted. The analysis of the first months of approbation of the standards showed the pedagogical community and the state that the initial ideology of the Federal state educational standards of education is to set an extensive list of competencies that the specialist should master before the end of the university is not implemented. Not only because of the unpreparedness of teachers, but, above all, because of the lack of transparent mechanisms for assessing competencies.

Competent approach to the construction of the educational process sets the conditions for designing master's programs:

- Construction of the basic educational program on a modular basis on the basis of new educational technologies;
- Achievement of close interrelations between theoretical training, social and practical knowledge, abilities and skills of expert analytical, design, organizational and consulting work in the scientific sphere.

The special importance of the organization of research activities lies precisely in the ability to link theoretical knowledge and practical research conducted by future masters in the framework of their professional activities.

NID is a mandatory section of the main educational program for the preparation of masters, the purpose of which is to form a wide range of competencies in accordance with the content of federal state educational standards for higher education that determine the basic requirements for the level and quality of master's training (hereinafter - GEF VO).

Along with these requirements, the developers of educational programs should take into account the regional specifics and expectations of future employers, presented to the professional portrait of their employee. I would also like to emphasize that this level of education is a base not only for training highly qualified specialists in certain fields of the economy, but also for future scientists who must have a level of training that is adequate to the constantly changing conditions in society.

According to GEF VO, one of the main objectives of the Master's program is to develop competences that are directly related to the performance of research work.

2 Theoretical Basis of the Research

Recently, due to the relevance of the issue under study, the number of publications and dissertations on this issue has significantly increased. A leading approach to solving the problem under discussion is a systematic approach. It is provides the formation, organization and implementation of the totality of professional competencies necessary for research and development. In addition, the following methods were used in the work: Comparison, theoretical and methodological analysis, competence-contextual approach and generalization of the experience of organization of higher professional education.

The study of this issue within the framework of the modern higher education school is based on the experience of foreign countries. In our opinion, similar positions for the purposes and the task of the research activity of masters can be seen in the works: Davies M. B., Gomm R., Lawrence A. Boland, Mansfield B., Marshall St., Kinuthia V., Taylor W. In the scientific works of Mansfield B. [1], it is noted that it is difficult to ensure the quality of education in the transition period and ways to overcome this problem. At the same time, the team of authors Marshall St., Kinuthia V., Taylor W. [2] Specifies ways of overcoming the knowledge gap and their improvement in the modern educational processes of universities.

The issues of implementing the master's degree programs are examined in the works of domestic authors, such as, E.A. Dereviantsenko, N.N. Komissarova, V.V. Lapteva, and E.I. Sakharchuk et al. Along with this, many researchers, such as A.A. Bulatbaeva, Y.I. Minyazhova, S.V. Osina, and Y.V. Solyannikov, focused in their work exclusively on the specifics of NID undergraduates. The development of the theoretical base, the availability of empirical material, as well as the studies carried out by us on this issue allow us to identify and formulate a number of shortcomings in the existing system of organizing and conducting research work with undergraduates of the specialty 38.04.01 "Economics".

The content of the research component in both Russian and foreign programs is determined by the head of the master's program, the supervisor of the master's thesis and the master student himself. The research topic should reflect the priorities of modern science, practice and vocational education.

Foreign practice shows that the research conducted at the initial stage of training in the magistracy is educational, but basic for the transition to the second year of study. A special (command) organization of the work of undergraduates is characteristic for the performance of some academic studies, when the research is conducted by a group of several people according to all standards of scientific research (relevance, subject, hypothesis, research tasks, etc.). At the same time, the results of the research are documented in an article, and in some cases - in a collective educational monograph, with all the requirements for the design of works of this kind. All this is typical for the educational process of Russian practice, but the lack of a clear understanding and implementation experience determines the main problem points.

Thus, one of the significant shortcomings of the currently existing Master's programs is their traditional, mainly teaching, character. In the process of training, students should develop not only the ability to freely navigate in scientific information materials, but also the ability to comprehend process and transform the content of these materials into a comprehensive system of knowledge and skills that will enable graduate students to solve the scientific, educational and professional tasks assigned to them. Unfortunately, practice shows that at the moment when master's programs are being implemented, there is not enough effort to develop research competencies and relevant practical skills for undergraduates.

3 Analysis of the Results of the Study

The content of the research activity of the undergraduate is determined in accordance with the profile of the Master's program, taking into account the subjects of scientific research of the departments, as well as the topic of the master's thesis of the student. Along with this, in practice, a number of problems arise that are related not only to the substantive and organizational aspects, but to the elementary lack of a structured presentation of research activities, methods of scientific cognition, structure and scientific apparatus of research activity, methods Performance of scientific research activities, etc.

According to our survey, 37% of the first year students of the Master's program in Economics are ready for active participation in research activities, while 16% of those surveyed are allowed to participate in research activities, but are not ready to do so This is due to a lack of experience of this type of activity and an understanding of the methods and specifics of the work ahead. Most of the masters of the first year of study (76% of respondents) are satisfied with the results of their research activities, but in the second year of education there is a significant decrease in this indicator - only 34% of respondents are satisfied with their results. Part of the decline in this indicator is due to the objective difficulties that the undergraduates have in the course of their research activities due to the insufficient level of competence at the previous level of training that would serve as a fundamental basis for carrying out research work.

The next important point, in our opinion, is that traditionally the work on the preparation of the master's thesis in some areas of training (for example, "Economics" or "Management") in Russian universities remains the main form of involvement of undergraduates in research activities. The difficulty lies in the fact that in these areas it is not easy to develop and implement other formalized types of research activities, such as the creation of scientific laboratories and the participation of undergraduates in the implementation of projects within the framework of these laboratories. Analysis of this problem showed that 35% of universities implement research activity as a learning task within a separate discipline, another 45% attempt to link the research activity of the master to the main scientific topic of the department or center, and only about 20% of universities are focused on a specific topic Master or its inclusion in projects close to his scientific and professional interests.

Another drawback is that the preparation of master's theses for master's theses begins too late, which negatively affects the theoretical and practical level, as well as the content of these dissertations.

For these reasons, the formation and development of the research component should be a fundamental approach and one of the main directions for the improvement of master's programs in the framework of professional activity. All this makes us talk about the extreme necessity of conducting multifaceted work in the structure of scientific, educational and production activities of undergraduates in modern conditions.

A review of existing works and studies on this topic, as well as an analysis of the public education programs of the magistracy and local documents of various educational institutions that were publicly available showed that the concept of NID in a master's degree is qualitatively different from the same concept in bachelor's programs, being a broader notion of synthesis Educational and research activities and directly research activities. Therefore, NID in the magistracy is on the one hand - the organized activity of undergraduates in mastering the methodology of scientific knowledge, on the other hand - directly organizing the scientific search for a master student.

The most important form of NID organization, contributing to its successful implementation, is a research seminar (hereinafter referred to as NIS), organized on each of the master's programs. The purpose of NIS (as a form of educational and research activities within the framework of NID) is to form the methodological foundations and methodological approaches to scientific research activities, as well as the optimal principles for organization of scientific work. In addition, this, in turn, is transformed into the goal of the NIS - to develop competencies and skills of master's work among the undergraduates, allowing them to coordinate their work and actions for the implementation of the master's thesis.

As a result, we can define the main tasks of the research seminar as follows:

- First, the organization and conduct of career guidance work with undergraduates, i.e. familiarizing them with the modern problems of the economy of the national economy, with a view to clarifying their theme of scientific research.

Secondly, the creation of conditions for the mastering of master's skills in academic research, including the preparation and conduct of applied research, participation in projects, obtaining scientific results, writing scientific papers.

- Thirdly, familiarity with the requirements for publications and speeches within the research topic, at the level of international and all-Russian conferences and scientific journals included in the list of VAK.
- Fourthly, the creation of conditions for the formation of skills of dialogue students and scientific discussion, preparation and presentation of research results.
- Fifthly, the study of the requirements of competitive selection, familiarity with the documentation on the design of scientific papers for contests or grants.

Consequently, the ultimate goal of the seminar will be the systematization of scientific work in the educational process and the involvement of undergraduates in the life of the scientific community.

Competent approach to the learning process allows you to set the following competencies within the training area 38.04.01 "Economics" [3]:

OK-3 - readiness for self-development, self-realization, use of creative potential;

OPK-1 - readiness for communication in oral and written forms in Russian and foreign languages for solving problems of professional activity;

PC-1 - the ability to generalize and critically evaluate the results obtained by domestic and foreign researchers, to identify promising areas, to compile a research program;

PC-2 - the ability to substantiate the relevance, theoretical and practical significance of the selected topic of scientific research;

PC-3 - the ability to conduct independent research in accordance with the developed program;

PC-4 - the ability to present the results of the study to the scientific community in the form of an article or a report.

The very process of NIS traditionally consists of lecture and seminar classes, as well as independent work of undergraduates. In our opinion, the main topics that allow formulating the declared competences and solving the tasks set are:

1. Science and scientific research. Science and its role in modern society. Basic concepts of scientific knowledge.
2. Methodology of scientific research. Types of scientific research. Levels of scientific research. Methodological, research and applied work.
3. Forms of scientific research. Goals and objectives of scientific research. Subject, object and subject of research.
4. The main stages of planning and presentation of the master's thesis. Requirements for the master's thesis, its structure and content of sections. Construction of theoretical and practical sections. Formulation of scientific conclusions.
5. Goverment management of scientific activity. Forms of organization of scientific activity on the levels of management.

4 Conclusions

In summary, the expected results (activity format) should be determined. From our point of view, the master must be prepared for the solution of professional tasks in accordance with the profile direction of the master's program and the types of professional activity. In particular, in the field of research and development, training should include:

- Formulation of research problems, processing, analysis and systematization of scientific information on the research topic;
- Definition of research tasks, development of conceptual models, work plans and programs for scientific research and methodological development, preparation of individual tasks for performers;
- Determination of the composition and operationalization of the main variables studied, selection of methods, planning and organization of conducting empirical studies, analysis and interpretation of their results, construction of mathematical models for the studied subject area;
- Preparation of scientific reports, reviews and publications based on the results of the research, planning, organization and psychological support of the implementation of the developments;
- Organization of scientific seminars, round tables, conferences and participation in their work.

In our work, we also proceeded from the hypothesis that the model of NID organization that we presented in our university functions more effectively only when certain pedagogical conditions are created [4]. One of the main conditions is not only the need to introduce the proposed model of NIS in the educational process, but also the activation of the involvement of all undergraduates in NID through the active implementation of innovative teaching technologies. The need to solve educational problems in the learning process, as well as to ensure the transformation of professional experience, determines the choice of pedagogical innovative technologies that allow active inclusion of undergraduates in NIDs, helping them to comprehend the practical importance of knowledge and skills that they acquire during NIDs, developing their desire to actualize knowledge and helping to accumulate experience in research activities, and so on. According to our experience, interactive and modular technologies are the most adequate for our goals. Based on this, we structured the contents of the training sessions on the basis of modular technologies. The main forms of organization of educational process on NIS are lectures with elements of problem, seminars with elements of the business game, seminar-discussions, etc. [5].

Thus, the master student will be involved in research processes within the framework of her research topic from the first year of her education, will form the necessary skills necessary for her scientific research work, and will be able to form an active scientific position using the results of the research conducted in professional managerial activities.

The materials of the article can serve as a basis for creating an effective system for organizing the research activities of masters and increasing the competitiveness of Russian education in the world community.

References

1. Mansfield, B.: Competence in transition. J. Eur. Ind. Training **28**(2/3/4), 296–309 (2004)
2. Marshall, St., Kinuthia, V., Taylor, W. (Eds.): Bridging the Knowledge Divide. Educational Technology for Development. USA. Information Age Publishing Inc., (2009). 413 p
3. Federal state educational standard of higher education level of higher education master's course of training 38.04.01 "ECONOMICS"
4. Strelkova, L.V., Makusheva, Y.A.: Experience in the use of active teaching methods in teaching economic disciplines of the profile "Economics of Enterprises and Organizations"/ "Innovative methods of teaching in higher education" project-oriented, problematic, search and other methods (Collected articles on the results of the UNN methodical conference 12–13 February 2014) (2014)
5. Nikulina, N.N., Shashkina, M.E.: On some organizational and methodological features of using the technique of "tuning" in master's programs/"Innovative methods of teaching in higher education" project-oriented, problematic, search and other methods (Collected articles on the results of the methodical conference of the UNN 10–12 February 2016) (2016)

Added Value as an Indicator of the Company's Economic and Social Development

Lev Y. Avrashkov[(✉)], Galina F. Grafova, Andrey V. Grafov,
Svetlana A. Shakhvatova, and Sergey M. Manasyan

Lipetsk Branch of the Russian Academy of National Economy
and Public Administration Under the President of the Russian Federation,
Lipetsk, Russian Federation
ekonomika310@mail.ru, grafova_gf@mail.ru,
grafav@mail.ru, ssha76@mail.ru, manasyan_sm@mail.ru

Abstract. The main purpose of economic development of economic entities at all levels and, first of all, of small or medium-sized businesses is the getting a profit, which is the basis for the formation of both explicit and implicit designated funds, namely the reserve fund, the production development fund, the collective development fund and fund of material incentive. Today, the formation of firms' funds for solving social problems is becoming increasingly important, the main one at the government level is improving the quality of life of the population of the country, which is primarily formed by the wages. Salary (payroll) and profit (the fund of material incentives) are the economic basis for the formation of the quality of life of workers. The amount of wages spent on the creation of products and profits of the enterprise from the sale of manufactured products is the added value or net production created by the firm for the analyzed period. The article considers the factors of the formation of added value as a result of the company's production and economic activities. Various variants of calculation of the added cost are considered depending on a level of rates of the tax to profit and deductions for social needs.

Keywords: Firm, salary, profit, sales revenue · Prime cost (cost)
Value added · Income tax · Social development of a company

1 Introduction

In the activities of economic entities of small and medium-sized businesses, the main goal is to achieve high economic results and profit. The profit of organizations and firms plays a significant role, since it is the main source of costs not only for financing current production and investment activities, but also for the social development of the firm.

Profit is the basis for the formation of a reserve fund and various funds whose main purpose is to promote the development of the firm, both in terms of investment and in terms of social development.

To date, only the standard for the formation of a reserve fund is established by legislation (for example, Article 35 of the Federal Law No. 208-FZ dated December 26,

© Springer International Publishing AG, part of Springer Nature 2018
E. G. Popkova (Ed.): HOSMC 2017, AISC 622, pp. 695–702, 2018.
https://doi.org/10.1007/978-3-319-75383-6_89

1995 (as amended on July 29, 2012) "On Joint Stock Companies" (http://www.consultant.ru/document/cons_doc_LAW_8743/) set a minimum reserve fund in the amount provided for the company's charter, but not less than 5% of its authorized capital). The norms of formation of other funds is practically not regulated by any legislative acts. The creation of such funds has the economic goal of increasing capitalization, financial stability and, consequently, increasing the market value of an economic entity. But more and more important is the formation of funds for solving social problems, the main of which on the national scale is to improve the quality of life of the population. Unfortunately, for today Russia according to the quality of life of the population according to various rating agencies is below most countries of the world community.

2 Theoretical Bases of Research

Firms and organizations of various forms of ownership are the primary and the main link in solving social problems related to improving the quality of life in the country. According to the Decree of the President of the Russian Federation of December 31, 2015. No. 683 "On the National Security Strategy of the Russian Federation" (http://www.consultant.ru/document/cons_doc_LAW_191669/), the strategic goals of ensuring national security in improving the quality of life of Russian citizens are the development of human potential, satisfaction of the material, social and spiritual needs of citizens, reducing the level of social and property inequality of the population, primarily due to growth of income.

This allows us to conclude that wages (payroll) and profit (the fund of material incentives) are the economic basis for shaping the quality of life of company employees.

The main factor determining the level of social security for employees of firms is the amount of their wages. Other payments, for example, material reward in the form of premiums from profits are an additional, but not the main factor in the growth of their social level. It should be noted that the law regulates only the minimum wage, which should be guided by the management of the firm as the lower boundary, while the upper limit of wages and the size of premiums from profits (the fund of material incentives) are not limited by any standards. At present, the minimum wage in the Russian Federation is set at RUB 7,800, or USD 132, which is less than the subsistence level of RUB 10,600 for 2017. This means that a working person does not have the ability not only to support a family, but also to meet elementary minimum needs. Russian economists link low wages in Russia to an insufficient level of labor productivity in comparison with developed countries.

The minimum wage at the base of the average monthly wage, the level of which in the Russian Federation also does not stand up to criticism. As can be seen from Table 1, even in the Central Federal District of the Russian Federation there is a significant gap in the level of average wages, from RUB 16,830 in the Orel region to RUB 66,880 in Moscow, that is about 4 times.

If you compare the Orel region with the Moscow region, then here the difference will be more than 2.5 times. As for the Lipetsk region, the gap in the average monthly

Table 1. Average monthly wages in the subjects of the Central Federal District of the Russian Federation in 2016 in RUB and USD.

Subjects Russian Federation	Average monthly salary, thousand RUB	Average monthly salary, $
Russian Federation	36.20	613.55
Central Federal District of Russia	43.78	742.03
Belgorod region	27.28	462.37
Voronezh region	26.07	441.86
Kursk region	22.77	385.93
Lipetsk region	24.64	417.62
Moscow region	42.46	719.66
Tambov region	21.45	363.55
Orel region	16.83	285.25
Moscow	66.88	1133.55

wage with the same indicator for the Central Federal District is 1.77 times, and in relation to Moscow, 2.7 times. It is difficult to assume that the differences in the level of labor productivity in the metropolitan region and in the provinces are so significant. The reason still seems to be that in Russia there is no thoughtful social policy in the context of its regional component in general and wages in particular.

A comparative analysis of the average wage across the regions of Russia continues to show a deepening trend towards widening the gap between the center and the province: money practically settles in Moscow and to a lesser extent the second unofficial capital of the Russian Federation - St. Petersburg. According to our estimates, the deformation of financial flows in the Russian Federation led to the fact that approximately 70% of the volume of financial resources is concentrated in Moscow, about 20% in St. Petersburg and the rest (10%) in other regions of the country.

In 2016, the salaries of Russians have grown significantly compared to 2015, on average in the country they have increased by 12%. However, this applies only to the denomination in national currency - the Russian ruble. If we take the same figure in US dollars, further subsidence is observed here. The average salary of Russians is still much lower than that in countries of the Baltic and former socialist camps, where wages, for example, in Estonia, Slovakia and Poland are already many times higher than in Russia, although the economic potential of these countries is not comparable to Russia (see Table 2).

As a result, at the present time, due to the low realization of the functions of wages, and, above all, its stimulating (motivating) function, more than 50% of the workers in the real sector of the economy in Russia barely make ends meet, practically live on hunger. That is why workers do not fully realize their physical and intellectual potential in the process of labor activity, which of course does not contribute to high labor productivity.

Table 2. Average monthly wages in the CIS countries and other countries of the world in 2016

Countries of the world	Average monthly salary, $	Average monthly salary, rub.
Norway	4650	274000
USA	4460	263000
Germany	4150	244000
France	1448	85000
Poland	1440	84000
Estonia	1240	73000
Slovakia	1050	61000
Romania	660	38000
Russia	613	36000
Kazakhstan	370	21000
Belarus	350	20000
Ukraine	200	11800

3   Research Methodology

The amount of wages spent for the creation of products, and the profits of the firm from the sale of the goods produced represent the added value or pure products created in the firm for the analyzed period.

The financial results of the company's production and economic activities are determined mainly by ordinary (core) activity, and the role of other types of economic activity is insignificant. Thus, net products (value added) can be calculated using the following formula:

$$\begin{aligned} VA_p &= (W + DSN) + (R - C) = (W + DSN) + [R - (MC + W + DSN + D)] \\ &= R - (MC + D) = (W + DSN) + P_s \end{aligned} \quad (1)$$

where

VA_p - net production (value added) from sales (from sales of products, works, services);

R - revenue (net) from sales (volume of sales, works, services), excluding VAT and other indirect taxes;

C - cost of sales, taking into account management and commercial expenses;

MC - material costs in the cost of goods sold (works, services);

D - depreciation in the cost of goods sold (works, services);

W - wages of workers in the cost price of the sold production (works, services);

DSN - deductions for social needs from the salary of employees;

P_s - profit on sales.

From the above formula, you can identify a close relationship between economic elements: wages with allowances for social needs (W + DSN) and profit from sales of P_s.

We accept the condition that the volume of products sold remains unchanged. Under this condition, an increase in the wages of workers will lead to a reduction in the mass of profit, and vice versa. Theoretically, to achieve maximum economic benefit in the form of profit can be with zero pay.

The patterns of social development of both the state and firms require the search for optimal options for the ratio of wages and profits.

Important in the formation of possible options is the impact of the state's tax policy on elements of net output (value added): for example, the wages of employees are affected by the level of the tax rate on individuals, deductions for social needs, and for profit - the level of the rate of income tax.

The current system of taxation is aimed at increasing the production, investment activities of the firm more than the social development of the firm's collective. Since even with the most favorable attitude of the company's management to the growth of the social development of the collective, an increase in wages is economically less profitable than an increase in the mass of profit.

This situation is caused by the fact that the level of the social contribution rate from the wages of employees (2017 - 30%, and if the enterprise is assigned a class of occupational risk, the amount of contributions for compulsory social insurance against occupational accidents and occupational diseases may be higher on the value of 0.2% to 8.5%, so the rate of social allocations may reach 38.5%) is significantly higher than the level of the income tax rate (2017 - 20%). This form of tax deductions for business is external costs, so it seems logical that economically competent management of the enterprise will seek options to reduce these total costs.

At the disposal of the firm remain the salary of employees and net profit, which should be attributed to the company's internal resources. In general, the net profit (NP) can be represented as follows:

$$NetP = P_b - \frac{T_p}{100} \cdot P_b \tag{2}$$

Where
P_b - profit before taxation;
T_p - the rate of the profit tax, %.

The normative level of tax deductions indicates that the increase in net profit by 1 rub. is associated with tax deductions at a ratio of $0.2 : 0.8 = 0.25$ RUB. with a tax rate on profit of 20%, while the level of deductions for social needs from one ruble of workers' salaries is 0.30 RUB at the rate of deductions for social needs 30%.

When choosing the option of formation of value added, the economic advantage of reducing the wage fund for a possible increase in the mass of profit is $0.30 - 0.25 = 0.05$ rubles. for one ruble of net output (added value). This advantage is achieved by reducing tax payments.

To increase the economic interest of the firm in increasing the wage fund, and to achieve an equilibrium balance of profits and wages in the net output (value added), it is first of all necessary to reduce the level of the social contribution rate from 30% to 25% with a 20% profit.

Thus, the priority of the formation of net products (added value) at the rate of allocations for social needs below the level of 25% will be an increase in the share of wages, and, conversely, with a level of social contributions exceeding 25%, the priority will be on the side of increasing the share of profits.

Equilibrium ratio in the formation of net production (value added) at the current stage of economic development is possible when establishing the following regulatory values of tax deductions:

– Income tax rate of 20%;
– The rate of deductions for social needs is 25%.

The level of tax payment standards will allow the company to conduct more than justified both social policy and industrial development policy.

It should be noted that for a number of years (2004–2009), which can be attributed to the most intensive period of development of the Russian economy, the rate of allocations to off-budget funds was 26.6%, which is close enough to the equilibrium level of 25%.

The current level of the general taxation system rate is 30%, and if not directly, then indirectly it promotes the spread of the shadow approach when determining the wage fund to pay its certain part informally, that is, to conceal funds to be transferred to extra-budgetary funds. The hidden fund of labor remuneration in 2016 amounted to 25.3% of the total amount of payment for employees according to the statistical yearbook "National accounts of Russia".

The size of wages and "mixed incomes" of Russians, not observed by direct statistical methods, reached 10.3 trillion rub. In relative terms, this is slightly less than the 2015 indicator (at that time it was 25.4%), but the informal wages fund has not fallen to this level since the beginning of the decade: in 2011, the size was 24.5%, after which it increased and remained at higher levels for four years. At the same time, in absolute terms, hidden wages even increased: a year ago, their size was 9.7 trillion rub.

Hidden wages and mixed incomes Rosstat considers with the balance method. Out of the expenses of Russians (including the increase in their financial assets, less liabilities), formally recorded revenues are deducted. The calculations are not broken down by industry, activities and territories. Peak of informal wages was reached in 2014, accounting for 28.2% of the total wage fund. Regarding GDP, the size of informal wages in 2016 was 12% (a year earlier it was 11.6%, in 2013 - 13.3%).

4   Analysis of Research Results

According to the above provisions, we will consider various options for the formation of added value from sales and net added value. In doing so, we will be guided by the provision that for all the different options the value of the net output (value added) remains unchanged (for example, RUB 1,580).

Table 3 presents the calculations for the proposed options.

When using the first option of forming value-added, the standards in 2017 are used: the income tax rate is 20%, the social contribution rate is 30%. The ratio of net added value remaining at the disposal of the enterprise is (1228: 1580) · 100 = 78% and tax

Table 3. Formation of value added, RUB thousand.

Indicators	Conventions	Option 1 P = 0.2; DSN = 30%	Option 2 P = 0; DSN = 30%	Option 3 P = 0.2; DSN = 1%	Option 4 P = 0.2; DSN = 25%
Sales revenue	S	2800	2800	2800	2800
Cost of sales, including	C	2400	2800	1220	2400
- material costs	MC	1000	1000	1000	1000
- salary of employees	SE	908	1216	0	944
- deductions for social needs	DSN	272	364	0	236
- depreciation	D	200	200	200	200
- other costs	OC	20	20	20	20
Revenue from sales	Rs	400 = 2800 − 2400	0	1580 = 2800 − 1220	400 = 2800 − 2400
Income taxes	I_t	80 = 400 * 0,2	0	316 = 1580 * 0,2	80 = 400 * 0,2
Net profit	Net_p	320 = 400 − 80	0	1264 = 1580 − 316	320 = 400 − 80
Value Added	VA_p	1580 = 908 + 272 + 320 + 80	1580 = 1216 + 364	1580 = 1580 + 0	1580 = 944 + 236 + 320 + 80
Net added value	Net_{AV}	1228 = 908 + 320	1216 = 1216 + 0	1264 = 0 + 1264	1264 = 944 + 320
Taxes and deductions	T	352 = 272 + 80	364 = 364 + 0	316 = 0 + 316	316 = 236 + 80

deductions (352: 1580) · 100 = 22%. In this case, this option is typical for the modern formation of added value and the company, when using this option, gives priority to investment development, but at the same time restrains the wages of employees, that is, the social development of the firm's staff.

In the second variant, the situation is considered where the firm does not have profit, that is, 100% of the formation of added value is carried out only at the expense of wages at the rate of allocations for social needs of 30%. The ratio of net added value (1216: 1580) · 100 = 77% and tax deductions (364: 1580) · 100 = 23%. In the real practice of Russian firms, the use of this option is quite a rare phenomenon, since it is focused solely on the social development of the firm. At the same time, investment activity is not supported (lack of profit). However, this option is the most attractive from the point of view of the formation of tax deductions (receipts to non-budgetary funds for social development).

In the third variant, a purely theoretical position is considered, in which there is no payroll. Thus, formation of 100% of added value is carried out only at the expense of profit at the rate of the profit tax of 20%. The ratio of net added value (1264: 1580) · 100 = 80% and tax deductions (316: 1580) · 100 = 20%. This option should be considered only in theory, since the absence of wages at the firm is not possible. However, hypothetically, this option has the greatest investment appeal: this option has the highest mass of profit.

In the fourth variant, the position proposed by the authors for the formation of added value is presented at an equilibrium value of allocations for social needs - 25% and income tax rates - 20%. Thus, the formation of added value is carried out both at the expense of wages, and at the expense of profit. The situation is logical when, at the equilibrium optimal value of the rates, the final results of the deductions are similar to the third option (VA_p – 1,264 thousand rubles and T - 316 thousand rubles). The implementation of this option in the practice of an operating firm is possible only if the equilibrium level of the income tax rate and social contribution rates are used. This option seems to be the most attractive for providing both the social, and innovation-investment policy of the firm.

5   Conclusions

To implement the most attractive option (option 4), it is necessary to observe the following ratio:

$$DSN = \frac{T_p}{100 - T_p}$$

The shift in the level of the rate of profit tax requires an adequate change in the level of the rate of deductions for social needs.

It is because of this position that the variant proposed by the authors is the least attractive for the structures forming off-budget funds. However, the amount of deductions to extra-budgetary funds depends more on the absolute value of wages, and then on the rate of deductions. Consequently, the higher the growth rate of the social development of a firm, the wage fund, the less the influence of the factor of the rate of allocations on social needs. The factor of reduction in the rate of deductions can be partially or completely compensated by an increase in the mass of the wages of workers, because of which the implementation of the principles of option four is most progressive for the economic and social development of the firm.

References

1. Rudolfs, B., Johnson, R.C.: Value-added exchange rates. In: NBER Working Paper No. 18498. National Bureau of Economic Research, Cambridge, Massachusetts (2012)
2. Gereffi, G., Humphrey, J., Sturgeon, T.: The Governance of global value chains. Rev. Int. Polit. Econ. **12**(1), 78–104 (2005)
3. Elms, D.K., Low, P. (eds.): Global value chains in a changing world. Fung Global Institute (FGI), Nanyang Technological University (NTU), and World Trade Organization (WTO). Printing by WTO Secretariat, Switzerland, p. 2 (2013). https://www.wto.org/english/res_e/booksp_e/aid4tradeglobalvalue13_e.pdf
4. SIA Alliance Media. https://www.Evrokatalog.eu
5. National Science Foundation. http://www.nsf.gov
6. Ya, A.L., Grafova, G.F., Grafov, A.V., Shakhvatova, S.A.: The Economics of Organizations (Firms). Textbook for Magisters, Moscow (2014)
7. Shahvatova, S.A., Avrashkov, L.Y., Grafova, S.A.: To the question of the interrelation between the indicators of the economic and social development of enterprises. The Auditor, vol. 10(236), pp. 86–90 (2014)
8. Average salary in 2016 by regions of Russia and other countries of the world. http://www.bs-life.ru

Intermediate Integration of Economic Disciplines in the System Military Engineers Training

Marina N. Gladkova(✉), Natalia S. Abramova, Olga G. Shagalova,
Oleg N. Abramov, and Elena I. Dvornikova

Tyumen Higher Military Engineering Command School Named After Marshal
Engineer Troops A.I. Proshlyakov, Tyumen, Russian Federation
glamarin@rambler.ru, ans.76@mail.ru,
sagal25@rambler.ru, aon.73@mail.ru, el.
dvornikova@mail.ru

Abstract. In the article, a systematic study of the effectiveness of the application of interdisciplinary integration of economic disciplines in the system of training of future military engineers is performed. The essence of the processes of integration and differentiation is revealed, as well as the objective necessity of applying these processes as a system for training future military engineers, so in the process of mastering economic content. It is theoretically justified that the structural integrity and continuity of the content of economic disciplines is achieved by the interdisciplinary integration of these disciplines. In addition, the implementation of interdisciplinary integration leads to the systematization of economic disciplines, and a unique cognitive result, the formation of an integral picture in the mind of a military engineer, which, as a result, leads to the formation of a qualitatively new type of knowledge, expressed in general scientific categories and concepts. The article develops an algorithm for interdisciplinary integration of economic disciplines, including the successive implementation of the following actions: The setting of a goal and the choice of a systematizing factor; The allocation of the functions of each academic discipline; Establishment of interrelation of economic disciplines; The definition of the functional dependence of economic disciplines; Development of a glossary; The definition of the system of abilities; Creation of a system of modules. The indicators of effectiveness of interdisciplinary integration of economic disciplines are defined: The quality of cadets' assimilation of the content of economic disciplines; The orientation of cadets to master military professional activities; Number of cadets who have creative methods of solving economic problems; Number of cadets engaged in research activities.

Keywords: Interdisciplinary integration · Economic disciplines
Military-professional training · Military engineer

1 Introduction

The implementation of a scientifically based training system for future military engineers differs from the professional training of specialists for scientific and technical, scientific, educational, professional, pedagogical, industrial, sociocultural and other fields of activity.

© Springer International Publishing AG, part of Springer Nature 2018
E. G. Popkova (Ed.): HOSMC 2017, AISC 622, pp. 703–709, 2018.
https://doi.org/10.1007/978-3-319-75383-6_90

The modern system of training future military engineers is a complex, multifaceted, multifunctional, integrative system that has been in the last decade in the process of continuous modernization, justified by the accelerated pace of scientific and technological progress, the development of technology and technology, the emergence of new means and objects of labor, the expansion of the spectrum of specializations, changes In the qualification requirements of engineers, etc. [1, 12, 13].

At the same time, integration processes in society, science and production, the introduction of high technology, the intensification and intellectualization of the work of a modern specialist require new, more efficient forms of integrating science, production and education, which justifies the choice of innovative type of development of educational institutions, including military- Engineering [6].

Theoretical basis of the research. The study showed that there is a development of a number of works on the integration of the content of training: The concept of realization of integrative and differentiated processes (A.P. Belyaeva, Yu.S. Tjunnikov, L.D. Fedotova, etc.); Integration of professional-pedagogical education (T.B. Vasilyeva, N.I. Vjunova, N.M. Zhukova, E.F. Zeer, M.G. Shalunova and others); Integration of the learning process (E.O. Galitskikh, V.V. Guzeev, K.K. Kolin, and others); Integration of the content of education (M.N. Berulava, K.Ya. Vazina, E.F. Zeer, Yu.N. Semin and others).

Methodological basis of the study. As the methodological foundations are integrative and differentiated approaches.

Integration processes include a broad penetration of the content characteristic of phenomena and the search for basic regularities, the implementation of universal scientific methods and means of pedagogical research, and therefore it must be studied both in general and in a narrow sense [7].

In the context of this study, by integration, in the broadest sense of the word, we mean bringing the content of economic and military-economic education to a single didactic form on the basis of scientific, technical, military-professional, socio-economic, psychophysiological and didactic communities; In the narrow sense of the word - bringing economic disciplines into a single educational complex [3].

At the same time, it is necessary to emphasize the dialectical unity of the processes of integration and differentiation.

Integration and differentiation of the economic component of military professional education is dictated by the need to transfer the continuous multi-level training of future military engineers (technician, bachelor, specialist, master), the integrity of the educational process and the specifics of the tasks to be solved, which in their essence have integrative-differentiated properties.

The content integrity, consistency and continuity of the content of economic disciplines with relative independence of each of them, the conditionality of all economic disciplines among themselves is achieved by the intersubject integration of these disciplines [4, 8].

Intersubject integration always has two sides - procedural and content: The first is to use the methods, means, and methods of some disciplines in the study of others; The second - in the field of intersection of the content of educational disciplines.

Based on this, the factors of intersubject integration may be different - and elements of content (meaningful education, concepts, events, problems, etc.), and some educational technologies (project method, business game, etc.).

In the study, we focused on the substantive component of interdisciplinary integration.

To the theoretical prerequisites for the structuring of the educational material, we classified the following initial assumptions [2, 5]:

- Establishing the correspondence of the content of the educational material to the current level of development of society, science, technology, production, education, defense industry;
- Taking into account the features of integrative and differentiated processes;
- Rational ratio of the invariant and variable part of the economic and military-economic content;
- The existence of links between economic disciplines.

Analysis of the results of the study. The analysis of the content of training military engineers allowed determining the content of economic disciplines is represented by three levels - general scientific, interdisciplinary and internal disciplinary [3].

The general scientific level of economic content implements a modern approach to presenting the content model for future military engineers, while new knowledge is derived from the knowledge of independent subsystems.

The interdisciplinary level of the content of economic disciplines is ensured by the integration of the component parts of the training material with the relationship between them through: Representation of the basic and variable parts of integral economic knowledge; Variative parts of integral economic knowledge; The development of a continuous hierarchy of goals and functions of economic disciplines.

The internal disciplinary level of content of economic disciplines provides, concretizes the essence and role of integration and differentiation of higher levels. It provides for bringing the integration processes from the general to the particular methodological laws and from the creation of a common model to the structure of individual economic disciplines.

The implementation of interdisciplinary integration leads to the systematization of economic disciplines and, as a consequence, to a unique cognitive result, the formation of an integral picture in the mind of a military engineer. This, of course, leads to the formation of a qualitatively new type of knowledge, which is expressed in the general scientific category and concepts.

Moreover, interdisciplinary integration always enriches intra-subject integration.

Based on the theoretical studies, we developed an algorithm for the interdisciplinary integration of economic disciplines, which assumes a consistent implementation of the following operations [2, 3]:

- The setting of a goal and the choice of a systematizing factor;
- The allocation of the functions of each academic discipline;
- Establishment of interrelation of economic disciplines;
- The definition of the functional dependence of economic disciplines;
- Development of a glossary;

– The definition of the system of abilities;
– Creation of a system of modules.

This algorithm is universal in nature, it does not depend on the program content of subjects and can be applied in the process of teaching to various disciplines [6].

The basic economic disciplines that took part in interdisciplinary integration were: Economic Theory, Marketing, Enterprise Economics, Management, Organization and Planning of Enterprise Production, Financial and Economic Activities in Military Units.

The aim of the work is to create a universal economic content in the system of training future military engineers [8].

And as a systematizing factor of interdisciplinary integration of economic disciplines is military-professional activity.

Having considered each economic discipline as an independent subsystem, their importance was determined in the system of training future engineers and the role in the formation of an integrated system of military professional activity.

Thus, the functions of economic disciplines were identified, which are the main components of the content of the economic component of the training of future military engineers [11].

Having established the functions of economic disciplines, we defined the logic of interaction of economic disciplines at their target and functional levels.

Further, a model of the object space was developed, in which the functional dependence of economic disciplines was reflected.

The next stage of the work was the creation of a single conceptual-terminological dictionary or glossary on economic disciplines.

To this end, it is necessary to differentiate the concepts used in the study of economic disciplines into basic and auxiliary ones. Basic concepts are those concepts that operate on two or more economic disciplines. Moreover, the use of basic concepts by various economic disciplines does not mean that in the study of one discipline, the interpretation of these concepts completely coincides with the disclosure of these definitions in the study of another discipline [9].

In the process of interdisciplinary integration of economic disciplines, there was an objective need for intra-subject integration of the content of these disciplines [8]. An algorithm for the intra-subject development of an integratively differentiated content of economic disciplines was developed:

(1) The choice of the systematizing factor of the intra-subject integration of the differentiation of the content of the academic discipline.
   (1) Construction of the domain system.
   (2) Development of subject abilities.
(2) Definition of the basic concepts of the academic discipline.
(3) Dosage of the content of the academic discipline.

By the educational block we mean the corresponding area of pedagogical discipline, which has a systemic character and relative independence in the whole educational material.

For the first stage of the selection of the training material, it is sufficient to determine which sections of the training material and in what order they are studied.

(4) Building a system of modules for the academic discipline. An objective necessity is to establish a connection between the individual concepts of a given piece of educational material and, accordingly, in the minds of future military engineers. Very often these connections do not lie on the surface, they are not seen without special logic-didactic analysis, but are inherent in the concepts under study.

A methodological tool that provides a structural unity of the content of economic discipline is the module.

As a basis, we assume that the module is primarily a means of systematizing the content of a person's self-development, based on the interaction of a person with any system. This requires a study of the structure of the system, its norms, ways of functioning and mastering these methods [2].

Thus, if in the economic discipline to single out the system and study it in the same sequence, as a result, the future military engineer will develop the necessary ability to successfully solve economic and military-economic problems.

Under the module, we will understand the methodological means of differentiating and systematizing the subject content, which is an invariant way of organizing the content of information exchange between people and the methodological guidance for its implementation, which guarantees the satisfaction of the need currently available to a person and determines the vector of a new, emerging Interest [3].

Thanks to the module, it became possible to turn sections and topics of the content of economic discipline into systems, which allows: Teacher dosage content, understand what information is discussed and for what purpose; Cadet - to realize that he "accepts" and why he needs it.

It was with the help of the module that we structured and systematized the content within each pedagogical discipline.

The effectiveness of the interdisciplinary integration of economic disciplines was confirmed by the results of experimental work, in which 200 cadets (100 people in the experimental and control groups) of the A.M. Proshlyakov Tyumen Higher Military Engineering Engineering School, trained in specialty 230502 - "Special Purpose Vehicles" (the experiment lasted from 2014 to 2017).

As indicators of the effectiveness of interdisciplinary integration of economic disciplines were selected:

1. The quality of cadets' assimilation of the content of economic disciplines;
2. The orientation of cadets to master military professional activities;
3. Number of cadets who possess non-standard ways and strategies for solving economic problems;
4. Number of cadets engaged in research activities.

To assess the effectiveness of the interdisciplinary integration of economic disciplines, the following were used: Observation of the educational process; Questioning of cadets; Conversation with cadets and teachers; Analysis, comparison, generalization of the obtained data, analysis of the results of current and boundary control, intermediate certifications and control works; Analysis of the structure and content of integrated educational and methodological support of the process of teaching economic disciplines; Number of cadets participating in research activities (conferences, competitions of scientific projects, etc.); Comparison of the results of training cadets in the control and experimental groups.

The effectiveness of the developed interdisciplinary integration of economic disciplines is confirmed by the following results of the shaping experiment:

– Growth of consciousness and strength of assimilation of economic knowledge at higher levels of generality and complexity of educational material (the achievement rate increased by 0.53);
– A significant increase in the number of cadets (by 33%), who possess non-standard methods and strategies for solving economic problems;
– An increase in the number of cadets involved in research and development (by 19%).

Thus, the results of experimental research show that the implementation of interdisciplinary integration of economic disciplines provides individualization of training future military engineers, raises the level and quality of economic knowledge, contributes to the formation and development of economic thinking.

References

1. Khizhnaya, A.V., Kutepov, M.M., Gladkova, M.N., Gladkov, A.V., Dvornikova, E.I.: Information technologies in the system of military engineer training of cadets. Int. J. Environ. Sci. Edu. **11**(13), 6238–6245 (2016)
2. Markova, S., Depsames, L., Tsyplakova, S., Yakovleva, S., Shherbakova, E.: Principles of building of objective-spatial environment in an educational organization. IEJME Math. Edu. **11**(10), 3457–3462 (2016)
3. Markova, S.M., Sedhyh, E.P., Tsyplakova, S.A.: Upcoming trends of educational systems development in present-day conditions. Life Sci. J. **11**(11s), 489–493 (2014)
4. Smirnova, Z., Vaganova, O., Shevchenko, S., Khizhnaya, A., Ogorodova, M., Gladkova, M.: Estimation of educational results of the bachelor's programme students. IEJME Math. Edu. **11**(10), 3469–3475 (2016)
5. Vaganova, O.I., Medvedeva, T.Y., Kirdyanova, E.R., Kazantseva, G.A., Karpukova, A.A.: Innovative approaches to assessment of results of higher school students training. Int. J. Environ. Sci. Edu. **11**(13), 6246–6254 (2016)
6. Gladkova, M.N.: Integrative-differentiated content of vocational education: Monograph. - N. Novgorod: VGIPA (2004)
7. Gladkova, M.N., Gladkov, A.V.: Implementation of the competence approach in the system of military-professional training of cadets. In: Proceedings of the II All-Russian Scientific and Practical Conference: Social and technical services: Problems and ways of development. - Nizhny Novgorod State Pedagogical University. K. Minin, pp. 14–17 (2015)
8. Gladkova, M.N., Luneva, Y.B.: The influence of economic training in the formation of a future military engineer. Collection of articles on the materials of the All-Russian scientific-practical conference "Innovative approaches to the solution of professional and pedagogical problems. - FGOU VO "Nizhny Novgorod State Pedagogical University named after K. Minin", pp. 200–203 (2016)

9. Ivanov, Y.M.: System approach to training a general engineer. Y.M. Ivanov. - Kiev: High school, p. 273 (1983)
10. Markova, S.M., Narkoziev, A.K.: Political theory of professional and pedagogical education. Bulletin of the University of Minin, No. 3(16), p. 2 (2016)
11. Markova, S.M., Tsyplakova, S.A.: Management of the pedagogical process as a system. School of the Future, No. 4, pp. 138–144 (2016)
12. Barber, M., Donneliy, K., Rizvi, S.: An avalanche is coming. Higher education and the revolution ahead. Institute for Public Policy Research (2013)
13. Hanushek, E.A.: The economic value of higher teacher quality. Working Paper No. 56, National Center for Analysis of Longitudinal Data in Education Research (2010). http://www.urban.org/UploadedPDF/1001507-Higher-Teacher-Quality.pdf. Accessed 03 Apr 2017

Human Resources in the Process
of Implementation of the Region's Economy
Innovational Potential

Oleg L. Goycher[✉], Roman V. Skuba, Olga S. Bugrova,
Maria I. Zakirova, and Vladimir E. Strelkov

Vladimir State University, Vladimir, Russian Federation
goll@pochta.ru, r_scuba@mail.ru,
olga_sergeevna89@inbox.ru, zakirova_maria@mail.ru,
strateg@vlsu.ru

Abstract. In the context of limited primary resources, the constant change in the internal and external factors that form them, the choice of the strategy for transforming Russia and its regions is directly related to determining the basis for innovative development and stimulating innovation potential. Human resources play the primary role in the implementation of the chosen strategy, both at the level of innovation creation and at the level of their implementation. From the level of competence of personnel depends the quality of innovation, and, consequently, the level of competitiveness of the organization, the economy of the region and the country in general.

Summarizing the research results of Russian and foreign scientists in the field of personnel management and innovation, based on their own research experience and data obtained, taking into account the current trends in the development of innovative management factors, a model for the participation of human resources in the process of realizing the potential of the regional economy was formed. Through this model, the interaction of its main elements - idea, technology of implementation, practical implementation, market development and regional support - is demonstrated in the process of shaping the development of the region's innovation strategy.

The combination of the procedure for interaction of human resources in the field of innovation with regional authorities at all levels of the innovation development process is reflected in the proposed model for implementing innovative initiatives with support at the regional level.

In general, the analysis of the main factors underlying the proposed models has made it possible to outline general provisions on the tasks of the territorial authorities in the practical implementation of tools to support innovative development.

Keywords: Human resources · Innovation · Innovation potential
Region · Economy of the region

© Springer International Publishing AG, part of Springer Nature 2018
E. G. Popkova (Ed.): HOSMC 2017, AISC 622, pp. 710–718, 2018.
https://doi.org/10.1007/978-3-319-75383-6_91

1 Introduction

The Russian economy at the present stage is in the conditions of limited financial resource potential caused by the slowdown in the development of the economies of the world, the sanctions policy of Europe and the United States. In such conditions, the traditional way of development of the country requires transformation, optimization of financial flows, transformation of sources of funds to the budget, strengthening of its internal potential. Thus, the heavy burden on the implementation of plans lies with the regions whose economy should become the catalyst that can stimulate the economy in general. Considering the vectors of the choice of the development strategy of the state and its regions, it should be said that the basis for it is structural reforms based on the implementation of the innovative potential of the territories. Thus, the main task of implementing the strategy is to identify the basis for innovation development [1], the creation and cultivation of innovations at different micro and macro levels, the formation of a system of stimulating innovation capacity, etc. In the practical process of implementing this type of strategy, it is necessary to use all types of resources, among which human resources play a primary role both in the procedures for the overall management of innovation development in the region and in the implementation of innovation potential at the level of enterprises and organizations. From the intellectual, productive, psychological qualities of man depends the implementation of both the multifaceted task of implementing the innovation strategy, and the objective creation of a "simple" innovation. In this way, we can talk about the need to create a unified approach to determining the place of human resources in the process of realizing the innovative potential of the regional economy.

2 Theoretical Basis of Research

The effectiveness of the functioning of socio-economic systems at all levels, including regional systems, depends more on the level of their provision with the necessary resources, which must meet certain quantitative and qualitative requirements. Among the entire list of resources, the most significant at the present stage are human resources.

The term "human resources" appeared relatively recently, in the 60–70's of 20 century, when it replaced the concept of "labor resources". This is related, in the opinion of Schultz [2], with the fact that it is the employees of the firm, its personnel, rather than equipment or real estate that form the competitive advantages, as previously thought.

In turn, labor resources are a relatively "narrow" category, which, as a rule, means "part of the population who is of working age and has physical and intellectual abilities for work" [3]. It is necessary to agree with the position of Shaburova [4], who in her monographic study defines "human resources as an able-bodied population, which is the material basis of human potential that characterizes the degree of development of the physical and spiritual abilities of people." The main difference between human resources and labor is the fact that the last are the usual resource of an enterprise that needs to be accounted for and can only be used in a certain way. Human resources include the personal characteristics of the employee, i.e. except ability and desire to work, also intellectual and psychological characteristics, which are the basis of competition of a new type, described in the writings of Schumpeter [5]. It is human who is the source of scientific knowledge, which is formalized and subsequently embodied in the form of an innovative product, goods or service.

Thus, it can be concluded that human resources are a strategic resource of socio-economic systems, since they are basis for the realization of creative potential, which is embodied in the form of innovations.

Generalizing most of the views of researchers in terms of the conceptual apparatus relating to innovation, we can distinguish a number of significant features of this term, namely:

1. Innovation is the result that can be formed in any sphere of the life activity of society and the market;
2. Innovation is a system that carries within itself and transforms all kinds of resources in the process of development;
3. Innovation at its core is the source of both evolutionary and revolutionary development of the territory's economy;
4. Innovation is, in the overwhelming majority, the result of intellectual, industrial, and so on human activity;
5. Innovation should ultimately represent, for commercialization purposes, a tangible asset demanded by the market;
6. Innovation at any stage of its formation and implementation is a high-risk asset;
7. Innovation, which is a source of advanced knowledge, with successful introduction into the market allows obtaining a cumulative effect for business, the scientific community, consumers, and for the region (country) in general.

In this way, it can be claimed that innovation is a multifaceted concept that has special features that are characteristic not only of the market, but of society as a whole, and for the region's economy that are of fundamental importance for development and increase of competitiveness level.

3 Research Methodology

The problems of human resources management are covered in the works of the following scientists: A.Y. Kibanova, B.M. Genkina, N.A. Volgin, S.V. Shekshni, A. P. Yegorshin, V.V. Novozhilova, T.I. Romanova, T.G. Vinichuk, A.V. Davydova, Y. A. Pikalina, S.V. Rachek, A.V. Shaburovoy, E.A. Sidenkova, E.A. Kosorukovoy. Including in the study of this problem, such prominent Russian scientists as Bazarov T. U., Vesnin V.R., Gagarinskaya G.P., Genkin B.M., Kibanov A.Y., Maslov V.I., Mitrofanova E.A., Odegov U.G., Shekshnya S.V and others.

With the development of innovations at the level of the region, studies of such scientists as Lenchuk E.B., Bortnik I.M., Mindeli L.E., and Golichenko O.G and others.

Development of the region's economy is devoted to the works of such authors as B.D. Babaeva, A.V. Bezgodova, E.T. Gaidar, S.U. Glazyeva, A.G. Granberg, S.D. Ilyenkova, L.V. Kantorovich, V. V. Klimanova, U.P. Morozova, E.G. Yasina and others.

In most cases, the calculation of human capital (the scientific potential of the territory) most often uses indicators related to the assessment of the number of scientific organizations and people engaged in research, as well as quantitative and qualitative characteristics of innovation activity. According to the methodology proposed by the National Research University "Higher School of Economics", when calculating the Russian regional innovation index (RRII), the specific weight of these people and the specificity of their employment are taken into account. It should be noted that in general, in the process of creating an innovative product, only people who create innovative ideas or produce the ultimate innovative product are considered. It does not take into account those human resources that participate in the remaining stages of the implementation of innovation activities, which determines the need for this study.

4 Analysis of Research Results

Consideration of the degree of participation of all party's in the innovation process is advisable to begin with the definition of the scientific potential of the analyzed territory. In this connection, parallels should be drawn between the total number of employed in the region and the number of specialists engaged in research and development (see Fig. 1).

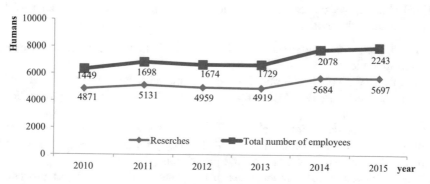

Fig. 1. Number of personnel engaged in scientific research and development in the Vladimir region

In the period from 2013, the total number of employees in the region is significantly reduced, by 2015, the economy of the Vladimir region employs 717.7 thousand people [6]. However, this does not affect the number of personnel engaged in research and development. In the presented structure it is worth noting the weak growth rates of the number of such a category of scientists, as researchers [7]. So, in 2013, the chain growth was 103.3%; In 2014 - 120.2% and in 2015 - 108.0%. Absolute growth in 2015 compared with the base growth in 2010 amounted to 794 people.

Further, it is necessary to consider the level of participation of human resources in the process of realizing the innovative potential of the regional economy (see Fig. 2).

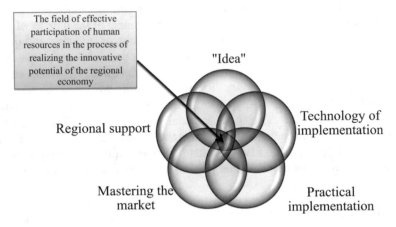

Fig. 2. Model of human resources participation in the process of realizing the innovation potential of the region's economy

The presented model suggests the interaction of human capital in the process of the stages of realizing the innovative potential of the regional economy. Achieving the necessary result on the ultimate implementation of innovation is possible only with the optimal participation of each of the subjects of this process. Let's consider in detail each element of the model:

- "Idea" implies the participation in this process of human resources that generate innovative ideas;
- "Technology implementation" implies the participation of all subjects of the innovation process, developing and describing the concept of implementing the proposed ideas;
- "Practical implementation" includes those human resources that relate to the objects of practical creation of an innovative product in the framework of scientific, practical or industrial institutions and productions;
- "Market development" unites those participants in the innovation process that contribute to the introduction of innovation into the market and introduction into the real sector;
- "Regional support" consists of human resources that provide all kinds of support for innovation by the regional authorities, including information, personnel, financial, infrastructure and administrative participation [8].

It should be noted that the model has a generalized character, i.e. participants are not identified with any specific sectors, government, regional authorities, etc.

The intersection of each two "petals" in particular is the area of finding between the participants of the compromise solutions for the creation of an innovative product. The goal is to achieve the field of efficiency of participation of labor potential, representing the final intersection of the areas of interaction of individual human resources.

It is useful to note that the size of each element depends on the degree of participation of human resources in the innovation process and can be changing. At the same time, the situation in a particular territory in which one of the model elements is missing is possible. The absence of an "idea" completely neutralizes the model, and, consequently, the whole process of its implementation. The presented model is the reference in the part of equal application of the forces of each of the subjects of this process.

Mathematically, this model can be represented in the form of the following interpretation:

$$HR = \sum_{i=0}^{n} HR_z(z_1 + z_2 + \ldots + z_n) + HR_t(t_1 + t_2 + \ldots + t_n) + HR_v(v_1 + v_2 +$$
$$\ldots + v_n) + HR_p(p_1 + p_2 + \ldots + p_n) + HR_r(r_1 + r_2 + \ldots + r_n) \to opt;$$
$$0 < HR_z \geq 0,2; 0 < HR_t \geq 0,2; 0 < HR_v \geq 0,2; 0 < HR_p \geq 0,2; 0 < HR_r \geq 0,2$$

$z, t, v, p, r > 0$

where: HR_z - *human resources that create the design of innovation;*

HR_t - *human resources, describing the technology of creation;*

ЧР$_v$ - *human resources that implement an innovative product at a specific object of innovation;*

HR_p - *Human resources that facilitate the entry of innovation into the market;*

HR_r - *human resources - representatives of regional authorities, contributing to the implementation of the process of creating an innovative product.*

z, t, v, p, r - *participation of each concrete subject of the innovation activity process.*

Each of the groups of innovation activity entities related to the model element is limited in total to 0.2, based on an equal application of their participation forces in the process of creating innovation. In Fig. 3 presents the detailed evaluation of the expert evaluation of the model described above. In this case, each of the blocks should not be less than or equal to zero, since in this case its participation in this process does not make sense.

Further, it is required to determine the place of the region in the process of creation and practical implementation of innovations. This can be represented in the form of a model that combines the procedure for the interaction of human resources at all levels of the innovation development process with instruments of support by regional authorities (Fig. 4).

The initial stage of the proposed model is the formation of the basic innovation plan. The significance of this stage is determined by the fact that the idea should not only contain a new view of the researcher, but also presume the initial inertia of its further commercialization.

The second stage involves transforming the idea into a visualized materialized model. The relevance of this stage is explained by the preparation of the technical and economic documentation of the embodiment of the idea in the project product.

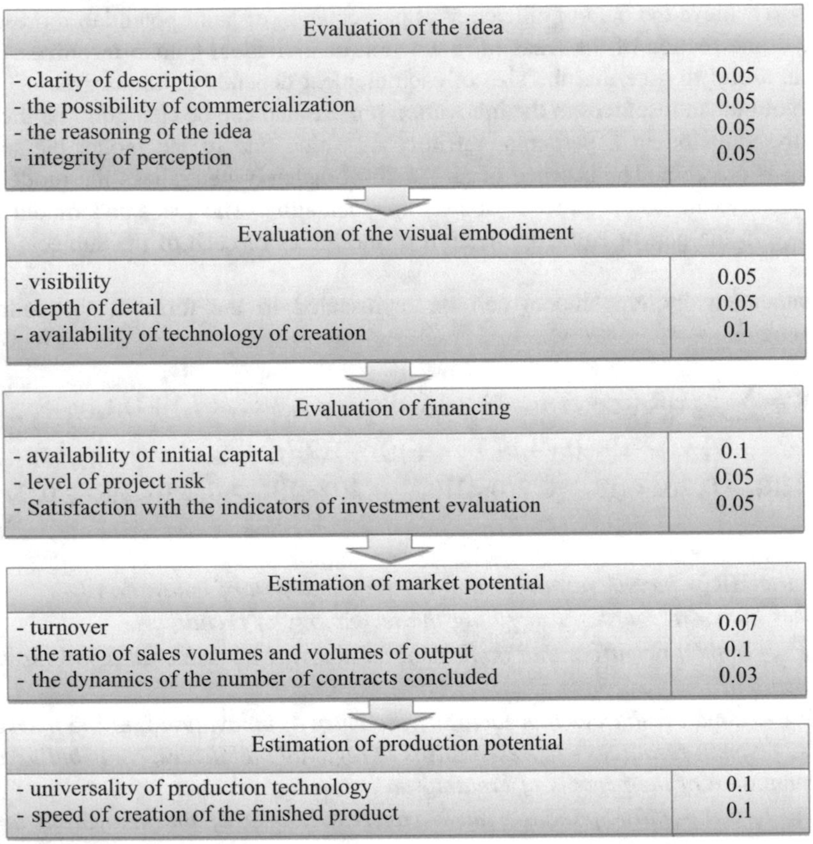

Evaluation of the idea	
- clarity of description	0.05
- the possibility of commercialization	0.05
- the reasoning of the idea	0.05
- integrity of perception	0.05

Evaluation of the visual embodiment	
- visibility	0.05
- depth of detail	0.05
- availability of technology of creation	0.1

Evaluation of financing	
- availability of initial capital	0.1
- level of project risk	0.05
- Satisfaction with the indicators of investment evaluation	0.05

Estimation of market potential	
- turnover	0.07
- the ratio of sales volumes and volumes of output	0.1
- the dynamics of the number of contracts concluded	0.03

Estimation of production potential	
- universality of production technology	0.1
- speed of creation of the finished product	0.1

Fig. 3. Algorithm for the practical assessment of the participation of human resources in the implementation of innovative initiatives (The weight of the criteria is determined on the basis of expert assessments)

At the next stage, there is a search for resource support for the implementation of an innovative product. The reasonableness of the use of this stage is because not every innovation was originally created with a project in the resource potential of implementation.

At the fourth stage of the implementation of innovative initiatives, the idea is supposed to move to the stage of the product, where the foundations and tools for the immediate commercial implementation of innovation are laid.

Despite the numerous features that distinguish traditional goods from innovative products, the result of all activities of institutions interested in the process is the immediate direct sale of finished goods, services, and works. In this sense, the innovation goes through the same stage as the classical product, which is represented in the fifth stage.

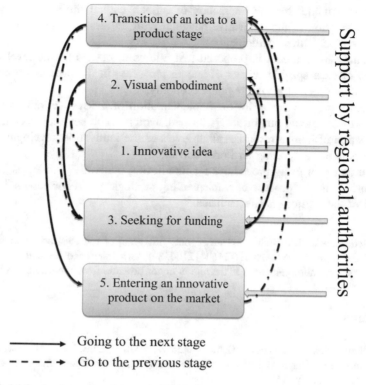

Fig. 4. Model for implementing innovative initiatives with support from regional authorities

The issue of participation of regional authorities in the process of implementing innovative initiatives should be separately mentioned. The authorities must take a very active part in each of the stages of innovation. The region is able to provide significant support for administrative, financial, information, legal, infrastructural, tools and resources, to create areas for direct interaction among participants in the innovation process, to ensure a more rapid agreement of the interests of participants, and, reducing bureaucratic barriers, to reduce the level of transaction costs. To implement the above instruments, regional authorities need constant monitoring of innovative processes occurring on the territory. It is also necessary to create conditions for optimizing the critical path, and if necessary, the region should expeditiously react to changes that occur in the process of implementing innovative initiatives.

5   Conclusions

The results presented in the research article include:

- Justification of the need to select an innovative vector for the development of the regional economy, taking into account the participation of human resources;

- Formed a model for the interaction of human capital during the stages of the implementation of the innovative potential with the designation of criteria for the effectiveness of their participation;
- A mathematical model is proposed that allows to determine the level of participation of each specific subject of human resources in the process of realizing the innovative potential of the regional economy;
- Formed a graphic model for the implementation of innovative initiatives with the support of regional authorities is formed, which combines the procedure for the interaction of human resources at all levels of the innovation development process with tools for supporting territorial education;
- The urgency of using procedures for monitoring innovative initiatives by regional authorities in the process of implementing strategic plans for economic development of the territory is substantiated.

Acknowledgments. The study was carried out with grant of the Russian Scientific Humanitarian Foundation No. AAAA-A16-116041210053-4 "An integrated approach to the implementation of innovation initiatives in the region under pressure from the external environment".

References

1. Grachev, S.A., Donichev, O.A., Zakirova, M.I.: Directions of transition from resource-dependent model of economy to innovation. Reg. Econ. Theor. Practic. **437**(2), 364–376 (2017). p. 15
2. Schultz, T.: Investment in Human Capital. NY (1971)
3. Beglaryan, A.G.: Problems of managing human resources. In: Modern High Technologies Regional Supplement No. 2, vol. 46, pp. 8–15 (2016)
4. Shaburova, A.V.: Management of the reproduction of high-quality labor resources of oil and gas producing enterprises in Western Siberia - Novosibirsk: SSGA, p. 313 (2014)
5. Schumpeter, J.: The theory of economic development. Capitalism, socialism, democracy, Eksmo (2007)
6. Labor resources [Electronic resource]. http://www.gks.ru/wps/wcm/connect/rosstat_main/rosstat/ru/statistics/wages/labour_force/. Accessed 05 Apr 2017
7. Science and innovation [Electronic resource]. http://www.gks.ru/wps/wcm/connect/rosstat_main/rosstat/en/statistics/science_and_innovations/science/. Accessed 03 Apr 2017
8. Goyher, O.L., Bugrova, O.S.: A model for monitoring innovation in the socio-economic system (on example of the Vladimir region. In: Economics and Entrepreneurship No. 10 (Part 2), pp. 219–223 (2016)
9. Egorova, A.O., Kuznetsov, V.P., Andryashina, N.S.: Methodology of formation and realization of competitive strategy of machine building enterprises. Eur. Res. Stud. J. **19**(2), 125–134 (2016)
10. Kuznetsov, V.P., Garina, E.P., Semakhin, E.A., Garin, A.P., Klychova, G.S.: Special aspects of modern production systems organization. Int. Bus. Manag. **10**(21), 5125–5129 (2016)
11. Garina, E.P., Kuznetsov, V.P., Lapaev, D.N., Romanovskaya, E.V., Yashin, S.N.: Formation of the production system elements and r&d product development processes in the early stages of the project. J. Appl. Econ. Sci. **48**(2), 538–545 (2017). Volume XII

Complex Evaluation of the Conditions of Formation of Regional Human Resources of Innovational Economy

Sergey A. Grachev[✉], Marina A. Gundorova, Oleg A. Donichev,
Denis Y. Freimovich, and Anna K. Holodnaya

Vladimir State University, Vladimir, Russian Federation
grachev-sa@yandex.ru, mg82.82@mail.ru,
kafedra-euii@mail.ru, fdu78@rambler.ru,
anya_nikolina@mail.ru

Abstract. The modern development of regional socio-economic systems and their stable functioning largely depend on the quality of human resources, expressed in the level of development, generation and application of new knowledge. It is these resources in the innovation economy that have a priority character and are the dominant factors of the reproductive and economic processes. In this connection, the article provides a comprehensive assessment of the use of regional human resources. During the research, the most significant factors that determine the development of the personnel potential of the territories are identified, which are grouped into groups of infrastructural, social and economic conditions. The work uses official Russian statistics for a long time interval and normalizes the initial values to ensure comparability of the data. The author's methodology also assumes the calculation of integral indicators by calculating the average geometric by groups of infrastructural, social and economic indicators. Methods of statistical and logical analysis, comparisons, tabular representation of data were used in the work. Also, to substantiate the author's conclusions, three-dimensional graphic models have been constructed that demonstrate and allow us to assess the real positions of the subjects of the Central Federal District of the Russian Federation in the space of the analyzed factors.

Keywords: Innovative economy · Human resources
Regional development indicators · Integrated assessment

1 Introduction

The progressive development of the country's innovative economy requires an objective analysis of its reproductive capabilities, coordinated management and proper provision of regional economic processes with all kinds of resources.

From the point of view of the resource approach, - one of the theoretical methods of modern economic theory - efficiency depends on the ability of the system to obtain scarce and valuable resources (in absolute or relative terms), their successful integration and management. Often, as resources that can provide strategic advantages, there are such things as Physical; Human; Financial; Organizational resources and other [8].

© Springer International Publishing AG, part of Springer Nature 2018
E. G. Popkova (Ed.): HOSMC 2017, AISC 622, pp. 719–726, 2018.
https://doi.org/10.1007/978-3-319-75383-6_92

It should be noted that the most important and valuable resources of the regional level are human. They directly affect the level of competitiveness of any innovative socio-economic system at the expense of the ability to generate structured scientific knowledge, which in the future are embodied in innovative products that ensure a high rate of profit in the market. Thus, for any region, the creation of the most favorable conditions for the development of human capital, which also contributes to raising the level and quality of life of the population, is an extremely important issue.

2   Theoretical Basis of the Research

There is a large number of definitions of the category "human capital", but there is no universal and single yet, and, according to some researchers, did not exist.

So, Suvorov and colleagues under human capital understand the "reserve" (potential) of the abilities, skills, and knowledge embodied in man [10].

Dolan and Lindsay's human capital is equated with mental abilities obtained in the stages of training or education, or through practical experience [3].

Critzky interprets human capital "as a universal form of economic life - the result of the historical movement of human society to the present state [5]."

Mahlup notes that unskilled labor should be distinguished from the qualified, which has become more productive due to investments, which increase the physical and mental ability of a person. These investments form human capital [1].

Schultz noted in his works that it is human abilities that are the engine of various socio-economic processes, which in turn bring income to the owner of these abilities [2].

In this way, under the category of "human capital" it is proposed to understand the reserve (potential) of knowledge, skills, competencies, intellectual and physical abilities belonging, including congenital, to a particular person, which he uses to create intellectual products, new technologies and products for provision of individual, social and economic well-being. They are formed, filled and developed on the basis of investments (both personal and investments of other persons) in the science, culture, medicine, sports and professional achievements of the individual to enhance his personal contribution to public welfare.

The most important place in the processes of "new industrialization" and import substitution is given to the development of science and education as a means of training qualified personnel for the renewed industry and improving the methods of managing human capital [4].

The use of real managerial innovations in the regions and the development of advanced market segments require an improvement in the quality of human resources. The new economy needs specialists whose effectiveness and effectiveness directly depend on their intellectual potential and the ability to apply it in practice. These are not just intellectual workers, they are carriers of intellectual capital. However, the concept of training such specialists on a mass scale has not yet been widely disseminated, and graduates of universities in the regions largely need retraining, on which companies spend significant funds. Advanced innovative and consulting centers retrain staff according to their methods, which allow them to "bring" specialists to the required level. The knowledge gained in the learning process is more systematized, not detached from reality and applicable in practice. The compactness of the information presented in such techniques facilitates the use of knowledge in management practice [7].

The sphere of education, including business schools and additional education centers, is one of the most important elements of the developed innovation infrastructure of regions. On the one hand, being a part of the most important social institution, business education ensures the formation and development of human intellectual capital as the basis for innovative transformation of society; on the other hand, it is business education that actively interacts with the institution of private business, no less important for achieving the objectives of innovation development with employers. Modern challenges facing the Russian economy require a radical restructuring of the business education system, since this direction of the formation of an innovative-oriented institutional environment directly affects the acceleration of economic growth in the regions and the increase of the country's competitiveness. Transformation of the business education system, which ensures its inclusion in the world market of educational services, suggests: Increase of its openness, unification of forms of training, introduction of institutional innovations, including those based on modern information technologies; Transition to the development of individual educational programs and the most flexible "client-oriented" schemes based on the credit-modular principle, competence approach; Increasing the requirements for quality control of education at all stages [6].

In addition, a special place in the innovation processes of building the sixth technological order should be given to infrastructural and socio-cultural development factors. Extremely strong inter-regional imbalances in a number of vital indicators, as evidenced by official statistics, cover the incentives for economic growth and hamper the dynamic development of new technologies and the creation of a full-fledged business environment on the territory of the Russian Federation. In this regard, the development of improved methods for studying the positions of the regions in terms of the complex of features and the justification of their reproductive capabilities by increasing the quality of the use of personnel reserves, especially in depressed and remote regions from the Federal Center, is becoming especially topical.

3 Methodology of the Research

A quantitative assessment of the conditions for the formation of human resources is proposed for three groups of indicators:

- Infrastructure;
- Social;
- Economic.

At the same time, the infrastructure complex can be diagnosed by the following factors:

- Number of general education organizations (at the beginning of the academic year);
- Number of educational organizations of higher professional education (at the beginning of the academic year);

- Number of hospital beds (at the end of the year, thousands);
- Number of gyms;
- The value of fixed assets (at the end of the year, at full accounting value, million rubles).

The social component is proposed to be expressed through:

- Termination of pregnancy (calculated for 100 deliveries);
- Morbidity per 1000 population;
- Number of registered crimes per 100, 000 people;
- The number of crimes committed by minors and with their complicity;
- The number of road accidents and injured in them per 100,000 people.

It is advisable to present the economic block with the following indicators:

- The Gini coefficient;
- Per capita income of the population;
- Average monthly nominal accrued wages of employees of organizations;
- The average size of the assigned pensions;
- Consumer spending on average per capita.

As an initial information for calculations, data from statistical collections "Regions of Russia: Social and economic indicators" for 2002–2016 [9]. The list of statistical indicators involved for a comprehensive assessment of the development of regional human resources allows us to characterize both the current situation and the dynamics of the processes taking place in the region. To ensure the possibility of comparing statistical data, originally expressed in different units of measurement, and their comparative evaluation, the procedure of index normalization was applied. The latter was realized by calculating the ratio of the achieved indicator for a specific period of time to the maximum result for the same period among all the analyzed subjects of the Federation (1).

$$X_{HopMi,j} = \frac{X_{ij}}{X_{\max j}} \tag{1}$$

where

- X_{norms} - the normalized value of the indicator of the i-th region for the j-th period;
- X_{ij} - the actual value of the i-th region indicator for the j-th period;
- $X_{\max j}$ - the maximum value of the indicator among the analyzed indicators for the j-th period.

To facilitate the ranking and evaluation process, a composite indicator of development for each component (infrastructure, social, economic) was defined. Calculation of the normalized indicators was carried out by calculating the geometric mean of the blocks.

An example of the results of the performed calculations for the social block is presented in Table 1.

Table 1. Consolidated indicators of development for the social block

Region/Territory[a]	2000	2001	2002	2003	2004	2005	2006	2007	2008	2009	2010	2011	2012	2013	2014	2015
1	0.450	0.503	0.459	0.480	0.464	0.471	0.446	0.438	0.436	0.413	0.392	0.403	0.395	0.416	0.379	0.395
2	0.584	0.595	0.509	0.546	0.545	0.600	0.593	0.596	0.626	0.636	0.635	0.608	0.576	0.645	0.596	0.635
3	0.697	0.781	0.733	0.784	0.755	0.773	0.795	0.826	0.822	0.812	0.796	0.823	0.788	0.826	0.792	0.792
4	0.523	0.540	0.494	0.529	0.555	0.603	0.594	0.649	0.596	0.590	0.562	0.567	0.586	0.639	0.625	0.640
5	0.642	0.625	0.640	0.678	0.611	0.643	0.654	0.672	0.626	0.629	0.629	0.631	0.623	0.623	0.584	0.582
6	0.637	0.619	0.625	0.648	0.614	0.664	0.618	0.632	0.668	0.642	0.645	0.627	0.677	0.681	0.731	0.711
7	0.553	0.596	0.509	0.530	0.491	0.547	0.585	0.558	0.576	0.527	0.517	0.484	0.462	0.513	0.508	0.530
8	0.486	0.493	0.479	0.504	0.519	0.574	0.610	0.616	0.610	0.559	0.562	0.589	0.602	0.622	0.582	0.564
9	0.528	0.529	0.482	0.485	0.465	0.513	0.517	0.534	0.574	0.558	0.550	0.568	0.523	0.569	0.518	0.545
10	0.735	0.723	0.729	0.756	0.735	0.735	0.766	0.759	0.725	0.743	0.716	0.701	0.695	0.724	0.661	0.631
11	0.544	0.536	0.533	0.554	0.569	0.576	0.583	0.629	0.664	0.630	0.607	0.587	0.624	0.639	0.624	0.643
12	0.514	0.530	0.491	0.516	0.463	0.532	0.534	0.526	0.523	0.483	0.484	0.492	0.527	0.571	0.539	0.529
13	0.645	0.661	0.621	0.584	0.565	0.617	0.642	0.656	0.662	0.672	0.639	0.640	0.626	0.682	0.578	0.554
14	0.502	0.523	0.530	0.503	0.488	0.552	0.563	0.580	0.586	0.572	0.535	0.526	0.553	0.588	0.540	0.559
15	0.617	0.696	0.647	0.662	0.655	0.720	0.750	0.796	0.797	0.729	0.713	0.678	0.677	0.698	0.682	0.709
16	0.537	0.571	0.512	0.535	0.559	0.562	0.594	0.588	0.614	0.612	0.606	0.571	0.543	0.581	0.554	0.561
17	0.766	0.756	0.719	0.753	0.744	0.789	0.756	0.776	0.768	0.728	0.698	0.674	0.671	0.653	0.599	0.679
18	0.415	0.471	0.473	0.500	0.445	0.476	0.490	0.477	0.467	0.440	0.404	0.395	0.403	0.450	0.422	0.407

[a]1 - Belgorod region, 2 - Bryansk region, 3 - Vladimir region, 4 - Voronezh region, 5 - Ivanovo region, 6 - Kaluga region, 7 - Kostroma region, 8 - Kursk region, 9 - Lipetsk region, 10 - Moscow region, 11 - Orel region, 12 - Ryazan region, 13 - Smolensk region, 14 - Tambov region, 15 - Tver region, 16 - Tula region, 17 - Yaroslavl region, 18 - Moscow.

4 Analysis of Research Results

When analyzing the components takes place, it should be noted that the direction of their most and least favorable values differs. Therefore, the best for the composite index on the social block is the minimum value, and for the economic and infrastructure - the maximum.

For a complex evaluation, a three-dimensional graphic model was constructed in which index values for the analyzed components were marked along the axes. Figure 1 shows the structure of the regions according to the level of favorable conditions for the formation of human resources as of 2000, and in Fig. 2 - as of 2015.

Analyzing the situation as of 2000, it should be noted that all the regions of the Central Federal District are fairly close in terms of the indicators being measured, but two subjects (Moscow region and Moscow) are statistical emissions, i.e. Function on trajectories different from other regions.

In 2015, the situation largely overlaps with the one discussed above, however Belgorod region is added to the regions-statistical emissions.

First of all, this is due to the fact that these subjects of the federation have normalized indicators close to the leading ones, i.e. aspire to a maximum (to 1) for the infrastructure and economic component and to a minimum - in the economic block.

If we consider the remaining regions of the CFD, then we should highlight the unevenness of their development in the context of specific blocks. So, according to the social component, the most optimal values in 2015 are Ryazan and Kostroma regions, and the least successful are Vladimir and Kaluga region. Leaders on the economic component are the Yaroslavl, Lipetsk and Kaluga regions, and the outsider regions are

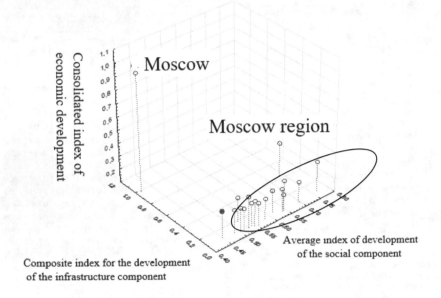

Fig. 1. Assessment of the conditions for the formation of human resources of the Central Federal District for 2000.

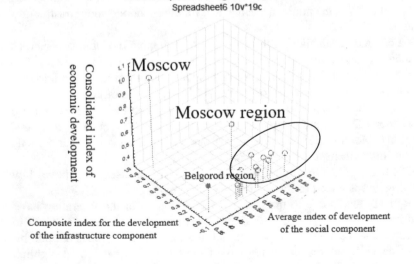

Fig. 2. Evaluation of the conditions for the formation of human resources of the Central Federal District for 2015.

Ivanovo and Kostroma regions. In the context of the infrastructure component, Tver, Smolensk and Bryansk regions have the best positions, and Kostroma, Ivanovo and Orel regions are among the lagging regions.

Thus, the least favorable conditions for the development of human resources are the Kostroma and Ivanovo regions, which belong to the outsider regions by two components, and the regions with the most optimal conditions include the Lipetsk and Tver regions.

5 Conclusions

As a result, the presented evaluation methodology allows to determine the positions of the regions on a new level of research on the set of conditions for the formation of human resources, to diagnose the nature of changes in a given time interval, and to identify subjects that require federal support for the development of the infrastructure and social environment and increase the efficiency of the use of the personnel reserve economy.

The computational algorithm developed in this work can be adapted to monitor the functioning of the knowledge economy of any region and does not exclude the possibility of expanding the data array.

The proposed set of tools can be used in the practical activities of the authorities, as well as for the training and retraining of specialists from the profile departments of federal, regional and municipal administrations.

References

1. Machlup, F.: Knowledge: Its Creation, Distribution, and Economic Significance. The Economics of Information and Human Capital, vol. 2, 419 p. Princeton University Press, Princeton (1984)
2. Schultz, T.W.: Investment in Human Capital. Am. Econ. Rev. 51(1), 1–17 (1961)
3. Dolan, E., Lindsay, J.: Market: Microeconomic Model, 477 p. (1992)
4. Donichev, O.A., Fraimovich, D.Y.: Effective management of the personnel of the organization. Bull. Vladimir State Univ. Named After Alexander Grigorievich Nikolai Grigorievich Stoletov. Ser. Econ. Sci. 3(9), 112–123 (2016). ISSN 2409-6210, p. 112
5. Kritsky, M.M.: Human Capital. Len. Univ., St. Petersburg (1991). p. 15
6. Lide, E.: Business education as an element of innovative economy. Probl. Manag. Theory Pract. 11, 101–107 (2011)
7. Otradnova, L.: Innovative approach to management training: stages and planes of management. Probl. Theory Pract. Manag. 4, 44–50 (2011)
8. Slavin, B., Soloviev, V.: Management of competences as resources. Probl. Manag. Theory Pract. 9, 72–78 (2015)
9. Statistical collections "Regions of Russia: Social and economic indicators" for 2002–2016. http://www.gks.ru/wps/wcm/connect/rosstat_main/rosstat/ru/statistics/publications/catalog/doc_1138623506156

10. Suvorov, A.V., Suvorov, N.V., Grebennikov, V.G., Ivanov, N.V., Boldov, O.N., Krasil-nikova, M.D., Bondarenko, N.V.: Approaches to measuring the dynamics and structure of human capital and assessing the impact of its accumulation on economic growth. Probl. Forecast. **3**, 3–17 (2014)
11. Kuznetsova, S.N., Garina, E.P., Kuznetsov, V.P., Romanovskaya, E.V., Andryashina, N.S.: Industrial parks formation as a tool for development of long-range manufacturing sectors. J. Appl. Econ. Sci. **XII**(2(48)), 391–402 (2017)
12. Garina, E.P., Kuznetsov, V.P., Egorova, A.O., Garin, A.P., Yashin, S.N.: Formation of the system of business processes at machine building enterprises. Eur. Res. Stud. J. **19**(2 Special Issue), 55–63 (2016)

Effects of Differences of Regional Economy's Dynamics: A Resource Aspect

Ilya V. Panshin$^{(\boxtimes)}$

Vladimir State University, Vladimir, Russian Federation
panshin@vlsu.ru

Abstract. The study is devoted to the analysis of the origin of the effects of the discrepancy between the dynamics of the productive and resource indicators of the regions in the context of the stagnation of the Russian economy in 2014–2016. During this period, multidirectional dynamics of the main indicators of economic growth was recorded, which is caused not only by market factors, but also by changes in the resource potential of the territories.

The modern principles of the "survival" of participants in economic relations (entrepreneurs, enterprises, regions) are formulated in the conditions of the crisis. The essence and systematization of the main features of the effects of discrepancy of dynamics in the analysis of regional socio-economic systems is determined. It is substantiated that the analysis of *decoupling* effects or the effects of divergence of dynamics is applicable to situations when the series of economic indicators that must have a correlation or fundamentally grounded dependence move either in different directions or in one direction, but it is incommensurable with other indicators of the intensity and range of the changes.

The analysis of the discrepancies in the dynamics of the main regional indicators of economic growth and consumption of resources (energy and investment) was carried out. It is revealed that resource differences in dynamics in different combinations of economic resources (labor, material, energy, investment, etc.) can also give the researcher useful information for optimizing the resource potential of the enterprise, industry, regional and national socio-economic system.

Keywords: Decoupling · Resource effect · Economic dynamic

JEL Codes: F 43 · R 11 · R 58

1 Introduction

Analysis of economic dynamics is one of the most common methods for studying the development of socio-economic systems, such as the national economy or the region one. In rapidly changing conditions and in view of the new challenges, the theory and methodology of investigating economic dynamics requires constant improvement.

In the context of the stagnation of the Russian economy in 2014–2016, due to both internal and external causes, import substitution, reduction of raw materials and energy dependencies, as well as modernization of manufacturing industries were declared the main imperatives of development. However, the sectoral and territorial structure of a

© Springer International Publishing AG, part of Springer Nature 2018
E. G. Popkova (Ed.): HOSMC 2017, AISC 622, pp. 727–735, 2018.
https://doi.org/10.1007/978-3-319-75383-6_93

significant part of regional economies turned out to be largely unprepared for such reforms. Some industries and enterprises showed positive dynamics of their economic performance, others, despite all the efforts, could not overcome the economic recession. These phenomena are difficult to explain by competitive struggle and the operation of other market mechanisms.

Preparation and adoption of decisions on neediness to modernize regional economies require new approaches to the analysis of economic dynamics. Deviations and disproportions in the trajectories of changes in economic indicators are some kind of signals for making managerial decisions. Thus, the resource intensity of economic processes does not always change proportionally and is commensurate with changes in the indexes of economic growth, which characterizes the emergence of decoupling effects. These *decoupling* effects in the analysis of the dynamics of indicators of use of resources (in particular energy and investment) and economic development become an independent indicator of the success of the ongoing reforms.

2 Theoretical Basis of the Research

The emergence of effects of decoupling in the dynamics of economic indicators connected with each other accompanies most economic, mathematical and statistical research. However, in most cases, analysts try to eliminate from the general aggregate the values of those indicators that do not fit into the logic of constructing a particular model, justifying it with either statistical errors or inaccuracies in data collection or random factors. Since one of the most important characteristics of the representativeness of the model is the level of correlation of the data included in it, then this approach seems quite justified. Excluding some of the indicators, the dynamics of which are at variance with economically justified trends, the analyst raises the correlation coefficient of his model's indicators and makes it more qualitative.

In reality, the discrepancy between the dynamics of a certain economic indicator and the logic of constructing a model can become the goal of independent research. One example of this approach is the analysis of *decoupling* effects or effects of discrepancy in dynamics. The English word "*decoupling*" in translation means "discrepancy, violation of synchronism, communication failure; Decoupling, decoupling, disengagement, disengagement." Thus, the analysis of *decoupling* effects or the effects of divergence of dynamics is applicable to situations where the series of economic indicators that must have a correlative or fundamentally grounded dependence move either in different directions or in one direction, but it is incommensurable with other indicators of the intensity and range of the changes. In the future, the terms "*decoupling*-effect" and "discrepancy effect" will be considered equivalent.

In general, the "divergence effect" in the analysis of economic growth is understood as a violation of synchronism in the trajectories of growth and decline in economies of countries and regions [2–7, 9]. So, Samarina believes that the essence of "decoupling" is to fix a situation where the processes that once demonstrate a certain conjugacy begin to change the development trajectories [7]. Zakharov believes that the effect of discrepancy arises when the achievement of economic progress is based on lower rates of resource consumption and a decrease in environmental degradation [5]. A similar interpretation is offered by Akulov: "… if, with positive dynamics of economic growth rates, the indicators of negative impact on the environment remain stable or even demonstrate a downward trend, there is a discrepancy effect" [2].

Regarding the problems of the nature management economy, decoupling *effects* originally characterized the phenomenon of divergence of trends in GDP growth and primary energy consumption in the countries of the Organization for Economic Co-operation and Development (OECD). With stable growth of GDP in the analyzed countries, the consumption of primary energy remained stable or even somewhat reduced [4]. This discrepancy is one of the signs of the intensification of the economy. In the future, the interpretation expanded, and *decoupling*-effects were understood as a mismatch in the growth rates of people's well-being and the dynamics of resource consumption. Special attention should be paid to studies of the effects of the discrepancy regarding the negative environmental impact of the economy on the environment [5, 9].

Among domestic and foreign scientists, Akulov A.O., Bashirova A.A., Zakharov V.M., Matveev I.E., Reznikov S.N., Samarina V.P., Stolbov M.I., Conrad E., Cassar L. F., Ward J.D., Sutton P.C., Werner A.D., Costanza R., Mohr S.H., Simmons C.T., Lopez J.A. et al. made the most significant contribution to the creation and development of the theory of the effects of divergence, an assessment of the relationship between environmental and economic development and resource saving, in particular energy conservation. However, in the works of the listed authors, the effects of discrepancy are mainly considered for comparing the dynamics of the development of developed and developing national economies, various sectors of the economy, and the environmental parameters of countries and regions [6–11]. The influence of the resource effects the discrepancy on regional and sectoral economic dynamics has not been adequately addressed and requires a deeper study.

Under the conditions of external economic pressure and taking into account the high degree of state participation in the economy of the Russian Federation, the laws of the market are being transformed. There are new economic principles of "survival" in a crisis, which include:

- participation in state and municipal purchases;
- access to budgetary support resources (grants, subsidies, subsidies);
- availability and accessibility of fuel and energy resources;
- Administrative support of business at the federal, regional and local levels, which ensures the preservation of the resource potential;
- gradual monopolization of domestic markets by large enterprises and their associations;
- creation of financial reserves to compensate for unplanned losses;
- minimization of participation in investment projects with high and medium risk levels.

Proceeding from the fact that regions, industries and even individual enterprises adapt to new conditions of management at different rates, the discrepancy in the dynamics of their main development indicators becomes even more significant.

The reasons for these imbalances are:

- significant differences in resource provision of territories both between regions and within the same region;

- different starting opportunities for the modernization of regional industries, sectors and enterprises;
- high differences in the level of flexibility of the applied technologies and adaptability of management systems at all levels.

The analysis of discrepancies in the economic dynamics of various socio-economic systems allows us to formulate the foundations of the economic concept of the effects of the divergence of economic dynamics. The main idea of which is the assertion that economic growth is possible without proportional growth in consumption of economic resources and negative effects on the environment.

Taking into account that the effects of the discrepancy between the dynamics of various economic indicators are to some extent present in a variety of studies, it is necessary to generalize some of their characteristics:

1. The presence of a close correlation or functional connection between the analyzed indicators. In this connection, pairs or groups of resource and performance indicators of economic systems, which are in objective functional dependence, are most often used to analyze the effects of divergence of dynamics.
2. The existence of external factors of influence, correcting the traditional economic model of the functioning of the analyzed socio-economic system. In the conditions of stagnation or crisis, the effects of the discrepancy of dynamics manifest themselves fully.
3. Multi-directional dynamics of the analyzed indicators, contradicting the relationship between indicators under normal conditions.
4. High level of differentiation of unidirectional changes in indicators.
5. High dependence on combinations of economic resources in the analysis of pair and group dynamics of economic indicators of several economic entities.

Thus, in our study, we will mainly consider the resource effects of the discrepancy, assuming a discrepancy between the economic dynamics of resource and performance indicators of socio-economic systems, such as the national economy, region, industry or a particular enterprise. A large number of types of economic resources and indicators characterize them, requires the concentration of our research on a limited list of resources used in the analysis of economic dynamics.

The purpose of the study **is to** determine the essence of the resource effects of discrepancies in the analysis of regional socio-economic systems, and to analyze their impact on economic development.

3   Research Methodology

In assessing *decoupling* effects, private indicators of resource intensity are widely used: Energy intensity, material intensity, water capacity, carbon intensity etc. [5]. Disproportionate to economic growth, the change in the indexes of resource intensity is evidence of the appearance in the economy of the resource effect of discrepancy.

In the theory and practice of economic research, a large number of methods for analyzing economic dynamics are used: Linear and nonlinear dynamic modeling, comparison method, graphical method, numerical methods, etc. Specialized methods for *analyzing decoupling* effects in the scientific literature are almost never found. In general, traditional economic-mathematical or graphical methods are used with a different interpretation of the obtained results.

The work of Bashirova A.A., which suggests calculating the decoupling factor for determining the relationship between changes in pressure indexes on the environment and indicators of economic development [3], can be recognized as interesting. Based on the calculation of this indicator, it was concluded that economic growth is possible without increasing environmental intensity and environmental damage. This corresponds to the concept of sustainable development, which involves meeting the growing needs of society while minimizing anthropogenic impact on the natural environment.

With respect to the analysis of the specific or multi-factor resource intensity of the development of regional economies, the formation of specialized methods for *decoupling* effects is in the process of elaboration and is the direction of further research.

4   Analysis of Research Results

One of the most common approaches to analyzing the resource intensity of the development of countries and regions is to study their energy intensity. This is due to the fact that the large volume and heterogeneity of the economic resources used significantly complicate the construction of economic models. Moreover, the construction of multi-factor models is a complex mathematical task, and the disparity of economic resources makes it very difficult to compare them. Therefore, attempts are often made by researchers to choose one priority resource, for example, energy, which makes the greatest contribution to economic development.

The convergence of the GDP trends of countries (GRP regions) with the dynamics of the use of energy resources is a direct characteristic of the extensive nature of the construction of the economic system. This situation leads to a high dependence of economies on the availability and cost of energy resources, and in the long term - loss of competitiveness and economic recession.

Solving this problem requires significant efforts to modernize regional economies, the widespread introduction of energy-saving technologies, increasing labor productivity, etc. The change in the energy intensity of a gross regional product can occur not only as a result of the implementation of special measures to save energy resources, but also because of other causes that determine the appearance of the effects of discrepancy:

- structural shifts in the regional economy
- economies of scale when opening new or expanding existing production facilities, increasing the utilization of existing production capacities;
- inaccuracies in the statistical accounting of energy consumption, due, for example, the need for the separation of costs for fuel and energy in the interregional transport system, etc.

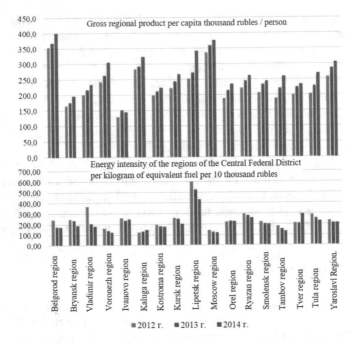

Fig. 1. Dynamics of energy intensity of GRP of the regions of the Central Federal District of Russia for 2012–2014. Source: Federal State Statistics Service.

The discrepancy in the trends of GDP dynamics (GRP) and volumes of consumption of energy resources is one of the target indicators for assessing the development of national and regional socio-economic systems.

According to the state report on energy conservation and increasing energy efficiency in the Russian Federation in 2014, high sensitivity to negative macroeconomic trends in non-energy-intensive sectors of the economy - the service sector, light industry - increased the pressure on the energy intensity of Russia's GDP [1].

Based on official statistics of the Federal State Statistics Service for 2012–2014. The regions of the Central Federal District of Russia demonstrated different dynamics of changes in the energy intensity of their GRP (Fig. 1).

The major part of the regions has managed to significantly reduce the energy intensity of its GRP even taking into account the slowdown in the growth of regional economies. The growth of GRP per capita was accompanied by a decrease in the specific energy intensity, which is evidence of the intensification of the region's economic development. The best indicator for the Vladimir, Kursk, Lipetsk and Tula regions. Other regions on the contrary showed an increase in GRP energy intensity with a decrease in its volume per capita (Ivanovo Region) and a significant lag in growth (Tver Region). The reason for this could be a negative divergence in the dynamics of the volumes of creation of a regional product and consumption of energy resources. Obviously, if the growth of production is ahead of consumption of energy resources, then the discrepancy in dynamics will be intense, the reverse situation is extensive.

Of course, such assessments are acceptable when eliminating the remaining parameters of the work of enterprises, such as personnel potential, raw materials production, etc.

In addition to energy resources, the effects of discrepancies in the dynamics of economic indicators can also be identified with respect to investment resources (Fig. 2).

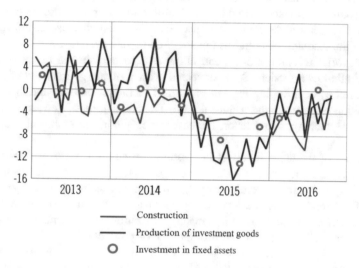

Fig. 2. Investment, construction and production of investment goods in the Russian Federation (an increase, in% to the corresponding period of the previous year). Source: Federal State Statistics Service, calculations of the Bank of Russia.

The investment resources necessary for the creation and development of any enterprise, both in the sphere of production and in the sphere of services, become a universal object of research, which makes it possible to compare the resource intensity of different regions and industries.

It can be seen from the graph that despite the high tightness of the connection between the dynamics of investment in fixed assets and the volume of production of investment goods, including construction materials, changes in the dynamics of construction volumes are insignificant and differ from the dynamics of the other two indicators. The reasons for the decline in investment in fixed assets in 2015 were the conservative investment policy of companies, the relatively high level of debt burden in a number of sectors, moderately stringent credit conditions, and uncertainty about the future dynamics of demand, whose contribution, however, by the end of 2016 gradually declined. The Russian construction industry, focused on deferred demand, reacted to crisis trends in the economy of the country in a much lesser degree in terms of volume indicators.

Based on the assessment of resource discrepancies, a primary diagnosis of the effectiveness of ongoing modernization reforms is possible. Anticipating the growth rates of consumption of economic resources over the dynamics of performance can also speak of the crisis trends manifested in the obsolescence of technology, increased wear and tear of equipment, a decrease in labor productivity and a low utilization of production capacities.

Special attention should be paid to the study of inter-resource discrepancies, manifested in a significant discrepancy in the dynamics of specific consumption of different types of economic resources. With regard to energy resources, changes in production technologies and in the structure of the product are possible. Other variants of inter-resource discrepancies in different combinations of production resources (labor, material, energy, investment, etc.) can also give the researcher useful information for optimizing the resource potential of the socio-economic system.

Multifactorial resource models can allow one to explain the discrepancy effects found in single-factor models. So, unexpected growth of energy intensity of production can be caused by outstripping rates of growth of labor productivity and investments in modernization of material and technical base. The acquisition of new, more automated equipment, replacing manual labor, increases energy consumption, but reduces labor costs.

Reduction of the areas involved in regional agricultural turnover does not necessarily lead to a decrease in yield. The compensatory effect of the discrepancy in dynamics, expressed in economic growth, can be obtained in this case by using more effective fertilizers, raising the level of mechanization, and due to favorable weather conditions. The extensive growth in capacity utilization, being generally positive, can lead to a decrease in net profit and profitability of production.

Managing the resource effects of the discrepancy in the dynamics of regional socio-economic systems in the process of their development is a complex multi-factor task. At the heart of its solution is the resource replacement methodology, based on the idea of optimizing resource proportions, the criterion of which is to maximize the cooperative effect from the joint use of a certain set of economic resources.

References

1. State report: On the state of energy conservation and energy efficiency in the Russian Federation in 2014. Ministry of Energy of Russia, Moscow (2015)
2. Akulov, A.O.: Decoupling effect in the industrial region (by the example of the Kemerovo region). Econ. Soc. Changes Facts Trends Forecast 4(28), 177–185 (2013)
3. Bashirova, A.A.: Determination of decoupling effect for problem territories in modern conditions (by the example of the NCFD subjects). Manage. Econ. Syst. 10(92) (2016)
4. Matveyev, I.E.: The effect of "decoupling" and renewable energy. Energy Fresh 3, 44–49 (2012)
5. Bobyleva, C.M., Zakharova, V.M. (eds.): Towards sustainable development in Russia. "Green" economy and modernization. Ecological and economic foundations of sustainable development, no. 60, p. 90. Institute for Sustainable Development (2012)
6. Reznikov, S.N.: The transformation of the vector and the model of the post-crisis development of the world economy: conceptual aspect. Integral 6, 39–41 (2011)
7. Samarin, V.P.: Decoupling effect as a criterion for sustainable regional development. In: Materials of the X International Scientific and Practical Conference "Actual problems of development of economic entities, territories and systems of regional and municipal management", Kursk, pp. 343–345, 28–30 May 2015

8. Stolbov, M.I.: Some results of the empirical analysis of the factors of the global crisis of 2008–2009. Issues Econ. **4**, 32–45 (2012)
9. Tereshina, M.V., Degtyaryova, I.N.: "Green growth" and structural changes in the regional economy: attempt of the theoretical and methodological analysis. Theor. Pract. Soc. Dev. **5**, 123–129 (2012)
10. Conrad, E., Cassar, L.F.: Decoupling economic growth and environmental degradation: reviewing progress to date in the small island state of Malta. Sustainability **6**, 6729–6750 (2014)
11. Ward, J.D., Sutton, P.C., Werner, A.D., Costanza, R., Mohr, S.H., Simmons, C.T.: Is decoupling GDP growth from environmental impact possible? PLoS ONE **11**(10), E0164733 (2016). https://doi.org/10.1371/journal.pone.0164733

Evaluation of the Digitalization Potential of Region's Economy

Tatyana O. Tolstykh[1]([⊠]), Elena V. Shkarupeta[2], Igor A. Shishkin[2],
Olga V. Dudareva[2], and Natalia N. Golub[2]

[1] Industrial Managment National University of Science and Technology
(MISIS), Moscow, Russia
tt400@mail.ru
[2] Voronezh State Technical University, Voronezh, Russia
79056591561@mail.ru, i-11223344@mail.ru,
dudarevaov@mail.ru, basl971@list.ru

Abstract. The technological, organizational and managerial changes related to
the spread of the digital economy, which take place today, covers all areas of
activity - economy, public administration, culture, health care, education,
transforming the everyday life of people and creating new ways of communication. The actual and potential effect of the development of the digital economy
bring great interest at the global, national and regional levels. Digitalization,
which we interpret as the use of the opportunities online and innovative digital
technologies by all participants in the economic system from individuals to large
companies and states, is a prerequisite for maintaining competitiveness for all
regions and countries. The article examines approaches to evaluation of the
digitalization potential of the regional economy by on example of the Central
Federal District of the Russian Federation using fuzzy-set methods through an
assessment of the digitalization potential of the region and the effectiveness of
the digitization potential. Regions show active involvement in the digital
economy as soon as the Internet infrastructure appears, although there are still
some inequalities in a number of additional indicators: it concerns the penetration and use of public services (gap with Moscow in 3–5 times), as well as the
activity of businesses in the use of digital opportunities (gap of 2–3 times). In
general, we see a huge under–utilized potential in the digitalization - of business,
regardless of regional specifics.

Keywords: Digital economy · Digital transformation · Digitalization
Potential · Region · Digitalization potential

1 Introduction

The main goal of any economy is structure of Gross Domestic Product and GDP by
itself, which allows ensuring competitiveness, sustainable and stable development. For
the digital economy phase, the goal is also to increase the share of the digital economy
in the total GDP in relation to the analog part of the economy. Since the growth of the
digital economy takes place 4–5 times more rapidly, this topic is extremely relevant.
The digital economy is characterized by an order of magnitude more rapid return of

© Springer International Publishing AG, part of Springer Nature 2018
E. G. Popkova (Ed.): HOSMC 2017, AISC 622, pp. 736–743, 2018.
https://doi.org/10.1007/978-3-319-75383-6_94

investment in specific projects and an order of magnitude higher profitability for individual projects that can be practically realized and, therefore, this issue affects the problem outside enterprises and organizations - where interstate associations and countries should direct their resources and including financial ones. Decisions in such a formulation of the task, as well as the goals set forth at the level of the digital transformation of the economic structure of Russia, cannot be imagined without a comprehensive analysis of the integrity of models of such transformation and consideration of the best world and domestic practices and remedies [1].

2 Theoretical Basis of the Research

Examples and analyzes of successful digital transformations are set forth in a significant number of published works on the digitization of the economy [2, 3, 5–10]. This accumulated experience of research shows that it is from the transformations or transformations of specific areas and businesses in the country that the volume of GDP of the digital economy is growing and developing and digital leaders are growing up.

The general pattern of digital economy projects is the orientation towards a specific consumer and the full use of information as a driving resource, taking into account the specific features of a particular consumer in a particular place, and making full use of digital transformation technologies for real business processes. In this way, these digital projects are characterized by very specific circumstances of their implementation in a specific place and only with the accumulation of economically positive results become the subject of standardization and other regulation.

Digital conversion is a difficult task. Countries that have reached the highest level of digital maturity had to solve complex cultural, organizational, technical problems, and only taking into account all these factors made these transformations successful. In order to become digital leaders today in specific sectors of the economy, priority digital projects that implement specific organizational teams will be singled out. Digital teams should focus on three key functional activities - developing digital strategy, managing digital activities through their national companies, and transforming them into operational excellence of their digital execution. Consequently, the creation of conditions for the emergence of digital leaders in Russia and other countries who is economic partner is also an urgent task [1].

3 Research Methodology

In the context of this article, it is necessary to assess the potential of Russia's digitalization through an assessment of the digitalization potential of each particular region and the degree of use digitalization in the context of the development strategy of the Russian Federation.

Let us formulate a methodology for assessing the potentials and effectiveness of digitalization of their use by regions.

The purpose of this technique is to obtain two complex estimates: The potential of digitalization and the effectiveness of use, and compare them.

Step 1. We define the classifier for estimating the digitalization potential of the region as a variety of the so-called "gray" Pospelov scale.

Step 2. The expert way determines a set of indicators characterizing the digitalization potential of the region. Depending on the achieved value of the indicator, the region occupies a certain place (rank) in this indicator, which we denote by the variable X_i ($i = 1, ..., n$), where n is the number of indicators.

Step 3. The effect of the indicator on the value of the digitization potential will be evaluated depending on rank, also using a linguistic variable.

Step 4. We define a term-set of five elements, i.e. $B = \{B_{i1}, B_{i2}, B_{i3}, B_{i4}, B_{i5}\}$, the following values of terms can be used depending on the values of ranks in the analyzed indicator (for example, in the ratio 1:2:4:2:1), the levels of indicators in the form of trapezoidal numbers:

B_{i1} "very low level X_i";

B_{i2} "low level X_i";

B_{i3} "middle level X_i";

B_{i4} "high level X_i";

B_{i5} "very high level X_i"

Step 5. We show the transition from the indicators characterizing the digitization potential $X = \{X_1, X_2, X_3, X_4, X_5, X_6\}$ to the statements about the magnitude of the digitization potential $G = \{G_1, G_2, G_3, G_4, G_5\}$.

Verification of this technique will be carried out using the example of assessing the digitalization potential of the regions of the Central Federal District of the Russian Federation.

Indicators characterizing the potential of digitalization of regions were obtained based on the data of the statistical collection of the Higher School of Economics for 2015 [4] (see Table 1):

X1   The level of digitalization of the local telephone network, %;

X2   The number of fixed broadband Internet subscribers per 100 people. Population;

X3   The number of mobile Internet access subscribers per 100 people of population;

X4   Average number of employees of ICT sector organizations, thousand people;

X5   Use of ICT in the business sector (in % of the total number of organizations of the business sector).

Indicators characterizing the effectiveness of using the digitalization potentials of the CFD regions were obtained on the basis of the data of the statistical collection of the Higher School of Economics for 2015 [4] (see Table 2):

X1   The volume of goods shipped (works performed, services) of own production by the ICT sector, billion rubles;

X2   The volume of goods shipped (works performed, services) of own production by the ICT sector organizations, as% of total volume;

X3   Investments in fixed assets of ICT sector organizations, billion rubles;

X4   Investments in fixed assets of ICT sector organizations, as % of total;

X5   Subscription fee for access to the Internet, rubles.

Table 1. Places of regions in the Central Federal District on the potential of digitalization of the economy for 2013

Rank	№	Region CFD	Indicators characterizing the digitalization potential of the region - place of the region among others					Average rank
			X1	X2	X3	X4	X5	
1	18	Moscow	1	1	1	1	1	1
2	10	Moscow region	5	18	1	2	5	6,2
3	17	Yaroslavl region	10	2	4	8	8	6,4
4	6	Kaluga region	9	6	7	6	6	6,8
5	12	Ryazan region	10	11	9	5	2	7,4
6	4	Voronezh region	7	5	16	3	7	7,6
7	3	Vladimir region	13	7	14	4	4	8,4
8	7	Kostroma region	2	4	6	18	17	9,4
9	16	Tula region	12	3	12	7	15	9,8
10	15	Tver region	6	15	3	13	14	10,2
11	13	Smolensk region	17	8	8	10	11	10,8
12	9	Lipetsk region	4	10	17	16	9	11,2
13	2	Bryansk region	8	11	15	11	12	11,4
14	14	Tambov region	3	9	18	9	18	11,4
15	1	Belgorod region	14	17	11	15	3	12
16	5	Ivanovo region	16	16	5	17	10	12,8
17	11	Orel region	15	13	10	12	16	13,2
18	8	Kursk region	18	14	13	14	13	14,4

Table 2. Places of the regions of the Central Federal District for the effectiveness of using the digitalization potential for 2013.

Rank	№	Region CFD	Indicators characterizing the digitalization potential of the region - place of the region among others					Average rank
			X1	X2	X3	X4	X5	
1	18	Moscow	1	7	1	2	1	2,4
2	6	Kaluga region	2	1	4	6	3	3,2
3	3	Vladimir region	4	3	5	1	10	4,6
4	10	Moscow region	3	16	2	4	5	6
5	12	Ryazan region	5	2	7	8	14	7,2
6	13	Smolensk region	9	4	12	12	2	7,8
7	17	Yaroslavl region	8	8	6	5	13	8
8	16	Tula region	7	11	9	13	4	8,8
9	4	Voronezh region	6	9	3	11	18	9,4
10	11	Orel region	15	6	14	9	7	10,2
11	15	Tver region	12	13	8	10	10	10,6
12	14	Tambov region	10	5	9	15	17	11,2
13	2	Bryansk region	11	10	12	14	12	11,8
14	7	Kostroma region	18	15	17	3	6	11,8
15	1	Belgorod region	13	17	11	16	7	12,8
16	5	Ivanovo region	17	12	15	6	15	13
17	8	Kursk region	14	14	15	17	9	13,8
18	9	Lipetsk region	16	17	17	18	16	16,8

4 Analysis of the Research Results

Stage I. Evaluation of digitalization potential of the regions of the Central Federal District of the Russian Federation

Step 1. We form the classifier in the form of a fuzzy linguistic description on the interval [0,1]. The universal set for the variable g is the interval [0,1], and the set of values of the variable g is the term-set $G = \{G_1, G_2, G_3, G_4, G_5\}$, where:

1. G_1 = The potential is extremely low;
2. G_2 = Potential is below average;
3. G_3 = Average potential;
4. G_4 = Above-average potential;
5. G_5 = High potential.

We compile a table of the membership functions of each term (Table 3), using the formula for the function of a trapezoidal fuzzy number $x = \{a_1, a_2, a_3, a_4\}$:

Table 3. The membership functions of subsets of the term-set g

The term G_k	The fuzzy set membership G
G_5 = "high potential" $G_5 \in [0; 0, 25]$	$\mu_5 = \begin{cases} 1, & åñëè\ 0 \leq g \leq 0,15 \\ 10(0,25-g) & åñëè\ 0,15 < g \leq 0,25 \end{cases}$
G_4 = "above-average potential" $G_4 \in (0, 15; 0, 45]$	$\mu_4 = \begin{cases} 1 - 10(0,25-g), & åñëè\ 0,15 < g \leq 0,25; \\ 1, & åñëè\ 0,25 < g \leq 0,35 \\ 10(0,45-g), & åñëè\ 0,35 < g \leq 0,45 \end{cases}$
G_3 = "medium potential" $G_3 \in (0, 35; 0, 65]$	$\mu_3 = \begin{cases} 1 - 10(0,45-g), & åñëè\ 0,35 < g \leq 0,45; \\ 1, & åñëè\ 0,45 < g \leq 0,55 \\ 10(0,65-g), & åñëè\ 0,55 < g \leq 0,65 \end{cases}$
G_2 = "below average potential" $G_2 \in (0, 55; 0, 85]$	$\mu_2 = \begin{cases} 1 - 10(0,65-g), & åñëè\ 0,55 < g \leq 0,65; \\ 1, & åñëè\ 0,65 < g \leq 0,75 \\ 10(0,85-g), & åñëè\ 0,75 < g \leq 0,85 \end{cases}$
G_1 = "the potential is extremely low" $G_1 \in (0, 75; 1]$	$\mu_1 = \begin{cases} 1 - 10(0,85-g), & åñëè\ 0,75 < g \leq 0,85; \\ 1, & åñëè\ 0,85 < g \leq 1 \end{cases}$

$$\mu(x) = \begin{cases} 0, & åñëè\ x < a_1; \\ \frac{x-a_1}{a_2-a_1}, & åñëè\ a_1 \leq x < a_2; \\ 1, & åñëè\ a_2 \leq x \leq a_3; \\ \frac{x-a_4}{a_3-a_4}, & åñëè\ a_3 < x \leq a_4; \\ 0, & åñëè\ x > a_4 \end{cases} \tag{1}$$

Step 2. The method of expert evaluation in Table 1 defines indicators in the form of indicators-ranks for 18 regions of the Central Federal District:

X1   A rank according to the indicator "Level of digitalization of local telephone network, %";

X2   Rank by the indicator "Number of subscribers of fixed broadband Internet access per 100 people of population";

X3   Rank by the indicator "Number of mobile Internet access subscribers per 100 people of population";

X4   a rank on an indicator "Average number of workers of the organizations of sector ICT, thousand people."

X5   rank of the indicator "Use of ICT in the business sector (in % of the total number of organizations of the business sector)."

Steps 3, 4. We define expertly the term-set $B = \{B_{i1}, B_{i2}, B_{i3}, B_{i4}, B_{i5}\}$:

$B_{i1} = (1, 5, 8, 11)$ – "very low level X_i";
$B_{i2} = (2, 9, 11, 13)$ – "low level X_i";
$B_{i3} = (3, 12, 14, 15)$ – the "average level X_i";
$B_{i4} = (4, 7, 12, 16)$ – "high level X_i";
$B_{i5} = (6, 10, 17, 18)$ – "very high level X_i".

The description of the state of the digitalization potential of the Voronezh region is characterized by two possible values: G5 ($\mu_5 = 0, 15$) = "high potential" is less significant than saying G4 ($\mu_4 = 0, 85$) = "above average".

Description of the status of digitalization potential of the Kostroma Region: G5 ($\mu_5 = 0, 5$) = "high potential" is also significant, as is the saying G4 ($\mu_4 = 0, 5$) = " above-average potential". If we compare the values of the digitalization potentials of the two regions, then on the G scale, the value of G5 for the Kostroma region will be larger than the G5 for Voronezh region.

Description of the state of digitalization potential of Tambov region is characterized by one meaning: G2 ($\mu_2 = 1, 0$) = "below average potential".

Stage II. Evaluation of the effectiveness of the digitization potential of the Central Federal District

Step 1. Similarly to the first stage, we form a classifier in the form of a fuzzy linguistic description on the interval [0,1].

Step 2. Ranking scores of performance indicators are shown in Table 2:

X_1 rank on the indicator "The volume of goods shipped (work and services performed) of own production by the ICT sector, billion rubles.";

X_2 rank according to the indicator "The volume of goods shipped (works performed, services) of own production by the ICT sector organizations, in% of the total volume";

X_3 rank on the indicator "Investments in fixed assets of ICT sector organizations, billion rubles.";

X_4 rank according to the indicator "Investments in fixed assets of ICT sector organizations, in% of total";

X_5 rank on the indicator "Subscription fee for Internet access, rubles."

Steps 3, 4. Let us evaluate the effectiveness of using the digitalization potential of the three regions through the ranks of indicators, term sets and the significance of the membership function.

We calculate the value of the membership function of the linguistic variable j = "the value of the effectiveness of using the digitalization potential" for each region of the CFD.

Conclusion: J5 $(\mu_5 = 0,55)$ - *the result of the digitization potential of the Voronezh region is excellent.*

Conclusion: J4 $(\mu_4 = 1,0)$ - *the result of the digitalization potential of the Kostroma region is good.*

Conclusion: J2 $(\mu_4 = 1,0)$ - *the result of the digitalization potential of the Tambov region is rather low with a tendency towards average.*

5 Conclusions

The results obtained should form the basis for a more detailed evolution of the digitalization potential of the regions.

A major achievement is the almost twofold reduction in the digital gap between Moscow and other regions. If in 2011 the gap between the digitization of Moscow and the average for regions was 2.6 times, today this figure has dropped to 1.35.

A key contribution to reducing the digital-gap was the development of-access infrastructures. Since 2011, this-indicator has increased by an average of 2.6 times in the regions.

The way of digital transformation requires a fundamental reorganization of the approaches of private business and the state interaction-, decision-making,- stimulation of innovations and the formation of a legislative environment, where every participant in the system has a significant role.

The concerted actions of all participants in the potential digital ecosystem will lead to a sustainable positive-result.

This will increase the share of the digital economy to 5.6% of GDP, as well as create large-scale inter sectoral effects and real value added in the industries up to 5–7 trillion rubles in year.

References

1. Kupriyanovsk, V., et al.: A holistic model of transformation in the digital economy-how to become digital leaders. Int. J. Open Inf. Technol. 5(1), 26–33 (2017)
2. Information-analytical report, Analysis of the world experience of industrial development and approaches to digital transformation of industry of the member states of the Eurasian Economic Union/Eurasian Economic Commission. Department of Industrial Policy, 116 p (2017)
3. Banke, B., Butenko, V., Kotov, I.: Russia online? Catch up cannot be left behind/The BosTon ConsulTing group, 56 p. (2017)
4. Abdrakhmanova, G.I., Gokhberg, L.M., Kovaleva, G.G.: Information society: development trends in the subjects of the Russian Federation. Issue 2: Statistical Collection/National Research Institute Higher School of Economics. NIUVSHE, Moscow (2015). 160 s
5. Schweer, D., Sahl, J.C.: The digital transformation of industry–the benefit for Germany. In: The Drivers of Digital Transformation. Springer International Publishing, pp. 23–31 (2017)

6. Grigore, G., Molesworth, M., Watkins, R.: New corporate responsibilities in the digital economy. In: Corporate Social Responsibility in the Post-Financial Crisis Era. Springer International Publishing, pp. 41–62 (2017)
7. Berdykulova, G.M.K., Sailov, A.I.U., Kaliazhdarova, S.Y.K., Berdykulov, E.B.U.: The emerging digital economy: case of Kazakhstan. Procedia Soc. Behav. Sci. **109**, 1287–1291 (2014). https://doi.org/10.1016/j.sbspro.2013.12.626. ISSN 1877-0428. http://www.sciencedirect.com/science/article/pii/S1877042813052658
8. Tolstykh, D.V., Shurshikova, G.V.: Methodology for the determination of complex estimates of the marketing potential of the territory and its use/economics and management of management systems, Voronezh, no. 3.1(5), p. 170 (2012)
9. Pfeffermann, N., Gould, J. (eds.): Strategy and communication for innovation: integrative perspectives on innovation in the digital economy. Springer (2017)
10. Cockayne, D.G.: Sharing and neoliberal discourse: the economic function of sharing in the digital on-demand economy. Geoforum **77**, 73–82 (2016). https://doi.org/10.1016/j.geoforum.2016.10.005. ISSN 0016-7185. http://www.sciencedirect.com/science/article/pii/S0016718516302305

Characteristics of the State of Russia's Labor Potential as a Component of Economy's Innovational Development

Nina A. Eldyaeva[1](✉), Elvira A. Yarnykh[1], Olga G. Lebedinskaya[1],
Sergey I. Kuzin[1], and Ekaterina S. Kovanova[2]

[1] G.V. Plekhanov Russian University of Economics,
Moscow, Russian Federation
{Eldyaeva.NA,Yarnykh.EA,Lebedinskaya.OG,
Kuzin.SI}@rea.ru
[2] Kalmyk State University Named After B.B. Gorodovikov,
Elista, Russian Federation
ekovanova@yandex.ru

Abstract. The article deals with the problems of measuring labor. In economic theory, there are still no well-founded and generally accepted definitions of the quality and efficiency of labor, their criteria and growth factors have not been established, there are no fixed relationships with the productivity of labor, the value of goods and other economic concepts and laws.

In most cases, the categories of labor are treated separately, in isolation, without necessarily taking into account their interdependence in real economic life, which indicates the non-systematic nature of their study, that is, there is no comprehensive, comprehensive approach to research. The article allows evaluating the factors influencing the level of labor potential. The strategy for the development of labor resources should be aimed at the formation of labor potential as the most important intellectual and professional resource of Russian society, which ensures effective social and economic development, high competitiveness and innovative development.

The article presents the main directions that make it possible to comprehensively assess the country's labor potential and set tasks that must be solved with a goal to overcome the current situation. Economic growth is impossible without improving the quality of labor.

Keywords: Labor · Labor potential · Labor statistics · Labor productivity
Working hours · Innovative development · Labor efficiency · Staffing

1 Introduction

In the innovative economy, the creative role of labor is enhanced and labor is much more than a simple source of material, spiritual goods and services. Human labor solves scientific, technical and socio-economic problems. However, labor not only makes it possible to solve problems and puts them. To solve labor problems, specialists of various profiles are working on it: Psych physiologists, economists and sociologists, organizers of production, etc. [1–3].

© Springer International Publishing AG, part of Springer Nature 2018
E. G. Popkova (Ed.): HOSMC 2017, AISC 622, pp. 744–750, 2018.
https://doi.org/10.1007/978-3-319-75383-6_95

Many categories and patterns of development of labor are defined, identified, and used in the practical activities of economic entities. However, there are many unsolved problems in the theory and practice of labor. On the one hand, a number of problems of labor and labor relations have not been studied for a long time by social and natural sciences. On the other hand, labor is a continuously developing psycho-physiological and socio-economic phenomenon. Identify the laws and patterns of development of labor - one of the most important tasks of science. From knowledge of laws - to management and acceleration of processes of development of work.

2 Theoretical Basis of the Research

Issues of accounting and analysis of the demographic and labor potential are devoted to the works of a number of foreign and domestic researchers, such as I.A. Aydrus, E.M. Andreev, J. Bourgeois-Pischa, O.I. Yevseyenko, S.I. Pirozhkov, E.V. Pismenaya, S.V. Ryazantsev, S.A. Sukneva, M.Y. Surmach, E. Filrose, L. Hersh, T.N. Shelekhova, D.M. Ediev, Thomas V. Malon, and others.

At the present stage of development of labor, the problems of effectiveness and quality are most urgent [2–8]. However, an analysis of the level of development of theoretical problems of quality and labor efficiency shows that economic science is not yet able to respond properly to the requests of practice in this field.

The amount of labor a worker spends daily has also great social significance. The magnitude of labor costs greatly affects the state of physical health, labor activity, the life expectancy of each person, etc. The insufficient theoretical elaboration of a number of basic problems of labor has led to the fact that in the economic literature there is still no common opinion on the fundamental concepts of labor. Such as the productivity of labor and the productive power of labor, the intensity of labor and its relation to the productivity of labor, simple and complex labor, the efficiency of labor and the quality of labor, etc. [3, 4].

Recently, Russia is moving to the development and implementation of long-term programs for economic development based on innovative solutions, which sets the task of scientific justification of approaches to information support in the field of labor management system. The system of indicators should reflect all ongoing processes, including those that can be a source of risks, losses and causes of non-fulfillment of adopted programs [3, 14, 15].

Consider the state of the information base for work.

Labor statistics changed simultaneously with the transformation of the entire economic system of Russia. As the I.I. Eliseeva notes, "in the 1970s, Soviet socio-economic planning was developing in the Soviet statistics aimed at intensifying and increasing the efficiency of the economy. Methods of performance analysis were developed, and they all relied on data available in the statistical reporting system and statistical publications. In recent years, we have lost almost all the indicators of resource efficiency: The statistics do not calculate the indicators of labor productivity, return on assets, energy and material consumption, etc." [2].

Currently, official statistics do not have a unified procedure for obtaining information from enterprises and organizations on the volume of output, it is replaced by the censor method in different statistical forms and at different intervals. Large and medium-sized organizations report on a monthly basis, small businesses - once a quarter on a selective basis. In order to obtain information on the volume of products, it is necessary to perform a number of adjustments using accounting information that is not harmonized with statistical accounting in many ways.

As a result of such changes, statistics almost lost data on the use of working time, and therefore, on the level of average hourly and average daily labor productivity, on the indexes of "tightness" of the working day and working period [2]. The indicator of labor productivity was almost the main qualitative indicator.

In the conditions of development of the innovation economy, a number of theoretical and methodical approaches to the use of traditional indicators and methods for their analysis need to be changed.

3   Research Methodology

The current level of Russia's labor potential does not correspond to the tasks of the country's economic growth. Its professional and qualification structure is not fully adequate to the needs of production. On the other hand, professional education does not have the proper level. Premature mortality and low level of public health also negatively affect labor potential. A fairly small salary of workers does not allow motivation for highly effective work [13]. In view of the low-quality workforce and the reduction in the number of the able-bodied population, there are risks to doing business and the competitiveness of domestic production is decreasing [9, 10].

Thus, the main factor in the growth of national GDP is the improvement of the quality of labor. Only investments in human capital bring a significant effect. According to the estimates of developed countries, one dollar invested in education yields a return of 2–10 dollars of net profit [11, 12].

According to the conducted researches, it is possible to formulate the following problems of development of labor potential of Russia in the beginning of the XXI century:

(1) *Depopulation*
    Depopulation in Russia has three main aspects.
    The essence of the *first aspect* is that the end of the twentieth century is characterized by a very low birth rate and a high mortality rate, which corresponds to the underdeveloped state. For the period 1992–2004. The population of Russia has decreased by 10.4 million people, which is comparable with the population of Yugoslavia, Belgium or Belarus. If we take into account the partial migration, the total loss was 4.45 million people.
    The second aspect is connected with long-term fundamental processes, and not with time, factors and the improvement of the social and economic situation in the country completely will not allow to get out of this state.
    The distinctive features of depopulation in Russia in comparison with the developed European countries, in which the population also decreases, characterize the third aspect. Depopulation results from a very low birth rate. The total

fertility rate for 2004 in Russia was 1.35 (for example, in Norway this coefficient is 1.73, in Sweden - 1.65, in France - 1.88, in the UK - 1.71. Russia also has a very high mortality rate.

In the absence of migration growth and the maintenance of the existing level of fertility and mortality by 2025, the population of Russia can reach 122.0 million people.

The consequence of depopulation is the aging of the labor potential, which entails an increase in the demographic burden on every able-bodied person.

(2) *Health status of the population*

One of the limitations of the country's economic growth is the health of the nation, mainly the able-bodied population. Statistical analysis showed that in 2009, the loss of resources due to the deterioration of workers' health amounted to 43% of GDP. Over the past 10 years due to premature mortality of the able-bodied population, the country's losses reached 1.5 trillion. Alternatively, 225 million person-years of active labor.

(3) *Staffing*

Deficiency of a labor generates personnel risks. In Russia, it is most difficult to ensure the production of workers' professions. Underfunding of qualified personnel of enterprises and organizations leads to a decline in production. This is especially true for enterprises with a high number of employees aged over 35–40 in the number of employees. Young people do not aspire to occupy vacant jobs at enterprises that are not engaged in production modernization, since very low wages and harsh working conditions.

(4) *Innovative way of development and changing the structure of demand for labor*

At present, the transition to an innovative development model is required, involving the introduction of new technology, research, the development of new types of competitive products, advanced technologies, which requires increased investment in improving production. This entails the renewal of the system of labor relations, the training of highly qualified specialists capable of generating new managerial decisions and using modern methods of work.

(5) *International migration and the "brain drain"*

Russia is one of the leading countries in the world for a "brain drain" along with countries such as South Africa, India, Poland, and the Philippines. The main factors affecting the international migration of labor resources are economic and non-economic.

Economic factors include:

– Different level of industrial production and standard of living;
– Different level of wages;
– The presence of organic unemployment;
– The functioning of international corporations and international capital flows.

About one third of talented Russian scientists work outside Russia. Consequently, the annual losses of the Russian economy amount to about 50–60 billion dollars US.

Branch specificity of the manifestation of the labor shortage. Despite the fact that in general, the labor deficit is not obvious in the economy, considering certain sectors, it is possible to observe significant distortions. For example, there is a significant decrease in the number of employees in key sectors of the economy. The most noticeable decrease in the number of workers occurred in science and scientific services, for the period 1990–2003. (2.6 times); Agriculture and forestry (1.3 times); Machine building and metalworking (2.1 times); Construction (in 1.8 times).

Professional qualification aspect of the deficit. At present, there is a problem of shortage of experienced and skilled workers of certain technical professions and qualifications, provided there are sufficiently well educated labor resources. This is due primarily to the reform of the system of primary professional education (PPE). In the system of PPE for the period 1990–2013. The output of qualified specialists decreased by more than 40%, while the number of students decreased by 15%.

In addition, most graduates of universities do not work for acquired specialty, which causes an additional shortage of staff. According to expert estimates, this figure is 50% for young engineers, 30–40% for graduates of medical schools, and 70% for graduates of agricultural institutes.

The age aspect of the deficit is the aging of the population at the working age. This also contributes to the reduction of human resources in the sectors of the economy. Moreover, against the background of a reduction in the number of all age groups, only the number of employed at the age of 45–54 is growing. Most of all, this reduction affected the age groups of up to 20 years, 30–39 years, 55–59 years.

Rosstat conducts a survey of the population on employment issues. Based on the data of this survey in 2016, the average age of people employed in the national economy has approached 43 years, and in some industries, for example, in science and scientific services, it reaches 48 years.

Extrapolation of the prevailing trends in the age structure of employed shows that the absolute number and share of employed at the age of 40–49 will continue to increase. For a whole group of ages (60–72, 20–29, 35–39 years), there will be a slight increase or stabilization of the values attained so far. However, there is a decrease in the number of employed in the age group 25–29 years.

Thus, if the prevailing trends persist in the long term, the number of the employed population of the most active age groups decreases.

It should be noted a weak influx of young personnel into a group of industries. This can be explained by the low level of wages in these industries and the prevailing system of preferences. From this, we can conclude that there is a structural deficit, which manifests itself in conditions of differentiation of wages.

The negative impact of structural problems of labor shortage on the demographic situation. Unemployment and low wages adversely affect the development of the family. Unemployment creates such social problems as drug addiction, crime, alcoholism, suicide. Low incomes entail a refusal to create a family, the birth of children or their delayed birth, divorce, abortion, etc. Unsecured work and harmful working conditions also lead to premature deaths, injuries and poor health.

The listed structural problems of the labor market entail a general decline in the economic growth rate and an increase in the resources needed to overcome these problems. This, in turn, leads to a reduction in spending on social needs and the implementation of programs to overcome the demographic crisis.

Thus, *it is necessary to talk not only about the impact of the demographic decline on the stagnation of the labor market, but also on the reverse process.*

4   Analysis of the Results

The study showed that at present Russia's labor potential does not correspond to the modern innovative way of its development. The professional qualification structure of the labor potential does not fully meet the needs of the economy. The quality of professional education is deteriorating. Premature mortality and poor health lead to large losses. There is no motivation for effective labor activity, since low wages do not stimulate the labor activity of workers. On the other hand, due to the inadequate quality of the workforce, the competitiveness of domestic goods and services is declining. In addition, a significant reduction in the working-age population creates additional risks for business and the economy as a whole.

Despite this, the achievement of economic growth is impossible without a sharp increase in the quality of labor. Therefore, the main factors for increasing national GDP should be the effective use of labor potential.

It should also be taken into account that as a result of the ongoing state policy of supporting the family, since 2014, the population has been growing and by the beginning of 2017 its level has reached 146,804,372 (according to the Federal State Statistics Service). Such growth is provided by the natural and migratory growth of the population, of which the migration gain was 95.5%.

5   Conclusions

In economic theory, there are still no well-founded and generally accepted definitions of the quality and efficiency of labor, their criteria and growth factors have not been established, there are no fixed relationships with the productivity of labor, the value of goods and other economic concepts and laws. Consequently, both in theory and in practice, labor efficiency is often identified with its productivity, the quality of products with the quality of labor, and so on.

In solving the problems of production and labor efficiency, the problem of the quantity of labor occupies one of the leading positions. The fact is that the effectiveness and efficiency of production and labor is achieved with the "optimal" quantity and quality of labor expended by the workers. Measuring and possibly quantifying this number in each workplace is an important task of science and practice.

References

1. Bashina, O.E., Yarykh, E.A.: The role and place of statistics in the economic development of society. Prob. Stat. **8**, 83–84 (2010)
2. Eliseeva, I.I.: Russian statistics at the present stage. Economics **2**, 75–92 (2011)
3. Zinchenko, A.P.: Problems of the development of statistics as a science. Prob. Stat. **8**, 67–70 (2011)
4. Eldiaeva, N.: Econometric methods in macroeconomic analysis: problems of constructing forecasting models. Vestnik Astrakhanskogo gos. Tech. Un-ta, Astrakhan, vol. 4, pp. 225–231 (2006)
5. Eldiaeva, N.: Statistical problems of construction of econometric models of market economy functioning. In: Scientific thought of the Caucasus: Application, vol. 8, pp. 156–162. Publishing House of the North Caucasian Scientific Center of Higher Education, Rostov-on-Don (2006)
6. Eldyaeva, N., Sibirskaya, E., Khokhlova, O., Lebedinskaya, O.: Statistical evaluation of the middle class in Russia. 125 49 Mediterr. J. Soc. Sci. 33 34 35 **6**(3, Suppl. 6), 125–135 (2015)
7. Eldyaeva, N., Sibirskaya, E., Yarnykh, E., Dubrova, T., Oveshnikova, L.: Development of entrepreneurial infrastructure of regional economy. Eur. Res. Stud. **18**(Special Issue), 235–258 (2015). Strategy of Systemic
8. Yarnykh, E.A., Oleynikov, B.I., Tsylina, I.T.: Statistical methods for the analysis of demographic processes within the region. In: 8th International Scientific Conference Science and Society, pp. 117–123, 24–29 November 2015
9. Malone, T.W.: The Future of Work/How the New Order of Business Will Share Your Organization, Your Management Style, and Your Life. Harvard Business School Press Boston, Massachusetts (2004)
10. Gerber, M.E.: The E-Myth Revisited Why Most Small Business Don not Work and What to Do About It. HarperBusiness, New York (2001)
11. Acemoglu, D.: Introduction to Modern Economic Growth. Princeton University Press, Princeton (2009)
12. Hall, R., Jones, C.I.: Why do some countries produce so much more output per worker than others? Q. J. Econ. **114**, 83–116 (1999)
13. Handbook of procedures relating to international labour conventions and recommendations. Intern Labour Office, Geneva (2012)
14. Sibirskaya, E., Mikheykina, L., Egorov, A., Safronova, A., Ivashkova, T.: Organization of favorable investment climate in the market of development and implementation of investment projects. Mediterr. J. Soc. Sci. **6**(36), 135–146 (2015)
15. Sibirskaya, E., Stroeva, O., Simonova, E.: The characteristic of the institutional and organizational environment of small innovative and big business cooperation. Procedia Econ. Financ. **27**, 1–736 (2015). pp. 507–515

Author Index

Printed in the United States
By Bookmasters